Renewable Energy

Its physics, engineering, use, environmental impacts, economy and planning aspects

Third Edition

Renewable Energy

Its physics, engineering, use, environmental impacts, economy and planning aspects

Third Edition

Bent Sørensen

Roskilde University
Energy & Environment Group, Institute 2,
Universitetsvej 1, P. O. Box 260
DK-4000 Roskilde, Denmark

ELSEVIER
ACADEMIC
PRESS

Amsterdam Boston Heidelberg London New York Oxford
Paris San Diego San Francisco Singapore Sydney Tokyo

Elsevier Academic Press

200 Wheeler Road, 6th Floor, Burlington, MA 01803, USA
525 B Street, Suite 1900, San Diego, California 92101-4495, USA
84 Theobald's Road, London WC1X 8RR, UK

This book is printed on acid-free paper. ⊗

Library of Congress Cataloging-in-Publication Data

Application submitted

British Library Cataloguing in Publication Data

A catalogue record for this book is available from the British Library

ISBN: 0-12-656153-2

For all information on all Academic Press publications
visit our Web site at www.academicpress.com

Layout and print-ready electronic-medium manuscript by author
Printed in the United States of America
04   05   06   07   08   09   9   8   7   6   5   4   3   2   1

Preface to third edition

The present edition has been updated in a number of renewable energy technology areas (Chapters 4 and 5), where progress has been made over the recent years. New solar simulation studies have been added to Chapter 6, and market considerations have been included in the overview in Chapter 1 and in discussing industry liberalisation in Chapter 7. The remarks on advanced subjects made in the preface to the second edition are still valid. A new growth area is fuel cells for stationary and mobile uses of hydrogen and other fuels. Only modest updates have been done in this area, as it is the subject of a new, companion book to be published about a year after this one (Sørensen: Hydrogen and Fuel Cells, Elsevier/Academic Press). Some older material has been omitted or tidied up, and manoeuvring through the book has been eased, both for reference and for textbook uses. The following diagrams may assist in following the tracks of interest through the book:

Topic-driven paths

Chapter	Wind	Solar power	Solar heat	Biofuels	Others
1	1.1	1.1	1.1	1.1	1.1
2	2.3.1 (end), 2.4.1, 2C	2.2.2, 2.4.1	2.2.2, 2.4.1	2.4.1	2.3.2, 2.4.1, 2B, 2D
3	3.2	3.1	3.1	3.6	3.3-3.5, 3.7
4	4.1.4, 4.3	4.1.5, 4.2.3	4.2.1-2, 4.6	4.8	4.1.3, 4.1.6, 4.4-5, 4.7, 4.8
5	5.1.2, 5.2.2	5.1.2, 5.2.2	5.1.1, 5.2.1		
6	6.2.5, 6.3.2, 6.4	6.2.4, 6.4	6.3.1, 6.4	6.2.7, 6.4	6.4
7	7.4.12-13, 7.5	7.4.12-13		7.4.12-13	
8	8	8	8	8	8

Course material uses

Chapter	Resource studies	Energy engineering	Energy planning	Energy economics	Energy & environment
1	1		1	1.1	1.2
2	2				2.4.1
3	3	as needed			3.4.2 (end)
4		4	as needed	4.8 (start)	4.5
5		5	as needed		
6			6	as needed	as needed
7			7	7	7.4
8	8	8	8	8	8

Gilleleje, October 2003, *Bent Sørensen*

Preface to second edition

When the first edition of *Renewable Energy* appeared in 1979, it was the first textbook and research monograph since the 1920s to deal with the renewable energy sources and systems at a scholarly level. Indeed, it was instrumental in establishing the now universally used term "renewable energy" for a new area of science, which emerged under names such as "regenerative energy" in Germany and "solar energy" in the United States of America. In many countries, renewable energy appeared in planning documents as "supplementary energy", based on a conviction by administrators, that this could never become a major source of energy. My suggestion in the journal *Science* (Sørensen, 1975b) that renewable energy could potentially become a hundred per cent solution was regarded as absurd by many. Things have changed today, where official energy plans of some countries call for over fifty per cent renewable energy coverage by year 2030 (Danish Department of Environment and Energy, 1996), where the best renewable energy technologies are already economically competitive relative to fossil options and where increased concern over greenhouse warming effects may well alter the perceived indirect costs of different energy solutions.

The structure of the first edition was determined by the aim of placing renewable energy on the academic agenda. It was my goal to show young scientists, engineers and future planners that renewable energy was at least as interesting and challenging as nuclear energy, and I tried to do this by showing the depth of problems to be solved using advanced techniques, shying no complication of quantum mechanics or non-linear mathematics. This was seen as positive by reviewers and colleagues, but may have limited the sales figures for the book! Today, the requirements are quite different: now many universities and polytechnic institutes have renewable energy courses in their curriculum, and the task at hand is to provide good teaching materials for the relevant levels of courses. Therefore, I have thoroughly revised the content and presentation in the second edition. The main sections of each chapter are now suited for introductory level study, with only very general prerequisites. Any topic requiring more background is deferred to special sections marked as **ADVANCED** topics at the top corner of each page. They can be added individually at the choice of the teacher, or they can be left for further study by the user of the book. My reflections on whether to separate elementary and advanced topics in two volumes or keep them together are as follows. Needing to go back to a topic for more detailed study, it is very convenient to be able to find it in a book that you have already worked with. The style and assumptions are known to you, and first of all, the book is on your shelf and need not be retrieved from somewhere else. Against the single volume solution speaks the book price for those who find it unlikely that they shall need more than the elementary sections. However, we are all surprised by the growth of our needs, and the price of this second edition is even below that of the first edition, thanks to modern preparation and printing methods.

Another issue is the arrangement of material, which I have basically kept as in the first edition: first describing the origin of renewable energy, then its disposition and availability at different geographical locations on Earth, then the techniques of energy conversion systems and systems suitable for each type of renewable energy, and finally the evaluation of the total system, in terms of economic and environmental impacts. The logic of this sequence is evident, but it means that someone wanting to know only about wind power will have to jump from chapter to chapter. This is made much easier in this edition, by the addition on each bottom left page references to previous and following sections dealing with the same form of renewable energy. As in the first edition, extensive references and an index are found at the end. The index also serves as a pointer to specialist words and concepts by giving the page where they are first explained. After the table of contents, a list of units and abbreviations is given.

The content has been revised in those areas where new advances have been made, notably in the sections on energy from biomass and on photovoltaic energy conversion, and in the economic chapter on life-cycle analysis. As in the first edition, emphasis is on basic principles. Fortunately, they do not wear much with time, and several sections needed only a light brush-up, sometimes with some tidying effort to keep the size down. However, new data available today has made it possible to improve many of the illustrations, notably in the area of global energy flows. At the end of each chapter, there are topics for discussion, including new ones. They are basically of two kinds: simple topics for classroom discussion and mini-project ideas that can serve as a basis for problem-oriented work extending from a few days to several months in duration. This is a reflection of the different style of teaching at different institutions, where small projects are often offered to individuals or groups of students for credit, with the indicated range of time devoted to each problem (and a corresponding difference in depth of investigation).

The Danish Energy Agency supported part of the work upon which the second edition updates are based. The author welcomes comments and suggestions, which may be addressed as indicated below.

Allerød, 1998, *Bent Sørensen*
email: bes@ruc.dk *or* novator@danbbs.dk
internet: http://mmf.ruc.dk/energy *or* http://www.danbbs.dk/~novator

Preface to first edition

Renewable energy is the collective name for a number of energy resources available to man on Earth. Their conversion has always played an important role for the inhabitants of the planet, and apart from a period of negligible length – relative to evolutionary and historical time scales – the renewable energy sources have been the only ones accessible to mankind.

Yet the study of renewable energy resources, their origin and conversion, may at present be characterised as an emerging science. During the past fifty years of scientific and technological revolution, much more effort has been devoted to the extraction and utilisation of non-renewable energy resources (fuels), than to the renewable ones. Only very recently have funds been made available to re-establish renewable energy research and development, and it is still unclear whether the technologies based on renewable energy sources will become able to constitute the backbone of future energy supply systems.

The purpose of the present book is to provide an incentive as well as a basis of reference for those working within the field of renewable energy. The discontinuity between earlier and present work on renewable energy, and the broadness of disciplines required for assessing many questions related to the use of renewable energy, have created a need for a comprehensive reference book, covering methods and principles, rather than specific engineering prescriptions of passing interest in a rapidly developing field.

A survey of renewable energy has to draw upon a large number of individual scientific disciplines, ranging from astrophysics and upper atmospheric science over meteorology and geology to thermodynamics, fluid mechanics, solid state physics, etc. Specialists in each discipline often use a vocabulary recognised only by insiders, and they rarely emphasise the aspects pertinent to renewable energy. I have attempted to use a common language throughout, and to restrict the prerequisites for understanding to a fairly elementary level (e.g. basic physics). However, this does not mean that I have avoided any degree of complication considered relevant, and the reader must be prepared to face a number of challenges.

I envisage my reader as a research worker or student working somewhere within the field of renewable energy. Such work is currently undertaken at universities, engineering schools and various offices and laboratories in the public or private sectors. However, since a substantial part of the book deals with *energy systems* comprising renewable energy elements, and with the management and economy of such systems, including environmental and social aspects, then I sincerely hope to attract also readers in the energy planning and management sectors, whether their concern is the physical planning and operation of energy supply systems, or the socio-economic assessment of such systems.

When used as a textbook, particular chapters may be more relevant than others. Cross-references are included in most cases where definitions or basic concepts have to be taken from a different chapter. Courses in engineering may place the emphasis around Chapter 4 (e.g. including Chapters 3 – 6), courses in "energy physics" or on energy in general may place more emphasis on Chapters 2 and 3, while courses on energy planning, systems aspects, and technological or economic assessments may find it natural to shift the emphasis to Chapters 6 and 7.

It should be stressed that the main purpose of the book is to provide general tools for treating problems relating to renewable energy. This is evident from the approach to energy conversion in Chapter 4 (stressing principles rather than describing individual pieces of equipment in detail), and from the treatment of supply systems in Chapter 6 (which contains no exhaustive reviews of possible system combinations, but illustrates basic modelling and simulation techniques by use of just a few, selected system examples). Energy storage and transmission (Chapter 5) are described in a highly condensed form, with the sole purpose of introducing the components for use in energy systems such as those discussed in Chapter 6.

I have been motivated to engage in work on renewable energy and to see the possibility of an increasingly important role played by the associated technologies by reflections which are largely summarised in Chapter 1, and which to some extent lie behind those amendments to conventional economic theory for application to long-term energy planning, proposed in Chapter 7. The subjective nature of a number of interpretations made in these two chapters is recognised, and an effort has been made to ban such interpretations from the remaining five chapters, so that readers disagreeing with my interpretations may still find the bulk of the book useful and stimulating.

I thank the following for reading and commenting on portions of the draft version of the manuscript: Niels Balling, Henning Frost Christensen, E. Eliasen, Frede Hvelplund, Johannes Jensen, Marshal Merriam, B. Maribo Petersen and Ole Ulfbeck.

Bent Sørensen, Allerød, January 1979

Contents

LIST OF CONTENTS

LIST OF CONTENTS

Units and conversion factors

Powers of 10

Prefix	Symbol	Value	Prefix	Symbol	Value
atto	a	10^{-18}	kilo	k	10^3
femto	f	10^{-15}	mega	M	10^6
pico	p	10^{-12}	giga	G	10^9
nano	n	10^{-9}	tera	T	10^{12}
micro	μ	10^{-6}	peta	P	10^{15}
milli	m	10^{-3}	exa	E	10^{18}

G, T, P, E are called milliard, billion, billiard, trillion in Europe, but billion, trillion, quadrillion, quintillion in the USA. M as million is universal.

SI units

Basic unit	Name	Symbol
length	metre	m
mass	kilogram	kg
time	second	s
electric current	ampere	A
temperature	Kelvin	K
luminous intensity	candela	cd
plane angle	radian	rad
solid angle	steradian	sr

Derived unit	Name	Symbol	Definition
energy	joule	J	$kg\ m^2\ s^{-2}$
power	watt	W	$J\ s^{-1}$
force	newton	N	$J\ m^{-1}$
electric charge	coulomb	C	$A\ s$
potential difference	volt	V	$J\ A^{-1}\ s^{-1}$
pressure	pascal	Pa	$N\ m^{-2}$
electric resistance	ohm	Ω	$V\ A^{-1}$
electric capacitance	farad	F	$A\ s\ V^{-1}$
magnetic flux	weber	Wb	$V\ s$
inductance	henry	H	$V\ s\ A^{-1}$
magnetic flux density	tesla	T	$V\ s\ m^{-2}$
luminous flux	lumen	lm	$cd\ sr$
illumination	lux	lx	$cd\ sr\ m^{-2}$
frequency	hertz	Hz	$cycle\ s^{-1}$

UNITS & CONVERSION FACTORS

Conversion factors

Type	Name	Symbol	Approximate value
energy	electron volt	eV	1.6021×10^{-19} J
energy	erg	erg	10^{-7} J (exact)
energy	calorie (thermochemical)	cal	4.184 J
energy	British thermal unit	Btu	1055.06 J
energy	Q	Q	10^{18} Btu (exact)
energy	quad	q	10^{15} Btu (exact)
energy	tons oil equivalent	toe	4.19×10^{10} J
energy	barrels oil equivalent	bbl	5.74×10^{9} J
energy	tons coal equivalent	tce	2.93×10^{10} J
energy	m^3 of natural gas		3.4×10^{7} J
energy	litre of gasoline		3.2×10^{7} J
energy	kilowatthour	kWh	3.6×10^{6} J
power	horsepower	hp	745.7 W
power	kWh per year	kWh/y	0.114 W
radioactivity	curie	Ci	3.7×10^{8} s^{-1}
radioactivity	becqerel	Bq	1 s^{-1}
radiation dose	rad	rad	10^{-2} J kg^{-1}
radiation dose	gray	Gy	J kg^{-1}
dose equivalent	rem	rem	10^{-2} J kg^{-1}
dose equivalent	sievert	Sv	J kg^{-1}
temperature	degree Celsius	°C	K − 273.15
temperature	degree Fahrenheit	°F	9/5 C+ 32
time	minute	m	60 s (exact)
time	hour	h	3600 s (exact)
time	year	y	8760 h
pressure	atmosphere	atm	1.013×10^{5} Pa
pressure	bar	bar	10^{5} Pa
mass	pound	lb	0.4536 kg
mass	ounce	oz	0.02835 kg
length	foot	ft	0.3048 m
length	mile (statute)	mi	1609 m
volume	litre	l	10^{-3} m^3
volume	gallon (US)		3.785×10^{-3} m^3

CHAPTER

1

PERSPECTIVES

1.1 Current penetration of renewable energy technologies in the marketplace

The penetration of renewable energy into the energy system of human settlements on Earth is from one point of view nearly 100%. The energy system seen by the inhabitants of the Earth is dominated by the environmental heat associated with the greenhouse effect, which captures solar energy and stores it within a surface-near sheet of topsoil and atmosphere around the Earth. Only 0.02% of this energy system is currently managed by human society, as illustrated in Fig. 1.1. Within this economically managed part of the energy sector, renewable energy sources currently provide about 25% of the energy supplied. As the figure indicates, a large part of this renewable energy is in the form of biomass energy, either in food crops or in managed forestry providing wood for industrial purposes or for incineration (firewood used for heat and cooking in poor countries or for mood-setting fireplaces in affluent countries, residue and waste burning in combined power and heat plants or incinerators). The additionally exploited sources of renewable energy include hydro, wind and solar. Hydropower is a substantial source, but its use is no longer growing due to environmental limits identified in many locations with potential hydro resources. Passive solar heating is a key feature of building design throughout the world, but active solar heat or power panels are still at a very minute level of penetration. Also, wind has both a passive and an active role. Passive use of wind energy for ventilation of buildings plays a significant role, and active power production by wind turbines is today a rapidly growing energy technology in many parts of the world. The highest penetration reaching nearly 20% of total electricity provided is found in Denmark, the country pioneering modern wind technology. Further renewable energy technologies, so far with small global penetration, include biofuels such as biogas and geothermal power and heat. As indicated in Fig. 1.1, the dominant energy sources are still fossil fuels, despite the fact that they are depletable and a cause of frequent national conflicts, due to the mismatch between their particular geographical availability and demand patterns.

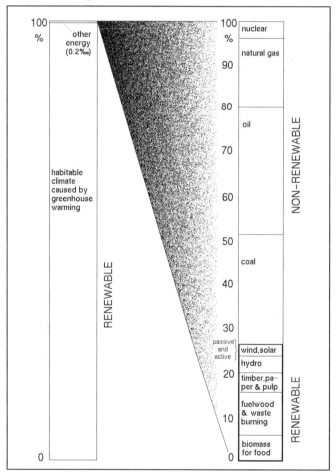

Figure 1.1. Renewable energy in the global energy system (Sørensen, 1992c).

From a business point of view, the total renewable energy flows, including free environmental heat, are, of course, not quite as interesting as the energy that can be traded in a market. Current renewable energy markets comprise both consumer markets and markets driven by government demonstration programmes and market-stimulating subsidy schemes. The reason for the initial support is partly industrial policy, aimed at getting new industry areas started, and partly a question of compensation for market distortions created by the fact that conventional energy industries are not fully paying for the negative environmental impacts caused by their products. This is a complex issue, partly because of the difficulty in exact determination of external costs and partly because most countries already levy taxation on energy products that may in part be contributing towards paying for the environmental damage, but often is just a government revenue not specifically used to offset the negative effects associated with using fossil or nuclear fuels (read more about these issues in Chapter 7).

The current penetration of active uses of renewable energy in national energy systems is growing, and Figures 1.2-1.14 show the values for the year 2000, which may serve as a reference year for assessing newer data. In cases where the growth rate is particular high, its annual value is mentioned in the caption to the figure showing the national distribution of markets.

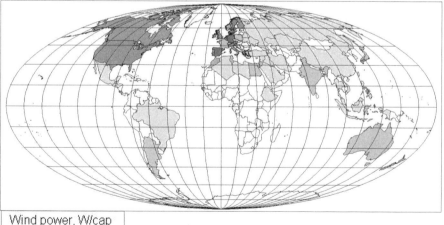

Wind power, W/cap

100	to	200
20	to	50
5	to	10
2	to	5
1	to	2
0.5	to	1
0.2	to	0.5
0.1	to	0.2
0.001	to	0.1

Figure 1.2. Wind power production. National average values for the year 2000 are shown, based upon BTM (2001) and an average capacity factor of 0.3. The world average for the year 2000 is 0.92 W/cap. The growth in cumulated installed capacity from 2000 to 2001 was 35% (BTM, 2002). Some observers expect the growth to slow during the following years, for economic and political reasons, but then to resume growth (Windpower Monthly, 2003).

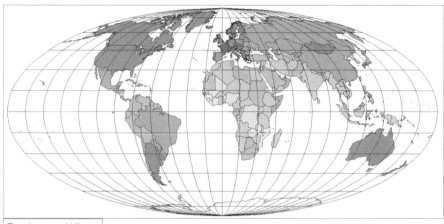

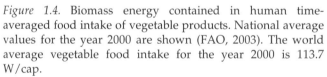

Food energy W/cap
Animal products

■ 65 to 75
■ 55 to 65
■ 45 to 55
■ 35 to 45
■ 25 to 35
■ 15 to 25
  5 to 15
  0.1 to 5

Figure 1.3. Biomass energy contained in human time-averaged food intake of animal products. National average values for the year 2000 are shown (FAO, 2003). The world average animal-based food intake for the year 2000 is 22.2 W/cap.

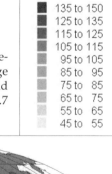

Food energy W/cap
Vegetable products

■ 135 to 150
■ 125 to 135
■ 115 to 125
■ 105 to 115
■ 95 to 105
■ 85 to 95
■ 75 to 85
  65 to 75
  55 to 65
  45 to 55

Figure 1.4. Biomass energy contained in human time-averaged food intake of vegetable products. National average values for the year 2000 are shown (FAO, 2003). The world average vegetable food intake for the year 2000 is 113.7 W/cap.

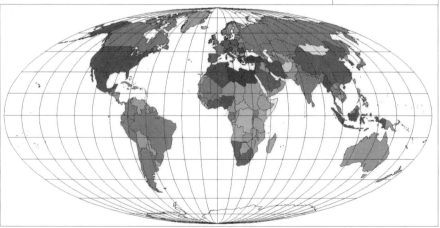

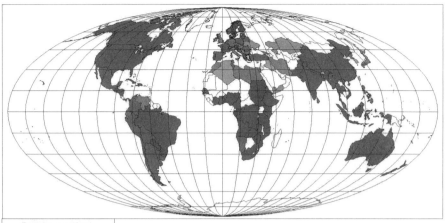

Solid woodfuel
W/cap

1000	to	2000
500	to	1000
200	to	500
100	to	200
50	to	100
20	to	50
10	to	20
5	to	10
2	to	5
1	to	2
0.5 to		1

Figure 1.5. Biomass energy contained in woodfuel. National average values for the year 2000 are shown (OECD/IEA, 2002a, 2002b). The world average woodfuel use in the year 2000, implied by the data, is 221.9 W/cap. No woodfuel use is reported for Russia. Some other countries are not covered by the data source.

Biomass waste
W/cap

100 to 200
50 to 100
20 to 50
10 to 20
5 to 10
2 to 5

Figure 1.6. Energy in biomass waste (refuse) utilised for power or heat production. National average values for the year 2000 are shown, based upon OECD/IEA (2002a, 2002b). The world average for the year 2000 is 3.7 W/cap.

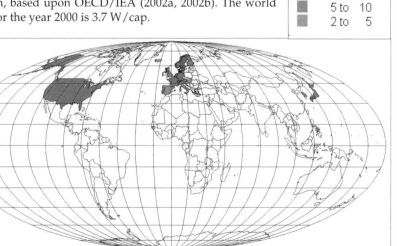

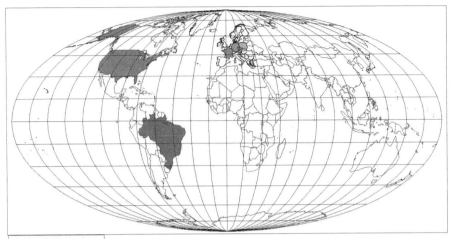

Liquid biofuels W/cap		
■ 50	to	100
■ 10	to	20
■ 5	to	10
■ 2	to	5
░ 0.5 to		1

Figure 1.7. Energy in liquid biofuels (presently ethanol and biodiesel). National average values for the year 2000 are shown, based upon EC-ATLAS (2003) and OECD/IEA (2002a, 2002b). The world average for the year 2000 is 2.3 W/cap.

Figure 1.8. Energy in biogas. National average values for the year 2000 are shown, based upon converting the number of units (FAO-Asia, 2003), assuming on average for each family unit a biogas production of half that of an optimally working unit fed manure from the equivalent of 2.5 cows and producing 1736 W of biogas. Additional data are from OECD/IEA (2002a, 2002b). The world average for the year 2000 is 2.8 W/cap.

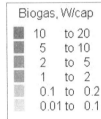

Biogas, W/cap	
■ 10	to 20
■ 5	to 10
■ 2	to 5
■ 1	to 2
░ 0.1	to 0.2
░ 0.01 to	0.1

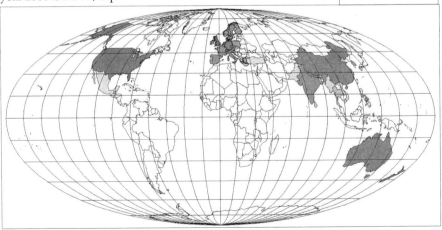

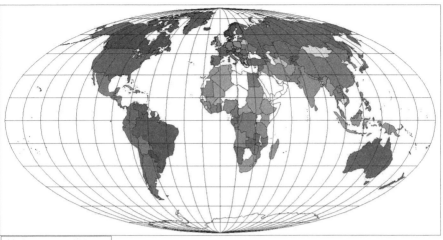

Hydro power, W/cap

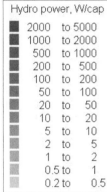

2000	to 5000
1000	to 2000
500	to 1000
200	to 500
100	to 200
50	to 100
20	to 50
10	to 20
5	to 10
2	to 5
1	to 2
0.5 to	1
0.2 to	0.5

Figure 1.9. Hydropower. National average values for the year 2000 are shown, based upon OECD/IEA (2002a, 2002b) and installed power data from Aqua-Media (1997) for countries not covered by IEA, using estimated capacity factors of 0.2 to 0.4. The world average for the year 2000 is 50 W/cap.

Figure 1.10. Tidal power. National average values for the year 2000 are shown, based upon OECD /IEA (2002a, 2002b). The world average for the year 2000 is 0.01 W/cap.

Tidal power, W/cap

1	to 2
0.1 to 0.2

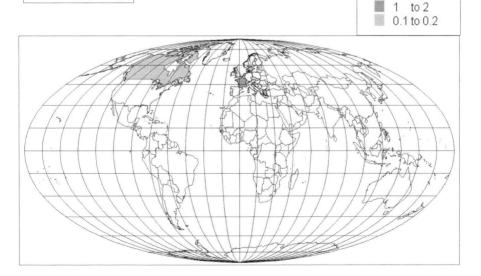

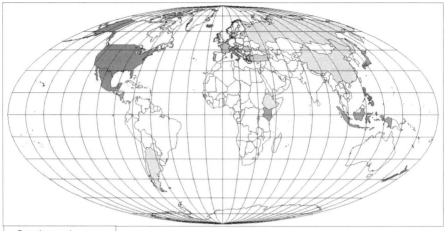

Geothermal power
W/cap

500	to 1000
50	to 100
20	to 50
10	to 20
5	to 10
2	to 5
1	to 2
0.5	to 1
0.2	to 0.5
0.1	to 0.2
0.01	to 0.1

Figure 1.11. Geothermal power. National average values for the year 2000 are shown, based upon either 10% of heat input given in OECD/IEA (2002a, 2002b) or 60% of installed power from Barbier (1999). The world average for the year 2000 is 9.3 W/cap.

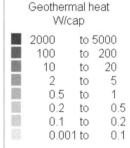

Geothermal heat
W/cap

2000	to 5000
100	to 200
10	to 20
2	to 5
0.5	to 1
0.2	to 0.5
0.1	to 0.2
0.001	to 0.1

Figure 1.12. Geothermal heat (mainly district heating). National average values for the year 2000 are shown, based upon OECD/IEA (2002a, 2002b). The world average for the year 2000 is 0.5 W/cap.

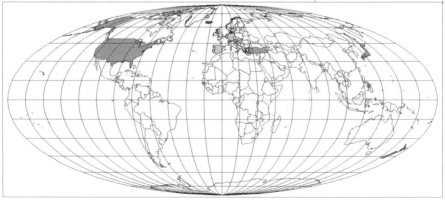

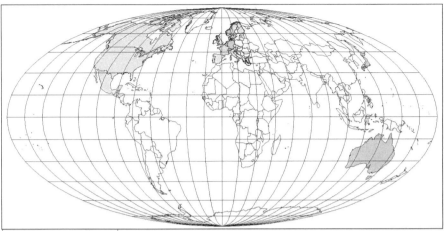

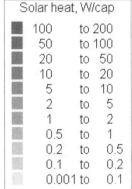

Figure 1.13. Solar power. National average values for the year 2000 are shown, based upon OECD/IEA (2002b) or an average energy production equal to 10% of the installed capacity given in IEA-PVPS (2001). The world average for the year 2000 is 0.007 W/cap. The growth rate from 2000 to 2001 was 35% (IEA-PVPS, 2002).

Figure 1.14. Solar heat. National average values for the year 2000 are shown, based upon IEA (2002). Both building-integrated and central district heating thermal systems are included. The world average for the year 2000 is 0.11 W/cap.

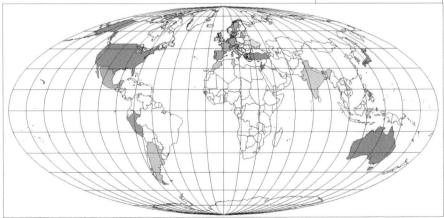

COUNTRY W/cap. or number	2000 Pop./1000	HYDRO	GEOTH POWR	GEOTH .HEAT	PV	SOLAR HEAT	TIDAL	WIND	BIO-RES	BIO SOLID	BIO LIQ	BIO-GAS	ANIM. FOOD	VEG. FOOD
Afghanistan	17270	3.6	0	0	0	0	0	0	0	0	0	0	11.67	62.86
Albania	3400	164.13	0	0	0	0	0	0	0	23.45	0	0	37.68	101.02
Algeria	30400	1.8	0	0	0	0	0	0.04	0	3.5	0	0	14.77	127.85
Andorra	70	0	0	0	0	0	0	0	0	0	0	0	44.41	121.07
Angola	13100	8.11	0	0	0	0	0	0	0	572.03	0	0	7.65	84.5
Anguilla	10	0	0	0	0	0	0	0	0	0	0	0	12	65
Antarctica	0	0	0	0	0	0	0	0	0	0	0	0	0	0
Antigua Barb.	70	0	0	0	0	0	0	0	0	0	0	0	38.69	77.34
Argentina	37000	89.05	0.01	0	0	0.22	0	0.13	0	105.93	0	0	48.43	105.62
Armenia	3800	38.46	0	0	0	0	0	0	0	0	0	0	15.01	79.08
Aruba	70	0	0	0	0	0	0	0	0	0	0	0	12	96.85
Australia	19160	100.06	0	0	0.15	6.8	0	0.47	0	349.36	0	9.57	50.8	102.95
Austria	8110	591.74	0.15	0.82	0.05	7.7	0	2.6	9.34	453.31	2.13	4.75	59.23	122.71
Azerbaijan	8000	21.59	0	0	0	0	0	0.05	0	0	0	0	17.53	101.99
Azores Port.	260	0	0	0	0	0	0	0	0	0	0	0	48.43	121.07
Bahamas	320	0	0	0	0	0	0	0	0	0	0	0	34.43	83.87
Bahrain	700	0	0	0	0	0	0	0	0	0	0	0	38.74	121.07
Bangladesh	131100	0.81	0	0	0	0	0	0	0	77.02	0	0	3.24	98.55
Barbados	260	0	0	0	0	0	0	0	0	0	0	0	33.95	112.4
Belarus	10000	0.2	0	0	0	0	0	0.04	0	131.54	0	0	38.79	101.74
Belgium	10250	5.06	0.01	0.13	0	0.26	0	0.6	15.94	33.96	0	3.89	54.29	124.94
Belize	250	0	0	0	0	0	0	0	0	0	0	0	29.06	110.8
Benin	6300	0	0	0	0	0	0	0	0	375.39	0	0	4.84	119.03
Bermuda	70	0	0	0	0	0	0	0	0	0	0	0	37	110.07
Bhutan	1840	64	0	0	0	0	0	0	0	0	0	0	4.84	96.85
Bolivia	8300	27.21	0	0	0	0	0	0	0	115.26	0	0	17.19	90.27
Bosnia Herzeg	4000	146.15	0.33	0	0	0	0	0	0	59.79	0	0	17.53	111.33
Botswana	1480	0	0	0	0	0	0	0	0	0	0	0	18.11	91.09
Brazil	170400	204.29	0	0	0	0	0	0.04	0	324.52	52	0	29.78	114.77
Br. Virgin Isl.	30	0	0	0	0	0	0	0	0	0	0	0	24.21	96.85
Brunei Daruss.	300	0	0	0	0	0	0	0	0	88.58	0	0	24.65	112.49
Bulgaria	8200	37.27	0	0	0	0	0	0	0	93.98	0	0	33.41	86
Burkina Faso	11100	0.7	0	0	0	0	0	0	0	0	0	0	5.62	105.42
Burundi	6030	0	0	0	0	0	0	0	0	0	0	0	1.74	75.98
Cambodia	6820	0.03	0	0	0	0	0	0	0	0	0	0	8.81	91.43
Cameroon	14900	26.75	0	0	0	0	0	0	0	444.07	0	0	6.44	102.76
Canada	30750	1331.4	0	0	0.02	0	0.13	1.36	0	484.27	0	0	45.42	108.28
Cape Verde	440	0	0	0	0	0	0	0.1	0	0	0	0	22.37	136.37
Cayman Isl.	30	0	0	0	0	0	0	0	0	0	0	0	24.21	96.85
Cen. African R	2790	1.7	0	0	0	0	0	0	0	0	0	0	9.49	84.7
Chad	6670	0	0	0	0	0	0	0	0	0	0	0	6.92	92.15
Chile	15200	143.35	0	0	0	0	0	0	0	369.75	0	0	30.22	109.35
China	1262500	20.13	0.02	0	0	0	0	0.08	0	225.66	0	5	28.23	118.45
Colombia	42300	86.69	0	0	0	0	0	0	0	165.22	0	0	20.63	105.18
Comoros	3800	171.33	29.37	0	0	5.59	0	0	0	87.41	0	0	4.46	80.44
Congo	3000	13.29	0	0	0	0	0	0	0	261.3	0	0	6.39	101.26
Cook Islands	20	0	0	0	0	0	0	0	0	0	0	0	24.21	96.85
Costa Rica	2500	74	0	0	0	0	0	6.1	0	0	0	0	25.18	109.64
Croatia	4400	154	0.3	0	0	0	0	0	0	111.73	0	0	24.94	95.3
Cuba	11200	1.19	0	0	0	0	0	0	0	347.58	0	0	17	107.17
Cyprus	800	0.17	0	0	0	59.79	0	0	0	16.61	0	0	46.15	111.62
Czech R	10270	19.54	0	0	0	0	0	0.05	0	41.14	5.95	4.66	40.53	109.78
Denmark	5340	0.5	0.02	0	0.03	1.99	0	131.5	145.55	220.94	8.46	17.17	63.24	101.21
Djibouti	70	0	0	0	0	0	0	0	0	0	0	0	12.83	86.44
Dominica	80	21	0	0	0	0	0	0	0	0	0	0	33.41	111.53
Dominican R	8400	11.07	0	0	0	0	0	0	0	215.11	0	0	16.46	96.17
Congo/Zaire	50900	12.27	0	0	0	0	0	0	0	355.26	0	0	2.28	71.04
Ecuador	12600	68.54	0	0	0	0	0	0	0	73.81	0	0	20.97	109.44
Egypt	64000	25.33	0	0	0	0	0	0.32	0	27.61	0	0	12.4	149.64
El Salvador	6300	21.09	14.34	0	0	0	0	0	0	291.04	0	0	14.67	106.54
Equat. Guinea	430	0	0	0	0	0	0	0	0	0	0	0	4.84	96.85
Eritrea	4100	0	0	0	0	0	0	0	0	165.27	0	0	4.99	75.64
Estonia	1400	0	0	0	0	0	0	0	0	474.51	0	0	42.52	120.97
Ethiopia	64300	2.89	0.04	0	0	0	0	0.03	0	359.95	0	0	5.08	92.88
Falkland Isl.	0	0	0	0	0	0	0	0	0	0	0	0	24.21	96.85
Fiji	730	23	0	0	0	0	0	0	0	0	0	0	26.97	111.57
Finland	5180	323.44	0	0	0.05	0	0	2.3	11.54	1659.77	0	4.62	55.79	100.48
France	60430	126.58	0.27	0	0.02	0.55	1.08	0.3	39.82	202.03	5.78	3.83	65.18	108.72
Fr. Guiana	130	0	0	0	0	0	0	0	0	0	0	0	24.21	96.85
Fr. Polynesia	210	0	0	0	0	0	0	0	0	0	0	0	40	98.16
Gabon	1200	66.43	0	0	0	0	0	0	0	1018.62	0	0	17.14	107.02
Gambia	690	0	0	0	0	0	0	0	0	0	0	0	5.67	114.14

Georgia	5000	132.86	0	0	0	0	0	0	0	18.6	0	0	19.23	97.63
Germany	82170	30.22	0	0.16	0.14	1.47	0	22.3	28.47	79.49	2.64	9.01	50.12	116.95
Ghana	19300	39.24	0	0	0	0	0	0	0	366.24	0	0	5.81	124.84
Gibraltar	30	0	0	0	0	0	0	0	0	0	0	0	4.84	121.07
Greece	10560	40.01	0	0.25	0	12.46	0	7.8	0	118.77	0	0.13	41.02	138.4
Greenland	50	200	0	0	0	0	0	0	0	0	0	0	48.43	96.85
Grenada	90	0	0	0	0	0	0	0	0	0	0	0	32.64	101.21
Guadeloupe	440	0	0	0	0	0	0	0	0	0	0	0	24.21	96.85
Guam	120	0	0	0	0	0	0	0	0	0	0	0	14.53	96.85
Guatemala	11400	23.31	0	0	0	0	0	0	0	454.54	0	0	10.07	95.06
Guinea	4050	6	0	0	0	0	0	0	0	0	0	0	3.87	110.07
Guinea Bissau	920	0	0	0	0	0	0	0	0	0	0	0	7.85	105.18
Guyana	870	0	0	0	0	0	0	0	0	0	0	0	20.29	104.75
Haiti	8000	3.32	0	0	0	0	0	0	0	252.44	0	0	6.44	93.12
Honduras	6400	39.44	0	0	0	0	0	0	0	276.11	0	0	16.66	99.32
HongKong	6800	0	0	0	0	0	0	0	0	9.77	0	0	38.74	96.85
Hungary	10020	1.99	0	0.66	0	0	0	0	2.65	47.21	0	0	53.9	113.56
Iceland	280	2595.6	623.51	2984.7	0	0	0	0	4.75	0	0	0	67.65	94.19
India	1015900	8.37	0	0	0	0.17	0	0.4	0	263.64	0	2.3	9.39	108.18
Indonesia	210400	4.93	1.44	0	0	0	0	0.02	0	299.95	0	0	5.67	134.87
Iran IR	63700	6.67	0	0	0	0	0	0.06	0	16.48	0	0	13.03	128.04
Iraq	23300	2.85	0	0	0	0	0	0.08	0	1.71	0	0	4.26	102.13
Ireland	3790	25.59	0	0	0	0	0	9.7	0	47.68	0	9.82	54.38	120.58
Israel	6200	0.2	0	0	0.007	127.72	0	0.08	0	0	0	0	31.96	140.53
Italy	57730	87.5	6.7	0	0.03	0.25	0	2.2	7.69	36.71	0	2.97	45.28	132.01
Ivory Coast	16000	12.46	0	0	0	0	0	0	0	350.43	0	0	4.75	120.68
Jamaica	2600	5.11	0	0	0	0	0	0	0	245.29	0	0	18.98	111.43
Japan	126920	78.55	3.01	2.33	0.25	8.41	0	0.34	10.56	47.64	0	0	27.55	106.2
Jordan	4900	0.3	0	0	0	17.62	0	0.08	0	0	0	0	15.79	117.34
Kazakhstan	14900	57.96	0	0	0	0	0	0.04	0	6.24	0	0	31.23	113.61
Kenya	30100	4.86	1.63	0	0	0	0	0	0	519.98	0	0	11.33	83.78
Kiribati	60	0	0	0	0	0	0	0	0	0	0	0	18.84	124.36
Korea	47280	9.7	0	0	0.008	1.18	0	0.06	41.84	4.83	0	1.1	22.47	127.31
Korea DPR	22300	109.03	0	0	0	0	0	0	0	59.58	0	0	5.96	99.81
Kuwait	2000	0	0	0	0	0	0	0	0	0	0	0	34.53	117.14
Kyrgyzstan	4900	319.96	0	0	0	0	0	0.02	0	0	0	0	26.73	112.3
Laos	2960	13	0	0	0	0	0	0	0	0	0	0	7.7	102.08
Latvia	2400	132.86	0	0	0	0	0	0	0	548.06	0	0	33.32	104.94
Lebanon	4300	12.36	0	0	0	2.16	0	0	0	40.17	0	0	19.56	133.22
Lesotho	1980	0.4	0	0	0	0	0	0	0	0	0	0	4.84	106.54
Liberia	2510	7	0	0	0	0	0	0	0	0	0	0	3.24	97.29
Libya ArabJam	5300	0	0	0	0	0	0	0.04	0	35.1	0	0	17.82	142.23
Liechtenstein	30	0	0	0	0	0	0	0	0	0	0	0	48.43	106.59
Lithuania	3700	10.77	0	0	0	3.59	0	0	0	226.23	0	0	34.09	113.12
Luxembourg	440	30.2	0	0	0	0	0	4.09	81.53	48.31	0	3.02	54.29	124.94
Macedonia FY	2000	66.43	1.99	0	0	0	0	0	0	139.51	0	0	24.21	121.31
Madagascar	8460	5	0	0	0	0	0	0	0	0	0	0	9.59	87.65
Malawi	9860	0	0	0	0	0	0	0	0	0	0	0	2.37	103.2
Malaysia	23300	34.21	0	0	0	0	0	0	0	144.27	0	0	27.41	113.95
Maldives	270	0	0	0	0	0	0	0	0	0	0	0	31.82	93.7
Mali	8070	1.1	0	0	0	0	0	0	0	0	0	0	4.84	106.25
Malta	400	0	0	0	0	0	0	0	0	0	0	0	44.21	127.36
Marshall Isl.	50	0	0	0	0	0	0	0	0	0	0	0	24.21	96.85
Martinique	450	0	0	0	0	0	0	0	0	0	0	0	24.21	96.85
Mauritania	2250	5	0	0	0	0	0	0	0	0	0	0	15.84	107.02
Mauritius	1220	11	0	0	0	0	0	0	0	0	0	0	20.68	123.87
Mexico	97220	38.94	6.94	0	0.014	0.59	0	0.02	0	109.81	0	0.08	28.23	125.04
Micronesia	0	0	0	0	0	0	0	0	0	0	0	0	24.21	96.85
Moldova Rep	4300	3.09	0	0	0	0	0	0	0	18.54	0	0	18.98	114.87
Monaco	30	0	0	0	0	0	0	0	0	0	0	0	45.28	108.72
Mongolia	2310	0.08	0	0	0	0	0	0	0	0	0	0	45.67	50.27
Morocco	28700	2.78	0	0	0	0.28	0	0.56	0	20.37	0	0	10.36	133.17
Mozambique	17700	45.04	0	0	0	0	0	0	0	496.18	0	0	2.32	90.99
Myanmar	47700	4.46	0	0	0	0	0	0	0	255.7	0	0.03	6	131.62
Namibia	1800	88.58	0	0	0	0	0	0	0	125.48	0	0	12.88	115.4
Nauru	10	0	0	0	0	0	0	0	0	0	0	0	19.37	96.85
Nepal	23000	8.09	0	0	0	0	0	0	0	388.77	0	3.7	7.75	110.22
Netherlands	15920	1	0	0	0.08	1.34	0	8.9	41.81	28.38	0	11.02	57.38	102.18
New Caledonia	170	94	0	0	0	0	0	0	0	0	0	0	31.72	101.02
New Zealand	3830	734.39	82.56	164.43	0	0	0	2.74	0	286.89	0	10.41	52.59	104.89
Nicaragua	5100	5.21	3.13	0	0	0	0	0	0	369.94	0	0	8.77	99.08
Niger	8650	0	0	0	0	0	0	0	0	0	0	0	5.42	95.74
Nigeria	126900	5.23	0	0	0	0	0	0	0	757.29	0	0	4.21	133.8
Niue	0	0	0	0	0	0	0	0	0	0	0	0	14.53	96.85

COUNTRY	POPULAT	HYDRO	GEO-P	GEO-H	PV	SOL-H	TIDAL	WIND	WASTE	B-SOLID	B-LIQ	B-GAS	ANIM.	VEG.
N Mariana Isl.	30	0	0	0	0	0	0	0	0	0	0	0	14.53	96.85
Norway	4490	3603.3	0	0	0.13	0	0	0.87	36.69	354.8	0	1.48	56.13	109.15
Oman	2400	0	0	0	0	0	0	0	0	0	0	0	33.9	96.85
Pakistan	138100	14.24	0	0	0	0	0	0	0	231.09	0	0.01	20.77	97.97
Palau Islands	20	0	0	0	0	0	0	0	0	0	0	0	19.37	96.85
Panama	2900	123.7	0	0	0	0	0	0	0	210.75	0	0	28.09	92.4
Papua N Guin.	4420	8	0	0	0	0	0	0	0	0	0	0	10.12	95.25
Paraguay	5500	1111.2	0	0	0	0	0	0	0	553.2	0	0	28.09	94.58
Peru	25700	71.86	0	0	0	2.74	0	0	0	115.29	0	0	16.8	110.31
Philippines	75600	11.78	17.57	0	0	0	0	0.02	0	167.66	0	0.01	17.14	98.06
Poland	38650	6.22	0	0	0	0	0	0.1	0	123.31	0.7	1.03	43.1	120.39
Portugal	10010	129.28	0.92	0.13	0.009	2.39	0	3.3	23.1	249.14	0	0.13	51.19	128.72
Puerto Rico	4940	4.6	0	0	0	0	0	0	0	0	0	0	48.43	121.07
Qatar	600	0	0	0	0	0	0	0	0	0	0	0	38.74	96.85
Reunion	650	38	0	0	0	0	0	0	0	0	0	0	29.06	96.85
Romania	22400	75.33	0.12	0	0	0	0	0	0	169.05	0	0	32.74	125.81
Russia Fed	185500	108	0.04	0	0	0	0	0.02	0	0	0	0	31.38	109.88
Rwanda	145600	128.76	0.05	0	0	0	0	0	0	63.97	0	0	2.47	98.11
Saint Lucia	150	0	0	0	0	0	0	0	0	0	0	0	32.11	105.33
San Marino	30	0	0	0	0	0	0	0	0	0	0	0	45.04	132.01
S Tome&Princ.	150	0	0	0	0	0	0	0	0	0	0	0	4.41	111.33
Saudi Arabia	20700	0	0	0	0	0	0	0.04	0	0	0	0	21.6	117.63
Senegal	9500	0	0	0	0	0	0	0	0	240.55	0	0	9.64	99.66
Seychelles	80	0	0	0	0	0	0	0	0	0	0	0	23.97	93.8
Sierra Leone	3480	0.2	0	0	0	0	0	0	0	0	0	0	3.15	87.12
Singapore	4000	0	0	0	0	21.26	0	0	0	0	0	0	29.06	96.85
Slovak R	5400	99.89	0	0	0	0	0	0.04	0	19.68	0	0	38.16	113.56
Slovenia	2000	219.23	0.66	0	0	0	0	0	0	305.59	0	0	45.08	108.33
Solomon IsL.	350	0	0	0	0	0	0	0	0	0	0	0	10.27	100
Somalia	10510	0	0	0	0	0	0	0	0	0	0	0	29.88	48.96
South Africa	42800	3.73	0	0	0	0	0	0	0	393.32	0	0	17.92	121.84
Spain	39930	81.19	0	0.2	0.02	1.03	0	21.3	7.75	130.47	0	3.79	44.41	117.92
Sri Lanka	19400	18.49	0	0	0	0	0	0	0	291.75	0	0.1	7.51	108.91
St Kitts&Nevis	50	0	0	0	0	0	0	0	0	0	0	0	36.17	93.85
St Vincent&Gr.	120	0	0	0	0	0	0	0	0	0	0	0	22.23	102.66
Sudan	31100	4.27	0	0	0	0	0	0	0	602.37	0	0	22.08	91.62
Suriname	420	166	0	0	0	0	0	0	0	0	0	0	17.68	110.75
Swaziland	700	12	0	0	0	0	0	0	0	0	0	0	18.89	107.99
Sweden	8870	1017.1	0	0	0.03	0.75	0	8.97	60.22	1155.48	0	4.19	49.83	100.78
Switzerland	7190	585.42	0	16.82	0.21	4.25	0	0.12	140.26	92.76	0	11.64	52.88	106.59
Syria Arab Rep	16200	65.61	0	0	0	0	0	0.08	0	0	0	0	19.85	127.26
Taiwan Teipei	22200	45.49	0	0	0	0	0	0.02	0	0.6	0	0	29.06	96.85
Tajikistan	6200	255.01	0	0	0	0	0	0	0	0	0	0	7.17	76.13
Tanzania UR	33700	7.49	0	0	0	0	0	0	0	567.73	0	0	5.91	86.39
Thailand	60700	11.38	0	0	0	0	0	0.04	0	312.13	0	0.15	13.75	107.6
Togo	4500	7	0	0	0	0	0	0	0	307.06	0	0	3.87	108.91
Tonga	100	0	0	0	0	0	0	0	0	0	0	0	29.06	96.85
Trinidad&Tob.	1300	0	0	0	0	0	0	0	0	30.66	0	0	21.16	113.32
Tunisia	9600	1.38	0	0	0	0.28	0	0.34	0	171.62	0	0	16.8	142.95
Turkey	66840	52.8	0.13	3.56	0	5.21	0	0.09	0	128.31	0	0.1	18.21	147.22
Turkmenistan	5200	0.4	0	0	0	0	0	0	0	0	0	0	21.94	107.6
Turks&Caicos I	10	0	0	0	0	0	0	0	0	0	0	0	29.06	96.85
Tuvalu	10	0	0	0	0	0	0	0	0	0	0	0	29.06	96.85
Uganda	16670	0	0	0	0	0	0	0	0	0	0	0	6.83	107.41
Ukraine	48500	26.85	0	0	0	0.03	0	0.04	0	7.12	0	0	28.33	110.7
U. Arab Emir.	2900	0	0	0	0	0	0	0	0	0	0	0	38.16	116.47
U. Kingdom	59760	9.76	0	0.02	0.003	0.24	0	2.13	6.23		0	17.88	48.52	112.93
United States	275420	103.04	6.09	2.5	0.05	7.32	0	2.84	34.29	290.69	15.55	15.91	50.51	92.16
Uruguay	3300	245.6	0	0	0	0	0	0	0	169.1	0	0	46.88	92.49
US Virgin Isl.	100	0	0	0	0	0	0	0	0	0	0	0	43.58	121.07
Uzbekistan	24800	27.32	0	0	0	0	0	0	0	0	0	0	20.97	93.85
Vanuatu	150	0	0	0	0	0	0	0	0	0	0	0	23.15	102.13
Vatican City	0	0	0	0	0	0	0	0	0	0	0	0	45.04	132.01
Venezuela	24200	296.47	0	0	0	0	0	0	0	29.65	0	0	17.19	92.06
Vietnam	78500	21.16	0	0	0	0	0	0.02	0	383.02	0	0.02	13.17	111.91
W Sahara	90	0	0	0	0	0	0	0	0	0	0	0	15.84	96.85
W Samoa	190	0	0	0	0	0	0	0	0	0	0	0	29.06	96.85
Yemen	17500	0	0	0	0	0	0	0	0	6.07	0	0	6.59	92.11
Serbia Monten	10600	130.36	0	0	0	0	0	0	0	31.34	0	0	48.23	76.22
Zambia	10100	88.14	0	0	0	0	0	0	0	674.84	0	0	4.5	88.09
Zimbabwe	12600	29.53	0	0	0	0	0	0	0	589.45	0	0	6.83	95.64

Table 1.1. Year 2000 renewable energy use (W/cap.), cf. Figs. 2.2–2.14.

As Table 1.1 shows, at a global average of 222 W/cap., the traditional use of biomass for combustion is still the dominating use of renewable energy, although it takes more efficient forms in many industrialised countries. Only slightly less (146 W/cap.) is the use of food energy in biomass of animal or vegetable origin (the nutrient value of food being, in any case, more than that of the energy it provides). Next comes hydropower (50 W/cap.) and then geothermal power, which only in part can be classified as renewable (as many steam reservoirs are exploited at a rate that will exhaust the reservoir over periods of decades). At the level of 1 W/cap., i.e. two orders of magnitude under the energy in food intake, one finds biomass waste (used for power or heat), biogas, liquid biofuels (used in the transportation sector), wind power and geothermal heat (used for district heating). At the bottom comes solar heat, tidal power and solar power, with the latter below 0.01 W/cap. However, the fastest growing markets are those of wind and solar power, with both currently adding 35% of installed power each year.

The market characteristics of the various renewable energy forms exhibit differences linked to the nature of each source. For food energy, the price is influenced by variations in production due to climatic variations, the choices made in regard to area use, livestock holdings, fish quotas and the competitive behaviour of the food processing and marketing industry. Yet the bulk prices of different commodities seem remarkably consistent with their energy content, varying only between some 70 US or euro-cents per kWh (heat value) and 200 c/kWh. Translating OECD data (OECD, 2002) to energy units, the current wholesale price of cereals such as rice or wheat are around 70 c/kWh, while the wholesale price of typical meat and diary products are about 100 c/kWh. Only specialised gourmet products obtain higher prices in the marketplace. Consumer retail prices are typically five times higher than the bulk prices just quoted. This is more than 30 times the current consumer price of a kWh of electricity produced from fossil fuels.

Wholesale market prices for biomass waste and fuelwood range from about 1 c/kWh (of "burning value", i.e. energy of combustion) in India (FAO-Asia, 2003) to 2 c/kWh in industrialised countries (e.g. straw, wood chips 1.6 c/kWh and wood pellets 1.9 c/kWh; Danish Energy Agency, 1996; Alakangas *et al.*, 2002). For comparison, the cost of coal before considering externalities is 0.5 c/kWh (Danish Energy Agency, 2002). The production cost of biogas is 3.6–7 c/kWh (Danish Energy Agency, 1992), while that of wind power is 3–7 c/kWh (depending on wind conditions) and that of photovoltaic solar power is 40–130 c/kWh (IEA-PVPS, 2002). The photovoltaic market enjoys substantial public start-up subsidies (often in the form of subsidising customer investments or offering attractive buy-back rates for excess solar power). This is the case in countries such as Germany and Japan, while in Switzerland, the market has largely been created by industries buying photovoltaic panels for reasons of aesthetics or image greening.

Hydropower costs 1–5 c/kWh, while coal- and gas-based power costs about 5 c/kWh to produce (Danish Energy Agency, 2002). To this comes distribution costs from centralised production units to the customers and, in many countries, taxes and environmental externality payments, leading to customer prices in excess of 14 c/kWh. As a result, wind power, being exempt from pollution and CO_2 fees, and biomass-based power are in many countries sold at prices very similar to that of fossil-based power. Also geothermal power is usually competitive with other forms of electricity, while the viability of geothermal heat depends on local costs of district heating distribution.

Current oil production costs vary from well under 1 c/kWh at some Middle East wells to near 2 c/kWh from off-shore facilities in the North Sea. The bulk sales price (currently – February 2003 – around 2 c/kWh) is not strongly coupled to production prices, but is determined by market and political considerations. Some countries are willing to wage war against oil-producing countries in order to control prices. Refined products such as gasoline are currently sold at prices around 4 c/kWh, with diesel fuel slightly lower, plus taxes and environmental fees where they apply (Danish Energy Agency, 2002; IEA 2002). Liquid biofuels have production costs of 3–7 c/kWh (ethanol from sugar cane the lowest, ethanol from sugar beet the highest, methanol from woody biomass at 4–5 c/kWh). Hydrogen from woody biomass is about 3 c/kWh (Turkenburg *et al.*, 2000). Natural gas market prices are currently 10% higher than those of oil (IEA, 2002).

Because of the cost of often advanced equipment, it is clear that prices of renewable energy only in particular cases can match those of fossil fuels, even with expected progress in technology and production facilities. The case for increasing the role of renewable energy sources is therefore linked to uncertainty of future fossil fuel prices (for political and resource depletion reasons) and increased awareness of the indirect costs of pollution caused by fossil and nuclear fuels, including in the fossil case emissions of substances contributing to excess greenhouse warming (cf. Chapter 7).

The source data for the Figs. 1.2–1.14 are given in tabular form in Table 1.1. It should be kept in mind that several of the numbers involve estimates and modelling, as direct energy production is not always monitored.

1.2 The energy scene – its history and present state

Taking now a scientific point of view, an issue more essential than the placement of renewable energy in the marketplace is its place within the physical universe. This view will be developed in the following, as a prerequisite for estimating the amounts of energy that can be extracted for use by the human society, at a rate that qualifies the process as renewable. Other views are the philosophical and the economic approaches, of which at least the latter will be taken up again in Chapter 7.

The speed of the Earth in its orbit around the Sun is about 3×10^4 m s^{-1}, corresponding to a kinetic energy of some 2.7×10^{33} J. The Earth further rotates around its axis with an angular velocity of about 7.3×10^{-5} rad s^{-1}, furnishing an additional kinetic energy of some 2.2×10^{29} J. The work required in order to pull the Earth infinitely far away from the Sun, against the gravitational attraction, is about 5.3×10^{33} J, and the corresponding work required to separate the Earth from its Moon is of the order of 8×10^{28} J.

These are some of the external conditions for our planet, spelt out in energy units. It is a little more difficult to obtain reliable estimates for the amount of energy residing within the Earth itself. The kinetic energy of molecular motion, i.e. heat energy, is of the order of 5×10^{30} J. This estimate represents the total heat energy, relative to the absolute zero temperature. It is extrapolated from the value 4×10^{30} J, given in section 3.5.2, for the heat energy in the interior of the Earth relative to the average surface temperature of 287 K.

The materials forming the Earth carry further energy, in addition to the heat energy corresponding to their temperature. About 10^{21} J is, on average, present as kinetic energy in the atmospheric and oceanic circulation (cf. section 2.4.1), and the potential energy of the continental height-relief, relative to sea-level, is about 2×10^{25} J, taking into account density variations in the crust (Goguel, 1976). Much larger amounts of energy are involved in the chemical and nuclear bindings, which determine the state and structure of matter. The carbon compounds of biological material provide an example of chemical energy. During earlier periods of the Earth's history, fossilisation of biological material created the deposits of coal, oil and natural gas, of which at least 10^{23} J is presently believed to be recoverable in a form suitable for fuel uses (see sections 2.4.1 and 4.8). Current standing crops of biomass correspond to an average of 1.5×10^{22} J (cf. sections 2.4.1 and 3.6).

Nuclear energy may be released in large quantities from nuclear reactions, such as fission of heavy nuclei or fusion of light nuclei. Except for spontaneously fissioning nuclear isotopes in the Earth's crust, which release about 4×10^{20} J y^{-1}, an initial amount of energy must be provided in order to get the energy-releasing fission or fusion processes going. Set-ups for explosive release of nuclear energy involving both types of processes are used for military purposes. As further discussed in section 3.7.3, only the fission process has as yet been demonstrated as a basis for controlled energy supply systems, and, with necessary additional improvements in the technology of fast breeder reactors, recoverable resources of nuclear fuels are estimated to be of the order of 10^{24} J. If fusion of deuterium nuclei to form helium nuclei could be made viable on the basis of deuterium present in sea water, this resource alone would amount to more than 10^{31} J (cf. section 3.7.3).

Energy conversion processes depleting certain materials of the Earth may be said to constitute irreversible processes. This is often true from a practical point of view, even if the reverse process may be theoretically possible.

The terms "energy use", "spending energy", etc., which are commonly used in energy literature as well as in everyday language, are of course imprecise expressions describing energy conversion processes. Such processes are in most cases associated with an increase in entropy. The entropy is a property of a system, which quantifies the "quality" of the energy contained in the system. The system may be, for example, an amount of fuel, a mass of air in motion, or the entire Earth–atmosphere system.

The entropy change for a process (e.g. an energy conversion process), which brings the system from a state 1 to a state 2, is defined by

$$\Delta S = \int_{T_1}^{T_2} T^{-1} dQ, \tag{1.1}$$

where the integral is over successive infinitesimal and reversible process steps (not necessarily related to the real process, which may not be reversible), during which an amount of heat dQ is transferred from a reservoir of temperature T to the system. The imagined reservoirs may not exist in the real process, but the initial and final states of the system must have well-defined temperatures T_1 and T_2 in order for (1.1) to be applicable.

Conversion of certain forms of energy, such as electrical or mechanical energy, among themselves may, in principle, not change entropy, but in practice some fraction of the energy always gets converted into heat. The energy processes characteristic of man's activities on Earth involve a series of successive conversion processes, usually ending with all the converted energy in the form of heat, radiated to space or released to the atmosphere, from which radiation of heat energy into space also takes place. The temperatures involved (the T_2) are typically 200–300 K. The processes involved will be discussed in more detail in Chapter 2.

Stored energy of any form, which may be converted to heat that is ultimately lost to space, could then be called a "non-renewable energy resource". The term a "renewable energy resource" is used for energy flows, which are replenished at the same rate as they are "used". The prime renewable energy resource is thus solar radiation intercepted by the Earth, because the Earth (i.e. the Earth–atmosphere system) re-radiates to space an amount of heat equal to the amount of solar radiation received (see Chapter 2). To utilise solar energy thus means converting it in a way convenient for man, but the net result is the same as if man had not interfered, i.e. ultimately to convert solar radiation into heat radiated to space. Such usage may involve a delay in returning the heat, either as a part of man's conversion scheme or by a natural process. For this reason energy stores, which are part of the natural process of converting solar energy into heat re-radiation, are also considered as "renewable energy resources".

"Renewable energy" is not narrowly defined here, and it may be taken to include the usage of any energy storage reservoir which is being "refilled" at rates comparable to that of extraction.

The amount of solar energy intercepted by the Earth and hence the amount of energy flowing in the "solar energy cycle" (from incident radiation flux via reflection, absorption and re-radiation to heat flux away from the Earth) is about 5.4×10^{24} J per year.

Energy fluxes of other than solar origin which occur naturally at the surface of the Earth are numerically much smaller. For example, the heat flux from the interior of the Earth through the surface is about 9.5×10^{20} J y^{-1} (cf. section 3.5.2), and the energy dissipated in connection with the slowing down of the Earth's rotation (due to tidal attraction by other masses in the solar system) is of the order of 10^{20} J y^{-1} (cf. section 2.3.2).

1.2.1 Man's energy history

The minimum energy requirement of man may be taken as the amount of "exchangeable" chemical energy that can be associated with the amount of food necessary to maintain life processes for someone performing a minimum of work and not losing weight. This minimum depends on the temperature of the surroundings, but for an adult man is generally considered to lie in the region of 60–90 W on average for extended periods, corresponding to $(6 \text{ to } 8) \times 10^6$ J day^{-1}. The total life requirements are, of course, more than energy, comprising adequate supplies of water, nutrients, etc.

In order to perform any (muscle) work not purely vegetative, additional energy must be supplied in the form of food, or energy stored in the body will become depleted. The efficiency in converting stored energy into work typically ranges from 5–50%, with the lower efficiencies being associated with activities involving large fractions of static conversion (e.g. carrying a weight, which requires the conversion of body energy even if the weight is not being moved). The percentage complementary to the efficiency is released as various forms of heat energy.

The maximum average rate of food energy intake that a human being can continue for extended periods is about 330 W, and the maximum average rate at which work can be delivered for extended periods is of the order of 100 W (Spitzer, 1954). During work periods, the "man-power" output level may be 300–400 W, and the maximum power which can be delivered by an adult male for a period of about a minute is roughly 2000 W.

Although it is not certain that the rates of energy conversion by the human body have remained constant during the evolution of man, it may be reasonable to assume that the average amount of "muscle power" used by the earliest members of the genus *Homo,* which evidence suggests lived some 4×10^6 years ago in Africa (Leakey, 1975), was of the order of 25 W.

The total energy flux received by an individual man in a food gathering or hunting society is then the sum of the energy in the food, averaging say 125 W, and the absorbed flux of radiation and heat from the surroundings, which may reach considerably larger values, but is highly dependent on clothing, climate and the nature of the surroundings (cf. e.g. Budyko, 1974). The outgoing energy flux again consists of heat and radiation fluxes, turnover of organic material, plus the amount of energy converted into work. For growing individuals, the net flux is positive and the mass of biological material increases, but also for adult individuals with zero net energy flux, new biomass continues to be produced to replace "respiration losses".

Man has successively developed new activities, which have allowed him to gain access to larger amounts of energy. Solar energy may have been used for drying purposes, and as soon as fires became available a number of activities based on firewood energy may have started, including heating, food preparation and process heat for tool making. The earliest evidence for fires used in connection with dwellings is from Hungary, 350 000–400 000 years ago (H. Becker, 1977, personal communication).

A good fire in open air, using some 10–50 kg of firewood per hour, may convert energy at a rate of 10^4–10^5 W, whereas indoor fires are likely to have been limited to about 10^3 W. Several persons would presumably share a fire, and it would probably not burn continuously at such a power level, but rather would be re-lit when required, e.g. from glowing embers. It is thus difficult to estimate the average fire energy per person, but it would hardly exceed 100 W in primitive societies. The efficiency of delivering energy for the desired task is quite low, in particular for open-air fires.

The next jump in energy utilisation is generally considered to have been associated with the taming of wild animals to form livestock and the introduction of agriculture. These revolutions have been dated to about 10^4 years ago for the Near East region (cf. e.g. DuRy, 1969), but may have been developed in other regions at about the same time, e.g. in Thailand and Peru (Pringle, 1998). This time corresponds to the ending of the last ice age (see Fig. 2.91), which may have caused changes in the low-latitude climate, including altered precipitation rates. The introduction of livestock would have promoted the tendency to settle at a given place (or vice versa), increasing in turn the requirement for food beyond the capacity of a hunting society. Agriculture was based at first on wild varieties of wheat, for example, and it is believed that artificial irrigation was necessary at many of the sites where evidence of agriculture (various tools) has been found. The power for water transport and, later, pumping would then be derived from suitable draught animals in the livestock pool, as a substitute for man's own muscle power. The transition from a hunting to an agricultural society, often called the *Neo-*

lithic or "new stone age", occurred several thousand years later in the temperate zones of northern America and Europe.

The creation of cultures of growing size and level of sophistication, leading to the formation of large cities, for example, at the rivers Euphrates, Tigris and the Nile, about 5000 years ago, witnessed a growing use of energy for ploughing, irrigation, grinding and transport (of food supplies and of materials, e.g. in connection with buildings and monuments), as well as the harvest of solar energy through agricultural crops. It is not known exactly how much of the physical work was performed by men and how much by animals, but it is likely that another 100–200 W was added to the average energy usage per capita in the most developed regions.

It is also important to bear in mind that there must have been large differences in energy use, both between different societies and between individuals within a given society. Throughout man's history (the meaning of "history" not being restricted to imply the presence of written records) there have been individuals whose access to energy was largely limited to that converted by their own bodies. Large regions in Asia and Africa today have an average energy spending per person, which is only a few hundred watts above the muscle power level (with firewood as an important source). This means that parts of the population today use no more energy than the average person during the Neolithic period.

The energy sources that have emerged so far are direct solar radiation, environmental heat, animal biomass, as well as primary (plant) biomass in the form of food and later as firewood, plus mechanical work from the muscle power of animals. In the Near East, oil was used for lighting, and bitumen had non-energy uses. Boat travel in the open sea (the Mediterranean) is believed to have started over 9000 years ago (Jacobsen, 1973), and there is evidence of wind energy utilisation by means of sails in Egypt about 4500 years ago (Digby, 1954). Per person, wind energy may not at this time have contributed a significant proportion of the total energy use in the Mediterranean region, but later, when trade became more developed (from about 4000 years ago), the total amount of energy spent on transportation on land and at sea constituted a less negligible share (maybe a few per cent) of the total amount of energy spent in the "developed regions" of the world at the time.

The building of houses in many cases implied the creation of a required indoor climate with utilisation of solar energy. In the low-latitude regions, structures of high heat capacities were employed in order to smooth out day-to-night temperature variations, and in many cases the houses were built partly underground, and the evaporation of soil moisture was utilised to create cool environments for living (during hot periods) and food storage (Bahadori, 1977). In regions with a colder climate, a number of insulating[*] building materials (e.g. roofs made of straw) were employed to reduce heat

---

[*] The term "insulating" is taken to include suppression of convective heat transfer.

losses, and heat production not involving fires was increased by keeping livestock within the living area of the houses, so as to benefit from their respirational heat release.

Water mills and windmills (e.g. the vertical axis panemone type probably derived from waterwheels, or the sail-wing type presumably copied from sail-ships) also played a role from a certain stage in development. The earliest mention of windmills in actual use is from India about 2400 years ago (Wulff, 1966). Considering its low efficiency and overall size, it is unlikely that wind power has at any time accounted for a large proportion of the average energy use. On the other hand, windmills and water mills offered the only alternative to muscle power for high-quality (i.e. low-entropy) mechanical energy, until the invention of the steam engine.

The industrial revolution 200–300 years ago was connected with placing at the disposal of man amounts of power capable of producing work far beyond his own muscle power. However, at that time firewood was barely a renewable resource in the developed regions of the world, despite quite extensive programmes to plant new forests to compensate for usage. The increase in energy usage made possible by the growing industrialisation did not really accelerate, therefore, before large amounts of coal became available as fuel. In the 20th century, the large growth in energy consumption was made possible by the availability of inexpensive fossil fuels: coal, natural gas and oil.

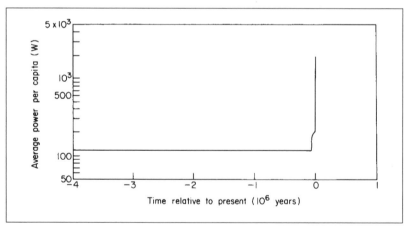

Figure 1.15. Trends in average rate of energy conversion per capita, not including fluxes associated with the local thermal environment.

An outline of the possible development in energy usage up to the present is presented in Figs. 1.15–1.17. Only over the past century or two have reliable world-wide data on energy usage been recorded, and even for this period the data comprise mainly direct use of commercial fuels, supplemented with incomplete information on biomass and other renewables. One reason for this is that it is more difficult to specify the remaining energy use, because e.g. solar collectors are often not individually monitored, local biomass use is not quantified in energy units, environmental heat gains vary from day to day, and so on. In Figs. 1.15–1.17, which are anyway only indicative, fuels are included in terms of their gross energy value, independently of end-use efficiency. The use of renewable energy flows, on the other hand, is given as an estimated net energy at the primary conversion stage, i.e. the energy in food intake rather than the total amount of energy absorbed by the plants or the total biomass of plants and animals. The environmental energy contribution to maintaining man's body temperature as well as the regulation of indoor climate by the choice of materials and building systems ("passive energy systems") are excluded.

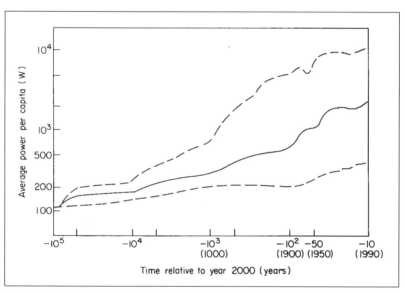

Figure 1.16. Trends in average rate (solid line) of energy conversion per capita, not including fluxes associated with the local thermal environment (same as Fig. 1.1, but on a logarithmic time scale). Dashed lines indicate the corresponding trends for the societies, which at a given time have the highest and the lowest average energy usage. For the more recent period, data from Darmstadter *et al.* (1971) and European Commission (1997) have been used, in a smoothed form.

Figure 1.15 shows the trend in average rate of energy conversion per capita, on a linear time scale, and Fig. 1.16 shows the same trend on a logarithmic time scale, extending backwards from the year 2000. Figure 1.16 also indicates the estimated spread in energy usage, with the upper curve representing the societies with highest energy use, at a given time, and the lower curve representing the societies with the lowest energy use. These curves, which do not reflect any great degree of accuracy, do not represent rigorous limits, and values outside the interval may certainly be appropriate for individuals of a given society – the very rich or the very poor.

The energy conversion rate corresponding to food only has been taken as 125 W throughout the time interval. The increase in energy usage from about -10^5 y is associated with access to fire. The amount of energy derived from fires depends on whether fires were used only for cooking or also for heating. The choice of the average curve further rests on the assumption that between -7×10^4 and -10^4 y (i.e. during the latest ice age, cf. Fig. 2.91) about half of the world population used fires for heating purposes.

In the time interval -10^4 to -10^3 y, human settlements developed into a variety of societies, some of which had a very high degree of organisation and urbanisation. The increase in energy usage was mainly associated with more systematic heating and cooking practices, with tool production (e.g. weapons) and with transportation (e.g. by riding or by draught animals). With increasing population density, materials which previously had been available in the immediate natural surroundings had to be transported from far away, or substitutes had to be manufactured; either way, additional energy had to be spent. In several of the societies in question, mechanical work was performed not only by animals but also by human slaves, so that the average per capita energy usage was less affected. The trends of the curves also reflect the differences in development characterising different geographical regions. Simultaneously with the culmination of the civilisations in Mesopotamia and Egypt, northern Europe and northern America entered the Neolithic period, with warm climatic conditions quite different from those of the preceding several thousand years.

During the last 1000 years, the increasing energy usage is, in part, due to the shift in population distribution towards higher latitudes, and to overall increased requirements for space heating in such regions (the "little ice age", cf. Fig. 2.91). It should also be mentioned that the efficiency of converting the energy of firewood (supplemented by animal dung and later by peat) into useful heat for cooking, craft work, hot water and space heating was quite low, for example in 16th-century Europe, but gradually improved as the 20th century approached (Bjørnholm, 1976). During the period 1500–1900, the curves are a result of this feature (in particular the early high maximum value attained for the most affluent societies) combined with increased energy demand (e.g. larger proportions of the population acquiring energy-demanding habits or lifestyles, such as taking hot baths, drinking hot beverages, washing clothes in hot water, etc.). The development in the last century is dominated by the energy consumption of the industrialised countries (industrial process heat, transportation, increased room temperature, refrigeration, lighting, etc.). During this period, the top curve in Fig. 1.16 represents the extravagant energy use of an average American, while the lowest curve represents the average energy use in the poor regions of Africa or India, including non-commercial fuels such as cow dung and stray wood (which used to be absent from official statistics, as first noted by Makhijani, 1977).

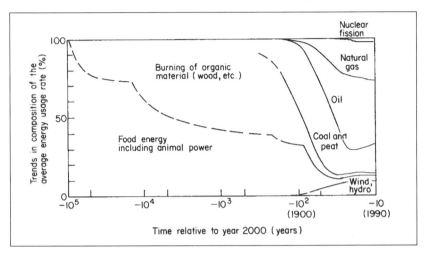

Figure 1.17. Trends in the distribution on different types of energy resources of the average rate of energy use. The most recent period is based on smoothed data from Darmstadter *et al.* (1971) and European Commission (1997), and the basis for the estimates pertaining to earlier periods is explained in the text. Needless to say, such estimates should be regarded as very tentative, and the definition of average use is itself uncertain, particularly for the early periods (e.g. the 20% contribution from fires 50 000 years ago depends sensitively on the fraction of the world population living in regions where space heating was desirable).

In Fig. 1.17, a sketch of the distribution of the energy consumption on different sources of energy is attempted. Again, only for the past century or two have actual data been used. The shape of the curve describing the diminishing share of food energy starting about 10^5 years ago is again dependent on the picture of emerging cultures and geographical distribution of the population, outlined above. It is clear, however, that the energy basis for human societies has been renewable energy sources until quite recently. Whether all the wood usage should be counted as renewable is debatable. Early agricultural practice (e.g. in northern Europe) involved burning forest areas for farming purposes and repeating the process in a new area after a few years, as the crop yield diminished owing to nutrient deficiency of the soil. Most forests not being converted into permanent agricultural land survived this exploitation, owing to the low population density and the stability of the soils originating from glacier deposits. Similar overuse, or overgrazing by livestock, would be (and was in fact) disastrous in low-latitude regions with a very shallow soil layer, which would simply be eroded away if the vegetation cover was removed (cf. section 2.4.2). Re-plantation of forests has been common in northern Europe during the last few centuries, but the strongly increasing demand for wood over the last century (not only for fuel purposes), as well as construction work associated with urbanisation, has led to an actual decrease in forest area in most parts of the world.

From the middle of the 19th century, the non-renewable fossil fuels have rapidly increased their share of the total energy usage, to the present 80–90 %. In the beginning, fossil fuels replaced wood, but they soon became the basis for exponential growth in energy use, associated with a number of novel energy-demanding activities. During the same period, usage of hydropower has increased, and recently nuclear fission power passed the 1% level. Growth has been interrupted by wars and periods of economic recession. The high dependence on non-renewable energy sources has developed over a very short period of time. The briefness of this era compared with the history of man on Earth stands out clearly on the linear scale used in Fig. 1.15.

1.3 The energy future and the role of renewable energy

Figure 1.16 shows a very large difference between the energy use of the leading countries or leading persons within countries, as compared with the least energy-using inhabitants of poor countries. This feature is currently changing, as the level of global interaction increases and every world citizen becomes aware of the kind of lifestyle that is "possible". However, the current development does not seem to indicate a diminishing ratio of the energy use of those using most and those using least energy. This is also true of other commodities related to living standard.

Energy use and resource depletion does not, of course, constitute the primary goals of any society or individual within a society. For example, average Europeans or Japanese use about half as much energy as the average North American, but have a living standard, which certainly is not lower than that of the North American citizens. This underlines the fact that the living standard and welfare depends on having primary (food, shelter, relations) as well as secondary standards of individual preference fulfilled and that this can be done in different ways with different implications for energy use.

The relationship between economic activities and social welfare has been debated for a considerable period of time, as has the possibility of physical limits to growth in material exploitation of the resources of a finite planet. The answer of conventional economists to this is that the inventiveness of man will lead to substitution of the materials threatened by exhaustion with others, in an ever-ongoing process. Recognising the finiteness of fossil and nuclear energy sources, this leads to the general prediction that renewable energy sources must take over at some stage, and the only debate is on how soon this will happen.

Most current geologists believe that oil and natural gas production will peak sometime in the next two decades. After that prices are bound to rise, thereby easing the introduction of alternative energy sources. Accepting a higher price of energy, it is also implied that energy must be used more efficiently, in order to prevent the belief that the higher energy cost slows down the development of human welfare.

This development in energy use is linked to another problem that may serve to accelerate the energy transition, namely, the increased awareness of the negative implications of environmental impacts of energy production and use. Early man was capable of causing environmental disturbance only on a very local scale. However, extensive burning of forests, for example, to provide land for agriculture, which would later be abandoned when over-exploitation diminished the crop or grazing yields, may have been instrumental in creating the desert and semi-desert regions presently found at low latitudes (Bryson, 1971). This is already an important example of a possibly man-made climatic change. Recently, man has reached a technological level enabling him to convert energy at rates which can be maintained over extended areas and which are no longer small compared to the energy fluxes of solar origin that are responsible for the climate.

The average heat flux of anthropogenic origin (i.e. from fossil fuels) in an industrial and urban area such as the Los Angeles Basin (about 10^{10} m^2) was estimated in 1970 to be 7 Wm^{-2} (Lees, 1970). The global average value around 1970 was 0.015 Wm^{-2} (see section 2.4.1), and the average solar flux absorbed by the Earth–atmosphere system is 240 Wm^{-2} (see Fig. 2.86). For comparison, a forest fire, which would burn down an area of fertile, tropical forests in one week, would release a heat flux of about 1000 Wm^{-2}. Yet the average heat flux from forest fires in all continental regions, the average being over several years, is less than the average anthropogenic heat flux. The nuclear weapons arsenal built up during the last 50 years is somewhere in the range of 10^4–10^5 megatons (Feld, 1976), with the higher figure corresponding to about 4.4×10^{20} J. If these weapons were detonated within a 24-hour interval, the average energy flux would be 5×10^{15} W, and if the target area were 10^{12} m^2, the average heat flux would be 5000 Wm^{-2}. The destructive effects would not be confined to those related to the immediate energy release. Radioactive contamination of the environment would cause additional death and decay, and would establish further mechanisms for climatic disturbance (e.g. destruction of the stratospheric ozone shield), in addition to the difficulty presented to human survival as the dominant species on our planet.

It has been suggested that fusion energy constitutes an alternative to renewable energy as the long-term solution. However, as elaborated in section 3.7.3, it is not clear at present whether fusion energy on Earth will ever become a feasible and practical source of controlled energy supply. It will create nuclear waste in amounts similar to those of fission technologies and will counteract the development towards decentralised technologies characterising the present trend. It is probably an exaggeration to imagine that the introduction of one kind of energy technology rather than another will determine or solve such institutional problems. What may be true, though, is that certain types of technology are more suitable for societies organised in a particular way and that the kind of technology imagined in connection with the use of certain renewable energy resources fits well both with the needs of sophisticated, decentralised societies based upon information technology and with the needs of the presently underprivileged regions.

Science and technology literature contains a range of suggestions for handling future energy demands. In the past, some of the technologies thus brought forward as "technically feasible" have actually been developed to commercial viability, and others not, for a variety of reasons. Renewable energy has over the last decades passed from the level of technical feasibility to a level of cautious introduction into the marketplace and not least into long-term government planning. One reason for the slow penetration is that some influential funding institutions, including the European Commission, have continued to use a large fraction of their R&D funds, as well as loan and aid money, on fission and fusion, ignoring the unattractiveness of the long-range radioactive waste problems and hoping to obtain short-term industry advantages in export of outdated fission technology to former East-block and developing nations. If funds were wholeheartedly aimed at a rapid transition from the fossil to the renewable era, the progress could be much faster. This has been demonstrated by a number of recent scenario studies, some of which will be described in Chapter 6. The general question of who controls technology development has been discussed by Elliott and Elliott (1976) and by Sørensen (1983; 2001a). During recent decades, a number of "grassroot" movements have advocated use of renewable energy, and it can be hoped that these preferences are preserved as that generation of people make their way into decision-making positions.

Renewable energy sources are typically characterised by a theoretical maximum rate at which energy may be extracted in a "renewable" mode, i.e. the rate at which new energy is arriving or flowing into the reservoirs associated with many of the renewable energy flows. In some cases, the additional loop on a given renewable energy cycle, caused by man's utilisation of the source, will by itself modify the rate at which new energy is arriving (for instance, utilisation of temperature differences in the oceans may alter surface evaporation rates and the velocities of ocean currents; in both cases the mechanisms for establishing the temperature differences may become altered, cf. section 3.5.1). The geothermal energy flux from the interior of the Earth is not a renewable resource, since the main part of the flux is associated with a cooling of the interior (cf. section 3.5.2). On the other hand, it is a very small fraction of the heat which is lost per year (2.4×10^{-10}), so for practical purposes geothermal energy behaves as a renewable resource. Only in case of overexploitation, which has characterised some geothermal steam projects, renewability is not ensured.

In Chapter 2, the nature and origin of renewable energy sources are discussed in what may resemble an odyssey through the sciences of astrophysics, atmospheric physics and chemistry, oceanography and geophysics. The importance of connecting all the pieces into an interlocking, overall picture becomes evident when the possible environmental impact of extended use of the renewable energy sources in the service of mankind is investigated in Chapter 7.

Chapter 3 provides, for each renewable energy source, an estimate of the size of the resource, defined as the maximum rate of energy extraction, which on an annual average basis will become renewed, independently of whether it is possible to extract such energy by known devices. Also issues of power density and variability are discussed in this Chapter.

Chapter 4 opens with some general features of energy conversion devices and then describes a number of examples of energy conversion equipment suitable for specific renewable energy sources.

Chapter 5 gives an overview of various methods of energy transport and storage, which, together with the energy conversion devices, will form the ingredients for total energy supply systems discussed in Chapter 6.

Chapter 6 discusses the modelling of the performance of individual renewable energy devices as well as whole systems and finally scenarios for the global use of renewable energy, with consideration of both spatial and temporal constraints in matching demand and supply.

In Chapter 7, renewable energy resources are first placed in the framework of current economic thinking, as a preliminary effort to quantify some of the considerations which should be made in constructing a viable energy supply system. Then follows a survey of indirect economic factors to consider, which leads to the description of the methodology of life-cycle analysis, which together with the scenario technique constitutes the package for an up-to-date economic analysis. It is then used in concrete examples, as a tool for systems assessment.

Chapter 8 offers some concluding remarks, notably on the renewable energy research and development areas most important for the near future.

1.4 Suggested topics for discussion

1.4.1
Find out what has happened since the year 2000 to the amounts of renewable energy used, based upon available statistical data, either globally or for your region.

1.4.2
Order the prices given on pages 13 and 14 according to the type of energy (heat, fuels for industry or transportation, electricity) and the type of customers (wholesale, retail). Try to find out what has happened to these prices since the year 2000 (e.g. using the annually updated sources such as OECD, IEA or Danish Energy Agency). Have the relative prices changed?

THE ORIGIN OF RENEWABLE ENERGY FLOWS

CHAPTER 2

In this chapter, renewable energy is followed from the sources where it is created – notably the Sun – to the Earth where it is converted into different forms, e.g. solar radiation to wind or wave energy, and distributed over the Earth–atmosphere system through a number of complex processes. Essential for these processes are the mechanisms for general circulation in the atmosphere and the oceans. The same mechanisms play a role in distributing pollutants released to the environment, whether from energy-related activities such as burning fossil fuels, or from other human activities. Because the assessment of environmental impacts plays an essential part in motivating societies to introduce renewable energy, the human interference with climate is also dealt with in this chapter, where it fits naturally.

2.1 Solar radiation

At present the Sun radiates energy at the rate of 3.9×10^{26} W. At the top of the Earth's atmosphere an average power of 1353 W m^{-2} is passing through a plane perpendicular to the direction of the Sun. As shown in Fig. 2.1, regular oscillations around this figure are produced by the changes in the Earth–Sun distance, as the Earth progresses in its elliptical orbit around the Sun. The average distance is 1.5×10^{11} m and the variation is $\pm 1.7\%$. Further variation in the amount of solar radiation received at the top of the atmosphere is caused by slight irregularities in the solar surface, in combination with the

Sun's rotation (about one revolution per month), and by possible timevariations in the surface luminosity of the Sun.

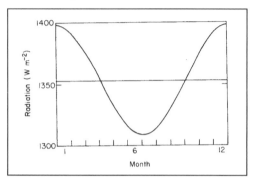

Figure 2.1. Yearly variations in the "solar constant" (the head-on solar radiation flux at the Earth's distance; NASA, 1971).

2.1.1 Energy production in the interior of stars like the Sun

The energy produced by nuclear reactions in the interior of the Sun must equal the amount of energy radiated from the surface, since otherwise the Sun could not have been structurally stable over long periods of time. Evidence for the stability of the Sun comes from several sources. Stability over a period of nearly 3×10^9 years is implied by the relative stability of the temperature at the Earth's surface (oxidised sediments and fossil remains indicate that water in its fluid phase has been present throughout such periods). Stability over an even longer time is implicit in our understanding of the evolution of the Sun and other similar stars. As an indication of this stability, Fig. 2.2 shows the variations in the radius of the Sun believed to have taken place since its assumed formation from clouds of dust and gas.

The conversion of energy contained in the atomic constituents of mainsequence stars such as the Sun from heat of nuclear reactions (which transforms hydrogen into helium) to radiation escaping from the surface is largely understood. The advanced topic section 2.A at the end of this chapter gives further details of the stellar evolution and a specific model for energy transport within the Sun. The basis for regarding such radiation as a renewable source is that it may continue essentially unaltered for billions of years. Yet there is also a possibility of tiny variations in solar energy production that may have profound implications for life on the planets encircling it.

2.1.2 Spectral composition of solar radiation

Practically all of the radiation from the Sun received at the Earth originates in the photosphere, a thin layer surrounding the convective mantle of high

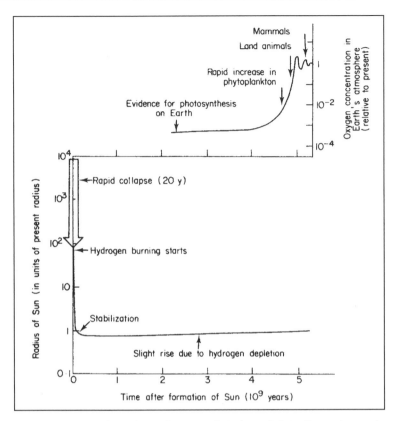

Figure 2.2. Variation in the solar radius as a function of time *(bottom)*, together with selected milestones in the development on Earth, associated with the building-up of oxygen in the Earth's atmosphere *(top)*. The rapid development of phytoplankton in the upper layers of the oceans at a relative oxygen concentration of 10^{-2} is associated with the formation of an ozone shield in the atmosphere, cutting down the ultraviolet part of the solar spectrum. When the oxygen level has reached 10^{-1}, the atmospheric ultraviolet absorption is strong enough to allow life on land. (Based on Herbig, 1967; Berkner and Marshall, 1970; Cloud and Gibor, 1970).

opacity. The depth to which a terrestrial observer can see the Sun lies in the photosphere. Owing to the longer path-length in the absorptive region, the apparent brightness of the Sun decreases towards the edges. The photosphere consists of atoms of varying degree of ionisation, plus free electrons. A large number of scattering processes take place, leading to a spectrum similar to the Planck radiation (see section 2.A) for a black body in equilibrium with a temperature $T \approx 6000$ K. However, this is not quite so, partly because of sharp absorption lines corresponding to the transitions between different electron configurations in the atoms present (absorption lines of over 60 elements have been identified in the solar spectrum), and partly because of the temperature variation through the photosphere, from around 8000 K near the convective zone to a minimum of 4300 K at the transition to the chromosphere (see Fig. 2.3). Yet the overall picture, shown in Fig. 2.4, is in fair agreement with the Planck law for an assumed effective temperature $T_{eff} \approx 5762$ K, disregarding in this figure the narrow absorption lines in the spectrum.

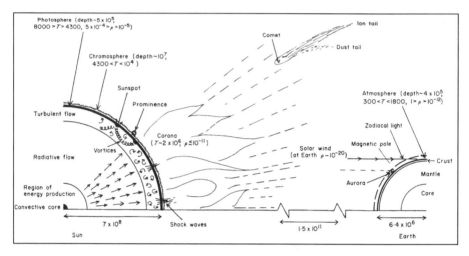

Figure 2.3. Schematic picture of solar layers, starting from the centre of the Sun to the left. The solar radius is defined by the bottom of the visible Sun. All distances are in metres, all temperatures T are in K and all densities ρ are in kg m^{-3}. The solar corona continues into an accelerating stream of particles, the solar wind. At the Earth's distance from the Sun (centre to the right, note the two changes of scale), the solar wind gives rise to magnetic induction and aurorae, and the extension of the corona is seen near the horizon as zodiacal light, in the absence of direct or scattered light from the solar photosphere. The solar wind is also responsible for the comet tails being directed away from the Sun (whereas the radiation pressure can only move the lightest material in the tail). The tails of a comet usually have an ion part and a dust part, with the latter moving more slowly and being deflected as a result of the Sun's rotation (a period of around 25 days). The inner part of the Earth is only sketched. The mantle is believed to consist of an outer part (silicates of Mg and Fe) and an inner part (oxides of Mg and Fe). Similarly, the core has an outer part (probably liquid FeS) and an inner part (a solid iron–nickel alloy).

The structure of the solar surface

The turbulent motion of the convective layer underneath manifests itself in the solar disc as a granular structure interpreted as columns of hot, vertical updrafts and cooler, downward motions between the grain-like structures (supported by observations of Doppler shifts). Other irregularities of the solar luminosity include bright flares of short duration, as well as the sunspots, regions near the bottom of the photosphere with lower temperature, appearing and disappearing in a matter of days or weeks in an irregular manner, but statistically periodic with an 11-year period. The sunspots first appear at latitudes at or slightly above 30°, reach maximum activity near 15° latitude, and end the cycle near 8° latitude. The "spot" is characterised by churning motion and a strong magnetic flux density (0.01–0.4 Wb m^{-2}), suggesting its origin from vorticity waves travelling within the convective layer, and the observation of reversed magnetic polarity for each subsequent 11-year period suggests a true period of 22 years.

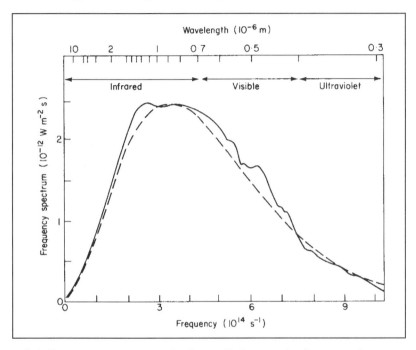

Figure 2.4. Frequency spectrum of solar radiation received on a unit area at the Earth's mean distance from the Sun, the area facing the Sun. Solid line: measured values (smoothed out over absorption lines, based on NASA, 1971). Dashed line: Planck law corresponding to an effective temperature of 5762 K, normalised to the experimental curve.

Above the photosphere is a less dense gas with temperature increasing outwards from the minimum of about 4300 K. During eclipses, the glow of this chromosphere is visible as red light. This is because of the intense H$_\alpha$ line in the chromospheric system, consisting primarily of emission lines.

After the chromosphere comes the corona, still less dense (of the order of 10^{-11} kg m^{-3} even close to the Sun), but of very high temperature (2×10^6 K, cf. Fig. 2.3). The mechanism of heat transfer to the chromosphere and to the corona is believed to be by shock waves originating in the turbulent layer

(Pasachoff, 1973). The composition of the corona (and chromosphere) is believed to be similar to that of the photosphere, but owing to the high temperature in the corona, the degree of ionisation is much higher, and e.g. the emission line of Fe^{13+} is among the strongest observed from the corona (during eclipses). A continuous spectrum (K-corona) and Fraunhofer absorption lines (F-corona) are also associated with the corona, although the total intensity is only 10^{-6} of that of the photosphere, even close to the Sun (thus the corona cannot be seen from the Earth's surface except during eclipses, owing to the atmospheric scattering of photospheric light). Because of the low density, no continuous radiation is produced in the corona itself, and the K-corona spectrum is due to scattered light from the photosphere, where the absorption lines have been washed out by Doppler shifts of random nature (as a result of the high kinetic energy).

The corona extends into a dilute, expanding flow of protons (ionised hydrogen atoms) and electrons, known as the solar wind. The increasing radial speed at increasing distances is a consequence of the hydrodynamic equations for the systems (the gravitational forces are unable to balance the pressure gradient; Parker, 1964). The solar wind continues for as long as the momentum flow is large enough not to be deflected appreciably by the magnetic fields of interstellar material. Presumably, the solar wind is penetrating the entire solar system.

Radiation received at the Earth

At the top of the Earth's atmosphere the solar wind has a density of about 10^{-20} kg m^{-3}, corresponding to roughly 10^7 hydrogen atoms per m^3. The ions are sucked into the Earth's magnetic field at the poles, giving rise to such phenomena as the aurorae borealis and to magnetic storms. Variations in solar activity affect the solar wind, which in turn affects the flux of cosmic rays reaching the Earth (many solar wind hydrogen ions imply larger absorption of cosmic rays).

Cosmic ray particles in the energy range 10^3–10^{12} MeV cross the interstellar space in all directions. They are mainly protons, but produce showers containing a wide range of elementary particles when they hit an atmosphere.

Figure 2.5 summarises a number of radiation sources contributing to the conditions at the Earth. Clearly, the radiation from the Sun dominates the spectral distribution as well the integrated flux. The next contributions, some six orders of magnitude down in the visible region, even integrated over the hemisphere, are also of solar origin, such as moonlight, airglow and zodiacal light (originating in the Sun's corona, being particularly visible at the horizon just before sunrise and just after sunset). Further down, in the visible regions of the spectrum, are starlight, light from our own galaxy and finally extra-galactic light.

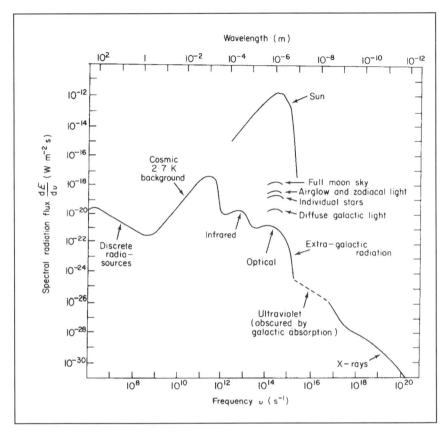

Figure 2.5. Frequency distribution of some contributions to the total radiation (integrated over the hemisphere). The radiation from the Sun and from extra-galactic sources is given outside the Earth's atmosphere. The relative intensity of some familiar light phenomena in the visible frequency region is indicated (some of these sources, e.g. the solar corona, also emit radiation in the ultraviolet and X-ray regions). In some regions of the spectrum, the information on the extra-galactic radiation is incomplete and the curve represents an estimate. (Based on NASA, 1971; Peebles, 1971; Allen, 1973; Longair and Sunyaev, 1969).

The peak in the spectral distribution of the extra-galactic radiation is in the microwave region. This is the universal background radiation approximately following a Planck shape for 2.7 K, i.e. the radiation predicted by the Big Bang theory (though probably consistent with other models) of the expanding universe (see section 2.A).

The main part of solar radiation can be regarded as unpolarised at the top of the Earth's atmosphere, but certain minor radiation sources, such as the light scattered by electrons in the solar corona, do have a substantial degree of polarisation.

2.2 Net radiation flux at the Earth

The solar radiation, which approximately corresponds to the radiation from a black body of temperature 6000 K, meets the Earth–atmosphere system and interacts with it, producing temperatures which at the Earth's surface typically vary in the range 220–320 K. The average (over time and geographical location) temperature of the Earth's surface is presently 288 K.

As a first approach to understanding the processes involved, one may look at the radiation flux passing through unit horizontal areas placed either at the top of the atmosphere or at the Earth's surface. The net flux is the sum (with proper signs) of the fluxes passing the area from above and from below. The flux direction towards the centre of the Earth will be taken as positive, consistent with reckoning the fluxes at the Sun as positive, if they transport energy away from the solar centre.

Since the spectral distributions of black-body radiation (see section 2.A) at 6000 and 300 K, respectively, do not substantially overlap, most of the radiation fluxes involved can be adequately discussed in terms of two broad categories, called short-wavelength (sw) and long-wavelength (lw) or thermal radiation.

2.2.1 Radiation at the top of the atmosphere

The flux of solar radiation incident on a surface placed at the top of the atmosphere depends on time (t) and geographical location (latitude ϕ and longitude λ) and on the orientation of the surface,

$$E_{0+}^{sw}(t,\phi,\lambda) = S(t)\cos\theta(t,\phi,\lambda).$$

Here $S(t)$ is the "solar constant" at the distance of the Earth (being a function of time due to changes in the Sun–Earth distance and due to changes in solar luminosity) and θ is the angle between the incident solar flux and the normal to the surface considered. The subscript "0" on the short-wavelength flux E^{sw} through the surface indicates that the surface is situated at the top of the atmosphere and "+" indicates that only the positive flux (in the "inward" direction) is considered. For a horizontal surface, θ is the zenith angle z, obtained by

$$\cos z = \sin\delta\sin\phi + \cos\delta\cos\phi\cos\omega,$$

where δ is the declination of the Sun and ω is the hour angle of the Sun (see Fig. 2.6). The declination is given approximately by (Cooper, 1969)

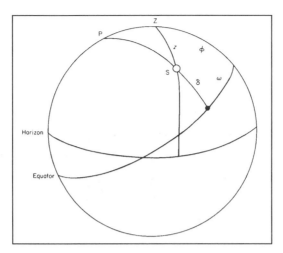

Figure 2.6. Coordinates in a geocentric picture. P represents the North Pole, Z is the zenith and ϕ is the geographical latitude. The unit radius great circle shown through P and Z is the median of the observer. S represents the direction to the Sun, and the solar coordinates are the declination δ and the hour angle ω. Also shown is the zenith angle z (equal to $\pi/2$ minus the height of the Sun over the horizon).

$$\delta = 0.4093 \sin\left(2\pi \frac{284 + day}{365} \right), \tag{2.1}$$

where *day* is the actual day's number in the year. The hour angle (taken as positive in the mornings) is related to the local time t_{zone} (in hours) by

$$\omega = 2\pi (12 - t_{zone})/24 - (\lambda - \lambda_{zone}) - TEQ. \tag{2.2}$$

Here λ_{zone} is the longitude of the meridian defining the local time zone (longitudes are taken positive towards east; latitudes and declinations are positive towards north; all angles are in radians). The correction TEQ (*"equation of time"*) accounts for the variations in solar time caused by changes in the rotational and orbital motion of the Earth, e.g. due to speed variations in the elliptical orbit and to the finite angle (obliquity) between the axis of rotation and the normal to the plane or orbital motion. The main part of TEQ remains unaltered at the same time in subsequent years and may be expressed as a function of the day of the year (see Duffie and Beckman, 1974),

$$
\begin{aligned}
TEQ = &- 0.0113 - 0.0019 \times day, \text{ if } day \le 20, \\
= &- 0.0227 - 0.0393 \cos(0.0357(day{-}43)), \text{ if } 20 < day \le 135, \\
= &- 0.0061 + 0.0218 \cos(0.0449(day{-}135)), \text{ if } 135 < day \le 240, \\
= &\; 0.0275 + 0.0436 \cos(0.0360(day{-}306)), \text{ if } 240 < day \le 335, \\
= &- 0.0020 \times (day{-}359), \text{ if } day > 335.
\end{aligned}
$$

The resulting daily average flux on a horizontal surface,

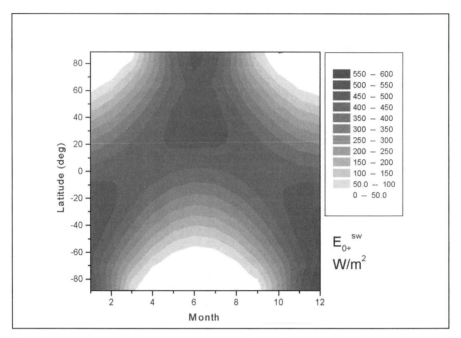

Figure 2.7. Daily average radiation, E_{0+}^{sw} (W m^{-2}), incident at the top of the Earth's atmosphere, as a function of time of the year and latitude (independent of longitude). Latitude is (in this and the following figures) taken as positive towards north (based on NCEP-NCAR, 1998).

$$E_{0+}^{sw} = \frac{1}{24} \int_{day} E_{0+}^{sw}(t, \phi, \lambda)\, dt,$$

which is independent of λ, is shown in Fig. 2.7. The maximum daily flux is about 40% of the flux on a surface fixed perpendicular to the direction of the Sun. It is obtained near the poles during midsummer, with light close to 24 hours of each day but a zenith angle of $\pi/2 - \max(|\delta|) \approx 67°$ (cf. Fig. 2.6, $|\delta|$ being the absolute value of δ). Daily fluxes near to 30% of the solar constant are found throughout the year at the Equator. Here the minimum zenith angle comes close to zero, but day and night are about equally long.

The disposition of incoming radiation

A fraction of the incoming solar radiation is reflected back into space. This fraction is called the albedo a_0 of the Earth–atmosphere system. The year and latitude–longitude average of a_0 is about 0.35. This is composed of about 0.2 from reflection on clouds, 0.1 from reflection on cloudless atmosphere (particles, gases) and 0.05 from reflection on the Earth's surface. Figure 2.8 gives the a_0 distribution for two months of 1997, derived from ongoing satellite

ADVANCED MATERIAL IN SECTION 2.B

measurements (cf. Wilson and Matthews, 1971; Raschke *et al.*, 1973; NCEP-NCAR, 1998).

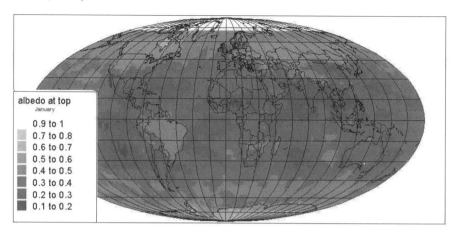

Figure 2.8a,b. Top of the atmosphere albedo for the Earth–atmosphere system, derived from satellite data for January (top) and July (bottom) 1997 (NCEP-NCAR, 1998)[*].

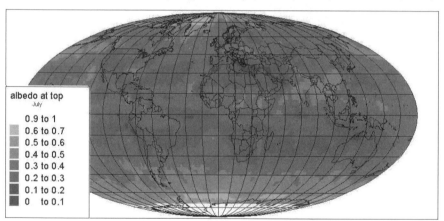

The radiation absorbed by the Earth–atmosphere system is

$$A_0 = E_{0+}^{sw}(1 - a_0).$$

Since no gross change in the Earth's temperature is taking place from year to year, it is expected that the Earth–atmosphere system is in radiation equilibrium, i.e. that it can be ascribed an effective temperature T_0, such that the black-body radiation at this temperature is equal to $-A_0$,

---

[*] These and most of the geographical figures in the following use the equal-area Mollweide projection, making it easy to compare quantities at different locations.

$$T_0 = \left(\frac{\frac{1}{4} S(1 - a_0)}{\sigma} \right)^{1/4} \approx 253 \ K.$$

Here S is the solar constant and $\sigma = 5.7 \times 10^{-8} \ W \ m^{-2} \ K^{-4}$ is Stefan's constant (cf. section 2.A). The factor $1/4$ comes because the incoming radiation on the side of the Earth facing the Sun is the integral of $S \cos\theta$ over the hemisphere, while the outgoing radiation is uniform over the entire sphere (by definition when it is used to define an average effective temperature). The net flux of radiation at the top of the atmosphere can be written

$$E_0 = E_{0+}^{sw}(1 - a_0) + E_0^{lw}, \tag{2.3}$$

where $E_0^{lw} = E_{0-}^{lw}$ on average equals $-\sigma T_0^4$ (recall that fluxes away from the Earth are taken as negative) and E_0 on average (over the year and the geographical position) equals zero.

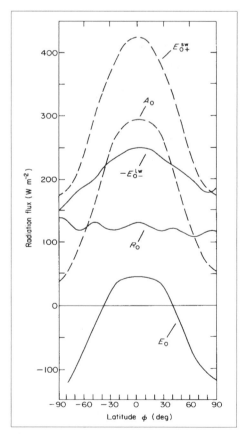

Figure 2.9. Latitude dependence (averaged over longitudes and over the year) of incident (E_{0+}^{sw}) and outgoing long-wavelength ($-E_{0-}^{lw}$) radiation at the top of the atmosphere, the amount of radiation absorbed (A_0) and reflected (R_0) by the Earth–atmosphere system, and the net radiation flux E_0 at the top of the atmosphere, in W m^{-2} (based on Sellers, 1965; Budyko, 1974).

The fact that T_0 is 34 K lower than the actual average temperature at the Earth's surface indicates that heat is accumulated near the Earth's surface, due to re-radiation from clouds and atmosphere. This is called the greenhouse effect, and it will be further illustrated below, when discussing the radiation fluxes at the ground.

The net radiation flux at the top of the atmosphere, E_0, is a function of time as well as of geographical position. The longitudinal variations are small. Figure 2.9 gives the latitude distribution of $\overline{E_0}$, $\overline{A_0}$ and $\overline{R_0}$, where the reflected flux is

$$R_0 = E_{0+}^{sw} a_0 \quad (= -E_{0-}^{sw}),$$

as well as $\overline{E_0^{lw}}$.

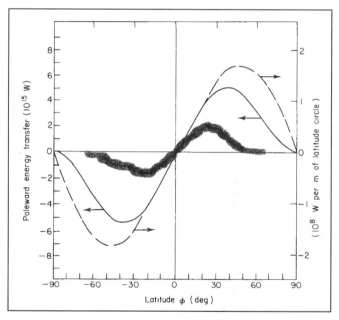

Figure 2.10. The poleward energy transfer required by the latitude variation of the net energy flux shown in Fig. 2.9. The dashed curve represents the same quantity divided by the length of the latitude circle, i.e. the energy transport required across a unit length of the latitude circle. These figures are annual and longitudinal averages (based on Sellers, 1965). The ocean transport contribution is indicated as a dark blurred region, based on measurements with estimated uncertainties (Trenberth and Solomon, 1994).

Although $\overline{R_0}$ is fairly independent of ϕ, the albedo $\overline{a_0}$ increases when going towards the poles. Most radiation is absorbed at low latitudes, and the net flux exhibits a surplus at latitudes below $40°$ and a deficit at higher absolute latitudes. Since no progressive increase in equatorial average temperatures, nor any similar decrease in polar temperatures, is taking place, there must be a transport of energy from low latitudes in the directions of both poles. The mechanisms for producing this flow are ocean currents and winds. The necessary amount of poleward energy transport, which is immediately given by the latitude variation of the net radiation flux (since radiation is the only mode of energy transfer at the top of the atmosphere), is given in Fig. 2.10, which also indicates the measured share of ocean transport. The total amount of energy transported is largest around $|\phi| = 40°$, but since the total circumference of the latitude circles diminishes as latitude increases, the maximum energy transport across a unit length of latitude circle is found at $|\phi| \approx 50°$.

Figures 2.11–2.12 give the time and latitude dependence at longitude zero of the long-wavelength re-radiation, E_{0-}^{lw} or E_0^{lw} (since the flux is only in the negative direction), and net flux, E_0, of the Earth–atmosphere system, as obtained from satellite measurements within the year 1997 (Kalnay *et al.*, 1996; NCEP-NCAR, 1998). A_0 may be obtained as the difference between E_{0+}^{sw} and E_{0-}^{sw}. The latter equals the albedo (Fig. 2.8) times the former, given in Fig. 2.7.

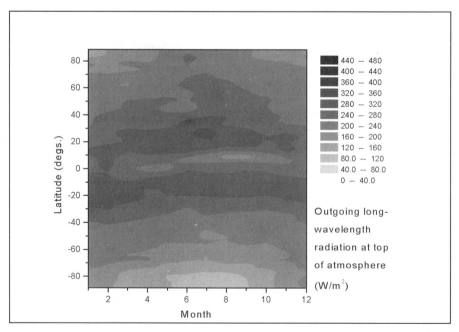

Figure 2.11. Long-wavelength radiation from the Earth–atmosphere system ($-E_{0-}^{lw}$).

2.2.2 Radiation at the Earth's surface

The radiation received at the Earth's surface consists of direct and scattered (plus reflected) short-wavelength radiation plus long-wavelength radiation from sky and clouds, originating as thermal emission or by reflection of thermal radiation from the ground.

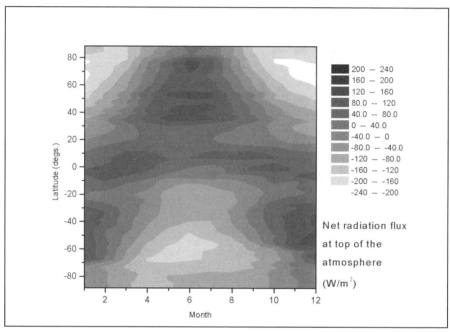

Figure 2.12. Net radiation flux E_0 at the top of the atmosphere (NCEP-NCAR, 1998).

Direct and scattered radiation

Direct radiation is defined as radiation which has not experienced scattering in the atmosphere, so that it is directionally fixed, coming from the disc of the Sun.

Scattered radiation is, then, the radiation having experienced scattering processes in the atmosphere. In practice, it is often convenient to treat radiation which has experienced only forward scattering processes together with the unscattered radiation, and thus define direct and scattered radiation as radiation coming from or not coming from the direction of the Sun. The area of the disc around the Sun used to define direct radiation is often selected in a way depending on the application in mind (e.g. the solid angle of acceptance for an optical device constructed for direct radiation measurement). In any case the "direct radiation" defined in this way will contain scattered radiation with sufficiently small angles of deflection, due to the finite solid angle of the Sun's disc.

The individual scattering and absorption processes will be discussed further below, but in order to give a first impression of the modifications in the spectral distribution, due to the passage through the atmosphere, Fig. 2.13 shows the amount of radiation surviving at sea-level for a clear day with the Sun in zenith. A large number of absorption lines and absorption bands can be seen in the low-frequency part of the spectrum (corresponding to wavelengths above 0.7×10^{-6} m). They are due to H_2O, CO_2, O_2, N_2O, CH_4 and other, minor constituents of the atmosphere. At higher frequencies the continuous absorption bands dominate, in particular those of O_3. Around 0.5×10^{-6} m, partial absorption by O_3 produces a dip in the spectrum, and below a wavelength of 0.3×10^{-6} m (the ultraviolet part of the spectrum) or a frequency above 9.8×10^{14} s^{-1}, the absorption by ozone is practically complete.

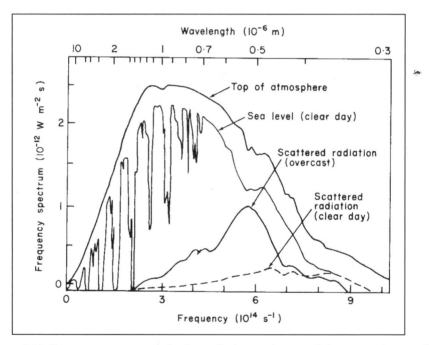

Figure 2.13. Frequency spectrum of solar radiation at the top of the atmosphere and at the surface (sea-level and minimum air mass, corresponding to the Sun in zenith on a clear day). Also shown are typical spectra of scattered radiation in the extremes of pure sky radiation (cloudless day) and pure cloud radiation (completely overcast). (Based on NASA, 1971; Gates, 1966.)

Figure 2.13 also shows the scattered part of the radiation on a clear day (dashed line). The scattered radiation is mainly composed of the most energetic (high frequency) radiation, corresponding to the blue part of the visible spectrum. The sky on a clear day is thus blue. The scattered light is also polarised. The radiation from clouds, i.e. from a completely overcast sky, is also shown in Fig. 2.13. It has a broad frequency distribution in the visible spectrum, so that the cloud light will appear white.

Disposition of the radiation at the Earth's surface

On average, roughly half of the short-wavelength radiation which reaches the Earth's surface has been scattered. Denoting the direct and scattered radiation at the Earth's surface D and d, respectively, and using the subscript s to distinguish the surface level (as the subscript "0" was used for the top of the atmosphere), the total incoming short-wavelength radiation at the Earth's surface may be written

$$E_{s+}^{sw} = D + d,$$

for a horizontal plane. The amount of radiation reflected at the surface is

$$R_s = E_{s+}^{sw} a_s \quad (= -E_{s-}^{sw}),$$

where the surface albedo, a_s, consists of a part describing specular reflection (i.e. reflection preserving the angle between beam and normal to the rejecting surface) and a part describing diffuse reflection. Here diffuse reflection is taken to mean any reflection into angles different from the one characterising beam reflection. An extreme case of diffuse reflection (sometimes denoted "completely diffuse" as distinguished from "partially diffuse") is one in which the angular distribution of reflected radiation is independent from the incident angle.

The total net radiation flux at the Earth's surface is

$$E_s = E_{s+}^{sw}(1 - a_s) + E_{s-}^{lw}, \tag{2.4}$$

where the long-wavelength net radiation flux is

$$E_s^{lw} = E_{s+}^{lw} + E_s^{lw},$$

in terms of the inward and (negative) outward long-wavelength fluxes.

Figure 2.14 gives the latitude distribution of the incoming short-wavelength flux E_{s+}^{sw} and its division into a direct, D, and a scattered part, d. The fluxes are year and longitude averages (adapted from Sellers, 1965). Similarly, Fig. 2.15 gives the total annual and longitude average of the net radiation flux at the Earth's surface, and its components, as function of latitude. The components are, from (2.4), the absorbed short-wavelength radiation,

$$A_s^{sw} = E_{s+}^{sw}(1 - a_s),$$

and the net long-wavelength thermal flux E_s^{lw} (Fig. 2.15 shows $-E_s^{lw}$). The figures show that the direct and scattered radiation are of comparable magnitude for most latitudes and that the total net radiation flux at the Earth's surface is positive except near the poles. The variations in the long-wavelength difference between incoming (from sky and clouds) and outgoing radiation indicate that the temperature difference between low and high latitudes is not nearly large enough to compensate for the strong variation in absorbed short-wavelength radiation.

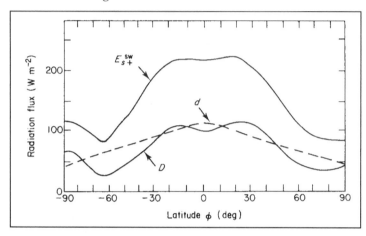

Figure 2.14. Incoming radiation fluxes on a horizontal plane at the Earth's surface. Averages over the year and over longitudes are shown of direct (D), scattered (d) and total radiation (E_{s+}^{sw}), as functions of latitude (adapted from Sellers, 1965).

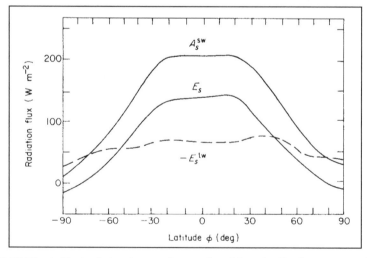

Figure 2.15. The latitude dependence of annual and longitudinal averages of net radiation flux on a horizontal plane at the Earth's surface (E_s) and its components, the amount of short-wavelength radiation absorbed (A_s^{sw}) and the net emission of long-wavelength radiation ($-E_s^{lw}$) (adapted from Sellers, 1965).

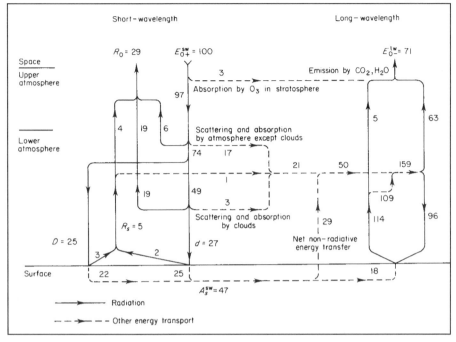

Figure 2.16. Schematic summary of the disposition of incoming radiation (E_{0+}^{sw}, put equal to 100 units) in the Earth–atmosphere system, averaged over time and geographical location. Symbols and details of the diagram are discussed in the text. (The data are adapted from Sellers, 1965; Budyko, 1974; Schneider and Dennett, 1975; Raschke *et al.*, 1973).

The conclusion is the same as the one drawn from the investigation of the Earth–atmosphere net radiation flux. A poleward transport of heat is required, either through the Earth's surface (e.g. ocean currents) or through the atmosphere (winds in combination with energy transfer processes between surface and atmosphere by means other than radiation).

Figure 2.16 summarises in a crude form the transfer processes believed to take place in the disposition of the incoming solar radiation. In the stratosphere, the absorption of short-wavelength radiation (predominantly by ozone) approximately balances the emission of long-wavelength radiation (from CO_2, H_2O, etc.). In the lower part of the atmosphere, the presence of clouds largely governs the reflection of short-wavelength radiation. On the other hand, particles and gaseous constituents of the cloudless atmosphere are most effective in absorbing solar radiation. All figures are averaged over time and geographical location.

Slightly over half the short-wavelength radiation which reaches the Earth's surface has been scattered on the way. Of the 27% of the solar radiation at the top of the atmosphere which is indicated as scattered, about 16% has been scattered by clouds and 11% by the cloudless part of the sky.

The albedo at the Earth's surface is shown in Fig. 2.17 for two months of 1997, based on a re-analysis of measured data, using climate modelling to ensure consistency. The model has not been able to achieve full consistency of fluxes, probably because of too high albedo values over ocean surfaces (Kalnay *et al.*, 1996). In Fig. 2.16 the average albedo is indicated as the smaller value $a_s \approx 0.12$. The albedo of a particular surface is usually a strongly varying function of the incident angle of the radiation. For sea water, the average albedo is about $a_s = 0.07$, rising to about 0.3 in case of ice cover. For various land surfaces, a_s varies within the range 0.05–0.45, with typical mean values 0.15–0.20. In case of snow cover, a_s lies in the range 0.4–0.95, with the smaller values for snow that has had time to accumulate dirt on its surface.

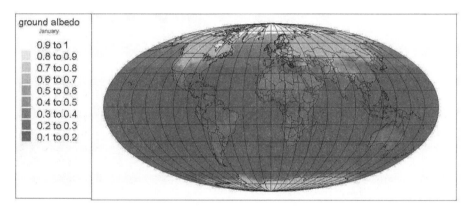

Figure 2.17a,b. Albedo at the Earth's surface, based on the analysis of ground-based and satellite measurements, for January (above) and July (below) of 1997. White areas: missing data, notably near poles (NCEP-NCAR, 1998).

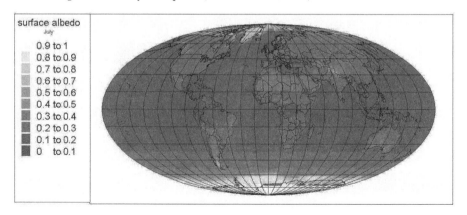

ADVANCED MATERIAL IN SECTION 2.B

As an illustration of the difficulties in estimating mean albedos, Fig. 2.18 shows the albedo of a water surface for direct sunlight, as a function of the zenith angle of the Sun. For a plane water surface and perpendicular incidence, the albedo implied by the air-to-water refraction index is 0.024. However, for ocean water the surface cannot be regarded as plane, and typical mean albedos (over the year) are of the order of 0.06, except at large latitudes, where the value may be greater. This estimate includes direct as well as scattered incident light, with the latter "seeing" a much more constant albedo of the water of about 0.09 (Budyko, 1974).

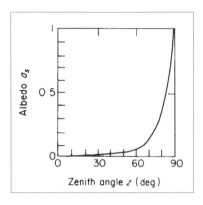

Figure 2.18. Albedo of water surface for direct radiation, as a function of zenith angle (i.e. angle between normal and the direction of radiation) (based on Budyko, 1974).

The total amount of direct and scattered light, which in Fig. 2.16 is given as reflected from the ground, is a net value representing the net reflected part after both the initial reflection by the ground and the subsequent back-reflection (some 20%) from the clouds above, and so on.

Disposition of radiation in the atmosphere

On average, only a very small fraction of the long-wavelength radiation emitted by the Earth passes unhindered to the top of the atmosphere. Most is absorbed in the atmosphere or by clouds, from which the re-radiation back towards the surface is nearly as large as the amount received from there.

The net radiation flux of the atmosphere,

$$E_a = E_a^{sw} + E_a^{lw} = A_a^{sw} + A_a^{lw} + S_a^{lw},$$

equals on average –29% of E_{0+}^{sw}. The short-wavelength radiation absorbed is $A_a^{sw} = E_s^{sw} = 21\%$, whereas the long-wavelength radiation absorbed and emitted in the lower atmosphere is $A_a^{lw} = 109\%$ and $S_a^{lw} = -159\%$, respectively. Thus, the net long-wavelength radiation is $E_a^{lw} = -50\%$.

The net radiation flux of the atmosphere is on average equal to, but (due to the sign convention adopted) of opposite sign to the net radiation flux at the Earth's surface. This means that energy must be transferred from the surface to the atmosphere by means other than radiation. Such processes are conduction and convection of heat, evaporation and precipitation of water and snow, etc. These processes produce energy flows in both directions, between surface and atmosphere, with a complex dynamic pattern laid out by river run-off, ocean currents and air motion of the general circulation. The average deviation from zero of the net radiation flux at the surface level gives a measure of the net amount of energy displaced vertically in this manner, namely, 29% of E_{0+}^{sw} or 98 W m^{-2}. As indicated in Fig. 2.14, the horizontal energy transport across a one metre wide line (times the height of the atmosphere) may be a million times larger. The physical processes involved in the transport of energy will be further discussed in section 2.3.

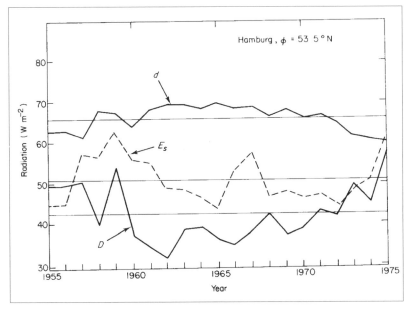

Figure 2.19. Variation of annual averages of direct and scattered radiation (solid, heavy lines) around their 21-year average value (thin, horizontal lines) for Hamburg. Also shown is the variation in net radiation flux (dashed line) around its 21-year average value (based on Kasten, 1977).

Annual, seasonal and diurnal variations in the radiation fluxes at the Earth's surface

In discussing the various terms in the radiation fluxes, it has tacitly been assumed that averages over the year or over longitudes or latitudes have a well-defined meaning. However, as is well known, the climate is not completely periodic with a one-year cycle, even if the solar energy input is so to a very large degree. The short-term fluctuations in climate seem to be very important, and the development over a time scale of a year depends to a large extent on the detailed initial conditions. Since, owing to the irregular short-term fluctuations, the initial conditions at the beginning of subsequent years are not exactly the same, the development of the general circulation and of the climate as a whole does not repeat itself with a one-year period. Although the gross seasonal characteristics are preserved as a result of the outside forcing of the system (the solar radiation), the year-average values of such quantities as mean temperature or mean cloudiness for a given region or for the Earth as a whole are not the same from one year to the next.

This implies that the components of the net radiation flux do not have the same annual average values from year to year. In particular, the disposition of incoming radiation as rejected, scattered, direct and absorbed radiation is very sensitive to variations in cloud cover, and, therefore, exhibits substantial changes from year to year. Figure 2.19 shows the variations of yearly means of direct and scattered radiation measured at Hamburg (Kasten, 1977) over a 20-year period. Also shown is the variation of the net radiation flux at the ground. The largest deviation in direct radiation, from the 20-year average, is over 30%, whereas the largest deviation in scattered radiation is 10%. The maximum deviation of the net radiation flux exceeds 20%.

Figures 2.20–2.21 give, for different seasons in 1997, the total short-wavelength radiation $(D+d)$ and the long-wavelength radiation fluxes (E_s^{lw}) at the Earth's surface, as a function of latitude and longitude.

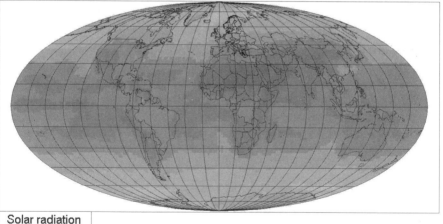

Solar radiation
annual mean W/m2

- 320 to 360
- 280 to 320
- 240 to 280
- 200 to 240
- 160 to 200
- 120 to 160
- 80 to 120

Figure 2.20. Annual average incoming short-wavelength radiation (E_{s+}^{sw}) on a horizontal surface at the Earth's surface, as a function of geographical position (W m^{-2}) (based on NCEP-NCAR, 1998). Seasonal variations are shown in Chapter 3, Figs. 3.1a–d.

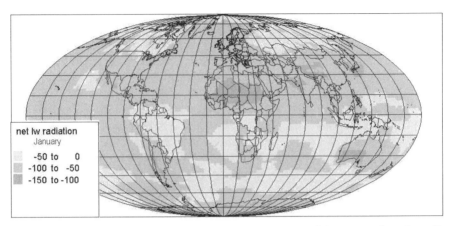

Figure 2.21a,b. Average net (i.e. downward minus upward) long-wavelength radiation flux (E_s^{lw}) in January (above) and July (below), at the Earth's surface, as a function of geographical position (W m^{-2})(based on NCEP-NCAR, 1998).

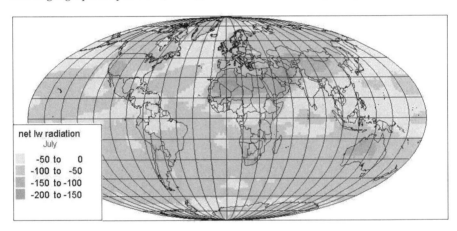

Figure 2.22 gives the results of measurements at a specific location, Hamburg, averaged over 19 years of observation (Kasten, 1977). Ground-based observations at meteorological stations usually include the separation of short-wavelength radiation in direct and scattered parts, quantities that are difficult to deduce from satellite-based data. Owing to their importance for solar energy utilisation, they will be further discussed in Chapter 3. The ratio of scattered to direct radiation for Hamburg is slightly higher than the average for the latitude of 53.5° (see Fig. 2.14), and the total amount of incident short-wavelength radiation is smaller than average. This may be due to the climate modifications caused by a high degree of urbanization and industrialisation (increased particle emission, increased cloud cover). On the other

hand, the amount of short-wavelength radiation reflected at the surface, R_s, exceeds the average for the latitude.

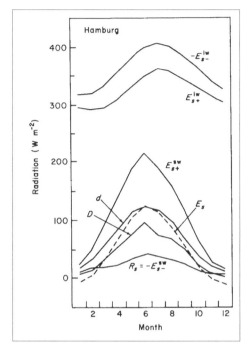

Figure 2.22. Components of the net radiation flux (E_s, dashed line) on a horizontal plane at the Earth's surface in Hamburg. Based on data for 19 consecutive years, the monthly averages have been calculated for direct (D), scattered (d) and total incident short-wavelength radiation (E_{s+}^{sw}), the rejected part ($R_s = -E_{s-}^{sw}$) and the flux of long-wavelength radiation received (E_{s+}^{lw}) and emitted (E_{s-}^{lw}) (based on Kasten, 1977).

Figures 2.23–2.25 give the variations with the hour of the day of several of the radiation contributions considered above, again based on long-term measurements at Hamburg (Kasten, 1977). In Fig. 2.23, the incident radiation at ground level is shown for the months of June and December, together with the direct and reflected radiation. The mean albedo is about 0.18 at noon in summer and about 0.3 in winter. In December, the amount of reflected radiation exceeds the amount of direct radiation incident on the plane. The long-wavelength components, shown in Fig. 2.24, are mainly governed by the temperature of the ground. The net outward radiation rises sharply after sunrise in June and in a weakened form in December. In June, the separate outward and inward fluxes have a broad maximum around 1400 h. The delay from noon is due to the finite heat capacity of the ground, implying a time lag between the maximum incident solar radiation and the maximum temperature of the soil. The downward flux changes somewhat more rapidly than the upward flux, because of the smaller heat capacity of the air. In December, the 24-h soil temperature is almost constant, and so is the outgoing long-wavelength radiation, but the re-radiation from the atmosphere exhibits a dip around noon. Kasten (1977) suggests that this dip is due to the dissolution of rising inversion layers, which during the time interval 1600–0800 h contributes significantly to the downward radiation. The hourly variation of the net total radiation flux is shown in Fig. 2.25 for the same two months. During daytime this quantity roughly follows the variation of incident radiation, but at night the net flux is negative, and more so during summer when the soil temperature is higher.

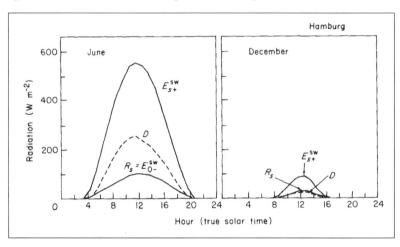

Figure 2.23. For the months of June and December, the total short-wavelength radiation on a horizontal plane at the Earth's surface in Hamburg is shown, and the direct and reflected radiation, for a given hour of the day. The fluxes are hourly averages based on 10 years of data, shown as functions of the hour of the day in true solar time (based on Kasten, 1977).

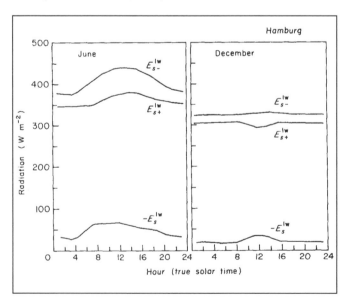

Figure 2.24. For June and December, the net long-wavelength radiation on a horizontal plane at the Earth's surface in Hamburg is shown, as well as the separate incoming and outgoing fluxes, for a given hour of the day. The fluxes are hourly averages based on 10 years of data, shown as functions of the true solar time hour (based on Kasten, 1977).

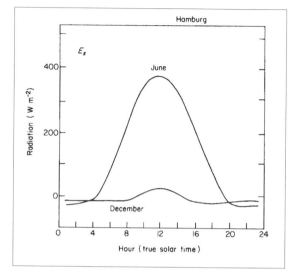

Figure 2.25. For the months of June and December the total net radiation flux on a horizontal plane at the Earth's surface in Hamburg is shown for a given hour of the day. The fluxes are hourly average values based on 10 years of measurements, shown as functions of the hour of the day in true solar time (based on Kasten, 1977).

Penetration of solar radiation

In discussing the solar radiation absorbed at the Earth's surface, it is important to specify how deep into the surface material, such as water, soil or ice, the radiation is able to penetrate. Also the reflection of sunlight may take place at a depth below the physical surface.

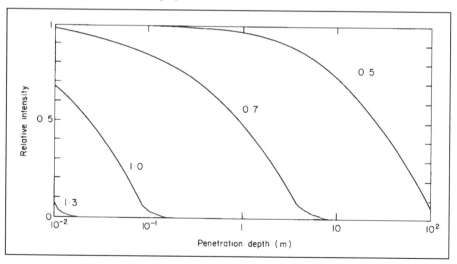

Figure 2.26. Penetration depth of monochromatic radiation through water as a function of wavelength of radiation (indicated on curves, in 10^{-6} m). The curves should only be considered rough indications of the behaviour (based on Sellers, 1965).

The penetration into soil, vegetation, sand and roads or building surfaces is usually limited to at most 1 or 2 mm. For sand of large grain size the penetration may reach 1–2 cm, in particular for the wavelengths above 0.5×10^{-6} m. For pure water the shorter wavelengths are transmitted more readily than the longer ones. Figure 2.26 shows calculated intensity variations as function of depth below water surface and wavelength. The availability of solar radiation (although only about 5% of that incident along the normal to the water surface) at ocean depths of nearly 100 m is, of course, decisive for the photosynthetic production of biomass by phytoplankton or algae. For coastal and inland waters the penetration of sunlight may be substantially reduced owing to dispersed sediments, as well as to eutrophication. A reduction in radiation intensity by a factor two or more for each metre penetrated is not uncommon for inland ponds (Geiger, 1961). The penetration of radiation into snow or glacier ice varies from a few tenths of a metre (for snow with high water content) to over a metre (ice).

2.3 Physical and chemical processes near the Earth's surface

A very large number of physical and chemical processes take place near the Earth's surface. No exhaustive description will be attempted here, but emphasis will be placed on those processes directly associated with or important for the possibility of utilising the renewable energy flows in connection with human activities.

In particular, this will lead to a closer look at some of the processes by which solar energy is distributed in the soil–atmosphere–ocean system, e.g. the formation of winds and currents. Understanding these processes is also required in order to estimate the possible disturbance of climate, which may follow certain types of interference with the "natural" flow pattern.

Although the processes in the different compartments are certainly coupled, the following subsections will first list a number of physical processes specific for the atmosphere, for the oceans and finally for the continents and then try to indicate how they can be combined in an attempt to account for the climate of the Earth.

2.3.1 The atmosphere

The major constituents of the present atmosphere are nitrogen, oxygen and water. For dry air at sea-level the nitrogen constitutes 78% by volume, oxygen 21% and minor constituents 1%, the major part of which is argon (0.93%)

and carbon dioxide (0.03%). The water content by volume ranges from close to zero at the poles to about 4% in tropical climates.

Typical variations of density, pressure and temperature, as functions of height, are shown in Fig. 2.27. Common names for the different layers are indicated. They are generally defined by turning points in the temperature profile ("pauses") or by the lower border of constant temperature regions if no sharp change in the sign of the temperature gradient takes place. Actual temperature profiles are not independent of latitude and season and do not exhibit sharp "turning points", as indicated in Fig. 2.28.

Estimates of the variation with height of the average abundance of some gaseous constituents of the atmosphere are shown in Fig. 2.29. The increasing abundance of ozone (O_3) between 15 and 25 km height is instrumental in keeping the level of ultraviolet radiation down, thereby allowing life to develop on the Earth's surface (cf. Fig. 2.2). Above a height of about 80 km, many of the gases are ionised, and for this reason the part of the atmosphere above this level is sometimes referred to as the ionosphere. Charged particles at the height of 60 000 km are still largely following the motion of the Earth, with their motion being mainly determined by the magnetic field of the Earth. This layer is referred to as the magnetosphere.

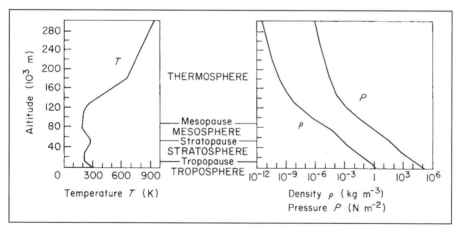

Figure 2.27. Altitude dependence of temperature, pressure and density for the US standard atmosphere (U.S. Government, 1962).

Particles in the atmosphere

In addition to gaseous constituents, water and ice, the atmosphere contains particulate matter in varying quantity and composition. A number of mechanisms have been suggested which produce particles varying in size from 10^{-9} to 10^{-4} m. One is the ejection of particles from the tops of the oceans, in the form of sea salt sprays. Others involve the transformation of gases into particles. This can occur as a result of photochemical reactions or as a result of adsorption to water droplets (clouds, haze), chemical reactions depending on the presence of water and finally evaporation of the water, leaving behind a suspended particle rather than a gas.

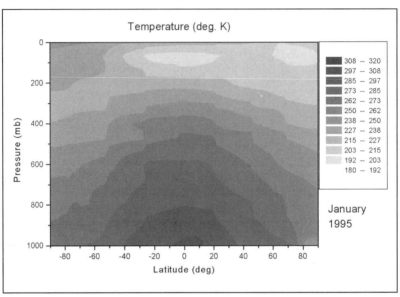

Figure 2.28a,b. Average temperature in the atmosphere as a function of height and latitude for January (above) and July (below) 1995, both at longitude zero (based on NCEP-NCAR, 1998).

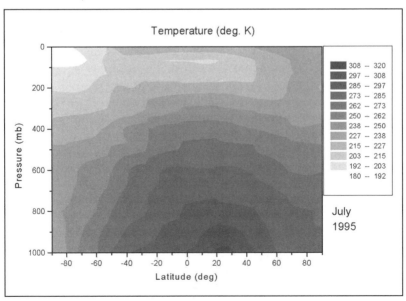

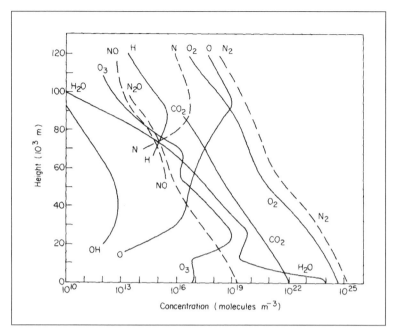

Figure 2.29. Estimated mean concentrations of some gaseous constituents of the atmosphere, as functions of height. (Based on Almquist, 1974).

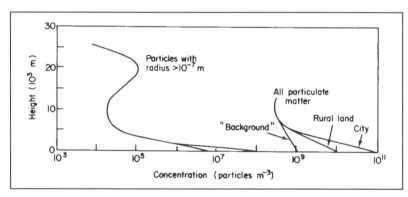

Figure 2.30. Trends of particle concentrations as functions of height, either for large particles only (left) or for all particles. The range of surface concentrations over different surfaces is indicated. "Background" refers to an extrapolated, unpolluted and clean surface, such as the Arctic ice covers. (Based on Craig, 1965; Wilson and Matthews, 1971).

Dust particles from deserts or eroded soil also contribute to the particle concentration in the atmosphere, as does volcanic action. Man-made contributions are in the form of injected particles, from land being uncovered (deforestation, agricultural practice), industry and fuel burning, and in the form of gaseous emissions (SO_2, H_2S, NH_3, etc.) which may later be transformed into particles by the above-mentioned processes or by catalytic transformations in the presence of heavy metal ions, for example.

Wilson and Matthews (1971) estimate that the yearly emission into or formation within the atmosphere of particles with a radius smaller than 2×10^{-5} m is in the range $(0.9–2.6) \times 10^{12}$ kg. Of this, $(0.4–1.1) \times 10^{12}$ kg is due to erosion, forest fires, sea salt sprays and volcanic debris, while $(0.3–1.1) \times 10^{12}$ kg is due to transformation of non-anthropogenic gaseous emissions into particles. Anthropogenic emissions are presently in the range $(0.2–0.4) \times 10^{12}$ kg, most of which is transformed from gaseous emissions.

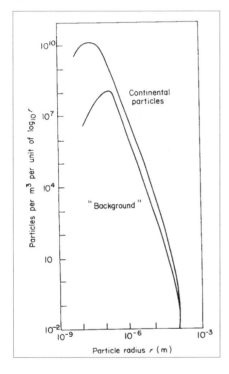

Figure 2.31. Size distribution of particles in the atmosphere. The estimated distribution over land is shown separately. The distribution over oceans resembles that of the clean "background" distribution, except in the radius interval 10^{-6}–10^{-5} m, where it approaches that for continental particles. (Based on Wilson and Matthews, 1971).

Figure 2.30 gives a very rough sketch of the height distribution of the particle number density. The distribution of large particles (radii in the range 10^{-7}–2×10^{-5} m) is based on measurements at a given time and place (based on Craig, 1965), while the total particle counts are based on extrapolations of the size distribution to very small particles (see Fig. 2.31). It is clear that the total number of particles depends sensitively on the density of very small particles, which on the other hand do not contribute significantly to the total mass of particulate matter. The concentrations of rural and city particles in the right–hand part of Fig. 2.30 correspond to an extrapolation towards smaller radii of the size distributions in Fig. 2.31, such that the number of particles per m^3 per decade of radius does not decrease rapidly towards zero.

Returning to mass units, the total mass of particles in the atmosphere can be estimated from the emission and formation rates, if the mean residence time is known. The removal processes active in the troposphere include dry deposition by sedimentation (important only for particles with a radius above 10^{-6} m), diffusion and impact on the Earth's surface, vegetation, etc., and wet deposition by precipitation. The physical processes involved in wet deposition are condensation of water vapour on particles, in-cloud adsorption of particles to water droplets (called rainout), or scavenging by falling rain (or ice) from higher-lying clouds (called washout). Typically, residence times for particles in the troposphere are in the range 70–500 h (Junge, 1963), leading to the estimate of a total particle content in the atmosphere of the order of 10^{14} kg.

Particles may also be removed by chemical or photochemical processes, transforming them into gases.

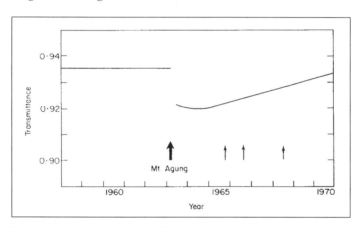

Figure 2.32. Transmittance of solar radiation in the direction of incidence, as measured in Hawaii near sunrise, during the periods before and after the volcanic eruption of Mt Agung in Bali. Other smaller volcanic eruptions in the following years are indicated by arrows. The curves have been smoothed over seasonal and other short-term variations, with the fluctuations being generally less than ±0.05 in the transmittance on clear days. (Based on Ellis and Pueschel, 1971).

Particle residence times in the stratosphere are much longer than in the troposphere. Sulphates (e.g. from volcanic debris) dominate, and the particles are spread rather homogeneously over each hemisphere, no matter where they are being injected into the stratosphere. Residence times are of the order of years, and the removal seems to be by diffusion into the troposphere at latitudes around 55°. This picture is based on evidence from major volcanic eruption and from detonation of nuclear weapons of megaton size.

Both of these inject particulate matter into the stratosphere. The particles in the stratosphere can be detected from the ground, because they modify the transmission of solar radiation, particularly close to sunrise or sunset, where the path-length through the atmosphere is at its maximum. Figure 2.32 shows atmospheric transmittance at sunrise, measured in Hawaii before and after the major volcanic eruption of Mt. Agung on Bali in 1963. The arrows indicate other eruptions which have taken place near the Equator during the following years and which may have delayed the return of the transmittance towards its value from before 1963. Figure 2.33 shows the ground deposition of radioactive debris (fallout), at a latitude of 56°N, after the enforcement of the test ban in 1963, which greatly reduced but did not fully end nuclear tests in the atmosphere. It is concluded that the amount of radioactivity residing in the stratosphere is reduced to half each year without new injection. Observations of increased scattering (resulting in a shift towards longer wavelengths) following the very large volcanic eruption of Krakatoa in 1883 have been reported (sky reddening at sunrise and sunset). It is believed that the influence of such eruptions on the net radiation flux has had significant, although in the case of Krakatoa apparently transient, effects on the climate of the Earth.

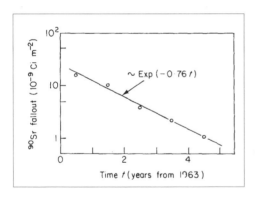

Figure 2.33. Yearly fallout of ^{90}Sr following the atmospheric bomb test ban in 1963, as measured in Denmark. (Based on Aarkrog, 1971).

Absorption and scattering in the atmosphere

Absorption and scattering of radiation take place throughout the atmosphere. The radiation may cause electrons in atoms to go from one orbital to another, or it may cause transition between vibrational or rotational levels in molecules. Generally, the fundamental energy differences associated with rotational excitations are smaller than those associated with vibrational excitations, which again are smaller than those associated with low-lying electron states. Only with a fairly small probability is the solar radiation capable of ionising atoms (expelling electrons) or dissociating molecules.

Since the wavelengths in the solar spectrum are much longer than the dimensions of atoms and molecules found in the atmosphere, the time-dependent fields of the atomic or molecular system can be approximated by the first few terms in a multipole expansion. In the dipole approximation, the probability of a transition between two levels i and f, accompanied by the emission or absorption of radiation with frequency equal to (or very close to)

$$v_{if} = |E_f - E_i|/h,$$

where $E_{i(f)}$ is the energy of level $i(f)$, is proportional to the square of the dipole transition moment d_{if} (an integral over the initial and final state quantal wavefunctions times the position vector; see e.g. Merzbacker, 1970). The absorption probability (or "rate") is further proportional to the spectral intensity at the required frequency.

"Rate of absorption" $\propto |d_{if}|^2 I(v_{if})$.

If the absorption and re-emission processes were sufficiently frequent, a state of thermal equilibrium would be reached, in the sense that the relative population of any two atomic or molecular levels would be given by the Boltzmann factor $\exp(-hv_{if}/(kT))$. If the radiation has a frequency distribution corresponding to the Planck law, the common temperature T will be that appearing in the Planck distribution for the radiation. In this way a single atom or molecule can be in equilibrium with radiation and can be ascribed the same temperature as that of the radiation.

On the other hand, the thermodynamic temperature is defined from the distribution of kinetic energy in the "external" motion of the atoms or molecules. This distribution is maintained by collisions between the atoms or molecules. If the frequency of collisions is comparable to that of interactions with the radiation, the kinetic and radiation temperatures will become similar (e.g. the radiative transfer section of the Sun).

Where collisions are much more frequent than interactions with the radiation field, the radiative energy absorbed by a given atomic or molecular transition will become shared among all the atoms or molecules, and if the corresponding emission spectrum can be described by the black-body law, the temperature entering should be the kinetic temperature, which need not be equal to the temperature of the radiation field (in fact, the kinetic temperature must be lower than the radiation temperature, if there are no other energy sources). Re-emission at frequencies corresponding to the same definite atomic or molecular transitions that caused absorption of radiation can be regarded as scattering of the incoming radiation.

In the Earth's atmosphere, the number of molecular collisions is generally large compared with the number of interactions with the solar radiation. Thus, the radiation which is not re-emitted as scattered light will be redistributed over a large number of thermal degrees of freedom, i.e. the absorbed light is transformed into heat. The population of various molecular levels will not be in equilibrium with the temperature of the solar radiation, and the spectrum of scattered light will not be of Planck form (cf. Fig. 2.13).

The scattering process may be viewed as the creation of a dipole field in the atom or molecule, as a result of the radiation field, and a subsequent emission of radiation, as is well known from an oscillating dipole. The angular distribution of scattered light is proportional to $(1+\cos^2\psi)$, where ψ is the angle between incident and scattered light.

Absorption processes in different frequency regions

Even in the thermosphere, most ultraviolet radiation with wavelength below 1.8×10^{-7} m is already being absorbed by N_2O and O_2. Owing to its low density, the thermosphere exhibits a rapid temperature response to variations in solar intensity, such as those caused by sunspots. However, for the same reason, these temperature changes have little or no effect on the lowerlying layers of the atmosphere. Figure 2.34 shows the penetration depth of ultraviolet radiation, confirming that the shortest wavelengths are being stopped at a height of 100 km or more.

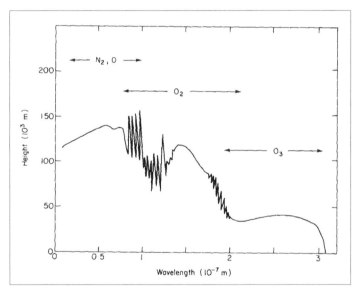

Figure 2.34. The altitude, as a function of wavelength, at which the intensity of solar radiation is reduced to 1/e. The main absorbing agents are indicated for different regions of the ultraviolet part of the spectrum. (Based on Ratcliffe, 1960).

This figure also shows that most of the radiation in the wavelength interval 1.8×10^{-7} to 3×10^{-7}m is absorbed in the mesosphere or the stratosphere. The agents of absorption are O_2 and in particular O_3. The rate of absorption by ozone peaks around 2.5×10^{-7}m wavelength, and the amount of energy converted to heat is sufficient to explain the peak in the temperature profile at about 50 km (see Fig. 2.28).

The ozone concentration is not uniform as a function of geographical position, as indicated by Fig. 2.35. Ozone is formed in the stratosphere by photodissociation of molecular oxygen,

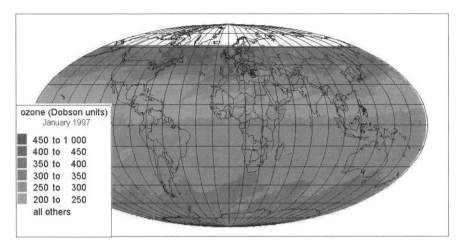

Figure 2.35a,b. Daily satellite measurements of ozone distributions for 1 January (above) and 1 April (below) 1997. The figures indicated are the volumes (in Dobson units, 10^{-5} m^3 m^{-2}) which the ozone would occupy at sea-level standard temperature and pressure, if all the ozone in a vertical column were put together. Areas obscured from the satellite are shown as white (NASA, 1998).

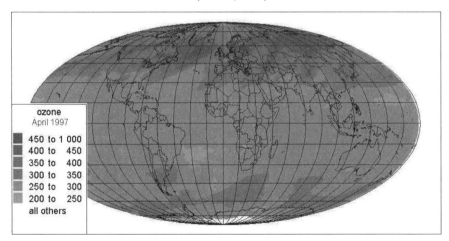

$$O_2 + \gamma \rightarrow 2O$$
$$O_2 + O \rightarrow O_3$$

(Chapman, 1943). Natural removal processes include

$$O_3 + \gamma \rightarrow O_2 + O$$
$$O_3 + O \rightarrow 2O_2$$
$$O_3 + NO \rightarrow NO_2 + O_2$$

(Hesstvedt, 1973).

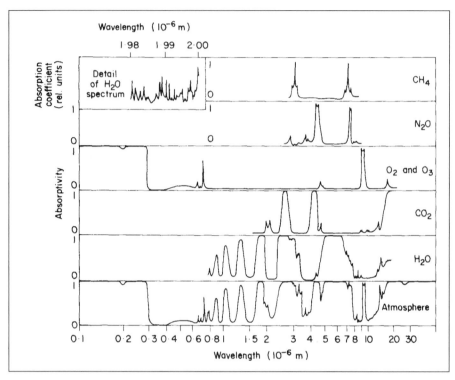

Figure 2.36. Spectral absorption efficiency of selected gaseous constituents of the atmosphere, and for the atmosphere as a whole (bottom) (from Fleagle and Businger, *An Introduction to Atmospheric Physics*, Academic Press, 1963).

The concentrations of nitrogen oxides in the stratosphere may be increased by a number of human activities. The ozone concentrations vary between years (Almquist, 1974), but anthropogenic emissions of fluorocarbons are believed to dominate the picture, as they have been demonstrated to cause increased ozone to accumulate near the poles (cf. Fig. 2.35).

While the ultraviolet part of the solar spectrum is capable of exciting electronic states and molecular vibrational–rotational levels, little visible light is absorbed in the atmosphere. The energy is insufficient for most electronic excitations, and few molecular bands lie in this frequency region. Of course, harmonics of lower fundamental bands can be excited, but the cross section for such absorption processes is very low.

In the infrared region, at wavelengths above 7×10^{-7} m, the fundamental vibrational and rotational bands of several molecules are situated. They give rise to clearly identifiable signatures of a large number of molecules, including H_2O, CO_2, N_2O, CH_4, CO, SO_2, H_2S, NO, NO_2 and NH_3. A few of these absorption spectra are shown in Fig. 2.36. Owing to the variations in water content, as well as the seasonal changes in the concentration of those molecules formed in biological cycles, the combined absorption spectrum is far from invariant.

Models for the description of scattered radiation

The cross section for scattering processes in the atmosphere is large enough to make multiple scattering important. The flux reaching a given scattering centre is thus composed of a unidirectional part from the direction of the Sun, plus a distribution of intensity from other directions. On the basis of scattering of radiation on atoms and molecules much smaller than the wavelength (the so-called Rayleigh scattering, described above), the intensity distribution over the sky may be calculated, including the direct (unscattered) part as well as the simply and multiply scattered parts. Figure 2.37 gives the result of such a calculation.

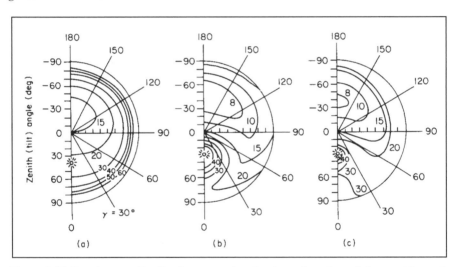

Figure 2.37. Luminance distribution over the sky (as a function of the zenith angle and azimuth relative to that of the Sun, for the direction of observation): (a) for pure Rayleigh atmosphere (multiple Rayleigh scattering on gases only), (b) for Mie atmosphere (including multiple Mie scattering on particles typical of continental Europe for a cloudless sky) and (c) measured on a clear day in the Swiss Alps. The three luminance distributions have been arbitrarily normalised to 15 at the zenith. The Sun's position is indicated by a circle (based on Möller, 1957).

Assuming the solar radiation at the top of the atmosphere to be unpolarised, the Rayleigh scattered light will be linearly polarised, and the distribution of light reaching the Earth's surface will possess a component with a finite degree of polarisation, generally increasing as one looks away from the direction of the Sun, until a maximum is reached 90° away from the Sun (Fig. 2.38).

The Rayleigh distribution of luminance (intensity) over the sky does not correspond to observations, even on seemingly clear days without clouds or visible haze (see Fig. 2.37). The reason for this is the neglect of scattering on particulate matter. It follows from the particle size distribution given in Fig. 2.31 that particles of dimensions similar to the wavelength of light in the solar spectrum are abundant, so that a different approach to the problem must be taken.

The theory of scattering of radiation on particles, taking into account reflection and refraction at the surface of the particle, as well as diffraction due to the lattice structure of the atoms forming a particle, was developed by Mie (1908). The cross section for scattering on a spherical particle of radius r is expressed as

$$\sigma = \pi r^2 K(y, n),$$

where the correction factor K multiplying the geometrical cross section depends on the ratio of the particle radius and the wavelength of the radiation through the parameter $y = 2\pi r/\lambda$ and on the refraction index (relative to air) of the material forming the particle, n.

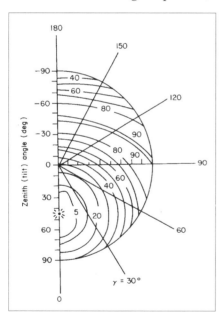

Figure 2.38. Distribution of polarisation over the sky (in %) for a typical clear sky atmosphere (based on Sekera, 1957).

Numerical calculations of the function K (Penndorf, 1959) are shown in Fig. 2.39 for fixed refraction index. The behaviour is very oscillatory, and the absorption may exceed the geometrical cross section by as much as a factor of four.

Angular distributions of Mie scattered light are shown in Fig. 2.40 for water particles (droplets) of two different size distributions representative for clouds and haze. The Rayleigh scattering distribution is shown for comparison. The larger the particles, the more forward peaked is the angular distribution, and the higher the forward to backward ratio.

In Fig. 2.37b, the results of a Mie calculation of the luminance distribution over the sky are presented. Here, a particle size distribution proportional to r^4 has been assumed, and the mean refraction index of the atmosphere has been fixed at $n = 1.33$ (Möller, 1957). Multiple scattering has been included as in the case of pure Rayleigh scattering, shown in Fig. 2.37a. For comparison, Fig. 2.37 also shows the results of measurements of luminance performed under the "clean air" conditions of a high-altitude Alpine site. The calculation using the Mie model is in qualitative agreement with the measured distribution, while the Rayleigh calculation is in obvious disagreement with the measurements.

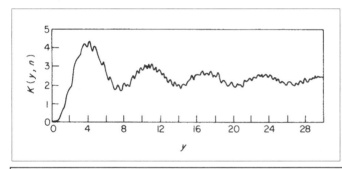

Figure 2.39. Cross-section function $K(y, n)$ for Mie scattering, as function of $y = 2\pi r/\lambda$ for fixed $n = 1.50$ (cf. text) (based on Penndorf, 1959).

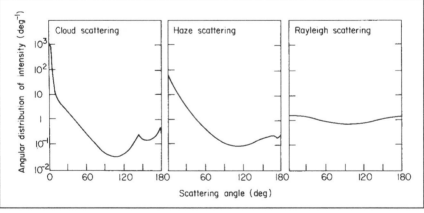

Figure 2.40. Angular distributions of radiation for simple (not multiple) scattering on water particles with different size distributions. Mie scattering theory was used in connection with the size distributions of clouds (comprising very large particles or droplets) and haze (smaller particle sizes but still comparable to *or* larger than the wavelength of solar radiation). For comparison, the angular distribution for Rayleigh scattering (particle sizes small compared to wavelength) is also shown. The intensity is averaged over different polarisation directions (based on Hansen, 1975).

The mean free path in the atmosphere, averaged over frequencies in the visible part of the spectrum, gives a measure of the visibility, giving rise to a clarity classification of the condition of the atmosphere (this concept can be applied to traffic control as well as astronomical observation, the main interest being on horizontal and vertical visibility, respectively). The atmospheric particle content is sometimes referred to as the turbidity of the air, with the small particles being called aerosols and the larger ones dust.

The types of stored energy

The net radiation fluxes discussed in section 2.2 showed the existence of energy transport processes other than radiation. Energy in the form of heat is transferred from the land or ocean surface by evaporation or conduction and from the atmosphere to the surface by precipitation, by friction and again by conduction in small amounts. These processes exchange sensible[*] and latent heat between the atmosphere and the oceans and continents. The exchange processes in the atmosphere include condensation, evaporation and a small amount of conduction. In addition, energy is added or removed by transport processes, such as convection and advection. The turbulent motion of the convection processes is often described in terms of overlaying eddies of various characteristic sizes. The advective motion is the result of more or less laminar flow. All motion in the atmosphere is associated with friction (viscosity), so that kinetic energy is constantly transformed into heat. Thus, the general circulation has to be sustained by renewed input of energy. The same processes are involved in the circulation in the oceans, but the quantitative relations between the different processes are different, owing to the different physical structure of air and water (density, viscosity, etc.). As mentioned earlier, the source of energy for the transport processes is the latitude variation of the net radiation flux. Additional transport processes may take place on the continents, including river and surface run-off as well as ground water flow. Direct heat transport in dry soil is very small, because of the smallness of the heat conductivity.

The amount of energy stored in a given volume of air, water or soil may be written

$$W^{stored} = W^{pot} + W^{kin} + W^{sens} + W^{lat}, \tag{2.5}$$

where W^{pot} is the geopotential energy, W^{kin} is the kinetic energy of external motion (flow), W^{sens} is the amount of sensible heat stored (internal kinetic motion) and W^{lat} is the amount of latent heat, such as the energies involved in phase changes (solid to quid, fluid to gaseous, or other chemical rearrangement)[**]. The zero point of stored energy is, of course, arbitrary. It may

---

[*] That is, due to heat capacity

[**] The sum of sensible and latent energy constitutes the total thermodynamic *internal energy* (cf. section 4.1).

be taken as the energy of a state where all atoms have been separated and moved to infinity, but in practice it is customary to use some convenient average state to define the zero of energy.

The geopotential energy may be written

$$W^{pot} = \rho(r)g(r)r, \tag{2.6}$$

where ρ is the density (of air, water or soil) and g is the gravitational constant at the distance r from the Earth's centre (eventually r is replaced by $z = r - r_s$ measured from the surface r_s). Both ρ and g further have a weak dependence on the geographical location (ϕ, λ). The kinetic energy W^{kin} receives contributions both from the mean flow velocity and from turbulent (eddy) velocities.

The sensible heat is

$$W^{sens} = \rho c_p T, \tag{2.7}$$

where the heat capacity c_p (at constant pressure) depends on the composition, which again may be a function of height and position. All quantities may also depend on time. The latent heat of a given constituent such as water may be written

$$W^{lat} = L_m(m_w + m_v) + L_v m_v, \tag{2.8}$$

where L_v (2.27 J kg^{-1} for water) is the latent heat of vaporisation and L_m (0.33 J kg^{-1} for ice) is the latent heat of melting. m_v is the mixing ratio (fractional content by volume) of water vapour and m_w is the mixing ratio of liquid water (applicable in oceans and soil as well as in the atmosphere). For other constituents it may be necessary to specify the chemical compound in which they enter in order to express the content of latent energy.

Total energy fluxes

As in the case of radiation, all relevant energy fluxes must now be specified, and the sum of radiation and non-radiation fluxes (E^{nonr}) gives the net total energy flux,

$$E^{total} = E^{rad} + E^{nonr} = E^{sw} + E^{lw} + E^{nonr}. \tag{2.9}$$

The sign convention introduced for radiative fluxes should still be applied.

Since many energy conversion processes (e.g. melting of snow and ice, run-off) take place at the surface of the Earth, the fluxes through strictly two-dimensional planes may vary rapidly for small vertical displacements of the defining plane and hence may be inconvenient quantities to work with. For this reason, the net total energy flux through such a plane will, in general, be different from zero, except possibly for year averages. For the upward boundary surface between Earth (continent or ocean) and atmosphere, the total net energy flux may be written

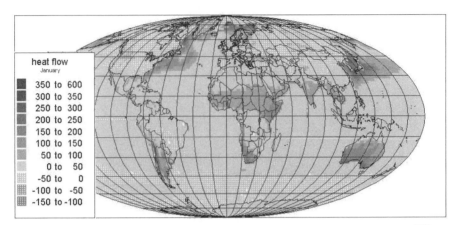

Figure 2.41a,b. Flux of sensible heat from the Earth's surface to the atmosphere $(-E_s^{sens})$ for January (above) and July (below) 1997, in W m^{-2} (based on NCEP-NCAR, 1998).

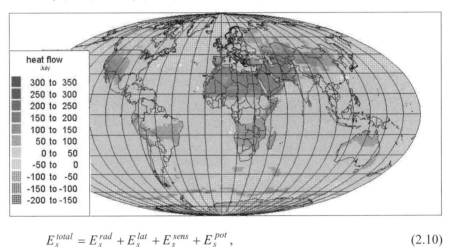

$$E_s^{total} = E_s^{rad} + E_s^{lat} + E_s^{sens} + E_s^{pot}, \tag{2.10}$$

The kinetic energy flow has been neglected, since the vertical velocity of matter at the Earth's surface is usually close to zero (an exception is volcanic matter being expelled into the atmosphere). Yet transport of matter does take place, usually by means of convective eddy flow with average vertical velocities that increase from zero as the height increases and by falling matter (including precipitation). The changes in geopotential energy involved in these flows are contained in the term E_s^{pot}. The radiation flux E_s^{rad} is given by (2.4). The latent energy flow, E_s^{lat}, is the rate of evaporation times the latent heat of vaporisation, L_v. Finally, the net flux of sensible heat, E_s^{sens}, includes all transfer of heat in both directions, such as transfer by convection, conduction and exchange of matter (e.g. precipitation or volcanic debris) with temperatures different from that of the surroundings.

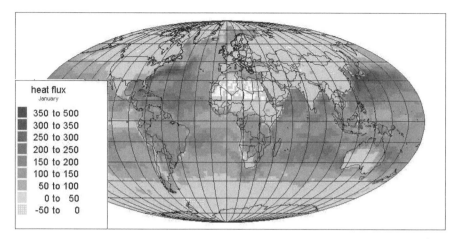

Figure 2.42a,b. Flux of latent heat from the Earth's surface to the atmosphere $(-E_s^{lat})$, i.e. L_v times the rate of evaporation, for January (above) and July (below) 1997, in W m^{-2} (based on NCEP-NCAR, 1998).

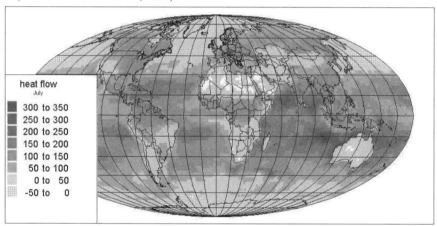

Figures 2.41 and 2.42 show the sensible and latent flux distributions for January and July 1997, and Fig. 2.43 shows the sum of the two fluxes, based on recent model evaluations within the reanalysis project of Kalnay *et al.* (1996). The analysis ensures balance between short- and long-wavelength radiation and non-radiative fluxes at the surface, and it is seen that there are significant areas where the latent and sensible heat fluxes are directed from the atmosphere towards the oceans or land masses. The sum of non-radiative fluxes behaves more smoothly than its components, indicating that the total energy surplus to be moved is split between evaporation and convective processes in ways depending strongly on local conditions (such as relative humidity).

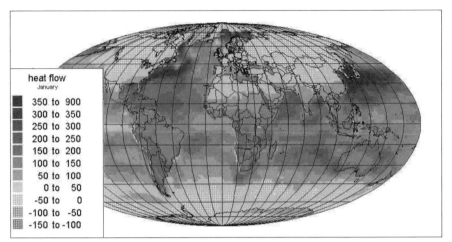

Figure 2.43a,b. Sum of non-radiative fluxes (i.e sensible and latent heat flows in Figs. 2.41–2.42) from the Earth's surface to the atmosphere for January (above) and July (below) 1997, in W m^{-2} (based on NCEP-NCAR, 1998).

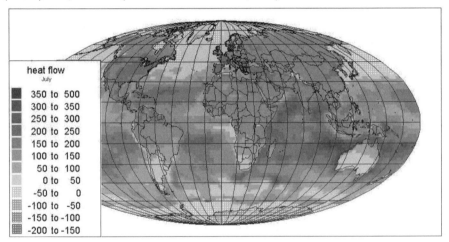

Energy transfer processes

Turning now to the behaviour of the amount of energy contained in a vertical column, it is convenient to consider first the atmospheric and the continent–ocean parts of the column separately. The depth of the continent–ocean column is taken as one at which no energy is transferred in either of the two vertical directions. This is an idealisation due to the neglect of outward transfer of heat produced in the interior of the Earth, which will be added in section 2.4.

Denoting the continent–ocean column by the suffix d, the change in stored energy may be expressed in the form

$$\frac{dW_d^{stored}}{dt} = E_s^{total} + F_d,\tag{2.11}$$

where F_d represents the horizontal net inflow of energy through the sides of the column considered. For a continental column, F_d would comprise the net surface run-off and ground water loss multiplied by the latent heat content L_w and the possible surplus of sensible heat over the surrounding soil, all taken with opposite sign because of the sign convention in (2.11).

Typically, F_d is much less important for continents than for oceans, where it comprises currents and other horizontal advection. The currents transport heat from the equatorial regions towards the poles, and in particular the North Atlantic (Gulf Stream) and Northern Pacific (Kuro Shio). According to Budyko (1974), the continental F_d is negligible in comparison. Also neglected in the above discussion of F_d is the addition or removal of potential energy by change of density (e.g. when a continental column loses part of its ground water during summer, or in an ocean column as a result of the different density of cold and warm water). According to (2.5)–(2.8), the change (2.11) in stored energy may involve a change in temperature, formation or melting of ice, etc.

Considering similarly a column in the atmosphere, bounded by the top of the atmosphere and the Earth's surface, the change in stored energy may be written

$$\frac{dW_a^{stored}}{dt} = E_0^{total} - E_s^{total} + F_a.\tag{2.12}$$

Since the flux of matter at the top of the atmosphere is negligible, E_0^{total} equals E_0^{raa}, which again equals zero if averaged over time and geographical position, but it depends on latitude and time as indicated in Fig. 2.12. The quantity F_a represents the energy gain of the column in question, as a result of horizontal transport by the general circulation of the atmosphere. The winds carry both sensible and latent heat, with the latter mainly in the form of water vapour. F_a may thus be written

$$F_a = F_a^{sens} + F_a^{lat} \approx F_a^{sens} + F_s^{lat} + L_v r,$$

where $-E_s^{lat}$ is the latent heat added to the atmosphere as a result of evaporation from the surface of the underlying ocean or continent, and r is the amount of precipitation, so that $L_v r$ represents the net transformation of latent heat (bound in water vapour) into sensible heat by condensation processes in the atmosphere. In this way $E_s^{lat} + L_v r$ represents the net vertical loss of latent heat for the atmospheric column. This may be equated to the gain by horizontal inflow, F_a^{lat}, provided the changes in stored, latent heat can be neglected, i.e. that $dW_a^{lat} / dt \approx 0$. Such an approximation is likely to be reasonable if annual averages are taken.

The net latitudinal transport in the atmosphere consists of the longitude-averaged value of F_a and the net transport by ocean currents, F_d, as shown schematically in Fig. 2.44. The quantities $-F_a$ and $-F_d$ express the transport of energy away from a given latitude region, in latitudinal direction. Being annual average values, the sum of $-F_a$ and $-F_d$ should equal the net energy flux at the top of the atmosphere, E_0^{rad}. In order to see this, (2.13) and (2.14) are added to give the change in stored energy for an entire vertical column of the Earth–atmosphere system, as a function of time,

$$\frac{d(W_a^{stored} + W_d^{stored})}{dt} = E_0^{rad} + F_a + F_d.$$

The annual average of the left-hand side must be approximately zero, since otherwise the mean temperature in a given latitude zone would continue to increase or decrease. However, there is a transport of energy from lower to higher latitudes, mediated by ocean currents and meridional wind transport, including processes such as evaporation, transport of water vapour and precipitation at another location. Details of the transport patterns can be deduced from data such as the ones presented in Figs. 2.41–2.43, in conjunction with the wind data presented in the following. Variations with time are substantial, including inter-annual differences. At latitudes higher than some 40–60°, there is a net inflow of latent heat (implying that precipitation exceeds evaporation) as well as a sensible energy gain.

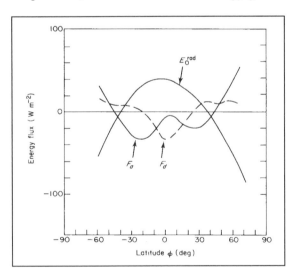

Figure 2.44. Schematic illustration of net energy flux components of the Earth-atmosphere system (longitude and time average). The net energy gains by a column at a given latitude, by transport of latent and sensible heat by atmospheric motion and ocean currents, respectively, are denoted F_a and F_d (based on Budyko; 1974).

Vertical transport in the Earth–atmosphere boundary layer

Returning to vertical transport, the important components of the net energy flux E_s^{total} at the Earth's surface, (2.10), are the net radiation flux, E_s^{rad}, the latent heat flux, E_s^{lat}, and the net sensible heat flux, E_s^{sens}. These were illustrated in Figs. 2.21 and 2.43.

The vertical transport of heat is predominantly achieved by turbulent motion. However, approaching the physical surface of the Earth, the turbulent air velocity must go to zero and, in a thin air layer close to the ground, no macroscopic motion takes place. This region, called the *laminar boundary layer* (typically only a fraction of a millimetre thick), allows vertical transport of heat, water vapour and horizontal momentum only by molecular diffusion processes. The laminar boundary layer provides a sort of coating of the surface, protecting it from the disruptive action of turbulent eddies. If the surface is rough, discontinuities may exist in the laminar boundary layer thickness, but a region of solely molecular processes will always exist.

A model of vertical transport near the ground is described and discussed in section 2.B.

The description of atmospheric motion

The external motion which takes place in the atmosphere (as opposed to the internal motion of molecules, defining the temperature) calls for mechanisms by which kinetic energy is created. Friction is an irreversible process by which kinetic energy is transformed into sensible heat, accompanied by an increase in entropy. External forces, such as the Coriolis force impressed by the rotation of the Earth, do not create or destroy kinetic energy. Neither do the processes by which latent heat is gained or lost. Thus, the only sources left to accomplish the maintenance of external kinetic motion despite the frictional damping are the potential energy and the energy of sensible heat, which can both be transformed into kinetic energy by adiabatic and reversible processes. The familiar examples of these transformations are gravitationally falling matter and buoyantly rising hot matter (or descending cool matter). In order to estimate the rate of creation and destruction of kinetic energy in the atmosphere, it is convenient first to collect the equations governing the general motion of the atmosphere.

These are the equation of motion,

$$\rho \frac{dv}{dt} = -\operatorname{grad} P + F_{viscous} + F_{ext},$$

where the frictional force $F_{viscous}$ depends on the internal properties of the fluid (the air), and the external force F_{ext} comprises the Coriolis force and the gravitational force, provided a coordinate system is employed which follows the rotation of the Earth; the equation of continuity,

$$\text{div}(\rho v) = -\frac{\partial \rho}{\partial t};$$

and the law of energy conservation (first law of thermodynamics),

$$\frac{dW^{total}}{dt} = \text{work plus heat added during time } dt.$$

The work performed on the system is that of the forces appearing on the right-hand side of the equation of motion, and the heat added is partly due to radiation, and partly to molecular conduction processes. W^{stored} contains kinetic, potential and sensible heat energy. The latent heat may be omitted in W^{stored}, provided any condensation or other process releasing latent heat is included in the heat added on the right-hand side, at the time when the latent heat is transformed into sensible heat.

The above equations would have to be complemented by the equation of state (e.g. the ideal gas law) and the source function specifying the external heat sources. These quantities are not independent of the constitution of the air, and in general the above equations are coupled to a large number of equations specifying the time development of the atmospheric composition (water vapour, ozone, etc.).

In practice, a simplified system of equations is used, based on an analysis of the relative importance of different terms, in particular of the forces, and depending on the scale of motion, which is considered interesting. The dominance of horizontal wind fields in the large-scale atmospheric motion makes it possible to simplify the vertical (z) component of the equation of motion, neglecting also the corresponding component of the viscous and Coriolis forces. This approximation leads to the hydrostatic equation,

$$\frac{\partial P}{\partial z} = -\rho g,$$

where g is the average acceleration of the Earth's gravity (neglecting anisotropy in the mass distribution inside the Earth and the latitude dependence associated with Coriolis corrections), $g \approx GM(r)/r^2 \approx 9.8$ m s^{-2}. This does not imply that terms in the equation of motion such as dw/dt, where w is the z-component of v, may be neglected in other circumstances, but only that they are small in comparison with the two terms kept in the equation expressing the hydrostatic approximation. This equation states that the pressure gradient at any infinitesimal horizontal area must balance the weight of the entire air column above it, just as if there were no motion in the atmosphere.

Time averaging

In order to describe the general circulation in the atmosphere, a time-averaging procedure is performed, over a time interval Δt which is small compared with the interesting large-scale circulation, but large compared with the bulk of turbulent eddy periods,

$$\overline{A} = \frac{1}{\Delta t} \int_{t_1}^{t_1 + \Delta t} A(t_2, ...) dt_2,$$

for any quantity A. The deviation of A from its time-averaged value is denoted $\tilde{A}$,

$$A = \overline{A} + \tilde{A}. \tag{2.13}$$

The equations describing the system may be simplified by further introducing the density weighted averages,

$$A^* = \overline{\rho A} / \overline{\rho}, \tag{2.14}$$

and the corresponding splitting of A,

$$A = A^* + A'. \tag{2.15}$$

One should note that $\overline{\tilde{A}} = 0$, but $\tilde{A}^* \neq 0$ and that $\overline{\rho A'} = \overline{\rho(A - A^*)} = 0$.

The basic equations for motion of the atmosphere in terms of time-averaged variables are derived in section 2.C, which also discusses the experimental evidence for expecting the splitting of atmospheric motion into large-scale average motion and small-scale turbulence to work under certain conditions. It is well known from the quality of meteorological forecasts that the period of predictability is of the order of 2–5 days. This means that the terms coupling the small- and the large-scale motion are small enough to be neglected for periods of a few days, but eventually grow (at speeds depending on the actual weather pattern) and make any forecast unreliable. This fundamental limitation on meteorological forecasting does not imply that climatic forecasting is impossible. The weather system is forced by the seasonally periodic incoming solar radiation, which makes the long-term behaviour (climate being defined as about 30-year averages of variables) stable and predictable, as a condition for maintaining an atmosphere at all. The boundary conditions of the system make overall stability co-exist with unpredictability on short-time or short-distance scales, a feature of quasi-chaotic systems discussed in more precise terms in section 2.3.3.

Features of the observed atmospheric circulation

For two seasons, Fig. 2.45 shows the longitude-averaged component of the large-scale wind velocity along the x-axis, i.e. parallel to the latitude circles. This component is called the zonal wind. At low latitudes, the zonal wind is directed towards the west, at mid-latitudes towards the east (reaching high speeds around 12 km altitude), and again towards the west near the poles.

In Fig. 2.46 the meridional circulation is exhibited, again for two different seasons and averaged over longitudes. The figure gives the streamfunction, related to the mass flux ρv^* by

$$\Psi(\phi,z) = \int \overline{\rho v}^* \cdot \boldsymbol{n} \, r_s \cos\phi \, \mathrm{d}\lambda \, \mathrm{d}s = 2\pi r_s \cos\phi \int_0^{(\phi,z)} \overline{\rho}[v^*] \cdot \boldsymbol{n} \, \mathrm{d}s,$$

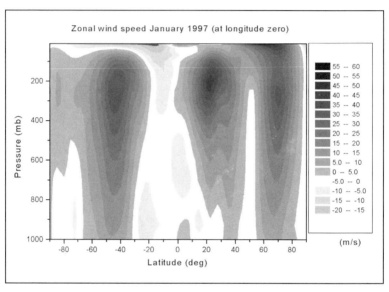

Figure 2.45a,b. Zonal winds (at longitude zero) for January and April of 1997, as functions of latitude and pressure (height) (based on NCEP-NCAR, 1998).

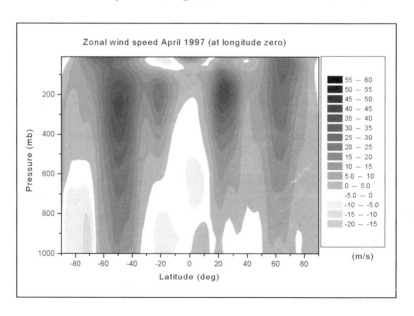

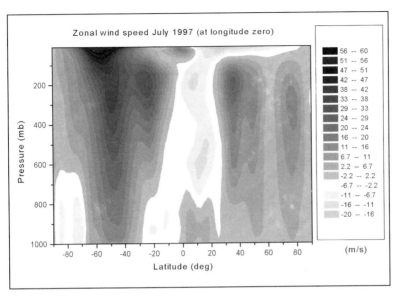

Figure 2.45c,d. Zonal winds (at longitude zero) for July and October of 1997, as functions of latitude and pressure (height) (based on NCEP-NCAR, 1998).

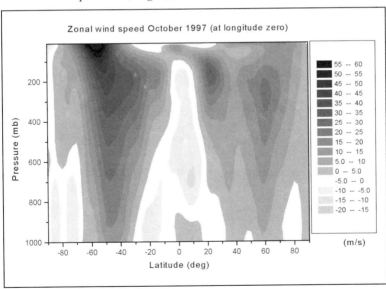

where *n* is normal to the area element $d\lambda\,ds$, and the integration path s in the (y,z) plane connects the surface $\Psi = 0$ with the point (ϕ, z) considered. "[ ]" denotes longitude average. The sign of Ψ along a closed loop streamline is positive if the direction of flow is towards the south at the ground and towards the north aloft. The loops are called meridional cell motion, and one notes a cell of large mass transport near the Equator. Its direction changes with season. On an average annual basis, there are two additional cells on each side of the Equator. They are called Hadley cells (Hadley, 1735), with the northern one being negative and the southern one being positive, according to the above sign convention. Near the poles there is an even weaker set of meridional cells without seasonal sign changes. 1997 data indicate that all the cells extend up to about 200 mb, in contrast to Fig. 2.46 (NASA, 1998). The cell structure is not entirely stable and features transient cells of shorter duration.

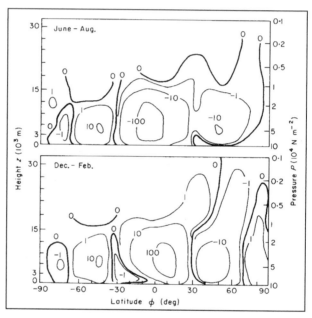

Figure 2.46. Summer (top) and winter (bottom) streamfunctions for the meridional mass transport (longitude average), given in units of 10^9 kg s^{-2}. Cells with a positive streamfunction have northward transport at top (based on Newell *et al.*, 1969).

Combining the standing cell picture with the mean zonal winds, one observes the following regularities near the Earth's surface. In a region near the Equator, the westward zonal winds combine with meridional transport towards the Equator (from both sides) to form the *trade winds* (blowing towards the southwest in the northern Hemisphere and towards northwest in the Southern Hemisphere). At mid-latitudes the eastward zonal winds combine with meridional cell motion away from the Equator to produce the *"westerlies"* (blowing towards the northeast in the Northern Hemisphere and towards southeast in the Southern Hemisphere). Further away from the Equator, in the polar regions, the wind directions are as for the trade winds.

The deviation of the wind field from the longitude average is not unimportant. A part of it is quite regular, in the form of standing horizontal cells, which are usually denoted (large-scale) eddies. They play an important role in transporting angular momentum towards the poles. Figure 2.47 shows the annual and longitude average of this transport.

Angular momentum is added to the atmosphere due to friction and drag forces between the atmosphere and the Earth's surface. Figure 2.48 gives a sketch of the torque acting on the atmosphere as a result of these forces. The main part of the torque is due to friction, but the contribution from drag forces is not negligible in mountain regions. The figure is based on a rough estimate by Priestley (1951), but it is consistent with the angular momentum transport in the atmosphere, as mentioned by Lorenz (1967).

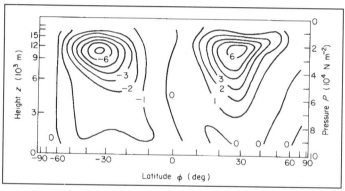

Figure 2.47. Northward transport of angular momentum by eddies, averaged over year and longitudes and expressed as 10^{18} kg m^2 s^{-1} per unit of pressure layer, i.e. per 10^4 N m^{-2}. Expressed per unit of height z, the magnitude of transport would increase less rapidly upwards towards the two maxima (based on Buch, 1954; Obasi, 1963; also using Lorenz, 1967).

The angular momentum added to the atmosphere at low latitudes is transported to a height of about 10 km by the meridional cell motion. Here it is taken over by the large-scale horizontal eddy motion, which brings it to mid-latitudes. Practically no transport across latitudes is performed by the meridional cell motion itself, as can be seen from Fig. 2.49. The absolute angular momentum (i.e. measured in a coordinate system not following the Earth's rotation) of a certain volume of the atmosphere is proportional to the tangential velocity and to the distance to the Earth's axis of rotation, $r \cos\phi$. If the same volume of air is moved polewards, the value of $r \cos\phi$ decreases, and the volume of air must either increase its tangential velocity or get rid of some of its angular momentum. Both happen in the mid-latitude atmosphere. The regions of increased tangential velocity are easily recognised at a height of about 12 km and latitudes of 30–50° (pressure about 200 mb, cf. Fig.

2.45). Some angular momentum is lost by internal friction in the atmosphere, and some is transported back down to Earth level, where it compensates for the angular momentum losses, occurring as a result of friction with the surface. The fact that the rotation of the Earth is not accelerating or decelerating shows that, on average, friction forces in opposite directions cancel each other out at low and mid-latitudes. At latitudes above roughly 60°, the transport of angular momentum reverses sign, becoming a net transport towards the Equator.

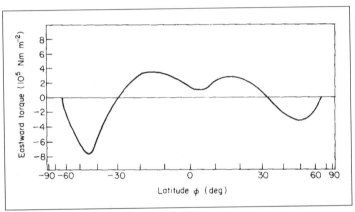

Figure 2.48. Estimate of the annual and longitude-averaged eastward torque acting on the atmosphere as a result of friction and momentum impact at the Earth's surface (based on Priestley, 1951, as quoted by Lorenz, 1967).

It has been pointed out by Lorenz (1967) that the transport of angular momentum by large eddies cannot be regarded as a simple diffusion process, since the corresponding "eddy diffusion coefficient" k would in that case have to be negative over a large part of the atmosphere. Indeed the transport of angular momentum has to be seen as an integral part of the general circulation, which cannot be considered to consist of just zonal and meridional motion. This is also inherent in the equation of motion written in terms of density averaged quantities (see section 2.C), which contains the horizontal eddy motion as a combination of the velocity vector components V_x^* and V_y.

Energy conversion processes and the separation of scales of motion

In order to include such essential properties as those mentioned above in a model of the circulation, it is important that the time interval Δt used in defining the average properties (2.13) and (2.14) be properly defined. This again poses the question of whether or not a rigorous division of atmospheric motion into large-scale and small-scale (turbulent eddy) motion is at all possible. This issue is dealt with in section 2.C, where it is seen that the wind velocity distribution exhibits two distinct categories, which can be denoted "macroscopic motion" and "microscopic motion" or turbulence. This allows averages to be consistently defined as containing only the macroscopic components of motion.

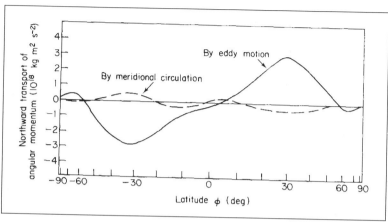

Figure 2.49. Northward transport of angular momentum averaged over year and longitude. The meridional contribution is the one carried by meridionally averaged winds, whereas the eddy contribution is obtained as the zonal average of the transport arising from the deviations from the zonal average of the wind velocity vector (thus including large-scale, horizontal eddies) (based on Buch, 1954; Obasi, 1963, as quoted by Lorenz, 1967).

In order to describe all the necessary transport of heat, any accurate model of the circulation of the atmosphere must include couplings to the heat transport within the ocean–continent system (ocean currents, rivers, run-off along the surface and to a lesser extent, as far as heat transport is concerned, ground water motion). Such coupled models will be considered below in section 2.3.2.

The kinetic energy of the atmospheric circulation is diminished by friction, leading to an irreversible transformation of kinetic energy into internal energy (heat). In order to make up for such frictional losses, new kinetic energy must be created in the atmosphere. This can be achieved essentially by two processes, both of which are reversible and may proceed adiabatically (details of the formulation are given in section 2.C). One is the conversion of potential energy into kinetic energy (by gravitational fall), and the other is the transformation of internal energy into kinetic energy by motion across a pressure gradient.

Creation and destruction of kinetic energy

Direct measurements of heat and temperature distributions (cf. Fig. 2.28 and section 2.C) allow an estimate of the generation of internal energy available for possible conversion into kinetic energy. With the use of wind data (e.g. Figs. 2.45–2.47), it is possible to obtain an estimate of the conversion of available energy and large-scale horizontal eddy motion into zonal motion. Also, the available energy may be split into a zonal part corresponding to the use of the zonal mean temperatures and a (large-scale) eddy part derived from the deviations of the temperature from longitudinal average (Lorenz, 1967; Newell et al., 1969). Separate estimates exist for the frictional losses from zonal and eddy motion as a function of height. An attempt to survey the main conversion processes (deriving poorly known quantities, such as the conversion of available eddy energy into kinetic eddy energy, from balance requirement) is given in Fig. 2.50, based on Lorenz (1967) and considering comments by Newell et al. (1969), who discusses the seasonal variations and uncertainties involved in the procedures used.

On an annual and global average basis, the creation of kinetic energy in the form of large-scale motion (V^*, w^*) amounts to 2.3 W m^{-2} or 0.7% of the solar radiation at the top of the atmosphere. For consistency, the frictional losses must be of equal magnitude, which is not quite consistent with direct estimates (4–10 W m^{-2}). Newell et al. argue that the value of about 2.3 W m^{-2} given in Fig. 2.50 is most likely to be correct.

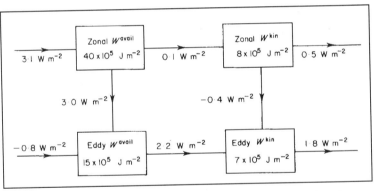

Figure 2.50. Estimated summary of the average energy (boxes) and main conversion processes in the atmosphere, for a vertical column, averaged over place and time. Boxes contain kinetic energy (W^{kin}) and the amounts of internal and potential energy available for conversion into kinetic energy (W^{avail}, "available" refers to energy above the reference state of the atmosphere). The compartments are further divided into the energy of zonally averaged motion and the average deviations from these, denoted eddy energy (based on Oort, 1964; Lorenz, 1967).

The process of creating kinetic energy may be intuitively interpreted in terms of air parcels of a temperature different from the ambient temperature. Such air parcels will rise (if they are hotter than the surroundings) or sink (if they are colder) and give rise to a redistribution of the surrounding air (i.e. to kinetic motion) for as long as temperature differences exist at the same altitude. More precisely, the change in kinetic energy is due to movement of the air parcels across horizontal or vertical pressure gradients, as directly implied by the equations given in section 2.C.

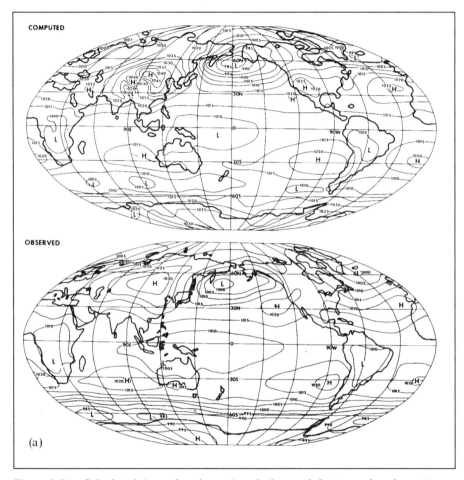

(a)

Figure 2.51a. Calculated (top of each pair) and observed (bottom of each pair) sea-level pressure in mb (1 mb being 10^2 N m^{-2}), (a, above) averaged over the period December–February (for *b*, see next page).

Models of general atmospheric circulation

Before going into the influence of the oceans and land masses on the atmospheric circulation, examples of early numerical simulation of the atmospheric circulation without including couplings to oceanic circulation will be presented (based on the work of Manabe *et al.*, 1974; Manabe and Holloway,

1975; Hahn and Manabe, 1975). They used a global, horizontal grid of about 250 km latitude and longitude steps and around ten vertical layers extending to a height of about 30 km, including equations of motion for wind, temperature and water vapour, and using the hydrostatic approximation and the ideal gas equation of state. Continental topography is modelled, whereas the oceans are prescribed seasonal temperatures as boundary conditions for the atmospheric integration. The radiation source term is calculated as a function of the seasonal variation in extraterrestrial radiation, the state variables plus ozone and carbon dioxide data. The hydrological cycle includes a model for evaporation, condensation and cloud distribution, as

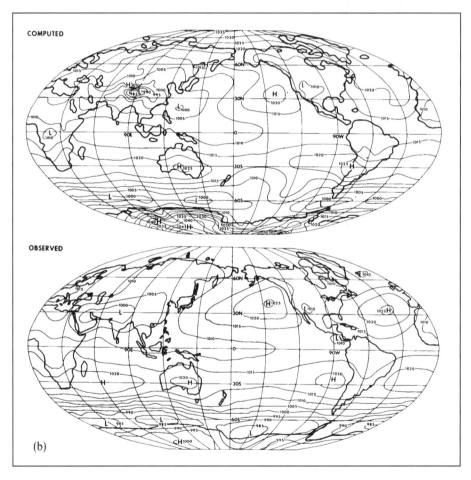

Figure 2.51b. Averaged over the period June–August (cf. *a* on previous page). From S. Manabe and J. Holloway (1975), *J. Geophys. Res.* **80**, 1617–1649, copyright of American Geophysical Union.

well as the processes involving ice and snow formation and accumulation as terrestrial snow cover or sea ice. Turbulent and sub-grid scale convection and transport of vapour have been modelled.

Figures 2.51a,b compare the average ground-level pressures during the periods December to February and June to August with observed values. Figure 2.52 gives the mean zonal wind for July as a function of latitude and height. The longitudinal averaging extends only over the interval 80°E to 95°E, whereas the observed distribution is for 100°E. Within the latitudes included in the figure, the agreement is fair, but in general the model is rather poor in predicting the longitude average of zonal winds in all height–latitude regions, in particular the magnitude and position of the strong mid-latitude jets observed at a height of about 12 km (an indication of the sensitivity of these features on the seasonal variation of radiation and on the model grid size may be inferred from a comparison with earlier work, e.g. Holloway and Manabe, 1971).

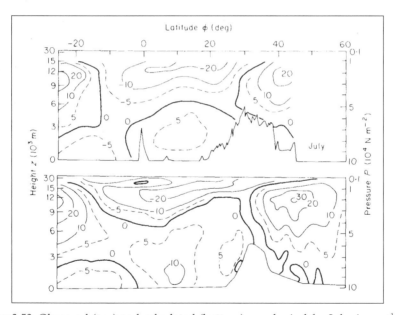

Figure 2.52. Observed (top) and calculated (bottom) zonal wind for July, in m s^{-1} (positive eastwards). The lower silhouette represents the mountain topography, an idealised one being assumed in the calculation. The observed wind field is for 100°E, whereas the calculated one has been averaged over the interval 80–95°E (based on Manabe *et al.*, 1974).

Figure 2.53 gives the streamfunction corresponding to the calculated meridional circulation for January and for July. Corresponding measured values, averaged over somewhat longer periods (three months), were shown in Fig. 2.46. In this case, the agreement is quite convincing, both in regard to the strength of the Hadley cells near the Equator and to the distribution of the weaker cells at higher latitudes. Figure 2.54 gives the northward transport of absolute angular momentum by eddies for July and January, averaged over longitudes. The corresponding measured values are found in Fig. 2.47, but averaged over the year. Figure 2.54 shows that the calculated transport is strongest in winter, and the average of January and July generally concurs with the value derived from measurements.

In Fig. 2.55, the annual average precipitation rates derived from the model calculation are compared with measured values. Apparently, agreement between the two is very good, over both oceans and continents. A comparison with earlier efforts by Manabe's group (e.g. those neglecting the seasonal variation in the extraterrestrial radiation) shows that the precipitation rates are very sensitive indicators of the appropriateness of the assumptions regarding the hydrological cycle. The more accurate presentation of measured data in Fig. 2.57 indicates detailed features not captured by the limited early atmospheric model results.

As suggested by Fig. 2.50, the formation of kinetic energy mainly involves the eddy motion (i.e. deviations from zonal mean winds). Figure 2.56 shows the calculated latitude variations of the main conversion branches: the transformation of available eddy energy into eddy kinetic energy and of zonal kinetic energy into eddy kinetic energy (which is mostly negative, i.e. the process goes the opposite way) and the dissipation of eddy kinetic energy through friction (transformation of large-scale eddies into small-scale eddies, cf. section 2.C, and the small-scale eddies eventually into heat).

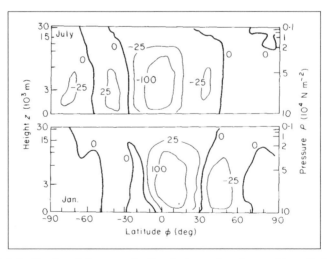

Figure 2.53. July (top) and January (bottom) streamfunctions of longitude-averaged meridional mass transport in units of 10^9 kg s^{-1} (based on Manabe and Holloway, 1975).

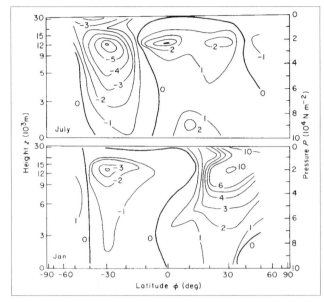

Figure 2.54. Northward transport of angular momentum by eddies, averaged over longitude, for July (top) and January (bottom), in 10^{18} kg m^2 s^{-2} per unit of pressure, 10^4 N m^{-2} (based on Manabe *et al.*, 1974).

Calculated height distributions indicate that the frictional dissipation is not confined to regions near the ground (Manabe *et al.*, 1974). Comparing the conversion rates with Fig. 2.50, it is evident that the order of magnitude is correct. The formation of eddy kinetic energy from eddy available energy and the frictional dissipation are both about 50% larger than the value deduced from observations, and the zonal to eddy kinetic energy is about three times too small. However, it should be remembered that there are no directly observed values for these conversions.

2.3.2 The oceans and continents

The water cycle

The processes by which water is transferred between different regions of the ocean–soil–atmosphere system may be described in much the same way as that used for the transfer of energy. Considering first a vertical column extending from the ocean–continent surface down to a level where no exchange of water takes place, the change in the amount of water present within the column, A_{wd}, can be written similarly to (2.11),

$$\frac{\mathrm{d}A_{wd}}{\mathrm{d}t} = E_{ws}^{total} + F_{wd}, \tag{2.16}$$

where the total water flux across the atmosphere-to-ocean/continent boundary is

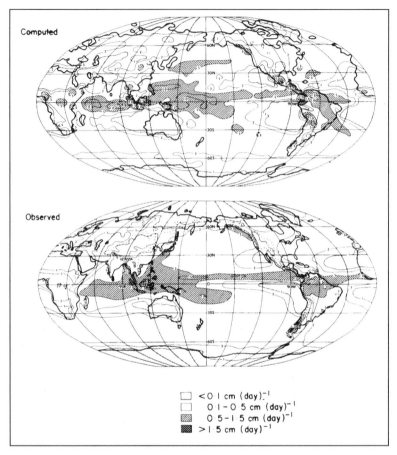

Figure 2.55. Computed (top) and observed (bottom) annual average rate of precipitation. The key to shading of different intervals is at the bottom, 1 cm/day being about 1.16×10^{-7} m³ water equivalent per m² and per second. (From S. Manabe and J. Holloway, 1975: *J. Geophys. Res.* **80**, 1617–1649, copyright of American Geophysical Union).

$$E_{ws}^{total} = r + d - e, \tag{2.17}$$

where r is the rate of precipitation, d is the formation of dew (usually negligible in comparison with the former) and e is the rate of evaporation. F_{wd} is the net inflow of water into the column due to advection, such as river or surface run-off and ground water motion on continents and ocean floors. In the oceans the net inflow is usually zero, except for density changes.

Figures 2.41 and 2.42 show the net upward flux of sensible and latent heat. The dominating upwards part of the latter is evaporation, whereas the precipitation r shown in Fig. 2.57 is only a part of the net sensible energy exchange. Assuming that dA_{wd}/dt is zero when averaged over the year, Sellers

(1965) has used (2.16) to calculate the average net inflow of water into a column, F_{wd}, shown in Fig. 2.58 along with longitudinal averages of r and e.

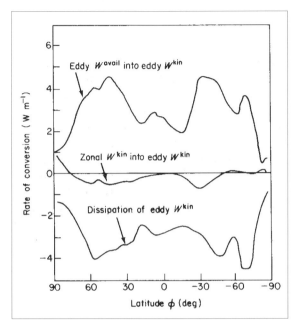

Figure 2.56. Calculated rates of eddy kinetic energy formation and destruction are shown, integrated over height and averaged over longitude and time. The middle curve is only approximately equal to the rate of conversion from zonal to eddy kinetic energy, since a contribution from the divergence of the eddy energy flux has been subtracted (based on Manabe *et al.*, 1974).

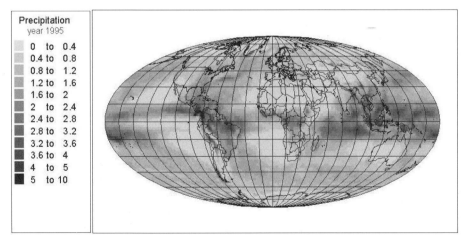

Figure 2.57. Annual precipitation 1995, m/y (based on NCEP-NCAR, 1998).

For a vertical column extending from the surface upward into the atmosphere, the change in the amount of water present can be written

$$\frac{dA_{wa}}{dt} = -E_{ws}^{total} + F_{wa},\qquad(2.18)$$

assuming that no water is leaving the column at the top of the atmosphere. If again dA_{wa}/dt is zero on an annual average basis, the transport of water horizontally through the atmosphere is given by

$$F_{wa} = -F_{wd}.$$

From the net influx into a vertical column one may calculate the required northward transport of water, which is very similar to the value estimated from measurements in the atmosphere, shown in Fig. 2.59. In order to construct an average water cycle, it is also necessary to split the ocean–continental transport term, F_{wd}, into its main components, i.e. the river and surface run-off (which can be estimated from observations as shown in Fig. 2.60), and the ground water flow (which may be calculated as a residue). This has been done in Fig. 2.61, based on Kalinin and Bykow (1969) and Lvovitch (1977). Also indicated are the total amounts of water stored in each compartment. Dividing these by the corresponding fluxes, one obtains the turnover time for water in each compartment, a quantity which is very large for ice caps and deep oceans.

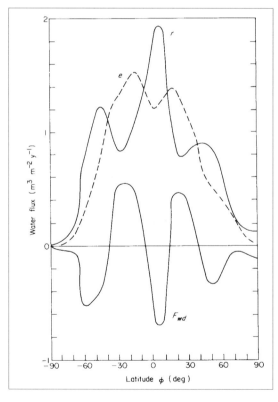

Figure 2.58. Latitude dependence (averaged over year and longitude) of water fluxes into a vertical column extending from the Earth's surface downwards: precipitation r, evaporation e, and net inflow into the column, F_{wd} (based on Sellers, 1965).

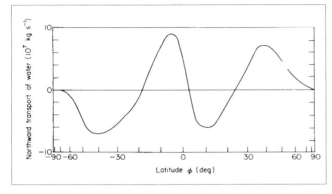

Figure 2.59. Northward transport of water, averaged over the year and summed over longitudes (computed from fluxes similar to those of Fig. 2.58, assuming the transport to be zero at the poles) (based on Peixoto and Crisi, 1965).

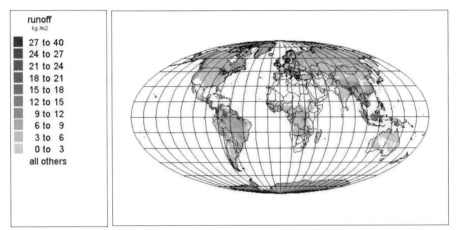

Figure 2.60. Accumulated run-off from land areas during 1997 (kg of water per m²) (NCEP-NCAR, 1998).

Evaporation processes

The evaporation e (2.17) from a given surface depends on several factors, such as the temperature and moisture of the surface; the temperature, humidity and wind velocity of the air above the surface; and the vertical gradients of these quantities. For soil evaporation, the moisture of the soil is considered to be of importance only if it is below a "critical" level. The physical parameter behind this "critical" value is likely to be the moisture tension, which plays the same role for evaporation, as does the surface tension for the evaporation from the surface of a fluid (e.g. a water surface). The evaporation from vegetation is further complicated by the plant' s ability to regulate the transportation of vapour from its interior, by opening and closing its pores (stomata). The release of water from plant-covered soil is called evapotranspiration (cf. Geiger, 1961; Sellers, 1965).

The vertical transport of moisture through the lower atmosphere is further described in section 2.B. The potential evaporation from land surfaces, inferred from measured data, is shown in Fig. 2.62 (Kalnay *et al.*, 1996). This is the evaporation that would take place if water transport from the surface were adequate. Evaporation from ocean surfaces has a dependence on wave and wind conditions, including the mechanism of whitecaps, from which water droplets evaporate before falling back into the sea. This evaporation process leaves salt particles that may serve as condensation nuclei for water vapour not carried away (up or to the side) quickly enough.

The state variables of the oceans

The state of the oceans, like that of the atmosphere, is given by the temperature T and the density $\rho = \rho_w$, but other variables are of great significance, primarily the salt concentration and, in regard to biomass production the amounts of dissolved oxygen and nutrients. The salinity S (salt fraction by mass) is necessary in order to obtain a relation defining the density, to replace the ideal gas law used in the atmosphere.

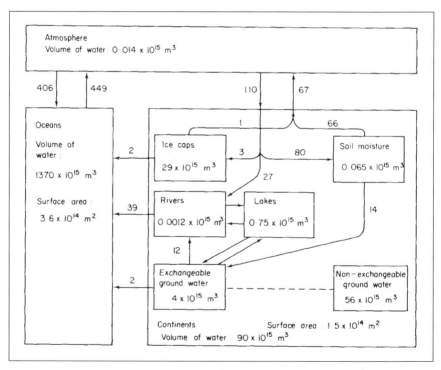

Figure 2.61. Schematic summary of the water cycle, including free water to a depth of about 5 km below the Earth's surface. The transfer rates, given at the arrows in 10^{12} $m^3 \ y^{-1}$, are averaged over time. For each compartment (box), the average water volume is indicated (based on Budyko, 1974; Kalinin and Bykow, 1969; Lvovitch, 1977).

Figures 2.63–2.65 show measured temperature and salinity distributions along sections through the three major oceans, at latitude circles at longitudes 30°W (Atlantic Ocean), 90°E (Indian Ocean) and 150°W (Pacific Ocean) (Levitus and Boyer, 1994).

Notable features of all three oceans are the large temperature gradients over the first 1000 m below the sea surface, observed at low latitudes. At latitudes of about 45°, the average temperature gradients become zero, and at higher latitudes they are generally small and irregular. The figures shown are annual average values, and the sea surface temperatures at the higher latitudes exhibit considerable seasonal variation, apart from regions covered by ice for most or all of the year. The regions in the upper water layers near the Equator generally have a high salinity, which can be understood in terms of the extensive evaporation, since most of the salt content is left behind by evaporation processes.

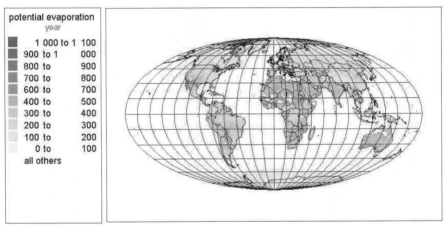

potential evaporation year	
1 000 to	1 100
900 to	1 000
800 to	900
700 to	800
600 to	700
400 to	500
300 to	400
200 to	300
100 to	200
0 to	100
all others	

Figure 2.62. Potential evaporation from land surfaces, expressed in energy units (by multiplication with L_v) and averaged over the year (NCEP-NCAR, 1998).

The asymmetry between the salinity distributions in the two hemispheres has to do with the geographical specificity of vertical mixing processes in the oceans. Convective overturn between the surface waters and deep sea waters (2000 m or more) takes place only in the Atlantic Ocean just south of Greenland and possibly in the Norwegian Sea (Reid, 1971). These regions are characterised by a small density increase as a function of depth. When cool and dense water is formed in these regions near the surface, such water can sink or mix downwards with little hydrostatic resistance. Also, in the Antarctic region, cool and dense water is formed in winter, but owing to substantial precipitation the salinity is low and, as seen from Figs. 2.63–2.65, the temperature and salinity in open sea remain fairly stratified. Only along the continental (Antarctic) shelf does the cool water sink down, in an essentially non-turbulent manner (primarily in the Weddell Sea). This accounts for the difference between the conditions (in particular the salinity) in the Southern Hemisphere and the Northern Hemisphere, which contains the only regions of deep convective mixing and thus the possibility of adding to the bottom some water that has recently been in strongly interactive contact with the atmosphere. In the Norwegian Sea, large masses of water have a composition revealing equilibrium between atmosphere and water. This water does not immediately sink to the bottom of the Atlantic Ocean, owing to the shallow sill separating the Norwegian Sea from the main part of the ocean (see Fig. 2.63). Instead, water from the Norwegian Sea is pressed through the nar-

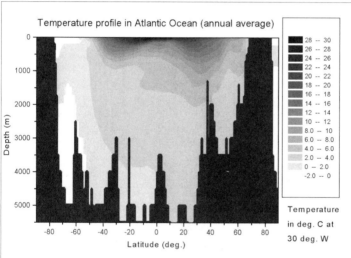

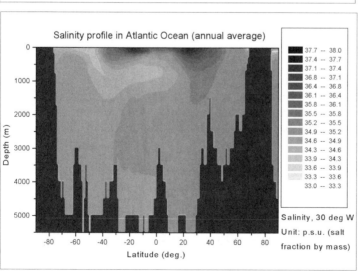

Figure 2.63a,b. Annual average temperature *T* (above) and salinity *S* (below) along a south–north section (30°W) through the Atlantic Ocean (based on Levitus and Boyer, 1994; data source: University of Columbia, 1998).

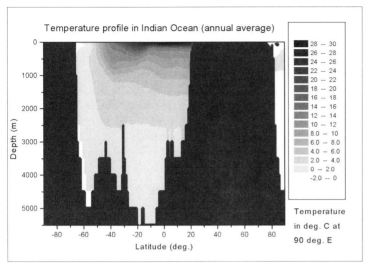

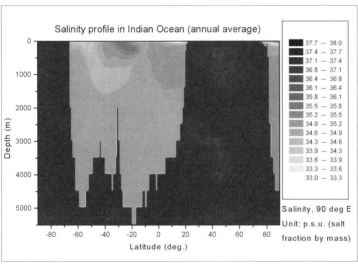

Figure 2.64a,b. Annual average temperature *T* (above) and salinity *S* (below) along a south–north section (90°E) through the Indian Ocean (based on Levitus and Boyer, 1994; data source: University of Columbia, 1998).

row passage in a highly turbulent mixing process, thus losing density and becoming unable to penetrate to the very bottom. It is therefore believed that the water near the bottom of the Atlantic Ocean is of Antarctic origin.

Scales of oceanic motion

The time scale of the oceanic circulation is very different from that of the atmosphere. A frequency spectrum analogous to the atmospheric one (cf. section 2.C) will reveal cycle times for the average circulation of a year or more, increasing with depth by about a factor of 10 and being another factor of 10 larger for motion in a vertical plane than for horizontal motion. Temperature adjustments in the upper ocean require times of the order of weeks.

As in the case of the atmosphere, it is thus possible to separate the general circulation from the short-term movements which, in the oceans, however, comprise not only quasi-random, turbulent motions (eddies), but also organized motion in the form of waves. As mentioned above, the wave motion may play an important role in the process of transferring mechanical energy between the atmosphere and the ocean through the wind stress, the magnitude of which may itself depend on the waves created.

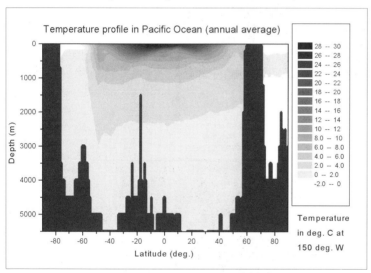

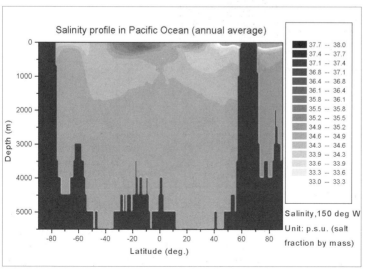

Figure 2.65a,b. Annual average temperature T (above) and salinity S (below) along a south–north section (150°W) through the Pacific Ocean (based on Levitus and Boyer, 1994; data source: University of Columbia, 1998).

To provide a rough picture of the classes of wave motion, Fig. 2.66 indicates the frequency spectrum of wave motion in the oceans (Roll, 1957). The harmonic waves in the spectral decomposition may be characterised by their wavelength, λ_w, the phase velocity, U_w, and the amplitude at the water surface, a. Figure 2.66 shows a time-averaged estimate of the distribution of amplitudes, a, on cycle times, $T_w = \lambda_w U_w^{-1}$. For waters of depth h much larger than the wavelength λ_w, harmonic waves governed by the gravity and surface tension forces are characterised by the relation

$$U_w = \left(\frac{g\lambda_w}{2\pi} + \frac{2\pi\gamma_w}{\lambda_w \rho_w} \right)^{1/2}, \tag{2.19}$$

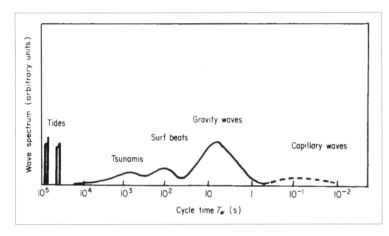

Figure 2.66. Sketch of the main features of the spectral decomposition of ocean wave amplitudes, averaged over time and position (based on Roll, 1957).

between wave velocity and wavelength, provided that the viscous forces in the water are neglected. The gravitational acceleration at the Earth's surface is $g = 9.81$ m s^{-1}, and the surface tension of water against air, γ_w, has a value of about 0.073 N m^{-1}. It follows from (2.19), which is shown graphically in Fig. 2.67, that U_w must exceed a minimum value of about 0.23 s^{-1} and that two values of λ_w are associated with each U_w above the minimum value. The branch of wavelengths smaller than about 0.017 m (corresponding to the minimum U_w) is called *capillary waves*, since they are mainly governed by the surface tension. The branch of wavelengths above $\lambda_w = 0.017$ m are the *gravity waves*, for which the term involving g in (2.19) dominates.

At cycle times around 2 min, pulsations in the amplitude of gravity waves formed earlier by interaction with the wind field may give rise to the surf beat phenomenon observed at shores (Sverdrup and Munk, 1947; Munk, 1980). The broad peak in the wave spectrum in Fig. 2.66, centred around $T_w = 20$ min, includes the flood-producing *tsunamis* occurring quite frequently, e.g. in the Pacific Ocean, as a result of sudden movements of the ocean floor caused by earthquakes (Neumann and Pierson, 1966).

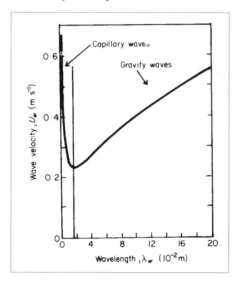

Figure 2.67. Theoretical relationship between wave (phase) velocity, U_w, and wavelength, λ_w, under the influence of gravity and surface tension, but neglecting viscosity (based on Roll, 1957).

The peaks at 12 and 24 h represent the tidal waves created by the time-dependence of the gravitational fields of various celestial bodies. The most important tides are caused by the Moon and the Sun in connection with the variations in distance from a given spot on the Earth's surface, resulting from the rotation of the Earth. Much longer tidal periods may be associated with orbital motion (lunar month, year, etc.). The theoretical description of waves and tides is discussed in section 2.D.

Joint models of general circulation in oceans and atmosphere

Early models coupling atmospheric and ocean circulation models did not include any detailed topography, had fixed amounts of cloud cover, and prescribed a radiation field corresponding to constant carbon dioxide levels, constant water vapour and constant ozone. Oceanic equations included temperature, salinity and ice, and the hydrological model was fairly detailed on precipitation, evaporation, soil moisture, snow cover, sea ice, melting processes and river run-off (e.g. assuming that when soil moisture exceeded a certain value, the excess would run off to the nearest ocean). The spatial grids were typically around 500 km by 500 km at the Equator, with around 10 vertical layers, and the time steps were of the order of a month for oceans, but of the order of 6 h for the atmosphere (Bryan, 1969; Manabe, 1971; Wetherald and Manabe, 1972).

As illustrated by e.g. the ocean salinity calculation (Fig. 2.68), the general results were quite realistic (compare with an average of the salinity data shown in Figs. 2.63–2.65). The models also allowed a first orientation into the structure of major current systems in the oceans (Fig. 2.69).

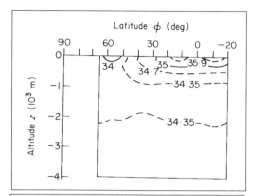

Figure 2.68. Average salinity (g kg^{-1}) calculated with an early joint ocean–atmosphere model without seasonal variations in radiation (based on Bryan, 1969).

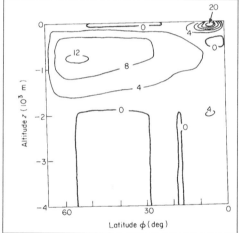

Figure 2.69. Streamfunction of meridional transport of water, averaged over year and longitude, calculated by a joint ocean–atmosphere model (unit: 10^9 kg of water per second). The sign convention is such that a meridional cell with positive streamfunction has northward transport at the top and southward below (based on Wetherald and Manabe, 1972).

Figure 2.70. Coupled ocean–atmosphere model results for ocean salinity, longitudinally averaged (England *et al.*, 1993).

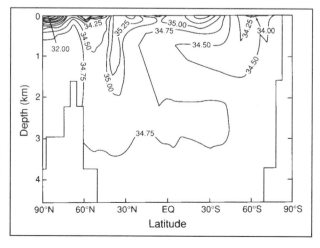

Over the years, models have been refined in terms of effects included, as well as in terms of the mesh sizes used. Some models replace the latitude–longitude compartments by a spectral model for the Fourier components of the variables [i.e. expanding all functions in terms of sine and cosine functions of a base frequency times n ($n = 1,...n_{max}$), with n_{max} typically around 30] (Manabe and Stouffer, 1988). This speeds up computing times, but once realistic topography is included, the advantage of this representation is diminished.

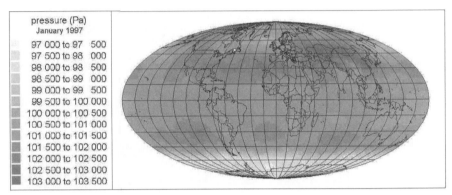

Figure 2.71a,b. Coupled ocean–atmosphere model results for mean sea-level pressure (Pa) in January 1997 (above) and July 1997 (below) HADCM2-SUL, the model used, which includes sulphate aerosols, has been integrated from pre-industrial times (Mitchell and Johns, 1997).

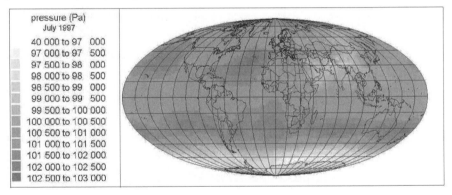

Current models include topography, a more complete water cycle with a cloud model, and a number of minor components in the atmosphere contributing to the radiation balance. The transfer of energy and momentum between ocean surfaces and the atmosphere has been a difficult process to model, and only since about 1995 has it been possible to formulate coupled ocean–atmosphere models not requiring some artificial correction for the mismatch between oceans and atmosphere to be introduced. This need was not only due to lack of knowledge of the processes involved, but also came from the different numerical treatment of ocean circulation and atmospheric circulation, notably the difference in time steps, and from having to start up the model by having it run for a period of time in order to gain stability (the finding that one could not model ahead from some known state of the atmosphere rests on the fact that input data were never fully consistent with the model structure and restrictions, so that an "artificial" initial state had to be created by running the model for several years, cf. the discussion in Gates et al., 1996). Current models use spatial grids of 30–100 km spacings and up to 30 atmospheric levels plus a similar number of ocean depth levels, with temporal grids of around half an hour for the atmosphere and an hour or more for the ocean part (see e.g. UK Meteorological Office, 1997).

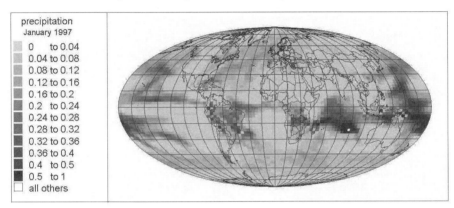

Figure 2.72a,b. Coupled ocean–atmosphere model results for total precipitation (m/y) in January 1997 (above) and July 1997 (below). The HADCM2-SUL model, which includes sulphate aerosols, has been integrated from pre-industrial times (Mitchell and Johns, 1997).

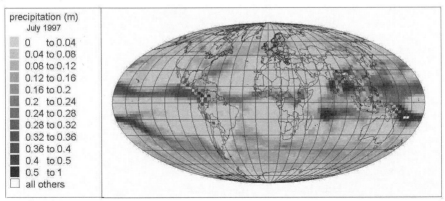

Models used near the year 2000 include the greenhouse effects of a number of gases as well as scattering and reflection effects of aerosols, such as those derived from sulphur dioxide emissions from human activity. Earlier models lumped these together as an effective change in solar radiation fluxes used in the models. Also, the interactions between the biosphere and the atmosphere and oceans are important for both heat and moisture fluxes, albedo and surface roughness experienced by winds. This means that the seasonal variation in standing crops as well as agricultural and forestry practices become important for climate modelling, along with direct anthropogenic interference through emissions of polluting substances to the atmosphere, in addition to natural processes such as volcanic eruptions. The human interference will be further discussed below.

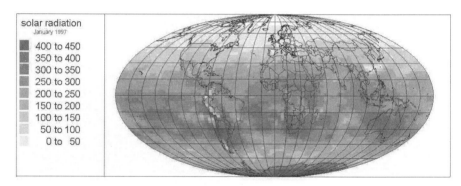

Figure 2.73a,b. Coupled ocean–atmosphere model results for surface solar radiation (W m^{-2}) on a horizontal surface in January 1997 (*a:* above) and April 1997 (*b:* below). The HADCM2-SUL model includes sulphate aerosols (Mitchell and Johns, 1997).

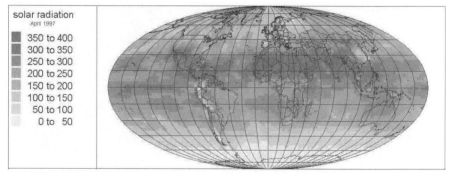

Figure 2.70 shows a newer model calculation of ocean salinity, considerably more accurate than the early model shown in Fig. 2.68. Figures 2.71–2.75 show mean sea-level pressure (i.e. pressure adjusted to zero height), precipitation, surface solar radiation, wind speed and temperature from a coupled ocean–atmosphere model including sulphate aerosol effects and greenhouse gases corresponding to 1997 levels. Where seasonal variation is important, the calculation is shown for two or four seasons. Measured data for comparison are contained in Fig. 2.76 for sea-level pressure, with Fig. 2.77

giving the pressure levels at the actual surface for comparison and Fig. 2.78 giving those for precipitation, while solar radiation on a horizontal plane is shown in Fig. 2.20 and seasonally in Fig. 3.1.

Figure 2.81 shows observed surface temperatures and Figs. 2.79 and 2.80 show wind speeds (w) and directions (d) at the surface (10 m). These two quantities are derived from the zonal (u) and meridional (v) components of the wind velocity,

$$w = (u^2 + v^2)^{1/2}, \quad d = \text{Atan}(v/u) \text{ or } \pm 90° \text{ (if } u=0 \text{ and } v>0 \text{ or } v<0).$$

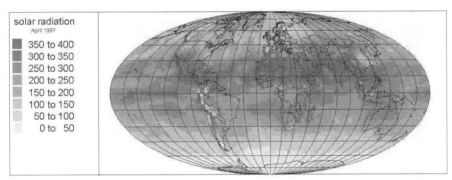

Figure 2.73c,d. Coupled ocean–atmosphere model results for surface solar radiation on a horizontal surface in July 1997 (*c*: above) and October 1997 (*d*: below). The model includes sulphate aerosols (Mitchell and Johns, 1997).

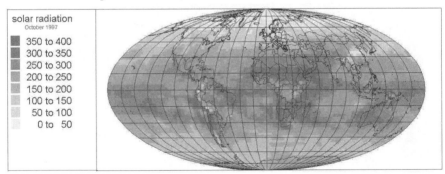

The sea-level pressure model results are generally consistent with the data, but detailed agreement is better on the Southern Hemisphere. On the Northern Hemisphere, the variations from North America over the North Atlantic Ocean to Europe are out of phase with data for January, as are the Europe to Asia variations in July. It should be kept in mind that the model does not aim at reproducing data for individual years, but the features of the Northern hemisphere measured pressures are fairly general, so a slight imbalance in the model treatment of oceans and land may be indicated.

The model results for precipitation rates are generally in better agreement with measured values, both for January and July of 1997. Only for an area

NE of Australia does the model predict much heavier rainfall than indicated by the observations.

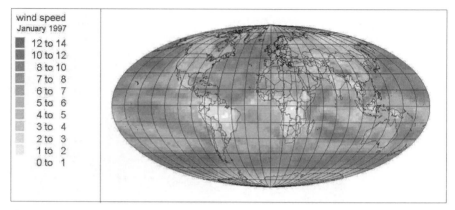

Figure 2.74a,b,c. Coupled ocean–atmosphere model results for surface wind speed (m s⁻¹, height 10 m), in January (*a:* above), April (*b:* below) and July 1997 (*c:* bottom). The HADCM2-SUL model includes sulphate aerosols (Mitchell and Johns, 1997).

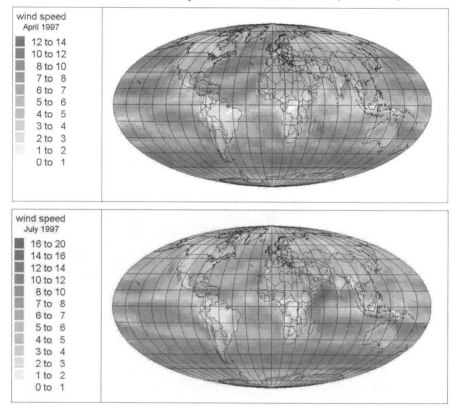

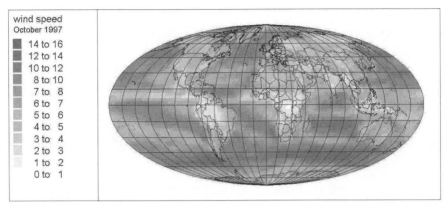

Figure 2.74d. Coupled ocean–atmosphere model results for surface wind speed (m s^{-1}, height 10 m) in October 1997 (*d:* below). The HADCM2-SUL model includes sulphate aerosols (Mitchell and Johns, 1997).

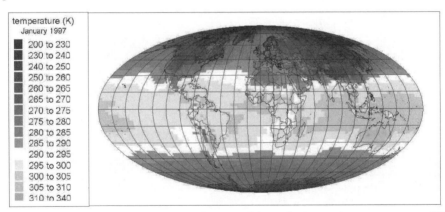

Figure 2.75a,b. Coupled ocean–atmosphere model results for surface temperature (K) in January (above) and July (below) of 1997. The HADCM2-SUL model includes sulphate aerosols (Mitchell and Johns, 1997).

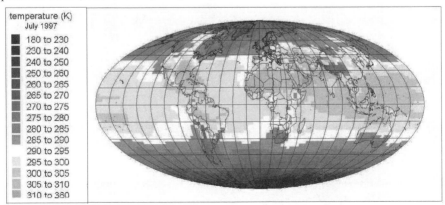

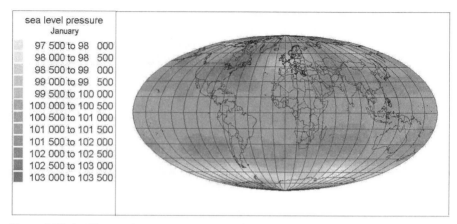

Figure 2.76a,b. Observed mean pressure, reduced to sea-level, for January (above) and July (below) of 1997 (NCEP-NCAR, 1998).

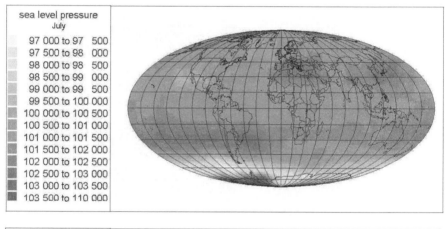

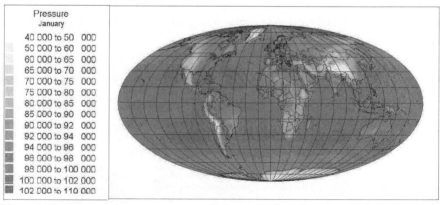

Figure 2.77a. Observed mean pressure (Pa) at actual surface height for January of 1995 (NCEP-NCAR, 1998).

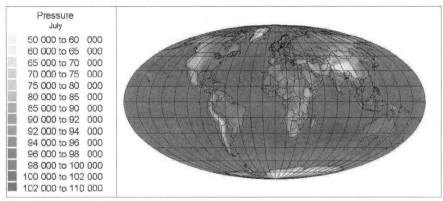

Figure 2.77b. Observed mean pressure (Pa) at actual surface height for July of 1995 (NCEP-NCAR, 1998).

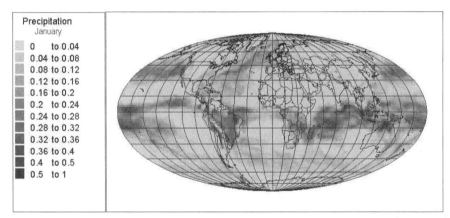

Figure 2.78a,b. Observed total precipitation (m per month) for January (above) and July (below) of 1995 (NCEP-NCAR, 1998).

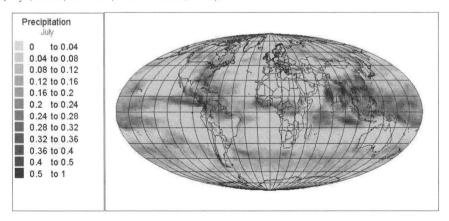

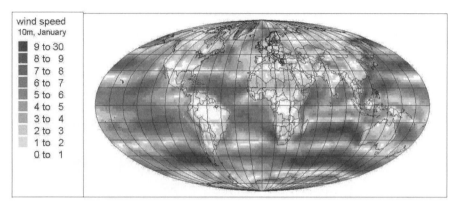

Figure 2.79a,b,c. Observed mean wind speed (m s^{-1}) at surface height (10 m) for January (above), April (below) and July (bottom) of 1997 (NCEP-NCAR, 1998).

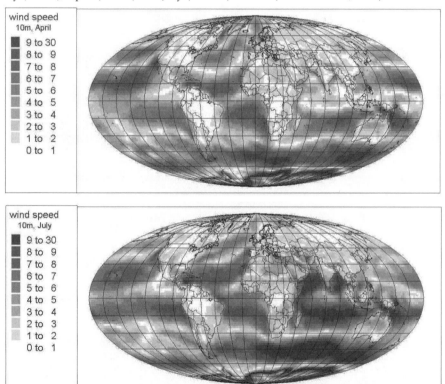

For wind speeds, the model calculations generally reproduce observed patterns, but the overall magnitude is on the low side, particularly for land locations in the Northern Hemisphere. This may be a problem with the data rather than with the calculation, because data are strongly influenced by

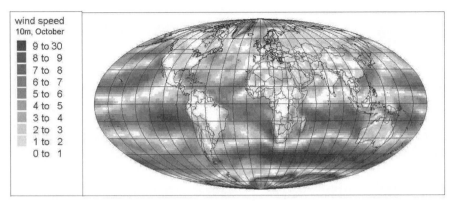

Figure 2.79d. Observed mean wind speed (m s^{-1}) at surface height (10 m) for October of 1997 (NCEP-NCAR, 1998).

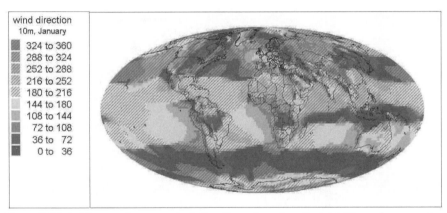

Figure 2.80a,b. Observed mean wind direction at surface (height 10 m, deg. counter-clockwise, 0° is west to east) for January (above) and April (below) of 1997 (NCEP-NCAR, 1998).

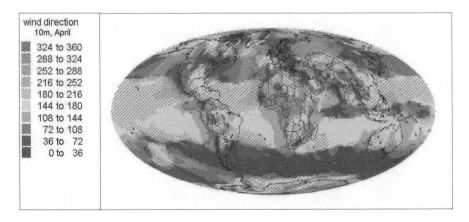

land-based measuring stations (which are particularly abundant for Northern Hemisphere continents) and, as many stations are placed in sheltered locations, they may not represent the true average wind speeds. This has implications for the use of meteorological data to forecast production from wind turbines, as further discussed in Chapter 3.

Figures 2.82 and 2.83 summarise calculated latitude distributions of the components of the net total energy flux, (2.10), and the rates of poleward energy transport for a seasonal model, but averaged over seasons. The fluxes compare well with those based on observations (Figs. 2.21, 2.41–2.43), while the sum of ocean and atmospheric transport should match that required by the net radiation flux at the top of the atmosphere (Figs. 2.7, 2.8).

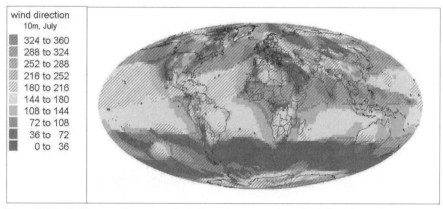

Figure 2.80c,d. Observed mean wind direction at surface (height 10 m, deg. counterclockwise, 0° is west to east) for July (above) and October (below) of 1997 (NCEP-NCAR, 1998).

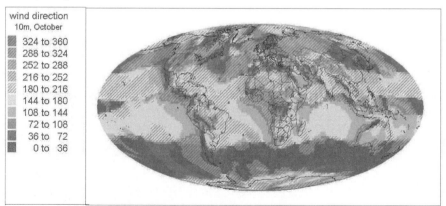

2.3.3 The climate

In the preceding subsections, the state of the atmosphere and the ocean–continent system has been described in terms of a finite set of state variables, such as temperature, pressure, salinity, moisture content of soil or air, ozone content, CO_2 content, etc. In addition, the motion of matter is described in terms of a velocity field, and the Earth–atmosphere system interacts with its surroundings through the exchange of radiation. The incoming radiation may at first be considered to constitute a fixed boundary condition, whereas the outgoing radiation may adjust itself as a result of the equations governing the system, i.e. the equations of motion, the rate equations for chemical processes, as well as the equations describing physical processes other than particle motion (e.g. the absorption and scattering of light).

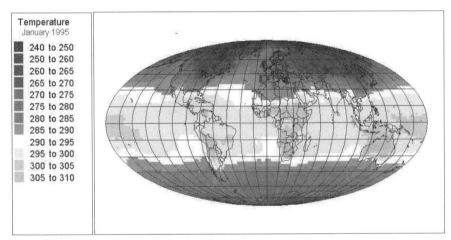

Figure 2.81a,b. Observed surface temperatures (in K at height 1000 mb) for January (above) and April (below) of 1995 (NCEP-NCAR, 1998).

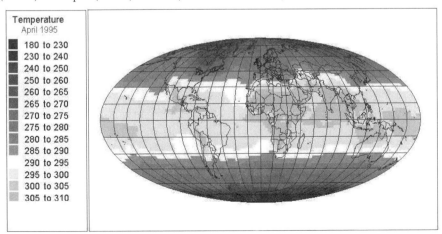

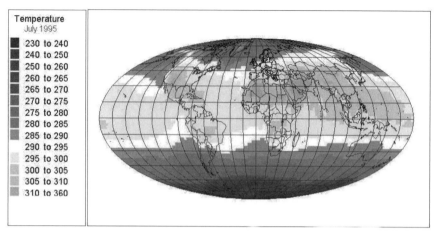

Figure 2.81c,d. Observed surface temperatures (in K at height 1000 mb) for July (above) and October (below) of 1995 (NCEP-NCAR, 1998).

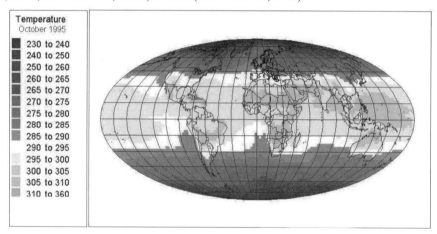

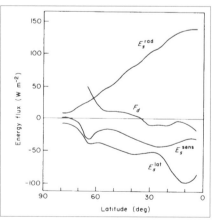

Figure 2.82. Annual and longitude average of calculated energy fluxes (in W m^{-2}) through a vertical column extending from the surface of the Earth downwards: net radiation, E_s^{rad}, net sensible energy flux, $E_{s\text{-}sens}$, net latent energy flux, E_s^{lat}, and net flux into the column, F_d, taken as the upward flux into the top ocean level of the joint ocean–atmosphere model, from deeper lying ocean levels (based on Wetherald and Manabe, 1972).

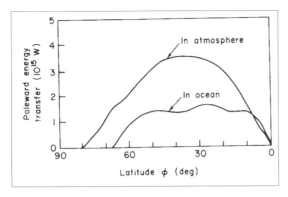

Figure 2.83. Poleward energy transfer (net) calculated from a joint ocean–atmosphere model (based on Wetherald and Manabe, 1972).

It has been indicated that a state of overall equilibrium prevails between the Earth–atmosphere system and its surroundings. This equilibrium is characterised by the absence of a net energy flux between the system and its surroundings, on a suitable average basis (involving at least annual and latitude–longitude averaging). The state of equilibrium may vaguely be described as "the climate of the Earth".

Climate variables

To arrive at a more accurate definition of climate, it is important to distinguish between micro- and macro-variables and to define the kind of averaging performed. From a physical point of view, the micro-variables may be thought of as variables describing the atomic or molecular state and motion, while macro-variables would be those associated with averages over statistical ensembles, e.g. pressure, temperature and wind velocity. However, this scale may still be considered microscopic from a meteorological point of view. Higher levels of macroscopic scale would thus be defined by further averaging, e.g. introducing a distinction between eddy motion and synoptic scale motion as discussed in section 2.C, and by making a distinction between zonal average circulation and deviations from it.

If the complete set of micro-variables is denoted

$$\{ x_j \mid j = 1,...,n \},$$

then the macro-variables corresponding to a given scale, s,

$$\{ \bar{x}_j^s \mid j = 1,...,n_s \},$$

are obtained by a suitable averaging over the micro-variables or functions of the micro-variables, e.g. time averages like

$$\bar{F}_j^s = \frac{1}{\Delta t_s} \int_{t_1}^{t_1 + \Delta t_s} F_j^s(\{x_i(t_2) \mid i = 1,...,n\}) \, \mathrm{d}t_2 ,$$

or by ensemble averages over N_s identical systems,

$$< F_j^s > = \frac{1}{N_s} \sum_{m=1}^{N_s} F_j^s(\{x_i \mid i = 1,...,n\})_m.$$

The number of macro-variables of scale s, n_s, may be smaller than the number of micro-variables, n, because not all variables survive the averaging procedure or because only a limited number are judged to be important (indeed, if the introduction of a macro-level description did not diminish the dimension, there would be little motivation for performing the averaging procedure at all). In general, it may be possible to define a very limited number of functions, K_j^s (to be denoted "climate functions"), which are *uniquely* determined from $\{x_j\}$ and depend only on the macro-variables of scale s,

$$K_k^s = G_k^s(\{x_j^s \mid j = 1,...,n_s\}),$$

or a similar expression involving ensemble averages, for $k = 1,..., m_s$.

The climate of scale s may then be defined as the set of functions $\{K_k^s \mid k = 1,..., m_s\}$. If, alternatively, a definition of climate is sought, which does not depend on whether unique definitions of the climate functions are possible, one might follow Lorenz (1968) in using the averages of the micro-variables directly as a "mathematical" climate, choosing $F_j^s(\{x_i\}) = x_j$,

$$\bar{x}_j^s = \frac{1}{\Delta t_s} \int_{t_1}^{t_1 + t_s} x_j(t_2) dt_2.$$

If $\{x_j\}$ are sufficiently well-behaved functions of time, these averages exist and may be called the mathematical climate of scale s. Letting Δt_s go towards infinity, one obtains what Lorenz calls "the climate of the variables x_j".

The equations, from which the time dependence of x_j may be determined (the micro-level equations of state, of motion, etc.), are often expressed as differential equations of the first order, and the time development in general depends on the initial conditions. This means that the averages $\bar{x}_j^s$ depend on the time t_1, in which case the equations are called *intransitive*, and the climate functions are not unique.

If, on the other hand, the averages $\bar{x}_j^s$ are independent of t_1, the equations are called *transitive*, and the mathematical climate is uniquely defined. A somewhat weaker condition would be obtained if only the climate functions, K_k^s, have to be uniquely defined, i.e. independent of initial conditions. As mentioned, the K_k^s then approximately (to the extent that these climate functions represent the solutions to the original micro-level equations) describe the actual climate of scale s. It is to be expected that the climate functions may gradually approach independence from initial conditions, as the time-averaging period increases. At $\Delta t_s = 1$ year, a certain stability is already obvious (due to the periodicity of radiative boundary conditions), but meteorologists normally define climate by requiring $\Delta t_s \approx 30$ years.

If time intervals of 30 years or longer are considered, it may no longer be possible to assume that the external boundary conditions remain constant. In that case one would talk about climate change, a subject which will be further dealt with in section 2.4.

Stability of climate

The uniqueness of climate functions of a given scale s implies that the state of the system is stationary. By stationary one should not infer that the time development has to be exactly periodic, but rather that the state of the system is periodic or constant, after fluctuations of frequency much higher than $(\Delta t_s)^{-1}$ have been filtered out (by the averaging procedure defining the scale s).

The climate may also be characterised by its response to a small perturbation of the state variables imposed at time t_1. If there is a finite probability that the climate functions will remain unaltered, the climate may be said to be stable, otherwise it is unstable. This follows directly from the independence (or dependence) on initial conditions at time t_1. A similar argument may be made for sequences of random perturbations, of closer resemblance to the actual "noise" introduced by fluctuations of the micro-variables around their average value (or more generally, fluctuations of time scales smaller than Δt_s).

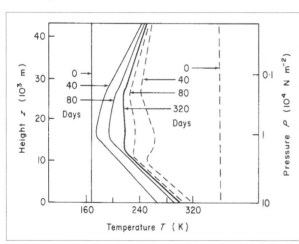

Figure 2.84. Approach towards equilibrium distribution of temperature for a one-dimensional model of the atmosphere with fixed external conditions. The initial condition is a uniform low (solid curves) or uniform high (dashed curves) temperature, and the different curves correspond to different periods of integration (based on Manabe and Strickler, 1964).

A transitive system only possesses one stable climate. If a system has more than one stable climate, it is by definition intransitive, and the interesting parameter is the magnitude of perturbation which would cause the system to make a transition from one stable climate to another. However, as pointed out by Lorenz (1968), it is not always necessary to induce a perturbation in order to make the system perform the transition. Some systems may remain in one stable state for a substantial time and then change into a different "stable" state, where they would again remain for a long time, and then eventually make the transition back, and so on, without requiring any external perturbation. Such systems (examples of which have been discussed, e.g. by Hide, 1958) may be called "near-in-transitive". Lorenz introduces this concept for systems which are strictly transitive in the limit $\Delta t_s \to \infty$, but for which the time averages over different finite time intervals may differ substantially. He has constructed numerical experiments with vacillating behaviour (periodic transition between two patterns of circulation), and suggests that the atmosphere may be such a quasi-intransitive system. This might explain some of the climatic changes in the past, without recourse to external causes.

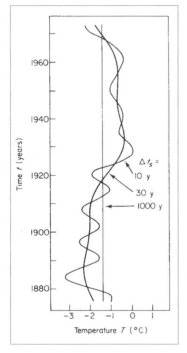

Figure 2.85. Time-averaged records of temperature from Godthaab, Greenland. The averaging period Δt_s is indicated for each curve (based on Dansgaard *et al.,* 1975).

The equations of section 2.C, used in sections 2.3.1 and 2.3.2 to describe the general circulation in the atmosphere, are in equilibrium calculations assumed to possess one stable solution, independent of initial conditions (at least over a substantial range). The way in which the equations were solved in early calculations (Manabe and Strickler, 1964) used this directly. Starting with simple and completely unrealistic conditions, such as a constant temperature throughout the atmosphere, the equations are first integrated over a long period of time, with fixed boundary conditions (such as radiation and ocean surface temperature). Then, after equilibrium has been reached, seasonal or other variations in the boundary conditions, as well as the ocean–atmosphere couplings, are introduced. After a while, the situation is again stationary. Figure 2.84 shows for a simple, one-dimensional model, how the stationary temperature profile is reached, starting from initial conditions of uniform high, or uniform low, temperature (Manabe and Strickler, 1964).

In more general cases, e.g. with time-dependent forces such as those associated with the solar radiation, one cannot expect the solutions to be independent of the initial conditions. Contemporary calculations use actual data as initial conditions, and provided these are of sufficient quality, the transient behaviour of the solutions should faithfully describe the effects of variations in external forcing of the climate system (IPCC, 1996b). This is important for calculations including greenhouse gas emissions into the atmosphere. Early calculations just aimed at deriving the equilibrium situation for a doubling of CO_2, whereas current models are capable of deriving the dynamic behaviour of climate during a period of time with changing emissions and changing response of the reservoirs included in the models.

Figure 2.85 indicates the variation in climate (represented by the single variable T = temperature) actually observed at a given location, using different averaging times Δt_s. It can be seen that the standard meteorological climate scale, 30 years, still leaves some structure in the climate function T (Dansgaard *et al.*, 1975).

The ergodic hypothesis

It is expected that, for a stationary climate, time averaging and the averaging over a statistical ensemble of identical "climate experiments" will yield the same value, when Δt_s and N_s become sufficiently large (in this case, the summation over the N_s experiments may be replaced by an integral over each variable times the probability density for that variable). This identity is known as the *ergodic hypothesis* (see e.g. Hinze, 1975),

$$\lim_{\Delta t_s = \infty} \bar{x}_j^s = \lim_{N_s = \infty} <x_j^s>.$$

Averaging over position vectors has been considered (e.g. in constructing zonal means), but these would equal the time or statistical averages at a fixed position only if the system were homogeneous, which the atmosphere is not. It should be clear that the macro-variables describing the climate need not be global averages, but may represent definite positions or zonal means. Also, the use of a large time-averaging interval, Δt_s, does not imply that the climate functions cannot comprise indicators of seasonal or diurnal changes, such as 30-year averages of May 18-humidity or of noon-pressure.

2.4 The energy cycle of the Earth

In this section an attempt is made to summarise the energy conversion processes taking place in the Earth–atmosphere system, for the more or less stationary state constituted by the present climate. In doing so, the energy flow is taken as a "common denominator" for the variety of physical and chemical processes by which matter is transformed or moved. There are certainly contexts in which it is necessary to follow the flow of matter directly, but only examples of this will be given.

The second question to be examined is the possible interference with this "natural" climate or energy flow, which is caused or may be caused by anthropogenic activity. The possibility of climatic changes resulting from the injection of CO_2 or particulate matter into the atmosphere has been widely discussed, and only a brief review will be given. However, the climatic modelling techniques utilised in the discussion of such effects may also prove valuable in discussing the impact of an increasing use of the renewable energy flows for human purposes, since the models deal with precisely those flows which would be redirected or displaced by renewable energy conversions.

2.4.1 The flows of energy and matter

The undisturbed energy cycle

Based on the material presented in sections 2.2 and 2.3, and in particular on the net radiation summary, Fig. 2.16, it is now possible to construct a schematic flow chart for the energy conversion and transport taking place in the Earth–atmosphere system (averaged over the year and geographical position). The result of such an attempt is shown in Fig. 2.86.

Different energy forms have been placed in different vertical columns, whereas the horizontal rows represent the different components of the Earth–atmosphere system. Compartments in which energy may become stored are represented by boxes, and the energy flows are indicated by lines with arrows, with the corresponding rates being given in units of terawatts (TW). The usefulness of diagrams of this type has been pointed out by King Hubbert (1971).

The first column contains the incoming short-wavelength solar radiation and its disposition into one part reflected into space and another part absorbed in various layers of the Earth–atmosphere system. The second column represents energy stored as (geo-)potential energy, latent energy of phase changes and other chemical (molecular and atomic) energy, plus nuclear energy.

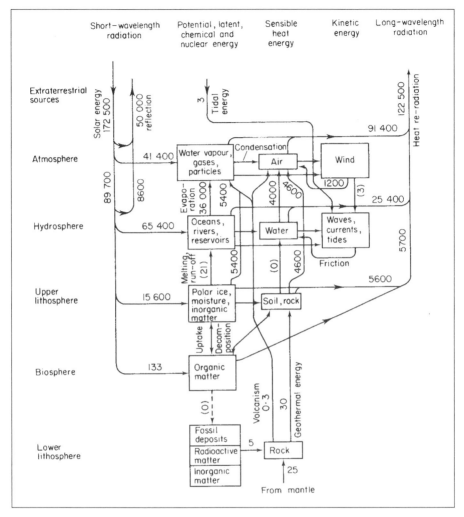

Figure 2.86. Schematic energy cycle without anthropogenic interference. The energy flows are in TW (10^{12} W). Numbers in parentheses are uncertain or rounded off (with use of Odum, 1972; Budyko, 1974; Gregg, 1973; Lorenz, 1967; Manabe, 1969).

Most of these energy quantities are bound within matter and do not participate in the undisturbed climatic processes. The third column represents storage of sensible heat, and the fourth column represents storage of kinetic energy in the general circulation of the atmosphere and the oceans. Certain compartments considered of little importance have been left out (e.g. kinetic energy of continents moving, tides of the solid Earth and movements of ground water and of animals, as well as sensible energy stored in the biosphere). The final column contains the long-wavelength re-radiation into space. According to the assumption of a stationary state, the net energy exchange between the Earth–atmosphere system and its surroundings is zero, which is the basis for talking about an "energy cycle": the outgoing flow can formally be joined to the incoming flow to form a closed cycle.

The top row contains the small contribution from gravitational attraction to other celestial bodies ("tidal energy"), in addition to the radiative fluxes.

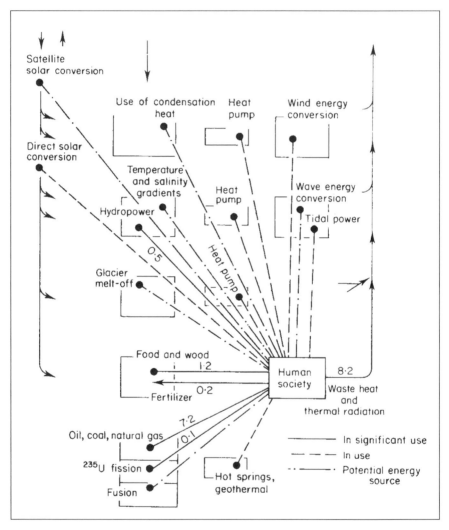

Figure 2.87. Past (1970s) and possible future human modification and points of interference with the energy cycle. Circles indicate possible sources of energy conversion for use in the human society, while a heavy arrow indicates return flows. The unit is TW, and the background box layout corresponds to the compartments in Fig. 2.86 (numerical entries based on Darmstadter *et al.*, 1971; European Commission, 1997).

In the atmosphere, 41 400 TW or about a quarter of the incoming radiation is absorbed. Based on this and a similar amount of energy from condensation of water vapour (36 000 TW of which is latent energy of vapour evaporated from the oceans, 5400 TW being from continents), the atmosphere maintains a strong interaction with the Earth' s surface, involving an intense exchange of long-wavelength radiation (see Fig. 2.16). The surface re-radiates a total of about 91 400 TW as long-wavelength radiation into space. The condensation of water vapour (and associated precipitation) does not take place over the regions where the evaporation took place, but transport processes generally have occurred. Yet only 1200 TW is used to sustain the atmospheric circulation (the "winds"), according to Fig. 2.50 (Lorenz, 1967). The net exchange of sensible energy between the atmosphere and the underlying surface amounts to 4000 TW from water surfaces and 4600 TW from continental surfaces, both in the direction towards the atmosphere.

The hydrosphere absorbs the major fraction of the incoming solar radiation, about 65 400 TW. This is more than expected from the fraction of surface area covered by oceans, but the average albedo of the oceans is smaller than the average albedo of the continents (cf. Fig. 2.17). The energy absorbed by the oceans is used to evaporate water and form sensible heat (36 000 TW), or it is re-radiated as long-wavelength radiation (25 400 TW).

According to Gregg (1973), the amount of energy transferred from the wind kinetic energy to oceanic circulation, by the surface stress, is around 3 TW. This implies that practically all the loss of kinetic energy goes into frictional heating. One would assume that the frictional losses are more pronounced over land than over ocean, due to the much larger roughness length (cf. section 2.B) and due to the impact against mountain topography. Although the frictional losses of kinetic energy are not confined to the surface layer, a substantial fraction of the losses takes place below a height of 1000 m (see e.g. the discussion of Newell *et al.*, 1969; another region of significant frictional losses is at the strong zonal winds around an altitude of 12 km). Since the heat arising from the frictional interaction is dispersed in the air at first, the 1200 TW have been transferred to the air sensible heat compartment in Fig. 2.86. If the heat is dispersed near the ground, it may, of course, contribute to heating the soil, but this process is listed as part of the (two-way) transfer between water or soil and air. Wave formation includes a transfer estimated at 0.3 TW from wind to currents (Bryan, 1969), but additional transfer may be directly to waves. A realistic treatment of wave formation will thus increase the estimated wind stress and hence the energy transfer above what is indicated in Fig. 2.86. Climate models often use the transfer as an adjustable parameter to compensate for numerical problems caused by the differences in the time scales used for the oceans and the atmosphere.

The lithosphere has been divided into an upper and a lower part. The upper part is characterised by having an active heat exchange with the atmosphere, whereas little heat transport takes place in the lower part, only the geothermal fluxes and a little transport by way of ground water flow. Of the 15 600 TW estimated to be absorbed at the continental surfaces, 5400 TW goes to evaporation and sublimation, and a net amount of 4600 TW is transferred to the atmosphere as sensible heat. Transfer from continents to oceans may take place by run-off if the water temperature differs from that of its environment. Such transfers are assumed to be negligible.

The polar ice is considered a part of the continents, and it is estimated that melting consumes 21 TW of the absorbed energy. The latent energy of melting is only formally transferred to the oceans, since it will normally not become available until the water has been evaporated into the atmosphere and condenses into snow.

The biosphere is presented as a separate compartment in Fig. 2.86, because of its importance in the following. The biomass production utilises 133 TW or 0.08% of the solar radiation at the top of the atmosphere, according to Odum (1972). The photosynthesis by phytoplankton and green algae in the top 100 m of the oceans (cf. Fig. 2.26) accounts for about a third of the biomass production. Almost as large a contribution comes from a much smaller area, namely, the wet tropical and subtropical forests (such as the region around the Amazon River). Grassland and pastures are responsible for roughly 10% of the biomass production.

The amounts of energy contained in each of the compartments considered in Fig. 2.86 are generally much harder to estimate reliably than are the conversion rates. If both are known, the turnover time for energy in a given compartment can be obtained by dividing the amount of stored energy by the conversion rate.

The amount of kinetic energy contained in the atmospheric circulation has been studied in some detail. From Fig. 2.50, the "available potential energy" (cf. section 2.3.1) in the atmosphere is 2.8×10^{21} J, and the kinetic energy of the circulation is 7.5×10^{20} J, corresponding to turnover times of 27.5 days and 7.4 days, respectively. The available potential energy is defined as the potential plus internal energy relative to a reference state of the atmosphere, and thus represents a measure of the non-radiative energy produced in a given region by absorbing radiation, by moving sensible heat into the region and by condensation of water vapour. The stored energy is denoted "available" if it can be converted into kinetic energy, and the reference state is one which cannot form any new kinetic energy (Lorenz, 1967; Newell et al., 1969). If the stored energy in the atmosphere is not taken to be the available energy, very different values could be obtained. For instance, the latent heat of vaporisation contained in all the water vapour present in the atmosphere in itself represents some 1.5×10^{22} J.

Much less is known about the magnitude of the energy stored in the oceans. An indication of the amount of energy stored in wave motion may be derived from the observed frequency distributions of wave height. In the North Atlantic ($\phi = 59°N$, $\lambda = 19°W$), the annual average wave amplitude is $<a> = 1.23$ m, and the average of the squared amplitude is $<a^2> = 2.13$ m (Salter, 1974). From the discussion in section 2.D, rigorously valid only for sinoidal waves, one then derives the average energy stored per m^2 ocean surface, $\overline{W}^{total} \approx 10^4$ J m^{-2}. (In Fig. 2.86, the wave compartment has been placed under "kinetic energy", although on average half the energy is in the form of potential energy, cf. section 2.D). If this amount of stored energy was found uniformly over the world's oceans, the total energy stored would be of the order of 10^{18} J. From similar order-of-magnitude arguments, based on measured velocities and estimated mass transports of surface currents (Sverdrup *et al.*, 1942), the kinetic energy of horizontal currents would be of the order of 10^{17} J, and the kinetic energy of the thermohaline (meridional) motion, based on the calculated circulation shown in Fig. 2.69, would be of the order of 10^{16} J. These rough estimates, which should be viewed with considerable reservation, would imply turnover times of the order of a few days for the surface wave motion, increasing to several years for the motion in the meridional cells. In the deep sea, which was not adequately described in the calculation referred to (Bryan, 1969), the turnover time may be several hundreds of years.

The amount of energy stored in the biosphere may be taken as the total biomass in energy units. Based on Odum (1972), the average biomass may be estimated to be about 1.5×10^{22} J, giving an average turnover time of 3.5 years. This average conceals a considerable spread, such values as a few weeks for phytoplankton and several hundreds of years for certain trees being included.

Man's interference with the energy cycle

Figure 2.87 attempts to summarise the modifications of the energy cycle caused by human society today and to identify a number of points on the flow diagram of Fig. 2.86 which could contribute to serving man's purpose. Some of these conversion methods are well known but only in modest use today. The compartments and main layout of Fig. 2.86 are repeated without text in Fig. 2.87, as a background for the lines connecting certain points, along the energy flows or at the energy stores, with the box representing human society.

The main source of present energy conversion in human society is fossil fuels, such as oil, coal and natural gas. This flow amounted to about 7.2 TW in the early 1970s, and about 12 TW in the late 1990s (European Commission, 1997), whereas the rate of forming new such fuel (peat, etc.) is negligible in comparison. Conversion of nuclear energy by induced fission of ^{235}U (plus other fissile material produced in the reactors) in a sustained neutron-rich environment presently accounts for about 0.7 TW (electric output), while hydropower converts about 0.5 TW of potential energy from the hydrological cycle (regulated through the formation of high-lying reservoirs).

The energy flow of organic matter from the biosphere to human society is about 1.2 TW, of which half is food. Not all the 0.6 TW transferred to human society as food actually serves as food intake, since part of it is lost in processing or is transferred to refuse. The other 0.6 TW of organic matter is mostly wood, but only a small fraction is burned directly. The rest goes into wood products or is lost in the production process. The amount of energy stored in wooden products is probably growing, although losses of stock take place when wooden (including paper) products are discarded and eventually burned.

Other conversion processes (indicated by dashed lines in Fig. 2.87) contribute less than 0.05 TW. They are geothermal energy (hot springs, etc.), tidal power, wind turbines, heat pumps and solar collectors (thermal) or solar cells. The possibility of an increasing role played by conversion schemes based on renewable energy flows is a major theme in the following chapters.

A minor theme is the study of those areas of the energy cycle which have not yet been utilised by man but for which conversion methods have been proposed. In several of these areas research is being carried out, although the practical viability of most of the proposed schemes remains to be demonstrated. The dot-and-dash lines in Fig. 2.87 indicate some such conversion possibilities, including wave energy, temperature and salinity gradients in the oceans, direct solar energy conversion outside the atmosphere, and hydropower stations in connection with glacier melt-off. Fusion of light elements to release nuclear energy is also included as an unproven process (that is, controlled), although its position might have been taken in the hydrosphere (deuteron–deuteron or deuteron–tritium processes) rather than in the lithosphere (bearing in mind that lithium enters the reaction chains primarily being studied today).

Energy conversion within human society produced about 8.2 TW of waste heat (in the early 1970s) in the form of sensible and latent energy emissions and long-wavelength radiation. The sensible energy emissions go into the hydrosphere (cooling water) or atmosphere, and the latent energy emissions (water vapour from cooling towers, etc.) go mostly into the atmosphere, where they add to the natural heat source function (see section 2.C). The long-wavelength radiation alters the net radiation flux at the level in question (most conversion presently being at ground level), which leads to a readjustment of the fluxes in both directions (upwards and downwards). Such changes have been observed locally ("heat islands" over metropolitan or heavily industrialised areas), whereas the evidence for global changes is at present inconclusive. The 8.2 TW constitutes 4.7×10^{-5} of the solar radiation at the top of the atmosphere. For comparison, the maximum relative change in the solar radiation at the top of the atmosphere, due to changes in the eccentricity of the Earth's orbit around the Sun during the past 500 000 years, has been 10^{-3} (Hays *et al.*, 1976).

Energy is stored in a number of products within the human society, and some are later exported, such as the 0.2 TW estimated to go back into the biosphere (agriculture) with fertilisers.

There is a net inflow of energy into human society, indicating a build-up of stored energy. The exact magnitude of the amount of energy stored is difficult to assess, since some of it is likely to be found in discarded products and other wastes, some of which may be on their way back into the biosphere or lithosphere, while other wastes cannot be expected to release stored energy in the near future (e.g. some manufactured products containing stored chemical energy). On the other hand, the stock of such energy (e.g. chemical) residing in the structures and materials forming our cities, transportation systems, etc. has certainly increased with time.

The amounts of non-renewable fuels expected to be recoverable with technology not too different from today's (proven plus estimated recoverable reserves) are of the order of 10^{22} J for oil and gas, 10^{23} J for coal and 10^{22} J for ^{235}U (Ion, 1975; Sørensen, 1999b). Ion states that not all the known coal reserves may be minable, and of the uranium estimate given here only about 10% has been verified today.

Matter cycles

Since the exchange of matter between the Earth–atmosphere system and its surroundings is small (cosmic particles are received and some atmospheric matter is probably lost to space), the matter cycles are essentially irrelevant when conserved quantities such as non-radioactive elements are considered. Normally, each element participates in several different chemical configurations during its cycle. Only in a few cases, such as the water cycle considered in Fig. 2.61, is it relevant to describe the flow of a chemical compound. Here the reason is that the gross flow of water in its different phases is little affected by chemical reactions between water and other elements, of which a number do take place in the atmosphere as well as in the lithosphere.

In most cases, a cycle would then be drawn for a single element, such as oxygen, carbon, or nitrogen. However, the flow of matter (the element) in itself does not give any indication of the processes involved, because it is often the chemical reactions which determine the flow rates in the system.

Of particular interest in connection with the renewable energy flows are the flows of those materials which participate in the biological processes depending on photosynthesis. Examples of such elements are carbon, nitrogen, phosphorus, potassium, oxygen and hydrogen, but a long list of other elements should be added to account for the structure of organic material, including the enzymes and hormones important for the control of biological processes.

The carbon cycle

Figure 2.88 shows the construction of a gross carbon cycle, based on a small number of highly aggregated compartments in the marine and terrestrial areas. The matter flow is expressed in units of 10^{12} kg of carbon per year, and the amounts of carbon stored in the compartments is expressed in units of 10^{12} kg of carbon.

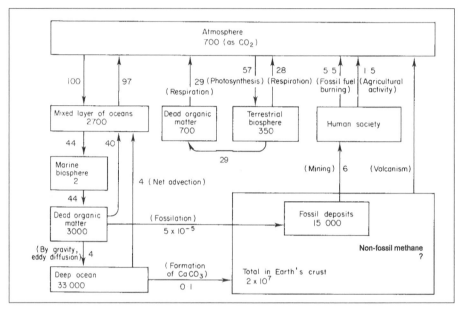

Figure 2.88. Schematic carbon cycle. The rates of transfer are given in 10^{12} kg of carbon per year, and the quantities stored are given in 10^{12} kg of carbon (based on Odum, 1972; Bolin, 1970; Wilson and Matthews, 1971; Gold and Soter, 1980).

The rates and stored quantities have been assumed to correspond to a stationary situation, before the interference by human society became noticeable. Over a longer period of historical development, as shown in Fig. 2.2 for oxygen, the photosynthetic conversion can no longer be assumed to be at a constant level. On this time scale, the most important processes adding or removing CO_2 from the atmosphere, aside from photosynthesis, are the weathering of limestone and the formation of such rock, by the combination of Ca^{2+} ions with CO_3^{2-} ions, and the exchange of CO_2 between the oceans and the atmosphere. The maximum CO_2 content of the atmosphere, which may have prevailed when there was no photosynthesis, is limited to about 1% (about 30 times the present level) by the $CaCO_3$ rate of formation (Rubey, 1951). Assuming the oceans to be present, the maximum amount of atmospheric oxygen would, in the absence of photosynthesis, be given by the rate of photodissociation of water in competition with the screening of this process by the oxygen already formed. The corresponding upper limit for atmospheric oxygen would be of the order of 0.02% (1000 times smaller than the present level) according to Berkner and Marshall (1970).

It follows that practically all the oxygen now present in the atmosphere has been formed by photosynthesis. The same seems to be true for the oxygen present in the lithosphere, and the ratio of carbon and oxygen in the atmosphere plus lithosphere (but excluding the hydrosphere, which contains oxygen in water) is very close to 0.5, indicating that all of it may have its origin in photosynthetic dissociation of CO_2 (Cloud and Gibor, 1970). A positive amount of CO_2 would have been transferred from oceans to atmosphere during most of the period in which photosynthesis has taken place outside the oceans (about 5×10^8 years).

Possible non-fossil carbon in the form of methane may be present in the Earth's crust (Gold and Soter, 1980), and it has been proposed that the sudden release of large amounts of methane could have been responsible for the climate change occurring 55 million years ago (Norris and Röhl, 1999).

Figure 2.88 indicates a surplus of CO_2 going into the atmosphere as a result of the burning of fossil fuels, the clearing of natural vegetation and the introduction of agricultural practices, which expose larger volumes of soil to contact with the air (deep ploughing), whereby they release more CO_2. Part of this surplus is absorbed by the oceans, but this process seems unable to keep pace with the rapid growth of CO_2 in the atmosphere, so the net result is that about half the extra CO_2 injected into the atmosphere is accumulating there. Also, the mixed upper layer of the oceans has a turnover time which is longer than the time scale of human interference with the carbon cycle, so the additional CO_2 absorbed by the oceans largely accumulates in the mixed layer, rather than reaching the long-term sink provided by the deep ocean.

The transfer rate of CO_2 across the atmosphere–ocean boundary, which in a stationary state would equal the ocean–atmosphere transfer rate, has been estimated on the basis of the depletion rate of ^{14}C injected into the atmosphere by nuclear weapons tests. Telegadas (1971) found that the effective half-life of the excess ^{14}C in the atmosphere was at an average value of 5.6 years during the period 1963–1966 and 8.2 years during the period 1966–1969. He interpreted the rising half-life as an approach to an equilibrium, in which re-emitted ^{14}C from biosphere and oceans must be subtracted from the fallout rate in order to give the effective half-life. The value $t_{1/2}^{eff} = 5.6$ y is thus an upper limit of the one-way fallout rate, which may become diminished if mechanisms of sufficiently short-time re-entry into the atmosphere are present. One such mechanism would be associated with the part of the biosphere respiration (cf. Fig. 2.88) derived from living plants or animals. The atmosphere–ocean transfer rate derived from these arguments is about 6×10^{13} kgC y^{-1}, but different interpretations of essentially the same data give transfer rates in the range $(3–14) \times 10^{13}$ kgC y^{-1} (Machta, 1971; Bolin, 1970; Fairhall, 1973). An average value of 9×10^{13} kg y^{-1} has been used in Fig. 2.88.

The rates of new fossilisation and formation of new rock (limestone) containing $CaCO_3$ have been estimated from the average history of deposits present today. Thus, the actual formation may have taken place over a shorter period, yielding a varying rate, which is larger than that indicated under certain circumstances, but zero during other periods.

The nitrogen cycle

Another biologically important element is nitrogen, the tentative cycle of which is depicted in Fig. 2.89. Nitrogen gas (N_2) is a major constituent of the atmosphere, but in order for it to become available for building biological material, the nitrogen must first be "fixed". By fixation is meant a process by which gaseous nitrogen (N_2) is transformed into a constituent of a compound, such as ammonia (NH_3):

Nitrogen fixation

N_2 + energy $(0.67 \times 10^6$ J mol$^{-1}) \rightarrow 2N,$

$2N + 3H_2 \rightarrow 2NH_3.$

The fixation may be accomplished by plants, but not all plants are suited to perform the process. Leafy vegetables are particularly suitable, and the increased cultivation of such vegetables by human activity is estimated to be responsible for a nearly 50% increase in the total fixation of nitrogen in the terrestrial biosphere (14×10^9 kg y^{-1}). The actual set of chemical reactions involved in the biological fixation is much more complicated than the summary equations given above, being combined with metabolic processes, by which the solar energy input is transformed into useful energy. Thus, the energy needed in the first step is derived from photosynthesis, while the second step, which is slightly exoergic, takes a different form, due to the absence of free hydrogen in practically all biological materials.

In the industrial nitrogen fixation process (the "Haber process"), the energy input may be from natural gas (methane) or another fossil fuel, which also provides the hydrogen for the second step. If renewable (other than biological) or nuclear energy is the starting point, hydrogen may be obtained by electrolysis. The rate of industrial fixation is presently around 30×10^9 kg y^{-1} (Delwicke, 1970). Most of the ammonia produced in this way is used to manufacture fertilisers, e.g. ammonium nitrate (NH_4NO_3) or urea ($CO(NH_2)_2$):

Fertilizer production

$NH_3 + 2O_2 \rightarrow HNO_3 + H_2O,$

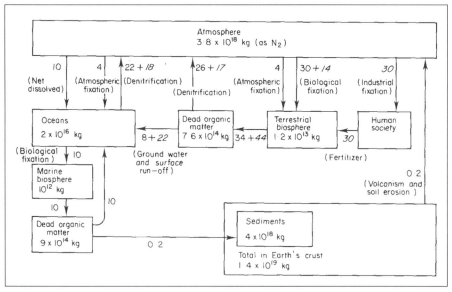

Figure 2.89. Schematic nitrogen cycle. The rates of transfer are given in 10^9 kg of nitrogen per year. Man-made contributions are shown in italics (with use of Delwicke, 1970).

$$HNO_3 + NH_3 \rightarrow NH_4NO_3,$$

or

$$2NH_3 + CO_2 \rightarrow CO(NH_2)_2 + H_2O.$$

The growth of plants which rely on nitrogen fixed by other plants can be greatly enhanced by administration of fertiliser. Nitrogen in fixed form appears to be a dominant limiting factor for such plants, but also the yield of plants which are themselves able to fix nitrogen can be enhanced by the addition of fertiliser. Since the fertiliser can replace the biological fixation, it saves the biosphere from spending the amount of energy which would otherwise be used to fix the corresponding amount of nitrogen. This is the basis for associating the use of fertiliser with an energy transfer from the human society to the biosphere, as shown in Fig. 2.87. To the extent indicated by this figure, one may say that plants are grown on fossil fuels rather than on solar energy.

A third source of fixed nitrogen, both for continents and oceans, is shown in Fig. 2.89 under the heading "atmospheric fixation". It is due to fixation within the atmosphere by ionisation of N_2 caused by radiation or lightning, and the nitrogen in fixed form is transferred to ocean or land surfaces by precipitation.

In a natural ecosystem which has reached a stationary state, no net removal of nitrogen from the atmosphere takes place. This is because the nitrogen fixed by the natural processes is later returned to the atmosphere by denitrification processes. The dead organic matter, in which the nitrogen may be bound in very complex compounds, is decomposed by bacteria which release the nitrogen in the form of ammonia ("ammonifying bacteria"). Another set of bacteria, called "nitrifying bacteria", oxidises ammonia to nitrite (NO_2^-) or nitrate (NO_3^-). Finally, the return of gaseous nitrogen, N_2, or N_2O to the atmosphere is achieved by "denitrifying bacteria", by highly exoergic processes based on sulphur or glucose, e.g.

Denitrification processes

$$C_6H_{12}O_6 + 6KNO_3 \rightarrow 6CO_2 + 3H_2O + 6KOH + 3N_2O,$$

$$5C_6H_{12}O_6 + 24KNO_3 \rightarrow 30CO_2 + 18H_2O + 24KOH + 12N_2.$$

The denitrification processes do not balance the sum of natural and man-made fixation processes, at the present level. The reason is not so much the time lag between the growing man-made nitrogen fixation and the formation of dead organic material, as the fact that only a fraction of the additional nitrogen in fixed forms adheres to the biological cycle all the way. A large quantity (in Fig. 2.89 estimated as 22×10^9 kg y^{-1}) leaves the terrestrial biosphere with ground water or surface run-off or is accumulated in soil (a build-up of 5×10^9 kg y^{-1} is indicated for the dead organic matter compartment on land). Some of the fertiliser administered to agricultural soils runs off with surface water even before it has been absorbed by plants, and some is carried away by ground water flows. Also, the conditions for a matching growth in denitrification do not appear to be fulfilled, for physical (the way in which the dead organic matter is deposited) or biological (the limiting factors governing growth of the population of denitrifying bacteria) reasons. The result is an accumulation of nitrogen in fixed forms in local water systems and in the oceans. According to the estimates made in Fig. 2.89, 4×10^9 kg is accumulated per year in the hydrosphere, and 9×10^9 kg y^{-1} is being removed from the atmosphere.

The above examples of matter cycles have illustrated some of the problems introduced by the rapidly increasing amount of interference, on the part of human society, with stationary cycles which have taken much longer to establish than the time scales of the changes now induced. It has also been demonstrated that the construction of a reasonable model of such cycles requires an intimate knowledge of all the chemical and physical processes of importance for the element considered.

2.4.2 Climate changes

The year-by-year fluctuations in climate parameters may appear as random fluctuations around an average climate when only a small number of years are considered. Over longer periods evidence for systematic changes has been found, and over the entire geological history it is quite certain that major changes in climate have taken place.

One important question is the extent to which the activities of human society inadvertently cause climatic changes that would not otherwise have occurred or speed up changes that would have happened anyway. In the longer term, the use of planned anthropogenic intervention may be considered, in order to avert natural climate changes that would limit or reduce the possibility of sustaining a human society of the type desired. At present, however, the consequences of human interference with climate, planned or inadvertent, cannot be predicted with a sufficient degree of certainty, so it would seem prudent to keep the amount of interference at the lowest possible level while more reliable information is being sought.

Energy use is in focus when addressing questions of anthropogenic interference with climate, as well as in the discussion of environmental pollution. The utilisation of renewable energy flows is often considered a non-polluting alternative (dams in connection with hydroelectric installations perhaps excepted), which may be true in terms of the absence of gaseous and particulate emissions. However, since these flows are integral parts of the climate, it is by no means certain that any utilisation – be it only a time delay in the return of long-wavelength radiation to space – could not have an effect on climate. In particular, agriculture is a way of utilising solar energy that involves changes of large land areas and almost certainly has had climatic impacts during the history of cultivation (cf. the discussion in Chapter 1).

Studies applying general circulation models have been used to address the question of the possible climatic impact of fuel burning with CO_2 emissions, of activities that may reduce this stratospheric ozone content, and general changes in net radiation, e.g. due to emissions altering cloud cover. Below, an outline will first be given of the climatic changes that have occurred in the past, inasmuch as they have been identified.

Climatic history

There is no detailed account of the early climate of the Earth before the Palaeozoic period starting about 6×10^8 years B.P. (before the present). Palaeontological evidence suggests that even the poles must have been ice-free more than 90% of the time since then. However, about 3×10^8 years B.P., a major glaciation took place at both poles, which at that time were covered by two vast continents, Laurasia in the north and Gondwana in the south. This period of glaciation, possibly with interruptions, lasted about 5×10^7 years and was followed by a warm period extending throughout the Mesozoic period (see Fig. 2.90). It is tempting to associate the cessation of this ice age with the continental drift, allowing warm ocean water to be circulated into the Arctic regions.

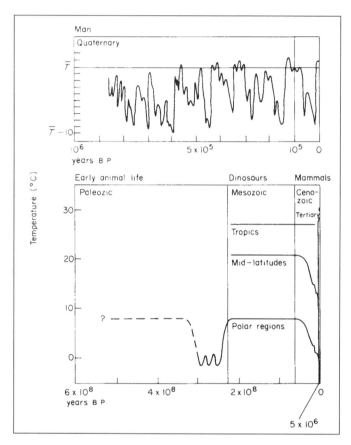

Figure 2.90. Trends in the surface temperature of the Earth. The lower figure gives gross estimates of temperature regimes during the eras of inhabited Earth for the tropical, mid-latitude and polar regions (based on paleontological evidence). Glaciations have taken place in the period between 3×10^8 and 2.5×10^8 years B.P., and again from 5×10^6 years ago at the South Pole, and from 2.5×10^6 years ago at the North Pole. The top figure shows, on an expanded scale, the trends of temperature variations during the last 10^6 years, believed to be representative for mid-latitudinal regions. The approximate temperature scale is given relative to the average temperature of the present (say last 10^3 years) climate, T. The temperature variations have been estimated on the basis of the extent of ice sheets (palaeontological data) and on cores drilled at ocean floors (^{18}O to ^{16}O ratio and fossilised organisms) (based on Wilson and Matthews, 1971; Shakleton and Opdyke, 1973).

From the Tertiary period (beginning about 6×10^7 years B.P.), a cooling in the mid-latitudes has been deduced, which finally led to new periods of glaciation in the Quaternary period (from about 10^6 years B.P.). A possible role of ocean currents and salinity changes has been suggested (Haug *et al.*, 1999).

During the last half of the Quaternary period, an oscillatory behaviour of the climate has been well established by studying layer sequences in ocean cores (Hays *et al.*, 1976), as well as evidence derived from estimated extents of the ice cover. The glaciation period started about 5×10^6 years B.P. in the Antarctic region, but 2–3 million years later in the Arctic region, presumably due to the distribution of continents at the time.

The fluctuating extent of the ice cover from the beginning of the Quaternary period to the present has been subjected to various kinds of smoothing and spectral analysis. Based on the ocean core data, Hays *et al.* (1976) found an overall frequency of the glaciations of about 10^{-5} y^{-1}. The shape of the temperature function is asymmetric, with a slow cooling from the interglacial maximum, taking about 9×10^4 y to reach the minimum (peak glaciation), followed by a rapid (10^4 years) rise of temperature back to the maximum value. The difference between maximum and minimum average temperature is believed to be about 10°C. Superimposed on the glaciation cycle are oscillations corresponding to a cycle time of about 4×10^4 years and another set, possibly with more than one component, of cycle times in the range $(1.9– 2.4) \times 10^4$ years.

Figure 2.91 shows estimated variations in the average temperature during the last 140 ky, based on ocean core data and (for the last 75 000 years) ice cores from Greenland (Dansgaard, 1977). The ice cores contain a more detailed record of climate, where seasonal variations can be followed. Each year corresponds to a depth interval of several centimetres, except for the very oldest (deepest) parts of the core. For the ocean cores one has to rely on estimated sedimentation rates. The temperature estimate is primarily based on the isotopic abundance of ^{18}O, which depends on the temperature prevailing in the atmosphere when the precipitation, which formed the ice, was falling. The reason is that $H_2^{18}O$ has a lower vapour pressure than $H_2^{16}O$, and thus the abundance of ^{18}O will be slightly higher in the precipitation than in the atmospheric vapour, implying a depletion of ^{18}O from the atmosphere. The lower the atmospheric temperature, the more ^{18}O has been removed from an air parcel having travelled for a while under the influence of condensation processes, so that a low ^{18}O content in the snow in Greenland indicates a low temperature throughout the "fetch region". It is clear that this type of temperature determination contains a number of uncertainties and that it is not possible to deduce the global average temperature from such data. However, the time sequence of warming and cooling trends should be very reliable, and comparisons with direct temperature measurements in recent periods indicate that these trends from the Greenland ice cores correlate well with the trends observed at northern mid-latitudes (Dansgaard *et al.*, 1975). Temperature scales similar to those indicated in Fig. 2.91 are suggested on the basis of other kinds of data, e.g. tree ring data, pollen records and fossil plankton data (see Bolin, 1974).

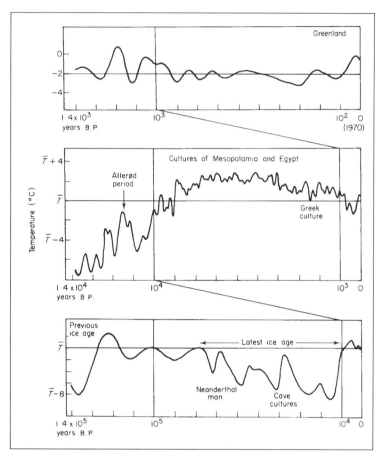

Figure 2.91. Trends in average surface temperature in recent times (last 1.4×10^5 years). The upper figures contain the most recent 1.4×10^4 and 1.4×10^3 years on expanded scales, and different smoothing times Δt_s, which "filters out" the short-term variations in temperature. The last 75 000 years are based on ice cores from drillings in Greenland, while the older periods are based on a number of determinations, including deep sea cores and pollen data. As in Fig. 2.90, a very tentative temperature scale relative to the average temperature of the most recent 10^3 years is indicated on the two lower figures. The top figure has a temperature scale derived from present-day measurements (Fig. 2.85), which should only be considered as representative for Greenland. The amplitudes of the temperature variations during the most recent couple of hundred years have been smaller at mid-latitudes than in Greenland (based on Bolin, 1974; Dansgaard *et al.*, 1975; Dansgaard, 1977).

It is shown how the latest ice age builds up over a period of at least 60 000 years and then disappears over something like 4000 years, with a number of fluctuations of appreciable amplitude superimposed on the main trend. During this glaciation period, man developed societies of substantial sophistication, including medical use of plants and the art of painting. Major urbanisations are found throughout the warm period starting about 10^4 years ago ("the climatic optimum"). During the most recent 10^3 years, the greatest change in climate has occurred in the present century, exhibiting a warming trend until about 1940, then a slight cooling trend until 1980, and finally more warming (IPCC, 1996a).

The 1880–1940 temperature rise from the Greenland data (Fig. 2.91 top) is somewhat larger than that observed at mid-latitudes, with the latter being 0.4°C according to IPCC (1996b).

Causes of climatic change

In the early history of the Earth, climate changes were integral parts of the evolution of continents and of the formation of the atmosphere and the oceans and their subsequent evolution. Major alterations in the surface structure, the formation of mountain chains and volcanic activity influenced climatic variables, and varying cloudiness and ozone content in the atmosphere modified the net radiation received at the surface, which itself, through changes in albedo, modified the radiation being reflected. In this way there are plenty of causes to choose from and to combine in order to account for the large-scale evolution of climate on Earth. As briefly mentioned above, the formation and dissolution of large continents near the poles may explain the late Palaeozoic glaciation and its subsequent disappearance. Superimposed on this large-scale trend of climatic behaviour shown in Fig. 2.90, short-term temperature excursions are likely to have taken place. Budyko (1974) argues that if the size and frequency of volcanic eruptions were stochastically distributed, then there would, during a time period as long as the Mesozoic era, be several occurrences of global dust-triggered temperature drops of perhaps 5 or 10°C, strong enough to last for several decades. The speculation is, then, that such episodes may have been responsible for the extinction of certain animal species (in particular species of dinosaurs, which appear to have had no heat adjustment mechanism). Of course, the durations of these temperature excursions were too short to leave any geological traces. Alternative speculations on the extinction of dinosaurs involve meteorite hits.

Considering now the Quaternary period, the physical state of the Earth (distribution of continents and oceans, composition of atmosphere) may be regarded as sufficiently stationary to exclude it as a likely cause of climatic changes. It is thus tempting to associate the periodic occurrence of ice ages and temperature oscillations (Fig. 2.90 top) with periodic variations in the external conditions, serving as boundary conditions for the climate equations. Such variations would primarily be variations in incident radiation, which could be caused, for example, by the periodic changes in the Earth's orbital parameters, due to many-body interactions in the planetary system.

Other suggestions invoke periodic passage through interplanetary dust (the difficulties of such theories have been pointed out by Dennison and Mansfield, 1976) or periodic changes in the Sun's radiation output (see Hammond, 1976), similar, but with different periods, to the well-known changes in magnetic activity of the Sun, e.g. the 11-year sunspot cycle. However, measured variations in solar output are much smaller than what seems needed, at least at present (IPCC, 1996b). Further, the oscillations in temperature during the Quaternary period might be interpreted in terms of Lorenz's model of near-intransitive, vacillating systems (see section 2.3.3), requiring no external causes.

The explanation in terms of orbital changes was first proposed by Milankovitch (1941). The frequency spectrum of such variations has a peak at 105×10^3 y (changes in the eccentricity of the Earth's orbit), 41 000 years (changes in obliquity, i.e. the angle between the equatorial and ecliptic planes) and a continuous shape with two peaks at about 23 000 and 19 000 years (changes in the longitude of the perihelion due to precession). The variations in obliquity and precession do not alter the total solar radiation received by the Earth, but lead to periodic variations at fixed geographical locations. The variation in eccentricity does alter the total amount of radiation, but by less than 0.1%. If this is considered negligible, the amount of radiation is unaffected, even locally, by the eccentricity variations.

In recent studies of ocean core samples, evidence for the presence of these spectral peaks, in the ^{18}O to ^{16}O ratio as well as in the sea surface temperature [estimated from the abundance of radiolarians (unicellular organisms) in the samples] has been strengthened (Hays *et al.*, 1976). The samples cover a time interval of 450 000 y back from the present (cf. Fig. 2.90). Whereas the periods around 4×10^4 years and around 2×10^3 years may then be interpreted in terms of variations in solar radiation, the coincidence of the main glaciation period and the period of eccentricity variations does not immediately suggest the nature of a possible causal relation. It is possible, however, that the 10^5 year period of climatic variations can be related to the amplitude of the precession components, since these amplitudes are directly proportional to the eccentricity. The coupling involved would have to be non-linear.

The white Earth state

It is important to realise that under present conditions, and probably throughout the Quaternary period, the most stable state of the Earth is one of complete glaciation (i.e. all oceans frozen, all continents snow covered). The explanation is of course that the albedo of such a "white Earth" would exceed 0.7 and thus reflect more than twice as much of the incident radiation as at present. The estimated global average surface temperature in this state is about $-70°C$ (Budyko, 1974). Model calculations indicate that the average temperature for the entire atmosphere would be about 20°C below the average surface temperature, but that stratospheric temperatures would be almost unchanged (Wetherald and Manabe, 1975).

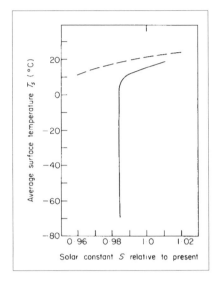

Figure 2.92. The calculated variation in globally averaged surface temperature, as a function of changes in the solar constant, relative to the present. The full curve is based on the one-dimensional model of Budyko (1974), while the dashed curve is based on a three-dimensional model of Wetherald and Manabe (1975).

Various attempts have been made to estimate the size of perturbation, which would trigger a transition from the present climate to a "white Earth". According to Budyko's (1974) simple, one-dimensional model, which parametrises the components of the net radiation flux and neglects seasonal variations and variations in cloudiness, a decrease as small as 1.6% in the solar constant would cause the ice caps to exceed a critical size, beyond which they would continue to grow until the entire Earth was covered. This process comes about as a result of the feedback loop: temperature decrease → increased glaciation → increased albedo → more temperature decrease. Budyko's model contains latitudinal transport of energy, in a highly idealised way, with parameters adjusted to reproduce present latitude dependence of temperature. In Fig. 2.92, the mean surface temperature derived from Budyko's model is compared with the results of Wetherald and Manabe, using the three-dimensional atmospheric circulation model with idealised topography, described in section 2.3.2, but without the oceanic circulation and without seasonal variations or detailed cloud modelling. The hydrological model is included, but, as the authors emphasise, the results concerning the relationship between solar radiation and temperature (or glaciation) may not be more reliable than those of Budyko, because his phenomenological parameters are fitted to (present) data, while Wetherald and Manabe approximately solve the true equations for a system which resembles but is not the actual Earth. As seen in Fig. 2.92, their average temperature for the Earth at present is about 6°C too high, and they predict a much slower dependence of the temperature on solar radiation, with no "white Earth" instability in the interval of solar constants considered, from 2% above the present actual value to 4% below. They warn that inclusion of the meridional transport by ocean currents may increase the sensitivity to the value of the solar constant. The extent of the maximum ice caps during the Quaternary period has exceeded 75% of the way to Budyko' s critical latitude of roughly 50°, which would have led to a white Earth.

From the above discussion, the long-term evolution of the Earth' s climate may be seen in a new light when one realises that an explanation is required to account for the presence of non-frozen water for more than 3×10^9 years, i.e. that the Earth could not have been in a stable state of complete glaciation during that time (or, if it was for shorter periods, an explanation is required to account for the large amount of energy necessary to melt the ice). The apparent contradiction is derived from the near certainty that the solar constant must have been increasing regularly from a value about 25% lower than the present value during the 4.5×10^9 years since the Sun became a main–sequence star. The gradual increase in solar luminosity is a result of the hydrogen depletion and radius increase, cf. Fig. 2.5. The magnitude of the relative increase in the solar constant, being thus about 0.05 per 10^9 years, is not strongly dependent on the particular solar model employed (Newman and Rood, 1977).

It is thus tempting to conclude that the change in solar constant must have been counteracted by a change in the Earth–atmosphere system, such that the average surface temperature has never been below the critical value leading to the white Earth, which again implies either that the Earth–atmosphere albedo was lower before or that the greenhouse effect was more pronounced. Even for an entirely ocean-covered Earth, the albedo decrease might not be sufficient, because increased cloudiness would counteract such a decrease. The greenhouse effect would be increased if the long-wavelength re-radiation from the atmosphere towards the Earth' s surface increased, as a result of an atmosphere of a composition different from the present one. The larger CO_2 content possible in the periods with little or no photosynthesis would act in this direction, but according to Sagan and Mullen (1972), even the maximum possible CO_2 content (in equilibrium with the CO_2 in oceans and the carbonate-containing rocks) would be insufficient. Sagan and Mullen suggest that a small amount of NH_3 (mixing ratio 10^{-5}) was present, which would effectively absorb radiation of wavelengths of around 10^{-5} m, leading to a surface temperature increase of as much as 70°C. At the same time, the atmospheric availability of NH_3 would enhance the chances of forming life, bypassing the step of nitrogen fixation. It may also be suggested that NH_3 was formed as a result of atmospheric fixation in periods of intense electrical activity in the atmosphere. If no new changes in the atmospheric composition take place, Sagan and Mullen estimate that the continued rise in the magnitude of the solar constant will lead to a 50°C increase in average temperature in another 4.5×10^9 years (this in itself may change the atmosphere into one like that of Venus today). A similar calculation for the Martian atmosphere suggests that in 4.5×10^9 years it will have a temperature much like the Earth' s today, which may serve as a consolation.

Man' s interference with climate

The first significant anthropogenic influence on climate is probably the one associated with extensive destruction of natural vegetation, primarily aimed at transforming it into arable land or pastures. Presently, about 10% of the continental area is cultivated and about 30% is occupied by grassland or pastures (Odum, 1972). The transformation of steppe or forest into arable land has a number of consequences. The components of the net energy flux (2.10) are altered. The net radiation flux is diminished owing to a (modest) increase in surface albedo. The albedos for steppe and coniferous forest (in non-Arctic regions) are about 0.13, for deciduous forest and tropical forest during the dry season 0.18, for tropical forest during wet season 0.24, for savannahs and semi-deserts 0.14 during the dry season and 0.18 during the wet season, and for deserts 0.28 (Budyko, 1974); the mean albedo for cultivated land and pastures is 0.20. Considering the types of transformation taking place, a reduction in net radiation ranging from 3 to 20% has been reported for the establishment of agricultural land in the temperate zones (Wilson and Matthews, 1971). Also, land reclaimed from the sea has a higher average albedo. The evapotranspiration and hence the latent heat flux away from the surface is smaller for grassland and slightly smaller for dry arable land than for forest, while it is slightly larger for wet arable land (rice paddies, etc.). The upward sensible heat flux is very small for wet arable land, medium for dry arable land and deciduous forest, and quite high for grassland and coniferous forest. Thus, the transformation of forest to arable land is likely to reduce considerably the Bowen ratio, defined as the ratio of sensible and latent heat fluxes,

$$\text{Bowen ratio} = E_s^{sens}/E_s^{lat}.$$

It is probable that man' s quest for agricultural land has had much larger unintended effects than the above-mentioned changes. It is believed that the savannah grasslands in the tropical regions are entirely man-made, since the natural vegetation of these semi-humid regions is dry, deciduous forest. The activities of man have greatly augmented the number of forest fires.

In the semi-arid regions, dry steppe is being transformed into semi-desert by overgrazing, which exposes the soil to wind erosion as well as to water erosion during the (short) rainy season. Calculations indicate that the albedo change due to overgrazing may induce severe drafts in such regions (Charney and Stone, 1975). Also, tropical forests have been and still are being destroyed by human activity. For instance, archaeological evidence suggests that the Rajputana desert in India was as fertile as a tropical forest only 1000 years ago, which implies that its Bowen ratio, which is now in the range 2–6, has decreased more than tenfold (Bryson, 1971).

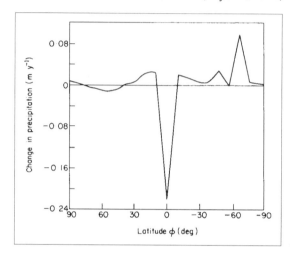

Figure 2.93. Calculated change in the rate of precipitation (averaged over longitudes), resulting from removing all tropical jungles (based on Potter *et al.*, 1975).

When a tropical forest is being removed, the soil is no longer able to retain its humidity. Then precipitation runs off along the surface, and the soil becomes still drier. The humus layer is washed away and a solid crust of clay is developed. Weathering of this dry soil may greatly increase the particle content of the atmosphere, and by reflecting incoming radiation this dust may reduce the net radiation at the surface. On the other hand, cloud cover is likely to become reduced. According to a computer simulation (Potter *et al.*, 1975), the surface temperature may drop by about 0.4°C, while the temperature at a height of 12–13 km may drop by 1°C (year average). Figure 2.93 shows the calculated change in zonal precipitation resulting from the assumed complete removal of the tropical rain forests. It is worth noting that the induced climate changes are global. The model does not include the influence of the dust particles from the barren soil on the rate of precipitation.

Urbanisation is another factor in climate modification due to anthropogenic land-use which may be significant, although the area involved is smaller than that of agricultural utilisation. Urbanisation usually entails an increase in temperature, a decrease in wind speed (through increased roughness length for the wind profile described in section 2.B) and, at least for societies relying on burning fossil fuels and on polluting industries, an increase in atmospheric content of particles and trace gases and an increase in cloudiness and precipitation, but less incoming radiation at the surface and less relative humidity of the atmosphere over the urban area.

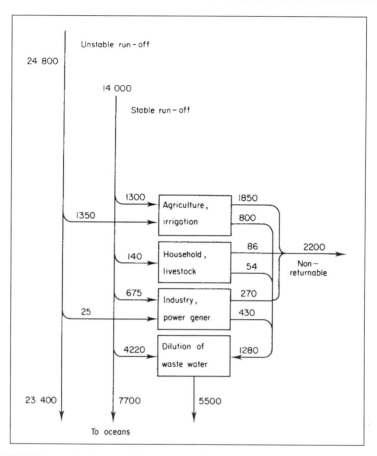

Figure 2.94. Schematic picture of human interference with river run-off. The rates are given in units of 10^9 m^3 of water per year. The "stable" part of the run-off is the part maintained by ground water flow plus the part associated with regulated reservoirs (based on Lvovitch, 1977).

Man's interference with the hydrological cycle (Fig. 2.61) has become quite significant. Lvovitch (1977) divides the total river run-off of 39×10^{12} m^3 water per year into a "stable" part, being maintained by ground water flows (12×10^{12} m^3 y^{-1}) or being regulated by reservoirs (2×10^{12} m^3 y^{-1}), plus a variable or "unstable" part (25×10^{12} m^3 y^{-1}). Anthropogenic interference with these flows is summarised in Fig. 2.94. It appears that the largest usage is for irrigation, then industry and then a small amount for household and livestock watering, but the significant measure for interference is probably the amount of water necessary to dilute the waste waters (which for some industrial enterprises may be very large), plus of course the amount of water really "used", i.e. left in an unretrievable form. As much as 45% of the stable run-off is being diverted through the human society, plus around 5% of the unstable run-off. Human extraction of ground water enhances the hydrological cycle, and significant changes in the net water flow at the surface are induced by the creation of artificial lakes or reservoirs and (working in the opposite direction) by drainage of swamps and other humid soils.

The amounts of latent energy moved by irrigation and subsequent evaporation and condensation are several times larger than the waste heat presently associated with direct anthropogenic energy conversion. But rather than changing the energy content of the atmosphere (as in urban "heat islands"), the effect of irrigation is to speed up the hydrological cycle. This could have climatic effects beyond the hydrological cycle which, as demonstrated e.g. by the model calculations described in section 2.3.1, is strongly coupled to the general circulation and temperature distribution.

The general circulation of the oceans has also been affected by man's activities. In this context, the regulation of rivers by building dams (in connection with hydropower) or canals (usually for transportation) may play a role by causing an "unnatural" timing of the peak run-off into the oceans or by diverting the run-off to other parts of the oceans than those originally receiving the river water. Few such planned engineering enterprises have been undertaken in open oceans, but a number of proposals have been put forward which could invoke large-scale climatic changes. An example is the proposal for building a dam across the Bering Strait and pumping water from the Arctic Ocean into the Pacific Ocean. Thereby, new warm water would be pulled into the Arctic region from the Atlantic Ocean, where it might melt the sea ice (the same might be achieved by spraying soot on the ice to reduce its albedo). Another example is the artificial upwelling of cold and nutrient-rich water into surface regions, with the purpose of increasing biomass production (e.g. in the tropical ocean surfaces, where the net radiation is abundant and the supply of nutrients is the main limiting factor) or energy production (thermal or salinity gradient conversion, see Chapter 4). Primary climatic effects might be increased fogginess in summer and increased heat flux in winter (Wilson and Matthews, 1971).

Concern has been expressed over the effect of oil spills (from tankers, offshore drilling, etc.) on the ocean–atmosphere interface (Wilson and Matthews, 1970). At large oil concentrations the heat fluxes may become impeded, but even at small concentrations wave motion is damped (particularly the short-wavelength wave components near the crests), and thus the transfer of kinetic energy from the atmosphere to the oceans may be diminished, with consequences for the circulation in the oceans as well as in the atmosphere.

Direct interference with the energy cycle takes place in connection with rain-making (seeding clouds with condensation or freezing nuclei, e.g. silver iodide crystals) or rain-preventing (hail suppression, hurricane control). The possible climatic effects have not been adequately studied, and they presumably depend strongly on the circumstances of a given operation, e.g.

whether the provoked precipitation is subsequently transferred by evapo-transpiration or by surface run-off.

Particle emissions (e.g. from industry) in a certain size range may provide an unintended source of cloud condensation nuclei. It is estimated (Hobbs *et al.*, 1974) that a relatively small number of suitable particles may affect the hydrological cycle in a globally significant way. All activities which may affect the cloud structure, cloud cover and cloud height have the potential of inducing climatic changes brought about by the influence on the radiation climate (short- and long-wavelength) and on the hydrological cycle.

Aerosols

Only particles with suitable properties, related to crystal structure for freezing nuclei and solubility for water condensation nuclei, can serve as cloud nuclei. The concentration of such particles is estimated as $(4\text{–}10) \times 10^{-5}$ m^{-3} over oceans and 4×10^{-4} m^{-3} over continents, their production rate being estimated as 4×10^{28} y^{-1}. Their average mass is 10^{-20} kg (Twomey and Wojciechowski, 1969). This is only a small fraction of the total particle content in the atmosphere (cf. Fig. 2.30), and the production rate is only about 0.1% of the total anthropogenic emissions (cf. section 2.3.1). The remaining particles influence climate primarily by scattering and absorption of radiation.

A layer of particles implies quite complex processes, in which radiation is (multiply) scattered, absorbed or transmitted, with each sub-process depending on the wavelength of radiation as well as on the properties of the particle (aerosol) on which the radiation is scattered. The net result is a modified albedo of the Earth–atmosphere system containing an aerosol layer, as well as a modification in the amount of upward radiation from the Earth's surface into space. The modified albedo is expected to be larger when the aerosol layer is present, leading to less radiation incident on the Earth's surface, which again constitutes a cooling effect. On the other hand, the presence of the aerosol layer is expected to decrease the amount of radiation from the Earth's surface back into space. In this way the temperature at the surface increases, implying a heating effect. It has been estimated (Reck *et al.*, 1974) that the cooling effect dominates if the albedo a_s of the Earth's surface is below 0.6 and that the heating effect dominates if a_s is above 0.6 (snow and ice), fairly independently of other parameters (note that albedos used here all refer to short-wavelength radiation).

Globally averaged, there is thus little doubt that an increase in atmospheric aerosol content will lead to a decrease in temperature, but in the polar region there will be a delicate balance between cooling and heating effects. The average albedo at high latitudes is about 0.6 for the combined Earth–atmosphere system (Raschke *et al.*, 1973), but higher for continental ice sheets (Greenland and Antarctic). At low latitudes, the surface albedo a_s is considerably smaller than the Earth–atmosphere albedo a_0, due to the influence of clouds, but in polar regions, a_s and a_0 are more similar. It may thus be expected that the addition of particles in Arctic regions may have a slight warming (ice melting) effect, while the effect at lower latitudes is definitely cooling.

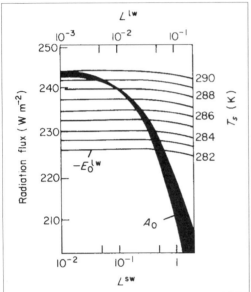

Figure 2.95. Absorption of short-wavelength radiation by the Earth–atmosphere system with an aerosol layer, as a function of the optical thickness L^{sw} of the aerosol layer in visible light. The absorbed flux A_0 is related to incoming radiation and albedo of the Earth–atmosphere system by $A_0 = E_{0+}^{sw}(1-a_0)$. The figure also shows the long-wavelength re-radiation into space, E_0^{lw}, as a function of the optical thickness L^{lw} of the aerosol layer in infrared light (top scale), for a range of surface temperatures T_s (right-hand scale) (based on Rasool and Schneider, 1971).

The magnitude of the cooling effect caused by increased aerosol load in the atmosphere was first studied by Rasool and Schneider (1971) and others at the same time. Figure 2.95 shows the amount of incoming (short-wavelength) radiation absorbed by the Earth–atmosphere system (including an aerosol layer), as function of the dimensionless optical thickness L of the aerosol layer (a measure of the amount of aerosols present),

$$L = \int \sigma_{abs} \, ds = \int l^{-1} \, ds = \int dI/I,$$

where σ_{abs} is the absorption cross section per unit volume, l is the mean free path of light due to aerosol absorption, and I is the intensity of the radiation. In all cases averages over wavelengths should be performed, but often the calculation simply uses one "characteristic" wavelength, in the present case $\lambda = 5.5 \times 10^{-7}$ m, corresponding to visible light. The corresponding optical thickness will be denoted L^{sw}. The interval of uncertainty indicated represents the lack of knowledge of the absorptive properties of the individual aerosols, the nature of which has not been specified. The calculation of Rasool and Schneider assumes that the aerosol layer lies between the surface and an altitude of 1000 m (cf. Fig. 2.30). It further assumes a 50% cloud cover and an aerosol size distribution proportional to r^{-4} for the (approximate) calculation of multiple Mie scattering (cf. section 2.3.1).

For the long-wavelength radiation, the Mie scattering cross section is much smaller than for wavelengths in the visible range, as seen from Fig. 2.39. Rasool and Schneider assume that the ratio of optical thicknesses for infrared (taken as $\lambda = 10^{-5}$ m) and visible radiation is

$$L^{lw}/L^{sw} = 0.108.$$

Thus, the infrared radiation to space is little affected by the presence of an aerosol layer, relative to the influence on the transmission of visible light. In Fig. 2.95, an L^{lw} scale is added at the top, and curves representing the long-wavelength radiation to space are drawn (thin lines) for different values of the surface temperature, T_s, i.e. the Planck radiation temperature of the upward radiation spectrum (section 2.A). The optical aerosol thickness at present is believed to be around 0.1. The figure indicates that in order to decrease the average surface temperature by one degree, the aerosol content would have to be increased by a factor of 1.5–2.0. For the average residence time around 15 days mentioned for atmospheric particles in section 2.3.1, the particle emission which increases the atmospheric load by a factor of 2 can be taken as twice the present. Thus, if 20% of the emissions (including emission of gases which later form particles) at present are of anthropogenic origin, then a doubling of the total amount would require a sixfold increase of the anthropogenic emissions.

The newest climate model calculations mentioned below under "carbon dioxide" all include a model for aerosol contents in the atmosphere.

Carbon dioxide

Gaseous emissions into the atmosphere may entail environmental side-effects and may affect human health. Such pollution capacities have been demonstrated, e.g. in connection with sulphur oxides from fossil fuel combustion and with photochemical smog. However, trace gases may also interfere directly with climate, by altering components of the radiation fluxes. As can be seen from Fig. 2.36, H_2O and CO_2 are very important absorbers in the infrared wavelength region, particularly for wavelengths above 10^{-5} m, where the peak of the long-wavelength radiation to space is situated. It is thus to be expected that an increase in the atmospheric content of these absorbers will lead to a higher temperature in the troposphere as well as on the Earth's surface. The present content of CO_2 is already big enough to imply a nearly total absorption at $\lambda = 1.5 \times 10^{-5}$ m. It is therefore unlikely that even a dramatic increase in CO_2 could have consequences similar to the addition of a small amount of NH_3 (temperature changes globally averaging something of the order of 10°C or more), discussed earlier in this section. However, a temperature change of just one or two degrees globally may have serious effects on climate, particularly if the local temperature change is amplified in sensitive regions, such as the polar ones.

As seen from Fig. 2.88, a little over half of the anthropogenic CO_2 emissions are accumulating in the atmosphere, but it is difficult to determine precisely the magnitude of future emissions, which would lead to a specific temperature increase, owing to the non-linear absorption response at the specific radiation wavelengths characterising each gas contributing to greenhouse warming (cf. Fig. 2.36).

Numerous calculations estimating the climatic effects of doubling the atmospheric CO_2 content have been performed over the years. An early model by Manabe and Wetherald (1967) used a global average one-dimensional (height) temperature distribution, taking into account both the changes in radiation components and the change in the turbulent (convective) fluxes from the surface, as a function of CO_2 content. Assuming a fixed relative humidity and cloudiness, a doubling of the CO_2 concentration in this calculation implied a temperature increase at the Earth's surface amounting to 2.4 K. A later three-dimensional equilibrium model used by Manabe and Wetherald (1975) employed the techniques described in section 2.3.1, but with the idealised topography discussed in section 2.3.2 in connection with the joint ocean–atmosphere model of the same authors. No seasonal variations of radiation were considered in that calculation, with no oceanic heat transport, and the cloud cover was fixed at 0.5. Test computations starting with substantially different initial conditions gave equilibrium temperature distributions that concurred with each other to better than about 1 K. Figure 2.96 shows the hemispherical longitude averages of temperature for the calculation with CO_2 content doubled, minus the corresponding temperatures for a reference calculation (CO_2 mixing ratio 4.56×10^{-7} by mass). In the stratosphere, the doubled CO_2 content leads to a net cooling, while the troposphere is heated. The temperature at mid-heights is increased by about 3 K, while the surface temperature is increased by 2–3 K, except in the polar region, where the temperature increases by up to 10 K. The sensitivity of the polar region is associated with the possibility of large albedo changes, primarily connected with ice melting. The model predicts a poleward movement of the border of the ice cap by about 10°, plus a reduction in the thickness of snow cover. The latter is due to reduced snowfall (in the polar region), while the total amount of precipitation increases at all latitudes, as a result of doubling CO_2. For this reason, the latent heat upward flux (evaporation) increases significantly, and the warming of the troposphere by a greenhouse effect is due to the combined absorption of radiation by water vapour and CO_2. The effective temperature T_0 of the entire Earth–atmosphere system is increased by roughly 0.6 K, but the authors warn that the strong interconnection between the CO_2 cycle and the hydrological cycle that the model has indicated may necessitate the inclusion of a predictive equation for the cloud cover in their model.

Finally, Fig. 2.97 gives the results of a recent modelling effort, using a grid with spatial resolution $2.5° \times 3.75°$ combined with 19 atmospheric height levels and 20 oceanic depth levels (Mitchell *et al.*, 1995; Mitchell and Johns, 1997). The model is transient, starting with realistic data from about the year 1860 and following the development under conditions of changing CO_2 emissions as well as taking into account the sulphate aerosol negative forcing. Model output for the present era is in substantial agreement with observed data, and the geographical distribution of calculated change between surface temperatures around 2055 and those characterising the 1990s are shown for each season in Fig. 2.97, averaged over period of 10 years.

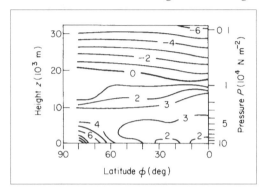

Figure 2.96. Calculated change in longitude averaged temperature distribution (K) in the atmosphere over the Northern Hemisphere, as a result of doubling the concentration of carbon dioxide (based on Manabe and Wetherald, 1975).

Compared to the earlier calculation, the only significant change is the cooling effect of particles in the atmosphere, which is strong at present and in the near future (giving a very complex temperature modification pattern), but diminishes in importance towards the end of the period because of assumed reductions in sulphate aerosol emissions from power plants and other activities, as environmental regulation introduced in many countries today is making itself felt globally. The average warming from the doubled CO_2 is about 2.5°C.

There is a continued effort to include more and more realistic effects in the climate simulation models. This includes differentiating between different residence time of different greenhouse gases (rather that lumping them together as an "effective CO_2 emission"); improving the cloud and generally water vapour modelling, the oceanic transport, and the transfer of heat and particles between oceans and atmosphere; and of course refining the grid resolution and accuracy of topography.

Support for the correctness of the calculated persistence in the atmosphere of greenhouse gases has recently come from the analysis of a climatic change that occurred 55 million years ago and is interpreted as caused by sudden release of methane from sedimentary hydrides. The residence time found by the sediment core study is 120 000 years (Norris and Röhl, 1999).

It is not the intention here to give an extensive list of possibilities for human society interfering with climate. In brief, it may be stated that life, and in particular human life, is sensitive to even very small changes in certain critical elements. Among these, a well-established case is that of ultraviolet radiation penetrating to the surface, which as discussed in section 2.3.1 depends on the presence of definite trace gases in the stratosphere (see Fig. 2.34). In particular, it has been suggested that the ozone concentration may be very sensitive to various human activities, such as aircraft combustion products (see e.g. Alyea *et al.*, 1975), chlorofluoromethanes (see e.g. Wofsy *et al.*, 1975) and nitrous oxide, N_2O, from increased use of fertilisers in agriculture (Johnston, 1977) (cf. section 2.4.1). In the two latter cases, the gases in question are very inert in the troposphere, but may become catalysts for chemical processes in the stratosphere by which ozone is depleted. The modelling of these changes, and their induced modifications of the radiative

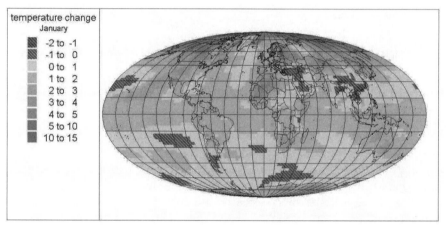

Figure 2.97a,b. January and April surface temperature changes for the mid-21st century, from a transient climatic model including both CO_2 increase (roughly doubling) and sulphate aerosol forcing (average 2050–2059 minus 1990–1999, in °C) (based on HADCM2-SUL; Mitchell and Johns, 1997; data source IPCC DDC, 1998).

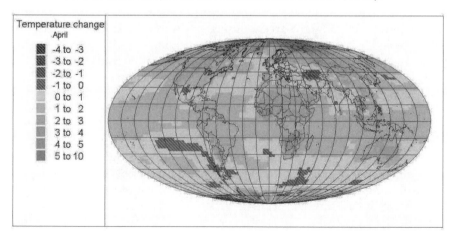

spectrum (Cutchis, 1974), was one of the first climate calculations undertaken. It involves a detailed knowledge of the chemistry and photochemistry of the atmosphere, together with the general atmospheric models describing the circulation and other physical processes, in particular the mixing and diffusion processes between the stratosphere and the lower atmosphere.

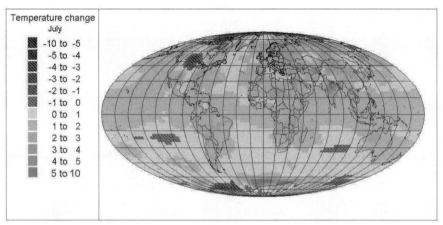

Figure 2.97c,d. July and October surface temperature changes for the mid-21st century, from a transient climatic model including both CO_2 increase (roughly doubling) and sulphate aerosol forcing (average 2050–2059 minus 1990–1999, in °C) (based on HADCM2-SUL; Mitchell and Johns, 1997; data source IPCC DDC, 1998).

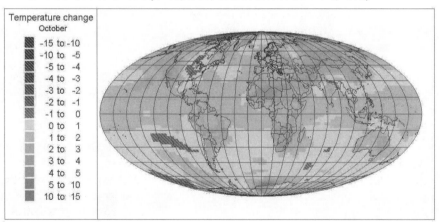

As a result of these calculations, international agreements on the reduction of emissions affecting the ozone content in the atmosphere have been reached, and similar agreements may be forthcoming regarding the curbing of greenhouse gas emissions. The very substantial economic implications involved are discussed in Chapter 7.

2.A Origins of stellar energy

The energy flows that are conventionally termed "renewable" are primarily associated with radiation from the Sun. This section presents a brief discussion of the origins of energy production in stars such as the Sun.

The birth and main-sequence stage of a star

Stars are believed to be created from particulate matter present in a forming galaxy. Contraction of matter is achieved by gravitational processes involving clouds of dust and gas, but counteracted by the increasing pressure of the gases as they are heated during contraction (cf. the equilibrium equations given below). However, once the temperature is high enough to break up molecules and ionise atoms, the gravitational contraction can go on without increase of pressure, and the star experiences a "free fall" gravitational collapse. If this is considered to be the birth of the star, the development in the following period consists of establishing a temperature gradient between the centre and the surface of the star.

While the temperature is still relatively low, transport of energy between the interior and the surface presumably takes place mainly through convection of hot gases (Hayashi, 1966). This leads to a fairly constant surface temperature, while the temperature in the interior rises as the star further contracts, until a temperature of about 10^7 K is reached. At this point the nuclear fusion processes start, based on a thermonuclear reaction in which the net result is an energy generation of about 25 MeV (1 MeV = 1.6×10^{-13} J) for each four hydrogen nuclei (protons) undergoing the chain of transformations resulting in the formation of a helium nucleus, two positrons and two neutrinos,

$$4\,{}^{1}_{1}\text{H} \rightarrow {}^{4}_{2}\text{He} + 2e^{+} + 2\nu + 26.2\,\text{MeV}.$$

After a while the temperature and pressure gradients have built up in such a way that the gravitational forces are balanced at every point and the gravitational contraction has stopped. The time needed to reach equilibrium depends on the total mass. For the Sun it is believed to have been about 5×10^7 years (Herbig, 1967). Since then the equilibrium has been practically stable, exhibiting smooth changes due to the hydrogen burning, such as the one shown in Fig. 2.2 for the radius. Before describing the conditions for the Sun's equilibrium state in a little more detail, a few words should be said about its subsequent development, which can be inferred from studies of stars which are more advanced in their evolution.

The red-giant and shell-burning stages

The hydrogen-burning stage will continue until the hydrogen in the centre of the star is nearly depleted. When this happens, the stability of the star is lost and it continues on a new evolutionary path. The stage of hydrogen burning in the core is projected to last about 10^{10} years for the Sun, implying that it is about halfway through this stage.

The diminishing number of protons in the centre of the star causes this region to contract slowly, and consequently, the temperature rises. The hydrogen burning is still taking place in a thick shell around the centre, and as a result of the rising temperature, the reaction rate increases and more energy is produced. In order to accommodate the increased flow of energy, the outer shell of the star has to expand. The radius may become some 50 times larger, and the surface temperature drops by about 1500 K, which causes the light emitted to change its spectrum in the direction of longer wavelengths. For this reason these stars are recognised as "red giants". Their luminosity is typically 1000 times that of the Sun (Iben, 1970).

As the temperature of the hydrogen-burning shell increases, the chain of hydrogen to helium sub-processes that dominates in stars like the Sun (the *pp*-chain described below) receives competition from another chain of reactions (the CNO-chain), which also results in a net hydrogen to helium conversion. Eventually, this process takes over. At this time, the conditions in the centre will allow the production of neutrino–antineutrino pairs, which escape from the star, carrying away large amounts of energy. The energy loss from the central region is compensated by a flow of matter inwards, mostly in the form of high-energy electrons. The central temperature rise is slowed down by the neutrino radiation, and it takes some 3×10^9 years (from the departure from the "main sequence" stage of hydrogen only burning in the core) until the temperature is sufficiently large to allow the burning of helium.

The burning of helium is achieved by the reaction

$$3 \, ^4_2\text{He} \rightarrow \, ^{12}_6\text{C} + \gamma + 7.28 \text{ MeV}.$$

The gamma-emission is associated with a definite decay process from an excited level of ^{12}C to its ground state, releasing 7.66 MeV of energy. Thus, the kinetic energy of the helium nuclei should furnish 0.38 MeV, which implies that the transition rate becomes significant only at temperatures above 10^8 K. Some of the carbon nuclei react with another helium nucleus to form oxygen,

$$^{12}_6\text{C} + \, ^4_2\text{He} \rightarrow \, ^{16}_8\text{O} + \gamma + 3.2 \text{ MeV}.$$

Since the temperature dependence of the helium-burning process is very strong just above 10^8 K, the star experiences a "helium flash" with a maximum temperature around 3×10^8 K. However, a new equilibrium is reached in which the helium burns stably in the centre at a temperature of 10^8 K, while hydrogen burns in a shell around the core, each contributing about half the energy production. This period may last of the order 10^8 years, and it is followed by alternating periods of instability and stability, as helium is depleted in the core and a helium-burning shell is formed.

Repeated contraction and temperature increase in the centre can ignite carbon- and oxygen-burning processes (10^9 K), resulting predominantly in the creation of $^{24}_{12}$Mg and $^{28}_{14}$Si. At a temperature around 3×10^9 K, gamma-rays cause the magnesium and silicon nuclei to evaporate a significant number of helium nuclei, which may be captured by other Mg, Si or heavier nuclei to successively create a number of nuclei in the mass range $A = 30$–60.

Equilibrium processes and supernovae explosions

Since the greatest amount of binding energy per nucleon (8.8 MeV) is found at $^{56}_{26}$Fe (see Fig. 3.82), there exist no exothermic reactions to burn iron, and hence further temperature rise of the star cannot be counterbalanced by radiative losses associated with further burning processes. The possibility exists, on the other hand, of establishing a stable, thermodynamic equilibrium with a temperature around 4×10^9 K and a density of around 10^{11} kg m^{-3}.

If net processes like

$$2\ ^{28}_{14}\text{Si} \rightarrow\ ^{56}_{26}\text{Fe} + 2e^- + 2\nu + 17.6\ \text{MeV}$$

were allowed to proceed until all the lighter elements had a chance to react, the abundance of any species ^A_Z(name) would be uniquely determined by the equations of thermodynamic equilibrium.

The abundance of elements and isotopes in the Sun' s planetary system suggests that the "equilibrium" processes that created the species around iron did not have the time required to reach a complete equilibrium, but that the material which did get through the Si to Fe type of process has, in fact, very nearly the distribution expected for a thermodynamic equilibrium (Clayton, 1968).

Assuming that the equilibrium processes are halted by a new phase of contraction and temperature increase, the fate of the newly created medium-heavy elements may be a dissociation or "melting" process of highly endothermic character,

$$^{56}_{26}\text{Fe} \rightarrow 13\ ^4_2\text{He} + 4n - 124.\ 3\ \text{MeV}.$$

This melting can take place at a temperature of about 7×10^9 K, leading to a rapid collapse of the star. Any lighter elements left in the outer regions of the star will start to burn explosively. It is possible that this is what happens in the observed supernovae outbursts, leading to the release of stellar matter into the interstellar space (Fowler and Hoyle, 1964). The ejection of material may be triggered by large quantities of neutrinos created in the central region. What is left behind in the centre after the supernova outburst will largely be neutrons, forming a dense "neutron star", unable to expand because most of the energy was carried away by the neutrinos (Chiu, 1964).

The formation of heavy elements

The evolution sketched above would explain the formation of elements up to the region of iron. Since the binding energy per particle then decreases as the weight of the element increases, heavier elements cannot be expected to form as a result of spontaneous fusion processes. More likely candidates are neutron capture processes, followed by β-decay processes. The source of the necessary neutrons is not entirely clear. Not until the carbon-burning stage do the main sequence of evolutionary nuclear reactions produce free neutrons. However, secondary processes during the CNO hydrogen-burning chain and during the helium-burning stage may be able to provide a sufficient flux of neutrons. These would be formed by (4He, n) reactions on nuclei such as ^{13}C, ^{16}O, ^{21}Ne or ^{22}Ne. The first three are found in small quantities for the CNO cycle (^{21}Ne only after proton capture and β-decay from ^{20}Ne), whereas the last is produced if CNO elements are present during the burning of helium,

$$^{14}_{7}N \text{ (end product of CNO chain)} + 2\,^{4}_{2}He \rightarrow\ ^{22}_{10}Ne + \gamma + e^+ + \nu.$$

For the synthesis of heavy elements to take place with neutrons originating from the helium–burning stage, elements up to the iron region must already be present in some quantity. It has been suggested that such conditions would be present in a second-generation star, the material of which is assembled from interstellar matter ejected from earlier stars, having already been through the carbon-burning stage (Clayton, 1968).

An alternative explanation for heavy element production is associated with the observed instabilities of stars in late burning stages. Such instabilities would occur in stars with carbon burning in the core and helium and hydrogen burning in subsequent outer shells, if hydrogen is periodically mixed into the helium- and carbon-enriched regions (Ulrich, 1973). The protons react with ^{12}C, forming ^{13}N, which subsequently β-decays into ^{13}C,

$$^{12}_{6}C + {}^{1}_{1}H \rightarrow\ ^{13}_{6}C + \gamma + e^+ + \nu.$$

Substantial amounts of neutrons would thus be formed by (4He, n) reactions, close to the carbon-burning regions where the target elements for the onset of the neutron capture processes are present.

The neutron capture processes may be fast or slow relative to the competing β-decay processes. The elements formed by the slow process (s-process) are close to the β-stability line (the nuclear mass and charge numbers A and Z for which neither electron nor positron emission is exothermic), whereas a succession of rapid neutron captures goes through very neutron-excessive nuclei, which eventually decay back towards the instability line, sometimes producing elements which could not be formed by the s-process. The rapid process is referred to as the r-process (Burbridge et al., 1957).

A comparison between the calculated abundance of species having long mean lifetime at their time of formation and their present abundance on Earth, in meteorites, etc., allows the establishment of a chronology or time scale. One example is offered by the uranium isotopes ^{235}U and ^{238}U, which are exclusively created by the r-process. If these isotopes were initially formed over a short period of time, their relative abundance would be

$$R_r = {}^{235}_{92}U / {}^{238}_{92}U = 1.45,$$

as compared with $R(t) = 7.3 \times 10^{-3}$ on Earth today (Seeger et al., 1965). The mean lives being $\tau(235) = 10^9$ y for ^{235}U and $\tau(238) = 6.5 \times 10^9$ y for ^{238}U, the time t since the r-process production of these isotopes would be

$$t = (\tau(235)^{-1} - \tau(238)^{-1})^{-1} \log(R(t)/R_r) = 6.3 \times 10^9 \text{ y}.$$

If, as suggested above, the r-processes take place periodically in multishell-burning stars, then the time from the onset of the r-process until now may be larger than t, whereas the time from the cease of production (due to supernova explosion) may be less than t. A precise determination of the age of the solar system based on such isotope ratios will involve not only a model for the time distribution of production processes, but also a model for the time distribution of supernova outbursts contributing to the formation of our solar system.

Equilibrium processes in the Sun

The present near-equilibrium state of the Sun may be described by a number of physical conditions. The first one is the condition of hydrostatic equilibrium, which states that the pressure gradient at every place must exactly balance the gravitational force. Assuming the Sun to be spherically symmetric, the condition may be written

$$dP/dr = - G M(r) \rho(r) / r^2, \tag{2.20}$$

where $M(r)$ is the integrated mass within a sphere of radius r, $\rho(r)$ is the local density and $G = 6.67 \times 10^{-11}$ m^3 kg^{-1} s^{-2} is the constant of gravitation.

The pressure $P(r)$ may be related to the local temperature $T(r)$ by an equation of state. The conditions everywhere in the Sun are close to those of an ideal gas, implying

$$P(r) = n(r) k T(r),$$

where $n(r)$ is the number of free particles per unit volume and $k = 1.38 \times 10^{-23}$ J K^{-1} is the Boltzmann constant. Alternatively, one may express n in terms of the gas constant, $\mathscr{R} = 8.315$ J K^{-1} mol^{-1}, the density and μ, the mean molecular weight (i.e. the mass per mole of free particles),

$$n(r) = \mathscr{R} \, \rho(r) / (k \, \mu).$$

For stars other than the Sun the ideal gas law may not be valid, as a result of electrostatic interactions among the charges (depending on the degree of ionisation, i.e. the number of free electrons and other "unsaturated" charges) and radiation pressure.

Cox and Giuli (1968) estimate the maximum pressure corrections to be 1.5% from the electrostatic interactions and 0.3% from the radiation pressure, calculated from the black-body radiation in the Sun' s centre. The mean molecular weight may be obtained from assuming the Sun to consist of hydrogen and helium in mass fractions X and $(1 - X)$. If all hydrogen and helium is fully ionised, the number of free particles is $2X + 3(1 - X)/4$ and the mass per mole of free particles is

$$\mu = 4 \, N_A \, M_p / (5X+3),$$

where $N_A = 6 \times 10^{23}$ is Avogadro' s number (the number of particles per mole) and 1.67×10^{-27} kg is the proton mass. The Sun' s hydrogen fraction X is about 0.7.

The potential energy of a unit volume at the distance r from the Sun' s centre, under influence of the gravitational forces, is

$$\varepsilon_{pot}(r) = -G \, M(r) \, \rho(r)/r,$$

while the kinetic energy of the motion in such a unit volume, according to statistical mechanics, is

$$\varepsilon_{kin}(r) = 1.5 \; n(r) \, k \, T(r).$$

Since ε_{pot} is a homologous function of order -1, the virial theorem of classical mechanics (see e.g. Landau and Lifshitz, 1960) states that for the system as a whole,

$$2 \, \bar{\varepsilon}_{kin} = - \, \bar{\varepsilon}_{pot}.$$

The bars refer to time averages, and it is assumed that the motion takes place in a finite region of space. Inserting the expressions found above, using $R = k \, N_A$ and integrating over the entire Sun and approximating $\int (M(r)/r) dM(r)$ by M/R (M and R being the total mass and radius), one finds an expression for the average temperature,

$$T = 4GMM_p / (3(5X + 3)kR).$$

An estimate for the average temperature of the Sun may be obtained by inserting the solar radius $R = 7 \times 10^8$ m and the solar mass $M = 2 \times 10^{30}$ kg, leading to

$T = 5 \times 10^6$ K.

The rigorousness of the hydrostatic equilibrium may be demonstrated by assuming a 1% deviation from equilibrium, implying an inward acceleration

$d^2r/dt^2 = \rho(r)^{-1}(GM(r)\,\rho(r)/r^2 - dP/dr) = -0.01\,GM(r)/r^2.$

The time needed for a 10% contraction ($\Delta r/R = -0.1$) under the influence of this acceleration is

$\Delta t = 7200$ s $= 2$ h,

at the solar surface $R = 7 \times 10^8$ m. The observed stability of the solar radius (see Fig. 2.2) demands a very strict fulfilment of the hydrostatic equilibrium condition throughout the Sun.

The energy transport equations

In order to derive the radial variation of the density and the temperature, energy transport processes have to be considered. Energy is generated by the nuclear processes in the interior. The part of the energy from nuclear processes which is carried by neutrinos would escape from a star of the Sun's assumed composition. One may express this in terms of the mean free path of neutrinos in solar matter (the mean free path being large compared with the Sun's radius) or in terms of the mass absorption coefficient,

$$\kappa = \frac{-dI}{\rho\,Ids} = \frac{1}{\rho}\sum_i \sigma_i,$$

with this quantity being very small for neutrinos in the Sun. I denotes the energy flux per unit solid angle in the direction ds, and s denotes the path length in the matter. The alternative way of writing the mass absorption coefficient involves the cross sections σ_i per volume of matter of all the absorption or other processes which may attenuate the energy flux.

The energy not produced in association with neutrinos is in the form of electromagnetic radiation (γ-rays) or kinetic energy of the particles involved in the nuclear reactions. Most of the kinetic energy is carried by electrons or positrons. The positrons annihilate with electrons and release the energy as electromagnetic radiation, and also part of the electron kinetic energy is gradually transformed into radiation by a number of processes. However, the electromagnetic radiation also reacts with the matter, notably the electrons, and thereby converts part of the energy to kinetic energy again. If the distribution of particle velocities locally corresponds to that predicted by statistical mechanics (e.g. the Maxwell–Boltzmann distribution), one says that the system is in local thermodynamic or statistical equilibrium.

Assuming that thermodynamic equilibrium exists locally, at least when averaged over suitable time intervals, the temperature is well defined, and the radiation field must satisfy the Planck law,

$$\frac{dF}{d\nu} = \frac{2h\nu^3}{c^2}\frac{1}{e^{h\nu/kT}-1}. \tag{2.21}$$

Here F is the power radiated per unit area and into a unit of solid angle, ν is the frequency of radiation, $h = 6.6 \times 10^{-34}$ J s is Planck's constant and $c = 3 \times 10^8$ m s^{-1} is the velocity of electromagnetic radiation (light) in vacuum.

The appropriateness of the assumption of local thermodynamic equilibrium throughout the Sun's interior is ensured by the shortness of the mean free path of photons present in the Sun, as compared to the Sun's radius (numerically the ratio is about 10^{-13}, somewhat depending on the frequency of the photons). The mean free path, l, is related to the mass absorption coefficient by

$l(\text{photons}) = 1/\rho\kappa.$

This relation as well as the relation between l and the cross sections of individual processes may be taken to apply to a specific frequency ν. κ is sometimes referred to as the opacity of the solar matter.

The radiative transfer process may now be described by an equation of transfer, stating that the change in energy flux, along an infinitesimal path in the direction of the radiation, equals the difference between emission and absorption of radiation along that path,

$dI(\nu)/(\rho\,ds) = e(\nu) - \kappa(\nu)\,I(\nu).$

Integrated over frequencies and solid angles, this difference equals, in conditions of energy balance, the net rate of local energy production per unit mass. In complete thermodynamic equilibrium one would have Kirchhoff's law, $e(\nu) = \kappa(\nu)I(\nu)$ (see e.g. Cox and Giuli, 1968), but under local thermodynamic equilibrium, $e(\nu)$ contains both true emission and also radiation scattered into the region considered. According to Cox and Giuli (1968), neglecting refraction in the medium,

$$e(\nu) = \kappa_t(\nu)\,dF/d\nu + \kappa_s(\nu)I_{av}(\nu), \qquad \kappa_t(\nu) + \kappa_s(\nu) = \kappa(\nu).$$

Here κ_t and κ_s are the true absorption and scattering parts of the mass absorption coefficient, and I_{av} is I averaged over solid angles. Cox and Giuli solve the equation of transfer by expressing $I(\nu)$ as a Taylor series expansion of $e(\nu)$ around a given optical depth (a parameter related to κ and to the distance r from the Sun's centre or a suitable path length s). Retaining only the first terms, one obtains, in regions of the Sun where all energy transport is radiative,

$$I(v) = \frac{dF(v)}{dv} - \frac{1}{\kappa(v)\rho} \frac{d}{ds} \frac{dF(v)}{dv} \cos\theta.$$

θ is the angle between the direction of the energy flux $I(v)$ and ds. The first term dominates in the central region of the Sun, but gives no net energy transport through a unit area. If the path s is taken as the radial distance from the centre, the net outward flux obtained by integrating $I(v) \cos\theta$ over all solid angles becomes

$$\int I(v)\cos\theta\, 4\pi\, d\cos\theta = -\frac{4\pi}{3\kappa(v)\rho} \frac{d}{dr} \frac{dF(v)}{dv}.$$

The luminosity $L(r)$ of a shell of radius r is defined as the net rate of energy flow outwards through a sphere of radius r, i.e. $4\pi r^2$ times the net outward flux. Introducing at the same time the temperature gradient, one obtains

$$L_v(r) = -\frac{(4\pi r)^2}{3\kappa(v)\rho}\left(\frac{\partial}{\partial T} \frac{dF(v)}{dv}\right) \frac{dT(r)}{dr}$$

for the spectral luminosity and

$$L(r) = -\frac{(4\pi r)^2\, 4\sigma T^3}{3\pi\kappa\rho} \frac{dT(r)}{dr} \tag{2.22}$$

for the total luminosity.

Here $\sigma = 5.7 \times 10^{-8}$ W m^{-2} K^{-4} is Stefan's constant in Stefan–Boltzmann's law for the total power radiated into the half-sphere, per unit area,

$$\iint \frac{dF(v)}{dv}\cos\theta\, 2\pi\, d\cos\theta\, dv = \sigma T^4.$$

From (2.22) it may be expected that dT/dr will be large in the central regions of the Sun, as well as near the surface, where the opacity κ is large. If dT/dr becomes sufficiently large, the energy cannot be transported away by radiation alone, and convective processes can be expected to play a role. A proper description of turbulent convection will require following all the eddies formed according to the general hydrodynamic equations. In a simple, approximate treatment, the convective transport is described in a form analogous to the first-order [in terms of $l(\text{photons}) = (\kappa\rho)^{-1}$] solution (2.22) to the equation of radiative transfer. The role of the mean free path for photons is played by a mixing length $l(\text{mix})$ (the mean distance that a "hot bubble" travels before it has given off its surplus heat and has come into thermal equilibrium with its local surroundings). The notion of "hot bubbles" is thus introduced as the mechanism for convective energy (heat) transfer, and the "excess heat" is to be understood as the amount of heat which cannot be transferred by adiabatic processes. The excess energy per unit volume, which is given away by a "hot bubble" traversing one mixing length, is

$$\rho c_P \left(\frac{\partial T}{\partial s} - \left(\frac{\partial T}{\partial s} \right)_{adiabatic} \right) l(\text{mix}),$$

where c_P is the specific heat at constant pressure. The energy flux is obtained by multiplying by an average bubble speed v_{conv} (see e.g. Iben, 1972):

$$F_{conv} \approx \rho c_P \left| \frac{\partial T}{\partial s} - \left(\frac{\partial T}{\partial s} \right)_{adiabatic} \right| l(\text{mix}) v_{conv}.$$

In this expression also the derivatives of the temperature are to be understood as average values. The similarity with the expression (2.22) for the radiative energy flux $F_{rad} = L(r)/(4\pi r^2)$ is borne out if (2.22) is rewritten slightly,

$$F_{rad} = -\frac{\partial P_{rad}}{\partial T} \frac{\partial T}{\partial r} l(\text{photons}) c,$$

where the radiation pressure P_{rad} equals one-third the energy density of black-body radiation,

$$P_{rad} = \frac{1}{3} a T^4,$$

and where use has been made of the relation $4\sigma = ac$ between Stefan's constant σ and the radiation constant a to introduce the photon velocity c (velocity of light) as an analogy to the average speed of convective motion, v_{conv}.

The radial bubble speed v_{conv} may be estimated from the kinetic energy associated with the buoyant motion of the bubble. The average density deficiency is of the order

$$\Delta \rho = \left(\frac{\partial \rho}{\partial r} - \left(\frac{\partial \rho}{\partial r} \right)_{adiabatic} \right) l(\text{mix}),$$

and the average kinetic energy is of the order

$$\tfrac{1}{2} \rho (v_{conv})^2 = \frac{1}{2} \frac{GM(r)}{r^2} \Delta \rho \, l(\text{mix})$$

(taken as half the maximum force, since the force decreases from its maximum value to zero, times the distance travelled). When p is inserted from the equation of state, one obtains the estimate for the bubble speed,

$$
v_{conv} = \frac{\rho c_P}{r} \left(\frac{GM(r)}{T} \left| \frac{\partial T}{\partial r} - \left(\frac{\partial T}{\partial r} \right)_{adiabatic} \right| \right)^{1/2}.
$$

The convective energy flux can now be written

$$
F_{conv} \approx \frac{\rho c_P}{r} \left(\frac{GM(r)}{T} \right)^{1/2} \left| \frac{\partial T}{\partial r} - \left(\frac{\partial T}{\partial r} \right)_{adiabatic} \right|^{3/2} (l(\text{mix}))^2. \tag{2.23}
$$

For an ideal gas, the adiabatic temperature gradient is

$$
\left(\frac{\partial T}{\partial r} \right)_{adiabatic} = \frac{c_P - c_V}{c_P} \frac{T}{P} \frac{dP}{dr},
$$

where c_V is the specific heat at fixed volume. For a monatomic gas (complete ionisation), $(c_P - c_V)/c_P = 2/5$.

The final equilibrium condition to be considered is that of energy balance, at thermal equilibrium,

$$
dL(r)/dr = 4\pi r^2 \rho(r) \, \varepsilon(\rho,T,X,...), \tag{2.24}
$$

stating that the radiative energy loss must be made up for by an energy source term, ε, which describes the energy production as function of density, temperature, composition (e.g. hydrogen abundance X), etc. Since the energy production processes in the Sun are associated with nuclear reactions, notably the thermonuclear fusion of hydrogen nuclei into helium, then the evaluation of ε requires a detailed knowledge of cross sections for the relevant nuclear reactions.

Before touching on the nuclear reaction rates, it should be mentioned that the condition of thermal equilibrium is not nearly as critical as that of hydrostatic equilibrium. Should the nuclear processes suddenly cease in the Sun's interior, the luminosity would, for some time, remain unchanged at the value implied by the temperature gradient (2.22). The stores of thermal and gravitational energy would ensure an apparent stability during part of the time (called the "Kelvin time") needed for gravitational collapse. This time may be estimated from the virial theorem, since the energy available for radiation (when there is no new energy generation) is the difference between the total gravitational energy and the kinetic energy bound in the thermal motion (i.e. the temperature increase associated with contraction),

$$
t_{Kelvin} = (\varepsilon_{gravit} - \bar{\varepsilon}_{kin})/L(R) = -(\varepsilon_{pot} + \bar{\varepsilon}_{kin})/L(R) = \bar{\varepsilon}_{kin}/L(R) = 3 \times 10^7 \text{ y}.
$$

On the other hand, this time span is short enough to prove that energy production takes place in the Sun's interior.

Nuclear reactions in the Sun

Most of the nuclear reactions taking place in the Sun involve two colliding particles and two, possibly three, "ejectiles" (outgoing particles). For example, one may write

$$A + a \rightarrow B + b + Q, \qquad \text{or} \qquad A(a, b)B,Q,$$

$$A + a \rightarrow B + b + c + Q, \qquad \text{or} \qquad A(a, bc)B,Q,$$

where Q is the energy release measured in the centre of mass coordinate system (called the "Q-value" of the reaction). Energy carried by neutrinos will not be included in Q. The main processes in the solar energy cycle are

$${}^1_1H \, ({}^1_1H, e^+ \nu) \, {}^2_1H, \qquad Q = 1.0 \text{ to } 1.4 \text{ MeV}, \qquad Q_{av} = 1.2 \text{ MeV},$$

$${}^2_1H \, ({}^1_1H, \gamma) \, {}^3_2He, \qquad Q = 5.5 \text{ MeV}.$$

It is highly probable (estimated at 91% for the present Sun) that the next step in the hydrogen-burning process is

Branch 1:

$${}^3_2He \, ({}^3_2He, 2{}^1_1H) \, {}^4_2He, \qquad Q = 12.9 \text{ MeV}.$$

This process requires the two initial ones to have taken place twice. The average total energy liberated is 26.2 MeV. In an estimated 9% of cases (Bahcall, 1969), the last step is replaced by processes involving ⁴He already present or formed by the processes of branch 1,

$${}^3_2He \, ({}^4_2He, \gamma) \, {}^7_4Be, \qquad Q = 1.6 \text{ MeV}.$$

The following steps may be either

Branch 2:

$${}^7_4Be \, (e^-, \nu) \, {}^7_3Li, \qquad Q = 0 \text{ or } 0.5 \text{ MeV}, \qquad Q_{av} = 0.1 \text{ MeV}$$

$${}^7_3Li \, ({}^1_1H, {}^4_2He) \, {}^4_2He, \qquad Q = 17.4 \text{ MeV},$$

or, with an estimated frequency among all the hydrogen-burning processes of 0.1%,

Branch 3:

$${}^7_4Be \, ({}^1_1H, \gamma) \, {}^8_5B, \qquad Q = 0.1 \text{ MeV},$$

followed by the decay processes

$$^8_5B \rightarrow {}^8_4Be + e^+ + \nu, \qquad Q = 4 \text{ to } 18 \text{ MeV}, \qquad Q_{av} = 10.8 \text{ MeV},$$

$$^8_4Be \rightarrow 2^4_2He \qquad Q = 0.1 \text{ MeV}.$$

The average Q-values for processes involving neutrino emission are quoted from Reeves (1965). The initial pp-reaction (^1_1H = hydrogen nucleus = p = proton) receives competition from the three-body reaction

$$^1_1H + e^- + {}^1_1H \rightarrow {}^2_1H + \nu.$$

However, under solar conditions only 1 in 400 reactions of the pp-type proceeds in this way (Bahcall, 1969).

In order to calculate the rates of energy production in the Sun, the cross sections of each of the reactions involved must be known. The cross section is the probability for the reaction to occur, under given initial conditions such as a specification of the velocities or momenta.

Among the above-mentioned reactions, those which do not involve electrons or positrons are governed by the characteristics of collisions between atomic nuclei. These characteristics are the action of a Coulomb repulsion of long range, as the nuclei approach each other, and a more complicated nuclear interaction of short range, if the nuclei succeed in penetrating their mutual Coulomb barrier. The effective height of the Coulomb barrier increases if the angular momentum of the relative motion is not zero, according to the quantum mechanical description. The height of the Coulomb barrier for zero relative (orbital) angular momentum may be estimated as

$$V_{coul} \approx Z_1 Z_2 e^2 / R(Z_2, A_2) \gtrsim 1 \text{ MeV},$$

where the radius of one of the nuclei has been inserted in the denominator (A_i is the mass of the nucleus in units of the nucleon mass; Z_i is the nuclear charge in units of the elementary charge e, and $e^2/4\pi\varepsilon_0 = 1.44 \times 10^{-15}$ MeV m). Thus, the thermal energy even in the centre of the Sun, $T(0) = 1.5 \times 10^7$ K,

$$\sigma(E) = S(E) \exp(-2\pi\eta(E))/E,$$

is far below the height of the Coulomb barrier (1 MeV = 1.6×10^{-13} J), and the nuclear reaction rate will depend strongly on the barrier penetration, i.e. the quantum tunnelling through the barrier. If the energy of the approaching particle does not correspond to a nuclear resonance (see Fig. 2.98), then the part of the cross section due to barrier penetration (and this will contain most of the energy dependence) can be estimated by a simple quantum approximation, called the WKB method (Merzbacher, 1970). In the absence of relative angular momentum, the nuclear reaction cross section may then be written, as a function of energy,

$$\sigma(E) = E^{-1} S(E) \exp(-2\pi\eta(E)),$$

where $S(E)$ is a slowly varying function of E, depending on the nuclear interactions, while the exponential function expresses the barrier penetration factor in terms of the Coulomb parameter

$$\eta(E) = 2\pi Z_1 Z_2 e^2 h^{-1} (\mu/2E)^{1/2},$$

with the reduced mass being

$$\mu = A_1 A_2 M_p / (A_1 + A_2).$$

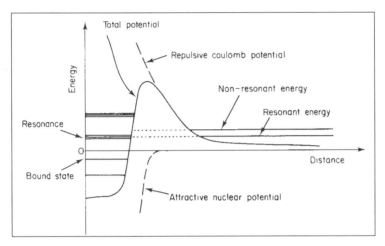

Figure 2.98. The nuclear potential seen by an impinging particle, a function of relative distance. A few nuclear states are indicated to the left of this highly schematic picture, in order to illustrate the concepts of resonant and non-resonant energies. The incident particle kinetic energy at solar temperatures is much lower than the height of the Coulomb barrier, so only quantum tunnelling can lead to a nuclear reaction.

The reactions involving electrons are governed not by nuclear (so-called "strong") interactions, but by the "weak interaction" type, which is available to all kinds of particles. The smallness of such interactions permits the use of a perturbational treatment, for which the cross section may be written (Merzbacker, 1970)

$$v\sigma = (2\pi)^2 h^{-1} |<H>|^2 \rho \text{ (final)}.$$

Here $v\sigma$ is the transition probability, i.e. the probability that a reaction will occur, per unit time and volume. If the formation of the positron–neutrino, electron–antineutrino or generally lepton–antilepton pair is not the result of a decay but of a collision process, then the transition probability may be split into a reaction cross section σ times the relative velocity, v, of the colliding particles [for example, the two protons in the $^1H(^1H, e^+v)^2H$ reaction]. ρ (final) is the density of final states, defined as the integral over the distribution of momenta of the outgoing particles. When there are three particles in the final state, this will constitute a function of an undetermined energy, e.g. the neutrino energy. Finally, $<H>$ is a matrix element, i.e. an integral over the quantum wavefunctions of all the particles in the initial and final states, with the interaction operator in the middle and usually approximated by a constant for weak interaction processes.

The reaction rate for a reaction involving two particles in the initial state is the product of the numbers n_1 and n_2 of particles 1 and of particles 2, each per unit volume, times the probability that two such particles will interact to form the specified final products. This last factor is just the product of the relative velocity and the reaction cross section. Since the relative velocity of particles somewhere in the Sun is given by the statistical distribution corresponding to the temperature, the local reaction probability may be obtained by averaging over the velocity distribution. For a non-relativistic gas, the relative kinetic energy is related to the velocity by $E = \frac{1}{2}\mu v^2$, and the velocity distribution may be replaced by the Maxwell–Bolzmann distribution law for kinetic energies (which further assumes that the gas is non-degenerate, i.e. that kT is not small compared with the thermodynamic chemical potential, a condition which is fulfilled for the electrons and nuclei present in the Sun),

$$<\sigma v> = (8/\pi\mu)^{1/2} (kT)^{-3/2} \int \sigma (E)\, E\, \exp(-E/kT)\mathrm{d}E.$$

The reaction rate (number of reactions per unit time and unit volume) becomes

$$P = n_1 n_2 <\sigma v> (1 + \delta_{12})^{-1},$$

where δ_{12} (=1 if the suffixes 1 and 2 are identical, otherwise 0) prevents double counting in case projectile and "target" particles are identical.

The number densities n (number per unit volume) may be re-expressed in terms of the local density $\rho(r)$ and the mass fraction of the ith component, X (e.g. the hydrogen fraction X considered earlier),

$$n_i = X_i\, \rho (r) / (A_i M_p).$$

The central part of the reaction rate, $<\sigma v>$, is shown in Fig. 2.99 for the branch 1 reactions of the pp-cycle, as well as for the $^3\mathrm{He}+^4\mathrm{He}$ reaction opening the alternative branches. The basic cross sections $S(0)$ used in the calculations are quoted in the work of Barnes (1971). The barrier penetration effects are calculated in a modified form, taking account of the partial screening of the nuclear Coulomb field by the presence of electrons. Such penetration factors are included for the weak interaction processes as well as for the reactions caused by nuclear forces.

One notes the smallness of the initial, weak interaction reaction rate, the temperature dependence of which determines the overall reaction rate above $T = 3 \times 10^6$ K. The four to five orders of magnitude difference in the rates of the ^{4}He+^{3}He and the ^{3}He+^{3}He processes account for the observed ratio of branch 2 (and 3) to branch 1, when the oppositely acting ratio of the ^{4}He to ^{3}He number densities is also included.

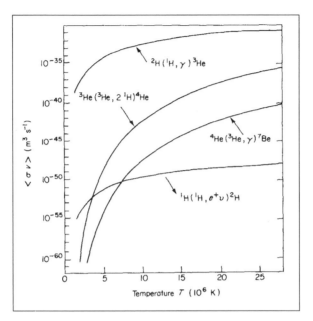

Figure 2.99. Reaction rates for various nuclear reactions of the *pp*-chain, as functions of temperature, and assuming that there is just one of each of the reacting particles per unit volume (denoted $<\sigma v>$) (based on Barnes, 1971).

The rate of energy production to be entered in the equation (2.24) for thermal equilibrium is obtained from the reaction rate by multiplying with the energy release, i.e. the Q-value. Since the energy production rate appearing in (2.24) is per unit mass and not volume, a further division by ρ is necessary,

$$\varepsilon(\rho,T,X_i) = \sum_j \varepsilon_j(\rho,T,X_i) = \sum_j n_1 n_2 <\sigma v> \frac{Q(j)}{\rho(1+\delta_{12})}. \qquad (2.25)$$

The sum over j extends over all the different branches; $Q(j)$ is the energy released in branch j; and the indices 1 and 2, for which n_1 and n_2 are taken and $<\sigma v>$ evaluated, may be any isolated reaction in the chain, since the equilibrium assumption implies that the number densities have adjusted themselves to the differences in reaction cross section. For decay processes, the reaction rate has to be replaced by the decay probability, in the manner described above.

Figure 2.100 shows the temperature dependence of the energy production rate for the *pp*-chain described above, as well as for the competing CNO-chain, which may come into play at temperatures slightly higher than those prevailing in the Sun' s interior. The curves should be regarded as typical for stars of composition similar to that of the Sun, but as (2.25) indicates, the energy production depends on the density as well as on the abundance of the reacting substances. Hence a change in composition or in central density will change the rate of energy production. For the *pp*-chain, Fig. 2.100 assumes a hydrogen fraction $X = 0.5$ and a density of 10^5 kg m^{-3}, believed to be reasonable for the centre of the Sun. For the CNO-chain, the additional assumptions have been made that the carbon, nitrogen and oxygen abundances are $X_C = 0.003$, $X_N = 0.001$ and $X_O = 0.012$ (Iben, 1972). These figures apply to the stellar populations similar to, but more advanced than, the Sun, whereas much smaller CNO-abundances and production rates are found for the so-called population-II stars. It should also be mentioned that, owing to the decrease in density and change in composition as one goes away from the centre of the star, the average energy production for the energy-producing region will be lower than the value deduced from the conditions near the centre.

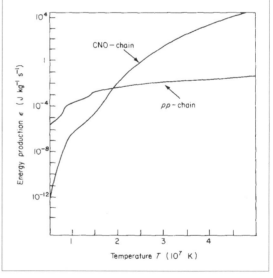

Figure 2.100. Rate of energy production by the *pp*- and CNO-chains, as a function of temperature, for a choice of composition (relative abundance of the constituents) typical for main-sequence stars such as the Sun (based on Iben, 1972).

A model of the solar processes

In order to determine the radial distribution of energy production, mass and temperature, the equilibrium equations must be integrated. Using the equation of state to eliminate $\rho(r)$ [expressing it as a function of $P(r)$ and $T(r)$], the basic equations may be taken as (2.20), (2.22) in regions of radiative transport, (2.23) in regions of convective transport, and (2.24), plus the equation defining the mass inside a shell,

$$\mathrm{d}M(r)/\mathrm{d}r \;=\; 4\pi r^2\, \rho(r) \;=\; 4\pi r^2\, \mu\, P(r)\,/\,(\mathscr{R} T(r)). \tag{2.26}$$

In addition to the equation of state, two other auxiliary equations are needed, namely, (2.25) in order to specify the energy source term and a corresponding equation expressing the opacity as a function of the state and composition of the gas. The solution to the coupled set of differential equations is usually obtained by smooth matching of two solutions, one obtained by integrating outward with the boundary conditions $M(0) = L(0) = 0$ and the other one obtained by inward integration and adopting a suitable boundary condition at $r = R$. At each step of the integration, the temperature gradient must be compared to the adiabatic gradient in order to decide which of the equations, (2.23) or (2.24), to use.

Figure 2.101 shows the results of a classical calculation of this type (Schwarzschild, 1958). The top curve shows the prediction at a convective region $0.85R < r < R$. The temperature gradient also rises near the centre, but not sufficiently to make the energy transport in the core convective. This may happen for slight changes in parameters, such as the initial abundances, without impinging on the observable quantities such as luminosity and apparent surface temperature. It is believed that the Sun does have a small, convective region near the centre (Iben, 1972). The bottom curves in Fig. 2.101 show that the energy production takes place in the core region, out to a radius of about $0.25R$, and that roughly half the mass is confined in the energy-producing region. Outside this region the luminosity is constant, indicating that energy is only being transported, not generated.

The time development of stars has been studied by use of models similar to the ones described here, relaxing the conditions of equilibrium and introducing the time variable explicitly. In this way a substantial understanding of the main burning stages has been achieved (Iben, 1972), as well as a detailed description of the synthesis of heavy elements (Clayton, 1968).

A very direct way to test the models of the conditions in the Sun' s interior would seem to be the measurement of the flux of neutrinos at the Earth. Since each of the elementary hydrogen-burning branches releases two neutrinos, the reaction rates of a given solar model can be translated into a flux of neutrinos at the Earth' s distance. The neutrino flux predicted from models similar to that of Fig. 2.101 is of the order of 10^{15} m^{-2} s^{-1}. Only one neutrino in 10^{11} is stopped within the Sun. This also explains the difficulty in constructing suitable detectors for use at the Earth.

Attempts to capture solar neutrinos by ^{37}Cl (forming radioactive ^{37}Ar, the decay of which can be detected) have not yet provided count rates significantly different from the background "noise" (the detecting system must be placed underground to avoid the formation of ^{37}Ar by other processes, such as collision with protons arising from the interaction between penetrating muons from cosmic rays and matter) (Bahcall and Davis, 1976). The ^{37}Cl system only captures neutrinos of energy above 0.8 MeV, i.e. a small fraction of all the neutrinos produced in the *pp*-chains. Although the predicted count rate depends significantly on such assumptions as the initial helium abundance in the Sun and on whether a convective core exists (or whether other processes have mixed the solar constituents at any time during its hydrogen-burning phase), most evaluations do predict the count rate in the ^{37}Cl detecting system exceeding the experimental "limit" by a factor two or more (Iben, 1972). Efforts are being made to improve the experimental system, and one should probably refrain from drawing any conclusions at present.

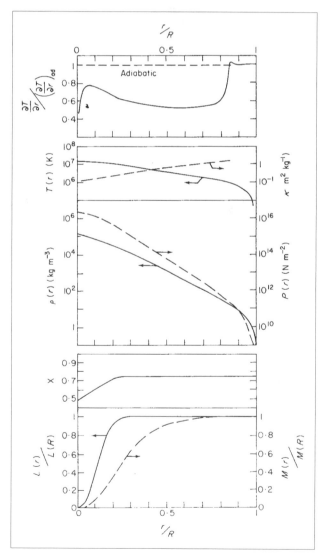

Figure 2.101. Radial model of the Sun. The solid curves represent (from top to bottom) temperature gradient relative to adiabatic, temperature, density, hydrogen abundance and relative luminosity. The dashed curves represent opacity, pressure and relative mass fraction inside a given radius. The opacity has not been calculated in the outer convective region (based on Schwarzschild, 1958).

One significant feature of the time-dependent models of the solar evolution is the presence of a substantial amount of helium (in the neighbourhood of 25%) before the onset of the hydrogen-burning process. At present about half the mass near the centre is in the form of helium (cf. Fig. 2.101). The initial abundance of helium, about 25%, is in agreement with observations for other stars believed to have just entered the hydrogen-burning stage. There is reasonable evidence to suggest that in other galaxies too, in fact everywhere in the universe, the helium content is above 25%, with the exception of a few, very old stars, which perhaps expelled matter before reaching their peculiar composition.

Cosmology

Two other distinct features stand out in the present observational knowledge of the universe as a whole. One is the uniform expansion indicated by the redshifts (Doppler shifts) of other galaxies; the other is the presence of cosmic background radiation, i.e. radiation not associated with identified objects, corresponding to the radiation from a black body of the temperature 2.7 K [found by comparing the spectral intensity with Planck's formula (2.21)].

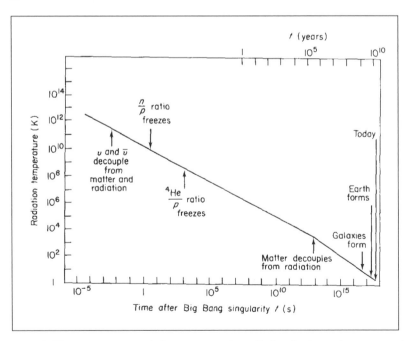

Figure 2.102. The temperature of electromagnetic radiation in the universe, as a function of time, according to the Big Bang model (based on Peebles, 1971; Webster, 1974).

The "Big Bang" model

The rate of expansion of the universe, given by the ratio of velocity and distance, indicates that – if backward extrapolation is permitted – all matter was together at a point about 5×10^{17} s or 15×10^9 years ago. The extrapolation hypothesis is not so daring as one might think, because the light from the faintest objects observed has travelled billions of years to reach us. The singularity or "Big Bang" theory is also consistent with the background temperature of 2.7 K. Extrapolating backwards yields an estimated radiation temperature as a function of the time after the Big Bang. Combining this with the density extrapolations it is then possible to estimate the state of the matter and set up a model for the types of reactions taking place at different times. Figure 2.102 gives an outline of such a model.

At temperatures above 10^{12} K, about 10^{-4} s after the singularity, most energy would be in the form of light ("Gamow' s fireball"). Matter in the form of particles may be present, either with a fixed number of nucleons (more generally baryons) or having been formed in particle–antiparticle pairs, in which case an explanation must be found for the present preponderance of particles in the known part of the universe. As the expansion continues and the temperature drops towards 10^{10} K (1 s after the singularity), processes like

$$\gamma + \gamma \leftrightharpoons e^+ + e^-$$
$$e^+ + e^- \leftrightharpoons v + v$$
$$e^+ + n \leftrightharpoons {}^1_1\text{H} + v$$
$$^1_1\text{H} + e^- \leftrightharpoons n + v$$
$$n \leftrightharpoons {}^1_1\text{H} + e^- + v$$

no longer have time to reach thermal equilibrium. Already below 10^{11} K, the neutrino mean free path becomes so large that most neutrinos or antineutrinos do not experience any further interaction with other particles or light quanta. Similarly, at 10^{10} K, the neutron equilibrium processes can no longer keep up with the expansion, so that the ratio of neutrons and protons freezes at the value implied by the Boltzmann distribution at $T \approx 10^{10}$ K and their energy difference $\Delta E = (M_n - M_p)c^2 = 1.3$ MeV,

$$X_n/X_p = \exp\left(-\Delta E/kT\right) \approx 0.2$$

(Peebles, 1971). As the temperature drops towards 10^9 K, fusion processes dominate the picture,

$$n + {}^1_1\text{H} \leftrightharpoons {}^2_1\text{H} + \gamma \quad ({}^1_1\text{H} \equiv p)$$

$$^2_1\text{H} + {}^2_1\text{H} \rightarrow {}^3_1\text{H} + {}^1_1\text{H}$$

$$^2_1\text{H} + {}^2_1\text{H} \rightarrow {}^3_2\text{He} + n$$

$$^3_1H + {}^1_1H \rightarrow {}^4_2He + \gamma$$

$$^3_2He + n \rightarrow {}^4_2He + \gamma.$$

Assuming at first that all the neutrons are transformed into helium (^{4}He), the mass fraction may be obtained from the number of neutrons bound in helium, $2X_n$, divided by the total number of nucleons,

$$X(^4He) = 2X_n / (X_p + X_n) = 0.33.$$

However, the first reaction, which forms the deuterons, ^{2}H, does not proceed entirely towards the right-hand side until the temperature reaches 10^9 K, 200 s after the Big Bang singularity. During these 200 s, about 30% of the neutrons have decayed into protons (plus electron and antineutrino). Replacing the neutron abundance in the above relation by $X_n = 0.7 \times 0.2 = 0.14$, the resulting helium abundance becomes

$$X(^4He) = 0.25$$

(cf. Peebles, 1971). Not much of this helium is lost by further nuclear reactions, since it so happens that there are no stable nuclei with 5 or 8 nucleons, so that any such nuclei formed by collisions between ^{4}He and p, n or other ^{4}He will soon again decay to the lighter species. Thus, the synthesis of nuclei by the Big Bang essentially ceases after 200 s, producing exactly the 25% helium abundance which is today observed as a minimum and which forms the initial condition for hydrogen burning in stars like the Sun.

The cosmic background radiation

For the following period of about 300 000 years the universe consisted of electromagnetic radiation, neutrinos and antineutrinos, protons, ^{4}He and enough electrons to balance the positive charges (plus minute fractions of ^{2}H, ^{3}He, ^{3}H, n, ^{7}Be, ^{7}Li and ^{6}Li). The universe expanded and cooled uniformly, ensured by frequent interaction between electromagnetic radiation and matter. However, when the temperature reached about 6000 K, the energy density of the radiation became smaller than that of matter, and the matter no longer got hit by photons frequently enough to assume the radiation temperature. Thus, electrons started to recombine with the ionised gas of matter, leading to a rapid decrease in the temperature of matter.

The photons, now de-coupled from matter, continued to cool, reaching today the observed value of 2.7 K. The matter, on the other hand, developed less regularly and eventually formed contracting clouds, from which galaxies were formed around 10^9 years after the singularity.

The Big Bang theory provides reasonably consistent connections between a number of astrophysical observations. Yet its implication of a point singularity in time and space need not be taken too seriously. The general lesson from past experience of singularities in physical theories seems to be that the appearance of singularities in a mathematical description of the physical world is only due to the use of oversimplified models, for convenience or because no better theory has yet been found. Whenever it became important to explore those regions of physical experience where singularities appeared to exist, the outcome was the formulation of a new theory which contained the old model as a special case, in the regions away from singularities where the old theory worked well. The more general applicability of the modified theories would remove the apparent singularities, but sometimes introduce other singularities to be removed by new generations of scientists. The Big Bang singularity may be no exception, and several attempts have been made to modify the behaviour of the universe near the singularity (see e.g. Hoyle, 1975; Alfven, 1965; Sørensen, 1987b).

2.B Vertical transport in or near the boundary layer

If the transport of sensible or latent energy is governed by the gradients of temperature and humidity, it may be described by the diffusion equation (cf. e.g. Sellers, 1965)

$$\frac{dT}{dt} = \frac{\partial}{\partial z}\left(k\frac{\partial T}{\partial z}\right), \tag{2.27}$$

where k is the diffusion coefficient and the z-axis is in the vertical direction. Considering a quasi-stationary situation (the partial derivatives of T with respect to time t as well as with respect to horizontal directions x and y being considered negligible), (2.27) may be integrated to give

$$k\frac{\partial T}{\partial z} = \text{constant} = \frac{E_s^{sens}}{\rho(0)c_P}. \tag{2.28}$$

The change in stored sensible heat in a small volume $dx\,dy\,dz$ close to the ground is $dW^{sens}/dt = \rho(0)\,dx\,dy\,dz\,c_P\,(dT/dt)$, and the sensible heat flux per unit area is thus

$$dE_s^{sens} = \frac{1}{dx\,dy}\frac{dW^{sens}}{dt} = \rho(0)c_P dz\frac{dT}{dt}.$$

Insertion of this into (2.27) explains the form of the constant used on the right-hand side of (2.28).

The vertical transport of water vapour through the laminar boundary layer may also be taken as a diffusion type process and is given by the equation analogous to (2.28),

$$k\frac{\partial q}{\partial z} = \frac{E_s^{lat}}{\rho(0)L_v},\tag{2.29}$$

where q is the specific humidity. The molecular diffusion coefficient in air is of the order of $k = 2 \times 10^{-5}$ m2 s^{-1}.

Considering the vertical change in the horizontal velocity u of the air, the expression analogous to (2.28) or (2.29),

$$\rho k\frac{\partial u}{\partial z} = \tau,\tag{2.30}$$

represents the shearing stress acting on the ground (or more generally on a horizontal unit area). τ has the dimension of force per unit area or pressure.

Above the laminar boundary layer, transport usually takes place by turbulent convection. The mass motion may be considered as consisting of overlaying eddies of varying size, eventually superimposed on a laminar flow. The fact that laminar flow is, in general, horizontal or nearly horizontal near the ground implies that all vertical transport is related to eddy motion.

It is now assumed (Rossby, 1932) that the expressions (2.28)–(2.30), based on molecular diffusion theory, can also be applied in the region of turbulent motion, provided that the coefficient k is given an effective value. This "*eddy diffusion parameter*" is no longer constant. It will depend on the quantity being transported and on the velocity field of the atmosphere, i.e. k will be a time-dependent function of the height z. The picture of turbulent transport given here is equivalent to the mixing length formulation used in the discussion in section 2.A of convective transport in stellar atmospheres. The eddy diffusion parameter k is related to the mixing length l by

$$k = l^2\frac{\partial u}{\partial z}.$$

If the temperature profile is approximately adiabatic,

$$\frac{dT}{dz} \approx \left(\frac{dT}{dz}\right)_{adiabatic} = \frac{c_P - c_V}{c_P}\frac{T}{P}\frac{dP}{dz}\tag{2.31}$$

(P is pressure, c_V is the heat capacity at constant volume), then it has been suggested that the mixing length be taken as proportional to or at least a linear function of z,

$$l = \kappa z \qquad \text{or} \qquad l = \kappa(z + z_0),\tag{2.32}$$

where κ close to 0.4 is von Kármáns constant (Prandtl, 1932). Using (2.32) together with the assumption that the shearing stress τ (2.30) is independent of z within the stable atmospheric layer, where the temperature gradient has been assumed to correspond to adiabaticity, then the horizontal velocity varies with height according to

$$u = \int_{z_0} \left(\frac{\tau}{\rho} \right)^{1/2} \frac{dz}{l} = \frac{1}{\kappa} \left(\frac{\tau}{\rho} \right)^{1/2} \log \left(\frac{l}{\kappa z_0} \right). \tag{2.33}$$

The lower limit of integration, z_0, is the height of the laminar boundary layer, for which the velocity u becomes zero if the first expression (2.32) for l is used, whereas for the second expression the velocity u becomes zero at the ground, $z = 0$. None of the expressions can be rigorously correct, since the second one, which gives the correct boundary condition $u(z = 0) = 0$, also predicts eddies which extend to the ground, $l(z = 0) \neq 0$, in contradiction with the basic assumptions regarding the laminar boundary layer. The first expression correctly predicts $l(0) = 0$, but cannot be used to estimate u if $z \leq z_0$. The coefficient $(\tau/\rho)^{1/2}$ has the dimension of velocity and is often referred to as the *friction velocity*.

If the eddy diffusion parameter k is left as a function of z [to be determined empirically rather than using (2.32)], and the heat flux E^{sens} is regarded as independent of z, then (2.28) may be integrated,

$$E_s^{sens} = \int_0^z \frac{dz}{k(z)} = \rho c_P (T(z) - T(0)).$$

According to Budyko (1974), the integral appearing on the left-hand side depends little on z, in contrast to k, so that the expression offers an empirical method of determining E_s^{sens} from temperature measurements. A similar expression for E_s^{lat} may be found from (2.29), again replacing the molecular diffusion coefficient k by the eddy diffusion parameter as a function of z.

Water transport

The transport of water vapour through the lower part of the atmosphere may be described by equation (2.29), which may be read as an expression of the potential evaporation from a "wet surface" (i.e. a surface of moisture exceeding the above-mentioned "critical" value), depending on the distribution of humidity q in the air above the surface, and on the diffusion coefficient $k(z)$, which again may depend on the wind velocity, etc.,

$$e = -\frac{E_s^{lat}}{L_v} = -\rho k \frac{\partial q}{\partial z}. \tag{2.34}$$

Within the laminar boundary layer, k may be considered constant, but in the turbulent layer k varies. If the integral

$$\int_0^z k^{-1} dz$$

can be considered as constant (Budyko, 1974), an integration of (2.34) shows that e varies linearly with the specific humidity q at some height z above the surface. On the other hand, if k is identified with the diffusion coefficient for horizontal wind velocity u (2.30), one obtains, similarly to (2.33),

$$\int_0^z k^{-1} dz = \kappa^{-1} \left(\frac{\tau}{\rho} \right)^{-1/2} \log\left(\frac{z}{z_0} \right) = \kappa^{-2} u^{-1} \left(\log\left(\frac{z}{z_0} \right) \right)^2$$

and

$$e \approx \rho \left(q(0) - q(z) \right) \kappa^2 u \left(\log\left(\frac{z}{z_0} \right) \right)^{-2}. \tag{2.35}$$

Several expressions of this structure or similar ones have been compared with measurements, where they are found to agree overall but not in specific detail (see e.g. Sellers, 1965; Sørensen, 1975a).

The evaporation from ocean surfaces is not independent of the occurrence of waves. At high wind speeds the surface tension is no longer able to hold the water surface together, and spray droplets are ejected into the atmosphere (whitecaps). Some of these droplets evaporate before they fall back into the sea, thereby providing a source of vapour, which does not follow the diffusion formulae given above.

Condensation of water vapour in the atmosphere may take place whenever the vapour pressure of water exceeds that of saturation at the prevailing conditions. However, if no condensation nuclei are present, the vapour may become supersaturated. One source of condensation nuclei, which is of particular importance over oceans, is the salt particles left over from the spray droplets discussed above, after the evaporation of the water. Typically, the mass of these salt particles is so small that gravitation can be disregarded, and the salt particles may remain suspended in the atmosphere for a considerable length of time. Other particles of suitable size and structure serve equally well as condensation nuclei, and globally the sea salt spray particles do not dominate the condensation processes (Junge, 1963).

The disposition of water and heat in soils

The penetration of water into soil and the movement of ground water depend strongly on local conditions, such as the sequence of soil types. Underground aquifers may transport water at moderate velocities, but the average turnover time even in the regions of active exchange is very slow, 200–300 years, as estimated from Fig. 2.61.

The ability of soils to store heat absorbed from solar radiation depends on the heat capacity of the soil, which may be written

$$C_s = C_s^{dry} m_s + C_w m_w,$$ (2.36)

in terms of the heat capacity of dry soil (C_s^{dry} typically in the range 2.0–2.5 $\times$ 10^6 J m^{-3} K^{-1}) and of water [$C_w \approx 4.2 \times 10^6$ J m^{-3} K^{-1} or roughly half of this value for ice (frozen soil)]. The mixing ratios (by volume) m_s and m_w are much more variable. For soils with different air content, m_s may be 0.2–0.6 (low for peat, medium for clay and high for sand). The moisture content m_w spans the range from zero to the volume fraction otherwise occupied by air, m_a (equal to about 0.4 for clay and sand, but 0.9 for peat) (Sellers, 1965). In temperate climates an average value of m_w for the upper soil layer is around 0.16 (Geiger, 1961).

Since the absorption of radiation takes place at the soil surface, the storage of heat depends on the competition between downward transport and long-wavelength re-radiation plus evapotranspiration and convective transfer. The thermal conductivity of the soil, $\lambda(z)$, is defined as the ratio between the rate of heat flow and the temperature gradient, at a given depth z,

$$E_s = \lambda(z) \frac{\partial T}{\partial z}.$$ (2.37)

(Note that by the sign convention previously adopted, z is negative below the soil surface.) Figure 2.103 gives a rough idea of the dependence of λ on the moisture content for different soil types.

If the downward heat transport can be treated as a diffusion process, equation (2.27) is valid. Taking the partial derivative of (2.37) with respect to depth ($-z$), and assuming that k and λ are independent of z (homogeneous soil),

$$\frac{\partial E_z}{\partial z} = \frac{\lambda}{k} \frac{dT}{dt}.$$

Together with the heat transport equation

$$\frac{\partial E_z}{\partial z} = C_s \frac{dT}{dt}$$

(cf. section 2.C) in the absence of heat sources, heat sinks and a fluid velocity in the soil), this implies the relation

$$k = \lambda/C_s.$$ (2.38)

Figure 2.104 shows an example of the variation in the monthly average temperature as a function of depth in the soil. The time variation diminishes with increasing depth, and the occurrence of the maximum and the minimum is increasingly delayed. These features are direct consequences of the diffusion approach, as one can easily see, e.g. by assuming a periodic temperature variation at the surface (Carslaw and Jaeger, 1959),

$$T(z = 0, t) = T_0 + T_1 \sin \omega t. \tag{2.39}$$

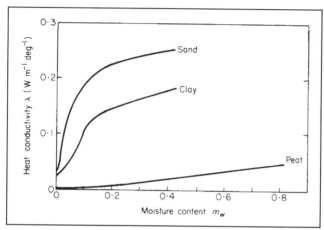

Figure 2.103. Heat conductivity, λ, for various soil types, as a function of water content by volume, m_w (based on Sellers, 1965).

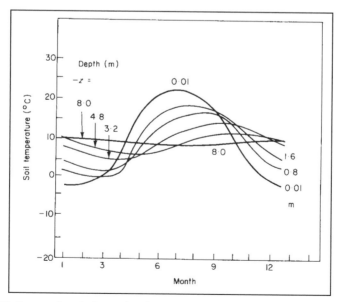

Figure 2.104. Seasonal variations of soil temperature at various depths measured at St Paul, Minnesota. In winter, the average air temperature falls appreciably below the top soil temperature ($z = -0.01$ m), due to the soil being insulated by snow cover (based on Bligh, 1976).

With this boundary condition, (2.27) can be satisfied by

$$T(z,t) = T_0 + T_1 \exp\left(z \left(\frac{\omega C_s}{2\lambda} \right)^{1/2} \right) \sin\left(\omega t + z \left(\frac{\omega C_s}{2\lambda} \right)^{1/2} \right), \quad (2.40)$$

and the corresponding heat flux is from (2.37)

$$E_z(t) = - T_1 (\omega C_s \lambda)^{1/2} \exp\left(z \left(\frac{\omega C_s}{2\lambda} \right)^{1/2} \right) \sin\left(\omega t + z \left(\frac{\omega C_s}{2\lambda} \right)^{1/2} + \frac{\pi}{4} \right). \quad (2.41)$$

The amplitude of the varying part of the temperature is seen from (2.40) to drop exponentially with depth (–z), as the heat flux (2.41) drops to zero. For a fixed depth, z, the maximum and minimum occur at a time which is delayed proportional to $|z|$, with the proportionality factor $(\omega C_s / 2\lambda)^{1/2}$. The maximum heat flux is further delayed by one-eighth of the cycle period (also at the surface).

The approximation (2.39) can be used to describe the daily as well as the seasonal variations in T and E_z. For the daily cycle, the surface temperature is maximum some time between 1200 and 1400 h. This temperature peak starts to move downward, but the highest flux away from the surface is not reached until 3 h later, according to (2.41). While the temperature peak is still moving down, the heat flux at the surface changes sign some 9 h after the peak, i.e. in the late evening. Next morning, 3 h before the peak temperature, the heat flux again changes sign and becomes a downward flux. Of course, the strict sine variation may only be expected at equinoxes.

Geothermal heat fluxes

In the above discussion, the heat flux of non-solar origin, coming from the interior of the Earth, has been neglected. This is generally permissible, since the average heat flux from the interior is only of the order of 3×10^{12} W, or about 8×10^{-2} W m^{-2}. Locally, in regions of volcanoes, hot springs, etc., the geothermal heat fluxes may be much larger. However, it has been estimated that the heat transfer through the Earth' s surface by volcanic activity only contributes under 1% of the average flux (Gutenberg, 1959). Equally small or smaller is the amount of heat transmitted by seismic waves, and most of this energy does not contribute any flux through the surface.

Although the distribution of heat generation within the solid Earth is not directly measurable, it may be estimated from assumptions on the composition of various layers of the Earth' s interior (cf. Lewis, 1974 and Fig. 2.3). Thus, one finds that most of the heat generation takes place within the rocks present in the Earth' s crust. The source of heat is the decay of radioactive elements, chiefly potassium, uranium and thorium. The estimated rate of generation (see e.g. Gutenberg, 1959) is roughly of the same order of magnitude as the outward heat flux through the surface, although there is considerable uncertainty about this.

It is believed that the temperature gradient is positive inwards (but of varying magnitude) all the way to the centre, so that the direction of heat transport is predominantly outwards, with possible exceptions in regions of convection or other mass movement.

A more detailed discussion of the nature of the geothermal energy flux is deferred to section 3.5.2.

Temperature gradients also exist in the oceans, but it will become clear that they are primarily maintained not by the heat flux from the Earth's interior, but by extraterrestrial solar radiation coupled with conditions in the atmosphere and in the surface layers of the continents.

Momentum exchange processes between atmosphere and oceans

The mixing and sinking processes sketched above constitute one of the two sources of oceanic circulation. The other one is the momentum exchange between the atmospheric circulation (winds) and the oceans. The interaction between wind and water is complex, because a large fraction of the kinetic energy is transformed into wave motion rather than directly into currents. Although the roughness length over water [z_0 of (2.32), (2.33)] is very small, a strong wind is capable of raising waves to a height of about 10 m.

The wind stress acting on the ocean surface may be taken from (2.33),

$$\tau = \rho_a \, \kappa^2 \, V(z_1)^2 \left(\log\left(\frac{z_1}{z_0}\right) \right)^{-2}, \tag{2.42}$$

where ρ_a is the density of air, κ is von Karman's constant and z_1 is a reference height. When the height variation of the horizontal wind velocity V is logarithmic as in (2.33), τ is independent of height. This is true for the lower part of the turbulent layer (the Prandtl layer), within which z_1 must be taken, and the value of τ found in this way will remain constant through the laminar boundary layer, i.e. the force exerted on the ocean surface can be determined from measurements in a height z_1 of, say, 10 m.

However, the approach may be complicated by the motion of the water surface. If waves are propagating along the water surface in the direction of the wind, one may argue that the stress should be proportional to the square of the relative velocity $(V - U_w)^2$, where U_w is the wave velocity, rather than to the square of V alone. Further, the roughness parameter z_0 may no longer be a constant, but it may depend on the wave velocity as well as on the wave amplitude. Given that the waves are formed as a result of the wind stress τ, this implies a coupled phenomenon, in which τ initiates wave motion, which again modifies τ as a result of a changed surface friction, etc. There is experimental evidence to suggest that z_0 is nearly constant except for the lowest wind velocities (and equal to about 6×10^{-3} m; Sverdrup, 1957).

The mechanism by which the surface stress creates waves is not known in detail, nor is the distribution on waves and currents of the energy received by the ocean from the winds, or any subsequent transfer (Pond, 1971). Waves may be defined as motion in which the average position of matter (the water "particles") is unchanged, whereas currents do transport matter. This may also be viewed as a difference in scale. Waves play a minor role in the atmosphere, but are important in the hydrosphere, particularly in the upper water layers, but also as internal waves. The reason is, of course, the higher density of water, which allows substantial potential energy changes to be caused by modest lifting heights.

2.C Modelling the atmosphere and the oceans

The basic equations in terms of time-averaged variables

As outlined in section 2.3, the averaging procedure leading to (2.15) may now be applied to the general Eulerian equation of transport, for any relevant quantity A,

$$\frac{\partial}{\partial t}(\rho A) + \mathrm{div}(v\rho A) + \mathrm{div}(s_A) = S_A. \tag{2.43}$$

This equation states that the change in A within a given volume, per unit time, is due to gains from external sources, S_A, minus the macroscopic and molecular outflow (divergence terms). The macroscopic outflow is due to the velocity field v, and the microscopic transport is described by a vector s_A, the components of which give the transport by molecular processes in the directions of the co-ordinate axes.

Applying the definitions (2.13)–(2.15), the time-averaged version of (2.43) becomes

$$\frac{\partial}{\partial t}(\overline{\rho} A^*) + \mathrm{div}(\overline{\rho} v^* A^* + \overline{\rho v' A'}) + \mathrm{div}(\overline{s}_A) = S_A. \tag{2.44}$$

If A is taken as unity, (2.44) becomes the equation of continuity (of mass),

$$\frac{\partial}{\partial t}\overline{\rho} + \mathrm{div}(\overline{\rho} v^*) = 0. \tag{2.45}$$

The density averaging method is essential for obtaining the same form as for the instantaneous values. If A is taken as a component of the velocity vector, (2.44) becomes [with utilisation of (2.45)]

$$\bar{\rho}\left(\frac{\partial v_i^*}{\partial t}+(v^*\cdot\mathrm{grad})v_i^*\right)+\overline{((\mathrm{div}v')\rho v_i')}=\sum_j\frac{\partial}{\partial x_j}\tau_{ij}+F_{ext,i}, \qquad (2.46)$$

The sources of velocity are the external forces F_{ext} (gravitational and Coriolis forces), and the molecular transfer of velocity (or momentum) is given in terms of the derivatives of the stress tensor τ_{ij},

$$-(s_{v_j})_i=\tau_{ij}=\left(-P+\left(\eta'-\frac{2}{3}\eta\right)\rho\,\mathrm{div}(v)\right)\delta_{ij}+\eta\rho\left(\frac{\partial v_i}{\partial x_j}+\frac{\partial v_j}{\partial x_i}\right),$$

where P is the thermodynamic pressure, η and η' are the kinematic and volume viscosities[□] and δ_{ij} is the Kronecker delta (Hinze, 1975). In the last term on the left-hand side of (2.46), the divergence operator div $=\Sigma_j\partial/\partial x_j$ is supposed to act on all three factors $v'_{\,j}\rho v'_{\,i}$ following in (2.46).

It is clear from the equation of motion (2.46) that the large-scale motion of the atmosphere, v^*, cannot generally be described by an equation that only depends on averaged quantities (for a discussion of early attempts to achieve such a description, see Lorenz, 1967). The divergence term on the left-hand side describes the influence of small-scale turbulent motion on the large-scale motion. It is usually referred to as the eddy transport term.

Apart from the well-known pressure gradient term, the contributions from the stress tensor describe the molecular friction forces. These may be important for calculations of energy transformations, but are often left out in calculations of general circulation (Wilson and Matthews, 1971). According to Boussinesq (1877), it may be possible to approximate the eddy transport term by an expression of the same form as the dynamic viscosity term [see (2.59)], introducing an effective "eddy viscosity" parameter similar to the parameter k considered above in connection with the turbulent transport of heat from the Earth' s surface to the boundary layer [i.e. the so-called *Prandtl layer*, extending about 50 m above the laminar boundary layer and defined by an approximately constant shear stress τ in expressions such as (2.33)].

The difference between the scales of vertical and horizontal motion makes it convenient to define separate velocities w and V,

$$v=(w\cdot e_z)e_z+(v-(v\cdot e_z)e_z)=we_z+V,$$

and split the equation of motion (2.46) into a horizontal part,

$$\bar{\rho}\left(\frac{\partial V^*}{\partial t}+(V^*\cdot\mathrm{grad})V^*+w^*\frac{\partial V^*}{\partial z}\right)+\frac{\partial}{\partial z}(\overline{\rho w'V'})=-\mathrm{grad}\,P-\rho f e_z\times V^*, \qquad (2.47)$$

and a vertical part, which as mentioned can be approximated by the hydro-static equation,

---

[□] The viscosities are sometimes taken to include the factor ρ.

$$\frac{\partial P}{\partial z} = -g\rho. \tag{2.48}$$

In (2.47), the molecular friction forces have been left out, and only the vertical derivative is kept in the eddy term. This term thus describes a turbulent friction, by which eddy motion reduces the horizontal average wind velocity V^*. The last term in (2.47) is due to the Coriolis force, and

$$f = 2\,\Omega \sin \phi,$$

where Ω is the angular velocity of the Earth's rotation and ϕ is the latitude.

As a lowest-order approximation to the solution of (2.47), all the terms on the left-hand side may be neglected. The resulting horizontal average wind is called the *geostrophic wind*, and it only depends on the pressure gradient in a given height and a given geographical position. The smallness of the left-hand terms in (2.47), relative to the terms on the right-hand side, is, of course, an empirical relationship found specifically in the Earth's atmosphere. It may not be valid for other flow systems.

The large-scale horizontal wind approaches the geostrophic approximation when neither height nor latitude is too small. At lower heights the turbulent friction term becomes more important (up to about 1000 m), and at low latitudes the Coriolis term becomes smaller, so that it no longer dominates over the terms involving local time and space derivatives.

In analogy to (2.46), the averaging procedure may now be applied to the general transport equation for some scalar quantity, such as the mixing ratio for some minor constituent of the atmosphere. Equation (2.46) then becomes

$$\overline{\rho}\left(\frac{\partial A^*}{\partial t} + (V^* \cdot \mathrm{grad})\,A^* + w^* \frac{\partial A^*}{\partial z} \right) + \frac{\partial}{\partial z}(\overline{\rho w' A'}) = S_A. \tag{2.49}$$

The source term on the right-hand side indicates whether the substance in question is created or lost, e.g. by phase changes or chemical reactions. S_A will, in general, depend on all the state variables of the atmosphere, the air velocity and the mixing ratios for other constituents. In an approximate manner, the source term may be taken to depend only on the average values of its parameters (Wilson and Matthews, 1971),

$$S_A = S_A(\rho, P, V^*, w^*, A_j^*).$$

The ideal gas approximation to the equation of state for the air reads, in averaged form,

$$P = \Re\,\rho\,T^*/\mu, \tag{2.50}$$

where $\Re/\mu \cong 280 \ \mathrm{m}^2\,\mathrm{s}^{-2}\,\mathrm{K}^{-1}$ up to a height z of about 90 km, above which level the molecular weight drops quickly. Eventually, a correction depending on humidity may be incorporated into (2.50).

The temperature is often replaced by a variable called the potential temperature, defined by

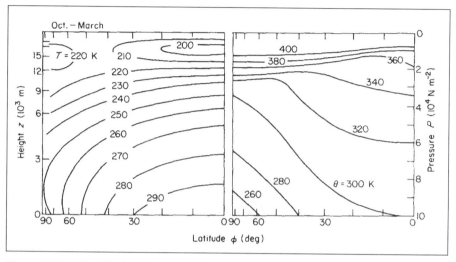

Figure 2.105. Distribution of seasonal (October–March) and longitudinal average of temperature T and potential temperature θ for the Northern Hemisphere (based on Lorenz, 1967).

$$\theta = T\left(\frac{P_0}{P}\right)^{(c_P - c_V)/c_P} \quad \text{or} \quad \theta^* = T^*\left(\frac{P_0}{P}\right)^{(c_P - c_V)/c_P} , \tag{2.51}$$

where the constant P_0 is usually taken as the standard sea-level pressure (10^5 N m^{-2}). The significance of the potential temperature is that it is constant during processes which do not change entropy. Figure 2.105 shows an example of latitude variations of θ^* and T^* based on the same data.

The first law of thermodynamics states that the change in internal energy per unit mass is given by

$$c_V \, dT = \rho^{-1} \, dW^{int} = -P \, d(\rho^{-1}) + Q \, dt, \tag{2.52}$$

where Q is the amount of heat added to the system (an air parcel of unit mass) per second. As in (2.47), the molecular friction (which also produces heat) will be left out. From the definition of specific entropy s, one obtains

$$T \, ds = c_P \, dT - \rho^{-1} \, dP,$$

and by introducing the potential temperature (2.51),

$$ds = c_P \, \theta^{-1} \, d\theta.$$

Inserting this into (2.52), the time derivative of θ is obtained,

$$\frac{d\theta}{dt} = \frac{Q}{c_P}\left(\frac{P_0}{P}\right)^{(c_P - c_V)/c_P} .$$

This is the source term to be entered on the right–hand side of the expression (2.49) for the averaged potential temperature,

$$\bar{\rho}\left(\frac{\partial\theta^*}{\partial t} + (\mathbf{V}^* \cdot \text{grad})\,\theta^* + w^*\frac{\partial\theta^*}{\partial z}\right)\frac{\partial}{\partial z}\overline{(\rho w'\theta')} = \frac{\overline{\rho Q}}{c_P}\left(\frac{P_0}{P}\right)^{(c_P - c_V)/c_P} . \qquad (2.53)$$

Again the last term on the left-hand side describes an amount of heat lost from a given volume by small-scale vertical eddy motion. The external heat sources are radiation and, according to the implicit way of treating latent heat, the heat of condensation of water vapour (plus possibly other phase change contributions),

$$\rho Q = R + C,$$

where R is the amount of heat added to a unit volume by radiation and C is the contribution from condensation processes.

For the approximations considered, a complete set of equations for the atmosphere is constituted by (2.45), (2.47), (2.48), (2.53) and a number of equations (2.49), the most important of which may be that of water vapour. Auxiliary relations include the equation of state, (2.50), and the equations defining the potential temperature (2.51), the heat sources and the source functions for those other constituents, which have been included in (2.49). The water vapour equation is important for calculating the condensation contribution to the heat source function Q. The calculation of the radiative contribution to Q involves, in addition to knowing the external radiation input from space and from continents and oceans as a function of time, a fairly detailed modelling of those constituents of the atmosphere which influence the absorption of radiation, e.g. ozone, water in all forms (distribution of clouds, etc.) and particulate matter.

The atmospheric heat source function

Alternatively, but with a loss of predictive power, one may take Q as an empirical function, in order to obtain a closed system of equations without (2.49), which – if the turbulent eddy term is also left out – allows a determination of the averaged quantities $\mathbf{V}^*$, w^*, $\bar{\rho}, \bar{P}$ and T^* or θ^*. Any calculation including the dissipation by turbulent eddies must supplement the equations listed above with either the equations of motion for the fluctuations $\mathbf{V}'$ and w', or (as is more common) by some parametrisation of or empirical approximation to the time averages $\overline{\rho w'\mathbf{V}'}$, $\overline{\rho w'\theta'}$ and eventually $\overline{\rho w' A'}$.

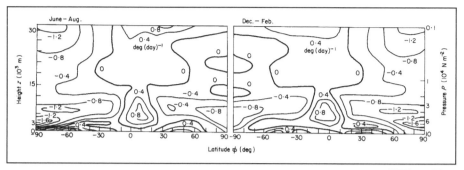

Figure 2.106. Height and latitude distribution of total diabatic heating $(\overline{\rho Q} / c_P \overline{\rho})$, averaged over longitude and season (left: June–August, right: December–February). The unit is K (day)$^{-1}$ (based on Newell *et al.*, 1969).

Figure 2.106 shows an empirical heat source function for two different seasons, but averaged over longitudes. This function could be inserted on the right-hand side of (2.53) to replace $\overline{\rho Q} / (c_P \overline{\rho})$ (the proper SI unit would be K s^{-1} and not K day^{-1} as used in the figure). The different contributions to the heat source function are shown in Figs. 2.107 and 2.108. Figure 2.107 shows the net radiation distribution, R, over the atmosphere (bottom) and the net condensation (condensation minus evaporation) of water vapour in the atmosphere, C (both in units of c_p). These two contributions constitute the truly external heat sources Q, but the total diabatic heating of the atmosphere illustrated in Fig. 2.106 also includes the addition of sensible heat by the turbulent eddy motion described by the last term on the left-hand side of (2.46). Figure 2.108 gives the magnitude of this "heating source" by the quantity

$$\frac{\partial}{\partial z}(\overline{\rho w' T'}) / \rho.$$

The diabatic heating contributions integrated over a vertical column through the atmosphere are indicated in Fig. 2.109. The difference between radiation absorbed and re-emitted from the atmosphere, E_a^{rad}, is negative and fairly independent of latitude. The heat of condensation, which is approximated by the precipitation rate r_a times the latent heat of vaporization L_v, varies substantially with latitude, depending primarily on the distribution of land and ocean surfaces. The heat added to the atmosphere by turbulent convection is equal to the net sensible heat flow at the Earth's surface, with opposite sign, i.e. $-E_s^{sens}$. The sum of the three contributions shown in Fig. 2.109 is not zero, but has to be supplemented by the transport of sensible heat in the atmosphere, between the different latitude areas. However, this is exactly what the equations of motion are supposed to describe.

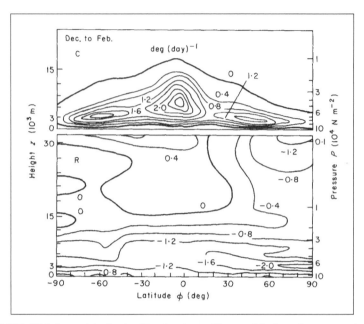

Figure 2.107. Height and latitude distribution of heat added to the atmosphere as a result of condensation of water vapour (C: top) and radiation (R: bottom). Both are in units of (c_p deg day^{-1}) and averaged over longitude and the period December–February (based on Newell *et al.*, 1969).

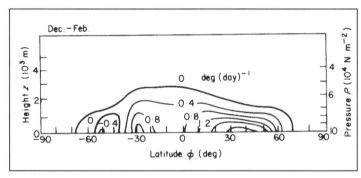

Figure 2.108. Height and latitude distribution of heat added to the atmosphere by turbulent transport of sensible heat from the Earth's surface, in units of (c_p deg day^{-1}) and averaged over longitude and the period December–February (based on Newell *et al.*, 1969).

Separation of scales of motion

The averaging procedure discussed in conjunction with (2.15) presumes that a "large" and a "small" scale of motion are defined. This may be properly done by performing a Fourier analysis of the components of the wind velocity vector, or of the wind speed $|v|$. It is convenient to consider the variance of the wind speed (because it is proportional to a kinetic energy), defined by $< \tilde{v} \cdot \tilde{v} >$. The brackets <> denote average over a statistical ensemble, corresponding to a series of actual measurements. Writing the spectral decomposition in the form

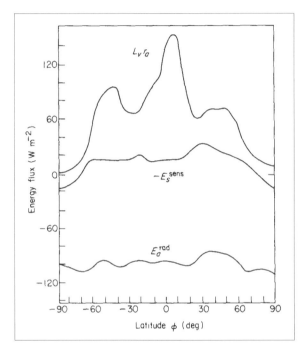

Figure 2.109. Components of the net energy flux passing the atmosphere, averaged over longitude and the year. E_a^{rad} is the net radiation flux, $L_v r_a$ is the heat gained by condensation of water vapour and $-E_s^{sens}$ is the turbulent flux of sensible heat from the Earth's surface to the atmosphere (i.e. the same quantities appearing in Figs. 2.107 and 2.108, but integrated over height). The deviation of the sum of the three quantities from zero should equal the heat gain from horizontal transport of sensible heat (F_a^{sens}) (based on Sellers, 1965).

$$\tfrac{1}{2}\langle \tilde{v} \cdot \tilde{v} \rangle = \int_0^\infty S(\omega)\,\mathrm{d}\,\omega = \int_0^\infty \omega\,S(\omega)\,\mathrm{d}(\log \omega),$$

one obtains a form-invariant spectral function $S(\omega)$, which is shown in Fig. 2.110, based on a modern version of a pioneering measurement effort made by van der Hoven (1957). The peaks exhibited by the spectrum vary in magnitude with the height of measurement (cf. Fig. 3.37).

A striking feature of the spectrum in Fig. 2.110 (and the analogues for other heights) is the broad gap between $\omega \approx 0.5$ h^{-1} and $\omega \approx 20$ h^{-1}. A large volume of measurements have confirmed that the existence of such a gap is an almost universal feature of the wind speed spectrum. Its significance is to provide a clear distinction between the region of large-scale motion ($\omega \lesssim 0.5$ h^{-1}) and the region of small-scale (eddy) motion ($\omega \gtrsim 5$ h^{-1}). The existence of the gap makes the time-averaging procedure in (2.13) easy and makes the exact choice of Δt insignificant over a reasonably large interval, so that the resulting large-scale motion is not critically dependent on the prescription for time averaging. One should be warned, however, not to conclude that the gap in Fig. 2.110 indicates that large-scale motion and small-scale motion are not coupled. It is not necessary for such couplings to involve all intermediate frequencies; on the contrary, distinct couplings may involve very different scales of motion in a straightforward way. Examples of this are evident in the equations of motion, such as (2.47), which allows kinetic energy to be taken out of the large-scale motion (V^*) and be put into small-scale motion ($\overline{\rho w'V'}$) or vice versa.

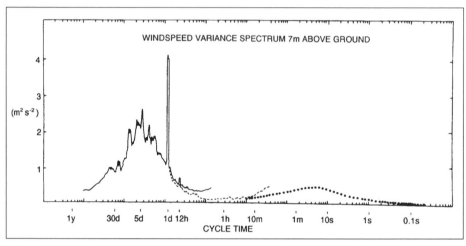

Figure 2.110. Wind speed variance spectrum, based on three different measurements (the main part from Petersen, 1974, cf. Sørensen, 1995a).

Thus small- and large-scale motions in the atmosphere are well-defined concepts, and no accurate model of the atmospheric circulation can be obtained without treating both simultaneously. Current modelling, both for short-term behaviour of the atmospheric circulation ("weather forecasts") and for long-term behaviour ("climate change"), uses averaged values of terms coupling the scales and thus effectively only includes large-scale variables. This is the reason for the poor validity of weather forecasts: they remain valid only as long as the coupling terms involving turbulent (chaotic) motion do not change. For climate modelling, the constraints offered by the system boundaries make the model results valid in an average sense, but not necessarily the detailed geographical distribution of the magnitude of climate variables. This also means that the stability of the atmosphere, which as mentioned in section 2.4 is not in its lowest energy state, cannot be guaranteed by this type of calculation and that the atmosphere could experience transitions from one relative stability state to another (as it has during its history, as evidenced by the transitions to ice ages and back).

Furthermore, since the motion within the atmosphere is not capable of carrying out all the necessary transport of heat (see Fig. 2.44), an accurate model of the circulation of the atmosphere is not possible without including the heat transport within the ocean–continent system (ocean currents, rivers, run-off along the surface and, to a lesser extent, as far as heat transport is concerned, ground water motion). Such coupled models will be considered below, after introducing the energy quantities of relevance for the discussion.

Energy conversion processes in the atmosphere

The kinetic energy of the atmospheric circulation can be diminished by friction, leading to an irreversible transformation of kinetic energy into internal energy (heat). In order to make up for frictional losses, new kinetic energy must be created in the atmosphere. This can be achieved essentially by two processes, both of which are reversible and may proceed adiabatically. One is the conversion of potential energy into kinetic energy (by gravitational fall), and the other is the transformation of internal energy into kinetic energy by motion across a pressure gradient. In terms of the averaged quantities (i.e. neglecting terms like $\overline{\rho V' \cdot V'}$), the kinetic, potential and internal energies may be written

$$W^{kin} = \tfrac{1}{2} \overline{\rho}(V^* \cdot V^* + w^* w^*),$$
$$W^{pot} = \overline{\rho} \, g z, \tag{2.54}$$
$$W^{int} = \overline{\rho} c_V T^*,$$

and the corresponding changes in time

$$\frac{dW^{kin}}{dt} = V^* \cdot \operatorname{grad} P - w^* \frac{\partial P}{\partial z} - w^* \overline{\rho} g - V^* \cdot \frac{\partial}{\partial z}(\overline{\rho w' V'}), \tag{2.55}$$

obtained from (2.46) by scalar multiplication with v^*,

$$\frac{dW^{pot}}{dt} = \overline{\rho} g w^* \tag{2.56}$$

[from the definition in (2.54)] and

$$\frac{dW^{int}}{dt} = -P \operatorname{div}(v^*) + \overline{\rho Q} + V^* \cdot \frac{\partial}{\partial z}(\overline{\rho w' V'}), \tag{2.57}$$

which follows from (2.52) by using

$$\rho \frac{d}{dt}\left(\frac{1}{\rho}\right) = \operatorname{div}(v)$$

and adding the heat gained by turbulent convection on the same footing as the external heat sources ρQ (radiation and gain of latent heat by condensation). As noted by Lorenz (1967), not all potential and internal energy is available for conversion into kinetic energy. The sum of potential and internal energy must be measured relative to a reference state of the atmosphere. Lorenz defines the reference state as a state in which the pressure equals the average value on the isentropic surface passing through the point considered [hence the convenience of introducing the potential temperature (2.51)].

Modelling the oceans

As mentioned in section 2.3, the state of the oceans is given by the temperature T, the density $\rho = \rho_w$ (or alternatively the pressure P), the salinity S (salt fraction by mass), and possibly other variables such as oxygen and organic materials present in the water. S is necessary in order to obtain a relation defining the density that can replace the ideal gas law used in the atmosphere. An example of such an equation of state, derived on an empirical basis by Eckart (1958), is

$$P = \frac{x_1(T,S)\rho}{1 - x_0\,\rho} - x_2(T,S), \tag{2.58}$$

where x_0 is a constant and x_1 and x_2 are polynomials of second order in T and first order in S^*.

The measured temperature and salinity distributions were shown in Figs. 2.63–2.65. In some studies the temperature is instead referred to surface pressure, thus being the potential temperature θ defined according to (2.51). By applying the first law of thermodynamics, (2.52), to an adiabatic process ($Q = 0$), the system is carried from temperature T and pressure P to the surface pressure P_0 and the corresponding temperature $\theta = T(P_0)$. The more complex equation of state (2.58) must be used to express $d(1/\rho)$ in (2.52) in terms of dT and dP,

$$d\left(\frac{1}{\rho}\right) = \frac{\partial X}{\partial P}\,dP + \frac{\partial X}{\partial T}\,dT,$$

where

$$X = \frac{x_1 + x_0(P + x_2)}{P + x_2} = \frac{1}{\rho}.$$

---

[*] $x_0 = 0.698 \times 10^{-3}$ m³ kg⁻¹,
$x_1 = (177950 + 1\,125\,T - 7.45\,T^2 - (380 + T)\,1000\,S)$ m² s⁻²,
$x_2 = (5890 + 38\,T - 0.375\,T^2 + 3000\,S)\,10^5$ N m⁻²,
where T should be inserted in °C (Bryan, 1969).

This is then inserted into (2.52) with $Q = 0$, and (2.52) is integrated from (T, P) to (θ, P_0). Because of the minimal compressibility of water (as compared to air, for example), the difference between T and θ is small.

Basic equations governing the oceanic circulation

Considering first a situation where the formation of waves can be neglected, the wind stress may be used as a boundary condition for the equations of motion analogous to (2.46). The "Reynold stress" eddy transport term may, as mentioned in connection with (2.46), be parametrised as suggested by Boussinesq, so that it gets the same form as the molecular viscosity term and – more importantly – can be expressed in terms of the averaged variables,

$$
\overline{(\text{div } v')\rho v'} \approx - k_z \overline{\rho} \frac{\partial}{\partial z} \frac{\partial}{\partial z} V^* - k_y \overline{\rho} \frac{\partial}{\partial y} \left(\text{grad } V_y^* + \frac{\partial}{\partial y} V^* \right)
$$

$$
- kx \overline{\rho} \frac{\partial}{\partial x} \left(\text{grad } V_x^* + \frac{\partial}{\partial x} V^* \right).
$$

(2.59)

Here use has been made of the anticipated insignificance of the vertical velocity, w^*, relative to the horizontal velocity, V^*. A discussion of the validity of the Boussinesq assumption may be found in Hinze (1975). It can be seen that (2.59) represents diffusion processes, which result from introducing a stress tensor of the form (2.30) into the equations of motion (2.46). In (2.59), the diffusion coefficients have not been assumed to be isotropic, but in practice the horizontal diffusivities k_x and k_y are often taken to be equal (and denoted K; Bryan, 1969; Bryan and Cox, 1967). Denoting the vertical and horizontal velocity components in the ocean waters w_w and V_w, the averaged equations of motion, corresponding to (2.47) and (2.48) for the atmosphere, become

$$
\frac{\partial \overline{P}_w}{\partial z} = - g \overline{\rho_w},
$$

(2.60)

$$
\frac{\partial V_w^*}{\partial t} + (V_w^* \cdot \text{grad}) V_w^* + w_w^* \frac{\partial V_w^*}{\partial z} - k_z \frac{\partial^2}{\partial z^2} V_w^* - K \left(\frac{\partial^2}{\partial x^2} + \frac{\partial^2}{\partial y^2} \right) V_w^*
$$

$$
- K \left(\frac{\partial}{\partial x} \text{grad } V_{wx}^* + \frac{\partial}{\partial y} \text{grad } V_{wy}^* \right) = \frac{1}{\rho} \text{grad } \overline{P}_w - f \mathbf{e}_z \times V_w^*,
$$

(2.61)

where the Coriolis parameter is

$$
f = 2\Omega \sin\phi + \frac{V_{wx}^*}{r_s} \tan\phi,
$$

in terms of the angular velocity of the Earth's rotation, Ω, plus the longitudinal angular velocity of the mean circulation, $V_{wx}{}^*(r_s \cos \phi)^{-1}$, with r_s being the radius of the Earth.

The boundary conditions at the ocean' s surface are

$$w_w^*(z = 0) = 0,$$

$$\overline{\rho_w k_z} \frac{\partial V_w^*(z = 0)}{\partial z} = \overline{\tau}, \qquad (2.62)$$

where τ is the wind stress (2.42). The boundary condition (2.62) [cf. (2.30)] reflects the intuitive notion that, for a small downward diffusion coefficient, k_z, a non-zero wind stress will lead to a steep velocity gradient at the ocean surface. Knowledge of k_z is limited, and in oceanic calculations k_z is often taken as a constant, the value of which is around 10^{-4} m^2 s^{-1} (Bryan, 1969).

If the density of water, ρ_w, is also regarded as a constant over time intervals much larger than that used in the averaging procedure, then the equation of continuity (2.45) simply becomes

$$\text{div } v_w^* = \frac{\partial V_{wx}^*}{\partial x} + \frac{\partial V_{wy}^*}{\partial y} + \frac{\partial w_w^*}{\partial z} = 0. \qquad (2.63)$$

Since the pressure, P_w in (2.61), is related to temperature and salinity through the equation of state, (2.58), a closed set of equations for the oceanic circulation must encompass transport equations analogous to (2.49) for the average values of temperature and salinity,

$$\frac{\partial A^*}{\partial t} + (V_w^* \cdot \text{grad})A^* + w_w^* \frac{\partial A^*}{\partial z} - k_z \frac{\partial^2}{\partial z^2} A^* - K\left(\frac{\partial^2}{\partial x^2} + \frac{\partial^2}{\partial y^2} \right)A^* = \frac{\overline{S_A}}{\rho_w}. \qquad (2.64)$$

The turbulent diffusion coefficients, k_z and K, being phenomenological quantities could eventually be chosen separately for temperature, $A^* = T_w^*$, and salinity, $A^* = S_w^*$, and not necessarily be identical to that used in (2.61) for momentum. If it is assumed that there are no sources or sinks of heat or salt within the ocean, the right-hand side of (2.64) may be replaced by zero. The exchange of heat and salt between the ocean and the atmosphere can then be incorporated into the boundary conditions, under which the two equations (2.64) are solved,

$$\overline{\rho_w k_z} \frac{\partial T_w^*(z = 0)}{\partial z} = -\frac{E_s^{total}}{C_w}, \qquad (2.65)$$

$$\overline{\rho_w k_z} \frac{\partial S_w^*(z = 0)}{\partial z} = S_w^*(z = 0)(e - r). \qquad (2.66)$$

Among the chemical reaction terms that have been suggested for inclusion as source terms S_A in (2.49) are e.g. the sulphate-forming reactions found in sea-salt spray particles (Laskin et al., 2003).

The surface net energy flux E_s^{total} [cf. (2.10)], the evaporation e and precipitation r are variables which link the ocean with the atmosphere, as well as the wind stress appearing in (2.62). In (2.65), C_w is the specific heat of water.

The above expressions do not include the contributions from ice-covered water. Bryan (1969) has constructed boundary conditions allowing for formation and melting of sea ice. Pack ice extending to a depth of less than 3 m is able to move, so Bryan adds a transport equation for such sea ice in the general form of (2.64) and with a suitable source term. More recent models include a specific ice cover model and take into account both melting and sublimation processes (see e.g. NCAR, 1997).

2.D Tides and waves

Consider a celestial body of mass M and distance R from the centre of the Earth. The gravitational force acting on a particle of mass m placed at distance r from the Earth's centre may in a coordinate system moving with the Earth be written

$$F^{tidal} = m\frac{d^2 r}{dt^2} = m\frac{d^2}{dt^2}(R'-R) = -mMG\left(\frac{R'}{R'^3} - \frac{R}{R^3}\right),$$

where R' is the position vector of the mass m relative to M. The force may be decomposed into a radial and a tangential part (along unit vectors e_r and e_t), introducing the angle θ between $-R$ and r and assuming that $|r|$ is much smaller than $|R|$ and $|R'|$ in order to retain only the first terms in an expansion of R'^{-2} around R^{-2},

$$F^{tidal} \approx \frac{2mMGr}{R^3}(\cos^2\theta e_r + \sin\theta\cos\theta e_t). \tag{2.67}$$

If the particle m is at rest within or near to the Earth, it is further influenced by the gravitational force of the mass of the sphere inside its radius, with the centre coinciding with that of the Earth. If the particle is following the rotation of the Earth, it is further subjected to the centrifugal force and, if it is moving relative to the Earth, to the Coriolis force. The particle may be ascribed a tidal potential energy, taken to be zero if the particle is situated at the centre of the Earth (Bartels, 1957),

$$W^{tidal} = \int_0^r F^{tidal} \cdot e_r dr = \frac{mMGr^2}{R^3}\cos^2\theta.$$

If the particle is suspended at the Earth's surface in such a way that it may adjust its height, so that the change in its gravitational potential, mgz, can balance the tidal potential, then the change in height, $z = \Delta r$, becomes

$$z = \frac{MGr^2}{gR^3} \cos^2 \theta.$$

Inserting the lunar mass and mean distance, this height becomes $z = 0.36$ m for $\cos^2\theta = 0$, i.e. at the points on the Earth facing the Moon or opposite (locations where the Moon is in zenith or in nadir). Inserting instead data for the Sun, the maximum z becomes 0.16 m. These figures are equilibrium tides which might be attained if the Earth's surface were covered with ocean and if other forces, such as those associated with the Earth's rotation, were neglected. However, since water would have to flow into the regions near $\theta = 0$ and $\theta = \pi$, and since water is hardly compressible, volume conservation would require a lowering of the water level at $\theta = \pi/2$.

The estimate of "equilibrium tides" only serves the purpose of a zero-order orientation of the magnitude of tidal effects originating from different celestial bodies. The actual formation of tidal waves is a dynamic process, which depends not only on the above-mentioned forces acting on the water particles, but also particularly on the topography of the ocean boundaries. The periods involved can be estimated by inserting into (2.67) the expression for the zenith angle θ of the tide-producing celestial body, in terms of declination δ and hour angle ω of the body,

$$\cos\theta = \sin\delta \ \sin\phi + \cos\delta \ \cos\phi \ \cos\omega$$

[cf. the discussion around (2.1) and (2.2)]. The tidal force will have components involving $\cos\omega$ and $\cos^2\omega = (\cos(2\omega)+1)/2$, or the corresponding sine functions, which proves the existence of periods for lunar tides equal to one lunar day (24 h and 50 min) and to half a lunar day. Similarly, the solar tides are periodic with components of periods of a half and one solar day. If higher-order terms had been retained in (2.67), periods of a third the basic ones, etc., would also be found, but with amplitudes only a small fraction of those considered here.

Owing to the inclination of the lunar orbit plane relative to that of the Earth–Sun system, the amplitudes of the main solar and lunar tidal forces (2.67) will only add in a fully coherent way every 1600 years (latest in 1433; Tomaschek, 1957). However, in case of incomplete coherence one can still distinguish between situations where the solar tidal force enhances that of the Moon ("spring tide") and situations where the solar tidal forces diminish the lunar tides ("neap tide").

Tidal waves lose energy due to friction against the continents, particularly at inlets with enhanced tidal changes in water level. It is estimated that this energy dissipation is consistent with being considered the main cause for the observed deceleration in the Earth's rotation (the relative decrease being 1.16×10^{-10} per year). Taking this figure to represent the energy input into tidal motion, as well as the frictional loss, one obtains a measure of the energy flux involved in the tides as equal to about 3×10^{12} W (King Hubbert, 1971).

Gravity waves in the oceans

The surface of a wavy ocean may be described by a function $\sigma = \sigma(x, y, t)$, giving the deviation of the vertical coordinate z from zero. Performing a spectral decomposition of a wave propagating along the x-axis yields

$$\sigma(x,t) = \int_{-\infty}^{\infty} \exp(-i(kx - \omega(k)t))S(k)dk, \tag{2.68}$$

where $S(k)$ represents the spectrum of σ. If ω is constant and positive, (2.68) describes a harmonic wave moving towards the right; if ω is constant and negative, the wave moves towards the left. By superimposing waves with positive and negative ω, but with the same spectral amplitude $S(k)$, standing waves may be constructed. If $S(k)$ in (2.68) is different from zero only in a narrow region of k-space, σ will describe the propagation of a disturbance (wave packet) along the ocean surface.

The wave motion should be found as a solution to the general equations of motion (2.60)–(2.63), removing the time averages and interpreting the eddy diffusion terms as molecular diffusion (viscous) terms, with the corresponding changes in the numerical values of $k_x = K$ into η_w (the kinematic viscosity of water, being around 1.8×10^{-6} m^2 s^{-1}). Owing to the incompressibility assumption (2.63), the equation of motion in the presence of gravity as the only external force may be written

$$\frac{\partial v_w}{\partial t} + (v_w \cdot \text{grad})v_w = -ge_z - \rho_w^{-1}\, \text{grad}\, P_w + \eta_w\, \text{div}\, \text{grad}\, v_w. \tag{2.69}$$

For irrotational flow, rot $v = 0$, the velocity vector may be expressed in terms of a velocity potential, $\phi(x, y, z, t)$,

$$v_w = \text{grad}\, \phi, \tag{2.70}$$

and the following integral to the equation of motion exists,

$$\frac{\partial \phi}{\partial t} + \tfrac{1}{2}v_w^2 + gz + \frac{P_w}{\rho_w} = 0, \tag{2.71}$$

where the constant on the right-hand side is only zero if proper use has been made of the freedom in choosing among the possible solutions ϕ to (2.70). In the case of viscous flow, rot v_w is no longer zero, and (2.70) should be replaced by

$$v_w = \text{grad}\, \phi + v_w^{rot}. \tag{2.72}$$

A lowest-order solution to these equations may be found using the infinitesimal wave approximation (Wehausen and Laitone, 1960) obtained by linearising the equations [i.e. disregarding terms like $\frac{1}{2}v_w^2$ in (2.71), etc.]. If the surface tension, which would otherwise appear in the boundary conditions at $z = 0$, is neglected, one obtains the first-order, harmonic solution for gravity waves, propagating in the x-direction,

$$\phi = a\, U_w e^{kz} \sin(k(x - U_w t)) \exp(-2k^2 \eta_w t), \tag{2.73}$$

$$\sigma = a \cos(k(x - U_w t)) \exp(-2k^2 \eta_w t)), \tag{2.74}$$

where the connection between wave velocity (phase velocity) U_w, wave number k and wavelength λ_w is

$$U_w = \left(\frac{g}{k}\right)^{1/2}; \qquad k = \frac{2\pi}{\lambda_w} \tag{2.75}$$

[cf. the relation in the presence of surface tension (2.19)]. No boundary condition has been imposed at the bottom, corresponding to a deep ocean assumption, $\lambda_w \ll h$, with h being the depth of the ocean. If the viscous forces are disregarded, the lowest-order solutions are similar to those of (2.73) and (2.74), but with the last exponential factor omitted. Improved solutions can be obtained by including higher-order terms in the equations of motion, by perturbative iteration starting from the first–order solutions given here. The second-order correction to ϕ is zero, but the wave surface (2.74) becomes (for $\eta_w \approx 0$)

$$\sigma^{(2)} = a \cos(k(x - U_w t)) + 0.5\, a^2 k \cos(2k(x - U_w t)).$$

The correction makes the wave profile more flat along the bottom (trough) and more steep near the top (crest). Exact solutions to the wave equations, including surface tension but neglecting viscous forces, have been constructed (see e.g. Havelock, 1919). An important feature of these solutions is a maximum value which the ratio between wave amplitude a and wavelength λ_w can attain, the numerical value of which is determined as

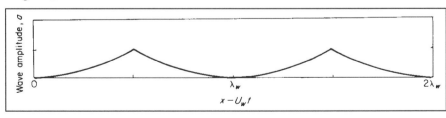

Figure 2.111. Wave profile for pure gravity wave of maximum amplitude-to-wavelength ratio in a coordinate system following the wave (based on Wehausen and Laitone, 1960).

$$ak = a\frac{2\pi}{\lambda_w} \le 0.89. \tag{2.76}$$

As already suggested by Stokes (1880), the "corner" occurring at the crest of such a wave would have an angle of 120°. Figure 2.111 indicates the form of the gravity wave of maximum amplitude-to-wavelength ratio. For capillary waves, the profile has opposite characteristics: the crest is flattened, and a corner tends to develop in the trough.

The formation and dissipation of energy in wave motion

The energy of wave motion is the sum of potential, kinetic and surface contributions (if surface tension is considered). For a vertical column of unit area, one has

$$W^{pot} = \tfrac{1}{2}\,\rho_w\, g\, \sigma^2, \tag{2.77}$$

$$W^{kin} = \tfrac{1}{2}\,\rho_w \int_{-\infty}^{0} \left(\left(\frac{\partial \phi}{\partial x}\right)^2 + \left(\frac{\partial \phi}{\partial z}\right)^2 \right) dz, \tag{2.78}$$

$$W^{surf} = \tfrac{1}{2}\,\gamma_w \left(\frac{\partial \sigma}{\partial x}\right)^2. \tag{2.79}$$

Thus, for the harmonic wave (2.73) and (2.74), the total energy, averaged over a period in position x, but for a fixed time t, and neglecting W^{surf}, is

$$W^{total} = \tfrac{1}{2}\,\rho_w\, g a^2 \exp(-4k^2 \eta_w t). \tag{2.80}$$

This shows that a wave which does not receive renewed energy input will dissipate energy by molecular friction, with the rate of dissipation

$$D = -\frac{d\overline{W}^{total}}{dt}\bigg|_{t=0} = 2\rho_w\, g a^2 k^2 \eta_w. \tag{2.81}$$

Of course, this is not the only mechanism by which waves lose energy. Energy is also lost by the creation of turbulence on a scale above the molecular level. This may involve interaction with the air, possibly enhanced by the breaking of wave crests, or oceanic interactions due to the Reynold stresses (2.59). Also, at the shore, surf formation and sand put into motion play a role in energy dissipation from the wave motion.

Once the wind has created a wave field, this may continue to exist for a while, even if the wind ceases. If only frictional dissipation of the type (2.81) is active, a wave of wavelength $\lambda_w = 10$ m will take 70 h to be reduced to half the original amplitude, while the time is 100 times smaller for $\lambda_w = 1$ m.

As mentioned above, the mechanisms by which a wave field is created by the wind field, and subsequently transfers its energy to other degrees of freedom, are not understood in detail. According to Pond (1971), about 80% of the momentum transfer from the wind may be going initially into wave formation, implying that only 20% goes directly into forming currents. Eventually, some of the energy in wave motion is transferred to the currents.

Early descriptions of wave formation by wind assumed that the energy transferred would be proportional to the wave velocity (most often taken as the phase velocity) and to the "steepness" of the wave, given by the ratio of height and wavelength, ka (see e.g. Neumann, 1949). The "steepness" would be interpreted as equivalent to the roughness length z_0 appearing in the logarithmic velocity law (2.33), but, as discussed in connection with (2.42), measurements did not support such a relation very well.

A fruitful way of looking at the complex wave fields found in the real ocean has been to consider the field of amplitudes (2.68) associated with gravity waves as a random quantity with definite statistical properties. For a stationary situation, these statistical properties may be taken as fixed, whereas the creation of waves by wind and other interaction phenomena may be described in terms of slowly varying statistical properties, using perturbation theory. For a first approximation, the wave amplitude field may be assumed to be Gaussian, so that the probability of finding a given amplitude σ is equal to (Pierson, 1955; see also Kinsman, 1965)

$$(2\pi\Sigma^2)^{-1/2} \exp(-\tfrac{1}{2}(\sigma - \sigma_0)^2/\Sigma^2).$$

The variance, Σ^2, may be expressed as

$$\tfrac{1}{2}\rho_w\Sigma^2 = \tfrac{1}{2}\rho_w g <\sigma(x,t)^2> = \int F(k)\,dk = \tfrac{1}{2}\ \overline{W}^{total}, \tag{2.82}$$

corresponding to the average potential energy or half the average total energy of the wave motion. The function $F(k)$ is called the "energy spectrum" (and the generalisation from the case of propagation along the x-axis considered here to a three-dimensional wavenumber k can readily be made). The Gaussian distribution contains waves with arbitrarily large amplitude (but correspondingly small probability), whereas real waves will break if the amplitude exceeds a certain value. Still, the Gaussian distribution is useful for the discussion of many properties of ocean waves, including their formation and growth.

The interactions important for the structure of the wave field are found to include couplings between different spectral components of the wave field, as well as couplings to mean and turbulent flows of wind and currents.

As a basis for constructing the perturbation expansion of the interactions, a set of "normal mode" solutions may be chosen, comprising the elementary (lowest-order) solutions of the wave equations, i.e. (2.74), in the absence of the friction factor, as well as harmonic solutions for the external fields, e.g. V

$- V^* = V'$ for the fluctuating wind velocity or the analogous fluctuating part of the water current. The couplings to the atmospheric fluctuations are often described not in terms of the velocity fluctuations, V', but in terms of pressure fluctuations, $\tilde{P} = P - \overline{P}$. The descriptions are equivalent due to the relation between thermodynamic pressure and velocity fluctuation (cf. section 2.3.1). The fluctuating fields, V' (or $\tilde{P}$) for wind and V'_w for currents, may be treated as random, e.g. assuming Gaussian distributions as used above for the lowest-order descriptions of the wave field.

Following Hasselmann (1967), the normal mode variables of the wave system itself are written

$$a_n(k) = 2^{-1/2}(p_n(-k) - i\omega_n(k)\,q_n(k)),$$

$$\tilde{a}_n(k) = 2^{-1/2}(p_n(-k) + i\omega_n(k)\,q_n(k)), \tag{2.83}$$

in terms of the normal mode frequencies, $\omega_n(k)$, where n enumerates the normal mode solutions, i.e. the solutions to the part of the classical Hamilton function, which is quadratic in the conjugate canonical variables $q_n(k)$ and (the momenta) $p_n(k)$. The coordinates $q_n(k)$ are the amplitudes of each normal mode [harmonic wave with $\sigma = \exp(ikx)$] in the lowest order solution. In this lowest order solution, the $a_n(k)$ are each proportional to the corresponding factor $\exp(-i\omega_n(k)t)$.

The external fields may, if they are known, be decomposed into variables $b_n(k)$, each of which is proportional to $\exp(-i\omega_n(k)t)$. In this case, the equations of motion for the unknown $a_n(k)$ (in the presence of couplings) are of the form

$$\frac{d}{dt}a_n(k) = -i\omega_n(k)\frac{\partial H}{\partial \tilde{a}_n(-k)} + G^{ext}, \tag{2.84}$$

where the Hamilton function H is a sum of terms with products of m normal mode variables $a_{n_i}(k_i)$ or $\tilde{a}_{n_i}(k_i)$, such that $m = 2$ is the uncoupled solution and $m > 2$ describes anharmonic wave–wave couplings. The external couplings G^{ext} are represented by a similar summation, where, however, each term may be a product of m' variables $a_{n_i}(k_i)$ or $\tilde{a}_{n_i}(k_i)$ and $(m-m')$ variables $b_{n_i}(k_i)$ or $b_{n_i}(k_i)$. The external couplings in general do not fulfil the symmetry relations characteristic of the intrinsic wave–wave couplings, and they cannot be derived from a Hamilton function.

Of course, seen from a broader perspective, the wave motion will influence the external fields, so that couplings both ways must be considered. In this case, $b_n(k)$ will also be unknown variables to be determined by the solution, and a Hamilton function may be constructed for the entire system of waves, currents and winds, but with a much greater dimensionality.

If the couplings are considered to be small, the solution for the wave system may be considered not to disturb the external fields, and the solution may be obtained in terms of a perturbation series,

$a_n(k) = {}^{(1)}a_n(k) + {}^{(2)}a_n(k) + \cdots,$ (2.81)

where each successive term may be evaluated by truncating the summations in the Hamilton function and G^{ext} terms of (2.80) at a given order m.

The energy spectrum is analogous to (2.85) (but for a single normal mode type n),

$F_n(k) = \tfrac{1}{2} <a_n(k)\, \tilde{a}_n(-k)>,$

involving the average of two a_n variables over a statistical ensemble. Hasselmann (1967) discusses the extent to which the energy spectra, defined this way, in turn determine the wave system.

The techniques described above are also being used extensively in the description of condensed matter and atomic nuclei.

In a number of simplified cases, perturbational solutions of the type (2.85) have been obtained, with the inclusion of anharmonic terms [also called "non-linear" terms, referring to the right-hand side of (2.81)] (Phillips, 1966; Hasselmann, 1962). Important studies of the wave formation by wind were made by Miles (1957) and by Phillips (1957). They formally correspond to the lowest-order treatment of two definite external fields in the above formalism.

In the model of Phillips (1957), the direct action of the fluctuating (turbulent) pressure field $\tilde{P}$ in the atmosphere on the ocean surface is assumed to create waves. Such resonant waves will start out with high frequencies (wavelengths around 10^{-2} m), corresponding to the turbulent mixing length in the atmosphere, of the small eddies present near the ocean surface. The growth of the wave spectrum is linear, since $\partial F(k)/\partial t$ is time independent and given by the (by assumption stationary and homogeneous) external $\tilde{P}$ field.

After a period with linear growth, the mechanism suggested by Miles (1957) is then believed to come into play. It is a coupling between the mean wind field V^* in the atmosphere and the wave field. Each wave component changes V^* slightly, inducing a pressure perturbation which acts back on the wave component and makes it grow.

The energy transfer may be understood in terms of a "critical layer" of air, covering the region from the (wavy) ocean surface to the height where the mean wind velocity profile $V(z)^*$, given by (2.33), for example, has reached the same speed as that of the surface waves, U_w (phase velocity). Within this "critical layer", energy is being removed from the mean wind flow by the Reynolds stresses [cf. (2.59)]. The rate of transfer may be derived from the last term in (2.55), after integration over z by parts,

$$\frac{dW^{kin}}{dt} = \int \overline{\rho w' V'}\, \frac{\partial V^*}{\partial z}\, dz.$$

One finds that the change in the wave spectrum, $\partial F(k)/\partial t$, implied by this mechanism is proportional to $F(k)$ itself, so that the wave growth becomes exponential.

The Phillips–Miles theory firstly of linear and then of exponentially growing spectral components of waves, under the influence of a stationary wind field, would be expected to be a reasonable first-order description of wave initiation, whereas the later "equilibrium situation" will require a balancing mechanism to carry away the additional energy and momentum transferred from wind to waves. Such mechanisms could be the wave–current interaction, or the dissipation of wave energy into turbulent water motion (an eddy spectrum), which might then in a secondary process transfer the momentum to currents. However, when detailed experimental studies of wave growth became available, it was clear that the actual growth rate was much larger than predicted by the Phillips–Miles theory (Snyder and Cox, 1966; Barnett and Wilkerson, 1967; Barnett and Kenyon, 1975).

Hasselmann (1967) has shown that in a systematic perturbation expansion of the solutions to (2.84), other terms than those considered by Phillips and Miles contribute to the same order of approximation. They represent couplings between the wave field and the fluctuating part of the wind field, such that two normal mode components of the wave field, n and n', are involved together with one component, $b_{n''}$, of the V'-field. Through its disturbance of the mean wind field V^*, the wave component n' interacts with the wind turbulence component n'', causing a pressure fluctuation which in turn increases the amplitude of wave component n. The energy transfer into component n is positive, while the transfer from component n' may have either sign, causing Hasselmann to speculate that this mechanism may be responsible both for the enhanced initial growth rate and also in part for the dissipation of wave energy at a later stage, balancing continued energy input from the wind.

In this model, the loss of energy from a wave field would not only go into oceanic turbulence and currents, but also into atmospheric turbulence. A similar mechanism is involved in breaking of waves [formation of whitecaps and plungers when a wave has developed to the size of maximum height-to-wavelength ratio (2.76), and more energy is added}. Although the breaking process primarily transfers energy from the waves into the atmosphere, it has been suggested that the breaking of waves pushes the streamlines of the air flow upwards, whereby the pressure field, which in Phillips' theory can transfer new energy to the wave field, gets more strength (Banner and Melville, 1976; Cokelet, 1977). Thus, it may be that the dissipative breaking process provides a key to the explanation of the fast growth of wave spectra, contrary to the predictions of the Phillips–Miles theory.

The energy flux associated with wave motion

Now let us reconsider a single, spectral component of the wave field. The energy flux through a plane perpendicular to the direction of wave propagation, i.e. the power carried by the wave motion, is given in the linearised approximation by (Wehausen and Laitone, 1960)

$$P = -\int_{-\infty}^{0} \rho_w \frac{\partial\phi}{\partial t}\frac{\partial\phi}{\partial x}dz - \gamma_w \frac{\partial\sigma}{\partial t}\frac{\partial\sigma}{\partial x}.$$
(2.86)

For the harmonic wave (2.73) and (2.74), neglecting surface tension, this gives

$$P = \tfrac{1}{4}\rho_w g \left(\frac{g}{k}\right)^{1/2} a^2,$$
(2.87)

when averaged over a period. Since $(g/k)^{1/2} = U_w$, this together with (2.84) for $q = 0$ gives the relationship

$$P = \tfrac{1}{2} U_w W^{total}.$$
(2.88)

Thus the energy is not transported by the phase velocity U_w, but by $\tfrac{1}{2}U_w$. For a spectral distribution, the power takes the form

$$P = \tfrac{1}{2}\rho_w g \int S(k)^2 U_g(k) \, dk,$$
(2.89)

where $U_g(k)$ is the group velocity $d(kU_w)/dk$ (equal to $\tfrac{1}{2}U_w$ for the ocean gravity waves considered above). It is no surprise that energy is transported by the group velocity, but (2.88) and (2.87) show that this is the case even if there is no group (or wave packet) at all, i.e. for a single, sinoidal wave.

Wind-driven oceanic circulation

The generation of currents by a part of the wind stress, or maybe as an alternative dissipation process for waves (Stewart, 1969), can be described by the equation of motion (2.61), with the boundary condition (2.62) eventually reducing $\bar{\tau}$ by the amount spent in the formation of waves. Considering the Coriolis and vertical friction terms as the most important ones, Ekman (1902) writes the solution in the form

$$V_{wx}^* = V_0^* e^{bz} \cos(bz + \alpha - \pi/4),$$

$$V_{wy}^* = V_0^* e^{bz} \sin(bz + \alpha - \pi/4),$$

where z is the directional angle of the surface stress vector (measured from the direction of the x-axis) and

$$V_0^* = |\tau|\rho_w^{-1}(fk_z)^{-1/2}; \qquad b = f^{1/2}(2k_z)^{-1/2}$$

(f is the Coriolis parameter). The surface velocity vector is directed 45° to the right of the stress vector, and it spirals down with decreasing length. The average mass transport is perpendicular to the wind stress direction, 90° to the right. The observed angle at $z \approx 0$ is about 10° rather than 45° (Stewart, 1969), supporting the view that more complicated processes take place in the surface region. The gross mass transport, however, is well established, e.g. in the North Atlantic Ocean, where water flows into the Sargasso Sea both from the south (driven by the prevailing eastward winds at the Equator) and from the north (driven by the mid-latitude "westerlies").

In the Sargasso Sea, the assembled water masses are pressed downwards and escape to the sides at a depth. All the water masses are, on average, following the rotation of the Earth, but the lateral squeezing process causes a change in rotational speed of the water involved, in particular a reduction in the vertical component of the angular speed vector. If the rotational speed of some mass of water does not fit with the rotation of the solid Earth at the latitude in question, the water mass will try to move to another latitude, which is consistent with its angular speed (unless continents prevent it from doing so). For this reason, the water masses squeezed out below the Sargasso Sea move towards the Equator. Later, they have to return northward (otherwise the polar part of the Atlantic Ocean would become emptied). Also, in this case the rotation has to be adjusted in order to conserve angular momentum. This change in rotation is achieved by frictional interaction with the western continents. Owing to the sense of the Earth's rotation, these return flows (e.g. the Gulf Stream) have to follow the west shores, whether the return flow is northward or southward.

A description has been provided above of how the wind drives the Ekman flow in the Atlantic Ocean, which again induces the major currents in that ocean. Similar mechanisms are present in the other oceans. Models of these flows, as well as the flows generated by sinking cold water (i.e. not by the winds), have been constructed and integrated along with the atmospheric models considered in section 2.3.1 (Manabe, 1969; Bryan, 1969; Wetherald and Manabe, 1972).

2.5 Suggested topics for discussion

2.5.1

How could one design instruments for measurement of the individual components of the net radiation flux, e.g. direct, scattered and reflected short-wavelength radiation, upward and downward directed long-wavelength radiation? (This question is probably most interesting for those who are not acquainted with such instruments, and who might wish to compare their proposals with instruments actually in use, cf. Thekaekara, 1976; Robinson, 1966; Sellers, 1965; Meinel and Meinel, 1976.)

2.5.2

Discuss the formulation of approximate equations for the flow of water in rivers (cf. e.g. Arnason, 1971).

2.5.3

Discuss the determinism of climate.

2.5.4

Discuss tides in the solid Earth and compare their properties with those in the ocean water and with those in the atmosphere. Construct, for instance, a simple model of a uniform interior of the Earth, a complete cover of ocean with fixed depth, and a uniform atmosphere of some effective height, and assume the densities for each of the three components to be constant (cf. e.g. Tomaschek, 1957; Kertz, 1957).

2.5.5

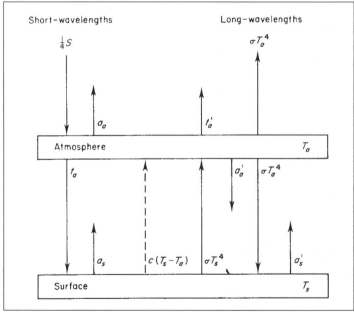

Figure 2.112. Crude model of the Earth–atmosphere system. The average incoming radiation is $S/4$, and the symbols a and t represent fractions being reflected or transmitted at the location indicated. Multiple reflections are not shown, but can be easily calculated. The constant σ in Stefan–Boltzmann's law has the value 5.7×10^{-8} W m^{-2} K^{-4}.

Construct a simple one-dimensional model of the greenhouse effect, containing two components: the Earth's surface of temperature T_s (variable), albedo a_s for short-wavelength radiation and a'_s for long-wavelength radiation, and an idealised atmosphere visualised as a layer (Fig. 2.112) of uniform temperature T_a (variable), short-wavelength albedo a_a, transmission coefficient t_a, long-wavelength albedo a'_a, and transmission coefficient t'_a. Both the surface and the atmosphere emit long-wavelength radiation as black bodies. All fluxes are treated as vertical.

With a definite assumption regarding the non-radiative transfer between surface and atmosphere, such as assuming the flux to be proportional to (T_s-T_a), a set of parameters may be set up such that the main features of the observed Earth–atmosphere system are reproduced (cf. Fig. 2.16). Now questions can be asked concerning the sensitivity of the model to the parameters entering into it. How does the surface temperature change if a_a is changed by ±10% (changing the albedo of the surface)? If the present energy use of the human society was to be covered by solar collectors of average efficiency 0.2, and a_s was changed from 0.15 to zero for the area covered by these collectors, then how much would the total albedo a_s of the Earth's surface become diminished? How does the surface temperature change if a_a is changed by ±10% (e.g. due to presence of aerosols)? How does the surface temperature change if a'_a is changed by ±10% (e.g. by the presence of carbon dioxide)?

2.5.6

In connection with the failure to detect neutrinos from the Sun in the expected quantity, it has been suggested that the energy production in the Sun's core by nuclear processes (and thereby the neutrino flux) may be variable and that the present production may be in a valley. How would this affect the radiation from the Sun's surface? Would it be possible, by suitable choice of a periodic variation in solar energy production, to account for some of the climatic changes of the past? (See Ulrich, 1975.)

THE INDIVIDUAL ENERGY SOURCES

3.1 Solar radiation

An assessment of the "magnitude" of solar radiation as an energy source will depend on the geographical location, including local conditions such as cloudiness, turbidity, etc. In section 2.2.2 a number of features of the radiation flux at a horizontal plane were described, such as spectral distribution, direct and scattered parts, geographical variations, and dependence on time, from annual to diurnal variations at a given location. The seasonal variation in solar radiation on a horizontal plane is shown in Fig. 3.1, corresponding to the annual average of Fig. 2.20.

For actual applications, it is often necessary to estimate the amount of radiation received by tilted or complexly shaped devices, and it is useful to look at relations which allow relevant information to be extracted from some basic measured quantities. For instance, radiation data often exist only for a horizontal plane, and a relation is therefore needed to predict the radiation flux on an arbitrarily inclined surface. In regions at high latitudes, directing solar devices towards the Equator at fairly high tilt angles actually gives an increase in incident energy relative to horizontally placed collectors.

Only the incoming radiation will be discussed in detail in this section, since the outgoing flux may be modified by the specific type of energy conversion device considered. In fact, such a modification is usually the very idea of the device. A description of some individual solar conversion devices will be taken up in Chapter 4, and their potential yield in Chapter 6, in terms of combined demand and supply scenarios.

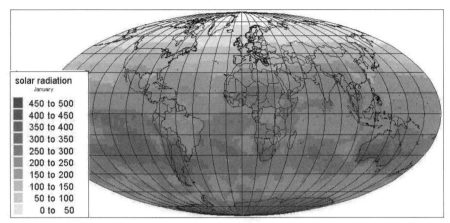

Figure 3.1a,b,c. Average short-wavelength solar radiation on a horizontal plane at the Earth's surface (W m⁻²), for the months of January (*a*, above), April (*b*, below) and July (*c*, bottom) of 1997 (NCEP-NCAR, 1998).

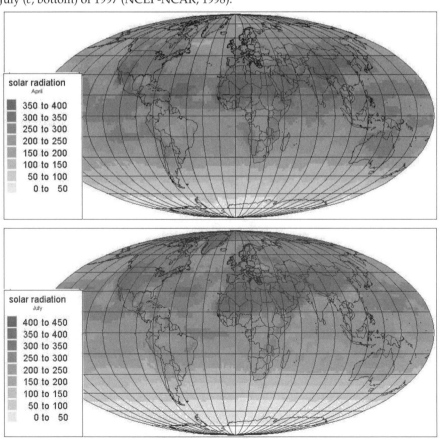

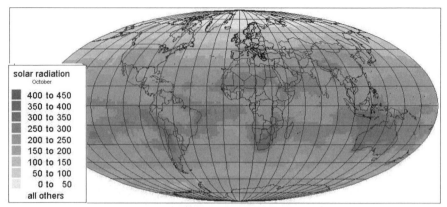

Figure 3.1d. Average short-wavelength solar radiation on a horizontal plane at the Earth's surface (W m^{-2}), for the month of October of 1997 (NCEP-NCAR, 1998).

3.1.1 Direct radiation

The inclination of a surface, e.g. a plane of unit area, may be described by two angles. The tilt angle, s, is the angle between vertical (zenith) and the normal to the surface, and the azimuth angle, γ, is the angle between the southward direction and the direction of the projection of the normal to the surface onto the horizontal plane; γ is counted positive towards east [in analogy to the hour angle (2.2)]. In analogy to the expression at the top of the atmosphere, given in section 2.2.1, the amount of direct radiation reaching the inclined surface characterised by (s, γ) may be written

$$D_{s,\gamma} = S_N \cos\theta, \tag{3.1}$$

where S_N is the "normal radiation", i.e. the solar radiation from the direction to the Sun. The normal radiation is entirely "direct" radiation, according to the definition of "direct" and "scattered" radiation given in section 2.2.2. The angle θ is the angle between the direction to the Sun and the normal to the surface specified by s and γ. The geometrical relation between θ and the time-dependent coordinates of the Sun, declination δ (2.1) and hour angle ω (2.2), is

$$\cos\theta = (SC - CSC)\sin\delta + (SS \sin\omega + (CC + SSC)\cos\omega)\cos\delta, \tag{3.2}$$

where the time-independent constants are given in terms of the latitude ϕ and the parameters (s, γ) specifying the inclination of the surface,

$$
\begin{aligned}
SC &= \sin\phi\cos s, \\
CSC &= \cos\phi\sin s\cos\gamma, \\
SS &= \sin s\sin\gamma. \\
CC &= \cos\phi\cos s, \\
SSC &= \sin\phi\sin s\cos\gamma.
\end{aligned}
\tag{3.3}
$$

For a horizontal surface ($s = 0$), θ equals the zenith angle z of the Sun, and (3.2) reduces to the expression for cos z given in section 2.2.1. Instead of describing the direction to the Sun by δ and ω, the solar altitude $h = \frac{1}{2}\pi - z$ and azimuth Az (conventionally measured as positive towards the west, in contrast to the hour angle) may be introduced. The two sets of coordinates are related by

$$\sin h = \sin \delta \sin\phi + \cos \delta \cos \phi \cos \omega,$$
$$\sin Az \cos h = - \cos \delta \sin \omega. \tag{3.4}$$

In the height–azimuth coordinate system, (3.2) may be written

$$\cos \theta = \sin h \cos s + \cos h \sin s \cos (Az + \gamma). \tag{3.5}$$

Again, the sign conventions for Az and γ are opposite, so that the argument of the last cosine factor in (3.5) is really the difference between the azimuth of the Sun and the projection of the normal to the surface considered. If cos θ found from (3.2) or (3.5) is negative, it means that the Sun is shining on the "rear" side of the surface considered. Usually, the surface is to be regarded as "one-sided", and cos θ may then be replaced by zero, whenever it assumes a negative value, in order that the calculated radiation flux (3.1) becomes properly zero.

If data giving the direct radiation flux D on a horizontal plane are available, and the direct flux impinging on an inclined surface is wanted, the above relations imply that

$$D_{s,\gamma} = D \cos \theta / \cos z.$$

It is evident that care should be taken in applying this relation when the Sun is near the horizon.

More reliable radiation fluxes may be obtained if the normal incidence radiation flux S_N is measured (as function of time) and (3.1) is used directly. S_N is itself a function of zenith distance z, as well as a function of the state of the atmosphere, including ozone mixing ratio, water vapour mixing ratio, aerosol and dust content, and cloud cover. The dependence on zenith angle is primarily a question of the path that the radiation has taken through the atmosphere. This path-length is shortest when the Sun is in zenith (often denoted "air mass one" for a clear sky) and increases with z, being quite large when the Sun is near the horizon and the path is curved due to diffraction. The extinction in the atmosphere is normally reduced at elevated locations (or low-pressure regions, notably mountain areas), in which case the effective air mass may become less than one. Figure 3.2 gives some typical variations of S_N with zenith angle for zero, small and heavy particle load ("turbidity"). Underlying assumptions are: cloudless sky, water vapour content throughout a vertical column equal to 0.02 m^3 m^{-2}, standard sea-level pressure and mean distance to the Sun (Robinson, 1966).

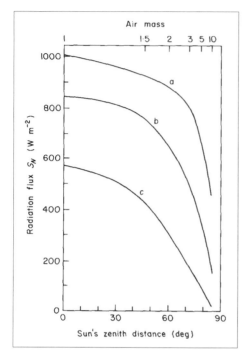

Figure 3.2. Normal incidence radiation as a function of zenith angle for a cloudless sky with different particle content: (a), hypothetical atmosphere with zero turbidity [$B = 0$, no aerosols but still water vapour and molecular (Rayleigh) scattering]; (b), clear sky ($B = 0.01$); (c), atmosphere with heavy aerosol pollution ($B = 0.4$). B is the turbidity coefficient B_λ (defined in the text) averaged over wavelengths. Sea-level pressure (10^5 N m^{-2}), a water content of 0.02 m^3 m^{-2}, an ozone content of 0.0034 m^3 m^{-2} (both referred to standard temperature and pressure) and mean Earth–Sun distance have been assumed. At the top, an approximate air mass scale is provided, relative to the one for a vertical path from sea-level (based on Robinson, 1966).

Since a complete knowledge of the state of the atmosphere is needed in order to calculate S_N, such a calculation would have to be coupled to the equations of state and motion discussed in section 2.3.1. Only some average behaviour may be described without doing this, and if, for example, hourly values of S_N are required in order to predict the performance of a particular solar energy conversion device, it would be better to use measured values of S_N (which are becoming available for selected locations throughout the world, cf. e.g. Turner, 1974) or values deduced from measurements for horizontal surfaces. Measuring techniques are discussed by Coulson (1975), amongst others.

Attempts to parametrise the solar radiation received at the ground are usually made for the global radiation (direct plus scattered, and for tilted planes also radiation reflected onto the surface), rather than separately for normal incidence and scattered radiation [cf. (2.12)].

Dependence on turbidity and cloud cover

The variability of S_N due to turbidity may be demonstrated by noting the range of S_N values implied by extreme high or low turbidities, in Fig. 3.2, and by considering the spread in mean daily turbidity values, an example of which is shown in Fig. 3.3. The turbidity coefficient B_λ for a given wavelength may be defined through an attenuation expression of the form

$$S_N(\lambda)/E_{0+}^{sw}(\lambda) \;=\; 10^{-m_r(\tau_\lambda^s + \tau_\lambda^a + B_\lambda)}.$$

Here m_r is the relative air mass (optical path-length in the atmosphere), τ^s_λ describes the scattering on air molecules and τ^a_λ is the absorption by ozone. In terms of the cross sections $\sigma_s(\lambda)$, $\sigma_a(\lambda)$ and $\sigma_p(\lambda)$ for scattering on air molecules, ozone absorption and attenuation by particles, one has

$$-m_r\,\tau^s_\lambda = \log_{10}\{\exp(\textstyle\int\sigma_s(\lambda)\mathrm{d}s)\},$$

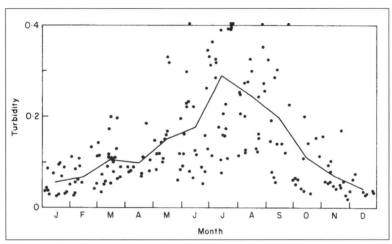

Figure 3.3. Turbidity coefficient B_λ at the wavelength $\lambda = 5 \times 10^{-7}$ m, as measured during 1972 at Raleigh (North Carolina) by Bilton *et at.* (1974). Daily means are indicated by dots, monthly means by the curve. Only days with a cloudless sky in the direction of the Sun have been included. Data points above $B_\lambda = 0.4$ are indicated at 0.4 (cf. the high turbidity curve of Fig. 3.2).

and similarly for τ^a_λ and B_λ. Figure 3.3 gives both daily and monthly means for the year 1972 and for a wavelength of 5×10^{-7} m (Bilton *et al.*, 1974). The data suggest that it is unlikely to be possible to find simple, analytical expressions for the detailed variation of the turbidity or for the solar fluxes, which depend on turbidity.

Another major factor determining the amount and distribution of different types of fluxes is the cloud cover. Both cloud distribution and cloud type are of importance. For the direct radiation flux the important questions are whether the path-line is obscured or not, and, if it is, how much attenuation is caused by the particular type of cloud. Figure 3.4 shows the total and scattered flux on a horizontal plane (and by subtraction the direct), for a clear sky and three different types of clouds, as a function of the zenith angle of the Sun. The cloud classification represents only a general indication of category. It is evident in this example that altocumulus and stratus clouds are almost entirely impermeable for direct radiation, whereas cirrus clouds allow the penetration of roughly half the direct radiation flux. Meteorological observations containing records of sunshine (indicating whether or not the direction to the Sun is obscured) and of cloud cover (as a percentage and identifying the cloud types and their estimated height) may allow a fairly reliable estimate of the fraction of direct radiation reaching a plane of given orientation.

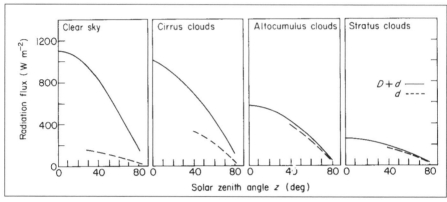

Figure 3.4. The influence of clouds on the radiation flux reaching a horizontal plane (total short-wavelength flux, $D + d$, and scattered flux alone, d) (based on measurements by J. Millard and J. Arvesen, as quoted in Turner, 1974).

3.1.2 Scattered radiation

The scattered radiation for a clear sky may be estimated from knowledge of the composition of the atmosphere, as described in section 2.3.1. The addition of radiation scattered from clouds requires knowledge of the distribution and type of clouds, and the accuracy with which a calculation of the scattered flux on a given plane can be made is rather limited. Even with a clear sky, the agreement between calculated and observed fluxes is not absolute, as seen, for example, in Fig. 2.37.

Assuming that the distribution of intensity, $S_{h,Az}^{scatt.}$, for scattered radiation as a function of the directional coordinates (h, Az) [or (δ, ω)] is known, then the total flux of scattered radiation reaching a plane tilted at an angle s and directed azimuthally, at an angle γ away from south, may be written

$$d_{s,\lambda} = \int S_{h,Az}^{scatt.} \cos \theta \, d\Omega = \int_0^{\pi/2} d(Az) \int_{h_{min}(Az)}^{\pi/2} dh \, S_{h,Az}^{scatt.} \cos \theta(h, Az) \cos h. \quad (3.6)$$

Here $h_{min}(Az)$ is the smallest height angle, for a given azimuth, for which the direction defined by (h, Az) is on the "front" side of the inclined plane. The unit solid angle is $d\Omega = \sin z \, dz \, d(Az) = -\cos h \, dh \, d(Az)$. For a horizontal plane $\theta(h, Az) = z = \frac{1}{2}\pi - h$.

If the scattered radiation is isotropic,

$$S_{h,Az}^{scatt.} = \text{constant} = S^{scatt.},$$

then the scattered radiation flux on a horizontal plane becomes

$$d = \pi S^{scatt.}, \tag{3.7}$$

and the scattered radiation flux on an arbitrarily inclined surface may be written

$$d_{s,\gamma} = d \cos^2 (s/2), \tag{3.8}$$

an expression that can also be derived by considering the fraction of the sky "seen" by the tilted surface.

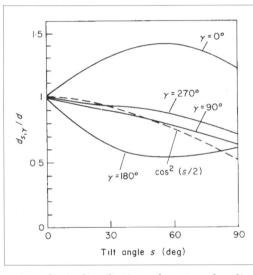

Figure 3.5. Ratio between scattered radiation flux on an inclined surface (tilt angle s, azimuth angle γ) and scattered radiation flux on a horizontal surface for a cloudless sky with solar height 15° (based on Kondratyev and Fedorova, 1976).

A realistic distribution of scattered radiation intensity is not isotropic, and as mentioned earlier it is not even constant for a cloudless sky and fixed position of the Sun, but depends on the momentary state of the atmosphere. Figure 3.5 shows the result of measurements for a cloudless sky (Kondratyev and Fedorova, 1976), indicating that the assumption of isotropy would be particularly poor for planes inclined directly towards or away from the Sun (*Az* equal to 0 or π relative to the solar azimuth). Robinson (1966) notes, from observations such as the one shown in Fig. 2.37, that the main differences between observed distributions of scattered radiation and an isotropic one are (a) increased intensity for directions close to that of the Sun and (b) increased intensity near the horizon. In particular, the increased intensity in directions near the Sun is very pronounced, as is also evident from Fig. 3.5, and Robinson suggests that about 25% of d, the scattered radiation on a horizontal plane, should be subtracted from d and added to the direct radiation, before the calculation of scattered radiation on an inclined surface is performed using the isotropic model. However, such a prescription would not be generally valid, because the increased intensity in directions near the Sun is a function of the turbidity of the atmosphere as well as of cloud cover.

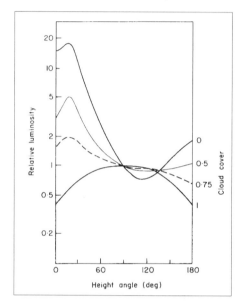

Figure 3.6. Luminance distribution along a great circle containing zenith as well as the direction towards the Sun, as a function of cloud cover. The distributions have been normalised at $h = 90°$. The data are based on several measurements with solar height 20° and mean turbidity, performed by Tonne and Normann (1960).

It is evident from Fig. 3.4 that the effect of clouds generally is to diminish direct radiation and increase scattered radiation, although not in the same proportion. Cirrus clouds and in particular altocumulus clouds substantially increase the scattered flux on a horizontal plane.

The radiation scattered by clouds is not isotropically distributed. Figure 3.6 gives the luminance distribution, i.e. the intensity of total radiation, along the great circle containing zenith as well as the direction to the Sun, the height angle of which was 20° at the time of measurement. For complete cloud cover the luminance distribution is entirely due to scattered radiation from the clouds, and it is seen to be maximum at zenith and falling to 0.4 times the zenith value at the horizon.

3.1.3 Total short-wavelength radiation

For inclined surfaces or surfaces surrounded by elevated structures, the total short-wavelength (wavelengths below, say, 3×10^{-6} m) radiation flux comprises not only direct and scattered radiation, but also radiation reflected from the ground or from the surroundings onto the surface considered.

Reflected radiation

The reflected radiation from a given area of the surroundings may be described in terms of an intensity distribution, which depends on the physical nature of the area in question, as well as on the incoming radiation on that area. If the area is specularly reflecting, and the incoming radiation is from a single direction, then there is also a single direction of outgoing, reflected radiation, a direction which may be calculated from the law of specular reflection (polar angle unchanged, azimuth angle changed by π). Whether the reflection is specular or not may not only depend on the fixed properties of the particular area, but also on the wavelength and polarisation of the incoming radiation.

The extreme opposite of specular reflection is completely diffuse reflection, for which by definition the reflected intensity is isotropic over the hemisphere bordered by the plane tangential to the surface at the point considered, no matter what the distribution of incoming intensity. The total (hemispherical) amount of reflected radiation is in this case equal to the total incoming radiation flux times the albedo a of the area in question,

$$R = aE_+,$$

where, for horizontal surfaces, $E_+ = D + d$ (for short-wavelength radiation).

In general, the reflection is neither completely specular nor completely diffuse. In this case, the reflected intensity in a given direction, e.g. specified by height angle and azimuth, $S_{\Omega_r}^{refl.}$, depends on the distribution of incident radiation intensities $S_{\Omega_r}^{inc.}$,[*]

$$S_{\Omega_r}^{refl.} = \int_{hemisphere} \rho_2(\Omega_r, \Omega_i) S_{\Omega_i}^{inc.} \cos\theta(\Omega_i)\, d\Omega_i. \tag{3.9}$$

Here $\rho_2(\Omega_r, \Omega_i)$ is called the bi-angular reflectance (Duffie and Beckman, 1974; these authors include an extra factor π in the definition of ρ_2). Dividing by the average incoming intensity,

---

[*] The intensity S_Ω is here defined as the energy flux passing through an infinitesimal area into an infinitesimal solid angle in the direction specified by Ω (e.g. h and Az). The infinitesimal area is perpendicular to the direction Ω, and the dimension of S_Ω is energy per unit time, unit area and unit solid angle. The energy flux passing through a unit area is found by multiplying S_Ω by $\cos\theta$, with θ being the angle between the direction Ω and the normal to the plane, and by $d\Omega$, and then integrating over the hemisphere above or below the unit area considered (giving directional fluxes E_- or E_+). These definitions are consistent with those used in sections 2.1 and 2.2, remembering that the incident solar radiation represents a limiting case of infinitesimal solid angle for the definition of intensity, because the rays from the Sun are treated as parallel, and recalling that several of the energy fluxes considered in section 2.2 were global averages.

$$\int S_{\Omega_i}^{inc.} \cos\theta(\Omega_i)\, d\Omega_i \Big/ \int \cos\theta(\Omega_i)\, d\Omega_i = E_+ / \pi,$$

a reflectance depending only on one set of angles (of reflected radiation) is defined,

$$\rho_1(\Omega\, r) = \pi\, S_{\Omega_r}^{refl.} / E_+. \tag{3.10}$$

In general, ρ_1 is not a property of the reflecting surface, since it depends on incoming radiation, but if the incoming radiation is isotrophic (diffuse or black-body radiation), the incoming intensity can be divided out in (3.10).

The total hemispherical flux of reflected radiation may be found by integration of (3.9),

$$R = \int_{hemisphere} S_{\Omega_r}^{refl.} \cos\theta(\Omega_r)\, d\Omega_r, \tag{3.11}$$

and the corresponding hemispherical reflectance,

$$\rho = R/E_+ = a, \tag{3.12}$$

is equal to the albedo defined above.

All of the above relations have been written without reference to wavelength, but they are valid for each wavelength λ, as well as for appropriately integrated quantities.

In considering the amount of reflected radiation reaching an inclined surface, the surrounding surfaces capable of reflecting radiation are often approximated by an infinite, horizontal plane. If the reflected radiation is further assumed to be isotropic of intensity $S^{refl.}$, then the reflected radiation flux received by the inclined surface is

$$R_{s,\gamma} = \pi S^{refl.} \sin^2(s/2). \tag{3.13}$$

The factor $\sin^2(s/2)$ represents the fraction of the hemisphere above the inclined surface from which reflected radiation is received. This fraction is evidently complementary to the fraction of the hemisphere from which scattered radiation is received and, therefore, equal to $1 - \cos^2(s/2)$ from (3.8). The albedo (3.12) may be introduced, noting from (3.11) that $R = \pi S^{refl.}$ for isotropic reflection,

$$R_{s,\gamma} = aE_+ \sin^2(s/2). \tag{3.14}$$

For short-wavelength radiation, the flux E_+ on a horizontal plane equals $D + d$, the sum of direct and scattered short-wavelength fluxes.

If the geometry of the reflecting surroundings is more complicated (than a plane), or if the reflected intensity is not isotropic, the calculation of the amount of reflected radiation reaching a given inclined surface involves an integration over the hemisphere seen by the inclined surface. Thus, for each direction of light reflected onto the inclined plane, a contribution to $S_{\Omega_r}^{refl.}$ is included from the first unobscured point on the line of sight capable of reflecting radiation. If semi-transparent objects (such as a water basin) are present in the surroundings, the integration becomes three dimensional, and the refraction and transmission properties of the partly opaque objects must be considered.

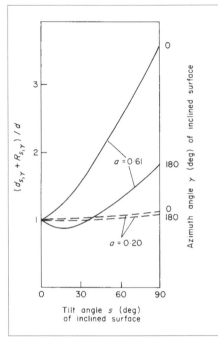

Figure 3.7. The ratio of scattered plus reflected radiation flux on inclined surfaces to that on a horizontal surface (scattered flux d only) for a clear sky with the Sun at height 34°. Measurements corresponding to two different ground albedos are depicted, based on summer and winter (snow-covered ground) conditions at a location in the Soviet Union (Kondratyev and Fedorova, 1976).

Figure 3.7 shows an example of the enhancement of indirect radiation that may result from increasing albedo of the surroundings – in this case it is due to snow cover in winter (Kondratyev and Fedorova, 1976). The rejected radiation on the inclined surface reaches a maximum value for the tilt angle $s = 90°$ (vertical).

Average behaviour of total short-wavelength radiation

The sum of direct, scattered and reflected radiation fluxes constitutes the total short-wavelength (sw) radiation flux. The total short-wavelength flux on a horizontal surface is sometimes referred to as the "global radiation", i.e. $D + d$ (if there are no elevated structures to reflect radiation onto the horizontal surface). For an inclined surface of tilt angle s and azimuth γ, the total short-wavelength flux may be written

$$E_{s,\gamma}^{sw} = D_{s,\gamma} + d_{s,\gamma} + R_{s,\gamma} ,\qquad(3.15)$$

with the components given by (3.1), (3.6) or (3.8), and (3.14) or a generalisation of it. The subscript "+" on E^{sw} has been left out, since the direction of the flux is clear from the values of s and γ (the E_- flux would generally require $s > 90°$).

Examples of the influence of clouds, and of solar zenith angle, on global radiation $E_{s=0,\gamma}^{sw}$ have been given in Fig. 3.4. Figure 3.8 illustrates the influence of cloud cover on the daily sum of total short-wavelength radiation received by inclined surfaces, relative to that received by a horizontal surface. For south-facing slopes the radiation decreases with increasing cloud cover, but for north-facing slopes the opposite takes place.

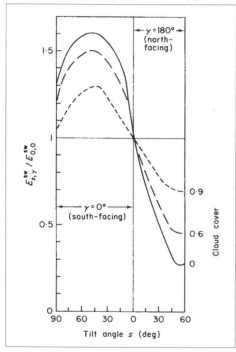

Figure 3.8. Daily sums of total short-wavelength radiation on an inclined plane, relative to that of a horizontal plane, for different cloud cover and as a function of tilt angle for north- and south-facing surfaces (based on measurements in the Soviet Union, Kondratyev and Fedorova, 1976).

Monthly averages of total short-wavelength radiation for different geographical locations are shown in Figs 3.9 and 3.10 for a horizontal surface. In Fig. 3.10, two sets of data are compared, each representing a pair of locations with the same latitude. The two at $\phi \approx 43°$N. correspond to a coastal and a continental site, but the radiation, averaged for each month, is nearly the same. The other pair of locations are at 35–36°N. Albuquerque has a desert climate, while Tokyo is near the ocean. Here the radiation patterns are vastly different during summer by as much as a factor two. The summer solar radiation in Tokyo is also smaller than in both the 43°N sites, presumably due to the influence of urbanisation (cf. section 2.4.2) and a high natural frequency of cloud coverage. Other average properties of radiation fluxes on horizontal surfaces were considered in section 2.2.2, notably for a latitude 53°N location.

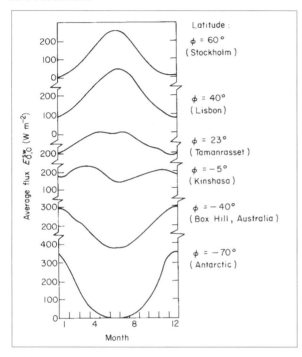

Figure 3.9. Smoothed variation with seasons of the average daily flux (24-h average) of total short-wavelength radiation on a horizontal plane for selected geographical locations (based on Flack and Morikofer, 1964).

Turning now to inclined surfaces, Fig. 3.11 shows the total short-wavelength radiation on a vertical plane facing south, west, east and north for a location at $\phi = 56°$N. The monthly means have been calculated from the hourly data of the Danish reference year (Andersen *et al.*, 1974), using the isotropic approximation (3.8) and (3.14) with an assumed albedo $a = 0.2$. The Danish reference year consists of selected meteorological data exhibiting "typical" fluctuations. This is achieved by selecting monthly sequences of actual data, with monthly averages of each of the meteorological variables in the vicinity of the 30-year mean. The different sequences, which make up the reference year, have been taken from different years. The variables pertaining to solar radiation are global radiation $(D + d)$, normal incidence radiation (S_N) and scattered radiation on a horizontal plane (d). Only global radiation has been measured over extended periods of time, but the two other variables have been constructed in such a way that the three variables, of which only two are independent, become reasonably consistent. Several other countries are in the process of constructing similar "reference years", which will allow an easy intercomparison of, for example, performance calculations of solar energy devices or building insulation prescriptions. Monthly averages of the basic data of the Danish reference year, $D + d$ and D, are shown in Fig. 3.12. Reference years have subsequently been constructed for a number of other locations (European Commission, 1985).

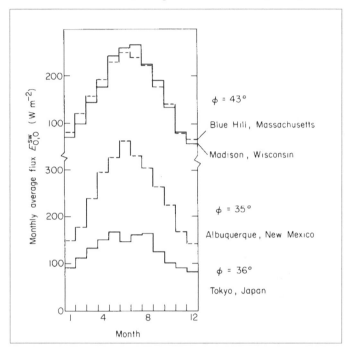

Figure 3.10.
Monthly average flux on a horizontal plane for pairs of locations with similar latitude but different climate [based on data from NOAA (US National Oceanic and Atmospheric Administration), quoted from Duffie and Beckman, 1974].

Figure 3.13 shows the composition of the total short-wavelength flux on a vertical, south-facing surface, in terms of direct, scattered and reflected radiation. It has been assumed that the albedo changes from 0.2 to 0.9 when the ground is covered by snow (a piece of information also furnished by the reference year). Snow cover is in this way responsible for the relative maximum in reflected flux for February. The proportion of direct radiation is substantially higher during winter for this vertical surface than for the horizontal surface (Fig. 3.12). Figure 3.14 gives the variation of the yearly average fluxes with the tilt angle s, still for a south-facing surface ($\gamma = 0$). The variation of monthly averages of the total flux with tilt angle for a south-facing surface is given in Fig. 3.15.

According to Fig. 3.14, the maximum yearly short-wavelength radiation in Denmark ($\phi = 56°N$) is obtained on a south-facing surface tilted about 40°, but the maximum is broad. The maximum direct average flux is obtained at a tilt angle closer to ϕ, which is clear because at the top of the atmosphere the maximum would be attained for s equal to the latitude plus or minus the

Sun's declination, the extremes of which are about ±23° at summer and winter solstices. From Fig. 3.15, one can see that the most constant radiation over the year is obtained for $s = 90°$. The December maximum value is for $s = 75°$, but the solar radiation on an $s = 90°$ surface is only 1–2% smaller.

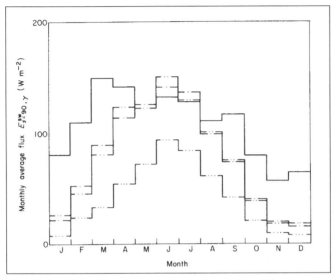

Figure 3.11. Monthly average flux on vertical surface with different azimuthal orientation, based on the Danish reference year (latitude 56°) and an assumed ground albedo of 0.2 (0.9 with snow cover). Key:——— $\gamma = 0°$ (south); —·—·— $\gamma = -90°$ (west); —··—··— $\gamma = 90°$ (east); —···—$\gamma = 180°$ (north). This and the following calculations assume that the scattered radiation is isotropic.

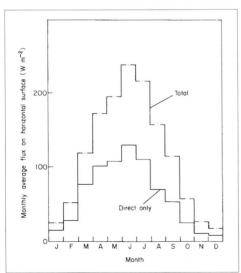

Figure 3.12. Monthly average short-wavelength flux on horizontal surface, and the direct part, D, based on the Danish reference year, $\phi = 56°N$ (Andersen *et al.*, 1974).

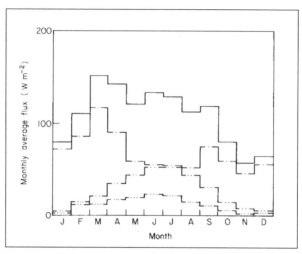

Figure 3.13. Components of monthly average fluxes on vertical, south-facing surface. Key: ———total short-wavelength radiation; —·—·— direct radiation $D_{s,\gamma}$; —··—··— scattered radiation $d_{s,\gamma}$; — ···— reflected radiation $R_{s,\gamma}$ (in all cases $\gamma = 0$). The calculation is based on the Danish reference year, $\phi = 56°N$, and an assumed albedo of 0.2 (0.9 with snow cover).

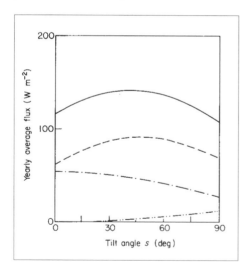

Figure 3.14. Components of yearly average fluxes on a vertical, south-facing surface. Key: ——— total short-wavelength radiation; ------ direct radiation $D_{s,\gamma}$; —·—·— scattered radiation $d_{s,\gamma}$; —··—··— reflected radiation $R_{s,\gamma}$ (in all cases $\gamma = 0$). The calculation is based on the Danish reference year, $\phi = 56°N$, and an assumed albedo of 0.2 (0.9 with snow cover).

3.1.4 Long-wavelength radiation

As suggested by Figs. 2.16 and 2.22, the long-wavelength radiation reaching a plane situated at the Earth's surface may be quite substantial, in fact, exceeding the short-wavelength radiation when averaged over 24 h. However, the outgoing long-wavelength radiation is (again on average) still larger, so that the net long-wavelength flux is directed away from the surface of the

Earth. The ground and most objects at the Earth's surface emit long-wavelength (*lw*) radiation approximately as a black body, while the long-wavelength radiation from the atmosphere often deviates substantially from any black-body spectral distribution.

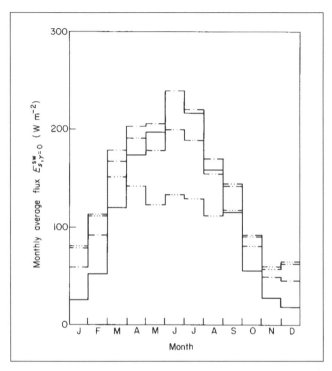

Figure 3.15. Monthly average of total short-wavelength radiation on inclined, south-facing surfaces, based on the Danish reference year, $\phi = 56°N$. Key:—— tilt angle $s = 0$; —·—· $s = 30°$; —··—·· $s = 60°$; — ··· — $s = 90°$.

In general, the emission and absorption of radiation by a body may be described by the spectral and directional emittance, $\varepsilon_\lambda(\Omega)$, and the corresponding absorptance, $\alpha_\lambda(\Omega)$. These quantities are related to $e(\nu)$ and $\kappa(\nu)$ of section 2.A, which may also depend on the direction Ω, by

$$e(\nu, \Omega) = \varepsilon_\lambda(\Omega)\, dF^{Planck} / d\nu, \qquad \kappa(\nu, \Omega) = \alpha_\lambda(\Omega),$$

for corresponding values of frequency ν and wavelength λ.

Thus $\varepsilon_\lambda(\Omega)$ is the emission of the body, as a fraction of the black-body radiation (2.21) for emission in the direction Ω, and $\alpha_\lambda(\Omega)$ is the fraction of incident flux, again from the direction specified by Ω, which is absorbed by the body.

Where a body is in complete thermodynamic equilibrium with its surroundings (or more specifically those with which it exchanges radiation), Kirchoff's law is valid (see e.g. Siegel and Howell, 1972),

$$\varepsilon_\lambda(\Omega) = \alpha_\lambda(\Omega). \tag{3.16}$$

However, since both $\varepsilon_\lambda(\Omega)$ and $\alpha_\lambda(\Omega)$ are properties of the body, they are independent of the surroundings, and hence (3.16) is also valid away from equilibrium. This does not apply if wavelength averages are considered. Then $\varepsilon(\Omega)$ is still a property of the surface, but $\alpha(\Omega)$ depends on the spectral composition of the incoming radiation, as follows from the definition of $\alpha(\Omega)$,

$$\alpha(\Omega) = \int \alpha_\lambda(\Omega) S_\lambda^{inc}(\Omega) \, d\lambda \bigg/ \int S_\lambda^{inc}(\Omega) \, d\lambda.$$

Further averaging over directions yields the hemispherical emittance ε and absorptance α, the former still being a property of the surface, the latter not.

Consider now a surface of temperature T_s and orientation angles (s, γ), which emit long-wavelength radiation in such a way that the hemispherical energy flux is

$$E_-^{lw,\,emission} = \varepsilon^{lw} \sigma T_s^4, \tag{3.17}$$

where σ is Stefan's constant (5.7×10^{-8} W m^{-2} K^{-4}). For pure black-body radiation from a surface in thermal equilibrium with the temperature T_s (forming an isolated system, e.g. a cavity emitter), ε^{lw} equals unity.

The assumption that $\varepsilon_\lambda = \varepsilon$ (emissivity independent of wavelength) is referred to as the "grey-body" assumption. It implies that the energy absorbed from the environment can be described by a single absorptivity $\alpha^{lw} = \varepsilon^{lw}$. Further, if the long-wavelength radiation from the environment is described as black-body radiation corresponding to an effective temperature T_e, the incoming radiation absorbed by the surface may be written

$$E_+^{lw,abs.} = \varepsilon^{lw} \sigma T_e^4. \tag{3.18}$$

If the part of the environmental flux reflected by the surface is denoted R (= $\rho^{lw} \sigma T_e^4$), then the total incoming long-wavelength radiation flux is $E_+^{lw} = R + E_+^{lw,abs}$, and the total outgoing flux is $E_-^{lw} = R + E_-^{lw,emission}$, and thus the net long-wavelength flux is

$$E^{lw} = \varepsilon^{lw} \sigma (T_e^4 - T_s^4). \tag{3.19}$$

The reflection from the environment of temperature T_e back onto the surface of temperature T_s has not been considered, implying an assumption regarding the "smallness" of the surface considered as compared with the effective surface of the environment. Most surfaces that may be contemplated at the surface of the Earth (water, ice, grass, clay, glass, concrete, paints, etc.) have long-wavelength emissivities ε^{lw} close to unity (typically about 0.95). Materials such as iron and aluminium with non-polished surfaces have low long-wavelength emissivities (about 0.2), but often their temperature T_s is higher than the average temperature at the Earth's surface (due to high absorptivity for short-wavelength radiation), so that (3.19) may still be a fair approximation, if T_s is chosen as an average temperature of physical surfaces. If the surface is part of an energy-collecting device, a performance evaluation will require the use of the actual temperature T_s with its variations (cf. Chapter 4).

The deviations of the long-wavelength radiation received from the environment from that of a black-body are more serious, leading both to a change in wavelength dependence and to a non-isotropic directional dependence. Decisive in determining the characteristics of this radiation component is the average distance, for each direction, to the point at which the long-wavelength radiation is emitted. Since the main absorbers in the long-wavelength frequency region are water vapour and CO_2 (cf. Fig. 2.36), and since the atmospheric content of water vapour is the most variable of these, then one may expect the long-wavelength flux to be primarily a function of the water vapour content m_v (Kondratyev and Podolskaya, 1953).

At wavelengths with few absorption bands the points of emission may be several kilometres away, and owing to the temperature variation through the atmosphere (see e.g. Figs. 2.27 and 2.28), one expects this component of T_e to be 20–30 K below the ambient temperature T_a at the surface. As humidity increases, the average emission distance diminishes, and the effective temperature T_e becomes closer to T_a. The temperature of the surface itself, T_s, is equal to or larger than T_a, depending on the absorptive properties of the surface. This is also true for physical surfaces in the surroundings, and therefore the "environment" seen by an inclined surface generally comprises partly a fraction of the sky with an effective temperature T_e below T_a and partly a fraction of the ground, possibly with various other structures, which has an effective temperature of T_e above T_a.

Empirical evidence for inclined surfaces

Figure 3.16 shows the directional dependence of long-wavelength radiation from the environment, averaged over wavelengths, for very dry and humid atmospheres. The measurements on which the figure is based (Oetjen et al., 1960) show that the black-body approximation is only valid for directions close to the horizon (presumably implying a short average distance to points of emission), whereas the spectral intensity exhibits deeper and deeper minima, in particular around $\lambda = 10^{-5}$ m, as the direction approaches zenith. Relative to the black-body radiation at ambient temperature, T_a, the directional flux in Fig. 3.16 starts at unity at the horizon, but drops to 79 and 56%, respectively, for the humid (Florida Beach region) and the dry atmospheres

(Colorado mountainous region). If an effective black-body temperature is ascribed, although the spectral distributions are not Planckian, the reduced T_e is 94.3 and 86.5% of T_a, corresponding to temperatures 27.5 and 37.9 K below ambient temperature. Although it is not clear whether they are in fact, Meinel and Meinel (1976) suggest that the two curves in Fig. 3.16 be used as limiting cases, supplemented with black-body emissions from ground and surrounding structures, which are characterised by ε^{lw} between 1.0 and 1.1, for the calculation of net long-wavelength radiation on inclined surfaces.

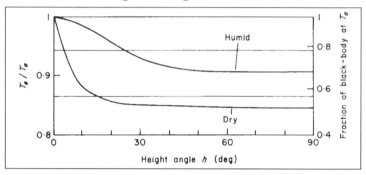

Figure 3.16. Variation in incoming long-wavelength radiation with height above the horizon (flat ground). The scale on the left is the ratio between the effective temperature of the long-wavelength radiation received and the ambient air temperature, whereas the scale on the right gives the average flux relative to that of black-body radiation at ambient temperature. The two curves represent measurements for a very humid and a very dry atmosphere; performed by Oetjen *et al.* (1960) and quoted by Meinel and Meinel (1976).

In calculations of the performance of flat-plate solar collectors it has been customary to use an effective environmental temperature about 6 K less than the ambient temperature T_a (Duffie and Beckman, 1974; Meinel and Meinel, 1976), but since the surfaces considered in this context are usually placed with tilt angles in the range 45–90°, a fraction of the hemisphere "seen" by the surface will be the ground (and buildings, trees, etc.). Thus, the effective temperature of the long-wavelength radiation received will be of the form

$$T_e = C\, T_{e,atmosphere} + (1 - C)\, T_{e,ground,} \tag{3.20}$$

where C is at most $\cos^2(s/2)$, corresponding to the case of an infinitely extended, horizontal plane in front of the inclined surface considered. For this reason T_e will generally not be as much below T_a as indicated by Fig. 3.16, but it is hard to see how it could be as high as $T_a - 6$, since $T_{e,ground}$ is rarely more than a few degrees above T_a. Silverstein (1976) estimates values of $T_e - T_a$ equal to −20 K for horizontal surfaces and −7 K for vertical surfaces.

Figure 3.17 shows measured values of the ratio of net long-wavelength radiation (3.19) on a black-painted inclined surface to that on the same surface in horizontal position. The quantity varying with angle of inclination is the effective black-body temperature (which could alternatively be calculated on the basis of curves such as those given in Fig. 3.16), $T_{e,s,\gamma}$, so that Fig. 3.17 can be interpreted as giving the ratio

$$(T_{e,s,\gamma}^4 - T_s^4) / (T_{e,0,0}^4 - T_s^4).$$

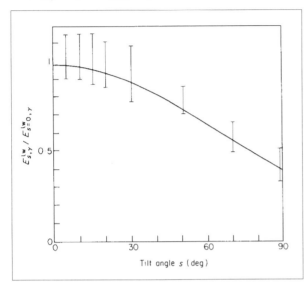

Figure 3.17. Ratio of net long-wavelength radiation on inclined surfaces and the corresponding flux for a horizontal surface. The error bars represent the spread of measured values. Although measurements for different azimuth angles γ have been aggregated, the position of data points within the error bars is not noticeably correlated with γ. The curve has been calculated with assumptions mentioned in the text (based on Kondratyev and Fedorova, 1976).

No significant differences are found between the data points for $\gamma = 0°$, $\pm 90°$ and $180°$, which are available for each value of the tilt angle s. The data are in agreement with the vertical to horizontal temperature difference $T_{e,s=90°} - T_{e,0}$ found by Silverstein and the calculation of Kondratyev and Podolskaya (1953) for an assumed atmospheric water mixing ratio of about 0.02 m^3 m^{-2} vertical column. However, the accuracy is limited, and it would be valuable to expand the experimental activity in order to arrive at more precise determinations of the effective environmental temperature under different conditions.

3.1.5 Variability of solar radiation

The fact that the amount of solar radiation received at a given location at the Earth's surface varies with time has been implicit in several of the data discussed in the preceding sections. The variability and part-time absence of solar radiation due the Earth's rotation (diurnal cycle) and orbital motion (seasonal cycle, depending on latitude) are well known and simple to describe, and so the emphasis here will be on a description of the less simple influence of the state of the atmosphere, cloud cover, etc.

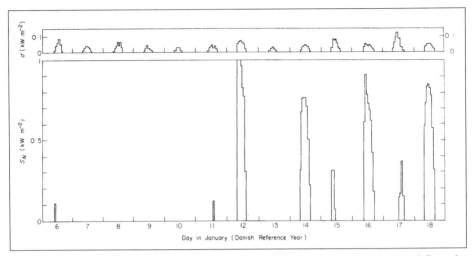

Figure 3.18. Hourly values of the normal incidence flux, S_N, and the scattered flux, d, on a horizontal plane for the Danish reference year, $\phi = 56°N$. Thirteen consecutive winter days are shown.

Most of the data presented in sections 3.1.1–3.1.3 have been in the form of averages over substantial lengths of time (e.g. a month), although some results were discussed in terms of instantaneous values, such as intensity as a function of solar zenith angle. In order to demonstrate the time structure of radiation quantities more clearly, Fig. 3.18 shows hourly averages of normal incidence radiation, S_N, as well as scattered radiation on a horizontal surface, d, hour by hour over a 13-day period of the Danish reference year (latitude 56°N). For this period in January, several days without any direct (or normal incidence) radiation appear consecutively, and the scattered radiation is quite low. On the other hand, very clear days occasionally occur in winter, as witnessed by the normal flux received on 12 January in Fig. 3.18. Fluctuations within each hour are also present, as can be seen from the example shown in Fig. 3.19. The data are for two consecutive days at a latitude of about 39°N (Goddard Space Flight Center in Maryland; Thekaekara, 1976), collected at 4-s intervals. The figure conveys the picture of intense fluctuations, particularly on a partially cloudy day.

The variability of incident solar energy may be displayed in various ways. Figure 3.20 gives the frequency distributions for the daily integrated radiation energy received by a horizontal plane for two different locations.

At Mauna Loa (Hawaii, at a height of 3400 m), the frequency curve peaks at about 6 kWh m^{-2} d^{-1}, while at Argonne (Illinois) the maximum is broad and extends from zero to about 2.5 kWh m^{-2} d^{-1}. The results are shown for two sampling periods: one year and several years. One sees that individual years may exhibit frequency distributions quite different from long-term averages.

Figure 3.21 gives, for the same two locations, some impression of the higher order structure of the fluctuations by trying to indicate the frequency of poor solar radiation days lumped together. For selected daily energy sums, the figure shows, for one year as well as for the average of several years, the number of occurrences of n consecutive days of daily solar radiation below the selected value. At Mauna Loa, the solar radiation falls below 1.742 kWh $m^{-2} d^{-1}$ only one day a year and never falls below 1.161 kWh $m^{-2} d^{-1}$, while at Argonne, 11 consecutive days with solar radiation below 1.742 kWh $m^{-2} d^{-1}$ on average occur once every year and isolated days with this condition occur nine times a year.

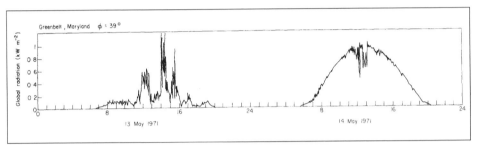

Figure 3.19. Two days of continuous record of total short-wavelength flux on a horizontal plane (based on Thekaekara, 1976).

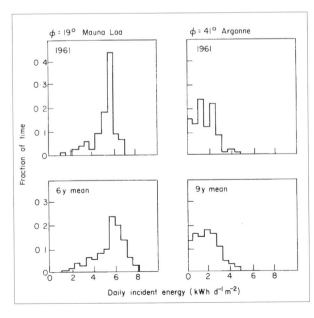

Figure 3.20. Frequency distribution of daily sums of total short-wavelength radiation flux on a horizontal plane for Mauna Loa (Hawaii, $\phi = 19°N$, elevation of measurement location: 3400 m) and Argonne (Illinois, $\phi = 41°N$). The upper curves are for 1961, while the frequency distributions shown in the lower row are based on several years of observation (based on Machta, 1976).

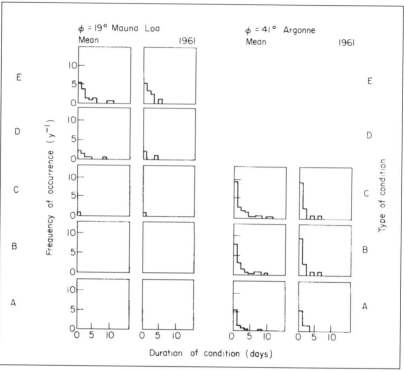

Figure 3.21. Occurrence of rows of consecutive days with daily sums of radiation on a horizontal plane below a certain value for two different locations. Five conditions are considered in rows from bottom to top: radiation in Wh m^{-2} d^{-1} below 581 (A), below 1161 (B), 1742 (C), below 3483 (D), and below 5225 (E). The histograms give the number of times per year, for which one of the conditions A–E persists during the number of days indicated on the abscissa. The two sets of columns correspond to mean frequencies for a number of years of observation (cf. Fig. 3.20) and to 1961 alone (based on Machta, 1976).

Geographical distribution of solar power

The geographical distribution of average incoming (short-wave) solar radiation on a horizontal plane was shown in Figs. 3.1a–d for each of the four seasons. The data exhibit considerable dependence on conditions of cloud cover and other variables influencing the disposition of radiation on its way from the top to the bottom of the atmosphere. These data form the basis for further analysis in terms of suitability of the power flux for energy conversion in thermal and electricity-producing devices (Chapter 4). Methods for estimating solar radiation on inclined surfaces from the horizontal data will be introduced in Chapter 6, along with an appraisal of the fraction of the solar resource that may be considered of practical use after consideration of environmental and area use constraints.

Power duration curves

The accumulated frequency distribution is called a "power duration curve", since it gives the fraction of time during which the energy flux exceeds a given value E, as a function of E. Figure 3.22a,b gives power duration curves for total and for direct radiation on a horizontal surface and two southward inclined surfaces for a location at latitude $\phi = 56°$N (Danish reference year, cf. European Commission, 1985). Since data for an entire year have been used, it is not surprising that the energy flux is non-zero for approximately 50% of the time. The largest fluxes, above 800 W m^{-2}, are obtained for only a few hours a year, with the largest number of hours with high incident flux being for a surface inclined about 45° (cf. Fig. 3.14). The shape of the power

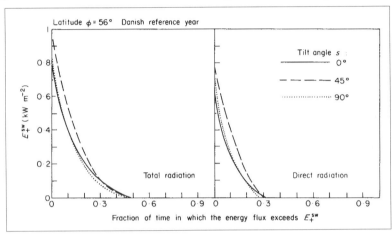

Figure 3.22a,b. One-year power duration curves of total (*a:* left) and direct (*b:* right-hand side) short-wavelength radiation on south-facing surfaces of three different inclinations, based on the Danish reference year, $\phi = 56°$N.

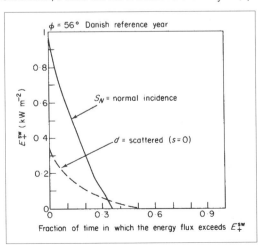

Figure 3.22c. One-year power duration curves of normal incidence radiation alone, and of scattered radiation alone, based on the Danish reference year ($\phi = 56°$N). The normal incidence curve would correspond to the radiation received by a fully tracking instrument. The curve for scattered radiation is for a horizontal surface.

duration curve may be used directly to predict the performance of a solar energy device, if this is known to be sensitive only to fluxes above a certain minimum value, say, 300 W m^{-2}. The power duration curves for direct radiation alone are shown on the right. They are relevant for devices sensitive only to direct radiation, such as most focusing collectors.

Figure 3.22c gives the power duration curves for normal incidence radiation, S_N, and for scattered radiation on a horizontal plane, d. The normal incidence curve is of interest for fully tracking devices, i.e. devices which are being continuously moved in order to face the direction of the Sun. By comparing Fig. 3.22c with Fig. 3.22a,b, it can be seen that the normal incidence curve lies substantially above the direct radiation curve for an optimum but fixed inclination ($s = 45°$). The power duration curve for normal incidence radiation is not above the curve for total radiation at $s = 45°$, but it would be if the scattered radiation received by the tracking plane normal to the direction of the Sun were added to the normal incidence (direct by definition) radiation. The power duration curve for scattered radiation alone is also shown in Fig. 3.22c for a horizontal plane. The maximum scattered flux is about 350 W m^{-2}, much higher than the fluxes received during the winter days shown in Fig. 3.18.

3.2 Wind

It follows from the discussion in section 2.4.1 that the kinetic energy content of the atmosphere, on average, equals about seven days of kinetic energy production or dissipation, also assuming average rates. Utilising wind energy means installing a device which converts part of the kinetic energy in the atmosphere to, say, mechanical useful energy, i.e. the device draws primarily on the energy stored in the atmospheric circulation and not on the energy flow into the general circulation (1200 TW according to Fig. 2.86). This is also clear from the fact that the production of kinetic energy, i.e. the conversion of available potential energy into kinetic energy, according to the general scheme of Fig. 2.50, does not take place to any appreciable extent near the Earth's surface. Newell *et al.* (1969) give height–latitude distributions of the production of zonal kinetic energy for different seasons. According to these, the regions of greatest kinetic energy, the zonal jet-streams at mid-latitudes and a height of about 12 km (see Fig. 2.45), are maintained from conversion of zonal available potential energy in the very same regions.

One of the most important problems to resolve in connection with any future large-scale energy extraction from the surface boundary region of the atmosphere's wind energy storage is that of the nature of the mechanism for restoring the kinetic energy reservoir in the presence of man-made energy extraction devices. To a substantial extent, this mechanism is known from the study of natural processes, by which energy is being removed from the kinetic energy storage in a given region of the atmosphere. The store of energy is replenished by the aforementioned conversion processes, by which kinetic energy is formed from the much larger reservoir of available potential energy (four times larger than the kinetic energy reservoir, according to section 2.4.1). The available potential energy is created by temperature and pressure differences, which in turn are formed by the solar radiation flux and the associated heat fluxes.

3.2.1 Wind velocities

The horizontal wind profile

The horizontal components of the wind velocity are typically two orders of magnitude larger than the vertical component, but the horizontal components may vary a great deal with height under the influence of frictional and impact forces on the ground (Fig. 2.48) and the geostrophic wind above, which is governed by the Earth's rotation (see section 2.3.1). A simple model for the variation of horizontal velocity with height in the lower (Prandtl) planetary boundary layer, assuming an adiabatic lapse rate [i.e. temperature gradient, cf. (2.31)] and a flat, homogeneous surface, was given in (2.33).

The zero point of the velocity profile is often left as a parameter, so that (2.33) is replaced by

$$u = \kappa^{-1} (\tau/\rho)^{1/2} \log((z + z_0 - d_0)/z_0) \tag{3.21}$$

The zero displacement, d_0, allows the zero point of the profile to be determined independently of the roughness length z_0, both being properties of the surface. This is convenient, for example, for woods and cities with roughness lengths from about 0.3 m to several metres, for which the velocity profile "starts" at an elevation (of maybe 10 or 50 m) above the ground. For fixed z_0 and d_0, a family of profiles is described by (3.21), depending on the one parameter $(\tau/\rho)^{1/2}$, called the friction velocity. For a given location it is essentially determined by the geostrophic wind speed, still assuming an adiabatic ("neutral") temperature profile.

If the atmosphere is not neutral, the model leading to (3.21) has to be modified. In fact, (2.33) may be viewed as a special case in a more general theory, due to Monin and Obukhov (1954) (see also Monin and Yaglom, 1965), according to which the wind speed may be expressed as

$$u(z) = \kappa^{-1} (\tau/\rho)^{1/2} (f(z/L) - f(z_0/L)).$$

Only in the case of an adiabatic lapse rate is the function f equal to a logarithm. This case is characterised by $z \ll L$, where the parameter L describing the atmosphere's degree of stability may be written

$$L = c_P \rho (\tau/\rho)^{3/2} T_0 / (g \kappa Q_0).$$ (3.22)

L is called the Monin–Obukhov length, g is the acceleration of gravity and T_0 and Q_0 are the temperature and heat flux at the surface. Thus, L is a measure of the transport of heat near the surface, which again determines the stability of the atmosphere. If the heat flux is sufficiently large and directed from ground to atmosphere (Q_0 and thus L negative), then transport of heat and matter is likely to be highly convective and unstable, so that there will be rapid mixing and therefore less variation of wind speed with height. On the other hand, if Q_0 and L are positive, the atmosphere becomes stable, mixing processes become weak and stratification becomes pronounced, corresponding to almost laminar motion with increasing wind speed upwards.

The connection between L and the temperature gradient becomes clear in the convective case, in which one can prove that z/L equals the Richardson number,

$$Ri = g \frac{1}{\theta} \frac{\partial \theta}{\partial z} \left/ \left(\frac{\partial u}{\partial z} \right)^2 \right. ,$$ (3.23)

where θ is the potential temperature (2.51). From the definition,

$$\frac{\partial \theta}{\partial z} = \frac{\theta}{T} \left(\frac{\partial T}{\partial z} - \frac{c_P - c_V}{c_P} \frac{T}{P} \frac{\partial P}{\partial z} \right),$$

which according to (2.31) is a direct measure of the deviation from adiabaticity. The adiabatic atmosphere is characterised by $Ri = 0$. The identity of z/L and Ri is easy to derive from (2.28), (2.30) and (2.32), but the linear relationship $l = \kappa z$ (2.32) may not be valid in general. Still, the identity may be maintained, if the potential temperature in (3.23) is regarded as a suitably defined average.

Figure 3.23 gives an example of calculated velocity profiles for neutral, stable and unstable atmospheres, assuming fixed values of z_0 (0.01 m, a value typical of grass surfaces) and $(\tau/\rho)^{1/2} = 0.5$ m s^{-1} (based on Frost, 1975). Many meteorological stations keep records of stability class, based on either observed turbulence, Ri or simply $\partial T/\partial z$. Thus, it is often possible with a certain amount of confidence to extrapolate wind speed measurements taken at one height (e.g. the meteorological standard of 10 m) to other heights that are interesting from the point of view of wind energy extraction.

The wind profiles also show that a wind energy conversion device spanning several tens of metres and placed in the planetary boundary layer is likely to experience substantially different wind velocities at different ends of the device (e.g. rotor blades).

The simple parametrisations of wind profiles considered above can be only expected to be valid over flat and homogeneous terrain. The presence of obstacles such as hills, vegetation of varying height and building structures may greatly alter the profiles and will often create regions of strong turbulence, where no simple average velocity distribution will give an adequate description. If an obstacle such as a hilltop is sufficiently smooth, however, and the atmospheric lapse rate neutral or stable, the possibility exists of increasing the wind speed in a regular fashion at a given height where an energy collecting device may be placed. An example of this is shown in Fig. 3.24, based on a calculation by Frost et al. (1974). The elliptical hill shape extends infinitely in the

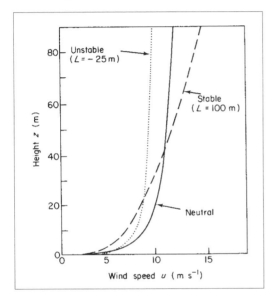

Figure 3.23. Wind speed profiles (i.e. variation with height) for a flat terrain characterised by a roughness length of 0.01 m and for a friction velocity of $(\tau/\rho)^{1/2} = 0.5$ m s^{-1}. The three different curves correspond to an atmosphere, which is neutral (has an adiabatic lapse rate), stable or unstable. The stability is expressed in terms of the Monin–Obukhov length (based on Frost, 1975).

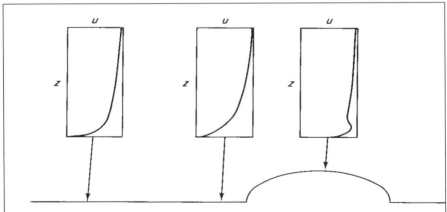

Figure 3.24. Wind speed profiles for a wind directed perpendicular to an elliptically shaped hill for two locations upwind, plus at the top of the hill. The profile far away from the disturbance corresponds to a surface of roughness length of 0.005 m, a friction velocity of 0.4 m s^{-1} and a neutral atmosphere (based on Frost et al., 1974).

direction perpendicular to the plane depicted. However, most natural hill shapes do not make a site more suitable for energy extraction. Another possibility of enhanced wind flow at certain heights exists in connection with certain shapes of valleys, which may act like a shroud to concentrate the intensity of the wind field, while still preserving its laminar features (see e.g. Frost, 1975). Again, if the shape of the valley is not perfect, regions of strong turbulence may be created and the suitability for energy extraction diminishes.

Only in relatively few cases can the local topography be used to enhance the kinetic energy of the wind at heights suitable for extraction. In the majority of cases, the optimum site selection will be one of flat terrain and smoothest possible surface in the directions of the prevailing winds. Since the roughness length over water is typically of the order 10^{-3} m, the best sites for wind energy conversion will usually be coastal sites with large fetch regions over water.

Wind speed data

Average wind speeds give only a very rough idea of the power in the wind (which is proportional to u^3) or the kinetic energy in the wind (which is proportional to u^3), owing to the fluctuating part of the velocity, which makes the average of the cube of u different from the cube of the average, etc. Whether the extra power of the positive speed excursions can be made useful depends on the response of the particular energy conversion device to such fluctuations.

For many locations, only average wind speed data are available, and since it is usually true in a qualitative way that the power which can be extracted increases with increasing average wind speed, a few features of average wind speed behaviour will be described.

In comparing wind speed data for different sites, the general wind speed–height relation (such as the ones shown in Fig. 3.23) should be kept in mind, along with the fact that roughness length, average friction velocity and statistics of the occurrence of different stability classes are also site-dependent quantities.

Figure 3.25 shows the seasonal variation of wind speed, based on monthly average values, for selected locations. Except for Risø and Toronto, the sites represent near optimum wind conditions for their respective regions. Not all parts of the Earth are as favoured with winds as the ones in Fig. 3.25. Figure 3.26 shows the variation of average wind speed throughout the day at two shoreline sites and one maritime site in Singapore. The average wind speed at the latter site is slightly over 3 m s^{-1}, and there is little seasonal variation. At the two land-based observational stations, the overall average speed is only around 2 m s^{-1}, with very little gain when the height increases from 10 to 65 m.

Global wind speed data were shown in Figs. 2.79-2.80. Figure 3.27 shows the levels of power in the wind at potential hub height for wind turbines (about 70 m), on a seasonal base, using data from NCEP-NCAR (1998). These data have been made more consistent by a re-analysis method (Kalnay *et al.*, 1996), in which general circulation models have been run to improve predictions in areas of little or poor data. The construction of power in the wind estimates from wind speed data is explained in section 6.2.5. It involves an averaging of circulation model data from height levels of 1000 and 925 mb (expressed as pressure levels), plus an empirical way of relating the average of the third power of the wind speed (the power) to the third power of the average wind speed. This procedure is necessarily very approximative. The main observation is that the highest power levels are found over oceans, including some regions close to shorelines. This points to the possibility of obtaining viable extraction of wind power from off-shore plants located at suitable shallow depths. It is also seen that seasonal variations are significant, particularly over large continents.

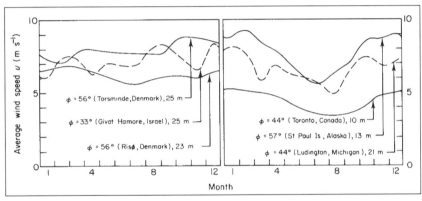

Figure 3.25. Seasonal variation of average wind speed for selected locations and heights. Torsminde data for 1961 from Jensen (1962), Givat Hamore data Nov. 1959 to Oct. 1960 from Frenkiel (1964), Risø 1958–1967 data from Petersen (1974), Toronto 1955–1972 data from Brown and Warne (1975), St. Paul Island 1943–1971 data from Wentink (1976), and the Ludington 1970, 1972 data from Asmussen (1975).

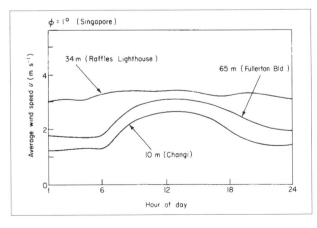

Figure 3.26. Variations of average wind speed with hour of the day for three Singapore stations. The data cover the period 1970–1974 (based on Nathan *et al.*, 1976).

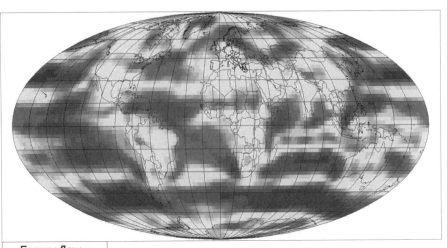

Energy flow W/m2	
■	1600 to 6000
■	800 to 1600
■	600 to 800
■	400 to 600
■	300 to 400
■	250 to 300
■	200 to 250
■	160 to 200
■	120 to 160
■	80 to 120
■	40 to 80
■	0 to 40

Figure 3.27a,b. Maps of wind power regimes for January and April 1997, based on NCEP-NCAR (1998). The power levels are estimated for a height of 70 m above ground, presently a typical hub height for wind turbines. The method of estimation is explained in section 6.2.5. [These and following area-based geographical information system (GIS) maps are from Sørensen and Meibom (1998), copyright B. Sørensen].

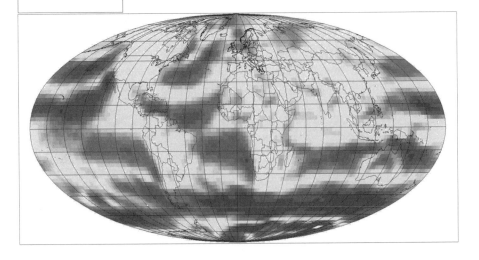

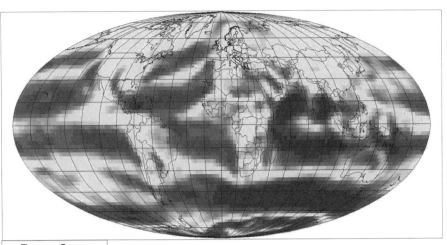

Energy flow W/m2	
■	1600 to 6000
■	800 to 1600
■	600 to 800
■	400 to 600
■	300 to 400
■	250 to 300
■	200 to 250
■	160 to 200
■	120 to 160
■	80 to 120
■	40 to 80
■	0 to 40

Figure 3.27c,d. Maps of wind power regimes for July and October 1997, based on NCEP-NCAR (1998). The power levels are estimated for a height of 70 m above ground, presently a typical hub height for wind turbines. The method of estimation is explained in section 6.2.5.

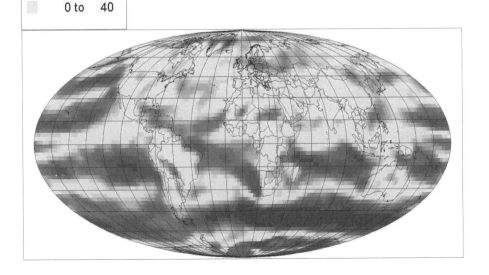

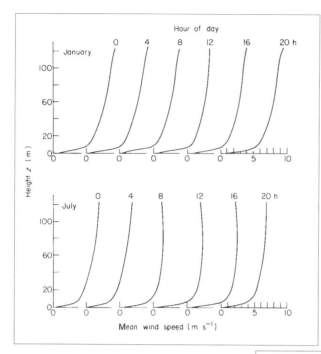

Figure 3.28. Variation of wind speed height profile with the hour of day at Risø, Denmark (56°N). Each curve has been averaged over all data from the period 1958–1967 for one winter or summer month (based on data from Petersen, 1974).

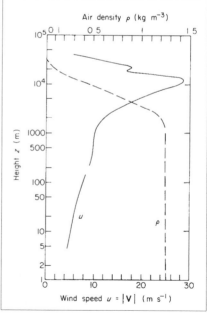

Figure 3.29. Annual average wind speed *vs.* height profile at latitudes 50–56°N. The lower part of the curve is based on Risø data (Petersen, 1974), and the upper part is based on a set of curves derived from German measurements (Hütter, 1976). The one matching the lower part of the curve has been selected. Also indicated (dashed line) is an average height profile of the air density, used in the construction of Figs. 3.30–3.33.

Figure 3.28 shows an example of measured wind profiles as a function of the hour of the day, averaged over a summer and a winter month. The tendency to instability due to turbulent heat transfer is seen during summer day hours (cf. the discussion above). Changes in wind direction with height have been measured (e.g. by Petersen, 1974). In the upper part of the planetary boundary layer this is simply the Ekman spiral (cf. section 2.3.2).

The continuation of the velocity profile above the Prandtl boundary layer may be important in connection with studies of the mechanisms of restoring kinetic energy lost to an energy extraction device (e.g. investigations of optimum spacing between individual devices). Extraction of energy in the upper troposphere does not appear attractive at the present technological level, but the possibility cannot be excluded. Figure 3.29 shows the 10-year average wind profile at Risø ($\phi = 56°$) matched to a profile stretching into the stratosphere, based on German measurements. The average wind speed near the tropopause compares well with the zonal means indicated in Fig. 2.45.

3.2.2 Kinetic energy in the wind

In analogy to Fig. 3.29, the height variation of the kinetic energy density is obtained in the form shown in Fig. 3.30. The density variation used in the expression (2.54) is the one indicated in Fig. 3.29, and the hourly data from Risø, Denmark, have been used to construct the kinetic energy at heights of 7–123 m, the average values of which form the lower part of the curve in Fig. 3.30. The fluctuations in wind speed diminish with height (cf. Fig. 3.36), and at $z = 123$ m the difference between the annual average of u^2 and the square of the average u is only about 1%. For this reason, the average wind speeds of Fig. 3.29 were used directly to form the average kinetic energy densities for the upper part of the figure. The kinetic energy density peaks some 3 km below the peak wind speed, owing to the diminishing air density. The curve from Fig. 3.30 has been redrawn in Fig. 3.31, on non-logarithmic scales. This illustrates the bend of the curve at a height of 1–2 km, where the near-logarithmic increase of wind speed with height is replaced by a much stronger increase.

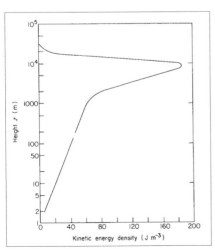

Figure 3.30. Annual average height profile of kinetic energy density at latitudes 50-56°N. The lower part of the curve is based on hourly data from Risø (Petersen, 1974), while the upper part has been calculated from the wind speeds of Fig. 3.29, neglecting any fluctuations. The contribution from the vertical component of the wind speed has been considered negligible.

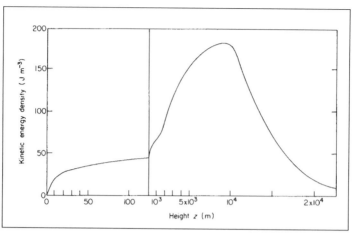

Figure 3.31. Same as Fig. 3.30, on a non-logarithmic scale (or rather two segments each with their linear scale).

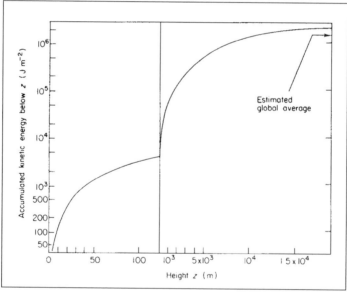

Figure 3.32. Accumulated kinetic energy below a given height, for latitudes around 50–56°N, obtained by integration of the curve in Fig. 3.31.

The kinetic energy density curve of Fig. 3.31 may readily be integrated to give the accumulated amount of kinetic energy below a given height, as shown in Fig. 3.32. This curve illustrates the advantage of extending the sampling area of a wind energy-collecting device into the higher regions of the atmosphere. The asymptotic limit of the accumulation curve is close to 2×10^6 J m^{-2} of vertical column. This order of magnitude corresponds to that of the

MORE ON WIND ENERGY IN 4.1.4, 4.3

estimate given in Fig. 2.50, of 8×10^5 J m^{-2} as a global average of zonal kinetic energy, plus a similar amount of eddy kinetic energy. The eddy kinetic energy, which is primarily that of large-scale eddies (any deviation from zonal means), may, in part, be included in the curve based on Fig. 3.29, which gives mean wind speed and not zonal mean. Since the data are for a latitude of 50–56°N, it may also be assumed that the kinetic energy is above the global average, judging from the latitude distribution of zonal winds given in Fig. 2.45.

3.2.3 Power in the wind

The energy flux passing through an arbitrarily oriented surface exposed to the wind is obtained by multiplying the kinetic energy density by $v \cdot n$, where v is the wind velocity and n is the normal to the surface. The energy flux (power) may then be written

$$E = \tfrac{1}{2} \rho u^3 \cos \theta, \tag{3.24}$$

where θ is the angle between v and n. If the vertical component of the wind velocity can be neglected, as well as the short-term fluctuations [termed v' according to (2.15)], then the flux through a vertical plane perpendicular to the direction of the wind becomes

$$E = \tfrac{1}{2} \bar{\rho}(V^*)^3, \tag{3.25}$$

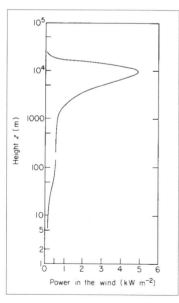

Figure 3.33. Annual average height profile of power in the wind, at latitudes 50–56°N. The lower part of the curve has been calculated from the Risø data of Petersen (1974), and the upper part is from the wind speeds of Fig. 3.29, neglecting any fluctuations.

where V^* is the average, horizontal wind velocity corresponding to a given time-averaging interval (2.13). If the wind speed or direction cannot be regarded as constant over the time interval under consideration, then an average flux (average power) may be defined by

$$E = \frac{1}{\Delta t} \int_{t_1}^{t_1 + \Delta t} \tfrac{1}{2} \rho v^3 \cos \theta \, \mathrm{d}t_2, \tag{3.26}$$

where both v and θ (and in principle ρ) may depend on the time integrand t_2. A situation often met in practice is one of an energy-collecting device, which is able to follow some of the changes in wind direction, but not the very rapid ones. Such a "yaw" mechanism can be built into the prescription for defining θ, including the effect of a finite response time (i.e. so that the energy-collecting surface is being moved towards the direction the wind had slightly earlier).

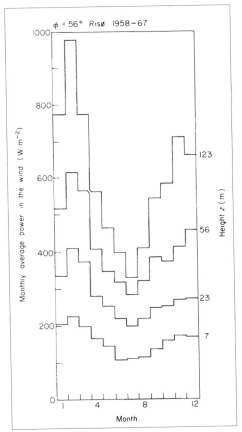

Figure 3.34. Monthly average power in the wind at Risø, Denmark, for different heights. Calculation based on ten years of data (Petersen, 1974).

Most wind data are available in a form where short-term variations have been smoothed out. For example, the Risø data used in the lower part of Fig. 3.29 are 10-min averages centred around every hour. If such data are used to construct figures for the power in the wind, as in Figs. 3.33–3.34 and the duration curves in Figs. 3.39–3.40, then two sources of error are introduced. One is that random excursions away from the average wind speed will imply a level of power, which on average is larger than that obtained from the average wind speed [owing to the cubic dependence in (3.24)]. The other is that owing to changes in wind direction the flux calculated from the average wind speed will not be passing through the same surface during the entire averaging period, but rather through surfaces of varying orientation. This second effect will tend to make the predicted power level an overestimate for any surface of fixed orientation.

Jensen (1962) overcame both of these difficulties by basing his measurements of power in the wind on a statistical ensemble of instantaneous measurements and by mounting unidirectional sensors on a yawing device which was chosen to respond to changing wind directions in a manner similar to that of the wind energy generators being considered. However, it is not believed that monthly average power or power duration curves will be substantially affected by using average data such as the ones from Risø (note that the ergodic hypothesis in section 2.3.3, relating the time averages to the statistical ensemble means, is not necessarily valid if the time averages are formed only over one 10-min interval for each hour).

In the upper part of Fig. 3.33 the power has been calculated from the average wind speeds of Fig. 3.29, using (3.25). The seasonal variations at lower heights are shown in Fig. 3.34. It is clear that the amount of seasonal variation increases with height. This indicates that the upper part of Fig. 3.33 may well underestimate the average of summer and winter energy fluxes.

Additional discussion of the power in the wind, at different geographical locations, may be found below in section 3.2.4. The overall global distribution of power in the wind at different seasons is shown in Fig. 3.27.

3.2.4 Variability in wind power

An example of short-term variations in wind speed at low height is given in Fig. 3.35. These fluctuations correspond to the region of frequencies above the spectral gap of Fig. 2.110. The occurrences of wind "gusts", during which the wind speed may double or drop to half the original value over a fraction of a second, are clearly of importance for the design of wind energy converters. On the other hand, the comparison made in Fig. 3.35 between two simultaneous measurements at distances separated horizontally by 90 m shows that little spatial correlation is present between the short-term fluctuations. Such fluctuations would thus be smoothed out by a wind energy conversion system, which comprises an array of separate units dispersed over a sufficiently large area.

The trends in amplitudes of diurnal and yearly wind speed variations are displayed in Fig. 3.36, as functions of height (Petersen, 1974). Such amplitudes are generally site-dependent, as one can deduce, for example, from Figs. 3.26 and 3.28 for diurnal variations and from Fig. 3.25 for seasonal variations. The diurnal amplitude in Fig. 3.26 diminishes with height, while the yearly amplitude increases with height. This is probably a quite general phenomenon when approaching the geostrophic wind, but the altitude dependence may depend on geographical position and local topography. At some locations, the diurnal cycle shows up as a 24-h peak in a Fourier decomposition of the wind speed data. This is the case at Risø, Denmark, as seen from Fig. 3.37, while the peak is much weaker at the lower height used in Fig. 2.110. The growth in seasonal amplitude, with height, is presumably determined by processes taking place at greater height (cf. Figs. 2.45 and 2.46) as well as by seasonal variations in atmospheric stability, etc.

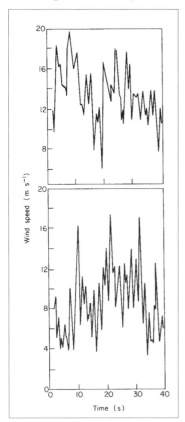

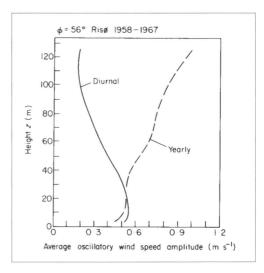

Figure 3.35 (left). Short-term variation in wind speed, from simultaneous records at two locations 90 m apart (measuring height probably in the range 5–10 m) (based on Banas and Sullivan, 1976).

Figure 3.36 (below). Height dependence of diurnal and yearly amplitude of oscillation of the wind speed at Risø. The average estimate is based on a Fourier decomposition of the data with no smoothing (Petersen, 1974).

The wind speed variance spectrum (defined in section 2.C in connection with Fig. 2.110) shown in Fig. 2.37 covers a frequency interval between the yearly period and the spectral gap. In addition to the 24-h periodicity, the amplitude of which diminishes with increasing height, the figure exhibits a group of spectral peaks with periods in the range 3–10 days. At the selected height of 56 m, the 4-day peak is the most pronounced, but moving up to 123 m, the peak period around 8 days is more marked (Petersen, 1974). It is likely that these peaks correspond to the passage time of typical meso-scale front and pressure systems.

In analysing the variability of wind speed and power during a certain period (e.g. month or a year), the measured data are conveniently arranged in the forms of frequency distributions and power duration curves, much in the same manner as discussed in section 3.1.5. Figure 3.38 gives the one-year frequency distribution of wind speeds at two Danish locations for a height of about 50 m. The wind speed frequency distribution (dashed curve) at Gedser (near the Baltic Sea) has two maxima, probably associated with winds from the sea and winds approaching from the land side (of greater roughness). However, the corresponding frequency distribution of power (top curve) does not exhibit the lower peak, as a result of the cubic dependence on wind speed. At the Risø site, only one pronounced maximum is present in the wind speed distribution. Several irregularities in the power distribution (which are not preserved from year to year) bear witness to irregular landscapes with different roughness lengths, in different directions from the meteorological tower.

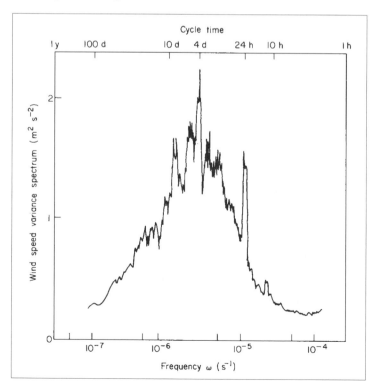

Figure 3.37. Variance spectrum of horzontal wind speeds at Risø ($\phi = 56°$ N) at a height of 56 m (based on ten years of observation; Petersen, 1974). The spectrum is smoothed by using frequency intervals of finite length (cf. Fig. 2.110 of section 2.C, where similar data for a height of 7 m is given).

Despite the quite different appearance of the wind speed frequency distribution, the power distribution for the two Danish sites peaks at roughly the same wind speed, between 10 and 11 m s^{-1}.

Power duration curves

On the basis of the frequency distributions of the wind speeds (or alternatively that of power in the wind), the power duration curves can be constructed, giving the percentage of time when the power exceeds a given value. Figures 3.39 and 3.40 give a number of such curves, based on periods of a year or more. In Fig. 3.39, power duration curves are given for four US sites which have been used or are being considered for wind energy conversion and for one of the very low-wind Singapore sites.

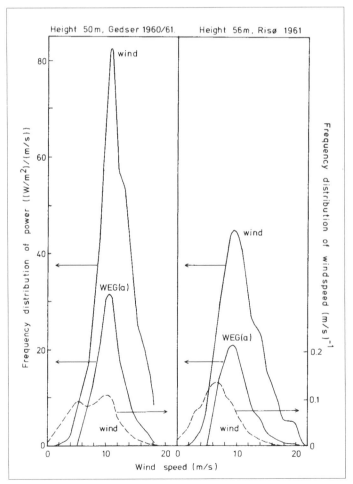

Figure 3.38. Frequency distribution of wind speed (right-hand scale) and of power for a height of about 50 m at two Danish sites. The middle curves give frequency distributions for the output of a typical wind energy generator (Sørensen, 1978a).

In Fig. 3.40, power duration curves are given for the two different Danish sites considered in Fig. 3.38, as well as for a site on the west coast of Sweden, at three different heights. These three curves for the same site have a similar shape, but in general the shape of the power duration curves in Figs. 3.39 and 3.40 depends on location. Although the Swedish Ringhals curves have non-negligible time percentages with very large power, the Danish Gedser site has the advantage of a greater frequency of medium power levels. This is not independent of conversion efficiency, which typically has a maximum as function of wind speed.

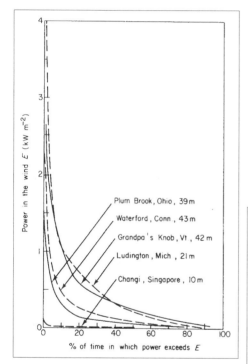

Figure 3.39 (left). One-year duration curves of power in the wind, for a number of locations and heights [Swanson *et al.*, 1975 (Plum Brook and Grandpa's Knob); Coste, 1976 (Waterford); Asmussen, 1975 (Ludington); Nathan *et al.*, 1976 (Changi)].

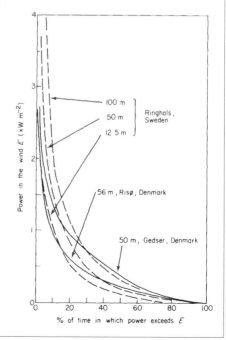

Figure 3.40 (right). One-year duration curves of power in the wind at Scandinavian sites (including different heights at the Ringhals site) [based on data from Ljungström, 1975 (Ringhals); Petersen, 1974 (Risø); Jensen, 1962 (Gedser)].

3.3 Ocean waves

The order of magnitude of the total energy in wave motion is about 10^{-3} of the total kinetic energy in the atmospheric wind systems, according to the rough estimate made in section 2.4.1 in connection with Fig. 2.86. The wave energy of about 10 kJ m^{-2} found as an annual average in the North Atlantic Ocean corresponds to the accumulated wind energy up to a height of about 200 m, according to Fig. 3.32. This implies that, although the amount of energy stored in waves is much smaller than the amount stored in wind, the wave energy may still be equivalent to the height-integrated fraction of wind energy accessible for practical use, at least at the current level of technology.

From the tentative estimates in section 2.4.1, the average turnover time for the energy in wave motion in the open ocean may be of the order of a few days. This time is consistent with average dissipation mechanisms, such as internal friction and weak wave–wave interactions, plus shoreline dissipation modes. The input of energy by the wind, on the other hand, seems to be an intermittent process which for extended intervals of time involves only slow transfer of energy between waves and turbulent wind components, or vice versa, and between wind and wave fields propagating in different directions (any angle from 0 to 2π). However, large amounts of energy may be transferred from wind to waves during short periods of time (i.e. "short periods" compared with the average turnover time). This suggests that the energy storage in waves may be varying more smoothly than the storage in wind (both waves and wind represent short-term stored solar energy, rather than primary energy flows, as discussed in connection with Fig. 2.86). As mentioned in section 2.3.2, the wave fields exhibit short-term fluctuations, which may be regarded as random. On a medium time scale, the characteristics of the creation and dissipation mechanisms may make the wave energy a more "dependable" energy source than wind energy, but on a seasonal scale, the variations in wind and wave energy are expected to follow each other (cf. the discussion topic in section 6.5.3).

3.3.1 Wave spectra

The energy spectrum $F(k)$ of a random wave field has been defined by (2.78). Since the wavelength (or wave number k) is very difficult to observe directly, it is convenient instead to express the spectrum in terms of the frequency,

$$v = \omega(k) / 2\pi = 1/T.$$

The frequency is obtained from the period, T, which for a harmonic wave equals the zero-crossing period, i.e. the time interval between successive passages of the wave surface through the zero (average) height, in the same direction. The spectral distribution of energy, or "energy spectrum" when expressed in terms of frequency, $F_1(\nu) = 2\pi F_1(\omega)$, is usually normalised to the total energy (Barnett and Kenyon, 1975), whereas the wavenumber-dependent spectrum $F(k)$, defined in section 2.3.2, is normalised to the average potential energy. Thus,

$$\int F_1(\omega)\, d\omega = W^{total} = 2 \int F(k)\, dk.$$

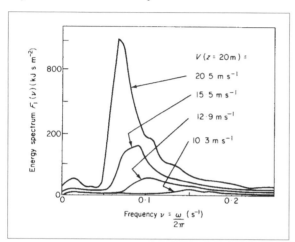

Figure 3.41. Energy spectrum of waves estimated to be "fully developed" for the Atlantic Ocean (data have been grouped according to the wind speed at a height of 20 m) (based on Moskowitz, 1964).

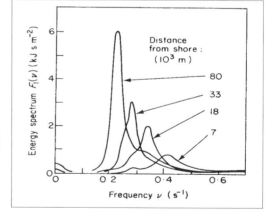

Figure 3.42. Fetch-limited energy spectrum of waves in the southern part of the North Sea. The wind is blowing from the continental shore (westward from Helgoland) (based on Hasselmann et al., 1973).

Figure 3.41 shows a set of measured energy spectra, F_1, based on data from the Atlantic Ocean (Moskowitz, 1964). The wave fields selected were judged to correspond to "fully developed waves", and data corresponding to the same wind speed at a height of 20 m were averaged in order to provide the spectra shown in the figure. It is seen that the spectral intensity increases and the frequency corresponding to the peak intensity decreases, with increasing wind speed.

Based on the similarity theory of Monin and Obukhov (1954), Kitaigorodskii (1970) suggested that $F_1(\omega)\,g^2\rho_w^{-1}\,V^{-5}$ (V being the wind speed) would be a universal function of $\omega V\,g^{-1}$, with both quantities being dimensionless. Based on the data shown in Fig. 3.41, which approximately satisfies Kitaigorodskii's hypothesis, Moskowitz (1964) suggested the following analytical form for the energy spectrum of fully developed gravity waves,

$$F_1(\omega) = 8.1 \times 10^{-3}\,\rho_w\,g^3\,\omega^5\,\exp(-0.74\,(V\{z=20\ \mathrm{m}\}\,\omega/g)^{-4}). \qquad (3.27)$$

The usefulness of this relation is limited by its sensitivity to the wind speed at a single height and by the difficulty of determining whether a given wave field is "fully developed" or not.

If the wave field is "fetch-limited", i.e. if the wind has only been able to act over a limited length, then the energy spectrum will peak at a higher frequency, and the intensity will be lower, as shown in Fig. 3.42. Hasselmann *et al.* (1973) have generalised (3.27) to situations in which the wave field is not necessarily fully developed, according to such data.

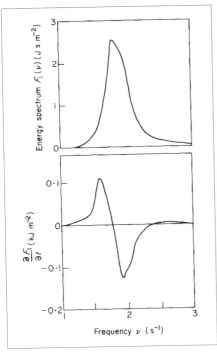

Figure 3.43. Simultaneous spectra of energy (top) and its time derivative (based on laboratory experiments by Mitsuyasu, 1968).

The position of the spectral peak will move downwards as a function of time owing to non-linear wave–wave interactions, i.e. interactions between

WAVE BASICS ARE IN 2.3.2, 2.D

different spectral components of the wave field, as discussed in connection with (2.80) (Hasselmann, 1962). This behaviour is clearly seen in the laboratory experiments of Mitsuyasu (1968), from which Fig. 3.43 shows an example of the time derivative of the energy spectrum, $\partial F_1 (v)/\partial t$. Energy is transferred from middle to smaller frequencies. In other experiments, some transfer is also taking place in the direction of larger frequencies. Such a transfer is barely present in Fig. 3.43. The shape of the rate-of-transfer curve is similar to the one found in later experiments for real sea (Hasselmann *et al.*, 1973), and both are in substantial agreement with the non-linear theory of Hasselmann.

Observations of ocean waves often exist in the form of statistics on the occurrences of given combinations of wave height and period. The wave height may be expressed as the significant height, H_s, or the root mean square displacement of the wave surface, H_{rms}. For a harmonic wave, these quantities are related to the surface amplitude, a, by

$$a^2 = 2\,(H_{rms})^2 = H_s^2\,/8.$$

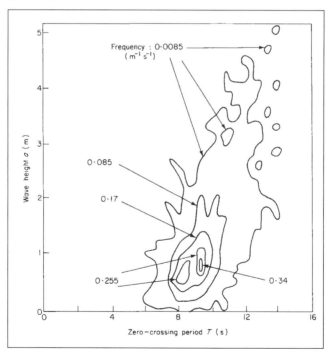

Figure 3.44. Frequency distribution of wave heights and zero-crossing periods for Station India (59°N, 19°W) in the North Atlantic [based on one year of observation by Draper and Squire (1967), quoted by Salter, 1974].

The period is usually given by the observed zero-crossing period. Figures 3.44 and 3.45 give examples of such measured frequency distributions, based on a year's observation at the North Atlantic Station "India" (59°N, 19°W, the data quoted by Salter, 1974) and at the North Sea Station "Vyl" (55.5°N, 5.0°E, based on data discussed in problem section 6.5.3). The probability of finding periods between 8 and 10 s and wave heights, a, between 0.3 and 1.5 m is quite substantial at the site "India", of the order of 0.3. Also the probability of being within a band of zero-crossing periods with a width of about 2 s and a centre which moves slowly upwards with increasing wave height is nearly unity (the odd values characterising the contour lines of Fig. 3.44 are due to the original definition of sampling intervals in terms of feet and H_s).

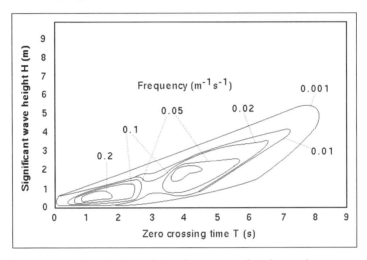

Figure 3.45. Frequency distribution of significant wave heights and zero-crossing periods for Station Vyl (55°N, 5°E) in the North Sea (based on one year of observation by Nielsen, 1977).

3.3.2 Power in the waves

The power of a harmonic wave, i.e. the energy flux per unit length of wave crest, passing through a plane perpendicular to the direction of propagation, is from (2.84)

$$P = \rho_w g\, U_w a^2/4 \;=\; \rho_w g^2 T a^2/(8\pi) = \rho_w g^2 a^2/(4\omega). \tag{3.28}$$

For the spectral distribution of energy given by the function F_1 (section 3.3.1), each component must be multiplied by the group velocity $\partial\,\omega(k)/\partial\,k$. Taking $\omega = (gk)^{1/2}$ for ocean gravity waves, the group velocity becomes $g/(2\omega)$, and the power becomes

$$P = \int \frac{\partial\omega}{\partial k} F_1(\omega)\,d\omega = \tfrac{1}{2} g \int \frac{F_1(\omega)}{\omega}\,d\omega. \tag{3.29}$$

Based on observed energy spectra, F_1, as a function of frequency or period $T = 2\pi/\omega$, the power distribution [(3.29) before integration] and total power may be constructed. Figure 3.46 gives the power distribution at the North Atlantic station also considered in Fig. 3.44. Using data for an extended period, the average distribution (labelled "year") has been calculated and, taking out data for the periods December–February and June–August, the curves labelled "winter" and "summer" have been constructed (Mollison *et al.*, 1976).

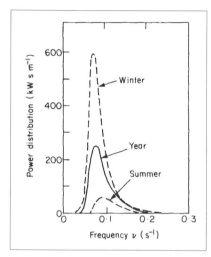

Figure 3.46 (left). Frequency distribution of wave power, based on one year of observations (full line) or summer or winter data only (dashed lines), for Station India (59°N, 19°W) in the North Atlantic. The yearly average power is 91 kW m^{-1} (based on Mollison *et al.*, 1976).

Figure 3.47 (below). Annual average wave power (in kW m^{-1}) for selected sites (based on United Kingdom Energy Technology Support Unit, 1976).

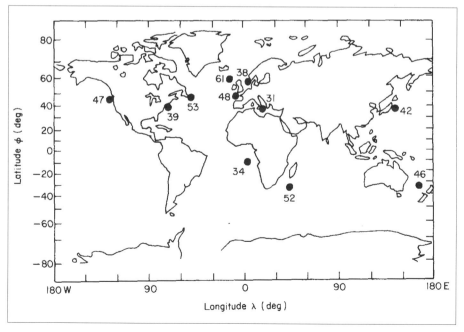

Compared with Fig. 3.34, for example, it is seen that the seasonal variations in wave power at the open ocean site considered in Fig. 3.46 are quite substantial and equivalent to the seasonal variations in wind power at considerable height (maybe around 100 m, considering that the roughness length over the ocean is smaller than at the continental site considered in Fig. 3.36).

Figure 3.47 summarises some of the data available on the geographical distribution of wave power. The figures given are yearly mean power at a specific location, and most of the sites chosen are fairly near to a coast, although still in open ocean. The proximity of a shore is considered important in connection with the potential for energy extraction from wave motion. Whether such a condition will be maintained depends on the development of suitable transmission methods for long-range energy transfer (see Chapter 5).

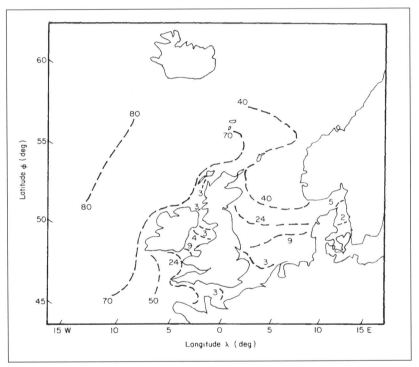

Figure 3.48. Contours of estimated equal annual average wave power in the North Sea, adjacent straits and a part of the North Atlantic Ocean (unit: kW m^{-1}) (based on United Kingdom Energy Technology Support Unit, 1976).

In Fig. 3.48 a more detailed map of wave power availability for the north European region is shown, based on the iso-power lines estimated in initial assessments of wave power in the United Kingdom, supplemented with estimates based on data for the waters surrounding Denmark. One might note the rapid decrease in power when passing the Hebrides in approaching the Scottish coast from the Atlantic and also when moving southwards through the North Sea.

The variability of wave power may be described in terms similar to those used for wind energy. Figure 3.49 shows the power duration curves for Station India in the North Atlantic, based on all year or the summer or winter periods only, as in Fig. 3.46. Again, the occurrence of periods of low and high power depends markedly on seasonal changes.

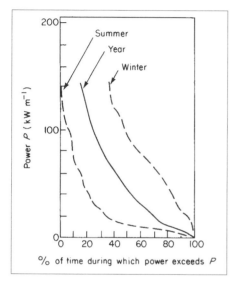

Figure 3.49. Time duration curves for wave power at Station India (59°N, 19°W) in the North Atlantic for the whole year (solid line) or only summer or winter (dashed lines) (based on Mollison *et al.*, 1976).

Waves in a climatic context

As in the case of wind energy, little is known about the possible impact of large-scale energy extraction from wave power. One may argue that the total amount of energy involved is so small compared to the energy exchanged by atmospheric processes that any climatic consequence is unlikely. However, the exchange of latent and sensible heat, as well as matter, between the ocean and the atmosphere may, to a large extent, depend on the presence of waves. In particular, the rates of transfer may be high in the presence of breaking waves, and the extraction of energy from the wave motion may prevent the waves from developing the wave profile to the point of breaking. Thus, a study of the environmental implications of wave energy utilisation, which seems to lend itself very naturally to computer simulation techniques, should be undertaken in connection with proposed energy extraction schemes. As suggested in previous sections, this is likely to be required for any large-scale use of renewable energy, despite the intuitive feeling that such energy sources are "nonpolluting". Yet the limited evidence available, mostly deriving from analogies to natural processes, does suggest that the results of more detailed analyses will be that renewable energy flows and stores can be utilised in quantities exceeding present technological capability, without worry about environmental or general climatic disturbances.

3.4 Water flows and tides

3.4.1 Ocean currents

The maximum speed in the centre of the Gulf Stream is about 2 m s^{-1}, corresponding to an energy density $\frac{1}{2}\,\rho_w\,V_w^2 = 2$ kJ m^{-3} and a power of $\frac{1}{2}\,\rho_w\,V_w^3 = 4$ kW m^{-2}. This power level, for example, approaches that of wave power at reasonably good sites, taken as power per metre of wave crest rather than per square metre perpendicular to the flow, as used in the above case. However, high average speed of currents combined with stable direction is found only in a few places. Figure 3.50 shows the isotachs in a cross section of the Gulf Stream, derived from a single set of measurements (in June 1938; Neumann, 1956). The maximum current speed is found at a depth of 100–200 m, some 300 km from the coast. The isotachs expand and become less regular further north, when the Gulf Stream moves further away from the coast (Niiler, 1977).

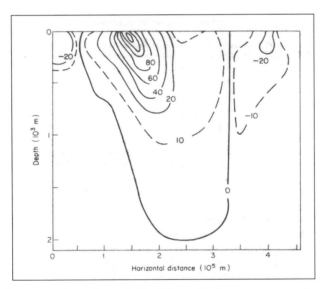

Figure 3.50. Contours of equal speed (in 10^{-2} m s^{-1}) along a cross section through the Gulf Stream from Bermuda to the US coast. The measurements were performed over a limited period of time (in June 1938, by Neumann, 1956), but they are consistent with other measurements in the same region (e.g. Florida Strait measurements by Brooks and Niiler, 1977).

Even for a strong current like the Gulf Stream, the compass direction at the surface has its most frequent value only slightly over 50% of the time (Royal Dutch Meteorological Institute, as quoted by Neumann and Pierson, 1966), and the power will, on average, deviate from that evaluated on the basis of average current speeds (as is the case for wind or waves) owing to the inequality of $<V^3>$ and $<V>^3$.

The geographical distribution of the strongest surface currents is indicated in Fig. 3.51. The surface currents in the Atlantic Ocean, along with the horizontal currents at three different depths, are sketched in Fig. 3.52. The current speeds indicated are generally decreasing downwards, and the preferred directions are not the same at different depths. The apparent "collision" of oppositely directed currents, e.g. along the continental shelf of Central America at a depth of around 4 km, conceals the vertical motion which takes place in most of these cases. Figure 3.53 gives a vertical cross section with outlined current directions. This figure shows how the oppositely directed waters "slide" above and below each other. The coldest water is flowing along the bottom, while the warmer water from the North Atlantic is sliding above the cold water from the Antarctic.

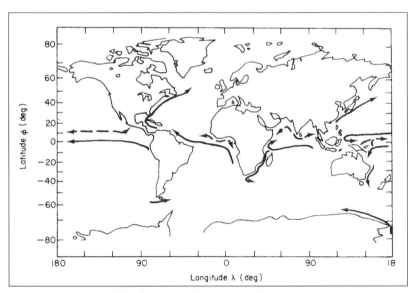

Figure 3.51. Indication of the location of strong surface currents (approximately defined as having average speeds above 0.5 m s⁻¹). Currents drawn with dashed lines are characterised by seasonal changes in direction (based on Defant, 1961).

Variability in current power

Although the general features of circulation in the open ocean were described in section 2.3.2, the particular topography of coastal regions may have an important influence on currents. As an example, water forced through a narrow

strait may acquire substantial speeds. The strait Storebælt ("Great Belt") between two Danish isles, which provides an outlet from the Baltic Sea, may
serve as an illustration. It is not extremely narrow (roughly 20 km at the
measurement site), but narrow enough to exhibit only two current directions,
separated by 180°. The currents may seem fairly steady, except for periods
when the direction is changing between north-going and south-going velocities, but when the energy flux is calculated from the third powers of the current speeds, the fluctuations turn out to be substantial. Figure 3.54, which
gives the power at 3-h intervals, during two weeks of January 1972, clearly illustrates this.

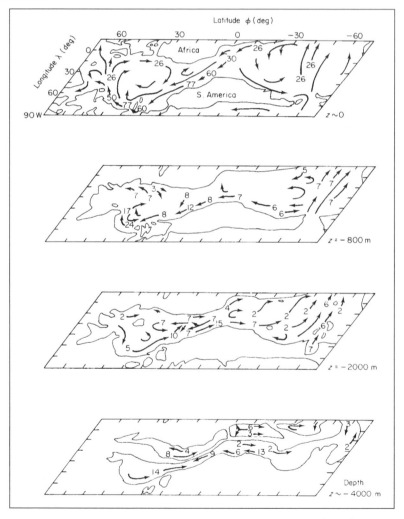

Figure 3.52. Indication of average horizontal current speeds (in 10^{-2} m s^{-1}) in the Atlantic Ocean for different depths (based on Defant, 1961).

BASIC MATERIAL IN SECTION 2.3.2

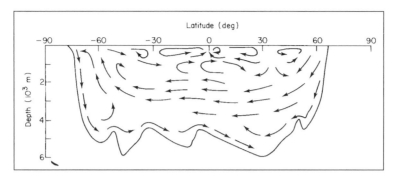

Figure 3.53. Indication of average current directions within a vertical cross section of the Atlantic Ocean (based on Neumann and Pierson, 1966).

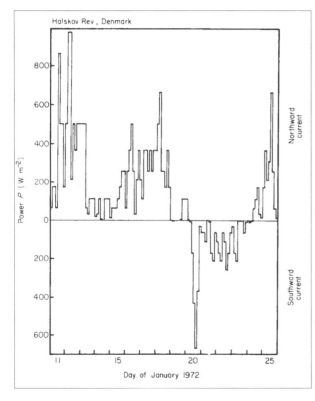

Figure 3.54. Power of surface current, based on observations made at 3-h intervals, for Halskov Rev, Denmark, during a 15-day period in 1972 (the strait is narrow enough to exhibit only two opposite current directions) (based on Danish Meteorological Institute, 1973).

Figure 3.55 shows, again for the Halskov Rev position in Storebælt, the variation in current speed with the hour of the day, based on one-month averages. A 12-h periodicity may be discerned, at least during January. This period is smaller than the one likely to be found in the open sea due to the motion of water particles in stationary circles under the influence of the Coriolis force [see (2.61)] and having a period equal to 12 h divided by sin ϕ (ϕ is the latitude).

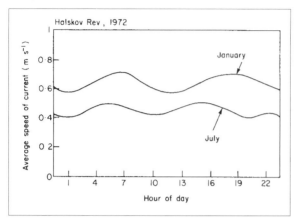

Figure 3.55. Dependence of average current speed on the hour of the day, for a summer and a winter month, for Halskov Rev, Denmark (based on Danish Meteorological Institute, 1973).

In Fig. 3.56, the frequency distributions of current speed and power are shown for a summer and a winter month. These curves are useful in estimating the performance of an energy extraction device, and they can be used, for example, to construct the power duration curve of the current motion, as shown in Fig. 3.57. This is the power duration curve of the currents themselves. That of an energy extracting device will have to be folded with the efficiency function of the device.

The peak in the frequency distribution of power is at a lower current speed for July than for January, and the average power is 93 W m^{-2} in July, as compared with 207 W m^{-2} in January (the corresponding kinetic energy densities are 138 and 247 J m^{-3}). This indicates that the fluctuations around the average values have a substantial effect on the energy and, in particular, on the power, because from the average current speeds, 0.46 m s^{-1} (July) and 0.65 m s^{-1} (January), the calculated kinetic energy densities would have been 108 and 211 J m^{-3}, and the power would have taken the values 50 and 134 W m^{-2}.

Few locations, whether in coastal regions or in open oceans, have higher average current speeds than the Danish location considered in Figs. 3.54–3.57, so that average power levels in the range 100–200 W m^{-2} are likely to be more representative than the 4000 W m^{-2} found (at least at the time of the measurement reported in Fig. 3.50) in the core of the Gulf Stream. This means that at many locations the power in currents is no greater than that found in the wind at quite low heights (cf. Fig. 3.34). Also, the seasonal variations and fluctuations are similar, which is not unexpected for wind-driven currents. For the currents at greater depths this may not hold true, partly because of the smoothing due to a long turnover time and partly because not all the deep sea motion is wind driven, but may also be associated with temperature and salinity gradients, as discussed in section 2.3.2.

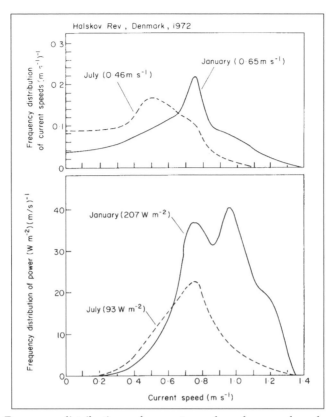

Figure 3.56. Frequency distributions of current speeds and power, based on a summer and a winter month (full and dashed curves, respectively), for Halskov Rev, Denmark. The data (taken from Danish Meteorological Institute, 1973) have been smoothed in calculating the distributions. The monthly average speed and power are indicated on the figure (in parentheses).

The power duration curves in Fig. 3.57 may be compared to those of wind power shown in Figs. 3.39 and 3.40. The fraction of time in which the monthly average power is available in the Halskov Rev surface current is 0.3 in both January and July. The overall average power of about 150 W m^{-2} is available for about 45% of the time in January, but only 17% of the time in July.

3.4.2 River flows, hydropower and elevated water storage

The kinetic energy of water flowing in rivers or other streams constitutes an energy source very similar to that of ocean currents. However, rather than being primarily wind driven or caused by differences in the state of the water masses themselves, the river flows are part of the hydrological cycle depicted in Fig. 2.61. Water vapour evaporated into the atmosphere is transported and eventually condensed. It reaches the ground as water or ice, at the elevation of the particular location. Thus, the primary form of energy is potential. In the case of ice and snow, a melting process (using solar energy) is usually necessary before the potential energy of elevation can start to transform into kinetic energy. The origin of many streams and rivers is precisely the ice melting process, although they are joined by ground water flows along their subsequent route. The area from which a given river derives its input of surface run-off, melt-off and ground water is called its "drainage basin".

The flow of water in a river may be regulated by means of dam building, if suitable reservoirs exist or can be formed. In this way the potential energy of water stored at an elevation can be transformed into kinetic energy (e.g. driving a turbine) at the times most convenient with respect to utilisation.

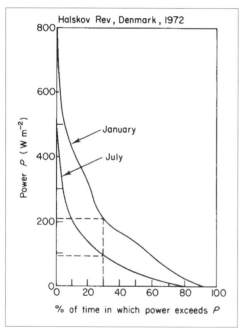

Figure 3.57. Power duration curves for power in currents at Halskov Rev, Denmark, based on a summer and a winter month of 1972. Thin, dashed lines indicate the monthly average power and the percentage of time during which it is available.

An estimate of the potential hydro-energy of a given geographical region could in principle be obtained by hypothetically assuming that all precipitation was retained at the altitude of the local terrain and multiplying the gravitational potential mg by the height above sea-level. According to Fig. 2.61, the annual precipitation over land amounts to about 1.1×10^{17} kg of water, and taking the average elevation of the land area as 840 m (Sverdrup et al., 1942), the annually accumulated potential energy would amount to 9×10^{20} J, corresponding to a mean energy flux (hydropower) of 2.9×10^{13} W.

Collection of precipitation is not usually performed as part of hydropower utilisation, but rather the natural processes associated with soil moisture and vegetation are allowed to proceed, leading to a considerable re-evaporation and some transfer to deeper lying ground water which eventually reaches the oceans without passing through the rivers (see Fig. 2.61). The actual run-off from rivers and overground run-off from polar ice caps comprise, on average, only about 0.36 of the precipitation over land, and the height determining the potential energy may be lower than the average land altitude, namely, that given by the height at which the water enters a river flow (from ground water or directly). The minimum size of stream or river which can be considered useful for energy extraction is, of course, a matter of technology, but these general considerations would seem to place an upper limit on the hydro-energy available of roughly 3×10^{20} J y^{-1}, corresponding to a power average of below 10^{13} W (10 TW).

If, instead of using average precipitation and evaporation rates together with average elevation, the geographical variations of these quantities are included, the result is also close to 3×10^{20} J y^{-1} or 10^{13} W. These figures are derived from the integral over all land areas,

$$\frac{\mathrm{d}W^{pot}}{\mathrm{d}t} = \int_{land\ area} (r-e)\,gz\,\mathrm{d}A,$$

where r and e are the rates of precipitation and evaporation (mass of water, per unit of area and time), g is the gravitational acceleration and z is the height above sea level. The observed annual mean precipitation and evaporation rates quoted by Holloway and Manabe (1971) were used in the numerical estimate.

Geographical distribution of hydropower resources

A different estimate of hydropower potential is furnished by counts of actual rivers with known or assumed water transport and falling height. According to such an estimate by the World Energy Conference (1974, 1995), the installed or installable hydro-generation capacity resource at average flow conditions may, in principle, amount to 1.2×10^{12} W, for both large installations and smaller ones and down to "micro-hydro" installations of around 1 MW. On the other hand, it is unlikely that environmental and other considerations will allow the utilisation of all the water resources included in the estimate. The World Energy Conference (1995) estimates 626 GW as a realistic reserve (including an already installed capacity producing, on average, 70 GW).

Figure 3.58 gives an idea of the geographical distribution of the hydro-power resources on a national basis. The largest remaining resources are in South America. The figures correspond to average flow conditions, and the seasonal variations in flow are very different for different regions. For example, in Zaire practically all the reserves would be available year round, whereas in the USA only 30% can be counted on during 95% of the year.

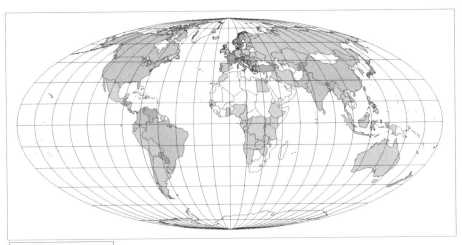

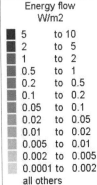

Energy flow W/m2	
■ 5	to 10
■ 2	to 5
■ 1	to 2
■ 0.5	to 1
■ 0.2	to 0.5
■ 0.1	to 0.2
■ 0.05	to 0.1
■ 0.02	to 0.05
■ 0.01	to 0.02
■ 0.005	to 0.01
■ 0.002	to 0.005
■ 0.0001 to	0.002
all others	

Figure 3.58 (above). Hydropower potential average production (W m⁻²), including existing plants, plants under construction, and planned and contemplated installations, both for large- and small-scale power schemes (based on data from World Energy Council, 1995).

Figure 3.59 (below). Seasonal variation in the power associated with the water flow into Norwegian hydropower reservoirs for a typical year (1985) and a "dry year" (1990) (based on Andersen, 1997; Meibom *et al.*, 1999).

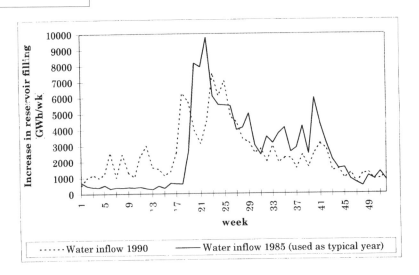

Figure 3.59 gives seasonal variations (for two years) in the flow into the existing hydropower reservoirs in Norway, a country where the primary filling of reservoirs is associated with the melting of snow and ice during the late spring and early summer months.

Environmental impact

The environmental impact of non-regulated hydro-generation of power is mainly associated with preventing the migration of fish and other biota across the turbine area, but the building of dams in connection with large hydro facilities may have an even more profound influence on the ecology of the region, in addition to introducing accident risks. For large reservoirs, there has been serious destruction of natural landscapes and dislocation of populations living in areas to be flooded. There are ways to avoid some of the problems. Modular construction, where the water is cascaded through several smaller reservoirs, has been used, e.g. in Switzerland, with a substantial reduction in the area modified as a result. The reservoirs need not be constructed in direct connection with the generating plants, but can be separate installations placed in optimum locations, with a two-way turbine that uses excess electric production from other regions to pump water up into a high-lying reservoir. When other generating facilities cannot meet demand, the water is then led back through the turbines. This means that although the water cycle may be unchanged on an annual average basis, considerable seasonal modifications of the hydrological cycle may be involved. The influence of such modifications on the vegetation and climate of the region below the reservoir, which would otherwise receive a water flow at a different time, has to be studied in each individual case. The same may be true for the upper region, for example, owing to increased evaporation in the presence of a full reservoir.

Although these modifications are local, they can influence the ecosystems with serious consequences for man. An example is provided by the building of the Aswan Dam in Egypt, which has allowed water snails to migrate from the Nile delta to the upstream areas. The water snails may carry parasitic worms causing schistosomiasis, and this disease has actually spread from the delta region to Upper Egypt since the building of the dam (Hayes, 1977).

It is unlikely that hydropower utilisation will ever be able to produce changes in the seasonal hydrological cycle, which could have global consequences, but no detailed investigation has yet been made. Such a study could proceed along the same lines as the investigation of the influence of deforestation, shown in Fig. 2.93.

3.4.3 Tides

The average rate of dissipation of tidal energy, as estimated from the slowing down of the Earth's rotation, is about 3×10^{12} W. Of this, about a third can be accounted for by frictional losses in definite shallow sea regions, bays and estuaries, according to Munk and MacDonald (1960).

In order to understand the concentration of tidal energy in certain coastal regions, a dynamic picture of water motion under the influence of tidal forces must be considered. The equations of motion for the oceans, (2.60) and (2.61), must be generalised to include the acceleration due to tidal attraction, i.e. F^{tidal}/m where an approximate expression for the tidal force is given in (2.67). Numerical solutions to these equations (see e.g. Nihoul, 1977) show the existence of a complicated pattern of interfering tidal waves, exhibiting in some points zero amplitude (nodes) and in other regions deviations from the average water level far exceeding the "equilibrium tides" of a fraction of a metre (section 2.D). These features are in agreement with observed tides, an example of which is shown in Fig. 3.60. Newer data based on satellite measurements can be followed in near-real time on the internet (NASA, 2004).

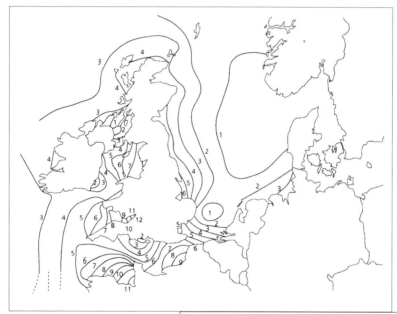

Figure 3.60. Tidal range H (difference between highest and lowest level in m) of the semi-diurnal lunar tides in the North Sea and adjacent straits (based on Kiilerich, 1965; Cavanagh et al., 1993).

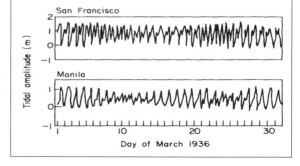

Figure 3.61. Examples of the time-development of tidal amplitudes at two different locations for the same month of 1936 (based on Defant, 1961).

The enhancement of tidal amplitudes in certain bays and inlets can be understood in terms of resonant waves. Approximating the inlet by a canal of constant depth h, the phase velocity of the tidal waves is $U_t = (gh)^{1/2}$ (Wehausen and Laitone, 1960), and the wavelength is

$$\lambda_t = T_t U_t,$$

where T_t is the period. For the most important tidal wave, T_t equals half a lunar day, and the condition for resonance may be written

$$L = i\,\lambda_t\,/4 = 0.25\,i\,T_t\,g^{1/2}\,h^{1/2}, \tag{3.30}$$

where i is an integer, so that the length L of the inlet is a multiple of a quarter wavelength. For $i = 1$, the resonance condition becomes $L = 34973h^{1/2}$ (L and h in metres). Bays and inlets satisfying this condition are likely to have high tidal ranges, with the range being the difference between highest and lowest water level. An example of this is the Severn inlet near Bristol in the UK, as seen from Fig. 3.60. Cavanagh, Clarke and Price (1993) estimate the total European potential to be 54 GW or about 100 TWh y^{-1}, of which 90% is in France and the UK.

As discussed in section 2.D, the tides at a given location cannot be expected to have a simple periodicity, but rather are characterised by a superposition of components with different periods, the most important of which being equal to one or one-half lunar or solar day. As a function of time, the relative phases of the different components change, leading to regularly changing amplitudes, of which two different patterns are shown in Fig. 3.61. The upper one is dominated by the half-day period, while the lower one is dominated by the full-day period.

If the water level at high tide, averaged over an area A, is raised by an amount H over the level at low tide, then the potential energy involved is

$$W^{tidal} = \rho_w H A g H,$$

and the maximum power which could be extracted or dissipated would, as an average over a tidal period T_t, be

$$P^{tidal} = \rho_w\,g\,A\,H^2\,T_t^{-1}. \tag{3.31}$$

Based on measured tidal ranges, and on an estimate of the area A of bay or inlet which could reasonably be enclosed by a barrage with the purpose of utilising the energy flow (3.31), a number of favourable sites have been identified, as shown in Fig. 3.62. These include only localities with considerable concentration of tidal energy, considering that if the tidal range decreases, the area to be enclosed in order to obtain the same power quadratically increases, and the length of barrage will have to be correspondingly greater. For the same reason, sites of adequate tidal range, but no suitable bay which could be enclosed by a reasonably small length of barrage have been excluded. Of course, the term "reasonable" rests on some kind of economic judgement, which may be valid only under given circumstances. It is estimated that 2–3 GW may be extracted in Europe, half of which at costs in the range 10–20 euro-cents (or US cents) per kWh (Cavanagh *et al.*, 1993, using costing methodology of Baker, 1987), and 20–50 GW in North America (Sorensen and MacLennan, 1974; Bay of Fundy Tidal Power Review Board, 1977).

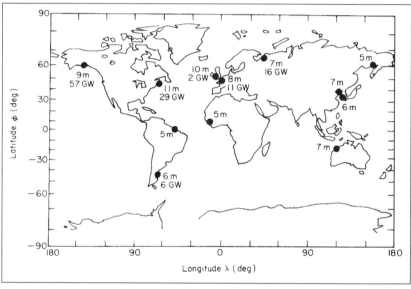

Figure 3.62. Tidal range for selected locations and the estimated average tidal power for each bay in a number of cases where utilisation has been proposed (based on King Hubbert, 1969; Gray and Gashus, 1972; Sorensen and MacLennan, 1974).

Environmental impacts may arise from utilisation of tidal power. When the La Rance tidal plant was built in the 1960s, the upper estuary was drained for water during two years, a procedure that would hardly be considered environmentally acceptable today. Alternative building methods using caissons or diaphragms exist, but in all cases the construction times are long and careful measures have to be taken to protect the biosphere (e.g. when stirring up mud from the estuary seabed). Generally, the coastal environment is affected by the building and operation of tidal power plants, both during construction and to a lesser extent during operation, depending on the layout (fish bypasses etc., as known from hydropower schemes). Some fish species may be killed in the turbines, and the interference with the periodic motion of bottom sand may lead to permanent siltation problems (and it has at la Rance).

The total estimated power of about 120 GW at the best sites throughout the world may be larger than what can be considered economic, but smaller than the amount of tidal energy actually available. It is still 12% of the above-

BASIC MATERIAL WERE IN 2.D

mentioned estimate of the total dissipation of tidal energy in the vicinity of land, and it is unlikely that all the coastal sites yielding a total of 1000 GW would be suitable for utilisation, so the order of magnitude is likely to be correct. The 2000 GW remaining relative to the tidal power derived from astronomical data (Munk and MacDonald, 1960) presumably becomes dissipated in the open ocean.

The maximal tidal power believed to be accessible, as well as the total resource estimate, is about 10% of the corresponding figures for hydropower, discussed in section 3.4.2. Tidal variations are coupled to river run-off and sea level rise due to global greenhouse warming (Miller and Douglas, 2004).

3.5 Heat flows and stored heat

A large fraction of the incoming solar radiation is stored as heat near the surface of the Earth. According to Fig. 2.16, on average 47% is absorbed by the oceans and continents. The more detailed picture in Fig. 2.86 shows that 38% is absorbed by the oceans, 9% by the continents and 24% by the atmosphere. Chapter 2 dealt with some of the ways in which this energy could be distributed and eventually dissipated. Previous sections dealt with the incoming radiation itself and with the kinetic energy in atmospheric and oceanic motion derived from the solar input by a number of physical processes. Storage in terms of potential energy has also been considered in connection with the processes of lifting water to a higher elevation, either in waves or in evaporated water, which may later condense and precipitate at a higher topographical level. Still, the energy involved in such kinetic and potential energy-creating processes is much less that the latent and sensible heat fluxes associated with the evaporation and condensation processes themselves and with the conversion of short-wavelength radiation to stored heat. The heat will be re-radiated as long-wavelength radiation in situations of equilibrium, but the temperature of the medium responsible for the absorption will rise, and in some cases the temperature gradients between the absorbing regions and other regions (deep soil, deep ocean, etc.), which do not themselves absorb solar radiation, cause the establishment of significant heat flows.

Utilisation of heat flows and stored heat may be direct if the desired temperature of use is no higher than that of the flow or storage. If this is not so, two reservoirs of different temperature may be used to establish a thermodynamic cycle yielding a certain amount of work, limited by the second law of thermodynamics. An alternative conversion scheme makes use of the heat pump principle by expending work added from the outside. Such conversion methods will be considered in Chapter 4, while the focus here will be on identifying those heat sources, which look most suitable for utilisation. The solar energy stores and flows will be surveyed in section 3.5.1, whereas section 3.5.2

will deal with the energy stored in the interior of the Earth and the corresponding geothermal flows.

3.5.1 Solar-derived heat sources

The ability of land and sea to act as a solar energy absorber and further as an energy store, to a degree determined by the local heat capacity and the processes, which may carry heat away, is of profound importance for the entire biosphere. For example, food intake accounts only for 25–30% of man's total income of energy during a summer day in central Europe (Budyko, 1974). The rest is provided by the body's absorption of solar energy. The biological processes that characterise the present biosphere have vanishing rates if the temperature departs from a rather narrow range of about 270–320 K. The present forms of life thus depend on the greenhouse effect maintaining temperatures at the Earth's surface in this range, at least during a fraction of the year ("growing season"), and it is hard to imagine life on the frozen "white Earth" (see section 2.4.2) which would result if the absorption processes were minimised.

Utilisation of heat stores and flows may then be defined as uses in addition to the benefits of the "natural" thermal regime of the Earth, but often no sharp division between "natural" and "artificial" uses can be drawn. For this reason, an assessment of the "magnitude" of the resource is somewhat arbitrary, and in each case it must be specified what is included in the resource base.

The total energy absorption rates are indicated in Fig. 2.86, and it is clear that the oceans are by far the most significant energy accumulators. The distributions of yearly average temperatures along cross sections of the main oceans are shown in Figs. 2.63–2.65.

The potential of a given heat storage for heat pump usage depends on two temperatures: that of the storage and the required usage temperature. This implies that no energy "amount" can be associated with the energy source, and a discussion of heat pump energy conversion will therefore be deferred until Chapter 4. An indication of possible storage temperatures may be derived from Figs. 2.63–2.65 for the oceans, Fig. 2.104 for a continental site, and Fig. 2.28 for the atmosphere.

For utilisation without addition of high-grade mechanical or electric energy, two reservoirs of different temperature must be identified. In the oceans, Figs. 2.63–2.65 indicate the presence of a number of temperature gradients, most noticeable at latitudes below 50° and depths below 1000–2000 m. Near the Equator, the temperature gradients are largest over the first few hundred metres, and they are seasonally stable, in contrast to the gradients in regions away from the Equator, which are largest during summer and approach zero during winter. At still higher latitudes, there may be ice cover during a part or all of the year. The temperature at the lower boundary of the ice, i.e. at the water–ice interface, is quite constantly equal to 271.2 K (the freezing point of sea water), and in this case a stable temperature difference can be expected between the water and the atmosphere, particularly during winter.

It follows from Fig. 2.86 that over half the solar energy absorbed by the oceans is used to evaporate water. Of the remaining part some will be transferred to the atmosphere as sensible heat convection, but most will eventually be re-radiated to the atmosphere (and maybe to space) as long-wavelength radiation. The length of time during which the absorbed energy will stay in the oceans, before being used in one of the above ways, determines the temperature regimes. At the ocean surface, residence times of a fraction of a day will be associated with the diurnal temperature variations, while residence times normally increase with depth and reach values of several hundred years in the deepest regions (section 2.3.2).

If oceanic temperature gradients are used to extract energy from the surface region, a cooling of this region may result, unless the currents in the region are such that the cooling primarily takes place elsewhere. A prediction of the climatic impact of heat extraction from the oceans thus demands a dynamic investigation, such as the one offered by the general circulation models discussed in section 2.3.2. The energy extraction would be simulated by an additional source term in (2.64), and attention should be paid to the energy loss from the ocean's surface by evaporation, which would decrease at the places with lowered temperatures. The extraction of energy from the oceans may for this reason lead to an increase of the total downward energy flux through the surface, as noted by Zener (1973). In addition to possible climatic effects, oceanic energy extraction may have an impact on the ecology, partly because of the temperature change and partly because of other changes introduced by the mixing processes associated with at least some energy conversion schemes (e.g. changing the distribution of nutrients in the water). Assuming a slight downward increase in the concentration of carbon compounds in sea water, it has also been suggested that artificial upwelling of deep sea water (used as cooling water in some of the conversion schemes discussed in Chapter 4) could increase the CO_2 transfer from ocean to atmosphere (Williams, 1975).

The power in ocean thermal gradients

In order to isolate the important parameters for assessing the potential for oceanic heat conversion, Fig. 3.63 gives a closer view of the temperature gradients in a number of cases with large and stable gradients. For the locations at the Equator, seasonal changes are minute, owing to the small variations in solar input. For the locations in Florida Strait, the profiles are stable, except for the upper 50–100 m, owing to the transport of warm water originating in the tropical regions by the Gulf Stream. The heat energy content relative to some reference point of temperature T_{ref} is $c_V(T - T_{ref})$, but only a fraction of this can be converted to mechanical work, and the direct use of water at temperatures of about 25°C above ambient is limited, at least in the region near the Equator. Aiming at production of mechanical work or electricity, the heat value must be multiplied with the thermodynamic efficiency (cf. Chapter 4), i.e. the maximum fraction of the heat at temperature T which can be converted to work. This efficiency, which would be obtained by a hypothetical conversion device operating in accordance with an ideal Carnot cycle, is (cf. section 4.1.1)

$$\eta_{Carnot} = (T - T_{ref})/T, \qquad (3.32)$$

where the temperatures should be in K.

Taking the reference temperature (the "cold side") as $T_{ref} = 6°C$, corresponding to depths of 380 m (Gulf Stream, 10 km from coast), 680 m (Gulf Stream, 50 km from coast), 660 m (Atlantic Ocean), 630 m (Pacific Ocean) and 1000 m (Indian Ocean) (see Fig. 3.63), one obtains the maximum energy which can be extracted from each cubic metre of water at temperature T (as a function of depth), in the form of mechanical work or electricity,

$$\eta_{Carnot} \, c_V \, (T - T_{ref}).$$

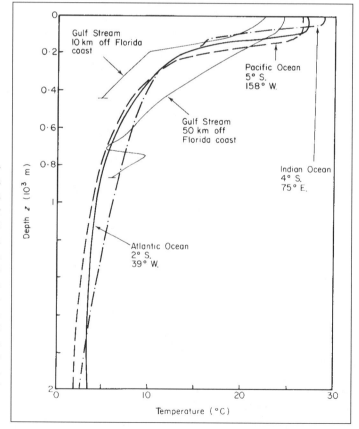

Figure 3.63. Temperature profiles for equatorial sites in the major oceans, as well as for the Gulf Stream at the Florida Strait. The profiles are derived from time-limited observations and may not represent annual averages (based on Neumann and Pierson, 1966; Sverdrup *et al.*, 1942).

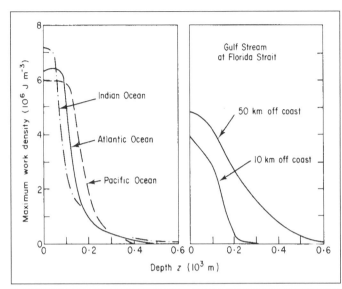

Figure 3.64. Maximum work density for the locations of Fig. 3.63, corresponding to an ideal Carnot machine extracting work from the temperature difference between the temperature at a given depth and a cold reservoir at constant temperature 6°C (the depth at which such cooling water might be collected can be found from Fig. 3.63).

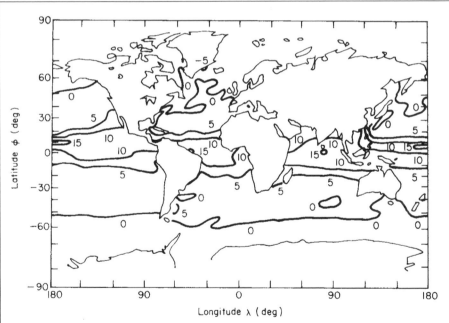

Figure 3.65. Average minimum temperature difference (°C) between sea surface and a depth of 200 m. "Average minimum" means that a smoothed curve for the seasonal variations in temperature difference has been used to determine the minimum (based on Putnam, 1953).

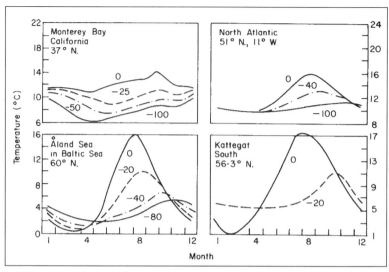

Figure 3.66. Trends of seasonal variations in water temperature at different depths (indicated in metres below surface) for selected locations in open ocean and near shore (based on Sverdrup *et al.*, 1942; Neumann and Pierson, 1966; Danish Meteorological Institute, 1973).

This quantity is shown in Fig. 3.64 for the sites considered in the preceding figure. It is seen that, owing to the quadratic dependence on the temperature difference, the work density is high only for the upper 100 m of the tropical oceans. Near the coast of Florida the work density drops more slowly with depth, owing to the presence of the core of the warm Gulf Stream current (cf. Fig. 3.50). Below the current the work density drops with a different slope, or rather different slopes, because of the rather complex pattern of weaker currents and counter-currents of different temperatures.

The power which could be extracted is obtained by multiplying the work density by the transfer coefficient of the conversion device, i.e. the volume of water from which a maximum energy prescribed by the Carnot value can be extracted per second. This quantity is also a function of T and T_{ref}, and losses relative to the ideal Carnot process are included in terms of a larger "effective" volume necessary for extracting a given amount of work. If the device were an ideal Carnot machine and the amount of water processed were determined by the natural flow through the device, then at a flow rate of 1 m s^{-1} (typical of the Gulf Stream current) the power extracted would be 3–4 MW m^{-2} facing the current flow direction, according to Fig. 3.64. In practice, as the treatment in Chapter 4 will show, only a fraction of this power will be obtained.

An indication of the global distribution of locations with large ocean temperature gradients may be obtained from Fig. 3.65, which gives the average temperature difference between the surface water and the water at a depth of

BASIC MATERIAL IS IN 2.3.1

200 m for the time of the year when this difference is smallest. As mentioned above, the seasonal variation is very small in the equatorial region, whereas the summer temperature difference is far greater than the winter temperature difference at mid- and high latitudes, as seen from Fig. 3.66.

The availability of currents, which could promote the flow through a temperature gradient-utilising device, may be estimated from the material in section 3.4.1, but it should be borne in mind that the currents are warm only after passage through the equatorial region. It then follows from Fig. 3.51 that most currents on the Northern Hemisphere are warm, while most currents in the Southern Hemisphere originate in the Antarctic region and are cold.

Temperature gradients in upper soil and air

The temperature gradient due to absorption of solar radiation in continental soil or rock has a diurnal and a seasonal component, as discussed in section 2.3.2. Transport processes in soil or rock are primarily by conduction rather than by mass motion (ground water does move a little, but speeds are negligible compared with the currents in the oceans), which implies that the regions which are heated are much smaller (typically depths of up to 0–7 m in the diurnal cycle and up to 15 m in the seasonal cycle). The seasonal cycle may be practically absent for locations in the neighbourhood of the Equator. An example of the yearly variation at fairly high latitude was given in Fig. 2.104. The picture of the diurnal cycle is very similar except for scale, as follows from the simple model given by (2.39)–(2.41). It follows that the potential for extracting mechanical work or electricity from solar-derived soil or rock temperature gradients is very small. Direct use of the heat is usually excluded as well, because the temperatures in soil or rock are mostly similar to or slightly

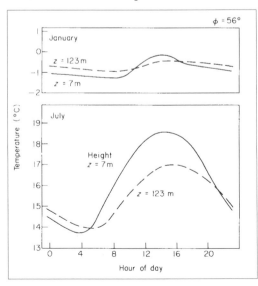

Figure 3.67. Variations of air temperature for two heights over ground and two months (January and July), as a function of the hour of the day. Ten years of observations from the meteorological tower at Risø, Denmark (Petersen, 1974) have been used to form the average trends.

lower than that of the air. An exception may be dry rock, which can reach temperatures substantially above the ambient after some hours of strong solar exposure. Extraction of heat from soil or rock by means of the heat pump principle is possible and will be discussed in the next chapter.

Temperature gradients in the atmosphere are rather small but of stable sign (except for the layer near the ground) up until the tropopause (see Fig. 2.28). At higher altitudes the gradient changes sign a few times (Fig. 2.27). The temperature gradient over the first few hundred metres above ground is largely determined by the sign of the net total radiation flux. This exhibits a daily cycle (cf. e.g. Fig. 2.25) characterised by downward net flux during daylight and upward net flux at night. An example of the corresponding diurnal variation in temperature at heights of 7 and 123 m is shown in Fig. 3.67 for January and July at a latitude 56°N location.

The density of air also decreases with height (Fig. 2.27), and the heat capacity is small (e.g. compared with that of water), so the potential for energy conversion based on atmospheric heat is small per unit of volume. Energy production using the atmospheric temperature gradient would require the passage of very large quantities of air through an energy extraction device, and the cold air inlet would have to be placed at an elevation of the order of kilometres above the warm air inlet. However, heat pump use of air near ground level is possible, just as it is for solar energy stored in the upper soil.

3.5.2 Geothermal flows and stored energy

Heat is created in some parts of the interior of the Earth as a result of radioactive disintegrations of atomic nuclei. In addition, the material of the Earth is in the process of cooling down from an initial high temperature or as a result of heat released within the interior by condensation and possibly other physical and chemical processes.

Regions of particularly high heat flow

Superimposed on a smoothly varying heat flow from the interior of the Earth towards the surface are several abnormal flow regions. The subsurface temperature exhibits variations associated with the magnitude of the heat flow and the heat capacity, or in general the thermal properties, of the material. The presence of water (or steam) and other molten material is important for the transportation and concentration of heat. It is the vapour-dominated concentrations of geothermal energy which have attracted most attention as potential energy extraction sources, up until the present. However, the occurrence of hot springs or underground steam reservoirs is limited to a very few locations. Reservoirs containing superheated water (brine) are probably more common, but are hard to detect from general geological data.

Some reservoirs attract special interest owing to chemical processes in brine–methane mixtures which release heat and increase the pressure, while the conductivity of such mixtures is low (so-called "geopressurised systems", cf. Rowley, 1977). Dry rock accumulation of heat is probably the most common type of geothermal heat reservoir (i.e. storage with temperature above average), but it does not lend itself so directly to utilisation because of the difficulties in establishing sufficiently large heat transfer surfaces. A number of high-temperature reservoirs are present in connection with volcanic systems in the form of lava pools and magma chambers.

Whereas the smoothly varying average heat flow from the interior of the Earth can be considered a renewable energy resource (see below), the reservoirs of abnormal temperature are not necessarily large in relation to possible usage and do not get renewed at rates comparable to possible extraction rates. It is estimated that electric power generation by use of geothermal steam at the locations where such steam is available will be possible only over a time of about 50 years. The total amount of geothermal heat stored in water or steam to a depth of 10 000 m is estimated to be 4×10^{21} J, of which some $(1–2) \times 10^{20}$ J is capable of generating steam of above 200°C (World Energy Conference, 1974). The same source estimates the total amount of energy stored in dry rocks to a depth of 10 km as around 10^{27} J. The steam of temperatures above 200°C represents an average power of 240×10^9 W for a period of 50 years.

It follows from the above that most of the abnormal geothermal heat reservoirs must be considered as non-renewable resources on the same footing as fossil and fissile deposits. However, the average geothermal heat flow also corresponds to temperature gradients between the surface and reachable interior of the Earth, which can be used for energy extraction, at least if suitable transfer mechanisms can be established. These may be associated with water flow in regions of high permeability, such that water which has been cooled by an energy extraction device may regain the temperature of the surrounding layer in a relatively short time.

The origin of geothermal heat

The radioactive elements mainly responsible for the geothermal heat production at present are ^{235}U (decay rate 9.7×10^{-10} y^{-1}), ^{238}U (decay rate 1.5×10^{-10} y^{-1}), ^{232}Th (decay rate 5.0×10^{-11} y^{-1}) and ^{40}K (decay rate 5.3×10^{-10} y^{-1}). These isotopes are present in different concentrations in different geological formations. They are more abundant in the granite-containing continental shields than in the ocean floors. Most of the material containing the radioactive elements is concentrated in the upper part of the Earth's crust. In the lower half of the crust (total depth around 40 km), the radiogenic heat production is believed to be fairly constant at a value of about 2×10^{-7} Wm^{-3} (Pollack and Chapman, 1977). The rate of radiogenic heat production at the top of the continental crust is typically at least ten times higher, but it decreases with depth and reaches the lower crust value approximately half way through the crust. Little is known about the radioactivity in the mantle (occupying the volume between the core and the crust) and in the core (the radius of which is about half the total radius), but from models of their composition it is believed that very little radioactive heat production takes place in the mantle or core.

From the radioactive decay of the above-mentioned isotopes, it can be deduced that the temperature of an average piece of rock near the surface of the crust must have decreased some 900°C during the last 4.5×10^9 years (i.e. since the Sun entered the main burning sequence; Goguel, 1976). At present, the radiogenic heat production accounts for an estimated 40% of the continental average heat flow at the surface. The rest, as well as most of the heat flow at the oceanic bottoms, may then be due to cooling associated with expenditure of stored heat.

In order to assess the nature and origin of the heat stored in the interior of the Earth, a model of the creation and evolution of the planet must be formulated. As discussed in section 2.A, the presence of heavy elements in the Earth's crust would be explained if the material forming the crust was originally produced during one or more supernovae outbursts. It is also consistent to assume that this material was formed over a prolonged period, with the last contribution occurring some 10^8 years before the condensation of the material forming the Earth (Schramm, 1974).

A plausible model of the formation of the planetary system assumes an initial nebula of dust and gases (including the supernova-produced heavy elements), the temperature of which would be increasing towards the centre. The Sun might then be formed by gravitational collapse (cf. section 2.A). The matter not incorporated into the "protosun" would be slowly cooling, and parts of it would condense into definite chemical compounds. The planets would be formed by gravitational accretion of matter, at about the same time as or slightly earlier than the formation of the Sun (Reeves, 1975).

One hypothesis is that the temperature at a given distance from the protosun would be constant during the formation of planets like the Earth ("equilibrium condensation model", see e.g. Lewis, 1974). As temperature declines, a sequence of condensation processes occur: at 1600 K oxides of calcium, aluminium, etc.; at 1300 K nickel–iron alloys; at 1200 K enstatite ($MgSiO_3$); at 1000 K alkali-metal-containing minerals (feldspar, etc.); at 680 K troilite (FeS). The remaining iron would be progressively oxidised and at about 275 K water ice would be formed. The assumption that the Earth was formed at a constant temperature of about 600 K would then imply that it would have a composition characteristic of the condensation state at that temperature.

Other models can be formulated which would also be relatively consistent with the knowledge of the (initial) composition of the Earth's interior. If, for example, the planet formation process was slow compared to the cooling of the nebula, then different layers of the planet would have compositions characteristic of different temperatures and different stages of condensation.

It is also possible that the creation of the Sun and the organisation of the primordial nebula did not follow the scheme outlined above, but were, for example, the results of more violent events such as the close passage of a supernova star.

If a constant temperature of about 600 K during the formation of the Earth is accepted, then the subsequent differentiation into crust, mantle and core (each of the two latter having two subdivisions) is a matter of gravitational settling, provided that the interior was initially in the fluid phase or that it became molten as a result of the decay of radioactive isotopes (which as mentioned must have been much more abundant at that time, partly because of the exponential decay of the isotopes present today and partly owing to now absent short-lived isotopes). A crust would have formed very quickly as a result of radiational cooling, but the presence of the crust would then have strongly suppressed further heat loss from the interior. The central region may have been colder than the mantle during the period of core formation by gravitational settling (Vollmer, 1977).

The gravitational settling itself is a process by which gravitational (potential) energy is transformed into other energy forms. It is believed that most of this energy emerged as heat, and only a smaller fraction as chemical energy or elastic energy. If the differentiation process took place over a short time interval, the release of 26×10^{30} J would have heated material of heat capacity similar to that of crustal rocks to a temperature of about 5000 K. If the differentiation was a prolonged process, then the accompanying heat would to a substantial extent have been radiated into space, and the temperature would never have been very high (Goguel, 1976). The present temperature increases from about 300 K at the surface to perhaps 4000 K in the centre of the core. During the past, convective processes as well as formation of steam plumes in the mantle are likely to have been more intense than at present. The present heat flow between the mantle and the crust is about 25×10^{12} W, while the heat flow at the surface of the crust is about 30×10^{12} W (Chapman and Pollack, 1975; Pollack and Chapman, 1977).

If heat conduction and convective processes of all scales can be approximately described by the diffusion equation (2.27), supplemented with a heat source term describing the radioactive decay processes, in analogy to the general transport equation (2.43), then the equation describing the cooling of the Earth may be written

$$\mathrm{d}T/\mathrm{d}t = \mathrm{div}(K \, \mathrm{grad} \, T) + S/\rho, \tag{3.33}$$

where K is the effective diffusion coefficient and S is the radiogenic heat production divided by the local heat capacity C. The rate of heat flow is given by

$$E^{sens} = \lambda \, \partial T / \partial r, \tag{3.34}$$

where the thermal conductivity λ equals KC, according to (2.38). If the temperature distribution depends only on the radial coordinate r, and K is regarded as a constant, (3.33) reduces to

$$dT/dt = K (\partial^2 T /\partial r^2 + 2 r^{-1} \partial T /\partial r) + S/\rho. \tag{3.35}$$

The simplified version is likely to be useful only for limited regions. The diffusion coefficient K is on average of the order of 10^{-6} m^2 s^{-1} in the crust and mantle, but increases in regions allowing transfer by convection. The thermal conductivity λ of a material depends on temperature and pressure. It averages about 3 W m^{-1} K^{-1} in crust and mantle, but may be much higher in the metallic regions of the core. The total amount of heat stored in the Earth (relative to the average surface temperature) may be estimated from $\int_{volume} \lambda K^{-1}(T - T_s) \, dx$, which has been evaluated as 4×10^{30} J (World Energy Conference, 1974). Using the average heat outflow of 3×10^{13} W, the relative loss of stored energy presently amounts to 2.4×10^{-10} per year. This is the basis for treating the geothermal heat flow on the same footing as the truly renewable energy resources. Also, in the case of solar energy, the energy flow will not remain unchanged for periods of the order billions of years, but will slowly increase, as discussed in section 2.4.2, until the Sun leaves the main sequence of stars (at which time the radiation will increase dramatically).

Distribution of the smoothly varying part of the heat flow

The important quantities for evaluating the potential of a given region for utilisation of the geothermal heat flow are the magnitude of the flow and its accessibility.

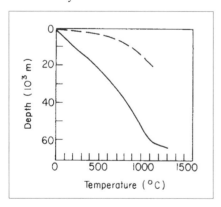

Figure 3.68. Temperature as a function of depth for a young and an old geological formation, representing average extremes of temperature profiles due to the smoothly varying part of the geothermal heat flow. The young formation (dashed line) corresponds to a mid-ocean ridge, and the old formation (solid line) corresponds to a Precambrian continental shield. The kink in the solid line may be associated with the onset of convective processes in the mantle (based on a model calculation, adjusted to observations, by MacGregor and Basu, 1974).

The temperature gradient and heat flow are usually large in young geological formations and, in particular, at the mid-oceanic ridges, which serve as spreading centres for the mass motion associated with continental drift. In the continental shields, the gradient and flow are much smaller, as illustrated in Fig. 3.68. Along sections from the mid-oceanic ridges to the continental shields the gradient may gradually change from one to the other of the two types depicted in Fig. 3.68. In some regions the temperature gradients change much more abruptly (e.g. in going from the Fenno-Scandian Shield in Norway to the Danish Embayment; cf. Balling, 1976), implying in some cases a high heat flux for land areas (e.g. northwestern Mexico, cf. Smith, 1974).

Maps of the geographical distribution of heat flow, both at the surface of the crust and at the surface of the mantle, have been prepared by Chapman and Pollack (1975; Pollack and Chapman, 1977) by supplementing available data with estimates based on tectonic setting and, for oceanic regions, the age of the ocean floor. The results, shown in Figs. 3.69 and 3.70, are contours of a representation in terms of spherical harmonic functions of latitude and longitude, $Y_{lm}(\phi, \lambda)$, of maximum degree $l = 12$. The advantage of this type of analysis is that oscillations of wavelength smaller than about 3000 km will be suppressed, so that the maps are presumably describing the smoothly varying average flow without perturbations from abnormal flow regions. A comparison, between the calculation shown in Fig. 3.69 and one in which the model predictions have been used also in regions where measurements do exist, has convinced Chapman and Pollack that little change will result from future improvements in the data coverage. The map showing the mantle flow, Fig. 3.70, is obtained by subtracting the contribution from the radiogenic heat production in the crust from the surface flow map. To do this, Pollack and Chapman have used a model for the continental regions in which the heat production decreases exponentially from its local value at the surface of the crust,

$$S = S_s \exp(-z/b),$$

where $b = 8.5$ km (if not measured), until it reaches the constant value $CS = 2.1 \times 10^{-7}$ W m^{-3} assumed to prevail throughout the lower crust. For oceanic regions, the difference between mantle and surface heat flows is estimated on the basis of the cooling of the oceanic crust (assumed to behave like a 6.5-km thick layer of basalt) with a fixed lower boundary temperature of 1200°C. The time during which such cooling has taken place (the "age" of the sea floor) is determined from the present surface heat flow (Pollack and Chapman, 1977).

It is evident that the mantle heat flow is very regular, with low continental values and increasing heat flow when approaching the oceanic ridges, particularly in the southern Pacific Ocean. The surface heat flow is considerably more irregular, in a manner determined by the composition of the crust at a given location, but the mean fluctuation around the average value of 5.9×10^{-2} W m^{-2} is only 22%. The surface heat flow has similar values if evaluated separately for continents and oceans (5.3 and 6.2×10^{-2} W m^{-2}), while the contrast is larger for the upper mantle flow (2.8 and 5.7×10^{-2} W m^{-2}, respectively).

The usefulness of a given flow for purposes of energy extraction depends on the possibilities for establishing a heat transfer zone of sufficient flux. As mentioned earlier, the most appealing heat transfer method is by circulating water. In this case the rate of extraction depends on the permeability of the geological material (defined as the fluid speed through the material, for a specified pressure gradient and fluid viscosity). The pressure is also of importance in determining whether hot water produced at a depth will rise to

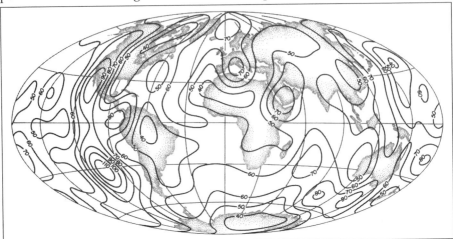

Figure 3.69. Surface heat flow contours (in 10^{-3} W m^{-2}) for the smoothly varying part of the geothermal flux, calculated on the basis of available data supplemented with a theoretical model. From H. Pollack and D. Chapman (1977), *Earth and Planetary Science Letters* **34**, 174–184, copyright by Elsevier Scientific Publ. Co., Amsterdam.

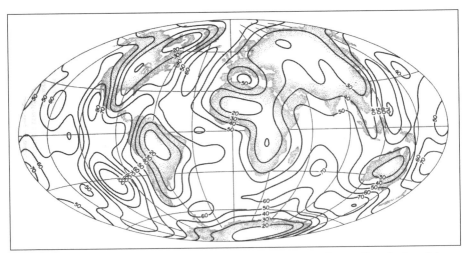

Figure 3.70. Heat flow contours at the top of the mantle (10^{-3} W m^{-2}), calculated from the surface fluxes shown in Fig. 3.69 by subtracting the contribution originating in the crust, for which a model has been constructed (see text). Ffrom H. Pollack and D. Chapman (1977), *Earth and Planetary Science Letters* **34**, 174–184, copyright by Elsevier Scientific Publ. Co., Amsterdam.

the top of a drill-hole unaided or will have to be pumped to the surface. High permeability is likely to be found in sand-containing sedimentary deposits, while granite and gneiss rock have very low porosity and permeability. Schemes for hydraulic fractionation or granulation caused by detonation of explosives have been proposed, with the purpose of establishing a sufficiently large heat transfer area, so that the vast areas of hot rock formations could become suited for extraction of geothermal energy, for example, by circulating water through such man-made regions of porous material.

3.6 Biological conversion and storage of energy

The amount of solar energy stored in the biosphere in the form of standing crop biomass (plants, animals) is roughly 1.5×10^{22} J (see section 2.4.1), and the average rate of production is 1.33×10^{14} W or 0.26 W m^{-2}. The residing biomass per unit area is small in oceans and large in tropical forests. The average rate of production on the land surface is 7.6×10^{13} W or 0.51 W m^{-2} (Odum, 1972), and over 0.999 of the standing crop is found on land.

Still larger amounts of dead organic matter are present on the Earth in various degrees of fossilation (see Fig. 2.88, expressed in carbon mass units rather than in energy units; 1 kg carbon roughly corresponds to 41.8×10^{6} J). Thus, the fossil deposits (varieties of coal, oil and natural gas) are estimated to be of the order of 6×10^{23} J. This may be an underestimate, in view of the floating distinction between low-concentration energy resources (peat, shale, tar sand, etc.) and non-specific carbon-containing deposits. A maximum of about 1×10^{23} J is believed to be recoverable with technology that can be imagined at present (cf. section 2.4.2), suggesting that the fossil energy resources are not very large compared with the annual biomass production. On the other hand, more sophisticated methods are generally required to convert fresh biomass into useful energy than is the case for the most pure and concentrated fossil fuel deposits. For bio-energy sources which have been in use for a long time, such as firewood and straw, an important requirement is often drying, which may be achieved by further use of direct solar radiation.

It should not be forgotten that plants and derived forms of biomass serve man in essential ways other than as potential energy sources, namely, as food, as raw material for construction and – in the case of green plants – as producers of atmospheric oxygen. The food aspect is partially an indirect energy usage, because man and other animals convert the energy stored in plants by metabolic processes, which furnish the energy for the life processes. It is also more than that, acting as a source of nutrients, vitamins, etc., which are required for reasons other than their energy content.

An assessment of the possible utilisation of bio-energy for purposes other than those associated with the life-cycles themselves must take into account the food requirements as well as other tasks performed by vegetation and animal stock, e.g. prevention of soil erosion, conservation of diversity of species and balanced ecological systems. Only energy uses which can be implemented in harmony with (maybe even in combination with) these other requirements can be considered acceptable. Although plant systems convert only a small fraction of the incident solar radiation, plans for energy usage should also consider the requirements necessary to avoid any adverse climatic impact. The management of extended areas of vegetation does hold possibilities of interfering with climate, e.g. as a result of the strong influence of vegetation on the water cycle (soil moisture, evaporation, etc.). Examples of such relations and actual climatic changes caused by removal of vegetation, overgrazing, etc. have been given in section 2.4.2.

Before the question of plant productivity in different geographical regions and under different conditions is considered, a brief survey of the biochemistry of photosynthetic processes in green plants will be outlined.

3.6.1 Photosynthesis

Mechanism of green plant photosynthesis

The cells of green plants contain a large number of chloroplasts in their cytoplasma. The interior of a chloroplast contains a liquid, the stroma, rich in dissolved proteins, in which floats a network of double membranes, the thylakoids. The thylakoid membranes form a closed space, containing chlorophyll molecules in a range of slightly different forms, as well as specific proteins important for the photosynthetic processes. In these membranes the photo-induced dissociation of water takes place,

$$2H_2O + light \rightarrow 4e^- + 4H^+ + O_2, \tag{3.36}$$

and the electrons are transferred, through a number of intermediaries, from the internal to the external side of the membrane. The transport system is depicted in Fig. 3.71, where the ordinate represents the redox potential [the difference in redox potential between two states roughly equals the amount of energy, which is released or (if negative) which has to be added, in order to go from one state to the other].

The first step in the reaction is the absorption of solar radiation by a particular chlorophyll pigment (a_{680}). The process involves the catalytic action of an enzyme, identified as a manganese complex, which is supposed to trap water molecules and become ionised as a result of the solar energy absorption. Thereby, the molecules (of unknown structure) denoted Q in Fig. 3.71 become negatively charged, i.e. electrons have been transferred from the water–manganese complex to the Q-molecules of considerably more negative redox potential (Kok et al., 1975; for a recently updated review, see the treatment given in Chapter 2 of Sørensen, 2004a).

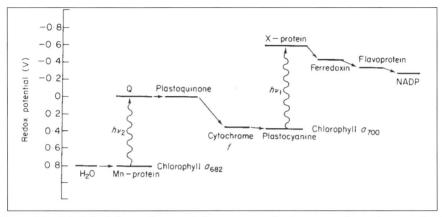

Figure 3.71. Model of the pathway of electron transfer in green plant photosynthesis (based on Trebst, 1974).

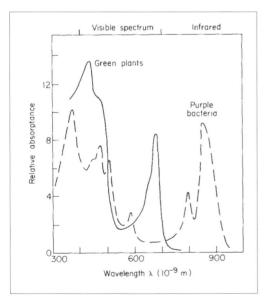

Figure 3.72. Spectrum of relative absorptance for a green plant and for a purple bacterium (based on Clayton, 1965).

The photosensitive material involved in this step is called "photosystem II". It contains the chlorophyll pigment a_{680} responsible for the electron transfer to the Q-molecules, where 680 is the approximate wavelength (in 10^{-9} m), for which the light absorptance peaks. It is believed that the initial absorption of solar radiation may be affected by any of the chlorophyll pigments present in the thylakoid and that a number of energy transfers take place before the particular pigment, which will transport an electron from H_2O to Q, receives the energy (Seliger and McElroy, 1965).

The gross absorptance as a function of wavelength is shown in Fig. 3.72 for green plants and, for comparison, a purple bacterium. The green plant spectrum exhibits a peak just above 400×10^{-9} m and another one a little below 700×10^{-9} m. The bacterial spectrum is quite different with a pronounced peak in the infrared region around 900×10^{-9} m. The positions of the peaks are somewhat different for the different types of chlorophyll (a, b, c,...), and they also move if the chlorophyll is isolated from its cellular surroundings. For a given chlorophyll type, different pigments with different absorption peak positions result from incorporation into different chemical environments (notably bonding to proteins).

Returning to Fig. 3.71, the electrons are transferred from the plastoquinone to cytochrome f by a redox reaction, which transfers a corresponding number of protons (H^+) across the thylakoid membrane. The electrons are further transferred to plastocyanine, which leaves them to be transported to another molecule of unknown structure, the X-protein. The energy required for this process is provided by a second photo-excitation step, in what is called "photosystem I", by means of the chlorophyll pigment a_{700} (spectral peak approximately at 700×10^{-9} m).

The electrons are now delivered to the outside region where they are picked up by ferredoxin, a small protein molecule situated at the outer side of the thylakoid membrane system. Ferredoxin may enter a number of different reactions, of which a very important one transfers the electrons via flavoprotein to nicotinamide–adenine dinucleotide phosphate (NADP), which reacts with protons to form NADPH$_2$, the basis for the carbon dioxide assimilation. The protons formed by (3.36) and by the cytochrome f–plastoquinone reduction are both formed inside the membrane, but they penetrate the thylakoid membrane by a diffusion process, regulated by the reaction (3.38) described below. The NADPH$_2$-forming reaction may then be written

$$2NADP + 4e^- + 4H^+ \rightarrow 2NADPH_2, \tag{3.37}$$

which describes the possible fate of electrons and protons formed by the process (3.36), after transport by the sequence shown in Fig. 3.71 and proton diffusion through the membrane.

The transport of protons from outside the thylakoid membrane to its inside space, by means of the plastoquinone–cytochrome f redox cycle, creates an acidity (pH) gradient, which in turn furnishes the energy necessary for the phosphorylation of adenosine diphosphate (ADP) into adenosine triphosphate (ATP),

$$ADP^{3-} + H^+ + HPO_4^{2-} \rightarrow ATP^{4-} + H_2O. \tag{3.38}$$

This process, which involves the action of an enzyme, stores energy in the form of ATP (4.8×10^{-20} J per molecule), which may later be used to fuel energy-demanding processes, e.g. in the chloroplast stroma or in the cytoplasm outside the chloroplast (see e.g. Douce and Joyard, 1977). It is the analogue of the energy-stocking processes taking place in any aerobic cell of plants and animals (i.e. cells using oxygen), as a result of the degradation of food (saccharides, lipids and proteins), and associated with expenditure of oxygen (the respiratory chain, the Krebs cycle; see e.g. Volfin, 1971). The membrane system involved in the case of food metabolism is the mitochondrion.

By means of the energy stored in ATP and the high reducing potential of $NADPH_2$, the CO_2 assimilation process may take place in the stroma of the chloroplasts independently of the presence or absence of solar radiation. The carbon atoms are incorporated into glyceraldehyde 3-phosphate, which forms the basis for synthesis of glucose and starch. The gross formula for the process leading to the synthesis of glyceraldehyde 3-phosphate in the chloroplasts (the Benson, Bassham and Calvin cycle; cf. Douce and Joyard, 1977) is

$$3CO_2 + 9ATP^{4-} + 6NADPH_2 + 5H_2O \rightarrow$$

$$POCH_2CH(OH)CHO^{2-} + 9ADP^{3-} + 8HPO_4^{2-} + 6NADP + 9H^+. \qquad (3.39)$$

The reactions (3.36) to (3.39) may be summarised,

$$3CO_2 + 2H_2O + HPO_4^{2-} + light \rightarrow POCH_2CH(OH)CHO^{2-} + 3O_2. \qquad (3.40)$$

Going one step further and including the synthesis of glucose, the classical equation of photosynthesis is obtained,

$$6H_2O + 6CO_2 + 4.66 \times 10^{-18} J \rightarrow C_6H_{12}O_6 + 6O_2, \qquad (3.41)$$

where 4.66×10^{-18} J is the net energy to be added by solar radiation.

Efficiency of conversion

The basis for energy utilisation has traditionally been the undifferentiated biomass produced. Other possibilities would include the direct dissociation of water by the action of sunlight (photolysis), or the formation of hydrogen rather than $NADPH_2$, after the initial oxygen formation (3.36) and electron transport to ferredoxin (Benemann and Weare, 1974; Mitsui and Kumazawa, 1977),

$$4e^- + 4H^+ \rightarrow 2H_2. \qquad (3.42)$$

In both cases the problem is to control the recombination process,

$$2H_2 + O_2 \rightarrow 2H_2O + 9.47 \times 10^{-19} J, \qquad (3.43)$$

so that this energy-releasing process takes place where desired and not immediately on the formation of hydrogen. The plants accomplish this by means of the thylakoid membrane and later by the chloroplast external membrane. Man-made processes may attempt to copy the membrane principle in various ways (Broda, 1975; Calvin, 1974, 1977; cf. Chapter 4), but it is not clear that present suggestions will allow separation of hydrogen and oxygen on a large scale. An advantage in this connection may lie in using the green plants to perform the first step (3.36) and transport electrons and protons (hydrogen ions) through their membranes, but to prevent the energy from being too deeply trapped in organic matter, for example, by adding a strongly reducing agent to ferredoxin, as well as suitable enzymes, in order to accomplish the reaction (3.42) at this stage.

The maximum theoretical efficiency of this process is the ratio between the heat release 9.47×10^{-19} J in (3.43) and the solar energy input required. The latter depends on the absorption properties of the plant and its chlorophyll molecules (cf. Fig. 3.72), as well as on the number of light quanta required for each molecular transformation (3.36). The energy, E, of each light quantum of wavelength λ may be found from

$$\lambda E_\lambda = hc = 1.986 \times 10^{-25} \text{ J m.}$$

The minimum number of quanta required to transport one electron as depicted in Fig. 3.71 is two, one of wavelength 680×10^{-9} m and the other of 700×10^{-9} m (although these need not be the quanta originally absorbed). Since (3.36) requires the transport of four electrons, the minimum requirement would be 8 quanta with a total energy of 2.3×10^{-18} J. Experimental estimates of the number of quanta needed typically give values between 8 and 10.

The efficiency of the photosynthetic process containing only the steps (3.36) and (3.42) may be written

$$\eta' = \eta_\lambda \, \eta_{geom} \, \eta_{chem} , \qquad\qquad (3.44)$$

where η_λ is the fraction of the frequencies in the solar spectrum (depending on cloud cover, etc.) that is useful in the photosynthetic process, η_{geom} is the geometrical efficiency of passing the incoming radiation to the chlorophyll sites (depending on penetration depth in leaves, on reflectance from outer and inner surfaces and on absorption by other constituents of the leaves), and η_{chem} is the efficiency of the photochemical reactions, the maximum value of which is given by

$$\eta_{chem} \leq 9.47 \times 10^{-19} / 2.30 \times 10^{-18} = 0.41.$$

This efficiency pertains to the amount of internal heat produced by (3.43). Only a part of this can be converted into useful work. This part is obtained by replacing in (3.43) the enthalpy change 9.47×10^{-19} J by the change in free energy, $\Delta G = 7.87 \times 10^{-19}$ J (cf. section 4.1.1). In this way the efficiency of the photochemical reaction becomes

$$\eta_{chem,free} \leq 0.34.$$

The efficiency η_λ associated with the chlorophyll absorption spectrum typically lies in the range 0.4–0.5 (Berezin and Varfolomeev, 1976), and the geometrical efficiency η_{geom} may be around 0.8 (for the leaf of a green plant, not including the reduction associated with the penetration of radiation through other vegetation, e.g. in a forest environment).

The overall maximum efficiency of about $\eta' \approx 0.14$ found here for a hypothetical hydrogen (H_2)-producing system is valid also for actual green plants which assimilate CO_2. Over extended periods of time, the biomass production efficiency will have to incorporate still another factor, η_{resp}, expressing the respiration losses associated with the life-cycle of the plant,

$$\eta = \eta' \, \eta_{resp}. \tag{3.45}$$

The respirative energy losses emerge as heat and evaporated water, at rates depending on temperature and wind speed (Gates, 1968). The value of η_{resp} is 0.4–0.5 for land plants and somewhat larger for aquatic plants and algae.

Actual plants may get close to the theoretical maximum efficiency, although the average plant does not. For the blue-green alga *Anacystis nidulans*, Goedheer and Hammans (1975) report an energy efficiency of $\eta_{chem} \approx 0.30$, based on 36 h of growth, including 6 h of irradiation, with a generous supply of nitrogen and other nutrients, as well as a CO_2-enriched atmosphere (i.e. 73% of the maximum energy efficiency calculated above).

It is estimated that, on average, each CO_2 molecule in the atmosphere becomes assimilated in a plant once every 200 years and that each O_2 molecule in the atmosphere is "renewed" through a plant once every 2000 years (Seliger and McElroy, 1965).

Bacterial photosynthesis

Several bacteria use solar radiation to dissociate a compound of the general form H_2X, with a net reaction scheme of the form

$$CO_2 + 2H_2X + \text{light} \rightarrow (CH_2O) + H_2O + 2X, \tag{3.46}$$

(Van Niel, 1941). Here (CH_2O) should be understood not as free formaldehyde, but as part of a general carbohydrate compound in analogy to the more precise equation (3.40). Actually, (3.46) was proposed to be valid for both green plant and bacterial photosynthesis, but there is no detailed analogy, since the bacterial photosynthesis has been found to take place in a single step, resembling the photosystem I of the green plant two-step process. Other photo-induced reactions do take place in bacteria, connected with the ATP formation, which in this case is not a side-product of the primary photosynthetic process.

The compound H_2X may be H_2S (sulphur bacteria), ethanol C_2H_5OH (fermentation bacteria), etc. Most photosynthetic bacteria are capable of absorbing light in the infrared region (wavelength $800–1000 \times 10^{-9}$ m). The role which the

NADP–NADPH$_2$ cycle (3.37) and (3.39) plays for green plant photosynthesis is played by NAD (nicotinamide-adenine dinucleotide)–NADH$_2$ for photosynthetic bacteria. The redox potential of NADH$_2$ is more or less the same as that of the initial compounds, e.g. H$_2$S, so practically no energy is stored in the process of bacterial photosynthesis (Hind and Olson, 1968).

3.6.2 Productivity in different environments

Ecological systems

The gross primary production of a plant or, in general, of an ecological system is the rate at which solar energy is assimilated, i.e. the total amount of energy produced by photosynthesis. The net primary production, on the other hand, is the difference between gross primary production and respiration. The respiration processes involve an increase in redox potential (oxidation), either by consumption of oxygen [aerobic respiration, the net result being equivalent to (3.43)] or by the action of some other agent of oxidation (anaerobic respiration; if the oxidant is an organic compound, the respiration process is called fermentation).

The primary producers are a part of an ecological system. Figure 3.73 gives a schematic view of the energy and matter flow through such a system. The primary chain comprises the primary producers capable of carrying out photosynthesis, plant-eating organisms (herbivores) and a chain of successive carnivorous predators (carnivores), of which some may also be plant eating (like man). Each compartment in the chain is called a trophic level, and the photosynthetic organisms are called autotrophs, while the consuming organisms further along the chain are called heterotrophs. Over short periods of time, withering and death of autotrophs can be neglected, and the net primary production available to the heterotrophic part of the community equals the gross primary production less the respiration of the autotrophs. Over longer periods, respiration, predation and death must be considered in order to describe the amounts of biomass in each compartment ("standing crop"). Generally, the biomass diminishes along the chain, but owing to the different average lifetime, which is often longest for the highest trophic levels of the food chain, the biomass at a given time may be maximal for, say, the second or third member of the chain. Also, the seasonal dependence of primary production in most geographical regions, in conjunction with an often short lifetime of the autotrophs, leads to a biomass distribution with much more long-range stability in the higher levels. The stability depends on the age of the ecosystem. A young system has fewer species and is more sensitive to external disturbances (climate variations, immigration of new predator species, etc.) which may destroy the food basis for the higher trophic levels. Old ecosystems are characterised by higher diversity and, hence, are better equipped to circumvent sudden changes in external conditions, at least for a while.

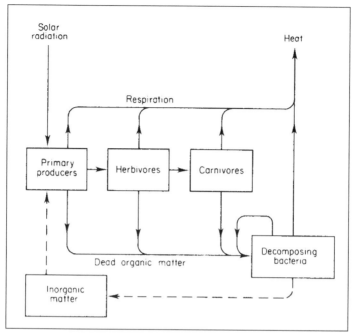

Figure 3.73. Model of an ecological system exhibiting flows of energy (solid lines, which in most cases also indicate the flow of organic matter) and of inorganic matter (dashed lines, inorganic matter includes CO_2, nutrients, etc.). Exchange of heat between compartments and their environment is not shown, neither is the possible presence of photosynthetic bacteria in the system. Some of the flows may be "delayed", e.g. man's harvesting of food crops and wood, which may go into a storage rather than being consumed immediately (i.e. transferred to the next compartment along the chain, "herbivores" in this example).

The dead organic matter may precipitate towards the sea floor or detritus layer of the land surface, where an environment suitable for sustaining a number of decomposing bacteria evolves. Thereby nutrients are transformed back to the inorganic form necessary for renewed uptake by autotrophs.

It follows from Fig. 3.73 that the flows of energy, which could be used for energy extraction serving human society, diminish along the food chain. This does not negate the fact that the concentration of energy flow may be higher and perhaps easier to utilise at one of the higher levels, as compared with the primary producers. For example, livestock leave a large fraction of their respiration heat and manure within their sheds, a spatial region much smaller than the size of a grazing range which may yield the same energy value.

Important factors in determining the net productivity of a given autotrophic system are solar radiation, temperature regime, water availability, climate in general, carbon dioxide and nutrient access, as well as the nature of the entire ecosystem, density and structure of plants in the neighbourhood, predation and harvesting by heterotrophs including man. The growth in biomass would be exponential if optimal conditions for photosynthetic production could be maintained indefinitely. Actual growth is stopped when one or more of the conditions can no longer be met, and the factor(s) responsible for halting the exponential growth is(are) called the limiting factor(s). A mature ecosystem may reach a stable situation in which the net production (and hence growth) of the community as a whole is zero.

Limiting factors

The effective area of a plant which is exposed to solar radiation depends on shadow effects caused by other vegetation, on the shape and arrangements of leaves or more generally of light-absorbing pigments. This, together with the loss caused by scattering and absorption on the leaf surface or other parts of the plant not involved in the light accumulation, accounts for the maximum geometrical efficiency in (3.44) being around 0.8. The effective area of fully developed terrestrial plant systems (forests, erect-leaved cereal crops) is large enough to intercept practically all incident radiation (Loomis and Gerakis, 1975), and most of the losses under such circumstances are due to reflection.

The availability of solar radiation itself is, of course, of basic importance. The geographical distribution of radiation has been discussed in connection with Fig. 2.20 and in section 3.1. The penetration of radiation through a depth of water is illustrated in Fig. 2.26. The corresponding distributions of productivity for typical coastal and oceanic regions of the North Atlantic are shown in Fig. 3.74. Total autotrophic plus heterotrophic biomass at these sites was found to be 167×10^3 J m^{-2} (coastal region) and 8×10^3 J m^{-2} (open ocean) (Odum, 1972).

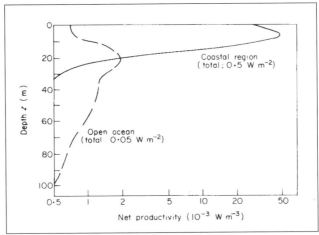

Figure 3.74. Net primary production as function of depth for two locations in the North Atlantic. The corresponding average biomasses (standing crops) are 167 kJ m^{-2} (coastal region) and 8 kJ m^{-2} (open ocean) (based on Odum, 1972).

Too high a solar intensity may diminish efficiency, if the chlorophyll pigments are unable to absorb it all or if the radiation destroys the chlorophyll. Also, the state of the photosynthetic systems as well as of the further energy conversion and transport system of the plant may play a role. Efficiency may be lowered by damage or ageing of such components (Wassink, 1975).

If changes do not occur too rapidly, many plants are able to adapt to external conditions of light, humidity etc., to a certain extent. The pathway of carbon synthesis described in section 3.6.1 involves a molecule (3.40) with three carbon atoms (C_3-pathway). Some plants are able to follow a different pathway, in which four carbon atoms are synthesised (C_4-pathway), notably as malic acid ($HO_2CCH:CHCO_2H$). In this case less CO_2 is needed in order to synthesise e.g. sugar, so that plant growth may occur more rapidly than for the C_3-plants, at least in some environments.

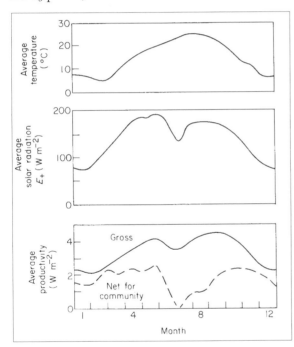

Figure 3.75. Trends of average temperature, solar radiation, gross and net production of biomass for an evergreen oak forest community at Minamata, Kyushu, Japan. The curves have been constructed from data collected in 1971/1972, and the conversion factor 1 kg dry matter = 18.8×10^6 J has been used (based on Kira, 1975).

The temperature plays an important role, both the temperature of air and soil for terrestial plants and the temperature of water for aquatic plants. The life cycles of aquatic plankton and algae are of relatively short duration, and for some species the biomass production is confined to a few weeks of the year. Many terrestrial plants are dormant during cold periods, and they are able to withstand low temperatures during winter. If, however, sudden frost occurs in the middle of a growth period, the tolerance is often low and damage may result. The number of days during the year with temperatures above certain values, depending on species, may be defined as the growing season, a concept which has been used extensively in estimates of average properties of productivity. A phenomenological treatment of the influence of thermal conditions on plant and animal life may be found in the work of Budyko (1974).

Figure 3.75 shows the trends of gross primary production of an evergreen oak forest, along with temperature and average solar radiation. The dashed line gives the net production of the community as a whole (trees and canopy), i.e. the surplus or growth rate, which is seen to be high most of the year except in summer and with peaks in spring and autumn. This is an example of a growing ecosystem. If the intervention of man can be achieved without changing the trophic relationships within the community, it would appear that a harvest (energy extraction) corresponding to the net production could be sustained, leaving a stable ecosystem with no growth. In this way man would enter as an additional loop in the system depicted in Fig. 3.73, and the ecosystem would provide a truly renewable energy source.

When the standing crop biomass is small, only a small fraction of the solar radiation is intercepted. Thus, the productivity initially increases with increasing biomass until a turning point of optimal growth conditions is reached, as seen in Fig. 3.76. After this point, the productivity slowly decreases towards equilibrium (zero).

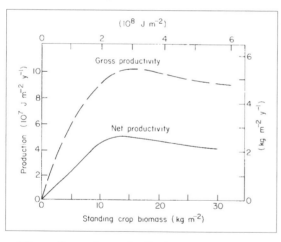

Figure 3.76. Trends of gross and net productivity as functions of standing crop biomass for a fir species (*Abies sachalinensis*) (based on Kira, 1975).

Water flows through all plants at average rates much higher than the rate at which water is dissociated by the first photosynthetic step (3.36). Extra water is required for the oxidation of the manganese enzyme involved in this process, as well as for the cyclic ADT–ATP and NADP–NADPH$_2$ processes, but these amounts of water are not consumed. The same is true for other metabolic processes in the plant, and water also performs physical tasks such as keeping the surfaces of membranes wet and transporting dissolved substances.

The overall photosynthetic process (3.41) requires one water molecule for each carbon atom assimilated, or 1.5 kg of water for each 1 kg of carbon or for each 42×10^6 J of biomass, corresponding to 2–3 kg of dry matter. The wet weight is typically in the range 4–10 times the dry weight, implying that the total water content of the plant is at least 4 times the amount fixed by photosynthesis and typically much higher.

The transport of water through a terrestrial plant is achieved by some kind of free-energy gradient, often referred to as the "water potential" [cf. (3.49) and e.g. Canny (1977)]. The corresponding force may thus be of a more general form than that associated with osmotic pressure. Loss of water by transpiration can be viewed as a local decrease in water potential near the area where water is leaving the plant. This means that a potential gradient has been established, which will attract water from other parts of the plant. Thereby the potential gradient moves through the plant and builds up in the root system, which then attracts water from the surrounding soil. If the water that can be drawn from the soil is insufficient, the plant will try to decrease its transpiration by contracting the stomata (openings at leaf surfaces). If the water deficit produces a strong and prolonged stress on the plant, the stomata will become tightly closed, in which case a subsequent situation of water saturation does not immediately lead to a complete resumption of the rate of photosynthesis at the previous level (as illustrated in Fig. 3.77).

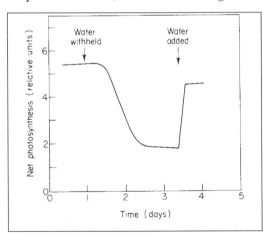

Figure 3.77. Effect on productivity of withholding water for 2.5 days. The experiment was performed on sunflower leaves under conditions of generous light (based on Slavik, 1975).

The turnover of water associated with a biomass production of 1 kg of dry matter is quoted as lying in the range 200–900 kg of er, with values around 500 kg being typical for grass and cereal crops (Geiger, 1961; cf. also Penman, 1970, which quotes 400 kg for food crops).

As an upper limit on the flow of water available to terrestrial plants, one may take the total land precipitation rate of 1.1×10^{17} kg of water per year (Fig. 2.61). Assuming the requirement of the plants to be 500 kg of water per kg of dry matter per year, one deduces a maximum sustainable production of 2.2×10^{14} kg of dry matter per year, or in energy units 4.0×10^{21} J y^{-1} (1.3×10^{14} W).

Since the total terrestrial gross primary production is presently 7.6×10^{13} W (Odum, 1972), and the net primary production may amount to about half this value, it is clear that a large fraction of the evapotranspiration (6.7×10^{16} kg of water per year according to Fig. 2.61) over land is due to plants. Accordingly, a substantial increase in terrestrial plant production would have to involve increased recycling of water on time scales shorter than one year. One might consider achieving this by artificial irrigation, using some of the water run-off from land areas (3.9×10^{16} kg of water per year according to Fig. 2.61). However, it is seen from Fig. 2.94 that about 70% of the water diverted to food crops is not immediately returned to the run-off streams. In view of the many other important functions performed by the rivers and streams responsible for the run-off, this suggests that there may be little room for expanding terrestrial biomass production on a world-wide scale if fresh-water resources alone are used for irrigation.

It would at least appear that schemes for ensuring quick recycling of the water added are as important as expansion of the irrigation potential and that really significant increases in biomass production must involve oceanic water resources, either by marine biomass production or by use of (eventually distilled) sea water for irrigation. Some terrestrial plants actually accept sea water irrigation, with some reduction in conversion efficiency (Epstein and Norlyn, 1977).

If the availability of CO_2 is the only limiting factor, it is reasonable to assume that the productivity is proportional to the amount of CO_2 entering the plant. The concentration X of CO_2 at the photosynthetic sites may be related to the atmospheric CO_2-concentration X_{atm}, e.g. by a linear relation

$$X = f X_{atm},$$

where f depends on properties of the plant (inside pressure, opening of stomata, etc.), as well as on the CO_2 eddy diffusion parameter in the air surrounding the plant. The atmospheric concentration, X_{atm}, decreases with height when no photosynthesis takes place (night condition), as shown, for example, in Fig. 2.29. However, during daytime in low wind conditions, the CO_2 concentration exhibits a minimum half way through the canopy of vegetation (Saeki, 1975).

If the proportionality factor between production P and CO_2 internal concentration X is g, when no other limiting factors are encountered, then the rate of primary production as function of incident radiation, E_+, may be written

$$\frac{1}{P} = \frac{1}{\eta' E_+} + \frac{1}{gX}, \tag{3.47}$$

where η' (3.44) is the efficiency at low incident radiation, when light is the only limiting factor. In (3.47), temperature requirements are supposed to be satisfied, and the water and nutrient supply are supposed to be adequate.

The state of the photosynthetic system within the plant may give rise to different efficiencies η_{chem} under different conditions. Basically, this can be understood in terms of competing pathways for the chain of molecular reactions involved. In the electron transport chain, Fig. 3.71, the Q to plastoquinone electron transport in photosystem II may receive competition by fluorescent de-excitation of the Q-molecule. A similar de-excitation may happen for the X-protein in photosystem I. In both cases, the energy captured by the chlorophyll pigments will become lost from the point of view of biomass production. This may be summarised by stating that under conditions which are not optimal more than 8 light quanta are required for each CO_2 molecule assimilated (the excess energy being lost as heat or radiation).

In aquatic environments, the rate of photosynthesis may depend on such factors as salinity and pH-value (acidity). In Sweden, it has been suggested that decreased productivity of coniferous forests is associated with increasing acidity of precipitation (and hence of soil) (Swedish Ministries of Foreign Affairs and of Agriculture, 1971).

Last, but not least, the role of nutrients as a limiting factor should be mentioned. Deprivation of nutrients decreases and eventually halts production. Transfer of plants to nutrient-free environments has shown effects of decreasing magnitude resulting from the deprivation of nitrogen, phosphorus and potassium in that order.

It follows from Figs. 2.88 and 2.89 that an average of 1.4×10^{-3} kg of nitrogen is fixed for each kg of carbon fixed in the terrestrial biosphere, in which there is about 0.0343 kg of N (kg of C)$^{-1}$ (the standing crop biomass shows a different relationship between N and C, owing to different turnover times). The amount of nitrogen with which the soil must be enriched in order to increase productivity, if N is a limiting factor, may be much larger, owing to the nature of the pathways of uptake. Delwicke (1970) quotes experiments in which 0.76 kg of N was needed in order to increase production by 1 kg of dry matter.

In evaluating the net energy yield of cultivated land or sea, non-solar energy subsidies must also be considered. These may be in the form of machinery and fuels for running the machinery (ploughing and harvesting tools). In present day practices of terrestrial agriculture, the largest energy subsidy in areas of intense farming is in the manufacture of fertilisers. In regions demanding irrigation, water pumping may expend substantial amounts of energy, whereas the amount of machinery employed typically is large only in regions of intense farming, so that the corresponding energy subsidy remains a fraction of that spent on fertiliser. If little machinery is used, more manpower is needed, combined with power provided by draught animals. Also, transport of harvested crops costs energy, particularly for remote fishing, and in the case of food crops, processing and packaging may in industrialised regions account for the largest part of the total energy subsidy. Food derived from animals also involves less efficient utilisation of primary energy, as discussed in connection with Fig. 3.73.

In the case of "energy crops", the size of energy subsidies in fertilisers may be of particular interest, in order to determine whether the extra primary production, i.e. solar energy utilisation, exceeds the energy subsidy. Using present-day manufacturing techniques, including the mining of phosphate rock, the energy needed to produce 1 kg of nitrogen in the form of fertiliser is about 10^8 J, the energy needed for 1 kg of phosphorus is about 1.4×10^7 J, and the energy for 1 kg of potassium is about 9×10^6 J (Steinhart and Steinhart, 1974; Blaxter, 1974). As an example of fertiliser usage in intense farming practice, the average use of fertiliser in Denmark in 1965–1966 amounted to 0.0114 kg of N per m^2, 0.003 58 kg of P per m^2 and 0.0109 kg of K per m^2 (Danish Statistical Survey, 1968), which with the above conversion figures correspond to a total subsidy of 1.3×10^6 J m^{-2} y^{-1} (0.04 W m^{-2}). The harvested yield averaged 0.5 kg of dry matter per m^2 and year. For the most common crop (barley), the net primary production would be about twice the amount harvested or 2×10^7 J m^{-2} y^{-1}, and the fertiliser energy subsidy would thus be 6.5%. By 1971, the fertiliser energy consumption had risen by 23%, but the harvest yield was unchanged per unit area (Danish Statistical Survey, 1972). This implies that, because of this farming practice, nutrients in a form suitable for uptake are being depleted from the soil in such a way that increasing amounts of fertiliser must be added in order to maintain a constant productivity.

A proposed scheme for "energy plantations" yielding biomass at the rate of about 15×10^7 J m^{-2} y^{-1} (e.g. based on sunflower) under (southern) US conditions anticipates the use of 4×10^6 J m^{-2} y^{-1} for fertilisers and 2×10^6 J m^{-2} y^{-1} for irrigation and other machinery, a total energy subsidy of 5% (Alich and Inman, 1976).

Productivity data

Data on productivity of different species under different climatic conditions are plentiful, but often hard to summarise because of the differences in techniques used and ways of representing the results. Most controlled experiments are performed under optimal conditions, and they do not reflect the average productivity of the particular plant and geographical region. On the other hand, statistical information for entire regions or countries does not usually convey the exact conditions of growth. In many cases, the productivity of a given crop depends on the history of previous uses of the land. For natural ecosystems there are other factors which make it difficult to compare data from different sources. The conditions are not always sufficiently well described by climatic zone and solar radiation data. For instance, the geology of the upper soil may vary over very short distances. Altogether, it is not reasonable to give more than broad ranges or limits inside which the productivity of a given plant in a given environment will normally lie, not excluding exceptional values outside the range given.

In order to partially eliminate the dependence on the amount of solar radiation, Fig. 3.78 indicates such ranges for the overall efficiency (3.45), representing the ratio between net primary production and total solar radiation. In this way, for example, the woods in different climatic regions (boreal coniferous, deciduous, temperate broad-leaved evergreen and subtropical or tropical rain forests) become represented by a fairly narrow range of efficiencies, despite the large variations in absolute productivity.

The range of efficiencies of natural plants goes from practically nothing to a little over 2%, the highest values being reached for tropical rain forests and algal cultures of coral reefs. Cultivated crops (terrestrial or marine) may reach some 4–5% under optimal conditions and nutrient subsidy. The optimum efficiency from the theoretical discussion of equations (3.44) and (3.45) is roughly

$$\eta = \eta_\lambda\, \eta_{geom}\, \eta_{chem}\, \eta_{resp} \leq 0.5 \times 0.8 \times 0.4 \times 0.6 \approx 0.1. \tag{3.48}$$

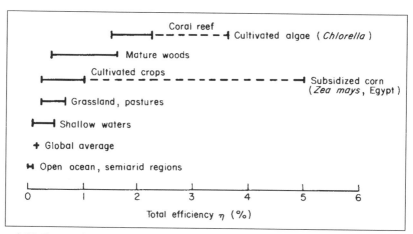

Figure 3.78. Intervals of typical values of total photosynthetic efficiency for different plants and communities (solid lines), as well as intervals of possible improvement for subsidised and optimised cultures (dashed lines). For non-perennial plants, the efficiency is an average over the growing season (constructed on the basis of information from Kira, 1975; Caldwell, 1975; Loomis and Gerakis, 1975).

Each of the conditions is difficult to achieve in practice, at least for extended periods. As mentioned earlier, mature ecosystems are characterised by diversity, which may not be reconcilable with maximum efficiency of primary production, and young ecosystems, such as non-perennial crops, need a growing period (cf. Fig. 3.76) before they reach full productivity, and thus their average efficiency on an area basis and over prolonged periods of time cannot be optimal.

Figure 3.79 gives a recent estimate of the geographical distribution of potential biomass production. These are net primary production data derived from the "Terrestrial Ecosystem Model (TEM)" of the Woods Hole group

(Melillo and Helfrich, 1998; Raich *et al.*, 1991; Melillo *et al.*, 1993; McGuire *et al.*, 1997; Sellers *et al.*, 1997). The assumption is that a mature ecosystem of natural vegetation has developed, and the model takes into account solar radiation, moisture, temperature, as well as access to water and nutrients. Not included here is global warming (increased CO_2), which could induce increased primary production in a fairly complex pattern and change the borders of natural vegetation zones, sometimes by several hundred kilometres (IPCC, 1996b). Use of these biomass data for energy scenarios is illustrated in Chapter 6. Seasonal variations of actual biomass land coverage, as deduced from satellite-based sensors, are regularly posted on the Internet (NOAA, 1998).

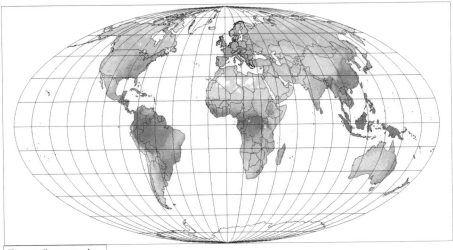

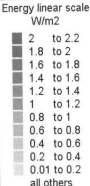

Energy linear scale W/m2

- 2 to 2.2
- 1.8 to 2
- 1.6 to 1.8
- 1.4 to 1.6
- 1.2 to 1.4
- 1 to 1.2
- 0.8 to 1
- 0.6 to 0.8
- 0.4 to 0.6
- 0.2 to 0.4
- 0.01 to 0.2
- all others

Figure 3.79. Annual average energy content in W m^{-2} of potential net biomass production in mature ecosystems (based on Melillo and Helfrich, 1998).

3.7 Other energy sources

The preceding parts of this chapter have been concerned with renewable energy flows and stores based on solar radiation, gravitation between celestial bodies, the mechanical energy involved in atmospheric and oceanic circulation, as well as a number of (sensible or latent) heat sources, the origin of which were solar or geological. It is believed that the most important such sources have been listed, but there could among those omitted be some that still have a certain amount of significance.

In the description of heat sources, in particular, the list is far from complete, considering that any difference in temperature constitutes a potential source of energy. It may not be practical to attempt to extract energy from the difference in air or soil temperature between the arctic and equatorial regions, but the corresponding difference in water temperature is, in fact, the basis for the temperature gradients in some oceanic regions (discussed in section 3.5.1), owing to the water transport by currents. One case in which a stable temperature difference might be usable in ways similar to those proposed for ocean temperature gradients is offered by the difference between the arctic air temperature and the water temperature at the base of the ice sheets covering arctic oceans. The average air temperature is about –25°C for the period October–March, but the temperature at the ice–water interface is necessarily close to 0°C (although not quite zero because of salinity- and pressure-induced depreciation of the freezing point, cf. Fig. 3.66).

It is natural to ask whether other energy forms present in the regime of the Earth, such as electromagnetic, earthquake, chemical (of which bio-energy has been treated in section 3.6) or nuclear energy, can be considered as energy resources and to what extent they are renewable. The rest of this section will briefly deal with examples of such energy forms.

3.7.1 Atmospheric electricity

The Earth is surrounded by a magnetic field, and from a height of about 80 km a substantial fraction of atmospheric gases is in an ionised state (cf. section 2.3.1). It is thus clear that electromagnetic forces play an important role in the upper atmosphere. Manifestations of electromagnetic energy in the lower parts of the atmosphere are well known in the form of lightning. Speculations on the possibility of extracting some of the electrical energy contained in a thunderstorm have appeared. An estimation of the amounts of energy involved requires knowledge of the electrical properties of the atmosphere.

In Fig. 3.80, calculated values of the electrical conductivity of the atmosphere are shown as a function of height for the lowest 200 km (Webb, 1976). It is shown that the conductivity is low in the troposphere, increases through the stratosphere and increases very sharply upon entering the ionosphere. Strong horizontal current systems are found in the region of high conductivity at a height of around 100 km (the "dynamo currents"). Vertical currents directed towards the ground are 6–7 orders of magnitude smaller for situations of fair weather, averaging about 10^{-12} A m^{-2} in the troposphere. Winds may play a role in generating the horizontal currents, but the energy input for the strong dynamo currents observed at mid-latitudes is believed to derive mainly from absorption of ultraviolet solar radiation by ozone (Webb, 1976; see also Fig. 2.16).

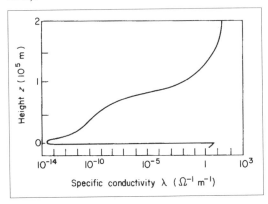

Figure 3.80. Electric conductivity of the atmosphere at noon on a clear day as a function of height (at about 10^5 m height, a substantial night-time reduction in λ takes place) (based on Webb, 1976).

The fair weather downward current of 10^{-12} A m^{-2} builds up a negative charge on the surface of the Earth, the total value of which on average is around 10^5 C. Locally, in the thunderstorm regions, the current is reversed and much more intense. Together these currents constitute a closed circuit, the flow through which is 500–1000 A, but in such a way that the downward path is dispersed over most of the atmosphere, while the return current in the form of lightning is concentrated in a few regions (apart from being intermittent, causing time variations in the charge at the Earth's surface). The power P associated with the downward current may be found from

$$P = \int I^2\, \lambda^{-1}\, dz,$$

where I ($\approx 10^{-12}$ A m^{-2}) is the average current and $\lambda(z)$ is the conductivity function (e.g. Fig. 3.80 at noon), or more simply from

$$P = IV = I\, Q/C,$$

where C is the capacitance of the Earth relative to infinity (V the corresponding potential difference) and Q ($\approx 10^5$ C) is its charge. The capacitance of a sphere with the Earth's radius r_s (6.4×10^6 m) is $C = 4\pi\, \varepsilon_0\, r_s = 7 \times 10^{-4}$ F (ε_0 being the dielectric constant for vacuum), and thus the average power becomes $P \approx 6 \times 10^{-5}$ W m^{-2}.

The energy stored in the charge Q of the Earth's surface, relative to a situation in which the charges were all moved to infinity, is

$$W_\infty^{electric} = \frac{1}{2} Q^2/C \approx 3 \times 10^{12} \text{ J}.$$

In "practice", the charges could only be moved up to the ionosphere (potential difference less than 10^6 V), reducing the value of the stored energy by almost two orders of magnitude. In any case, the charge of the Earth, as well as the trophospheric current system, constitutes energy sources of very small magnitude, even if the more concentrated return flux of lightning could be utilised (multiplying the average power estimate by the area of the Earth, one obtains a total power of 3×10^{10} W).

3.7.2 Salinity differences

Useful chemical energy may be defined as energy that can be released through exotermic chemical reactions. In general, chemical energy is associated with chemical bindings of electrons, atoms and molecules. The bindings may involve overlapping electron wavefunctions of two or more atoms, attraction between ionised atoms or molecules, and long-range electromagnetic fields created by the motion of charged particles (notably electrons). In all cases, the physical interaction involved is the Coulomb force. Examples of chemical energy connected with molecular binding structure have been given in section 3.6 (bio-energy sources, including fossil fuels).

The organisation of atoms or molecules in regular lattice structures represents another manifestation of chemical bindings. Some substances possess different crystalline forms, which may exist under given external conditions. In addition to the possibility of different solid phases, phase changes associated with transitions among solid, liquid and gas phases all represent different levels of latent energy. Examples in which such latent energy differences have been considered as potential energy sources have been mentioned, for example, in section 3.4.2.

Solutions represent another form of chemical energy, relative to the pure solvent. The free energy of a substance with components $i = 1, 2,...$, there being n_i mol of the ith component, may be written

$$G = \sum_i n_i \mu_i, \tag{3.49}$$

where μ_i is called the "chemical potential" of component i. For a solution, μ_i can be expressed in the form (see e.g. Maron and Prutton, 1959)

$$\mu_i = \mu_i^0 + \mathscr{R} T \log (f_i x_i), \tag{3.50}$$

where $\mathscr{R}$ is the gas constant (8.3 J K^{-1} mol^{-1}), T is the temperature (K) and $x_i = n_i/(\Sigma_j n_j)$ the mole fraction. μ_i^0 is the chemical potential that would correspond to $x_i = 1$ at the given pressure P and temperature T, and f_i is the "activity coefficient", an empirical constant which approaches unity for "ideal solutions", an example of which is the solvent of a very dilute solution (whereas, in general, f_i cannot be expected to approach unity for the dissolved component of a dilute solution).

It follows from (3.49) and (3.50) that a solution represents a lower chemical energy than the pure solvent. The most common solution present in large amounts on the Earth is saline ocean water. Relative to this, pure or fresh water such as river run-off represent an elevated energy level. In addition, there are salinity differences within the oceans, as shown in Figs. 2.63–2.65.

Taking the average ocean salinity as about 33×10^{-3} (mass fraction), and regarding this entirely as ionised NaCl, $n_{Na+} = n_{Cl-}$ becomes about 0.56×10^3 mol and $n_{water} = 53.7 \times 10^3$ mol, considering a volume of 1 m^3. The chemical potential of ocean water, μ, relative to that of fresh water, μ^0, is then from (3.50)

$$\mu - \mu^0 = \mathscr{R} \, T \log x_{water} \approx -2\mathscr{R} \, T \, n_{Na+} / n_{water}.$$

Consider now a membrane which is permeable for pure water but impermeable for salt (i.e. for Na$^+$ and Cl$^-$ ions) as indicated in Fig. 3.81. On one side of the membrane, there is pure (fresh) water, on the other side saline (ocean) water. Fresh water will flow through the membrane, trying to equalise the chemical potentials μ^0 and μ initially prevailing on each side. If the ocean can be considered as infinite and being rapidly mixed, then n_{Na+} will remain fixed, also in the vicinity of the membrane. In this case each m^3 of freshwater penetrating the membrane and becoming mixed will release an amount of energy, which from (3.49) is

$$\delta G = \sum_i (n_i \delta \mu_i + \mu_i \delta n_i) \approx n_{water} (\mu^0 - \mu) \approx 2\mathscr{R} \, T n_{Na^+}. \tag{3.51}$$

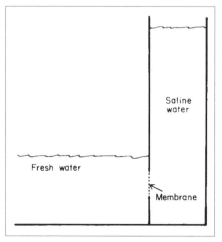

Figure 3.81. Schematic picture of an osmotic pump. In order to mix the fresh water penetrating the semi-permeable membrane in the direction towards the right and to maintain the salinity in the salt water compartment, new saline water would have to be pumped into the salt water compartment, and water motion near the membrane would have to be ensured.

For a temperature $T \approx 285$ K (considered fixed), $\delta G \approx 2.65 \times 10^6$ J. The power corresponding to a fresh-water flow of 1 m^3 s^{-1} is thus 2.65×10^6 W (cf. Norman, 1974). The world-wide run-off of about 4×10^{13} m^3 y^{-1} (Fig. 2.61) would thus correspond to an average power of around 3×10^{12} W.

The arrangement schematically shown in Fig. 3.81 is called an osmotic pump. The flow of pure water into the tube will ideally raise the water level in the tube, until the pressure of the water head balances the force derived from the difference in chemical energy. The magnitude of this "osmotic pressure", P^{osm}, relative to the atmospheric pressure P_0 on the fresh water surface, is found from the thermodynamic relation

$$V \, dP - S \, dT = \Sigma_i n_i \, d\mu_i,$$

where V is the volume, S is the entropy and T is the temperature. Assuming that the process will not change the temperature (i.e. considering the ocean a large reservoir of fixed temperature), insertion of (3.51) yields

$$P^{osm} = \delta P \approx n_{water} \, V^{-1} \, \delta\mu_{water} \approx 2\mathscr{R} \, T \, n_{Na+} \, V^{-1}. \tag{3.52}$$

Inserting the numerical values of the example above, $P^{osm} = 2.65 \times 10^6$ N m^{-2}, corresponding to a water-head some 250 m above the fresh water surface. If the assumption of fixed mole fraction of salt in the tube is to be realised, it would presumably be necessary to pump saline water into the tube. The energy spent for pumping, however, would be mostly recoverable, since it also adds to the height of the water-head, which may be used to generate electricity as in a hydropower plant.

An alternative way of releasing the free energy difference between solutions and pure solvents is possible when the dissolved substance is ionised (the solution is then called electrolytic). In this case direct conversion to electricity is possible, as further discussed in Chapter 4.

3.7.3 Nuclear energy

The atomic nuclei (consisting of protons and neutrons) carry the bulk of the mass of matter on Earth as well as in the known part of the present universe (cf. section 2.A). By a nucleus is usually understood a bound system of Z protons and N neutrons (unbound systems, resonances, may be formed under laboratory conditions and they are observed in cosmic ray showers for short periods of time). Such a nucleus contains an amount of nuclear binding energy, given by

$$E_{Z,N} - (NM_n + ZM_p) \, c^2 = -B,$$

with the difference between the actual energy $E_{Z,N}$ of the bound system and the energy corresponding to the sum of the masses of the protons and neutrons if these were separated from each other. It thus costs energy to separate all the nucleons, and this is due to the attractive nature of most nuclear forces.

However, if the binding energy B (i.e. the above energy difference with opposite sign) is evaluated per nucleon, one obtains a maximum around ^{56}Fe (cf. section 2.A), with lower values for both lighter and heavier nuclei. Figure 3.82 shows the trends of $-B/A$, where $A = Z + N$ is the nucleon number. For each A, only the most tightly bound nucleus $(Z, A - Z)$ has been included, and only doubly even nuclei have been included (if Z or N is odd, the binding energy is about 1 MeV lower).

This implies that nuclear material away from the iron region could gain binding energy (and thereby release nuclear energy) if the protons and neutrons could be re-structured to form ^{56}Fe. The reason why this does not happen spontaneously for all matter on Earth is that potential barriers separate the present states of nuclei from that of the most stable nucleus and that very few nuclei are able to penetrate these barriers at the temperatures prevailing on Earth.

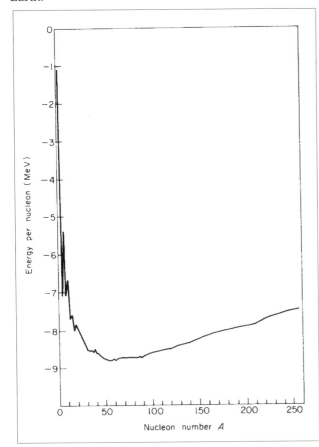

Figure 3.82. Trends of nuclear binding energy of bound nuclei (taken per nucleon) relative to a hypothetical situation in which all nucleons are separated (see text) (based on Bohr and Mottelson, 1969).

A few nuclei do spontaneously transform to more tightly bound systems (the natural radioactive nuclei mentioned in section 3.5.2), but the rate at which they penetrate the corresponding "barriers" is low, since otherwise they would no longer be present on Earth now, some 5×10^9 years after they were formed. As mentioned in section 3.5.2, these nuclei are responsible for 40% of the average heat flow at the surface of continents and contribute almost nothing to the heat flow at the ocean floors.

As schematically illustrated in Fig. 3.83, the barrier which a nucleus of $A \approx$ 240 must penetrate in order to fission is typically a very small fraction (2–3%) of the energy released by the fission process. This barrier has to be penetrated by quantum tunnelling. The width of the barrier depends on the state of the initial nucleus. It is smaller if the nucleus is in an excited state rather than in its ground state. Some heavy isotopes, such as ^{235}U, easily absorb neutrons, forming a compound system with dramatically reduced fission barrier and hence with dramatically increased fission probability. This process is called "induced fission", and it implies that by adding a very small amount of energy to a "fissile" element such as ^{235}U, by slow-neutron bombardment, a fission energy of some 200 MeV can be released, mostly in the form of kinetic energy of the products, but a few per cent usually occurring as delayed radioac-

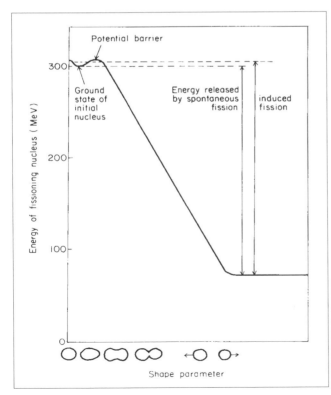

Figure 3.83. Schematic view of a nuclear fission process described by a "one-dimensional" shape parameter (axially symmetric shapes are indicated along the lower boundary of the figure). The zero point of the energy scale on the ordinate corresponds to the energy of a (fractional) number of ^{56}Fe nuclei with mass identical to that of the fissioning nucleus.

tivity of unstable fragments. An example of binary fission, i.e. with two end-nuclei plus a number of excess neutrons (which can induce further fission reactions), is

$$_{92}^{235}\text{U} + n \rightarrow \, _{54}^{142}\text{Xe} + _{38}^{92}\text{Sr} + 2n.$$

The probability of finding an asymmetrical mass distribution of the fragments is often larger than the probability of symmetric fission. The reason why further fission processes yielding nuclei in the $_{26}^{56}\text{Fe}$ region do not occur is the high barrier against fission for nuclei with $Z^2/A \lesssim 30$ and the low probability of direct fission of, say, uranium into three fragments closer to the iron region (plus a larger number of neutrons).

The amount of recoverable fissile material in the Earth's crust would correspond to a fission energy of about 10^{22} J (Ion, 1975). However, other heavy elements than those fissioning under slow-neutron bombardment can be made fissile by bombardment with more energetic neutrons or some other suitable nuclear reaction, spending an amount of energy much smaller than that which may later be released by fission. An example is $_{92}^{238}\text{U}$, which may absorb a neutron and emit two electrons, thus forming $_{94}^{239}\text{Pu}$, which is a "fissile" element, capable of absorbing neutrons to form $_{94}^{240}\text{Pu}$, the fission cross section of which is appreciable. Including resources by which fissile material may be "bred" is this way, a higher value may be attributed to the recoverable energy resource associated with nuclear fission. This value depends both on the availability of resource material such as ^{238}U (the isotope ratio ^{238}U to ^{235}U is practically the same for all geological formations of terrestrial origin) and on the possibility of constructing devices with a high "breeding ratio". Some estimates indicate a 60-fold increase over the energy value of the "non-breeder" resource (World Energy Conference, 1974).

As indicated in Fig. 3.82, the energy gain per nucleon which could be released by fusion of light elements is several times larger than that released by fission of heavy elements. A number of possible fusion processes have been mentioned in section 2.A. These reactions take place in stars at high temperature. On Earth, they have been demonstrated in the form of explosive weapons, with the temperature and pressure requirements being provided by explosive fission reactions in a blanket around the elements to undergo fusion. Controlled energy production by fusion reactions such as $^2\text{H}+^3\text{H}$ or $^2\text{H}+^2\text{H}$ is under investigation, but at present the necessary temperatures and confinement requirements have not been met. Theoretically, the fusion processes of elements with mass below the iron region are, of course, highly exenergetic, and energy must be added only to ensure a sufficient collision frequency. In practice, it may take a while to realise devices for energy extraction for which the net energy gain is at all positive.

In theory, the nuclear fusion energy resources are much larger than the fission resources, but in any case only a fraction of the total nuclear energy on

Earth (relative to an all-iron state) could ever become energy resources on a habitable planet. In principle, the nuclear energy resources are clearly non-renewable, and the recoverable amounts of fissionable elements do not appear large compared with the possible requirements of man during the kind of time span for which he may hope to inhabit the planet Earth. However, it cannot be denied that the fusion resources might sustain man's energy expenditure for a long time, if, for example, the reaction chain entirely based on naturally occurring deuterium could be realised [cf. the cosmological reactions occurring some 300 s after the singularity in the Big Bang theory (section 2.A), at temperatures between 10^9 and 10^8 K]:

$$\textstyle{}^2_1\text{H} + {}^2_1\text{H} \rightarrow {}^1_1\text{H} + {}^3_1\text{H} + 3.25 \text{ MeV},$$

$$\textstyle{}^2_1\text{H} + {}^2_1\text{H} \rightarrow {}^1_0 n + {}^3_2\text{He} + 4.0 \text{ MeV},$$

$$\textstyle{}^2_1\text{H} + {}^3_1\text{H} \rightarrow {}^1_0 n + {}^4_2\text{He} + 17.6 \text{ MeV},$$

$$\textstyle{}^2_1\text{H} + {}^3_2\text{He} \rightarrow {}^1_1\text{H} + {}^4_2\text{He} + 18.3 \text{ MeV}.$$

The abundance of ^{2}H in sea water is about 34×10^{-6} (mass fraction). The potential nuclear energy released by one of the deuterium–to–helium fusion chains is thus 10^{13} J m^{-3} of sea water, or over 10^{31} J for all the oceans (an energy storage equivalent to the entire thermal energy stored in the interior of the Earth, cf. section 3.5.2).

The prime environmental concern over utilisation of fission or fusion energy is the inherent change in the radioactive environment. The fragments formed by fission reactions cover a wide range of nuclear isotopes, most of which are unstable and emit nuclear radiation of gamma or particle type (see e.g. Holdren, 1974). Also, the fusion devices are likely to produce large amounts of radioactivity, because the high-energy particles present in the reaction region may escape and experience frequent collisions with the materials forming the "walls", the confinement, and thereby induce nuclear reactions. The range of radioactive elements formed by fusion reactions can be partially optimised from an environmental point of view (minimising the production of the most biologically hazardous isotopes) by choosing appropriate materials for the confinement, but the choice of materials is also limited by the temperature and stability requirements (see e.g. Post and Ribe, 1974).

Large-scale implementation of fission- or fusion-based energy conversion schemes will raise the question of whether it will be possible safely to manage and confine the radioactive "wastes" that, if released to the general environment, could cause acute accidents and long-range alterations in the radiological environment to which man is presently adapted. These dangers exist equally for the use of nuclear reactions in explosive weapons, in addition to the destructive effect of the explosion itself, and in this case no attempt is even made to confine or control the radioactive material.

In addition to possible health hazards, the use of nuclear energy may also have climatic consequences, e.g. associated with the enhanced heat flow (during nuclear war; cf. Feld, 1976) or with the routine emissions of radioactive material (e.g. ^{85}Kr from the uranium fuel cycle; cf. Boeck, 1976).

In summary, nuclear energy production based on the existing once-through reactors constitutes a parenthesis in history, given the very limited amounts of fissile resources available for this mode of operation, comparable at most to those of oil. Nuclear fusion research has been ongoing for more than 50 years, so far with little success. Commercialisation is still predicted to happen some 50 years into the future, just as it was at any earlier stage.

3.8 Suggested topics for discussion

3.8.1

Discuss the reflected flux of solar radiation on an inclined plane, in front of which a horizontal mirror of given dimensions has been placed (cf. e.g. Seitel, 1975).

3.8.2

Which data on solar radiation, wind, waves, etc. are available in your region (e.g. check with the local meteorological service)?

3.8.3

Discuss the cooling of the Earth in terms of simplified models such as (a) a uniform sphere of constant heat capacity and diffusion coefficient, initially at a uniform temperature, (b) adding radiogenic heat production in the crust, by constraining the surface temperature to be the sum of two terms, one entirely due to the heat flow from the interior, and the other term exponentially decaying, such that the drop in 4.5×10^9 years becomes 900 K.

Use present surface heat flow and temperature (average values) as boundary conditions, and discuss the long-range development of the thermal conditions in the interior of the Earth (cf. e.g. Goguel, 1976). Recent models suggest a core temperature of nearly 4000 K (Hofmeister, 1999).

3.8.4

Consider a compartment model for biological energy production and transfer within a closed ecosystem of the general form shown in Fig. 3.73. Assume, for instance, that the limiting factor governing the rate of primary production is solar radiation, such that the production becomes proportional to a schematic solar input, constant or with only seasonal sinusoidal variations [cf. (3.47)]. Think of a food chain, in which you may find data on the rates and biomass levels appearing in Fig. 3.73 or some modification of it, and try to determine conditions for stability (no net production of the community as a whole).

If you have a computer at hand, you might try to set up a numerical simulation model for the system, assuming each rate (time derivative of a compartment level) to be a linear combination of all the compartment levels (some coefficients may be zero, of course). In this case you may also be able to follow the growth phase of the system, and attempt to judge whether the parameter values chosen are reasonable (cf. e.g. Odum, 1972, where examples of parameter values may also be found; Patten, 1971, 1972; Sørensen, 1975a).

3.8.5

Use the current biomass production data (available on the Internet) mentioned in connection with Fig. 3.79 to estimate possible bio-energy sources present in your region and their seasonal distribution. Does the seasonality have implications for energy use? Compare your results with the model considered in Chapter 6.

THE ENERGY CONVERSION PROCESSES

4.1 Basic principles of energy conversion

A large number of energy conversion processes take place in nature, some of which were described in Chapters 2 and 3. Man is capable of performing a number of additional energy conversion processes by means of various devices invented during the history of man. Such devices may be classified according to the type of construction used, according to the underlying physical or chemical principle, or according to the forms of energy appearing before and after the action of the device. In this chapter, a survey of conversion methods, which may be suitable for the conversion of renewable energy flows or stored energy, will be given. A discussion of general conversion principles will be made below, followed by an outline of engineering design details for specific energy conversion devices, ordered according to the energy form being converted and the energy form obtained. The collection is necessarily incomplete and involves judgement about the importance of various devices.

4.1.1 Conversion between energy forms

For a number of energy forms, Table 4.1 lists some examples of energy conversion processes or devices currently in use or contemplated, organised according to the energy form emerging after the conversion. In several cases more than one energy form will emerge as a result of the action of the device, e.g. heat in addition to one of the other energy forms listed. Many devices also perform a number of energy conversion steps, rather than the single ones given in the table. A power plant, for example, may perform the conversion process chain between energy forms: chemical → heat → mechanical → electrical. Diagonal transformations are also possible, such as conversion of mechanical energy into mechanical energy (potential energy of elevated fluid → kinetic energy of flowing fluid → rotational energy of turbine) or of heat into heat at a lower temperature (convection, conduction). A process in which the only change is that heat is transferred from a lower to a higher temperature is forbidden by the second law of thermodynamics. Such transfer can be established if at the same time some high-quality energy is degraded, e.g. by a heat pump (which is listed as a converter of electrical into heat energy in Table 4.1, but is further discussed in the heat conversion section, 4.6.1).

Initial energy form	Converted energy form				
	Chemical	Radiant	Electrical	Mechanical	Heat
Nuclear					Reactor
Chemical			Fuel cell, battery discharge		Burner, boiler
Radiant	Photolysis		Photovoltaic cell		Absorber
Electrical	Electrolysis, battery charging	Lamp, laser		Electric motor	Resistance, heat pump
Mechanical			Electric generator, MHD	Turbines	Friction, churning
Heat			Thermionic & thermoelectric generators	Thermodynamic engines	Convector, radiator, heat pipe

Table 4.1. Examples of energy conversion processes listed according to the initial energy form and one particular converted energy form (the one primarily wanted).

The efficiency with which a given conversion process can be carried out, i.e. the ratio between the output of the desired energy form and the energy input, depends on the physical and chemical laws governing the process. For the heat engines, which convert heat into work or vice versa, the description of thermodynamic theory may be used in order to avoid the complication of a microscopic description on the molecular level (which is, of course, possible, e.g. on the basis of statistical assumptions). According to thermodynamic theory (again the "second law"), no heat engine can have an efficiency higher than that of a reversible Carnot process, which is depicted in Fig. 4.1, in terms of different sets of thermodynamic state variables,

$(P, V) = $ (pressure, volume),
$(T, S) = $ (absolute temperature, entropy),

and

$(H, S) = $ (enthalpy, entropy).

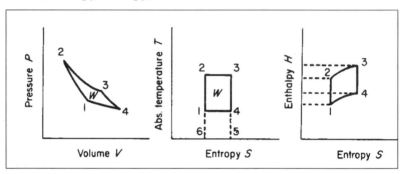

Figure 4.1. The cyclic Carnot process in different representations. Traversing the cycle in the direction $1 \to 2 \to 3 \to 4$ leads to the conversion of a certain amount of heat into work (see text for details).

The entropy was defined in (1.1), apart from an arbitrary constant fixed by the third law of thermodynamics (Nernst's law), which states that S can be taken as zero at zero absolute temperature $(T = 0)$. The enthalpy H is defined by

$$H = U + PV, \tag{4.1}$$

in terms of P, V and the internal energy U of the system. According to the first law of thermodynamics, U is a state variable given by

$$\Delta U = \int dQ + \int dW, \tag{4.2}$$

in terms of the amounts of heat and work added to the system [Q and W are not state variables, and the individual integrals in (4.2) depend on the paths of integration]. The equation (4.2) determines U up to an arbitrary constant, the zero point of the energy scale. Using the definition (1.1),

$$dQ = T\, dS$$

and

$$dW = - P\, dV,$$

both of which are valid only for reversible processes The following relations are found among the differentials:

$$dU = T\,dS - P\,dV,$$
$$dH = T\,dS + V\,dP. \tag{4.3}$$

These relations are often assumed to have general validity.

If chemical reactions occur in the system, additional terms $\mu_i\,dn_i$ should be added on the right-hand side of both relations (4.3), in terms of the chemical potentials, which were discussed briefly in section 3.7.2.

For a cyclic process such as the one shown in Fig. 4.1, $\int dU = 0$ upon re-turning to the initial locus in one of the diagrams, and thus according to (4.3) $\int T\,dS = \int P\,dV$. This means that the area enclosed by the path of the cyclic process in either the (P, V) or the (T, S) diagram equals the work $-W$ per-formed by the system during one cycle (in the direction of increasing num-bers on Fig. 4.1).

The amount of heat added to the system during the isothermal process 2-3 is $\Delta Q_{23} = T(S_3 - S_2)$, if the constant temperature is denoted T. The heat added in the other isothermal process, 4-1, at a temperature T_{ref} is $\Delta Q_{41} = -T_{ref}(S_3 - S_2)$. It follows from the (T, S) diagram that $\Delta Q_{23} + \Delta Q_{41} = -W$. The efficiency by which the Carnot process converts heat available at temperature T into work, when a reference temperature of T_{ref} is available, is then

$$\eta = \frac{-W}{\Delta Q_{23}} = \frac{T - T_{ref}}{T}. \tag{4.4}$$

The Carnot cycle (Fig. 4.1) consists of four steps: 1-2, adiabatic compres-sion (no heat exchange with the surroundings, i.e. $dQ = 0$ and $dS = 0$); 2-3, heat drawn reversibly from the surroundings at constant temperature (the amount of heat transfer ΔQ_{23} is given by the area enclosed by the path 2-3-5-6-2 in the (T, S)-diagram); 3-4, adiabatic expansion; and 4-1, heat given away to the surroundings by a reversible process at constant temperature $[|\Delta Q_{41}|$ equal to the area of the path 4-5-6-1-4 in the (T, S)-diagram].

The (H, S)-diagram is an example of a representation in which energy dif-ferences can be read directly on the ordinate, rather than being represented by an area.

It requires long periods of time to perform the steps involved in the Car-not cycle in a way that approaches reversibility. As time is important for man (the goal of the energy conversion process being power rather than just an amount of energy), irreversible processes are deliberately introduced into the thermodynamic cycles of actual conversion devices. The thermodyna-mics of irreversible processes are described below using a practical ap-proximation, which will be referred to in several of the examples to follow. Some devices of minor importance will be deferred to the advanced section

4.A at the end of this chapter, and readers without specific interest in the thermodynamic description may go lightly over the formulae.

Irreversible thermodynamics

The degree of irreversibility is measured in terms of the rate of energy dissipation,

$$D = T \, dS/dt, \tag{4.5}$$

where dS/dt is the entropy production of the system while held at the constant temperature T (i.e. T may be thought of as the temperature of a large heat reservoir, with which the system is in contact). In order to describe the nature of the dissipation process, the concept of "free energy" may be introduced (cf. Prigogine, 1968; Callen, 1960).

The free energy, G, of a system is defined as the maximum work that can be drawn from the system under conditions where the exchange of work is the only interaction between the system and its surroundings. A system of this kind is said to be in thermodynamic equilibrium if its free energy is zero.

Consider now a system divided into two subsystems, a small one with extensive variables (i.e. variables proportional to the size of the system) U, S, V, etc. and a large one with intensive variables T_{ref}, P_{ref}, etc., which is initially in thermodynamic equilibrium. The terms "small system" and "large system" are meant to imply that the intensive variables of the large system (but not its extensive variables U_{ref}, S_{ref}, etc.) can be regarded as constant, regardless of the processes by which the entire system approaches equilibrium.

This implies that the intensive variables of the small system, which may not even be defined during the process, approach those of the large system when the combined system approaches equilibrium. The free energy, or maximum work, is found by considering a reversible process between the initial state and the equilibrium. It equals the difference between the initial internal energy, $U_{init} = U + U_{ref}$, and the final internal energy, U_{eq}, or it may be written (all in terms of initial state variables) as

$$G = U - T_{ref} S + P_{ref} V, \tag{4.6}$$

plus terms of the form $\Sigma \mu_{i,ref} n_i$ if chemical reactions are involved, and similar generalisations in case of electromagnetic interactions, etc.

If the entire system is closed it develops spontaneously towards equilibrium through internal, irreversible processes, with a rate of free energy change

$$\frac{dG}{dt} = \frac{d}{dt} (U_{init} - U_{eq}(t)) = \left(\frac{\partial}{\partial S(t)} U_{eq}(t) \right) \frac{dS(t)}{dt},$$

assuming that the entropy is the only variable. $S(t)$ is the entropy at time t of the entire system, and $U_{eq}(t)$ is the internal energy that would be possessed by a hypothetical equilibrium state defined by the actual state variables at time t, i.e. $S(t)$ etc. For any of these equilibrium states, $\partial U_{eq}(t)/\partial S(t)$ equals T_{ref} according to (4.3), and by comparison with (4.5) it is seen that the rate of dissipation can be identified with the loss of free energy, as well as with the increase in entropy,

$$D = -dG/dt = T_{ref}\, dS(t)/dt. \tag{4.7}$$

For systems met in practice, there will often be constraints preventing the system from reaching the absolute equilibrium state of zero free energy. For instance, the small system considered above may be separated from the large one by walls keeping the volume V constant. In such cases the available free energy (i.e. the maximum amount of useful work that can be extracted) becomes the absolute amount of free energy, (4.6), minus the free energy of the relative equilibrium which the combined system can be made to approach in the presence of the constraint. If the extensive variables in the constrained equilibrium state are denoted U^0, S^0, V^0, etc., then the available free energy becomes

$$\Delta G = (U - U^0) - T_{ref}(S - S^0) + P_{ref}(V - V^0), \tag{4.8}$$

eventually with the additions involving chemical potentials, etc. In the form (4.6) or (4.8), G is called the Gibbs potential. If the small system is constrained by walls, so that the volume cannot be changed, the free energy reduces to the Helmholtz potential $U - TS$, and if the small system is constrained so that it is incapable of exchanging heat, the free energy reduces to the enthalpy (4.1). The corresponding forms of (4.8) give the maximum work that can be obtained from a thermodynamic system with the given constraints.

A description of the course of an actual process as a function of time requires knowledge of "equations of motion" for the extensive variables, i.e. equations which relate the currents such as

$$J_s = dS/dt \text{ (entropy flow rate) or } J_Q = dQ/dt \text{ (heat flow rate)},$$

$$J_m = dm/dt \text{ (mass flow rate) or } J_\theta = d\theta/dt \text{ (angular velocity)}, \tag{4.9}$$

$$J_q = dq/dt = I \text{ (charge flow rate or electrical current), etc.}$$

to the (generalised) forces of the system. As a first approximation, the relation between the currents and the forces may be taken as linear (Onsager, 1931),

$$J_i = \sum_j L_{ij}\, F_j. \tag{4.10}$$

The direction of each flow component is J_i / J_i. The arbitrariness in choosing the generalised forces is reduced by requiring, as did Onsager, that the dissipation be given by

$$D = -dG/dt = \Sigma_i J_i \cdot F_i. \tag{4.11}$$

Examples of the linear relationships (4.10) are Ohm's law, stating that the electric current J_q is proportional to the gradient of the electric potential ($F_q \propto$ grad ϕ), and Fourier's law (3.34) for heat conduction or diffusion, stating that the heat flow rate $E^{sens} = J_Q$ is proportional to the gradient of the temperature.

Considering the isothermal expansion process required in the Carnot cycle (Fig. 4.1), heat must be flowing to the system at a rate $J_Q = dQ/dt$, with $J_Q = LF_Q$ according to (4.10) in its simplest form. Using (4.11), the energy dissipation takes the form

$$D = T \, dS/dt = J_Q F_Q = L^{-1} J_Q^2.$$

For a finite time Δt, the entropy increase becomes

$$\Delta S = (dS/dt) \, \Delta t = (LT)^{-1} J_Q^2 \, \Delta t = (LT\Delta t)^{-1} (\Delta Q)^2,$$

so that in order to transfer a finite amount of heat ΔQ, the product $\Delta S \, \Delta t$ must equal the quantity $(LT)^{-1} (\Delta Q)^2$. In order that the process approaches reversibility, as the ideal Carnot cycle should, ΔS must approach zero, which is seen to imply that Δt approaches infinity. This qualifies the statement made in the beginning of this subsection that, in order to go through a thermodynamic engine cycle in a finite time, one has to give up reversibility and accept a finite amount of energy dissipation and an efficiency which is smaller than the ideal one (4.4).

Efficiency of an energy conversion device

A schematic picture of an energy conversion device is shown in Fig. 4.2, sufficiently general to cover most types of converters in practical use (Angrist, 1976; Osterle, 1964). There is a mass flow into the device and another one out from it, as well as an incoming and outgoing heat flow. The work output may be in the form of electric or rotating shaft power.

It may be assumed that the converter is in a steady state, implying that the incoming and outgoing mass flows are identical and that the entropy of the device itself is constant, i.e. that all entropy created is being carried away by the outgoing flows.

From the first law of thermodynamics, the power extracted, E, equals the net energy input,

$$E = J_{Q,in} - J_{Q,out} + J_m (w_{in} - w_{out}). \tag{4.12}$$

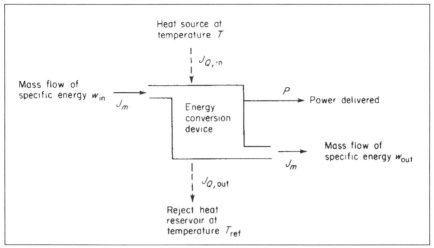

Figure 4.2. Schematic picture of an energy conversion device with a steady-state mass flow. The sign convention is different from the one used in (4.2), where all fluxes into the system were taken as positive.

The magnitude of the currents is given by (4.9), and their conventional signs may be inferred from Fig. 4.2. The specific energy content of the incoming mass flow, w_{in}, and of the outgoing mass flow, w_{out}, are the sums of potential energy, kinetic energy and enthalpy. The significance of the enthalpy to represent the thermodynamic energy of a stationary flow is established by Bernoulli's theorem (Pippard, 1966). It states that for a stationary flow, if heat conduction can be neglected, the enthalpy is constant along a streamline. For the uniform mass flows assumed for the device in Fig. 4.2, the specific enthalpy, h, thus becomes a property of the flow, in analogy with the kinetic energy of motion and, for example, the geopotential energy,

$$w = w^{pot} + w^{kin} + h. \tag{4.13}$$

The power output may be written

$$E = -J_\theta \cdot F_\theta - J_q \cdot F_q, \tag{4.14}$$

with the magnitude of currents given by (4.9) and the generalised forces given by

$$F_\theta = \int r \times dF_{mech}(r) \qquad \text{(torque)},$$
$$F_q = -\text{grad}(\phi) \qquad \text{(electric field)} \tag{4.15}$$

corresponding to a mechanical torque and an electric potential gradient. The rate of entropy creation, i.e. the rate of entropy increase in the surroundings of the conversion device (as mentioned, the entropy inside the device is constant in the steady-state model), is

$$dS/dt = (T_{ref})^{-1} J_{Q,out} - T^{-1} J_{Q,in} + J_m (s_{m,out} - s_{m,in}),$$

where $s_{m,in}$ is the specific entropy of the mass (fluid, gas, etc.) flowing into the device, and $s_{m,out}$ is the specific entropy of the outgoing mass flow. $J_{Q,out}$ may be eliminated by use of (4.12), and the rate of dissipation obtained from (4.7),

$$D = T_{ref} \, dS/dt =$$
$$J_{Q,in} (1 - T_{ref}/T) + J_m (w_{in} - w_{out} - T_{ref} (s_{m,in} - s_{m,out})) - E = \max(E) - E. \quad (4.16)$$

The maximum possible work (obtained for $dS/dt = 0$) is seen to consist of a Carnot term (closed cycle, i.e. no external flows) plus a term proportional to the mass flow. The dissipation (4.16) is brought in the Onsager form (4.11),

$$D = J_{Q,in} F_{Q,in} + J_m F_m + \mathbf{J}_\theta \cdot \mathbf{F}_\theta + \mathbf{J}_q \cdot \mathbf{F}_q, \quad (4.17)$$

by defining generalised forces

$$F_{Q,in} = 1 - T_{ref}/T,$$
$$F_m = w_{in} - w_{out} - T_{ref} (s_{m,in} - s_{m,out}) \quad (4.18)$$

in addition to those of (4.15).

The efficiency with which the heat and mass flow into the device is converted to power is, in analogy to (4.4),

$$\eta = \frac{E}{J_{Q,in} + J_m w_{in}}, \quad (4.19)$$

where the expression (4.16) may be inserted for E. This efficiency is sometimes referred to as the "first law efficiency", because it only deals with the amounts of energy input and output in the desired form and not with the "quality" of the energy input related to that of the energy output.

In order to include reference to the energy quality, in the sense of the second law of thermodynamics, account must be taken of the changes in entropy taking place in connection with the heat and mass flows through the conversion device. This is accomplished by the "second law efficiency", which for power-generating devices is defined by

$$\eta^{(2.law)} = \frac{E}{\max(E)} = -\frac{\mathbf{J}_\theta \cdot \mathbf{F}_\theta + \mathbf{J}_q \cdot \mathbf{F}_q}{J_{Q,in} F_{Q,in} + J_m F_m}, \quad (4.20)$$

where the second expression is valid specifically for the device considered in Fig. 4.2, while the first expression is of general applicability, when $\max(E)$ is taken as the maximum rate of work extraction permitted by the second law of thermodynamics. It should be noted that $\max(E)$ depends not only on the system and the controlled energy inputs, but also on the state of the surroundings.

Conversion devices for which the desired energy form is not work may be treated in a way analogous to the example in Fig. 4.2. In the form (4.17), no distinction is made between input and output of the different energy forms. Taking, for example, electrical power as input (sign change), output may be obtained in the form of heat or in the form of a mass stream. The efficiency expressions (4.19) and (4.20) must be altered, placing the actual input terms in the denominator and the actual output terms in the numerator. If the desired output energy form is denoted W, the second law efficiency can be written in the general form

$$\eta^{(2.\ law)} = W / \max\ (W). \tag{4.21}$$

For conversion processes based on other principles than those considered in the thermodynamic description of phenomena, alternative efficiencies could be defined by (4.21), with max(W) calculated under consideration of the non-thermodynamic types of constraints. In such cases, the name "second law efficiency" would have to be modified.

4.1.2 Thermodynamic engine cycles

A number of thermodynamic cycles, i.e. (closed) paths in a representation of conjugate variables, have been demonstrated in practice. They offer examples of the compromises made in modifying the "prototype" Carnot cycle into a cycle that can be traversed in a finite amount of time. Each cycle can be used to convert heat into work, but in traditional uses the source of heat has mostly been the combustion of fuels, i.e. an initial energy conversion process, by which high-grade chemical energy is degraded to heat at a certain temperature, associated with a certain entropy production.

Figure 4.3 shows a number of engine cycles in (P, V)-, (T, S)- and (H, S)-diagrams corresponding to Fig. 4.1.

The working substance of the Brayton cycle is a gas, which is adiabatically compressed in step 1-2 and expanded in step 3-4. The remaining two steps take place at constant pressure (isobars), and heat is added in step 2-3. The useful work is extracted during the adiabatic expansion 3-4, and the simple efficiency is thus equal to the enthalpy difference $H_3 - H_4$ divided by the total input $H_3 - H_1$. Examples of devices operating on the Brayton cycle are gas turbines and jet engines. In these cases, the cycle is usually not closed, since the gas is exhausted at point 4 and step 4-1 is thus absent. The somewhat contradictory name given to such processes is "open cycles".

The Otto cycle, presently used in a large number of automobile engines, differs from the Brayton cycle in that steps 2–3 and 4–1 (if the cycle is closed) are carried out at constant volume (isochores) rather than at constant pressure.

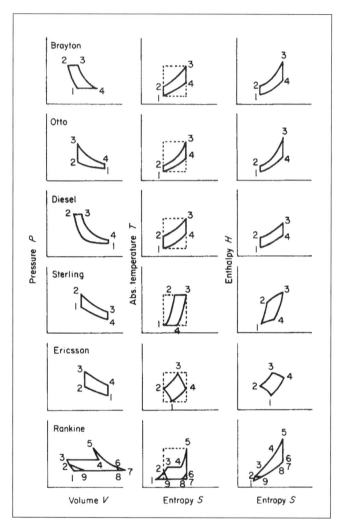

Figure 4.3. Examples of thermodynamic cycles in different representations. For comparison, the Carnot cycle is indicated in the (*P, S*)-diagram (dashed lines). Further descriptions of the individual cycles are given in the text (cf. also section 4.6.2 for an alternative version of the Ericsson cycle).

The Diesel cycle (e.g. common in ship and lorry engines) has step 2-3 as isobar and step 4-1 as isochore, while the two remaining steps are approximately adiabates. The actual designs of the machines, involving turbine wheels or piston-holding cylinders, etc., may be found in engineering textbooks (such as Hütte, 1954).

Closer to the Carnot ideal is the Stirling cycle, involving two isochores (1-2 and 3-4) and two isotherms.

The Ericsson cycle has been developed with the purpose of using hot air as the working fluid. It consists of two isochores (2-3 and 4-1) and two curves somewhere between isotherms and adiabates (cf. e.g. Meinel and Meinel, 1976).

The last cycle depicted in Fig. 4.3 is the Rankine cycle, the appearance of which is more complicated owing to the presence of two phases of the working fluid. Step 1-2-3 describes the heating of the fluid to its boiling point. Step 3-4 corresponds to the evaporation of the fluid, with both fluid and gaseous phases being present. It is an isotherm as well as an isobar. Step 4-5 represents the superheating of the gas, followed by an adiabatic expansion step 5-7. These two steps are sometimes repeated one or more times, with the superheating taking place at gradually lowered pressure, after each step of expansion to saturation. Finally, step 7-1 again involves mixed phases with condensation at constant pressure and temperature. The condensation often does not start until a temperature below that of saturation is reached. Useful work is extracted during the expansion step 5-7, so the simple efficiency equals the enthalpy difference $H_5 - H_7$ divided by the total input $H_6 - H_1$. The second law efficiency is obtained by dividing the simple efficiency by the Carnot value (4.4), for $T = T_5$ and $T_{ref} = T_7$.

Thermodynamic cycles such as those of Figs. 4.1 and 4.3 may be traversed in the opposite direction, thus using the work input to create a low temperature T_{ref} (cooling, refrigeration; T being the temperature of the surroundings) or to create a temperature T higher than that (T_{ref}) of the surroundings (heat pumping). In this case step 7-5 of the Rankine cycle is a compression (8-6-5 if the gas experiences superheating). After cooling (5-4), the gas condenses at the constant temperature T (4-3), and the fluid is expanded, often by passage through a nozzle. The passage through the nozzle is considered to take place at constant enthalpy (2-9), but this step may be preceded by undercooling (3-2). Finally, step 9-8 (or 9-7) corresponds to evaporation at the constant temperature T_{ref}.

For a cooling device the simple efficiency is the ratio of the heat removed from the surroundings, $H_7 - H_9$, and the work input, $H_5 - H_7$, whereas for a heat pump it is the ratio of the heat delivered, $H_5 - H_2$, and the work input. Such efficiencies are often called "coefficients of performance" (COP), and the second law efficiency may be found by dividing the COP by the corresponding quantity ε_{Carnot} for the ideal Carnot cycle (cf. Fig. 4.1),

$$\varepsilon_{Carnot}^{cooling} = \frac{\Delta Q_{14}}{W} = \frac{T_{ref}}{T - T_{ref}}, \tag{4.22a}$$

$$\varepsilon_{Carnot}^{heatpump} = \frac{\Delta Q_{32}}{W} = \frac{T}{T - T_{ref}}. \tag{4.22b}$$

In practice, the compression work $H_5 - H_7$ (for the Rankine cycle in Fig. 4.3) may be less than the energy input to the compressor, thus further reducing the COP and the second law efficiency, relative to the primary source of high-quality energy.

4.1.3 Thermoelectric and thermionic conversion

If the high-quality energy form desired is electricity, and the initial energy is in the form of heat, there is a possibility of utilising direct conversion processes, rather than first using a thermodynamic engine to create mechanical work and then in a second conversion step using an electricity generator.

Thermoelectric generators

One direct conversion process makes use of the thermoelectric effect associated with heating the junction of two different conducting materials, e.g. metals or semiconductors. If a stable electric current, I, passes across the junction between the two conductors A and B, in an arrangement of the type depicted in Fig. 4.4, then quantum electron theory requires that the Fermi energy level (which may be regarded as a chemical potential μ_i) is the same in the two materials ($\mu_A = \mu_B$). If the spectrum of electron quantum states is different in the two materials, the crossing of negatively charged electrons or positively charged "holes" (electron vacancies) will not preserve the statistical distribution of electrons around the Fermi level,

$$f(E) = (\exp((E - \mu_i)/kT) + 1)^{-1}, \tag{4.23}$$

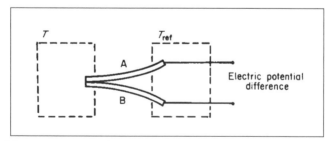

Figure 4.4. Schematic picture of a thermoelectric generator (thermocouple). The rods A and B are made of different materials (metals or better p- and n-type semiconductors).

With E being the electron energy and k being Boltzmann's constant. The altered distribution may imply a shift towards a lower or a higher temperature, such that the maintenance of the current may require addition or removal of heat. Correspondingly, heating the junction will increase or decrease the electric current. The first case represents a thermoelectric generator, and the voltage across the external connections (Fig. 4.4) receives a term in addition to the ohmic term associated with the internal resistance R_{int} of the rods A and B,

$$\Delta\phi = -IR_{int} + \int_{T_{ref}}^{T} \alpha \, dT',$$

The coefficient α is called the Seebeck coefficient. It is the sum of the Seebeck coefficients for the two materials A and B, and it may be expressed in terms of the quantum statistical properties of the materials (Angrist, 1976). If α is assumed independent of temperature in the range from T_{ref} to T, then the generalised electrical force (4.15) may be written

$$F_q = R_{int}J_q - \alpha T F_{Q,in}, \tag{4.24}$$

where J_q and $F_{q,in}$ are given in (4.9) and (4.18).

Considering the thermoelectric generator (Fig. 4.4) as a particular example of the conversion device shown in Fig. 4.2, with no mass flows, the dissipation (4.11) may be written

$$D = J_Q F_Q + J_q F_q.$$

In the linear approximation (4.10), the flows are of the form

$$J_Q = L_{QQ} F_Q + L_{Qq} F_q,$$
$$J_q = L_{qQ} F_Q + L_{qq} Fq,$$

with $L_{Qq} = L_{qQ}$ because of microscopic reversibility (Onsager, 1931). Considering F_Q and J_q (Carnot factor and electric current) as the "controllable" variables, one may solve for F_q and J_Q, obtaining F_q in the form (4.24) with $F_Q = F_{Q,in}$ and

$$L_{qq} = (R_{int})^{-1}; \qquad L_{qQ} = L_{Qq} = \alpha T/R_{int}.$$

The equation for J_Q takes the form

$$J_Q = CTF_Q + \alpha T J_q, \tag{4.25}$$

where the conductance C is given by

$$C = (L_{QQ}L_{qq} - L_{Qq}L_{qQ})/(LqqT).$$

Using (4.24) and (4.25), the dissipation may be written

$$D = CTF_Q^2 + R_{int} J_q^2, \tag{4.26}$$

and the simple efficiency (4.19) may be written

$$\eta = \frac{-J_q F_q}{J_Q} = \frac{-J_q(R_{int}J_q - \alpha TF_Q)}{CTF_Q + \alpha TJ_q}. \tag{4.27}$$

If the reservoir temperatures T and T_{ref} are maintained at a constant value, F_Q can be regarded as fixed, and the maximum efficiency can be found by variation of J_q. The efficiency (4.27) has an extremum at

$$J_q = \frac{CF_Q}{\alpha}\left(\left(1+\frac{\alpha^2 T}{R_{int}C}\right)^{1/2}-1\right),$$

(4.28)

corresponding to a maximum value

$$\max(\eta) = F_Q \frac{(1+ZT)^{1/2}-1}{(1+ZT)^{1/2}+1},$$

(4.29)

with $Z = \alpha^2(R_{int}C)^{-1}$. Equation (4.29) is accurate only if the linear approximation (4.10) is valid. The maximum second law efficiency is obtained from (4.29) by division by F_Q [cf. (4.20)].

The efficiencies are seen to increase with temperature, as well as with Z. Z is largest for certain materials (A and B in Fig. 4.4) of semiconductor structure and small for metals as well as for insulators. Although R_{int} is small for metals and large for insulators, the same is true for the Seebeck coefficient α, which appears squared. C is larger for metals than for insulators. Together, these features combine to produce a peak in Z in the semiconductor region. Typical values of Z are about 2×10^{-3} (K)$^{-1}$ at $T = 300$ K (Angrist, 1976). The two materials A and B may be taken as a p-type and an n-type semiconductor, which have Seebeck coefficients of opposite signs, so that their contributions add coherently for a configuration of the kind shown in Fig. 4.4.

Thermionic generators

Thermionic converters consist of two conductor plates separated by vacuum or by a plasma. The plates are maintained at different temperatures. One, the emitter, is at a temperature T large enough to allow a substantial emission of electrons into the space between the plates due to the thermal statistical spread in electron energy (4.23). The electrons (e.g. of a metal emitter) move in a potential field characterised by a barrier at the surface of the plate. The shape of this barrier is usually such that the probability of an electron penetrating it is small until a critical temperature, after which it increases rapidly ("red-glowing" metals). The other plate is maintained at a lower temperature T_{ref}. In order not to have a build-up of space charge between the emitter and the collector, atoms of a substance such as caesium may be introduced in this area. These atoms become ionised near the hot emitter (they give away electrons to make up for the electron deficit in the emitter material), and for a given caesium pressure the positive ions exactly neutralise the space charges of the travelling electrons. At the collector surface, recombination of caesium ions takes place.

The layout of the emitter design must allow the transfer of large quantities of heat to a small area in order to maximise the electron current responsible for creating the electric voltage difference across the emitter–collector system, which may be utilised through an external load circuit. This heat transfer can be accomplished by a so-called "heat pipe" – a fluid-containing pipe which allows the fluid to evaporate in one chamber when heat is added. The vapour then travels to the other end of the pipe, condenses and gives off the latent heat of evaporation to the surroundings, where after it returns to the first chamber through capillary channels, under the influence of surface tension forces.

The description of the thermionic generator in terms of the model converter shown in Fig. 4.2 is very similar to that of the thermoelectric generator. With the two temperatures T and T_{ref} defined above, the generalised force F_Q is defined. The electrical output current, J_q, is equal to the emitter current, provided that back-emission from the collector at temperature T_{ref} can be neglected and provided that the positive-ion current in the intermediate space is negligible in comparison with the electron current. If the space charges are saturated, the ratio between ion and electron currents is simply the inverse of the square root of the mass ratio, and the positive-ion current will be a fraction of a per cent of the electron current. According to quantum statistics, the emission current (and hence J_q) may be written

$$J_Q = AT^2 \exp(-e\,\phi_e/(kT)), \tag{4.30}$$

where ϕ_e is the electric potential of the emitter, $e\phi_e$ is the potential barrier of the surface in energy units, and A is a constant (Angrist, 1976). Neglecting heat conduction losses in plates and the intermediate space, as well as light emission, the heat $J_{Q,in}$ to be supplied to keep the emitter at the elevated temperature T equals the energy carried away by the electrons emitted,

$$J_{Q,in} = J_q\,(\phi_e + \delta + 2kT/e), \tag{4.31}$$

where the three terms in brackets represent the surface barrier, the barrier effectively seen by an electron due to the space charge in the intermediate space, and the original average kinetic energy of the electrons at temperature T (divided by e), respectively.

Finally, neglecting internal resistance in plates and wires, the generalised electrical force equals the difference between the potential ϕ_e and the corresponding potential for the collector ϕ_c,

$$-F_q = \phi_c - \phi_e\,, \tag{4.32}$$

with insertion of the above expressions (4.30) to (4.32). Alternatively, these expressions may be linearised in the form (4.10) and the efficiency calculated exactly as in the case of the thermoelectric device. It is clear, however, that a linear approximation to (4.30), for example, would be very poor.

4.1.4 Turbines and other flow-driven converters

A turbine is a device delivering rotational shaft power on the basis of some other type of mechanical energy. If the temperature of the surroundings is regarded as fixed, the simple model in Fig. 4.2 allows the energy dissipation (4.17) to be written

$$D = J_m F_m + J_\theta F_\theta, \tag{4.33}$$

since from (4.18) $F_{Q,in}$ is zero, and no electrical output has been considered in this conversion step. The output variables are the angular velocity of the shaft, J_θ (4.9), and the torque acting on the system, F_θ (4.15), while the input variables are the mass flow rate, J_m (4.9), and the generalised force F_m given in (4.18). The specific energy contents w_{in} and w_{out} are of the form (4.13), corresponding to e.g. the geopotential energy of a given water head,

$$w_{in}^{pot} = g\,\Delta z, \qquad\qquad w_{out}^{pot} = 0, \tag{4.34}$$

the kinetic energy of the working fluid,

$$w_{in}^{kin} = \tfrac{1}{2} u_{in}^2, \qquad\qquad w_{out}^{kin} = \tfrac{1}{2} u_{out}^2, \tag{4.35}$$

and the enthalpy connected with the pressure changes,

$$h_{in} = P_{in} / \rho_{in}, \qquad\qquad h_{out} = P_{out} / \rho_{out}, \tag{4.36}$$

where the internal energy term in (4.1), assumed constant, has been left out, and the specific volume has been expressed in terms of the fluid densities ρ_{in} and ρ_{out} at input and output.

If a linear model of the Onsager type (4.10) is adopted for J_m and J_θ and these equations are solved for J_m and F_θ, one obtains

$$J_m = L_{m\theta}\, J_\theta / L_{\theta\theta} + (L_{mm} - L_{m\theta} L_{\theta m} / L_{\theta\theta})\, F_m,$$
$$-F_\theta = -J_\theta / L_{\theta\theta} + L_{\theta m} F_m / L_{\theta\theta}. \tag{4.37}$$

The coefficients may be interpreted as follows: $L_{m\theta} / L_{\theta\theta}$ is the mass of fluid displaced by the turbine during one radian of revolution, $(L_{mm} - L_{m\theta} L_{\theta m} / L_{\theta\theta})$ is a "leakage factor" associated with fluid getting through the turbine without contributing to the shaft power, and finally, $L_{\theta\theta}^{-1}$ represents the friction losses. Insertion into (4.33) gives the linear approximation for the dissipation,

$$D = (L_{mm} - L_{m\theta} L_{\theta m} / L_{\theta\theta})\, (F_m)^2 + (J_\theta)^2 / L_{\theta\theta}. \tag{4.38}$$

An ideal conversion process may be approached if no heat is exchanged with the surroundings, in which case (4.19) and (4.12) give the simple efficiency

$$\eta = (w_{in} - w_{out}) / w_{in}. \tag{4.39}$$

The second law efficiency in this case is, from (4.20), (4.14) and (4.12),

$$\eta^{(2.\ law)} = (w_{in} - w_{out}) / (w_{in} - w_{out} - T_{ref}(s_{m,in} - s_{m,out})). \tag{4.40}$$

The second law efficiency becomes unity if no entropy change takes place in the mass stream. The first law efficiency (4.39) may approach unity if only potential energy change of the form (4.34) is involved. In this case $w_{out} = 0$, and the fluid velocity, density and pressure are the same before and after the turbine. Hydroelectric generators approach this limit if working entirely on a static water head. Overshot waterwheels may operate in this way, and so may the more advanced turbine constructions, if the potential to kinetic energy conversion (in penstocks) and pressure build-up (in the nozzle of a Pelton turbine and in the inlet tube of many Francis turbine installations) are regarded as "internal" to the device (cf. section 4.5). However, if there is a change in velocity or pressure across the converter, the analysis must take this into account, and it is no longer obvious whether the first law efficiency may approach unity or not.

Free stream flow turbines

Consider, for example, a free stream flow passing horizontally through a converter. In this case, the potential energy (4.34) does not change and may be left out. The pressure may vary near the converting device, but far behind and far ahead of the device the pressure is the same if the stream flow is free. Thus,

$$w = w^{kin} = \tfrac{1}{2}\,(u_x^2 + u_y^2 + u_z^2) = \tfrac{1}{2}\,\boldsymbol{u} \cdot \boldsymbol{u},$$

and

$$w_{in} - w_{out} = \tfrac{1}{2}\,(\boldsymbol{u}_{in} - \boldsymbol{u}_{out}) \cdot (\boldsymbol{u}_{in} + \boldsymbol{u}_{out}). \tag{4.41}$$

This expression and hence the efficiency would be maximum if $\boldsymbol{u}_{out}$ could be made zero. However, the conservation of the mass flow J_m requires that u_{in} and u_{out} satisfy an additional relationship. For a pure, homogeneous streamline flow along the x-axis, the rate of mass flow is

$$J_m = \rho\, A_{in}\, u_{x,in} = \rho\, A_{out}\, u_{x,out}, \tag{4.42}$$

in terms of areas A_{in} and A_{out} enclosing the same streamlines, before and after the passage through the conversion device. In a more general situation, assuming rotational symmetry around the x-axis, there may have been induced a radial as well as a circular flow component by the device. This situation is illustrated in Fig. 4.5. It will be further discussed in section 4.3, and the only case treated here is the simple one in which the radial and tangential components of the velocity field, u_r and u_t, which may be induced by the conversion device, can be neglected.

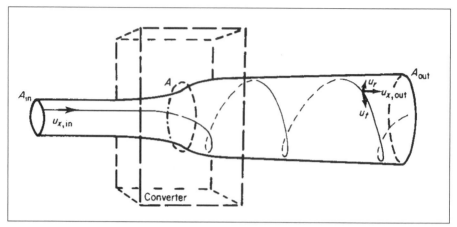

Figure 4.5. Schematic picture of a free stream flow converter or turbine. The incoming flow is a uniform streamline flow in the *x*-direction, while the outgoing flow is allowed to have a radial and a tangential component. The diagram indicates how a streamline may be transformed into an expanding helix by the device. The effective area of the converter, *A*, is defined in (4.44).

The axial force ("thrust") acting on the converter equals the momentum change,

$$F_x = J_m \left(u_{x,in} - u_{x,out} \right). \tag{4.43}$$

If the flow velocity in the converter is denoted *u*, an effective area of conversion, *A*, may be defined by

$$J_m = \rho A u_x. \tag{4.44}$$

according to the continuity equation (4.42). Dividing (4.43) by ρA, one obtains the specific energy transfer from the mass flow to the converter, within the conversion area *A*. This should equal the expression (4.41) for the change in specific energy, specialised to the case of homogeneous flows u_{in} and u_{out} along the *x*-axis,

$$u_x \left(u_{x,in} - u_{x,out} \right) = \tfrac{1}{2} \left(u_{x,in} + u_{x,out} \right) \left(u_{x,in} - u_{x,out} \right)$$

or

$$u_x = \tfrac{1}{2} \left(u_{x,in} + u_{x,out} \right). \tag{4.45}$$

The physical principle behind this equality is simply energy conservation, and the assumptions so far have been the absence of heat exchange [so that the energy change (4.12) becomes proportional to the kinetic energy difference (4.41)0 and the absence of induced rotation (so that only *x*-components of the velocity needs to be considered). On both sides of the converter, Bernoulli's equation is valid, stating that the specific energy is constant along a streamline (Bernoulli's equation for an irrotational flow is given in section 2.D, with ϕ = constant defining the streamlines). Far from the converter, the pressures are equal but the velocities are different, while the velocity just in front of or behind the converter may be taken as u_x, implying a pressure drop across the converter,

$$\Delta P = \tfrac{1}{2} \rho \left(u_{x,in} + u_{x,out}\right) \left(u_{x,in} - u_{x,out}\right). \tag{4.46}$$

The area enclosing a given streamline field increases in a continuous manner across the converter at the same time as the fluid velocity continuously decreases. The pressure, on the other hand, rises above the ambient pressure in front of the converter, then discontinuously drops to a value below the ambient one, and finally increases towards the ambient pressure again, behind ("in the wake of") the converter.

It is customary (see e.g. Wilson and Lissaman, 1974) to define an "axial interference factor", a, by

$$u_x = u_{x,in} \left(1 - a\right), \tag{4.47}$$

in which case (4.45) implies that $u_{x,out} = u_{x,in} \left(1 - 2a\right)$. With this, the power output of the conversion device can be written

$$E = J_m \left(w_{in} - w_{out}\right) = \rho A \left(u_{x,in}\right)^3 2a \left(1 - a\right)^2, \tag{4.48}$$

and the efficiency can be written

$$\eta = E / \left(J_m w_{in}\right) = 4a \left(1 - a\right). \tag{4.49}$$

It is seen that the maximum value of η is unity, obtained for $a = \tfrac{1}{2}$, corresponding to $u_{x,out} = 0$. The continuity equation (4.42) then implies an infinite area A_{out}, and it will clearly be difficult to defend the assumption of no induced radial motion.

In fact, for a free stream device of this type, the efficiency (4.49) is of little relevance since the input flux may not be independent of the details of the device. The input area A_{in}, from which streamlines would connect with a fixed converter area A, could conceivably be changed by altering the construction of the converter. It is therefore more appropriate to ask for the maximum power output for fixed A, as well as fixed input velocity $u_{x,in}$, this being equivalent to maximising the "power coefficient" defined by

$$C_p = E / \left(\tfrac{1}{2} \rho A \left(u_{x,in}\right)^3\right) = 4a \left(1 - a\right)^2. \tag{4.50}$$

The maximum value is obtained for $a = 1/3$, yielding $C_p = 16/27$ and $u_{x,out} = u_{x,in}/3$. The areas are $A_{in} = (1 - a)A = 2/3\,A$ and $A_{out} = (1 - a)A/(1 - 2a) = 2A$, so in this case it is not unlikely that it may be a reasonable approximation to neglect the radial velocity component in the far wake.

The maximum found above for C_p is only a true upper limit with the assumptions made. By discarding the assumption of irrotational flow, it becomes possible for the converter to induce a velocity field, for which rot(u) is no longer zero. It has been shown that if the additional field is in the form of a vortex ring around the converter region, so that it does not contribute to the far wake, then it is possible to exceed the upper limit power coefficient 16/27 found above (cf. section 4.3.4).

Magnetohydrodynamic converters

For the turbines considered above, it was explicitly assumed that no heat was added. Other flow-type converters are designed to receive heat during the process. An example of this is the gas turbine, which was described in section 4.1.2 from the point of view of thermodynamics. The gas turbine (Brayton cycle) allows the conversion of heat into shaft power, but it may equally well be viewed as the successive conversion of heat into kinetic energy of flow and of the kinetic energy of flow into shaft power.

The magnetohydrodynamic converter is another device converting heat into work, but delivering the work directly as electrical power without intermediate steps of mechanical shaft power. The advantage is not in avoiding the shaft power to electrical power conversion, which can be done with small losses, but rather in avoiding a construction with moving parts, thereby permitting higher working temperatures and higher efficiency. The heat added is used to ionise a gas, and this conducting gas ("plasma") is allowed to move through an expanding duct, upon which an external magnetic field B is acting. The motion of the gas is sustained by a pressure drop between the chamber where heat is added and the open end of the expanding duct. The charged particles of velocity u in the plasma are subjected to a Lorentz force

$$F = \rho_{el}\, u \times B, \tag{4.51}$$

where the direction of this induced force is perpendicular to B and u, but opposite for the positive atoms and for the negatively charged electrons. Since the mass of an electron is much smaller than that of an atom, the net induced current will be in the direction given by a negative value of ρ_{el}. Assuming a linear relationship between the induced current J_{ind} and the induced electric field $E_{ind} = F/\rho_{el} = u \times B$, the induced current may be written

$$J_{ind} = \sigma u \times B, \tag{4.52}$$

where σ is the electrical conductivity of the plasma. This outlines the mechanism by which the magnetohydrodynamic (MHD) generator converts kinetic energy of moving charges into electrical power associated with the induced current J_{ind} across the turbine. A more detailed treatment must take into account the contributions to the force (4.51) on the charges, which arise from the induced velocity component J_{ind} / ρ_{el}, as well as the effect of variations (if any) in the flow velocity u through the generator stage (see e.g. Angrist, 1976).

The generator part of the MHD generator has an efficiency determined by the net power output after subtraction of the power needed for maintaining the magnetic field B. Only the gross power output can be considered as given by (4.12). Material considerations require that the turbine be cooled, so in addition to power output there is a heat output in the form of a coolant flow, as well as the outgoing flow of cooled gas. The temperature of the out-flowing gas is still high (otherwise recombination of ions would inhibit the functioning of the converter), and the MHD stage is envisaged as being followed by one or more conventional turbine stages. It is believed that the total power generation in all stages could be made to exceed that of a conversion system based entirely on turbines with moving parts, for the same heat input.

Very high temperatures are required for the ionisation to be accomplished thermally. The ionisation process can be enhanced in various ways. One is to "seed" the gas with suitable metal dust (sodium, potassium, caesium, etc.), for which case working MHD machines operating at temperatures around 2500 K have been demonstrated (Hammond *et al.*, 1973). If the heat source is fossil fuel, and particularly if it is coal with a high sulphur content, the seeding has the advantage of removing practically all the sulphur from the exhaust gases (the seeding metals are rather easily retrieved and must anyway be recycled for economic reasons).

4.1.5 Photovoltaic conversion

Conversion of radiant energy (light quanta) into electrical energy can be achieved with the use of semiconductor materials, for which the electron excitation caused by impinging light quanta has a strongly enhancing effect on the conductivity.

It is not sufficient, however, that electrons are excited and are able to move more freely, if there is no force to make them move. Such a force would arise from the presence of a gradient of electrical potential, such as the one found in a p–n junction of doped semiconductor materials (a p–n junction is a junction of a p-type and an n-type semiconductor, as further described below). A p–n junction provides an electrical field which will cause the electrons excited by radiation (such as solar) to move in the direction from the p-type to the n-type material and cause the vacancies (holes) left by the excited electrons to move in the opposite direction. If the electrons and holes reach the respective edges of the semiconductor material, the device is capable of delivering electrical power to an external circuit. The motion of electrons or holes receives competition from recombination processes (electrons being recaptured into vacancies), making such factors as overall dimensions and electron mobility in the material used of importance.

The p–n junction

An essential constituent of photovoltaic cells is the p–n junction. A refresher on the semiconductor physics needed for understanding the p–n junction is given in section 4.A at the end of this chapter. When a p-type and an n-type semiconductor are joined so that they acquire a common surface, they are said to form a p–n junction. This will initially cause electrons to flow in the n to p direction because, as seen in Fig. 4.118 of section 4.A, the electron density in the conduction band is higher in n-type than in p-type material and because the hole density in the valence band is higher in the p-type than in the n-type material (the electron flow in the valence band can also be described as a flow of positive holes in the direction p to n).

This electron flow builds up a surplus of positive charge in the n-type material and a surplus of negative charge in the p-type material, in the neighbourhood of the junction (mainly restricted to distances from the junction of the order of the mean travelling distance before recombination of an electron or a hole in the respective materials). These surplus charges form a dipole layer, associated with which is an electrostatic potential difference, which will tend to hinder any further unidirectional electron flow. Finally, an equilibrium is reached in which the potential difference is such that no net transfer of electrons takes place.

Another way of stating the equilibrium condition is in terms of the Fermi energy [cf. Fig. 4.118 and (4.23)]. Originally, the Fermi energies of the p- and n-type materials, μ_p and μ_n, are different, but at equilibrium $\mu_p = \mu_n$. This is illustrated in Fig. 4.6, and it is seen that the change in the relative positions of the conduction (or valence) bands in the two types of material must equal the electron charge, $-e$, times the equilibrium electrostatic potential.

The number of electrons in the conduction band may be written

$$n_c = \int_{E_c}^{E'_c} n'(E) f(E)\, dE,$$

(4.53)

where E_c and E_c' are the lower and upper energy limit of the conduction band, $n'(E)$ is the number of quantum states per unit energy interval (and, for example, per unit volume of material, if the electron number per unit volume is desired), and finally, $f(E)$ is the Fermi–Dirac distribution (4.23). If the electrons in the conduction band are regarded as free, elementary quantum mechanics gives (see e.g. Shockley, 1950)

$$n'(E) = 4\pi h^{-3} (2m)^{3/2} E^{1/2},$$

(4.54)

where h is Planck's constant and m is the electron mass. The corrections for electrons moving in condensed matter, rather than being free, may to a first approximation be included by replacing the electron mass by an "effective" value.

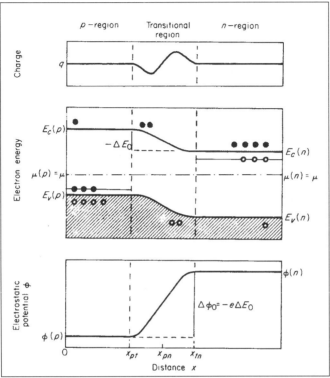

Figure 4.6. Schematic picture of the properties of a p–n junction in an equilibrium condition. The x-direction is perpendicular to the junction (all properties are assumed to be homogeneous in the y- and z-directions). The charge (top) is the sum of electron charges in the conduction band and positive hole charges in the valence band, plus charge excess or defect associated with the acceptor and donor levels. In the electron energy diagram (middle), the abundance of minority charge carriers (closed circles for electrons, open circles for holes) is schematically illustrated. The properties are further discussed in the text.

If the Fermi energy if not close to the conduction band,

$$E_c - \mu \gg kT,$$

the Fermi–Dirac distribution (4.23) may be replaced by the Boltzmann distribution,

$$f_B(E) = \exp(-(E - \mu)/kT). \tag{4.55}$$

Evaluating the integral, (4.53) then gives an expression of the form

$$n_c = N_c \exp(-(E_c - \mu)/kT). \tag{4.56}$$

The number of holes in the valence band is found in an analogous way,

$$n_v = N_v \exp(-(\mu - E_v)/kT), \tag{4.57}$$

where E_v is the upper limit energy of the valence band.

The equilibrium currents in a p–n junction such as the one illustrated in Fig. 4.6 can now be calculated. Considering first the electron currents in the conduction band, the electrons thermally excited into the conduction band in the p-region can freely flow into the n-type materials. The corresponding current, $I_0^-(p)$, may be considered proportional to the number of electrons in the conduction band in the p-region, $n_c(p)$, given by (4.56),

$$I_0^-(p) = \alpha N_c \exp\left(-(E_c(p) - \mu(p))/kT\right), \tag{4.58}$$

where the constant α depends on electron mobility in the material and on the electrostatic potential gradient, grad ϕ. The electrons excited into the conduction band in the n-type region will have to climb the potential barrier in order to move into the p-region. The fraction of electrons capable of doing this is given by a Boltzmann factor of the form (4.56), but with the additional energy barrier $\Delta E_0 = -\Delta \phi_0/e$ ($-e$ being the electron charge),

$$n_c(n) = N_c \exp(-(E_c(n) - \mu(n) - \Delta E_0)/kT).$$

Using $-\Delta E_0 = E_c(p) - E_c(n)$ (cf. Fig. 4.6) and considering the current $I_0^-(n)$ as being proportional to $n_c(n)$, the corresponding current may be written

$$I_0^-(n) = \alpha' N_c \exp\left(-(E_c(n) - \mu(n))/kT\right), \tag{4.59}$$

where α' depends on the diffusion parameter and on the relative change in electron density, $n_c^{-1}\text{grad}(n_c)$, considering the electron motion against the electrostatic potential as a diffusion process. The statistical mechanical condition for thermal equilibrium demands that $\alpha = -\alpha'$ (Einstein, 1905), so (4.58) and (4.59) show that the net electron current,

$$I_0^- = I_0^-(p) + I_0^-(n),$$

becomes zero precisely when

$$\mu(p) = \mu(n),$$

which is then the condition for thermal equilibrium. The same is true for the hole current,

$$I_0^+ = I_0^+(p) + I_0^+(n).$$

If an external voltage source is applied to the p–n junction in such a way that the n-type terminal receives an additional electrostatic potential $\Delta\phi_{ext}$ relative to the p-type terminal, then the junction is no longer in thermal equilibrium, and the Fermi energy in the p-region is no longer equal to that of the n-region, but satisfies

$$\mu(p) - \mu(n) = e^{-1}\Delta\phi_{ext} = \Delta E_{ext} \qquad (4.60)$$

if the Boltzmann distributions of electrons and of holes are to maintain their shapes in both p- and n-regions. Similarly $E_c(p) - E_c(n) = -(\Delta E_0 + \Delta E_{ext})$, and assuming that the proportionality factors in (4.58) and (4.59) still bear the relationship $\alpha = -\alpha'$ in the presence of the external potential, the currents are connected by the expression

$$\Gamma(n) = -\Gamma(p)\exp(\Delta E_{ext}/kT).$$

The net electron current in the conduction band then becomes

$$\Gamma = \Gamma(n) + \Gamma(p) = -\Gamma(p)\,(\exp(\Delta E_{ext}/kT) - 1). \qquad (4.61)$$

For a positive $\Delta\phi_{ext}$, the potential barrier which electrons in the n-region conduction band (see Fig. 4.6) have to climb increases and the current $\Gamma(n)$ decreases exponentially (ΔE_{ext} negative, "reverse bias"). In this case, the net current Γ approaches a saturation value equal to $\Gamma(p)$, according to (4.61).

For negative $\Delta\phi_{ext}$, (positive ΔE_{ext}, "forward bias"), the current $\Gamma(n)$ increases exponentially with the external potential. In both cases $\Gamma(p)$ is assumed to remain practically unchanged, when the external potential of one or the other sign is applied, considering that $\Gamma(p)$ is primarily limited by the number of electrons excited into the conduction band in the p-type material, a number which is assumed to be small in comparison with the conduction band electrons in the n-type material (cf. Figs. 4.6 and 4.118).

The contributions to the hole current, I^+, behave similarly to those of the electron current, and the total current I across a p–n junction with an external potential $\Delta\phi_{ext} = -e\,\Delta E_{ext}$ may be written

$$I = I^- + I^+ = -I(p)\,(\exp(\Delta E_{ext}/kT) - 1). \qquad (4.62)$$

The relationship between current and potential is called the "characteristic" of the device, and the relation (4.62) for the p–n junction is illustrated in Fig. 4.7 by the curve labelled "no light". The constant saturation current $I(p)$ is sometimes referred to as the "dark current".

Solar cells

A p–n junction may be utilised to convert solar radiation energy into electric power. A solar cell is formed by shaping the junction in such a way that, for example, the p-type material can be reached by incident solar radiation, e.g. by placing a thin layer of p-type material on top of a piece of n-type semiconductor. In the dark and with no external voltage, the net current across the junction is zero, as was shown in the previous subsection, i.e. the intrinsic potential difference $\Delta\phi_0$ is unable to perform external work.

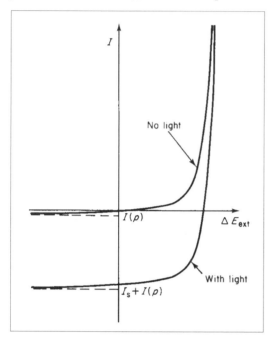

Figure 4.7. Characteristics (i.e. current as a function of external voltage) of a *p–n* junction, in the dark and with applied light. The magnitude of the short-circuit current, I_s, is a function of light intensity and spectral distribution.

However, when irradiated with light quanta of an energy $E_{light} = h\nu = hc/\lambda$ (*h* is Planck's constant, *c* is the velocity of light and ν and λ are the frequency and wavelength of radiation), which is larger than the energy difference between the conduction and valence band for the *p*-type material,

$$E_{light} \geq E_c(p) - E_v(p),$$

then electrons may be photo-excited from the valence band into the conduction band. The absorption of light quanta produces as many holes in the valence band of the *p*-type material as electrons in the conduction band. Since in the dark there are many fewer electrons in the *p*-type conduction band than holes in the valence band, a dramatic increase in the number of conduction band electrons can take place without significantly altering the number of holes in the valence band. If the excess electrons are sufficiently close to the junction to be able to reach it by diffusion before recombining with a hole, then the current in this direction exceeds $I_0(p)$ of (4.58) by an amount I_s, which is the net current through the junction in case of a short-circuited external connection from the *n*-type to the *p*-type material. The photo-induced current is not altered if there is a finite potential drop in the external circuit, since the relation between the current (4.62) and the external potential drop $e \, \Delta E_{ext}$ was derived with reference only to the changes in the n-region.

An alternative n–p type of solar cell may consist of a thin n-type layer exposed to solar radiation on top of a p-type base. In this case, the excess holes in the n-type valence band produce the photo-induced current I_s.

The total current in the case of light being absorbed in the p-type material and with an external potential drop is then

$$I = I_s - I(p) \, (\exp(- \Delta \, \phi_{ext} / kT) - 1). \tag{4.63}$$

The short-circuit current I_s depends on the amount of incident light with frequencies sufficient to excite electrons into the conduction band, on the fraction of this light actually being absorbed, and on the conditions for transporting the excess electrons created in the conduction band, in competition with electron–hole recombination processes. I_s may be written as the sum of a conduction and a diffusion type current, both related to the number of excess electrons in the conduction band, n_c^{ind}, induced by the absorption of light,

$$I_s = e(m_c E_e n_c^{ind} + k_c \, dn_c^{ind}/dx), \tag{4.64}$$

where e is the numerical value of the electron charge (1.6×10^{-19} C), m_c is the mobility of conduction band electrons [e.g. 0.12 $m^2 V^{-1} s^{-1}$ for silicon (Loferski, 1956), the dependence on the degree of doping being displayed in Fig. 4.8], E_e is the local electrical field, k_c is the diffusion constant [cf. section 2.B and (3.33); e.g. $k_c = 10^{-3}$ $m^2 s^{-1}$ (Loferski, 1956)], and x is the depth below the solar cell surface, assumed to be the only significant co-ordinate (as in Fig. 4.6).

The excess electron number density, $n_c^{ind}(x)$, at a depth x, changes when additional electrons are photo-excited, when electrons are carried away from x by the current I_s, and when electrons recombine with holes,

$$\frac{\partial \, n_c^{ind}(x)}{\partial t} = \int \sigma(v) n_{ph}(v) \exp(-\sigma(v)x) \, dv + \frac{I}{e} \frac{\partial I_s}{\partial x} - n_c^{ind}(x) \frac{1}{\tau_c}. \tag{4.65}$$

Here $\sigma(v)$ is the cross section for absorption of light quanta ("photons") in the p-type material, and $n_{ph}(v)$ is the number of photons at the cell surface ($x = 0$) per unit time and unit interval of frequency v. The absorption cross section is zero for photon energies below the semiconductor energy gap, $hv < E_c(p) - E_v(p)$, i.e. the material is transparent to such light. The most energetic light quanta in visible light could theoretically excite more than one electron per photon (e.g. 2–3 in Si with an energy gap slightly over 1 eV), but the probability for exciting just one electron to a higher energy is higher, and such a process is usually followed by a transfer of energy to other degrees of freedom (e.g. lattice vibrations and ultimately heat), as the excited electron approaches the lower part of the conduction band, or as the hole left by the electron de-excites from a deep level to the upper valence band. Thus, in practice the quantum efficiency (number of electron–hole pairs per photon) hardly exceeds one.

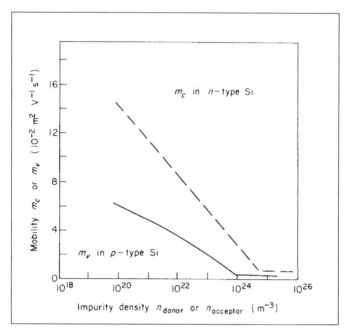

Figure 4.8. Mobility of minority carriers in Si at room temperature (about 300 K), extrapolated from measurements (Wolf, 1963). The mobility plotted is the "conduction mobility", equal to the conductivity divided by the number of minority carriers and by the electron charge. The attenuation of the flow of actual carriers by recombination effects (trapping) is not considered.

The last parameter introduced in (4.65), τ_c, is the average lifetime of an electron excited into the conduction band, before recombination [τ_c may lie in the interval 10^{-11} to 10^{-7}, with 10^{-9} being a typical value (Wolf, 1963)]. The lifetime τ_c is connected to the cross section for recombination, σ_c, and to the mean free path l_c of electrons in the conduction band by

$$l_c = \sigma_c^{-1} = v_c \tau_c N_a,$$

where v_c is the average thermal velocity of the electrons, $v_c = (2kT/m)^{1/2}$ (m being the electron mass, k being Boltzmann's constant and T being the absolute temperature) and N_a is the number of recombination centres ("acceptor impurities", cf. Fig. 4.118).

The boundary conditions for solving (4.65) may be taken as the absence of excess minority carriers (electrons or holes) at the junction $x = x_{pn}$,

$$n_c^{ind}(x_{pn}) = 0,$$

and a prescribed (material dependent) excess electron gradient at the surface $x = 0$. This gradient, $(dn_c^{ind}/dx)|_{x=0}$, is often expressed in terms of a surface recombination velocity, s_c, through (4.64) by writing the left-hand side

$$I_s = s_c n_c^{ind}(0).$$

Typical values of s_c are of the order of 10^3 m s^{-1} (Wolf, 1963, 1971).

For n–p type solar cells, expressions analogous to the above can be used.

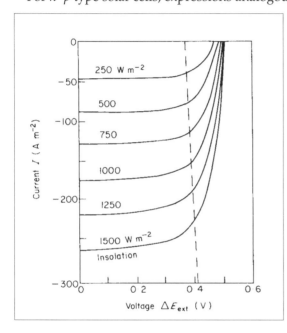

Figure 4.9. Characteristics of Cu$_2$S–CdS solar cell at 300 K for different intensities of incident radiation with typical clear sky solar frequency distribution. The points of intersection with the dashed line give the voltage and current leading to maximum power output for given solar radiation (based on Shirland, 1966).

Once $n_c^{ind}(x)$ has been found, I_s can be calculated. Figure 4.7 shows an example of the total current through a p–n junction, as a function of applied voltage but for a fixed rate of incoming solar radiation on the p-type surface. The short-circuit current I_s increases linearly with intensity of light, if the spectral composition is kept constant, for the entire interval of intensities relevant for applications of solar cells at or near Earth. This is illustrated in Fig. 4.9 for a solar cell based on a p–n heterojunction, with the p-type material being Cu$_2$S and the n-type material CdS.

For an open-ended solar cell (i.e. no external circuit), the difference in electrical potential between the terminals, $V_{oc} = \Delta\phi_{ext}(I=0)$, is obtained by putting I equal to zero in (4.63),

$$V_{oc} = kTe^{-1} (\log(I_s/I(p)) + 1). \tag{4.66}$$

The amount of electrical power, E, delivered by the irradiated cell to the external circuit is obtained by multiplication of (4.63) by the external voltage,

$$E = (\Delta\phi_{ext})I = \Delta\phi_{ext}\,(I_s - I(p)(\exp(-e\Delta\phi_{ext}/kT) - 1)). \qquad (4.67)$$

From $\partial E/\partial\,(\Delta\phi_{ext}) = 0$, the external voltage V_{opt} may be found, which leads to the maximum value of power, E_{max}. In the situations of interest, V_{opt} is a slowly varying function of the amount of incident radiation, as illustrated by Fig. 4.9. The corresponding current may be denoted I_{opt}.

The efficiency of solar cell radiant-to-electrical energy conversion is the ratio of the power E delivered and the incident energy, denoted E_+^{sw} as in sections 3.1 and 2.2.2 (eventually for a tilted orientation of the solar cell), $\eta = E/E_+^{sw}$. In terms of the flux of photons of given frequency incident on the solar cell [introduced in (4.65)], the non-reflected energy flux at the surface may be written (a is the albedo of the cell surface)

$$E_+^{sw}\,(1-a) = \int_0^\infty h\nu n_{ph}(\nu)\,d\nu, \qquad (4.68)$$

where h is Planck's constant. For a given semiconductor material, the maximum fraction of the energy (4.68), which can be absorbed, is

$$\int_{h\nu=E_c(p)-E_v(p)}^\infty h\nu n_{ph}(\nu)\,d\nu.$$

The part of the integral from zero up to the energy gap (i.e. the part above a certain wavelength of light) constitutes a fundamental loss. The same can be said of the energy of each light quantum in excess of the semiconductor energy gap $E_c(p) - E_v(p)$, assuming a quantum efficiency of at most one, i.e. that all such quanta are indeed absorbed (which may not be true if their energy is, say, between the upper limit of the conduction band and the lower limit of the following band) and that all excess energy is spent in exciting lattice degrees of freedom (vibrational phonons) that do not contribute to the photovoltaic process. In that case the energy flux available for photoconversion is only

$$(E_c(p) - E_v(p)) \int_{h\nu=E_c(p)-E_v(p)}^\infty n_{ph}(\nu)\,d\nu = E^{avail}. \qquad (4.69)$$

Further losses in addition to reflection and insufficient or excess photon energy may be associated with imperfections in the junction materials or in the current extraction system, causing heat formation or light re-emission rather than electrical power creation. Both heat creation (in the lattice) and re-radiation may take place in connection with the recombination of photo-excited electrons and holes. Since many of these processes are highly temperature dependent, the maximum efficiency that can be obtained in practice is also temperature dependent. Examples of maximum theoretical efficiencies, as well as those obtained in practice, will be given in section 4.2.3.

Rather than being p- and n-doped materials of the same elemental semiconductor, the solar cell junction may be based on different materials ("heterojunction") or on a metal and a semiconductor ("Schottky junction").

The discussion of individual types of solar cells is given below in section 4.2.3, after presenting a few more general energy conversion methods.

4.1.6 Electrochemical conversion

Electrochemical energy conversion is the direct conversion of chemical energy, i.e. free energy of the form (4.8) into electrical power or vice versa. A device that converts chemical energy into electric energy is called a *fuel cell* (if the free energy-containing substance is stored within the device rather than flowing into the device, the name "primary battery" is sometimes used). A device that accomplishes the inverse conversion (e.g. electrolysis of water into hydrogen and oxygen) may be called a *driven cell*. The energy input for a driven cell need not be electricity, but could be solar radiation, for example, in which case the process would be photochemical rather than electrochemical. If the same device can be used for conversion in both directions, or if the free energy-containing substance is regenerated outside the cell (energy addition required) and recycled through the cell, it may be called a *regenerative* or *reversible fuel cell* and, if the free energy-containing substance is stored inside the device, a *regenerative* or *secondary battery.*

The basic ingredients of an electrochemical device are two electrodes (sometimes called anode and cathode) and an intermediate electrolyte layer capable of transferring positive ions from the negative to the positive electrode (or negative ions in the opposite direction), while a corresponding flow of electrons in an external circuit from the negative to the positive electrode provides the desired power. Use has been made of solid electrodes and fluid electrolytes (solutions), as well as fluid electrodes (e.g. in high-temperature batteries) and solid electrolytes (such as ion-conducting semiconductors). A more detailed treatise of fuel cells may be found in Sørensen (2004a).

Fuel cells

The difference in electric potential, $\Delta\phi_{ext}$, between the electrodes (cf. the schematic illustration in Fig. 4.10) corresponds to an energy difference $e\Delta\phi_{ext}$ for each electron. The total number of electrons which could traverse the external circuit may be expressed as the product of the number of moles of electrons, n_e, and Avogadro's constant N_A, so the maximum amount of energy emerging as electrical work is

$$\Delta W^{(elec)} = n_e\, N_A\, e\, \Delta\phi_{ext},$$

(4.70)

where $\mathscr{F} = N_A e = 96\ 400$ C mol^{-1} (Faraday's constant) is sometimes introduced. This energy must correspond to a loss (conversion) of free energy,

$$- \Delta G = \Delta W^{(elec)} = n_e \mathscr{F} \Delta \phi_{ext},$$ (4.71)

which constitutes the total loss of free energy from the "fuel" for an ideal fuel cell. This expression may also be derived from (4.8), using (4.2) and $\Delta Q = T \Delta S$, because the ideal process is reversible, and $\Delta W = -P \Delta V + \Delta W^{(elec)}$.

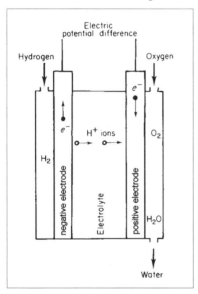

Figure 4.10. Schematic picture of a hydrogen–oxygen fuel cell. The electrodes are in this case porous, so that the fuel gases may diffuse through them.

Figure 4.10 shows an example of a fuel cell, based on the free energy change $\Delta G = -7.9 \times 10^{-19}$ J for the reaction

$$2H_2 + O_2 \rightarrow 2H_2O,$$

[cf. (3.42)]. Hydrogen gas is led to the negative electrode, which may consist of a porous material, allowing H^+ ions to diffuse into the electrolyte, while the electrons enter the electrode material and may flow through the external circuit. If a catalyst (e.g. a platinum film on the electrode surface) is present, the reaction at the negative electrode

$$2H_2 \rightarrow 4H^+ + 4e^-$$ (4.72)

may proceed at a much enhanced rate (see e.g. Bockris and Shrinivasan, 1969). Gaseous oxygen (or oxygen-containing air) is similarly led to the positive electrode, where a more complex reaction takes place, the net result of which is

$$O_2 + 4H^+ + 4e^- \rightarrow 2H_2O.$$ (4.73)

This reaction may be built up by simpler reactions with only two components, such as oxygen first picking up electrons or first associating with a hydrogen ion. Also, at the positive electrode the reaction rate can be stimulated by a catalyst. Instead of the porous material electrodes, which allow direct contact between the input gases and the electrolyte, membranes can be used (cf. also Bockris and Shrinivasan, 1969) like those found in biological material, i.e. membranes which allow H^+ to diffuse through but not H_2, etc.

The drop in free energy (4.71) is usually considered to be mainly associated with the reaction (4.73), expressing G in terms of a chemical potential (3.49), e.g. of the H^+ ions dissolved in the electrolyte. Writing the chemical potential μ as Faraday's constant times a potential ϕ, the free energy for n moles of hydrogen ions is

$$G = n\mu = n\mathscr{F}\phi = nN_A\, e\phi. \tag{4.74}$$

When the hydrogen ions "disappear" at the positive electrode according to the reaction (4.73), this chemical free energy is converted into the electrical energy (4.70) or (4.71), and since the numbers of electrons and hydrogen ions in (4.73) are equal, $n = n_e$, the chemical potential μ is given by

$$\mu = \mathscr{F}\phi = \mathscr{F}\Delta\phi_{\text{ext}}. \tag{4.75}$$

Here ϕ is the quantity usually referred to as the electromotive force (e.m.f.) of the cell, or "standard reversible potential" of the cell, if taken at standard atmospheric pressure and temperature. From the value of ΔG quoted above, corresponding to -2.37×10^5 J per mole of H_2O formed, the cell e.m.f. becomes

$$\phi = -\Delta G / n\mathscr{F} = 1.23 \text{ V}, \tag{4.76}$$

with $n = 2$ since there are two H^+ ions for each molecule of H_2O formed. The chemical potential (4.75) may be parametrised in the form (3.50), and the cell e.m.f. may thus be expressed in terms of the properties of the reactants and the electrolyte [including the empirical activity coefficients appearing in (3.50) as a result of generalising the expression obtained from the definition of the free energy, (4.6), assuming P, V and T to be related by the ideal gas law, $PV = \mathscr{R}T$, valid for one mole of an ideal gas (cf. e.g. Angrist, 1976)].

The efficiency of a fuel cell is the ratio between the electrical power output (4.70) and the total energy lost from the fuel. However, it is possible to exchange heat with the surroundings, and the energy lost from the fuel may thus be different from ΔG. For an ideal (reversible) process, the heat added to the system is

$$\Delta Q = T\,\Delta S = \Delta H - \Delta G,$$

and the efficiency of the ideal process thus becomes

$$\eta^{ideal} = -\Delta G / (-\Delta G - \Delta Q) = \Delta G / \Delta H. \tag{4.77}$$

For the hydrogen–oxygen fuel cell considered above, the enthalpy change during the two processes (4.72) and (4.73) is $\Delta H = -9.5 \times 10^{-19}$ J or -2.86×10^5 J per mole of H_2O formed, and the ideal efficiency is

$$\eta^{ideal} = 0.83.$$

There are reactions with positive entropy change, such as $2C+O_2 \rightarrow 2CO$, which may be used to cool the surroundings and at the same time create electric power with an ideal efficiency above one (1.24 for CO formation).

In actual fuel cells, a number of factors tend to diminish the power output. They may be expressed in terms of "expenditure" of cell potential fractions on processes not contributing to the external potential,

$$\Delta \phi_{ext} = \phi - \phi_1 - \phi_2 - \phi_3 - \dots,$$

where each of the terms $-\phi_i$ corresponds to a specific loss mechanism. Examples of loss mechanisms are blocking of pores in the porous electrodes, e.g. by piling up of the water formed at the positive electrode in the process (4.73), internal resistance of the cell (heat loss) and the building up of potential barriers at or near the electrolyte–electrode interfaces. Most of these mechanisms limit the reaction rates and thus tend to place a limit on the current of ions that may flow through the cell. There will be a limiting current, I_L, beyond which it will not be possible to draw any more ions through the electrolyte, because of either the finite diffusion constant in the electrolyte, if the ion transport is dominated by diffusion, or the finite effective surface of the electrodes at which the ions are formed. Figure 4.11 illustrates the change in $\Delta \phi_{ext}$ as a function of current,

$$I = \Delta \phi_{ext} R_{ext},$$

expressed as the difference between potential functions at each of the electrodes, $\Delta \phi_{ext} = \phi_c - \phi_a$. This representation gives some indication of whether the loss mechanisms are connected with the positive or negative electrode processes, and it is seen that the largest fraction of the losses is connected with the more complex positive electrode reactions in this example. For other types of fuel cells it may be negative ions that travel through the electrolyte, with corresponding changes in characteristics.

It follows from diagrams of the type shown in Fig. 4.11 that there will be an optimum current, usually lower than I_L, for which the power output will be maximum,

$$\max(E) = I^{opt} \Delta \phi_{ext}^{opt}.$$

Dividing by the rate at which fuel energy ΔH is added to the system in a steady-state situation maintaining the current I^{opt}, one obtains the actual efficiency of conversion maximum,

$$\max(\eta) = I^{opt} \, \Delta\phi_{ext}^{opt} / (dH/dt). \tag{4.78}$$

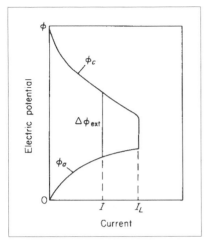

Figure 4.11. Fuel cell negative electrode potential ϕ_a and positive electrode potential ϕ_c as a function of current. The main cause of the diminishing potential difference $\Delta\phi_{ext}$ for increasing current is at first incomplete electrocatalysis at the electrodes, for larger currents also ohmic losses in the electrolyte solution, and finally a lack of ion transport (based on Bockris and Shrinivasan, 1969).

The potential losses within the cell may be described in terms of an internal resistance, the existence of which implies a non-zero energy dissipation, if energy is to be extracted in a finite time. This is the same fundamental type of loss as that encountered in the case of solar cells (4.67) and in the general thermodynamic theory described in section 4.1.1.

Other electrochemical conversion schemes

A driven-cell conversion based on the dissociation of water may be accomplished by electrolysis, using components similar to those of the hydrogen–oxygen fuel cell to perform the inverse reactions. Thus, the efficiency of an ideal conversion may reach 1.20, according to (4.78), implying that the electrolysis process draws heat from the surroundings.

By combining a fuel cell with an electrolysis unit, a regenerative system has been obtained, and if hydrogen and oxygen can be stored, an energy storage system or a "battery" results. The electric energy required for the electrolysis need not come from the fuel cell, but may be the result of an intermittent energy conversion process (e.g. wind turbine, solar photovoltaic cell, etc.).

Direct application of radiant energy to the electrodes of an electrochemical cell has been suggested, aiming at the achievement of the driven-cell process (e.g. dissociation of water) without having to supply electric energy. The electrodes could be made of suitable p- and n-type semiconductors, and the presence of photo-induced electron–hole excitations might modify the electrode potentials ϕ_a and ϕ_c in such a way that the driven-cell reactions become thermodynamically allowed. Owing to the electrochemical losses discussed above, additional energy would have to be provided by the solar radiation.

If the radiation-induced electrode processes have a low efficiency, the overall efficiency may still be higher for photovoltaic conversion of solar energy into electricity followed by conventional electrolysis (Manassen *et al.*, 1976). In recent years, several such photoelectrochemical devices have been constructed. Some of these will be described below in section 4.2.3.

4.2 Conversion of solar radiation

4.2.1 Heat generation

Conversion of solar energy to heat requires a light-absorbing material, a *collector*, which is able to distribute the absorbed radiant energy over internal degrees of freedom associated with kinetic energy of motion at the molecular level (e.g. lattice vibrations in case of a solid). The Earth and its atmosphere are examples of such collectors, as discussed in sections 2.2 and 2.3. Absorption of solar energy will raise the temperature of the collector or transfer energy to a reservoir, if the collector is connected to one. The collector will also emit radiation, and it may lose heat energy by conduction and convection processes. The frequency spectrum of the emitted radiation will correspond to the Planck spectrum (section 2.A) for the collector temperature T_c, if the collector is in a state allowing the definition of a thermodynamic temperature.

Man-made collectors may try to achieve a large absorption by minimising reflection and transmission and to achieve small losses, e.g. by operating the collector at temperatures not much above ambient air temperatures or, if higher load temperatures are required, by reducing the heat loss rates, e.g. by suitable transparent covers and insulation.

One may distinguish between "passive" and "active" systems, according to whether energy is specifically added (pumps, etc.) in order to bring the collector heat gain to the load areas or not. A passive system need not be characterised by the absence of definite heat flow paths between collectors and load areas, but such flows should be "natural", i.e. they should not depend on other energy inputs provided by man. There may be borderline cases in which the term "natural circulation" would be difficult to define.

Examples of passive solar heat systems are ordinary windows in buildings which transmit a large fraction of the solar radiation (if the angle of incidence is small, i.e. if the direction of the incident light does not make a small angle with the pane). The room behind the window may, to a large extent, act like a black body, absorb practically all of the radiation transmitted

MORE ON PV CELLS IN SECTION 4.2.3

through the window and re-emit only a small fraction again to the outside (providing that the total window area of the room is not large).

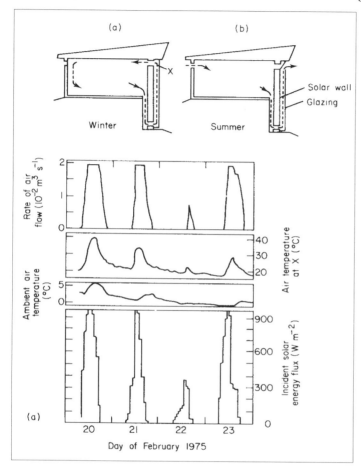

Figure 4.12. Solar wall type passive heating and cooling system. Top cross sections show air flows during (a) winter and (b) summer. Curves below show, for a selected period of a few days operation, the air flow-rate and temperature [at the location X in (a)], ambient temperature outside the house, and solar radiation. The inside wall temperature remains above 20°C during the period considered. The house is situated at 43°N latitude (based on Trombe, 1973; Stambolis, 1976).

Another kind of passive solar heat system uses the heat capacity of walls facing the sun during the daytime. The walls absorb radiation and accumulate it (the heat capacity of building materials increases roughly in proportion to mass), and at night they lose heat to their colder surroundings, including the inside area which is thus heated. The wall's heat capacity also serves to cool the building during at least the first part of the daytime, if the wall temperature after the night's cooling off is lower than the next day's ambient temperature. More elaborate versions of such solar wall systems, directing the natural convection according to conditions (night/day, summer/winter), are shown in Figs. 4.12 and 4.13. These systems require a little "active" help in opening and closing vents, shutters or covers. Being dependent on a daily cycle, these systems are of most interest in climatic regions where the daily solar input is substantial also during the cold season.

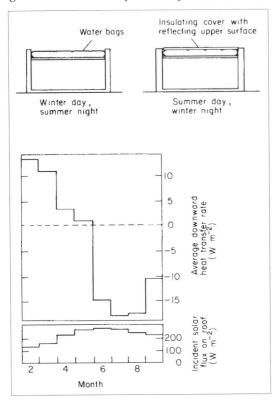

Figure 4.13. Solar roof-type passive heating and cooling system. The top cross sections show operation of cover panel at day and night, during summer and winter. Below are heat transfer rates through the roof (positive downward) and solar radiation, both averaged month by month during an 8-month period. The house is situated at 35°N latitude (based on Hay and Yellot, 1972; Stambolis, 1976).

Figure 4.14. Schematic cross section through a solar pond (cf. Tabor, 1967). The temperature profile may have a secondary peak within the top hundredths of a metre, due to absorption of ultraviolet radiation in this layer, and depending on the re-radiation conditions (effective sky temperature relative to temperature of pond surface layer).

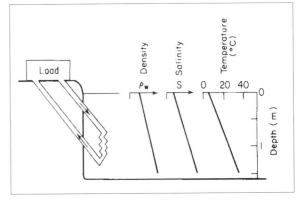

Greenhouses are also passive solar-collecting systems, as are water heaters based on water bags (e.g. placed on roofs) or on flat-plate collectors (see below) delivering heat to water, which is transferred by natural circulation to a storage tank lying higher than the collector. A saline pond can play the role of the collector, with a heat exchanger at the bottom of the pond transferring the collected energy to a working fluid, which can be circulated to the load area by natural or forced circulation. The solar pond itself (see Fig. 4.14) contains water with a high content of dissolved salts, causing the formation of a salinity and density gradient, preventing the mixing of surface and bottom water. The water absorbs some solar radiation, but if the pond is shallow most of the absorption takes place at the bottom, thereby creating a stable temperature gradient increasing towards the bottom, because heat is transferred upwards only by slow processes.

Flat-plate collectors

The term "flat-plate collector" is used for absorbers with a generally flat appearance, although the collecting surface need not be flat in detail (it might have V-shaped carvings or even focusing substructure). The side of the collector facing the sun may have a cover system (e.g. one or more layers of glass), and a mass flow J_m^c of some fluid (e.g. water or air) passes the absorber and is supposed to carry the heat, which is transferred from the absorber plate to the mass flow, to the load area (i.e. the place of actual usage) or to some temporary energy storage. The general layout is shown in Fig. 4.15.

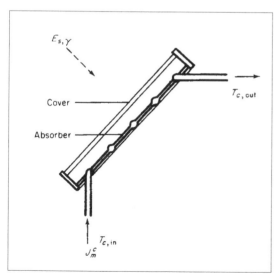

Figure 4.15. Example of flat-plate solar collector. The performance may not be given simply by the net incident radiation flux $E_{s,\gamma}$ (s is tilt angle and γ is azimuth angle), because the transmission–absorption product of the cover system may be different for different components of the incident radiation.

The absorber is characterised by an absorptance, $\alpha_\lambda(\Omega)$, which may depend on wavelength and on the direction of incident light (cf. section 3.1.4). In many situations, it may be assumed in a first approximation that the absorptance is independent of direction and of wavelength ("grey surface"). For a black painted surface, this α may be around 0.95, but, if the surface is structureless, the assumption that α is independent of direction breaks down when the angle between the normal to the surface and the direction of incidence exceeds about 60°. Towards 90°, $\alpha_\lambda(\Omega)$ actually approaches zero (Duffie and Beckman, 1974). According to (3.16), the emittance ε equals the absorptance (both are properties of the surface in the grey-surface approximation), and high absorptance thus implies high emittance for all wavelengths, including those characterising the thermal emission from the absorber of temperature T_c. In order to reduce this loss, use is made of surface treatments causing the emittance to assume two distinct values, a high one for the short wavelengths of the solar spectrum (implying a high solar absorptance α^{sw}) and a low one (α^{lw}) for the longer wavelengths characterising the emission from typical absorber temperatures, assumed to have a spectrum approximately equal to the black-body spectrum for that temperature. Such surfaces are called *selective surfaces,* and their wavelength-dependent absorptance/emittance may resemble that shown in Fig. 4.16, exhibiting the regions that can be characterised by $\alpha^{sw} \approx 0.95$ and $\alpha^{lw} \approx 0.1$. A further discussion of selective surface technologies may be found in Meinel and Meinel (1976).

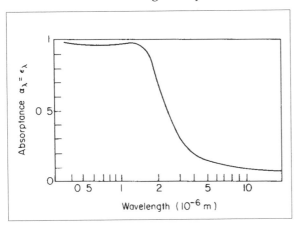

Figure 4.16. Spectral absorptance (or emittance) of a commercial "black-chrome" type of selective surface (based on Masterson and Seraphin, 1975).

As indicated in Fig. 4.15, the absorber may be covered by material that reduces heat losses and at the same time transmits most of the incoming radiation. The cover may consist, for example, of one or more layers of glass. The transmittance of glass depends on the type of glass and in particular on minority constituents such as Fe_2O_3. Figure 4.17 gives an example of the wavelength dependence of the transmittance, τ_λ, through ordinary window glass, defined as the fraction of incident light which is neither reflected [cf. (3.12)] nor absorbed,

$$\tau_\lambda = 1 - \rho_\lambda - \alpha_\lambda. \tag{4.79}$$

For a collector of the type considered, multiple reflections may take place between the absorber and the cover system as well as within the cover system layers, if there are more than one. The reflections are mostly specular (angle of exit equal to angle of incidence) at glass surfaces, but mostly diffuse (hemispherically isotropic) at the absorber surface. Thus, the amount of radiation absorbed by the absorber plate, in general, cannot be calculated from knowledge of the total incoming flux but must be calculated for each incident direction from which direct, scattered or reflected light is received, as well as for each wavelength. The total absorbed flux is then obtained by integration.

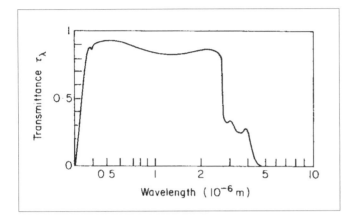

Figure 4.17. Spectral transmittance, $(1 - \rho_\lambda - \alpha_\lambda)$, of 2.8×10^{-3} m thick glass layer with a Fe_2O_3 content of 0.1 (based on Dietz, 1954).

In many cases a sufficiently good approximation can be obtained by considering only the two wavelength intervals, denoted "short" and "long" wavelengths above (the dividing wavelength being about 3×10^{-6} m), in view of the near-constancy of the absorptance and transmittance of the relevant materials (cf. Figs. 4.16 and 4.17). The net short-wavelength energy gain by the collector may then be written

$$E_c^{sw} = A \int E_{c+}^{sw}(\Omega)\, P^{t.a.}(\Omega)\, d\Omega, \qquad (4.80)$$

in terms of the incoming flux from a given direction Ω, $E_{c+}^{sw}(\Omega)$, and the "transmission–absorption product", $P^{t.a.}(\Omega)$, describing the fraction of incident short-wavelength radiation from the direction Ω, which gets transmitted through the cover system and gets absorbed by the absorber plate, the area of which is denoted A. It has been assumed that A serves as a simple proportionality constant, implying that incident radiation as well as transmission–absorption properties are uniform over the entire collector area. For a cover system consisting of N layers, e.g. of glass, characterised by a refraction index relative to air, n (~1.5 for glass), and an extinction coefficient, x, such that the reduction in intensity of radiation from traversing a distance d through the material is $\exp(-xd)$, the total thickness of each cover layer being L, then $P^{t.a.}(\Omega)$ may be approximately written (see e.g. Duffie and Beckman, 1974)

$$P^{t.a.}(\Omega) = \frac{1-\rho}{1+(2N-1)\rho} e^{-xNL\cos\theta} \frac{\alpha^{sw}}{1-(1-\alpha^{sw})\rho_d(N)}, \qquad (4.81)$$

where θ is the polar angle between the incident direction Ω and the normal to the collector surface, and ρ is the reflectance of one cover layer, given by the Fresnel formula in case of unpolarised light,

$$\rho = \frac{1}{2}\left(\frac{\sin^2(\theta'-\theta)}{\sin^2(\theta'+\theta)} + \frac{\tan^2(\theta'-\theta)}{\tan^2(\theta'+\theta)}\right),$$

with the polar angle of refraction, θ', given by

$$\theta' = \text{Arcsin}\left(\frac{\sin\theta}{n}\right).$$

The factorisation of (4.81) into a factor describing the multiple reflection through the N-layer cover system in the absence of absorption, times a factor describing the attenuation of the intensity from passage through glass, and a final factor describing the absorption in the absorber plate, after multiple reflections back and forth between the plate and cover system, is a valid approximation for most practical applications, owing to the smallness of the extinction product xL. The final factor in (4.81) may be rewritten as

$$\frac{\alpha^{sw}}{1-(1-\alpha^{sw})\rho_d} = \alpha^{sw}\sum_{i=1}^{\infty}((1-\alpha^{sw})\rho_d)^i,$$

revealing that it is the sum of terms corresponding to i reflections back and forth between the plate and cover (considering only the lower cover surface), and assuming the non-absorbed fraction, $(1-\alpha^{sw})$, from each impact on the absorber plate to be diffusely reflected from the cover inside surface with a reflectance ρ_d different from that of specular reflection, ρ. In order to make up for processes in which light reflected back from the absorber gets transmitted to and reflected from cover layers other than the lowest one, and then eventually reaches back to become absorbed by the absorber plate, effective values of ρ_d depending on N may be used, as indicated in (4.81). Duffie and Beckman (1974) suggest that the diffuse reflectance ρ_d may be approximated by the specular reflectance for an incident polar angle of 60°, using values of 0.16, 0.24, 0.29 and 0.32 for $N = 1,2,3$ and 4.

For the direct part (3.1) of the incident radiation flux (3.15), only one direction Ω is permitted [given by (3.2)], and the integration in (4.81) should be left out. For specular reflection on to the collector, the same is true in simple cases (cf. Seitel, 1975), but for complicated geometry of the specularly reflecting objects in the surroundings more than one direction may have to be taken into account. In general cases of reflection on to the collector, as well as for scattered radiation from the atmosphere [see (3.6)], the integration in (4.80) would generally have to be kept, but approximations such as multiplying the total scattered *or* reflected flux at the outside collector surface, (3.6) or (3.8) and (3.11) or (3.14), by the transmission–absorption product for an "average" angle of incidence, such as $\theta = 60°$, are sometimes used, permitting the total short-wavelength gain by the collector plate to be written

$$E_c^{sw} = A(D_{s,\gamma}P^{t.a.}(\theta) + (d_{s,\gamma}+R_{s,\gamma})<P^{t.a.}>), \qquad (4.82)$$

e.g. with $<P^{t.a.}> = P^{t.a.}(\theta = 60°)$.

The net long-wavelength energy gain by the collector is the difference between the long-wavelength radiation received from the surroundings, and the long-wavelength radiation emitted by the collector system. Considering the temperature T_c of the absorber plate as constant, the thermal radiation from the plate is of the form (3.17), and the net gain of an absorber without cover would be

$$E_c^{lw}(N{=}0) = A\varepsilon^{lw}\sigma(T_e^{4} - T_c^{4}), \qquad (4.83)$$

where the temperature T_e of the environment, in general, is lower than the ambient air temperature, T_a, as discussed in section 3.1.4.

Finally, sensible heat may be exchanged between the collector and its surroundings through conduction and convection processes. The back and edge heat transfers are usually expressed in the same way as heat losses from buildings,

$$E_{back}^{sens} = -A\,U_{back}\,(T_c - T_b), \qquad (4.84)$$

where U_{back} is a constant depending on the insulating properties of the materials separating the back side of the absorber plate from the environment of temperature T_b, which may be equal to the ambient temperature, T_a, if the collector is mounted freely or equal to the indoor temperature, T_L, if the collector forms an integral part of a building wall or roof. U_{back} may be assumed to include the "edge" heat losses from the sides of the collector, mainly of conduction type.

The net exchange of sensible energy at the front of the collector depends on the number of cover layers and on the wind speed, V, since the convection of heat is largely limited by the rate at which warm air at the top of the front cover is removed. In the absence of cover, the front (or top) exchange of heat may be written

$$E_{back}^{sens}(N{=}0) = -A f_1 (T_c - T_a), \qquad (4.85)$$

where f_1 is a polynomial in V with empirical coefficients, as suggested by Duffie and Beckman (1991),

$$f_1 = max\ (5.0;\ 3.95\ V[\text{m s}^{-1}]^{\,0.6})\ [\text{W K}^{-1}\text{m}^{-2}].\tag{4.86}$$

With glass covers, the expressions (4.83) and (4.85) must be modified, for example, by explicitly considering the heat transfer equations connecting each layer and solving for steady-state solutions with constant (but different) temperatures of each layer. Klein (1975) has parametrised these solutions in an explicit form, given by

$$E_c^{lw} = \frac{A\sigma(T_e^4 - T_c^4)}{(\varepsilon_p^{lw} + c_1 N(1 - \varepsilon_p^{lw}))^{-1} + N(2 + (f_2 - 1)/N) - N)/\varepsilon_g^{lw}},\tag{4.87}$$

where ε_p^{lw} and ε_g^{lw} are the long-wavelength emittances of the absorber plate and of the cover glasses, $c_1 = 0.05$ and

$$f_2 = (1 + 0.089\,f_1 - 0.1166\,f_1\,\varepsilon_p^{lw})\,(1 + 0.07866N),\tag{4.88}$$

with f_1 inserted from (4.86) the units indicated (Duffie and Beckman, 1991). The relation replacing (4.85) is

$$E_{top}^{sens} = -A(T_c - T_a) \cdot$$

$$\left\{ \left(\frac{NT_c}{f_3} f_4^{-c_2} + f_1^{-1} \right)^{-1} + \sigma(T_c + T_e)(T_c^2 + T_e^2) \left((\varepsilon_p^{lw} + 0.00591Nf_1)^{-1} + \frac{2N + f_2 - 1 + 0.133\varepsilon_p^{lw}}{\varepsilon_g^{lw}} - N \right)^{-1} \right\}.$$

$$\tag{4.89}$$

Here $c_2 = 0.43(1 - 100/T_c)$ and

$$f_3 = 520\ [\text{K}]\ (1 - 0.000051s^2),\tag{4.90}$$

with s being the tilt angle of the collector or 70°, whichever is smaller. Further,

$$f_4 = (T_c - T_a)/(N + f_2),\tag{4.91}$$

unless this expression becomes smaller than 1, in which case f_4 should rather be equal to 1. All temperatures in (4.87) and (4.89) should be inserted in K. The parametrisations are clearly phenomenological, and substantial changes between editions of the Duffie and Beckman book, notably affecting winter results, emphasise the *ad hoc* nature of the procedure.

The total net energy gain of the absorber plate of the collector is the sum of the contributions specified above,

$$E_c^{gain} = E_c^{sw} + E_c^{lw} + E_{top}^{sens} + E_{back}^{sens}.\tag{4.92}$$

Stalled and operating collector

If the mass flow $J_m^c = dm/dt$ through the solar collector is zero, the entire energy gain (4.92) will be used to raise the temperature of the collector. This situation of a "stalled collector" is of considerable interest for determining the maximum temperature which the materials used in constructing the collector must be able to withstand, for example, in a situation of pump failure or complete loss of circulating fluid. Denoting the heat capacity of the "dry" collector C', taken per unit area of collector, the equation determining the time development of the stalled-collector plate temperature becomes

$$AC' \, dT_c/dt = E_c^{sens}(T_c). \tag{4.93}$$

As T_c rises, the negative terms in (4.92) increase, and if the incident energy can be regarded as constant, an equilibrium situation will result in which the heat losses exactly balance the gains and the temperature thus remains constant,

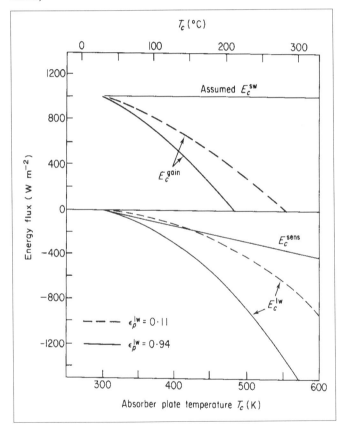

Figure 4.18. Energy flux components for a stalled collector, as a function of plate temperature T_c for a selective and a non-selective absorber surface (dashed and solid lines). There are two cover glasses of emittance $\varepsilon_g^{lw} = 0.88$, there is no wind (the curves are sensitive to wind speed changes), and all environmental temperatures have been taken as 300 K. The plate absorptance is $\alpha^{sw} = 0.94$, and the collector is placed horizontally. The curve E_c^{sens} represents the sum of fluxes reaching the top and back side of the collector (both positive towards the collector).

$$E_c^{gain}(T_{c,max}) = 0. \tag{4.94}$$

The determination of $T_{c,max}$ and the magnitude of the individual terms in (4.92), as a function of T_c, are illustrated in Fig. 4.18 for a high incoming flux E_c^{sw}, environmental temperatures $T_e = T_a = T_L = 300$ K, and no wind, based on the explicit expressions (4.84)–(4.92). Two collectors are considered, one with selective surface ($\varepsilon_p^{lw} = 0.11$) and one with non-selective black surface ($\varepsilon_p^{lw} = \alpha^{sw} = 0.94$). Both have two layers of cover glass. The corresponding equilibrium temperatures are approximately $T_{c,max} = 550$ K and $T_{c,max} = 480$ K. If wind were present, the convective heat loss (4.89) would be much higher and the maximum temperature correspondingly lower.

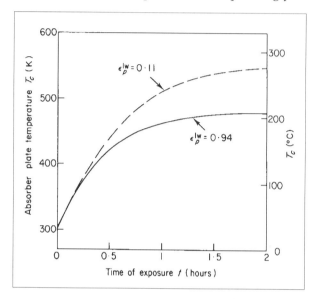

Figure 4.19. Plate temperature for a stalled, dry collector as a function of time. The incident flux is constant $E_c^{sw} = 1000$ W m^{-2}, and the heat capacity C' in (4.93) has been taken as 10^4 J m^{-2} K^{-1}, corresponding to a fairly lightweight construction. Other parameters are as in Fig. 4.18. About 7 min are required to reach operating temperatures (some 50 K above ambient). The time scale is proportional to C', so results for other values of C' are easily inferred.

The time required to reach a temperature T_c close to $T_{c,max}$ can now be calculated from (4.93) by integration, as illustrated in Fig. 4.19. While the maximum temperature is independent of the heat capacity of the collector, the time scales linearly with C'. The value $C' = 10^4$ J m^{-2} K^{-1} used in Fig. 4.19 corresponds to a fairly light collector (e.g. absorber plate of extruded aluminium), in which case the asymptotic temperature region is reached after about 1 h of constant high incoming radiation. The shortness of this period makes the assumption of a constant E_c^{sw} acceptable. The amount of time needed to reach operating temperatures of about 50°C is roughly 7 min. It would double if the ambient temperature were at the freezing point of water and would further increase if the radiation were lower than the 1000 W m^{-2} assumed in the example used in Figs. 4.18 and 4.19. A short response time of the collector is also of relevance in situations of variable radiation (e.g. caused by frequent cloud passage).

In case of an operating collector, J_m^c is no longer zero and a fluid carries energy away from the absorber (cf. Fig. 4.15). The temperature of the outgoing fluid, $T_{c,out}$, will in this case be higher than that of the incoming fluid, $T_{c,in}$, and it is no longer consistent to assume that the plate temperature T_c is uniform over the plate. The change in plate temperature must be at least as large as the change in fluid temperature, between the points where the fluid is receiving its first and last heat transfer contribution from the absorber plate. Because of the non-linear relation between plate temperature and heat loss, it is not correct to evaluate the losses at the mean plate temperature, T_c. Still, this is often done in approximate calculations, choosing

$$\overline{T}_c = \frac{1}{2}(T_{c,in} + T_{c,out}), \tag{4.95}$$

and assuming further that the average fluid temperature is also the average plate temperature. Incomplete heat transfer between absorber plate and fluid channels, as well as the non-linearity of loss functions, may be simulated by adding a constant term to (4.95), or by using a factor larger than ½. The transfer of heat to the fluid may be effectuated by passing the fluid along the absorber plate (used e.g. in most collectors with air as a working fluid) or by the fluid's passage through pipes in highly heat-conducting contact with the absorber plate (most often used when the working fluid is water, oil, etc.). In the latter case, the non-uniform heat extraction constitutes a further cause of non-uniform plate temperature (the temperature gradients towards the pipe locations being, of course, the reason for obtaining a heat flow from any plate position to the relatively small area of contact with the fluid-carrying pipes).

If the fluid of fixed inlet temperature $T_{c,in}$ is allowed to pass only once through the collector (e.g. for hot water production), and if the simplifying assumptions mentioned above are made, then the equation replacing (4.93) for an operating collector will be

$$E_c^{gain}(T_c) = AC' \, d\overline{T}_c/dt + J_m^c \, C_p^c \, (T_{c,out} - T_{c,in}), \tag{4.96}$$

or in a steady state situation just

$$E_c^{gain}(\overline{T}_c) = J_m^c \, C_p^c \, (T_{c,out} - T_{c,in}).$$

C_p^c is the heat capacity of the fluid flowing through the collector. This equation determines $T_{c,out}$, but the left-hand side depends on $T_{c,out}$, through (4.95) or some generalisation of this relation between the flow inlet and outlet temperatures and the effective average plate temperature. Therefore, it is conveniently solved by iteration,

$$T_{c,out}^{(i+1)} = T_{c,in} + E_c^{gain} \, (\frac{1}{2} \, (T_{c,in} + T_{c,out}^{(i)})) \, / \, (J_m^c C_p^c). \tag{4.97}$$

A better approximation for the transfer of heat from the absorber plate to the fluid channels may be obtained along the lines described below for the general problem of heat exchange encountered in several places with application of the solar collectors in heat supply systems, for example, involving a heat storage which is not part of the collector flow circuit.

Heat exchange

A situation like the one depicted in Fig. 4.20 is often encountered in energy supply systems. A fluid is passing through a reservoir of temperature T_3, thereby changing the fluid temperature from T_1 to T_2. In order to determine T_2 in terms of T_1 and T_3, in a situation where the change in T_3 is much smaller than the change from T_1 to T_2, the incremental temperature change of the fluid by travelling a short distance dx through the pipe system is related to the amount of heat transferred to the reservoir, assumed to depend linearly on the temperature difference,

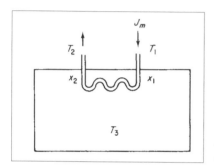

Figure 4.20. Heat exchanger, an idealised example of a well-mixed T_3-reservoir.

$$J_m C_p^{fluid}\, dT^{fluid}/dx = h'\, (T_3 - T^{fluid}).$$

Integrating from T_1 at the inlet $(x = x_1)$ gives

$$T^{fluid}(x) - T_3 = (T_1 - T_3)\exp(- h'(x - x_1)/(J_m C_p^{fluid})), \tag{4.98}$$

where h' is the heat transfer per unit time from a unit length of the pipe for a temperature difference of one unit. The heat transfer coefficient for the entire heat exchanger is

$$h = \int_{x_1}^{x_2} h'\, dx,$$

which is sometimes written $h = U_h A_h$, with U_h being the transfer coefficient per unit area of pipe wall and A_h being the effective area of the heat exchanger. For $x = x_2$, (4.98) becomes (upon re-ordering)

$$(T_1 - T_2) = (T_1 - T_3)(1 - \exp(- h/(J_m C_p^{fluid}))). \tag{4.99}$$

Flat-plate collector with heat storage

The general layout of a flat-plate solar collector with heat storage is shown in Fig. 4.21. There are two heat exchangers on the collector side and two on the load side. If the actual number is less, for example, if the storage is in terms of a fluid that can be circulated directly through the collector, then the corresponding heat transfer coefficient [h in (4.99)] should be taken as infinitely large. While the relationship between temperatures at every heat exchanger may be taken from (4.99), the net energy transfer to the storage is associated with a change in the storage temperature (the average value of which is $\bar{T}_s$), which may be obtained from the equation accounting for the relevant energy fluxes, assuming the storage temperature to be uniform,

$$SC^s \, dT_s/dt = J_m^c C_p^c(T_{c,out} - T_{c,in}) - J_m^L C_p^L(T_{L,in} - T_{L,out}) - h_s(\bar{T}_s - T_0). \qquad (4.100)$$

Here S is the storage volume (or mass), C^s is the heat capacity on a volume (or mass) basis of the material used for storage (assuming for simplicity storage in the form of sensible heat), superscripts c and L distinguish the flows in the collector and the load circuits, and finally, T_0 is the temperature of the surroundings of the storage volume (e.g. soil in the case of underground storage), to which the storage is losing energy through its walls. (4.100) must be supplemented by (4.96) and a similar equation for the load area, e.g. of the form

$$LC^L \, dT_L/dt = J_m^L C_p^L (T_{L,in} - T_{L,out}) - h_L (\bar{T}_L - T_0), \qquad (4.101)$$

assuming that the load area is a building of volume (or mass) L and average temperature $\bar{T}_L$, which may lose energy to surroundings of ambient air temperature T_a. If the load is in the form of water heating, it may be described by a term of the same form as the first one on the right-hand side of (4.101), but with $T_{L,out}$ being the water inlet temperature and $T_{L,in}$ being the water temperature of the water supplied by the system for usage.

Together with the four heat exchanger equations (4.99) with h_c, h_{cs}, h_{sL} and h_L (see Fig. 4.21), (4.96), (4.100) and (4.101) constitute seven equations for determining the seven temperatures appearing in Fig. 4.21. If the load circuit is not activated, the four equations for $T_{c,in}$, $T_{c,out}$, $\bar{T}_s$ and $\bar{T}_c$ remain. Instead of the heat exchanger equation (4.99) for the collector, which for $h_c \to \infty$ gives $T_{c,out} = \bar{T}_c$ owing to the assumption that the reservoir of temperature $\bar{T}_c$ is large compared with the amounts of heat exchanged, relations such as (4.95) may be used, or a more elaborate calculation taking into account the geometry of the collector and fluid pipes or channels may be performed.

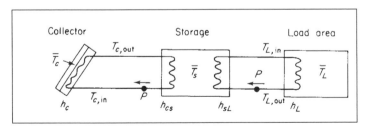

Figure 4.21. Solar heating system with storage. h denotes a heat exchanger, P denotes a circulation pump, and the Ts are temperatures.

The efficiency of the collector itself may be defined as

$$\eta_c = J_m^c C_p^c \, (T_{c,out} - T_{c,in}) / E_{s,\gamma}^{sw}, \tag{4.102}$$

where the numerator may be obtained from (4.96) and the denominator from (3.15). Only if the heat capacity of the collector is neglected, or if the collector is operating at a constant temperature, can the collector heat output (to the fluid) be taken as the collector gain $E_c^{gain}(T_c)$ from (4.92) directly.

If an energy store is used in connection with the collector, the relevant efficiency is that of delivering the heat to the storage, which is formally the same as (4.102). In practice, however, the introduction of an energy storage may completely alter the efficiency by changing $T_{c,in}$, which again causes a change in $T_{c,out}$. When the storage temperature, T_s, is low, the average collector plate temperature $\bar{T}_c$ will be fairly low and the loss terms correspondingly small. Since $T_{c,in}$ will always be higher than T_s (but not always higher than $\bar{T}_s$, since often the fluid circuit is made to leave the storage at a place of low temperature, if T_s is not uniform), the accumulation of stored energy and associated increases in T_s and $T_{c,in}$ will cause the collector loss terms to increase, implying a diminishing efficiency. Examples of this behaviour will be given in Chapter 6, in the discussion of simulations of complete solar heating systems.

Another important factor in determining the efficiency of energy collection is the rate of fluid circulation through the collector circuit, J_m^c. Figure 4.22 gives an example of the performance of the same system with two different values of J_m^c. With the low fluid velocity, the exit temperature from the collector, $T_{c,out}$, is about 45°C above the inlet temperature, but the energy transferred to the storage is much less than in the case of a higher fluid velocity, causing a smaller difference between $T_{c,out}$ and $T_{c,in}$, but larger net collector gain, owing to the smaller loss terms. The optimum fluid flow rate for a given heat exchange capability h_{cs} (assuming the heat transfer at the collector to be nearly perfect) may be found by calculating the collection efficiency as a function of J_m^c or, as a more generally applicable criterion, by calculating the coverage of the actual energy demand which is sought covered by the solar heat system. If a given solar heat system is considered, the percentage of the total heat load over an extended period, such as a year, may be evaluated as function of J_m^c, as done in Fig. 4.23.

If 100% coverage is aimed at, the parameter to be optimised may be taken as the minimum size (e.g. collector area A) which will allow the full load to be covered at any time. The example shown in Fig. 4.23 relates to the heating and hot water requirements of a one-family dwelling at 56°N latitude. For each value of the heat exchange coefficient h_{cs} between the collector circuit and the storage, the optimum flow rate is determined. With increasing h_{cs}, the region of "acceptable" values of J_m^c becomes larger, the increased J_m^c indicated for large h_{cs} does not significantly improve the performance of the system, and it will soon become uneconomical due to the energy spent in pumping the fluid through the circuit (the energy required for the pump is negligible for low fluid speeds, but the pipe resistance increases with the fluid speed, and the resistance through the heat exchanger increases when h_{cs} does). The results of Fig. 4.23 may then be interpreted as suggesting a modest increase in fluid velocity, from about 0.1 to about 0.2 kg (or litre) s^{-1}, when the heat exchange coefficient is increased from 400 to 4000 W K^{-1} (for the particular system considered).

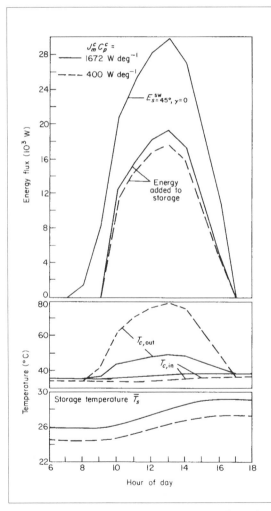

Figure 4.22. Single-day performance of a solar heating system, based on a simulation model using the Danish Reference Year (latitude 56°N). The day is 8 February, the collector area $A = 40$ m^2, and the storage volume $S = 25$ m^3 (water equivalent). The collector has one glass cover layer and a selective surface ($\varepsilon_c^{lw} = 0.11$), and the heat transfer coefficient $h_{cs} = 4000$ W K^{-1}. Two different flow rates are considered for the collector circuit, $J_m^c C_p^c = 1672$ or 400 W K^{-1} corresponding to 0.4 or 0.096 kg of water per second.

The optimisation problem outlined in this example is typical of the approach to the concept of an "efficiency" for the total solar heat system, including storage and load sections. Because the time distribution of the load is, in general, barely correlated with that of the solar energy collection (in fact, it is more "anti-correlated" with it), an instantaneous efficiency defined as the ratio between the amount of energy delivered to the load area and the amount of incident solar energy is meaningless. A meaningful efficiency may be defined as the corresponding ratio of average values, taken over a sufficient length of time to accommodate the gross periodicity of solar radiation and of load, i.e. normally a year,

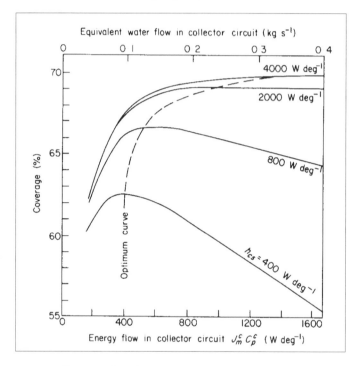

Figure 4.23. Annual average coverage of heating and hot water load for a one-family dwelling (average load 1.91 kW) under conditions set by the Danish Reference Year (56°N) (cf. section 6.3.1) as a function of water flow in collector circuit and heat exchanger transfer coefficient. The collector is as described in the caption to Fig. 4.22 and placed on a south-facing roof with tilt angle $s = 45°$.

$$\bar{\eta}_{system} = \bar{c} / \bar{E}_{s,\gamma}^{sw}, \tag{4.103}$$

where c is the average load covered by the solar energy system and $\bar{E}_{s,\gamma}^{sw}$ is the average incident short-wavelength radiation on the surface of the collector (its cover if any). Conventionally, the long-wavelength or other heat contributions to the incident energy flux are not included in the efficiency calculations (4.103) or (4.102).

Concentrating collectors and tracking systems

Various advanced versions of the flat-plate collector are conceivable in order to increase the amount of absorbed energy and decrease the amount of heat lost. The incident average flux may be increased by replacing the fixed installation (of tilt and azimuth angles s and γ) by a movable one, which can be made to track the direction towards the Sun. Fully tracking (i.e. in both height and azimuth angle) devices may be mounted in a manner similar to the one used for (star-)tracking astronomical instruments, and incompletely tracking systems (e.g. only following the azimuth angle of the Sun) might combine a simpler design with some improvement in energy gain. An idea of the gain made possible by tracking the Sun can be obtained by comparing Fig. 3.19 with Fig. 3.20. The scattered and (diffusely) rejected fluxes do not change much on average by going from a fixed to a tracking collector, but the direct part (3.1) becomes replaced by the normal incidence flux S_N. For a flat-plate collector, the maximum average gain from using a fully tracking system (less than a factor 2) would rarely justify the extra cost of the tracking system which, at least at present, exceeds the cost of doubling the collector area.

Other possible improvements of the flat-plate collector system include coating the cover layers to reduce reflection losses on the outside and increase reflection (back on to the absorber) on the inside surfaces. The mainly convective heat loss E_{top}^{sens} may be reduced by installation of shields that diminish the wind velocity outside the top cover layer. The strong dependence of convection losses on wind speed is clear from (4.89). The convective losses may be more radically diminished by evacuating the space between the absorber plate and the cover system. In vacuum, heat transport can take place only by radiation. The evacuation of the substantial space between absorber and cover typical of collector designs of the type shown in Fig. 4.15 is not practical, and systems using evacuation-type insulation would be built in a different way, e.g. with units of evacuated cylinders or spheres containing the absorbers and possibly concentrating equipment.

Focusing, or more generally concentrating, devices constitute a separate class of solar collectors. These are necessary if very high temperatures are required, but may, in principle, also be considered for heating purposes involving modest temperature rises over the ambient. For a flat-plate collector without any of the special loss-reducing features mentioned above, the net gain is a rapidly decreasing function of the plate temperature, as seen in Fig. 4.18. Since the losses are proportional to the absorber area, there is a special interest in reducing this area relative to the collector area A determining the incident radiation energy.

One extreme is a point-focusing device, such as the parabolic reflectors in Fig. 4.24 or the lenses in Fig. 4.26. If the absorber is placed in the focus on the symmetry axis and its dimension corresponds to that of the image of the Sun when the Sun is in the direction of the symmetry axis ("optical axis"), then direct radiation will be accepted, but only as long as the optical axis is made to track the Sun. Imperfections and, in the case of lenses, the wavelength dependence of the refraction index may further enlarge the image dimensions. The absorber dimension would have to be chosen so that it covered the bulk of the image under most conditions.

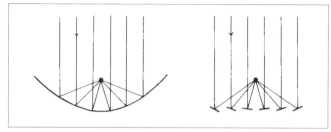

Figure 4.24. Parabolic- and Fresnel-type reflectors, e.g. for use in heliostats and tower absorbers, respectively.

The total short-wavelength radiation received by the absorber, E_c^{sw}, may be written in the same form as for the flat-plate collector, (4.80) or (4.82), but with a transmission–absorption product, $P^{t.a.}(\Omega)$, suitable for the focusing geometry rather than (4.81). If reflections and re-absorptions at the absorber surface are neglected, the transmission–absorption product for a focusing device may be written as the product of the absorptance α^{sw} of the absorber surface and a transmission function $t^{sw}(\Omega)$, which expresses the fraction of the incident short-wavelength radiation from the direction Ω that reaches the absorber surface (directly or after one or more reflections or refractions within the device),

$$P^{t.a.}(\Omega) = \alpha^{sw}\, t^{sw}(\Omega). \tag{4.104}$$

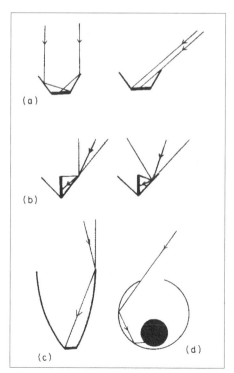

(a)

(b)

(c)

(d)

Figure 4.25. Examples of concentrating collectors.

For the parabolic reflector (Fig. 4.24) and the lens (Fig. 4.26), the idealised transmission function is zero, unless the absorber is placed exactly in the focal point corresponding to incoming radiation from the direction Ω. Thus, either the collector has to fully track the Sun or the absorber has to be moved to the focal point corresponding to a given direction to the Sun. For scattered radiation, which by definition does not come from the direction of the Sun (the same being normally the case for reflected radiation), $t^{sw}(\Omega)$ is zero, and such radiation will not be available for these devices. As a compromise, the acceptance angle may be increased at the expense of a high concentration ratio. Some scattered and reflected radiation will be accepted, and complete tracking may not be necessary.

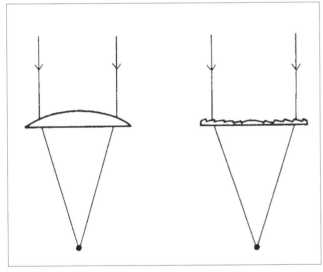

Figure 4.26. Simple Fresnel-type lenses.

Figure 4.25 gives a number of examples of such devices. The inclined "booster" mirrors along the rim of a flat-plate collector (Fig. 4.25a) result in a modest increase in concentration for incident angles not too far from the normal to the absorber plate, but a decrease (due to shadows) for larger incident angles. The V- or cone-shaped collector with the absorber in the shape of a wall or an upright line (cylinder) in the middle (Fig. 4.25b) has an acceptance angle depending on the distance from the convergence point of the cone. At this point, the acceptance angle equals the opening angle, but at points further out along the cone sides the acceptance angle is smaller (depending on the length of the absorber line, or wall height) relative to that of the cone. A fairly high concentration can be achieved with a cusp construction (Fig. 4.25c), comprising two parabolic sections, each of which has its focus at the point of intersection between the opposite parabolic section and the absorber. Figure 4.25d shows a "trapping" device consisting of a cylinder (or sphere) with a window, and an absorber displaced from the centre. The dimensions of the absorber are such that a large fraction of the radiation passing through the window will be absorbed after (multiple) reflections on the reflecting inside walls of the cylinder. Also, for the long-wavelength radiation emitted by the absorber, the reflecting walls will lead a substantial fraction back to the absorber, so that the loss-to-gain ratio is improved, even though there is no actual concentration (if the dimension of the absorber is as large as that of the window).

Energy collection from focusing systems

A survey of a number of optically focusing systems for solar energy collection may be found in Meinel and Meinel (1976).

In the absence of cover systems, the long-wavelength emission from a focusing system may be written in the form (4.83),

$$- E_c^{lw} = A_a \varepsilon^{lw} \sigma (T_c^4 - T_e^4), \tag{4.105}$$

where A_a is the absorber area.

Similarly, the sensible heat exchange with the surroundings of ambient temperature T_a may be given in analogy to (4.84) and (4.85),

$$E_c^{sens} = - A_a U (T_c - T_a), \tag{4.106}$$

where U may depend on wind speed in analogy to (4.86). The collector gain is now given by $E_c^{gain} = E_c^{sw} + E_c^{lw} + E_c^{sens}$, where the first term is proportional to A, while the two last terms are proportional to A_a. Assuming now a fluid flow J_m^c through the absorber, the expression for determination of the amount of energy extracted becomes [cf. (4.96)]

$$J_m^c C_p^c (T_{c,out} - T_{c,in}) = E_c^{gain}(\overline{T}_c) - A_a C' \, dT_c / dt$$
$$= E_c^{sw} + E_c^{lw}(\overline{T}_c) + E_c^{sens}(\overline{T}_c) - A_a C' \, d\overline{T}_c / dt, \tag{4.107}$$

where $A_a C'$ is the heat capacity of the absorber and the relation between $\overline{T}_c$ and $T_{c,out}$ may be of the form (4.97) or (4.99). Inserting (4.82) with use of (4.104), (4.105) and (4.106) into (4.107), it is apparent that the main additional parameters specifying the focusing system are the area concentration ratio,

$$X = A/A_a, \tag{4.108}$$

and the energy flux concentration ratio,

$$C^{flux} = A t^{SW}(\Omega) / A_i, \tag{4.108'}$$

where t^{SW} is given in (4.104) and A_i is the area of the actual image (which may be different from A_a). Knowledge of A_i is important for determining the proper size of the absorber, but it does not appear in the temperature equation (4.107), except indirectly through the calculation of the transmission function t^{SW}.

For a stalled collector, $J_m^c = 0$ and d $\bar{T}_c/dt = 0$ give the maximum absorber temperature $T_{c,max}$. If this temperature is high, and the wind speed is low, the convective heat loss (4.108) is negligible in comparison with (4.105), yielding

$$\sigma (T_{c,max}^4 - T_c^4) = X D \, \alpha^{sw} \, \varepsilon^{lw},$$

where D is the incident short-wavelength flux in the direction of the Sun [cf. (4.82)], assuming the device accepts only the flux from this direction and assuming perfect transmission to the absorber ($t^{SW} = 1$). With the same conditions, the best performance of an operating collector is obtained from (4.108'), assuming a steady-state situation. The left-hand side of (4.108') is the amount of energy extracted from the collector per unit of time, E^{extr}:

$$E^{extr} = A\alpha^{SW} D - A_a\varepsilon^{lw} \, \sigma (\bar{T}_c^4 - T_c^4), \tag{4.109}$$

and the efficiency corresponding to the idealised assumptions

$$\eta_c^{ideal} = \frac{E^{extr}}{AD} = \alpha^{SW} \left(1 - \frac{\varepsilon^{lw}\sigma}{X\alpha^{SW}D}(\bar{T}_c^4 - T_e^4) \right). \tag{4.110}$$

This relation is illustrated in Fig. 4.27, assuming $D = 800$ W m^{-2}, $T_c = 300$ K and $\alpha^{SW} = 0.9$ independent of absorber temperature. It is clear that the use of a selective absorber surface (large ratio $\alpha^{SW}/\varepsilon^{lw}$) allows a higher efficiency to be obtained at the same temperature and that the efficiency drops rapidly with decreasing light intensity D. In case the total short-wavelength flux on the collector is substantially different from the direct component, the actual efficiency relative to the incident flux $E_{s,\gamma}^{SW}$ of (3.15),

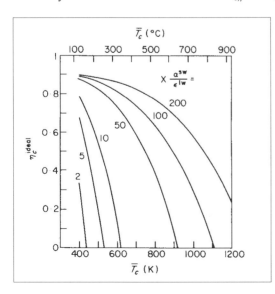

Figure 4.27. Ideal collector efficiency for concentrating collectors, evaluated for a direct radiation incident flux of $D = 800$ W m^{-2}, an environmental temperature $T_e = 300$ K and $\alpha^{SW} = 0.9$.

$$\eta_c = E^{extr} / E_{s,\gamma}^{SW},$$

may become far less than the "ideal" efficiency (4.110).

4.2.2 Applications for cooling, pumping, etc.

A variety of applications of solar energy thermal conversion have been considered, in addition to heat use. Living comfort and food preservation require cooling in many places. As in the case of heating, the desired temperature difference from the ambient is usually only about 10–30°C. Passive systems such as those depicted in Figs. 4.12 and 4.13 may supply adequate cooling in the climatic regions for which they were designed. Radiative cooling is very dependent on the clearness of the night sky (cf. section 3.1.4). In desert regions with a clear night sky, the difference between the winter ambient temperature and the effective temperature of the night sky has been used to freeze water. In Iran, the ambient winter temperature is typically a few degrees Celcius above the freezing point of water, but the temperature T_e of the night sky is below 0°C. In previous centuries, ice (for use in the palaces of the rich) was produced from shallow ponds surrounded by walls to screen out the daytime winter Sun, and the ice produced during the night in such natural icemakers ("yakhchal", see Bahadori, 1977) could be stored for months in very deep (say, 15 m) underground containers.

Many places characterised by hot days and cool nights have taken advantage of the heat capacity of thick walls or other structures to smooth out the diurnal temperature variations. If additional cooling is required, active solar cooling systems may be employed, and a "cold storage" may be introduced in order to cover the cooling need independent of the variations in solar radiation, in analogy to the "hot storage" of solar heating systems.

The solar cooling system may consist of flat-plate collectors delivering the absorbed energy to a "hot storage", which is then used to drive an absorption cooling cycle (Fig. 4.28), drawing heat from the "cool storage", to which the load areas are connected (see e.g. Wilbur and Mancini, 1976). In principle, only one kind of storage is necessary, but with both hot and cold storage the system can simultaneously cover both heating needs (e.g. hot water) and cooling needs (e.g. air conditioning).

The absorption cooling cycle (Fig. 4.28) is achieved by means of an absorbent–refrigerant mix, such as LiBr–H_2O or H_2O–NH_3. The lithium–bromide–water mix is more suitable for flat-plate solar collector systems, giving higher efficiency than the water–ammonia mix for the temperatures characteristic of flat-plate collectors. LiBr is hygroscopic, i.e. it can mix with water in any ratio. The solar heat is used in a "generator" to vaporise some water from the mix. This vapour is led to a condenser unit, using a coolant flow, and is then expanded to regain a gaseous phase, whereby it draws heat from the area to be cooled, and is returned to an "absorber" unit. Here it becomes absorbed in the LiBr–H$_2$O mix with the help of a coolant flow. The coolant inlet temperature would usually be the ambient one, and the same coolant passes through the absorber and the condenser. The coolant exit temperature would then be above ambient, and the coolant cycle could be made closed by exchanging the excess heat with the surroundings (e.g. in a "cooling tower"). The refrigerant-rich mix in the absorber is pumped back to the generator and is made up for by recycling refrigerant-poor mix from the generator to the absorber (e.g. entering through a spray system). In order not to waste solar heat or coolant flow, these two streams of absorbent–refrigerant mix are made to exchange heat in a heat exchanger. The usefulness of a given absorbent–refrigerant pair is determined by the temperature dependence of vaporisation and absorption processes.

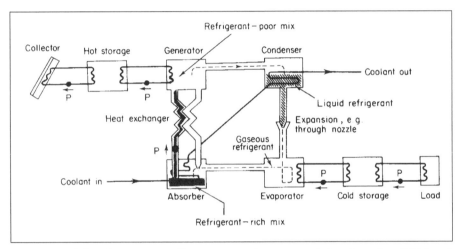

Figure 4.28. Solar absorption cooling system using a mix of absorbent (e.g. LiBr) and refrigerant (e.g. H$_2$O). P denotes pump. The coolant may be recycled after having rejected heat to the environment in a cooling tower.

In dry climates a very simple method of cooling is to spray water into an air-stream (evaporative cooling). If the humidity of the air should remain unchanged, the air has first to be dried (spending energy) and then cooled by evaporating water into it, until the original humidity is again reached.

In principle, cooling by means of solar energy may also be achieved by first converting the solar radiation to electricity by one of the methods described in section 4.2.3 and then using the electricity to drive a thermodynamic cooling cycle, such as the Rankine cycle in Fig. 4.2.

The same applies if the desired energy form is work, as in the case of pumping water from a lower to a higher reservoir (e.g. for irrigation of agricultural land). In practice, however, the thermodynamic cycles discussed in section 4.1.2 are used directly to make a pump produce mechanical work on the basis of the heat obtained from a solar collector. Figure 4.29 shows two types of devices based on the Stirling cycle, using air or another gas as a working fluid (cf. Fig. 4.2). In the upper part of the figure, two ball valves ensure an oscillatory, pumping water motion, maintained by the tendency of the temperature gradient in the working fluid (air) to move air from the cold to the hot region. The lower part of the figure shows a "free piston Stirling engine", in which the oscillatory behaviour is brought about by the presence of two pistons of different mass, the heavy one being delayed by half an os- cillatory period with respect to the lighter one. The actual water pumping is made by the "membrane-movements" of a diaphragm, but if the power level is sufficient the diaphragm may be replaced by a piston-type pump. For Stirling cycle operation, the working fluid is taken as a gas. This is an effi- cient cycle at higher temperatures (e.g. with use of a focusing solar collector), but if the heat provided by the (flat-plate) solar collector is less than about 100°C, a larger efficiency may be obtained by using a two-phase working fluid (corresponding to one of the Rankine cycles shown in Fig. 4.2).

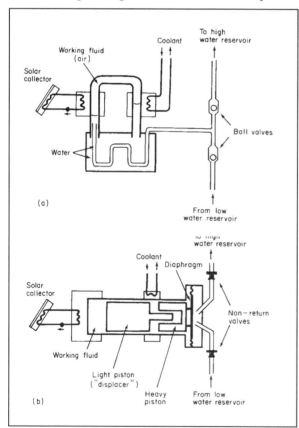

Figure 4.29. Solar water pumps based on Stirling cycles. In (a), air and water should not mix. The ball valves produce cyclic op- eration. In (b), the light and heavy pistons should be out of phase by a half cycle (based on West, 1974; Beale 1976).

If focusing solar collectors are used, the diaphragm may be replaced by a more efficient piston pump.

Two Rankine-type solar pumps, based on a fluid-to-gas and gas-to-fluid cycle, are outlined in Fig. 4.30. The one shown in the upper part of the figure is based on a cyclic process. The addition of heat evaporates water in the container on the left, causing water to be pumped through the upper one-way valve. When the vapour reaches the bottom of the U-tube, all of it moves to the container on the right and condenses. New water is drawn from the bottom reservoir, and the pumping cycle starts all over again. This pump is intended for wells of shallow depth, below 10 m (Boldt, 1978).

The lower part of Fig. 4.30 shows a pump operated by a conventional Rankine engine, expanding the working fluid to gas phase passing through a turbine, and then condensing the gas using the pumped water as coolant, before returning the working fluid to the heat exchanger receiving solar absorbed heat. For all the pumps based on a single thermodynamic cycle, the

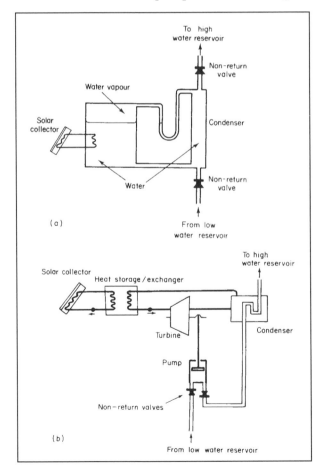

Figure 4.30. Solar water pumps based on Rankine cycles. In (a), the formation of vapour, its transfer through the U-tube and condensation produces a cyclic pressure–suction variation, which draws water from the lower reservoir and delivers water to the high reservoir (based on Boldt, 1978). In (b), a turbine-type pump is shown, the working fluid of which may be an organic compound (cf. Meinel and Meinel, 1976).

maximum efficiency (which cannot be reached in a finite time) is given by
(4.4) and the actual efficiency is given by an expression of the form (4.19).

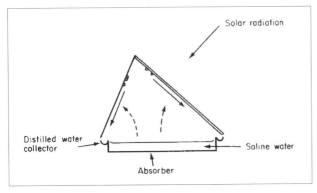

Figure 4.31. Solar still.

Distillation of saline water (e.g. sea water) or impure well water may be
obtained by a solar still, a simple example of which is shown in Fig. 4.31. The
side of the cover system facing the Sun is transparent, whereas the part
sloping in the opposite direction is highly reflective and thus remains at ambient temperature. It therefore provides a cold surface for condensation of
water evaporated from the saline water surface (kept at elevated temperature due to the solar collector part of the system). In more advanced systems,
some of the heat released by condensation (about 2.3×10^6 J kg^{-1}, which is
also the energy required for the vaporisation) is recovered and used to heat
more saline water (Talbert *et al.*, 1970).

4.2.3 Solar electricity generation

Conversion of solar radiation into electric energy may be achieved either in
two steps by, for example, first converting radiation to heat, as described in
section 4.2.1, and then heat into electricity, using one of the methods described in section 4.1.2 or 4.1.3, or alternatively by direct conversion of radiation into electricity, using the photovoltaic conversion scheme discussed
in section 4.1.6. Two-step conversion using chemical energy rather than heat
as the intermediate energy form is possible, for instance, by means of the
photogalvanic conversion scheme briefly mentioned in section 4.1.6. Examples of the layout of systems capable of performing such conversions are
given below, and their possible performance is discussed.

Photo-thermoelectric converters

A two-step conversion device may be of the general form shown in Fig. 4.32.
The solar collector may be of the flat-plate type or may perform a certain
measure of concentration, requiring partial or full tracking of the Sun. The
second step is performed by a thermodynamic engine cycle, for example, a

Rankine cycle with expansion through a turbine as indicated in the figure. Owing to the vaporisation of the engine-cycle working fluid in the heat exchanger linking the two conversion steps, the heat exchange performance does not follow the simple description leading to (4.99). The temperature behaviour in the heat exchanger is more like the type shown in Fig. 4.33. The fluid in the collector circuit experiences a uniform temperature decrease as a function of the distance travelled through the heat exchanger, from x_1 to x_2. The working fluid entering the heat exchanger at x_2 is heated to boiling point, after which further heat exchange is used to evaporate the working fluid (and perhaps superheat the gas), so that the temperature curve becomes nearly flat after a certain point.

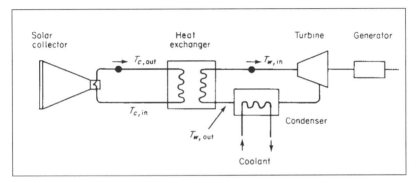

Figure 4.32. Photo-thermoelectric generator shown as based on concentrating solar collectors.

The thermodynamic engine cycle (the right-hand side of Fig. 4.32) can be described by the method outlined in section 4.1.1, with the efficiency of heat to electricity conversion given by (4.19) [limited by the ideal Carnot process efficiency (4.4)],

$$\eta_w = E^{electric} / J_{Q,in}{}^w \leq (T_{w,in} - T_{w,out}) / T_{w,in},$$

where $E^{electric} = -J_q F_q$ is the electric power output [cf. (4.14)] and

$$J_{Q,in}{}^w = J_m{}^c C_p{}^c (T_{c,out} - T_{c,in})$$

Is the rate of heat transfer from the collector circuit to working fluid circuit in the heat exchanger. The right-hand side is given by (4.107) for a concentrating solar collector and by (4.96) and (4.92) for a flat-plate collector. The overall conversion efficiency is the product of the efficiency η_c of the collector system and η_w,

$$\eta = \eta_c \eta_w.$$

The determination of the four temperatures $T_{c,in}$, $T_{c,out}$, $T_{w,in}$ and $T_{w,out}$ (cf. Fig. 4.32) requires an equation for the collector performance, for the heat transfer to the collector fluid, for the heat transfer in the heat exchanger and for the processes involving the working fluid of the thermodynamic cycle. If the collector performance is assumed to depend only on an average collector temperature, $\bar{T}_c$ [e.g. given by (4.95)], and if $T_{w,in}$ is considered to be roughly equal to $\bar{T}_c$ (Fig. 4.33 indicates that this may be a fair first-order approximation), then the upper limit for η_w is approximately

$$\eta_w \lesssim (\bar{T}_c - T_{w,out}) / \bar{T}_c .$$

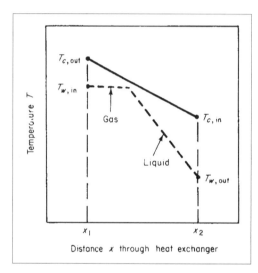

Figure 4.33. Temperature behaviour in heat exchanger for two-phase working fluid of temperature T_w and the fluid in the collector circuit of temperature T_c (based on Athey, 1976).

Taking $T_{w,out}$ as 300 K (determined by the coolant flow through the condenser in Fig. 4.32), and replacing η_c by the idealised value (4.110), the overall efficiency may be estimated as shown in Fig. 4.34, as a function of $\bar{T}_c$. The assumptions for η_c^{ideal} are as in Fig. 4.27: no convective losses (this may be important for flat-plate collectors or collectors with a modest concentration factor) and an incident radiation flux of 800 W m^{-2} reaching the absorber (for strongly focusing collectors this also has to be direct radiation). This is multiplied by η_w^{ideal} (the Carnot limit) to yield η_{total}^{ideal} depicted in Fig. 4.34, with the further assumptions regarding the temperature averaging and temperature drops in the heat exchanger which allowed the introduction of $\bar{T}_c$ as the principal variable both in η_w^{ideal} and in η_c^{ideal}.

In realistic cases, the radiation reaching the absorber of a focusing collector is perhaps half of the total incident flux $E_{s,\gamma}^{SW}$ and η_w is maybe 60% of the Carnot value, i.e.

$$\eta_{total} \approx 0.3 \, \eta_{total}^{ideal}.$$

This estimate may also be valid for small concentration values (or rather small values of the parameter $X\alpha^{sw}/\varepsilon^{lw}$ characterising the curves in Figs. 4.34 and 4.27), since the increased fraction of $E_{s,\gamma}^{sw}$ being absorbed is compensated for by high convective heat losses. Thus, the values of η_{total}^{ideal} between 6 and 48%, obtained in Fig. 4.34 for suitable choices of $\bar{T}_c$ (this can be adjusted by altering the fluid flow rate J_m^c), may correspond to realistic efficiencies of 2–15%, for values of $X\alpha^{sw}/\varepsilon^{lw}$ increasing from 1 to 200.

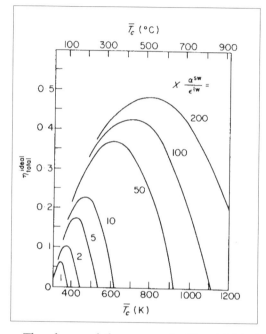

Figure 4.34. Ideal overall efficiency of photo-thermoelectric converter, based on the assumptions underlying Figs. 4.27 and 4.33 and with the Rankine cycle efficiency replaced by an ideal Carnot cycle efficiency.

The shape of the curves in Fig. 4.34 is brought about by the increase in η_w^{ideal} as a function of $\bar{T}_c$, counteracted by the accelerated decrease in η_c^{ideal} as a function of $\bar{T}_c$, which is seen in Fig. 4.27.

Photo-thermoelectric conversion systems based on flat-plate collectors have been discussed, for example, by Athey (1976); systems based on solar collectors of the type shown in Fig. 4.25d have been considered by Meinel and Meinel (1972). They estimate that a high temperature on the absorber can be obtained by use of selective surfaces, evacuated tubes (e.g. made of glass with a reflecting inner surface except for the window shown in Fig. 4.25d), and a crude Fresnel lens (see Fig. 4.26) concentrating the incoming radiation on the tube window. Molten sodium is suggested as the collector fluid. Finally, fully tracking systems based on the concept shown on the right-hand side of Fig. 4.24 ("power towers") have been suggested, e.g. by Teplyakov and Aparisi (1976) and by Hildebrandt and Vant-Hull (1977).

Photovoltaic converters

A photovoltaic converter consists of a number of solar cells connected in a suitable way, plus eventually some auxiliary equipment such as focusing devices in front of the cells and tracking systems. The maximum efficiency of a solar cell is given by the ratio of the maximum power output (4.67) and the incident radiation flux,

$$\max(\eta) = \max(E) / E_{s,\gamma}^{sw}. \tag{4.111}$$

It is smaller than unity for a number of reasons. Firstly, as discussed in section 4.1.5, radiation of frequency below the semiconductor band gap is not absorbed. Secondly, according to (4.69), the excess energy of radiation with frequencies above the semiconductor band gap is not available to the photovoltaic conversion process. This loss would be small if the solar spectrum was peaked across the band gap, but most semiconductor gaps only correspond to a limited part of the broad solar spectrum (cf. Fig. 2.13).

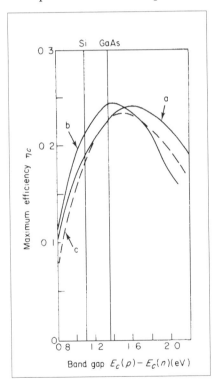

Figure 4.35. Early calculation of maximum efficiency for simple *p*-on-*n* solar cell; (a) outside the Earth's atmosphere (E^{sw} = 1350 W m^{-2}); (b) at the Earth's surface under standard conditions (E^{sw} = 890 W m^{-2}, air mass one, water content 0.02 m^3 m^{-2} and major absorption bands included in the calculation); (c) overcast condition (E^{sw} = 120 W m^{-2}) (based on Loferski, 1956).

Thirdly, as seen e.g. in Fig. 4.9, the maximum power output is less than the maximum current times the maximum voltage. This reduction in voltage, necessary in order to get a finite current, is analogous to the necessity of a finite dissipation term in a thermodynamic engine in order to get energy out in a finite time (cf. section 4.1.1). The expression (4.67) does not fit measured power values in detail, and it has been suggested that a second exponential term be added in the current–voltage relation (4.63), of similar form but with $\Delta\phi_{ext}$ replaced by $\frac{1}{2}\Delta\phi_{ext}$ (Sah *et al.*, 1957). The origin of such a term is thermal generation and recombination of carriers in the vicinity of the junction. Fourthly, the external potential (times the electron charge) has a maximum value (4.66), which is smaller than the semiconductor gap $E_c(p) - E_c(n)$, since it equals the difference in Fermi level between the *p*- and *n*-regions [cf. (4.60) and section 4.A]. This loss may be diminished by increasing the number of impurities (and hence the number of minority carriers) in both the *p*- and the *n*-regions. However, if these are increased above a certain level, the increased recombination probability offsets the voltage gain.

Figure 4.35 shows an early example of calculated maximum efficiency as a function of the semiconductor band gap, including the above-mentioned losses, for radiation conditions corresponding to the top of the Earth's atmosphere (a), for a clear sky day at the Earth's surface with average atmospheric absorption and scattering conditions (b), and for a situation with cloud-covered sky (c). It is shown that for common solar cell materials such as Si or GaAs, the maximum efficiency is larger for the spectral composition of clear-day solar radiation at ground level than for the solar spectrum not disturbed by the Earth's atmosphere. The performance for scattered solar radiation (overcast case) is not substantially impaired.

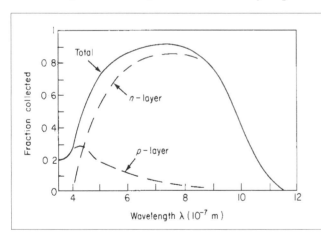

Figure 4.36. Spectral collection efficiency for a simple *p*-on-*n* solar cell. The curve labelled "total" is based on measurements, and the individual contributions from the *p*-layer (thickness 5×10^{-7} m) and the *n*-layer have been calculated. The total thickness is about 4.5×10^{-4} m (based on Wolf, 1963).

The wavelength dependence of the collection efficiency (i.e. the efficiency including the first two loss terms discussed above, but not the last two), is shown in Fig. 4.36 for a simple silicon cell consisting of a thin (5×10^{-7} m) *p*-layer on top of an *n*-layer base (of thickness 4.5×10^{-4} m). The total efficiency curve is based on measurements, whereas the separate contributions from the *p*- and *n*-layers have been calculated (Wolf, 1963). Short wavelengths are absorbed in the *p*-layer and give rise to a current of electrons directed towards the junction, but the bulk of the solar wavelengths are not absorbed until the photons reach the *n*-layer base. They give rise to a hole current towards the junction. The efficiency is not independent of cell temperature, as indicated in Fig. 4.37. The currently dominating silicon material exhibits an absolute loss in efficiency of 0.4–0.5% for each °C of temperature rise. Figure 4.38 adds more recent data on the temperature dependence of different types of solar cells, based upon measurements.

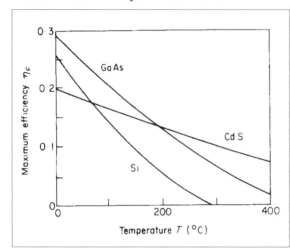

Figure 4.37 (left). Calculated temperature dependence of solar cell efficiency (based on Wysocki and Rappaport, 1960).

Figure 4.38 (below). Solar cell efficiency as function of operating temperature, normalised to a typical 25°C efficiency for each cell type. From Sørensen (2000b), based on Ricaud (1999), Dutta *et al.* (1992), Wysocky and Rappaport (1960), Yamamoto *et al.* (1999), Rijnberg *et al.* (1998).

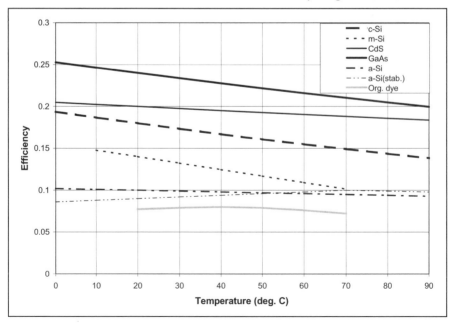

The data shown in Fig. 4.38 has, for each type of solar cell, been normalised to a typical absolute efficiency for current commercial or near-commercial versions of the type of device in question. The early theoretical calculations (Wysocki and Rappaport, 1960) are largely confirmed by current measurements, and the mechanisms are thus well understood, at least for conventional photovoltaic devices. The temperature dependence is chiefly due to band-gap effects, which explains why the slope of the crystalline silicon (c-Si) and multicrystalline silicon (m-Si) are identical (Yamamoto *et al.*, 1999). In other words, the grain boundaries do not give rise to additional temperature effects. Cd-S cells have a lower but still significant temperature gradient, whereas the temperature effect for amorphous silicon cells and organic dye-sensitised TiO_2 cells is very small.

The temperature effect is negative with increasing working temperature for all devices except two: the organic cells show a maximum near 40°C (Rijnberg *et al.*, 1998) and the amorphous silicon-hydrogen cells (a-Si) show a reversal of temperature trends after annealing (Dutta *et al.*, 1992). This positive temperature coefficient only persists until the un-degraded efficiency is reached, and it requires annealing as opposed to light soaking treatment, which causes the development of a stronger negative temperature coefficient. The modest temperature dependence is conveniently modelled by a power expansion of the efficiency,

$$\eta = \eta(298K) + a\,(T - 298K) + b\,(T - 298K)^2. \tag{4.112}$$

The operating temperature dependence of the solar energy to electricity conversion efficiency suggests that cooling the cell by extracting heat may improve the electric performance of the cell and thereby pay for some of the extra expense of the heat extraction equipment. This is further discussed in a case study in section 6.3.1. Typical operating temperatures for un-cooled cells are about 50°C. Fig. 4.38 shows that improvement is indeed obtained for e.g. crystalline or multicrystalline silicon photovoltaic (PV) cells, but not notably for dye-sensitised cells or amorphous PV cells. On the other hand, in order to make use of the heat it should preferably be collected at higher temperatures, which would indicate that the best solutions are those with little operating temperature effect of the electricity yields. This trade-off is further illustrated in the simulation models of section 6.3.1.

In cells currently produced, each loss factor is carefully minimised, and the resulting efficiencies have increased over time, for all types of solar cells, as discussed below.

Monocrystalline silicon cells

The photovoltaic cell principles described in section 4.1.5 and above form the basis for monocrystalline cells, which are cells constructed from single crystals, usually in the form of ingots sliced into a number of cells.

A number of improvements have brought the cell efficiency of state-of-the-art monocrystalline silicon cells up to about 25%. The light capture is improved through trapping structures that minimise reflection in directions not benefiting the collection area and by back–side designs reflecting light rays back into the active areas (see e.g. Fig. 4.39). The doping degree is altered near electrodes (n^+ and p^+ areas), and a thin oxide layer further helps to prevent electrons reaching the surface rather than the electrode (this process being termed "passivation"). Further, top electrodes may be buried in order not to produce shadowing effects for the incoming light (Green, 1992).

Figure 4.40 shows the measured characteristics, i.e. current as function of voltage, for a cell of the type shown in Fig. 4.39.

Simulation of the light trapping and electron transport processes in one, two or three dimensions has helped in selecting the best geometry and degree of doping (Basore, 1991; Müller *et al.*, 1992). Figure 4.41 gives an example of one-dimensional simulation of the variation in cell performance as function of doping degree for a cell of the kind shown in Fig. 4.39.

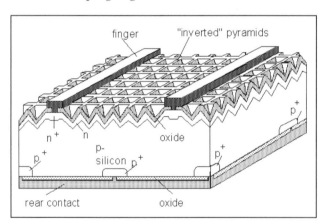

Figure 4.39. Structure of a monocrystalline silicon cell with passivated emitter and a locally diffused rear structure ("PERL"), used to obtain a 23% module efficiency (from Green et al., 1998, used with permission).

Figure 4.40. Current-voltage curve for PERL cell similar to that of Fig. 4.39 (but with cell efficiency 22%), as measured at the Sandia Laboratories (USA) at 1006 W m^{-2} airmass 1.5 simulated radiation, for a 4 cm^2 cell. Key findings are V_{oc} = 696 mV (4.66), I_s = 160 mA (4.64) and a fill factor of 0.79 (Wenham *et al.*, 1995b). The fill factor is the ratio between the area under the curve and $I_s V_{oc}$.

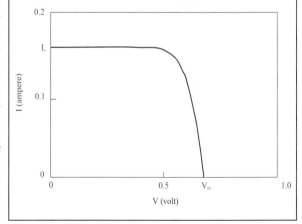

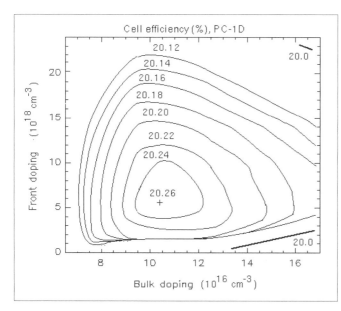

Figure 4.41. Calculated efficiency as function of doping parameters for a simplified silicon cell of the type depicted in Fig. 4.39. The one-dimensional finite–element model used is described in the text (Sørensen, 1994c).

The model takes into account the 10 μm facet depth and uses a curved one-dimensional path through the cell. The two most important doping parameters (impurities per unit volume) are the uniform bulk doping of the *p*-material and the *n*-doping at the front, assumed to fall off as an error function, thereby simulating the average behaviour of near-electrode sites and sites away from electrodes. Transport and recombination are calculated in a finite-element model (Basore, 1991). The back–side doping is kept at 2×10^{19} cm^{-3}.

Multicrystalline cells

Another technology for producing solar cells uses multicrystalline (sometimes referred to as "polycrystalline") materials, instead of the single-crystal materials. Multicrystalline materials consist of small domains or grains of crystalline material, randomly oriented relative to each other. The crystal grains in multicrystalline materials sustain conductivity in the same way as single crystals do, but the transport of electrons across grain boundaries induces losses, reduces conductivity and thus makes the cells less efficient. On the other hand, they can be produced by simpler methods than those needed for monocrystals, e.g. by evaporating suitable coatings onto a substrate. This field is in rapid development, as it is becoming possible to deposit only a few atomic layers onto a substrate, and with suitable techniques (such as using magnetic fields to align grains) it may soon be possible in this way to form near-monocrystalline layers without having to grow crystals.

It was initially believed that the additional losses at grain boundaries would necessarily make the efficiency of multicrystalline cells substantially lower than what could be obtained by crystalline materials. Actually, the difference has narrowed as a result of better understanding of the options for optimising performance of complex cell structures. One problem has been the damage inflicted upon multicrystalline cells by attempting to copy to them some of the efficiency-improving techniques that have worked well for monocrystalline cells (surface texturing, rear passivation by oxide layers). Yet, etching of inverted pyramids on the surface of multicrystalline cells has improved efficiency considerably (Stock *et al.*, 1996), and recently, less damaging honeycomb texture patterns have brought the efficiency up to 20% (Zhao *et al.*, 1998). This is a trend likely to induce the long-predicted change from expensive ingot-grown monocrystalline cell materials to deposition techniques for multicrystalline materials on suitable backings, much more suited for mass production and price reduction efforts. However, the development away from single-crystalline solar cell materials is slower than anticipated because of the higher maturity of the crystalline industry processes. Figure 4.42 shows the structure of the 20% efficient multicrystalline cell, obtaining over 90% absorption of incoming radiation. The advantages of thin-film multicrystalline solar cells over monocrystalline ones would on basic principles seem to more than compensate for the remaining 5% efficiency difference. This does not exclude that crystalline and multicrystalline technologies will continue to co-exist in the marketplace for a while.

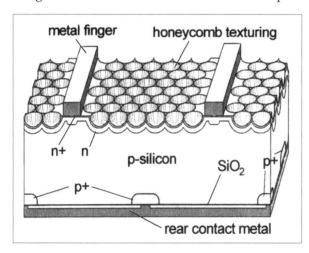

Figure 4.42. Texturing used to raise multicrystalline cell efficiency to 20% in a PERL-type cell (from Zhao *et al.*, 1998, used with permission).

Stacked cells

Instead of basing a solar cell on just a single *p–n* junction, it is possible to stack several identical or different cells on top of each other. The use of different cell materials aims at capturing a wider range of frequencies than possible with a single junction. In this case materials of different band gaps will be stacked (Yazawa *et al.*, 1996; Takamoto *et al.*, 1997). In the case of stacked identical cells, the aim is to be able to use lower quality material (e.g. the mentioned thinly sprayed crystalline cells in contrast to ingot-grown ones) and still get an acceptable overall efficiency by stacking several layers of low individual efficiency (Wenham *et al.*, 1995a). This concept is illustrated in Fig. 4.43, where a calculation performed for low-quality multicrystalline silicon finds a broad maximum of efficiency for about six layers (Green, 1994).

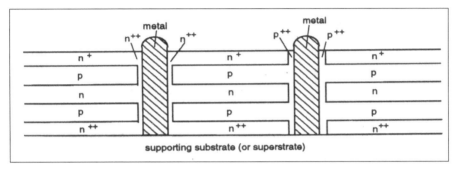

Figure 4.43. Concept of a multilayer thin-film silicon solar cell (Wenham *et al.*, 1995a).

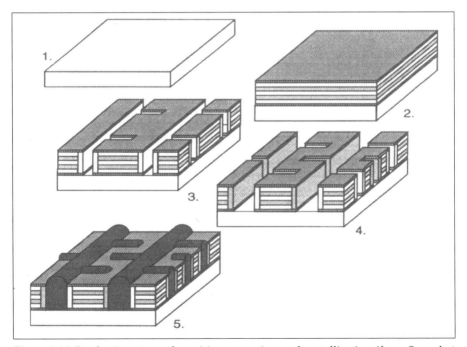

Figure 4.44. Production steps: deposition, grooving and metallisation (from Sproul *et al.*, 1995, used with permission)

Figure 4.44 shows the industrial process that may be used for producing such multilayer cells, using vapour deposition to deposit each layer and laser grooving to make room for electrodes. Module assembly uses automatic series connection achieved by the layout shown in Fig. 4.45. With six junctions of low-quality material (grain sizes above 3 µm), a cell efficiency of 15.2% has been measured (for a small specimen; Sproul *et al.*, 1995). The dependence on the number of layers is weak for the optimum bulk doping concentration of about 10^{18} cm^{-3} and a total thickness of around 10 µm. Efforts to scale the processes up to industrial scale production levels have not succeeded so far, due to difficulties in controlling the transport of electrons from the layers of low-quality silicon to the electrodes.

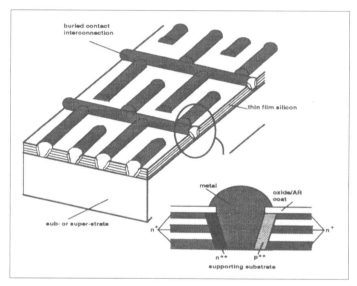

Figure 4.45. Design of cells for easy electrical interconnection between adjacent cells in a module (from Wenham *et al.*, 1995a, used with permission).

Amorphous cells

Amorphous semiconductor materials exhibit properties of interest for solar cell applications. While elemental amorphous silicon has a fairly uniform energy distribution of electron levels, composite materials have been constructed which exhibit a pronounced energy gap, i.e. an interval of energy essentially without any states, as in a crystal. Spear and Le Comber (1975) first produced such an amorphous material, which was later proved to be a silicon–hydrogen alloy, with the distribution of energy states shown in Fig. 4.46 as curve a. Ovshinsky (1978) produced a silicon–fluorine–hydrogen alloy with further depression of gap states (Fig. 4.46, curve b). The gap is about 1.6 eV wide and thus should be more favourable with respect to the solar energy spectrum than the 1.1-eV gap of crystalline silicon. Furthermore, doping (introduction of boron or phosphorus atoms) has proved possible, so that *p*-

and n-type amorphous semiconductors can readily be made, and a certain amount of "engineering" of materials with exactly the desired properties with regard to gap structure, doping efficiency, conductivity, temperature sensitivity and structural stability (lifetime) can be performed.

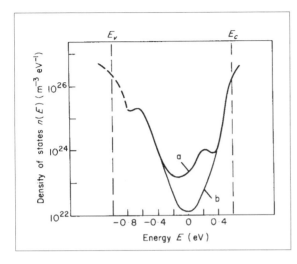

Figure 4.46. Density of electron states as a function of excitation energy for silicon-based amorphous materials: (a) silicon–hydrogen alloy; (b) silicium–fluorine–hydrogen alloy (based on Spear and Le Comber, 1975; Ovshinsky, 1978).

A theoretical description of the gap occurrence and electron conduction in amorphous materials was first presented by Street and Mott (1975) and followed up by Fritsche (1977) and Pfister and Scher (1977). The basis is the occurrence in amorphous material of defects of particular affinity to attract or reject electrons (e.g. lone pair sites), and the transport of electrons is thought of as a quasi-random "hopping" between sites, some of which are capable of "trapping" an electron for a shorter or longer period of time. Abundance of broken or "dangling" bonds may give rise to serious trapping problems (low conductivity), and the success obtained by incorporating hydrogen seems due to its occupying and neutralising such sites.

Not very long after the theoretical description of amorphous solar cells, Japanese scientists succeeded in creating designs of such cells that were suited for industrial production (Hamakawa *et al.*, 1981) and soon found a market in powering calculators and similar small-scale devices, where the cell cost was relatively unimportant. A typical structure of a commercial amorphous cell is illustrated in Fig. 4.47. Band gaps in the range from 1.0 to 3.6 eV can be engineered with different silicon alloys (SiGe, Si, SiC), and such cells may be stacked to obtain a broader frequency acceptance (Ichikawa, 1993; Hamakawa, 1998).

However, the simplest version has just one type of material: an intrinsic layer of an a-Si:H compound is the main area of light absorption, and adjacent p- and n-type layers ensure the transport to the electrodes, of which the front one is made of a transparent material. The whole structure is less than

1 μm thick and is deposited onto a plastic backing material. Maximum efficiencies of around 13% have been demonstrated (Fig. 4.48), but one problem has persisted: because the structure of the material is without order, bombardment with light quanta may push atoms around, and the material degrades with time. Current practice is to degrade commercial cells before they leave the factory, thereby decreasing the efficiency by some 20%, but in return obtaining reasonable stability over a 10-year period under average solar radiation conditions (Sakai, 1993). Several layers of p-, i- and n-layers may be stacked, and the highest efficiency is obtained by replacing the amorphous n-layers by a multicrystalline pure Si or silicon compound layer (Ichikawa, 1993; Ma *et al.*, 1995).

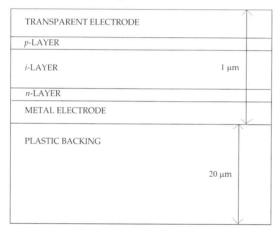

Figure 4.47. Structure of the most common a-Si cell. Several *pin*-layers may be stacked.

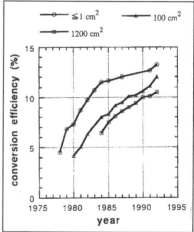

Figure 4.48. Development in a-Si solar cell efficiency for different cell area (before degradation; Sakai, 1993).

Other materials and other thin-film cells

Use of materials from the chemical groups III and V, such as GaAs, CdS and CdTe, instead of silicon allows better engineering of band gaps in crystalline solar cells to suit particular purposes and brings forward new properties suitable for certain tasks (notably space applications and use in concentrating collectors). Important considerations in selecting materials include temperature dependence where crystalline silicon cell efficiency drops rather fast with the increasing temperature likely to prevail (despite possible active cooling) for cells operating at high levels of solar radiation (see Figs. 4.37 and 4.38).

The GaAs band gap of 1.43 eV is well suited for the solar spectrum, and with a tandem cell of GaAs plus $GaInP_2$ an efficiency of over 30% has been reached (Bertness et al., 1994; Deb, 1998). At the moment these cells are expensive and are mainly used in space. However, thin-film versions may be developed as in the Si case, as they already have for CIS cells (copper-indium-diselenide). The highest efficiency obtained so far is about 17% for a Cu (In,Ga) Se_2 structure (Tuttle et al., 1996).

Among a range of nontraditional designs of solar cell is the use of spherical droplets. This idea grew out of an idea of reusing scrap material from the microelectronics industry, but it also may increase acceptance of light from different directions and reduce reflection that otherwise would have to be dealt with by surface texturing (Maag, 1993; Drewes, 2003; ATS, 2003).

Organic and other photoelectrochemical solar cells

Photoelectrochemistry is an area of confluence between solar cell technology, discussed here, and battery or fuel cell technology, discussed in section 4.7 below. Organic solar cells are a special kind of photoelectrochemical (PEC) devices that try to take advantage of inexpensive organic materials as opposed to the more expensive metals and doped semiconducting materials used in the photovoltaic devices discussed above. Suitable organic materials can, as discussed for biological photosynthesis in section 3.6, trap sunlight and convert radiation into other forms of energy. It has been attempted to copy this process in various ways. Calvin (1974) considered a double membrane that would separate the ionised reactants of a photo-excitation process,

$$A + h\nu \rightarrow A^* ; \qquad A^* + B \rightarrow A^+ + B^- .$$

In addition, a transport system is needed to get the ions to an electrode. No practical version of this idea has been produced. The same is the case for a concept aimed at both removing CO_2 from the atmosphere and at the same time producing methanol (Jensen and Sørensen, 1984, pp. 217–218). The absorption of solar radiation is used to fix atmospheric carbon dioxide to a ruthenium complex [Ru], which is then heated with water steam,

$$[Ru]CO_2 + 2H_2O + 713 \text{ kJ mol}^{-1} \rightarrow [Ru] + CH_3OH + \tfrac{1}{2}O_2.$$

One scheme that has been realised is the attachment of a ruthenium complex as a dye to TiO_2, which may then transport electrons formed by photo-excitation in the Ru-complex to an electrode. The process is similar to the dye-excitation processes used in conventional photographic prints and was first proposed by Moser (1887). He called the dye substance enhancing the absorption of solar radiation above what can be achieved by the TiO_2 a "sensitiser". The Ru-complex is restored by a redox process in an electrolyte joining the other electrode, with use of a platinum catalyst as indicated in Fig. 4.49. Because a monolayer of even a very efficiently absorbing dye will absorb less than 1% of incoming solar radiation, the dye absorption layer is made three dimensional by adhering the dye to a nanostructured TiO_2 network of nodules, as first demonstrated by Tsubomura *et al.* (1976). They also introduced the liquid electrolyte, capable of penetrating into the cavities in the nanostructured sensitised titanium dioxide and providing the necessary contact for transfer of the replacement electron to the dye. The effective absorption surface may be increased by three orders of magnitude and provides an overall cell efficiency to about 10%. It should still be possible for the TiO_2 nodules to transfer the absorbed electron to the back electrode through a series of transport processes.

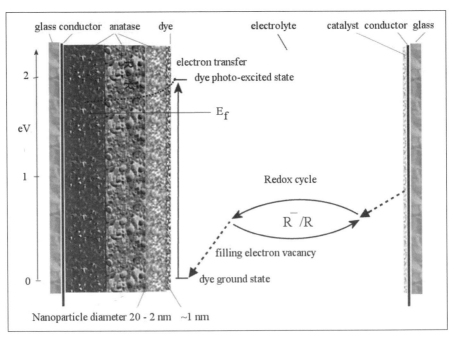

Figure 4.49. Layout of photoelectrochemical solar cell with indication of energy levels (E_f is the Fermi level of the semiconductor material). Solar radiation is absorbed in a dye layer, creating an excited electron state, from which an electron is transferred to the semiconductor at left and replenished from the counter electrode through a redox cycle in an electrolyte to the right (Sørensen, 2003a).

The material presently favoured for the anode nanoparticles is TiO_2 in the form of anatase (Fig. 4.50). Compared to the other forms of titanium dioxide (rutile and brookite), anatase better accommodates the dye molecules and forms nodules rather than flakes. The large-side dimension of the unit cell shown in Fig. 4.50 is about 0.2 nm. Several other semiconductor materials have been investigated, but so far anatase has shown the best overall properties and is used fairly universally. Electron transport through the anatase layers follows conventional solid state physics, except for the issue of nodule coherence. A simple modelling effort has successfully described the transport as random walk, rather than governed by hopping models (Nelson *et al.*, 2001).

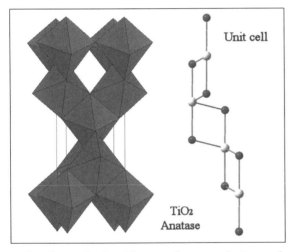

Figure 4.50. Anatase structure (first determined by Horn *et al.*, 1970) and unit cell (Sørensen, 2003a).

On the cathode side, a redox couple is used to supply the electron to replace the one being excited in the dye and transferred to the anatase before it decays back to the dye ground state. The electrolyte is typically acetonitrile (C_2H_3N), and the redox couple is iodine/tri-iodine (I^-/I_3^-), which has been used rather exclusively since the work of Tsubomura *et al.* (1976). This does not seem ideal, as the difference between the anatase Fermi level and the I^-/I_3^- chemical potential, which determines the cell open circuit voltage, is only about 0.9 eV, as compared with typical dye excitation energies of 1.8 eV. Many efforts have been directed at finding more appropriate redox shuttle systems, but so far none have shown overall properties making them preferable to the I^-/I_3^- couple (Wolfbauer, 1999). Electrolyte and redox couple integrity and lifetimes are of concern. A comparison to batteries is appropriate, and battery lifetimes are rarely as long as desired for solar cells that may be incorporated directly into building components and structures.

A further energy loss takes place at the cathode, where application of a catalyst is required in order to obtain the desired rate of electron transfer from electrode to electrolyte. As in batteries and fuel cells, traditionally preferred catalysts are based on platinum, but alternatives are under investigation. Generally speaking, the use of liquid electrolytes and catalysts is undesirable, and the much slower electron transfer through the electrolyte and its redox couple (as compared with the semiconductor transport) is likely to be the overall limiting factor for current in the device. However, the reason for this choice is also obvious. The cell is produced by deposition of anatase layers on the anode and subsequent annealing, processes requiring temperatures well over 100°C. The dye is then applied, either by a soaking or by a flushing process, creating the huge intrinsic surface for solar collection. Typical dye melting points are 80-100°C, so applying a second semiconductor material (if one with appropriate properties could be found) from the other side at appropriate temperatures would destroy the cell. More gentle application not requiring high temperatures is not likely to allow the surface contact area between dye and semiconductor to be equally large on both sides.

An alternative would be to find another material not requiring high temperatures for penetrating into the cavity structure of the initial semiconductor plus dye layers. Possible candidates would be conducting polymers or the ion-carrying polymers used in fuel cells. Actual achievements of 2-3% energy conversion efficiency have been obtained with two types of polymer systems. One uses a gel network polymer as electrolyte (Ren *et al.*, 2001). The other is a type of plastic solar cell, where the already known ability of ^{60}C-molecules to absorb solar radiation (Sariciftci *et al.*, 1992) is used to create a fairly large absorption area of ^{60}C sensitiser imbedded in a suitable polymer (Shaheen *et al.*, 2001; Yu *et al.*, 1995).

The choice of sensitiser is ideally based upon fulfilment of requirements including at least the following:

• high absorption capability over the range of spectral frequencies characteristic of sunlight

• energetically suitable excited states

• good attachment to semiconductor nanoparticles, that ensures rapid electron transfer (in competition with de-excitation and back-transfer from semiconductor surface to dye sensitiser)

• easily accepting replacement electron from electrolyte

• dye lifetime consistent with stipulated device life

The search for optimised sensitisers has usually focused on a particular family of molecules. For example, one group (O'Regan and Grätzel, 1991; Nazeeruddin *et al.*, 1993, 2001; Shklover *et al.*, 1998) has looked at metal complexes based on ruthenium polypyridines, meticulously synthesising one variant after the other, adding rings, thiocyanate ligands and carboxylate groups in different combinations. The size of the molecule, in combination with its excitation spectrum, determines the frequencies of solar radiation that can be absorbed and the associated cross sections. The "black dye" (1 ruthenium atom, 3 pyridine rings, 3 thiocyanate ligands and 3 carboxylate groups) has led to the currently highest overall conversion efficiency of 10% for laboratory cells (area about 10^{-4} m²). For comparison, an efficiency of 5% is claimed for a large cell (of the order of 1 m²) in industrial production (STI, 2002). An earlier favourite was the "N3 dye" (Ru, 2 bipyridine rings, 2 thiocyanate ligands and 4 carboxylate groups). It is particularly willing to transfer an excited electron to an anatase surface, a fact that has been attributed to its attachment to the anatase surface by two carboxylate binding sites at approximately the same spacing as the "indents" in one anatase surface. However, the light absorption stops below 800 nm, implying smaller efficiency for many potential real-life collector sites. Figure 4.51 compares spectral sensitivities of the two dyes mentioned above, plus the coumarin-derivative organic dye considered in the following.

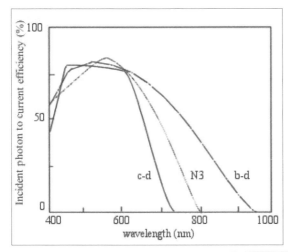

Figure 4.51 (left). Spectral sensitivity [c-d: coumarin derivative, N3 and b-d (black dye) ruthenium complexes] (based upon Hara *et al.*, 2001; Nazeeruddin *et al.*, 2001).

Figure 4.52 (below). The structure of N3 ($RuS_2O_8N_6C_{26}H_{24}$, below left) and b-d ($RuS_3O_6N_6C_{21}H_{22}$, below right) ruthenium sensitisers synthesised by Nazee-ruddin *et al.* (1993; 2001). A modest structure optimisation has been performed (Sørensen, 2004b).

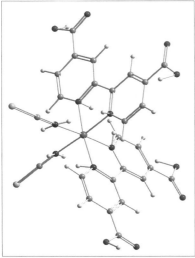
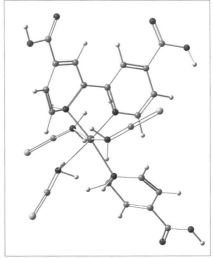

Figure 4.52 gives the molecular structure of the purple N3 and the black ruthenium dye (Nazeeruddin *et al.* 1993, 2001). Several variants have been studied (see e.g. Shklover *et al.*, 1998; Zakeeruddin *et al.*, 1997).

The structure of modified sensitiser molecules is often roughly given by that of known components plus some general rules of thumb. Measurements of spectra such as the nuclear magnetic resonance (NMR) spectra will help determine the positions of specific atoms (e.g. hydrogen atoms), but not always in a unique way. The addition of new features to a dye molecule in order to enhance solar absorption or help electron transfer out of the molecule may give rise to structures not amenable to simple guesses, e.g. due to isomerism or for other reasons such as energy surfaces in configuration space with more than one minimum.

Quantum mechanical modelling of the molecular structure* will lead to theoretical predictions of likely structures, as outcomes of optimisation studies. These involve following a path of steepest descent in the potential energy surface, eventually using second-order derivatives in order to stabilise the sizes of jumps made for each of the iterations (Sørensen, 2003a).

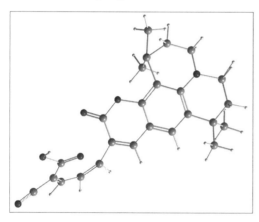

Figure 4.53 Optimised structure of coumarin-derivative organic dye (gross formula $H_{26}C_{25}N_2O_4$) synthesised by Hara *et al.* (2001) and yielding overall efficiencies of around 6% when used in PEC solar cells (Sørensen, 2003a).

Figure 4.53 shows the result of a theoretical optimisation (Sørensen, 2001a; 2003a) for a modification of a coumarin dye, being part of a family of molecules studied by Hara *et al.* (2001) as potential inexpensive, purely organic candidates for dyes to use with TiO_2 semiconductor nanoparticles in solar cells.

---

* So-called *ab initio* calculations in quantum chemistry involve solving the Schrödinger equation under several simplifying assumptions: nuclear motion is neglected, the basis functions used are linear combinations of Gaussian functions for each atom, and far fewer than needed for completeness. Interactions are first treated by the Hartree-Fock method, implying that each electron is assumed to move in the mean field from all the other particles. Higher electron correlations are added in a perturbative way or by the phenomenological density functional theory.

Figure 4.54 shows the molecular orbits near the Fermi level for the coumarin-derivative dye, obtained from a quantum mechanical Hartree-Fock self-consistent field calculation of the best ground state configuration (using software of Frisch *et al.*, 1998 and including a basis of 641 molecular states or about 3 for each physical orbit). It is seen that moving an electron from the highest occupied orbit (HOMO) to the lowest unoccupied one (LUMO) involves reducing the electron density in the coumarin part of the molecule and increasing it along and particularly at the end of the "arm" attached to the molecule. However, at this level of approximation, the energy difference (LUMO minus HOMO) is still nearly 8 eV, as opposed to an experimental value of 1.8 eV (Hara *et al.*, 2001). The second unoccupied orbit (Fig. 4.54 bottom) is quite different in nature.

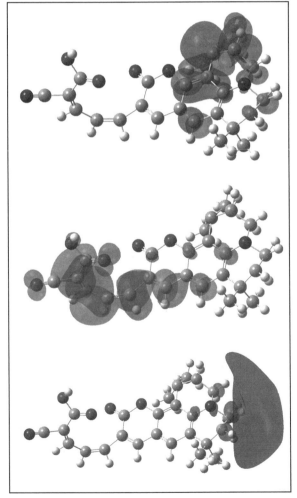

Figure 4.54. Electron density of HOMO (top), LUMO (middle) and second unoccupied molecular orbital (bottom) for coumarin-derivative dye, based on self-consistent field (SCF) calculation (Sørensen, 2001a, 2003a).

In order to estimate more realistically the energy of the first excited state of the coumarin-derivative dye molecule, a number of calculations have been performed (Sørensen, 2003a). Interactions not included in the self-consistent ground state calculation may be added in a time-dependent Hartree-Foch calculation for one or more excited states (TDHF; Casida *et al.*, 1998), and further improvement is obtained by performing the TDHF calculation on top of ground state calculations including further interaction, such as including exchange forces in a density functional method (Kohn and Sham, 1965; Becke, 1993). Figure 4.55 shows the excitation energy of the lowest spin-0 and spin-1 excited states using successive improvements in the sophistication of the interactions included and in the number of basis states used for the calculations (Sørensen, 2003b).

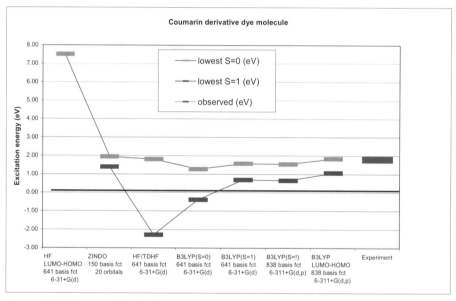

Figure 4.55. Measured (right-hand side) and calculated excitation energies of first excited states of spin 0 or 1 for the coumarin-derivative molecule shown in Fig. 4.53. The abscissa indicates acronyms for the type of calculation made, with complexity increasing from left to right (except the last, which by comparison with the first column shows the correlations built into the molecular orbits calculated with inclusion of exchange forces). It is seen that two of the calculations give unphysical states of negative energy, indicating the fragile nature of the approximations that effectively use other electron interactions for the excited state calculation than for the ground state calculation (because at a realistic level of approximation, the same approach will not yield both). The problem is solved in the subsequent columns by using basis states more appropriate for the spin-1 calculations (Sørensen, 2003b).

The fact that the first excited state comes down from an initial 8 eV molecular orbital energy difference to the observed 1.8 eV indicates that it must contain a substantial amount of correlations, or in other words that this excitation must comprise collective involvement of a large number of electrons.

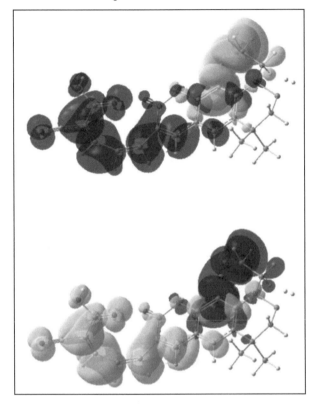

Figure 4.56. Calculated electron density difference between the first excited singlet and ground state of coumarin-derivative dye (top: positive values enhanced; below: negative values enhanced; Sørensen, 2001a, 2003a).

This is borne out in Fig. 4.56, showing the electron density difference between the first singlet excited state of the coumarin-derivative, and the ground state, using the large basis TDHF calculation. The excited state is made up by a dozen significant molecular orbital (MO) excitation pairs. The density difference shows the expected effect, already surmised from the HF ground state MO's, that the excitation moves electron density from the coumarin core to the peripheral arm added to the molecule. It would then be natural to assume that this is from where the transfer to the anatase surface, to which the dye adheres, takes place. This interpretation is supported by the large dipole moment found in the calculations (13.7 debyes).

The precise attachment of the dye to the anatase surface might be investigated by a combined optimisation of the dye plus a chunk of semiconductor, with the distance and rotation angles of the dye relative to the surface as parameters (cf. Fig. 4.57). A recent study of a similar material has revealed the nature of surface distortion and surface Ti and O molecules (Erdman *et al.*, 2002). The distortion only penetrates one layer down, in contrast to the over

10 nm thick space charge regions in surfaces of solids. The conclusion from the present calculation is that both dye and surface are modified, with the dye "arm" being bent to better attach to the surface, and the surface atoms to accommodate the dye particle (but even in the absence of the dye, the surface of the lattice structure would be different from the regular interior used as starting point for the optimisation).

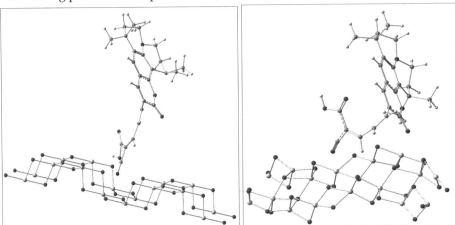

Figure 4.57. Attachment of coumarin-derivative dye to and modification of anatase surface. Left: initial guess, right: optimised (the depth of TiO₂ surface layer distortion cannot be determined by modelling just the top layer; Sørensen, 2003a, 2004b).

A number of further "handles" on the quantum chemical calculations is the comparison of measured spectra (IR, NMR, etc.) with those predicted by the calculations. The understanding of complex molecules has progressed rapidly as a result of the combination of spectral studies in the laboratory with quantum model calculations.

The photoelectrochemical dye and nanostructure technique has several applications beyond the possibility of forming solar cells. Among these are smart windows (Bechinger *et al.*, 1996; Granqvist *et al.*, 1998; STI, 2002), energy storage (Hauch *et al.*, 2001; Krasovec *et al.*, 2001), environmental monitors (Kamat and Vinodgopal, 1998; Yartym *et al.*, 2001), hydrogen production (Khaselev and Turner, 1998; Luzzi, 1999; Mathew *et al.*, 2001), computer and TV screens also suitable for outdoor use (using the dye as light emitter rather than absorber; Rajeswaran *et al.*, 2000; Tang, 2001; Müller *et al.*, 2003), and three-dimensional data storage (Dwayne-Miller *et al.*, 2001).

Module construction

Individual cells based on monocrystalline materials have areas of typically 100 cm² (limited by techniques for growing ingots of monocrystalline material). Multicrystalline and photoelectrochemical cells, where semiconductor material is deposited on a backing template (usually glass), may have larger cell size, and for amorphous cells, there are essentially no limitations. Amorphous cells have been produced on rolls of flexible plastic backing materials, with widths of 1–2 m and rolls of any length. The same may also become possible for other thin-film types, such as spray-deposited multicrystalline materials.

It is customary to assemble cells into modules by a mixture of parallel and series connections, so that the resulting voltages become suitable for standard electricity handling equipment, such as inverters transforming the DC currents into AC currents of grid specifications and quality. Alternatively, microprocessor inverters may be integrated into each module or even into each cell in order to minimise transport losses. In recent years, specific inverters optimised for solar cell applications have been produced, with an inverter efficiency increase from 90% to some 98% (IEA, 1999).

The solar cell technology is then characterised by two main solar radiation conversion efficiencies: the efficiency of each cell and the efficiency of the entire module sold to the customer. The latter is currently about 5% lower than the former, notably because of phase mismatch between the individual cell current components, but this does not need to be so, and the difference between the two efficiencies is expected to diminish in the future.

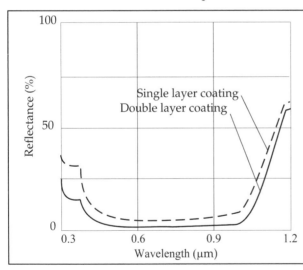

Figure 4.58. Front reflection for cell of the type shown in Fig. 4.39 (single layer anti-reflection coating) and the same with double layer coating (Zhao *et al.*, 1995).

Optical subsystem and concentrators

As indicated by the devices shown in Figs. 4.39 and 4.42, optical manipulation of incoming solar radiation is in use for non-concentrating solar cells. They serve the purpose of reducing the reflection on the surface to below a few per cent over the wavelength interval of interest for direct and scattered solar radiation, as seen from Fig. 4.58. The collection efficiency is far better than the one shown on Fig. 4.36, exceeding 90% for wavelengths between 0.4 and 1.05 µm, and exceeding 95% in the interval 0.65–1.03 µm.

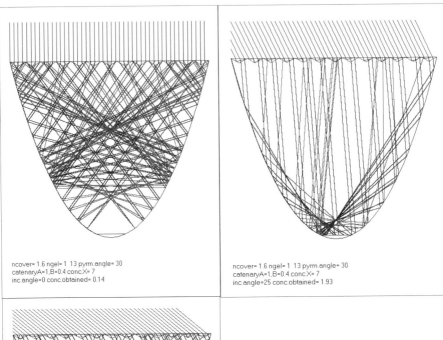

ncover= 1.6 ngel= 1 13 pyrm.angle= 30
catenaryA=1,B=0.4 conc.X= 7
inc.angle=0 conc.obtained= 0.14

ncover= 1.6 ngel= 1 13 pyrm.angle= 30
catenaryA=1,B=0.4 conc.X= 7
inc.angle=25 conc.obtained= 1.93

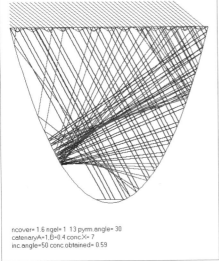

ncover= 1.6 ngel= 1 13 pyrm.angle= 30
catenaryA=1,B=0.4 conc.X= 7
inc.angle=50 conc.obtained= 0.59

Figure 4.59. Two-dimensional ray-tracing through a concentrator with glass cover textured with 30° inverted pyramids, for incident angles 0°, 25° and 50°. Only for incident angles around 25° does the absorber at the bottom receive more light than would have hit it if there had been no concentrator at all.

For large-factor concentration of light onto a photovoltaic cell much smaller than the aperture of the concentrator, the principles mentioned for thermal systems (towards the end of section 4.2.1) apply unchanged. Most of these concentrators are designed to focus the light onto a very small area, and thus tracking the Sun (in two directions) is essential, with the implications that scattered radiation largely cannot be used and that the expense and maintenance of non-stationary components have to be accepted.

One may think that abandoning very high concentration would allow the construction of concentrators capable of accepting light from most directions (including scattered light). However, this is not so easy. Devices that accept all angles have a concentration factor of unity (no concentration), and even if the acceptance angular interval is diminished to, say, 0–60°, which would be suitable because light from directions with a larger angle is anyway reduced by the incident cosine factor, only a very small concentration can be obtained (examples such as the design by Trombe are discussed in Meinel and Meinel, 1976).

The difficulty may be illustrated by a simple two-dimensional ray-tracing model of an (arbitrarily shaped) absorber, i.e. the PV cell, sitting inside an arbitrarily shaped concentrator (trough) with reflecting inner sides, and possibly with a cover glass at the top, that could accommodate some texturing or lens structure. The one component that unfortunately is not available is a semi-transparent cover that fully transmits solar radiation from one side and fully reflects the same wavelengths of radiation hitting the other side. With such a cover, nearly 100% of the radiation coming from any direction could reach the absorber, the only exception being rays being cyclically reflected into paths never reaching the absorber.

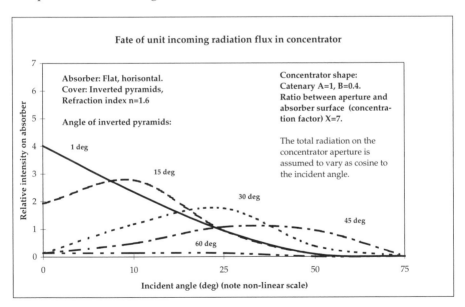

Figure 4.60. Intensity of rays on the absorber surface relative to that experienced in the absence of the concentrator (times cosine of incident angle, in order to reflect the different intensity of rays from different directions) for the $X = 7$ concentrator with a refracting, inverted pyramid cover as illustrated in Fig. 4.59.

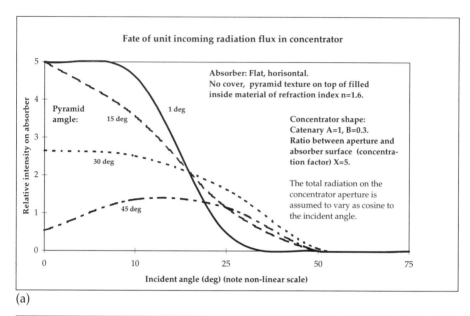

(a)

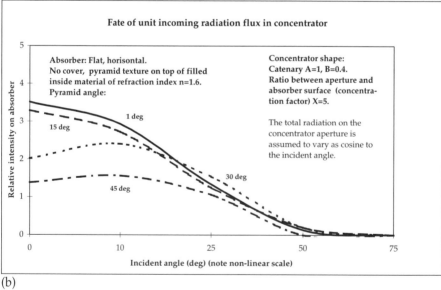

(b)

Figure 4.61a-c. Intensity of rays on the absorber surface relative to that experienced in the absence of the concentrator (times cosine of incident angle, in order to reflect the different intensity of rays from different directions) for different depths of the X = 5 concentrator without cover, but with a refracting inside material having pyramid texture on top, as illustrated in Fig. 4.63. (*a* and *b* above, *c* on next page).

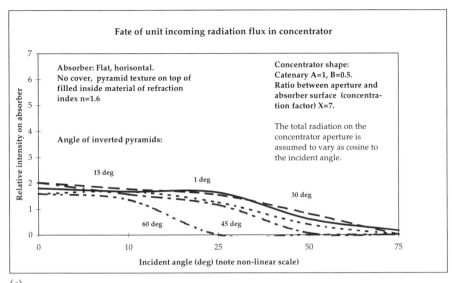

(c)

Figure 4.61c. See preceding page for details.

Figure 4.59 shows some ray-tracing experiments for a catenary-shaped concentrator holding a flat absorber at the bottom, with a concentration factor $X = 7$ (ratio of absorber and aperture areas, here lines). The cover has an inverted pyramid shape. This is routinely used for flat-plate devices (see Fig. 4.39), with the purpose of extending the path of the light rays inside the semiconductor material so that absorption is more likely. For the concentrator, the purpose is only to make it more likely that some of the rays reach the absorber, by changing their direction from one that would not have led to impact onto the absorber. However, at the same time, some rays that otherwise would have reached the absorber may now become diverted. Thus, not all incident angles benefit from the concentrator, and it has to be decided whether the desired acceptance interval is broad (say, 0–60°) or can be narrowed, e.g. to 0–30°. As illustrated, the concentration of rays towards the centre absorber is paid for by up to half of the incident rays being reflected upwards after suffering total reflection on the lower inside of the cover glass.

The result of varying some of the parameters is shown in Fig. 4.60. An inverted pyramid angle of 0° corresponds to a flat-plate glass cover, and the figure shows that in this situation the concentration is limited to incident angles below 20°, going from a factor of four to unity. With increasing pyramid angle, the maximum concentration is moved towards higher incident angles, but its absolute value declines. The average concentration factor over the ranges of incident angles occurring in practice for a fixed collector is not even above unity, meaning that a PV cell without the concentrating device would have performed better.

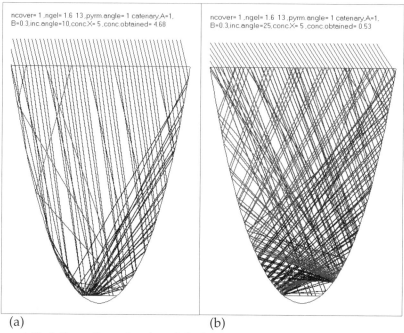

ncover= 1 ,ngel= 1.6 13 ,pyrm.angle= 1 catenary.A=1,
B=0.3,inc.angle=10,conc.X= 5 ,conc.obtained= 4.68

ncover= 1 ,ngel= 1.6 13 ,pyrm.angle= 1 catenary.A=1,
B=0.3,inc.angle=25,conc.X= 5 ,conc.obtained= 0.53

(a)   (b)

Figure 4.62a,b. Two-dimensional model of concentrator without cover but filled with a refracting material with slight pyramid texture at top. Incident angles are 10° and 25°.

Another idea is to direct the cover pyramids upwards, but to avoid "stretching out" the rays again at the lower glass boundary by having the entire trough filled with the refractive substance (which might be a gel or plastic material, textured at the upper surface). This is illustrated in Figs. 4.61a–c for an $X = 5$ concentrator. For a deep catenary-shaped collector, all rays up to 10° may be led to the absorber, but in this case (Figs. 4.61a and 4.62), incident angles above 25° are not accepted. To get reasonable acceptance, pyramid angles around 30° would again have to be chosen, and Figs. 4.61a–c show that the concentration penalty in trying to extend the acceptance interval to angles above 35° (by decreasing the depth of the concentrator, see Fig. 4.63) is severe. The best choice is still the deep catenary shape, with maximum concentration between 2 and 3 and average concentration over the interesting interval of incident angles (say, to 60°) no more than about 1.5. This is better than the inverted pyramid structure in Fig. 4.59, but it is still doubtful whether the expense of the concentrator is warranted. The collector efficiency as measured relative to total aperture area is the ratio of absorber intensity and X, i.e. considerably less than the cell efficiency.

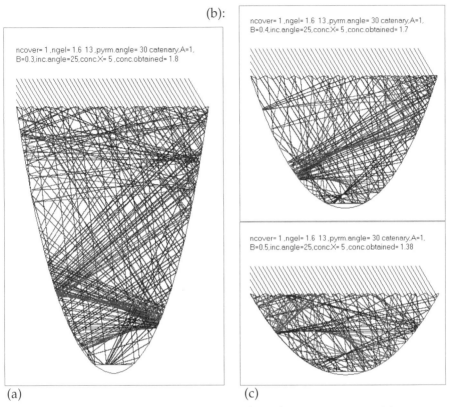

(b):

Figure 4.63a-c. Two-dimensional ray-tracing through a concentrator without cover, but filled with a refracting material, having its top pyramid textured, for an incident angle of 25° and three catenary depths.

Generation 24 h a day at constant radiation intensity could be achieved by placing the photovoltaic device perpendicular to the Sun at the top of the Earth's atmosphere. If it were placed in a geosynchronous orbit, it has been suggested that the power generated might be transmitted to a fixed location on the surface of the Earth after conversion to microwave frequency (see e.g. Glaser, 1977).

Use of residual energy

Given that practical efficiencies of photovoltaic devices are in the range of 10–30%, it is natural to think of putting the remaining solar energy to work. This has first been achieved for amorphous cells, which have been integrated into window panes, such that at least a part of the energy not absorbed is passed through to the room behind the pane. Of course, conductors and other non-transparent features reduce the transmittance somewhat. The same should be possible for other thin-film photovoltaic materials, including the emerging multicrystalline silicon cells.

Another possibility, particularly for PV panels not serving as windows, is to convert the solar energy not giving rise to electricity into heat (as some of it in actuality already is). One may think of cooling the modules of cells by running pipes filled with water or another suitable substance along their bottom side, carrying the heat to a thermal store, or alternatively by an air flow above the collector (but below a transparent cover layer). The energy available for this purpose is the incoming radiation energy minus the reflected and the converted part. Reflection from the back side of the semiconductor material, aimed at increasing the path-length of the photons in the material, could be chosen to optimise the value of the combined heat and power production, rather than only the power production. Examples of combined cycle solar panels are modelled in section 6.3.

For concentrator cells, active cooling is needed in any case, because of the temperature dependence of the characteristics of the photovoltaic process (cf. Fig. 4.37).

The maximum electrical efficiency of a single junction photovoltaic device implied by semiconductor physics is around 40% (see discussion earlier in section 4.2.3, pp. 384-386). If reflections can be minimised and many different light absorption materials are stacked, the theoretical Carnot efficiency (4.4) for the temperature of the Sun relative to a reference temperature at the surface of the Earth, i.e. about 95%, may be approached. Considerations of power flow optimisation diminish this limit further. Honsberg *et al.* (2001) and Green (2001) find a thermodynamical value of some 87% for an infinite stack of cells, to be used as a starting point for discussing the losses deriving from semiconductor physical arguments.

For a solar cell with electric efficiency below 40%, the possible associated heat gain is of the order of magnitude another 40% (cf. the simulation results in section 6.3.1).

4.3 Conversion of wind energy

4.3.1 Conversion of wind flow

Conversion of wind energy into linear motion of a body has been utilised extensively, particularly for transportation across water surfaces. A large sail-ship of the type used in the 19th century would have converted wind energy at peak rates of a quarter of a megawatt or more.

The force on a sail or a wing (i.e. profiles of negligible or finite thickness) may be broken down into a component in the direction of the undisturbed wind (drag) and a component perpendicular to the undisturbed wind direction (lift). When referring to an undisturbed wind direction it is assumed that a uniform wind field is modified in a region around the sail or the wing, but that beyond a certain distance such modifications can be disregarded.

In order to determine the force components, Euler's equations (cf. section 2.C) may be used. If viscous and external forces are neglected, and the flow is assumed to be irrotational (so that Bernoulli's equation is valid) and steady (so that the time-derivative of the velocity potential vanishes, cf. section 2.D), then the force on a segment of the airfoil (sail or wing) may be written

$$\frac{dF}{dz} = \oint_C P n \, ds = -\tfrac{1}{2}\rho \oint_C (v \cdot v) n \, ds.$$

Here dz is the segment length (cf. Fig. 4.64), C is a closed contour containing the airfoil profile, n is a unit vector normal to the contour [in the (x, y)-plane] and ds is the path-length increment, directed along the tangent to the contour, still in the (x, y)-plane. Taking advantage of the fact that the wind velocity v approaches a homogeneous field W (assumed to be along the x-axis) far from the airfoil, the contour integral may be reduced and evaluated (e.g. along a circular path),

$$dF/dz = \rho W \Gamma \, e_y, \tag{4.113}$$

$$\Gamma = \oint_C v \cdot ds \approx \pi c W \sin \alpha. \tag{4.114}$$

Here e_y is a unit vector along the y-axis, c is the airfoil chord length and α is the angle between the airfoil and W. In the evaluation of the circulation Γ, it has been assumed that the airfoil is thin and without curvature. In this case c and α are well defined, but in general the circulation depends on the details of the profile, although an expression similar to the right-hand side of (4.114) is still valid as a first approximation, for some average chord length and angle of attack. Equation (4.113) is known as the theorem of Kutta (1902) and Joukowski (1906).

The expressions (4.113) and (4.114) are valid in a co-ordinate system fixed relative to the airfoil (Fig. 4.64), and if the airfoil is moving with a velocity U, the velocities v and W are to be interpreted as relative ones, so that

$$W = u_{in} - U, \tag{4.115}$$

if the undisturbed wind velocity is u_{in}.

The assumption that viscous forces may be neglected is responsible for obtaining in (4.113) only a lift force, the drag force being zero. Primitive sailships, as well as primitive windmills, have been primarily aimed at utilising the drag force. It is possible, however, with suitably constructed airfoils, to make the lift force one or two orders of magnitude larger than the drag force and thereby effectively approach the limit where the viscous forces and hence the drag can be neglected. This usually requires careful "setting" of the airfoil, i.e. careful choice of the angle of attack, α, and in order to study operation at arbitrary conditions the drag component should be retained.

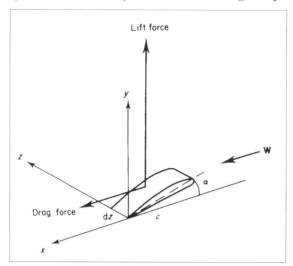

Figure 4.64. Forces on an air-foil segment.

It is customary to describe the drag and lift forces on an airfoil of given shape, as a function of α, in terms of two dimensionless constants, $C_D(\alpha)$ and $C_L(\alpha)$, defined by

$$dF_x/dz = \tfrac{1}{2}\,\rho\,C_D W^2 c,$$

$$dF_y/dz = \tfrac{1}{2}\,\rho\,C_L W^2 c. \tag{4.116}$$

The constants C_D and C_L are not quite independent of the size of the system, which is not unexpected since the viscous forces (friction) in air contribute most to turbulent motion on smaller scales (cf. the discussion in section 2.C). Introducing the Reynolds number,

$$\text{Re} = Wc/\eta,$$

where η is the kinematic viscosity of air defined in section 2.C as a measure of the ratio between "inertial" and "viscous" forces acting between airfoil and air, the α-dependence of C_D and C_L for fixed Re, as well as the Re-dependence for the value of α which gives the highest lift-to-drag ratio, L/D = C_L/C_D, may look as shown in Figs. 4.65 and 4.66. The contours of these "high lift" profiles are indicated in Fig. 4.65.

Assuming that C_D, C_L and W are constant over the area $A = \int c\,dz$ of the airfoil, the work done by a uniform (except in the vicinity of the airfoil) wind field u_{in} on a device (e.g. a ship) moving with a velocity U can be derived from (4.116) and (4.115),

$$E = \boldsymbol{F} \cdot \boldsymbol{U}.$$

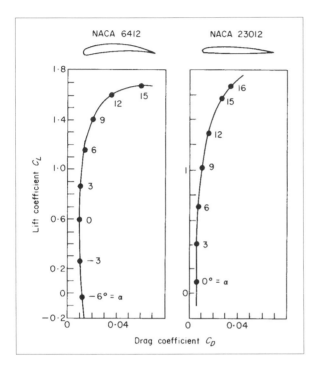

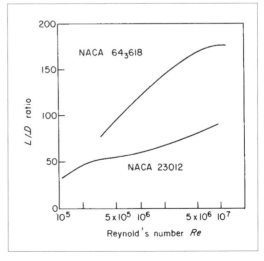

Figure 4.65. Lift and drag forces as a function of the angle of attack for two NACA airfoils (National Advisory Committee for Aeronautics; cf. e.g. Betz, 1959). The Reynolds number is Re $= 8 \times 10^6$.

Figure 4.66. Reynolds number dependence of the lift-to-drag ratio, defined as the maximum value (as a function of the angle of attack) of the ratio between the lift and drag coefficients C_L and C_D (based on Hütter, 1977).

The angle β between u_{in} and U (see Fig. 4.67) may be maintained by a rudder. The power coefficient (4.50) becomes

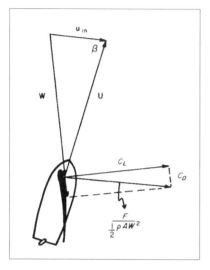

Figure 4.67. Velocity and force components for a sail-ship.

$$C_p = f\,(C_L \sin \beta - C_D\,(1 - \sin^2 \beta + f^2 - 2f \cos \beta)^{\frac{1}{2}})\,(1 + f^2 - 2f \cos \beta)^{\frac{1}{2}},$$

with $f = U/u_{in}$. For $C_L = 0$ the maximum C_p is $4C_D/27$, obtained for $f = 1/3$ and $\beta = 0$, whereas the maximum C_p for high lift-to-drag ratios L/D is obtained for β close to $\frac{1}{2}\pi$ and f around $2C_L/(3C_D)$. In this case, the maximum C_p may exceed C_L by one to two orders of magnitude (Wilson and Lissaman, 1974).

It is quite difficult to maintain the high speeds U required for optimum performance in a linear motion of the airfoil, and it is natural to focus the attention on rotating devices, in case the desired energy form is shaft or electric power and not propulsion. Wind-driven propulsion in the past (mostly at sea) has been restricted to U/u_{in}-values far below the optimum region for high L/D airfoils (owing to friction against the water), and wind-driven propulsion on land or in the air has received little attention.

4.3.2 Propeller-type converters

Propellers have been extensively used in aircraft to propel the air in a direction parallel to that of the propeller axis, thereby providing the necessary lift force on the aeroplane wings. Propeller-type rotors are similarly used for windmills, but here the motion of the air (i.e. the wind) makes the propeller, which should be placed with its axis parallel to the wind direction, rotate, thus providing the possibility of power extraction. The propeller consists of a number of blades which are evenly distributed around the axis (cf. Fig. 4.68), with each blade having a suitable aerodynamic profile usually designed to produce a high lift force, as discussed in section 4.3.1. If there are two or more blades, the symmetrical mounting ensures a symmetrical mass distribution, but if only one blade is used it must be balanced by a counterweight.

Theory of non-interacting streamtubes

In order to describe the performance of a propeller-type rotor, the forces on each element of blade must be calculated, including the forces produced by the direct action of the wind field on each individual element as well as the forces arising as a result of interactions between different elements on the same or on different blades. Since the simple airfoil theory outlined in section 4.3.1 deals only with the forces on a given blade segment, in the absence of the other ones and also without the inclusion of "edge effects" from "cutting out" this one segment, it is tempting as a first approximation to treat the different radial segments of each blade as independent. Owing to the symmetrical mounting of blades, the corresponding radial segments of different blades (if more than one) may for a uniform wind field be treated together, considering as indicated in Fig. 4.68 an annulus-shaped "*stream-tube*" of flow, bordered by streamlines intersecting the rotor plane at radial distances r and $r+dr$. The approximation of independent contributions from each streamtube, implying that radially induced velocities (u_r in Fig. 4.5) are neglected, allows the total shaft power to be expressed as

$$E = \int_0^R \frac{dE}{dr}\,dr,$$

where dE/dr depends only on the conditions of the streamtube at r. Similar sums of independent contributions can in the same order of approximation be used to describe other overall quantities, such as the axial force component T and the torque Q (equal to the power E divided by the angular velocity Ω of the propeller).

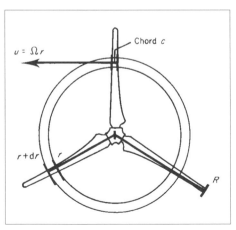

Figure 4.68. Definition of the streamtubes and blade integration variable for a propeller-type rotor.

In Fig. 4.69, a section of a wing profile (blade profile) is seen from a direction perpendicular to the cut. The distance of the blade segment from the axis of rotation is r, and its velocity $r\Omega$ is directed along the y-axis. This defines a coordinate system with a fixed x-axis along the rotor axis and moving y- and z-axes such that the blade segment is fixed relative to this coordinate system.

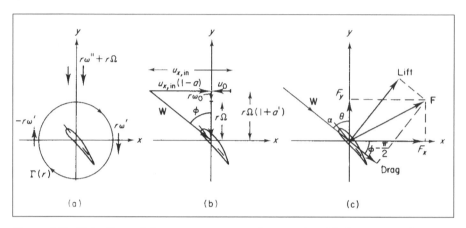

Figure 4.69. Velocity and force components of a rotating blade segment in a coordinate system following the rotation (which is in the direction of the y-axis): (a) determination of induced angular velocities; (b) determination of the direction of the apparent wind velocity; (c) determination of force components along x- and y-axes.

In order to utilise the method developed in section 4.3.1, the apparent wind velocity W to be used in (4.116) must be determined. It is given by (4.115) only if the velocity induced by, and remaining in the wake of, the device, $u^{ind} = u_{out} - u_{in}$ (cf. Fig. 4.5), is negligible. Since the radial component u_r of u^{ind} has been neglected, u^{ind} has two components, one along the x-axis,

$$u^{ind}_x = u_{x,out} - u_{x,in} = -2au_{x,in}$$

[cf. (4.47)], and one in the tangential direction,

$$u^{ind}_t = \omega^{ind} r = 2a'\Omega\, r,$$

when expressed in a non-rotating coordinate system (the second equality is defining a quantity a', called the "tangential interference factor").

W is determined by the air velocity components in the rotor plane. From the momentum considerations underlying (4.47), the induced x-component u_0 in the rotor plane is seen to be

$$u_0 = \tfrac{1}{2}u^{ind}_x = -au_{x,in}. \tag{4.117}$$

In order to determine $r\omega_0$, the induced velocity of rotation of the air in the rotor plane, one may use the following argument (Glauert, 1935). The induced rotation is partly due to the rotation of the wings and partly due to the non-zero circulation (4.114) around the blade segment profiles (see Fig. 4.69a). This circulation may be considered to imply an induced air velocity component of the same magnitude, $|r\omega'|$, but with opposite direction in front of and behind the rotor plane. If the magnitude of the component induced by the wing rotation is called $r\omega''$ in the fixed coordinate system, it will be $r\omega'' + r\Omega$ and directed along the negative y-axis in the coordinate system following the blade. It has the same sign in front of and behind the rotor. The total y-component of the induced air velocity in front of the rotor is then $-(r\omega'' + r\Omega + r\omega')$, and the tangential component of the induced air velocity in the fixed coordinate system is $-(r\omega'' - r\omega')$, still in front of the rotor. But here there should be no induced velocity at all. This follows, for example, from performing a closed line integral of the air velocity v along a circle with radius r and perpendicular to the x-axis. This integral should be zero because the air in front of the rotor has been assumed to be irrotational,

$$0 = \int_{circle\ area} \text{rot}\ v\ dA = \oint v \cdot ds = -2\pi r(r\omega'' - r\omega'),$$

implying $\omega'' = \omega'$. This identity also fixes the total induced tangential air velocity behind the rotor, $r\omega^{ind} = r\omega'' + r\omega' = 2r\omega''$, and in the rotor plane (note that the circulation component $r\omega'$ is here perpendicular to the y-axis),

$$r\omega_0 = r\omega'' = \tfrac{1}{2}r\omega^{ind} = \Omega\ ra'. \tag{4.118}$$

The apparent wind velocity in the co-ordinate system moving with the blade segment, W, is now determined as seen in Fig. 4-69b. Its x-component is u_x, given by (4.47), and its y-component is obtained by taking (4.118) to the rotating coordinate system,

$$W_x = u_x = u_{x,in}(1 - a),$$

$$W_y = -(r\omega_x + r\Omega) = -r\Omega\ (1 + a'). \tag{4.119}$$

The lift and drag forces (Fig. 4.69c) are now obtained from (4.118) (except that the two components are no longer directed along the coordinate axes), with values of C_D, C_L and c pertaining to the segments of wings at the streamtube intersecting the rotor plane at the distance r from the axis. As indicated in Figs. 4.69b and c, the angle of attack, α, is the difference between the angle ϕ, determined by

$$\tan \phi = -W_x / W_y, \tag{4.120}$$

and the pitch angle θ between blade and rotor plane,

$$\alpha = \phi - \theta. \tag{4.121}$$

Referred to the coordinate axes of Fig. 4.69, the force components for a single blade segment become

$$F_x = \tfrac{1}{2}\,\rho\,cW^2\,(C_D(\alpha)\,\sin\phi + C_L(\alpha)\,\cos\phi),$$

$$F_y = \tfrac{1}{2}\,\rho\,cW^2\,(-C_D(\alpha)\,\cos\phi + C_L(\alpha)\,\sin\phi), \tag{4.122}$$

and the axial force and torque contributions from a streamtube with B individual blades are given by

$$dT/dr = BF(r),$$

$$dQ/dr = BrF_y(r). \tag{4.123}$$

Combining (4.123) with equations expressing momentum and angular momentum conservation for each of the (assumed non-interacting) streamtubes, a closed set of equations is obtained. Equating the momentum change of the wind in the x-direction to the axial force on the rotor part intersected by an individual streamtube [i.e. (4.43) and (4.44) with $A = 2\pi r$], one obtains

$$dT/dr = -A\rho\,u_x u^{ind}_x = 2\pi r \rho\,u_{x,in}\,(1-a)\,2au_{x,in}, \tag{4.124}$$

and similarly equating the torque on the streamtube rotor part to the change in angular momentum of the wind (from zero to $r \times u^{ind}_t$), one gets

$$dQ/dr = A\rho\,u_x u^{ind}_t = 2\pi r^2 \rho\,u_{x,in}\,(1-a)\,2a'\Omega\,r. \tag{4.125}$$

Inserting $W = u_{x,in}(1-a)/\sin\phi$ or $W = r\Omega\,(1+a')/\cos\phi$ (cf. Fig. 4.69c) as necessary, one obtains a and a' expressed in terms of ϕ by combining (4.122)–(4.126). Since, on the other hand, ϕ depends on a and a' through (4.120) and (4.119), an iterative method of solution should be used. Once a consistent set of (a, a', ϕ)-values has been determined as function of r [using a given blade profile implying known values of θ, c, $C_D(\alpha)$ and $C_L(\alpha)$ as function of r], either (4.123) or (4.124) and (4.125) may be integrated over r to yield the total axial force, total torque or total shaft power $E = \Omega\,Q$.

One may also determine the contribution of a single streamtube to the power coefficient (4.50),

$$C_p(r) = \frac{\Omega\,dQ/dr}{\tfrac{1}{2}\,\rho\,u^3_{x,in}\,2\pi r} = 4a'\,(1-a)\left(\frac{r\Omega}{u_{x,in}}\right)^2. \tag{4.126}$$

The design of a rotor may utilise $C_p(r)$ for a given wind speed $u_{x,in}^{design}$ to optimise the choice of blade profile (C_D and C_L), pitch angle (θ) and solidity

$(Bc/(\pi r))$ for given angular velocity of rotation (Ω). If the angular velocity Ω is not fixed (as it might be by use of a suitable asynchronous electrical generator, except at start and stop), a dynamic calculation involving $d\Omega/dt$ must be performed. Not all rotor parameters need to be fixed. For example, the pitch angles may be variable, by rotation of the entire wing around a radial axis, such that all pitch angles $\theta(r)$ may be modified by an additive constant θ_0. This type of regulation is useful in order to limit the C_p-drop when the wind speed moves away from the design value $u_{x.in}^{design}$. The expressions given above would then give the actual C_p for the actual setting of the pitch angles given by $\theta(r)_{design} + \theta_0$ and the actual wind speed $u_{x,in}$, with all other parameters left unchanged.

In introducing the streamtube expressions (4.124) and (4.125), it has been assumed that the induced velocities and thus a and a' are constant along the circle periphery of radius r in the rotor plane. As the model used to estimate the magnitude of the induced velocities (Fig. 4.69a) is based on being near to a rotor blade, it is expected that the average induced velocities in the rotor plane, u_0 and $r\omega_0$, are smaller than the values calculated by the above expressions, unless the solidity is very large. In practical applications, it is customary to compensate by multiplying a and a' by a common factor, F, less than unity and a function of B, r and ϕ (see e.g. Wilson and Lissaman, 1974).

Furthermore, edge effects associated with the finite length R of the rotor wings have been neglected, as have the "edge" effects at $r = 0$ due to the presence of the axis, transmission machinery, etc. These effects may be described in terms of "trailing vortices" shed from the blade tips and blade roots and moving in helical orbits away from the rotor in its wake. Vorticity is a name for the vector field rot$v$, and the vortices connected with the circulation around the blade profiles (Fig. 4.69a) are called "bound vorticity". This can "leave" the blade only at the tip or at the root. The removal of bound vorticity is equivalent to a loss of circulation I (4.114) and hence a reduction in the lift force. Its dominant effect is on the tangential interference factor a' in (4.125), and so it has the same form and may be treated on the same footing as the corrections due to finite blade number. (Both are often referred to as "tip losses", since the correction due to finite blade number is usually appreciable only near the blade tips, and they may be approximately described by the factor F introduced above.) Other losses may be associated with the "tower shadow", etc.

Model behaviour of power output and matching to load

A calculated overall C_p for a three-bladed, propeller-type wind energy converter is shown in Fig. 4.70, as a function of the tip-speed ratio,

$$\lambda = \Omega R / u_{x,in}$$

and for different settings of the overall pitch angle, θ_0. It is clear from (4.126) and the equations for determining a and a' that C_p depends on the angular velocity Ω and the wind speed $u_{x,in}$ only through the ratio λ. Each blade has been assumed to be an airfoil of the type NACA 23012 (cf. Fig. 4.65), with chord c and twist angle θ changing along the blade from root to tip, as indicated in Fig. 4.71. A tip-loss factor F has been included in the calculation of sets of corresponding values of a, a' and ϕ for each radial station (each "streamtube"), according to the Prandtl model described by Wilson and Lissaman (1974). The dashed regions in Fig. 4.70 correspond to values of the product aF (corresponding to a in the expression without tip loss) larger than 0.5. Since the axial air velocity in the wake is $u_{x,out} = u_{x,in}(1 - 2aF)$ [cf. (4.47) with $F = 1$], $aF > 0.5$ implies reversed or re-circulating flow behind the wind energy converter, a possibility which has not been included in the above description. The C_p-values in the dashed regions may thus be inaccurate and presumably overestimated.

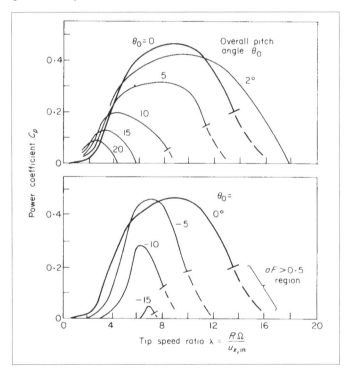

Figure 4.70. Dependence of power coefficient on tip-speed ratio for different settings of the overall pitch angle. The calculation is based on the wing geometry shown in Fig. 4.71 and the NACA 23012 profile of Fig. 4.65.

Figure 4.70 shows that for negative settings of the overall pitch angle θ_0, the C_p distribution on λ-values is narrow, whereas it is broad for positive θ_0 of modest size. For large positive θ_0, the important region of C_p moves to smaller λ-values. The behaviour of $C_p(\lambda)$ in the limit of λ approaching zero is important for operating the wind energy converter at small rotor angular velocities and, in particular, for determining whether the rotor will be self-starting, as discussed in more detail below.

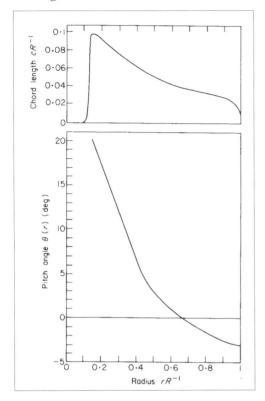

Figure 4.71. Chord variation and twist along the blades used for the calculations in this section.

For a given angular speed Ω of the rotor blades, the power coefficient curve specifies the fraction of the power in the wind which is converted, as a function of the wind speed $u_{x,in}$. Multiplying C_p by $\frac{1}{2} \rho A u_{x,in}^3$, where the rotor area is $A = \pi R^2$ (if "coning" is disregarded, i.e. the blades are assumed to be in the plane of rotation), the power transferred to the shaft can be obtained as a function of wind speed, assuming the converter to be oriented ("yawed") such that the rotor plane is perpendicular to the direction of the incoming wind. Figures 4.72 and 4.73 show such plots of power E, as functions of wind speed and overall pitch angle, for two definite angular velocities ($\Omega = 4.185$ and 2.222 rad s^{-1}, if the length of the wings is $R = 27$ m). For the range of wind speeds encountered in the planetary boundary layer of the atmosphere (e.g. at heights between 25 and 100 m), a maximum power level around 4000 W per average square metre swept by the rotor (10 MW total) is reached for the device with the higher rotational speed (tip speed $R\Omega = 113$ ms^{-1}), whereas about 1000 W m^{-2} is reached by the device with the lower rotational velocity ($R\Omega = 60$ m s^{-1}).

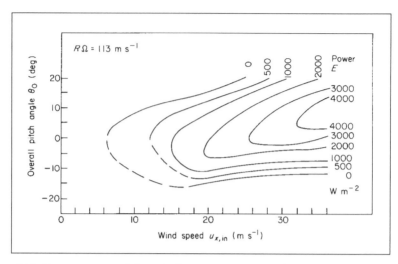

Figure 4.72. Power output of the wind energy converter described in Figs. 4.70 and 4.71, for a tip speed of 113 m s⁻¹, as a function of wind speed and overall pitch angle. The actual Reynolds number is in this case about 5×10^6 from blade root to tip, in reasonable consistency with the value used in the airfoil data of Fig. 4.65. Dashed lines indicate regions where a [defined in (4.119)] times the tip speed correction, F, exceeds 0.5 for some radial stations along the blades.

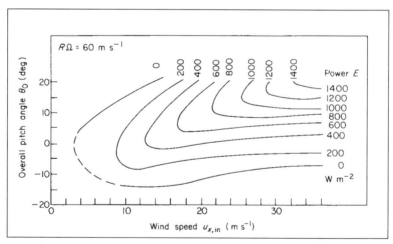

Figure 4.73. Same as Fig. 4.72, but for a tip speed of 60 m s⁻¹.

The rotor has been designed to yield a maximum C_p at about $\lambda = 9$ (Fig. 4.70), corresponding to $u_{x,in} = 12.6$ and 6.6 m s⁻¹ in the two cases. At wind speeds above these values the total power varies less and less strongly, and for negative pitch angles it even starts to decrease with increasing wind speeds in the range characterising "stormy weather". This phenomenon can be used to make the wind energy converter self-regulating, provided that the "flat" or "decreasing" power regions are suitably chosen and that the constancy of the angular velocity at operation is assured by, for example, coupling the shaft to a suitable asynchronous electricity generator (a generator allowing only minute changes in rotational velocity, with the generator angular speed Ω_g being fixed relative to the shaft angular speed Ω, e.g. by a constant exchange ratio $n = \Omega_g / \Omega$ of a gearbox). Self-regulating wind energy converters with fixed overall pitch angle θ_0 and constant angular velocity have long been used for AC electricity generation (Juul, 1961).

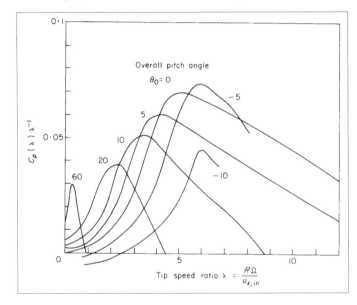

Figure 4.74. Variation of power coefficient over tip-speed ratio, as a function of tip-speed ratio, for the converter considered in the previous figures.

According to its definition, the torque $Q = E/\Omega$ can be written

$$Q = \tfrac{1}{2} \rho \, \pi R^3 \, u_{x,in}^2 \, C_p(\lambda) \, / \, \lambda, \qquad (4.127)$$

and Fig. 4.74 shows C_p/λ as a function of λ for the same design as the one considered in Fig. 4.70. For pitch angles θ_0 less than about $-3°$, C_p/λ is negative for $\lambda = 0$, whereas the value is positive if θ_0 is above the critical value of about $-3°$. The corresponding values of the total torque (4.127), for a rotor with radius $R = 27$ m and overall pitch angle $\theta_0 = 0°$, are shown in Fig. 4.75, as a function of the angular velocity $\Omega = \lambda u_{x,in}/R$ and for different wind speeds. The small dip in Q for low rotational speeds disappears for higher pitch angles θ_0.

The dependence of the torque at $\Omega = 0$ on pitch angle is shown in Fig. 4.76 for a few wind speeds. Advantage can be taken of the substantial increase in starting torque with increasing pitch angle, in case the starting torque at the pitch angle desired at operating angular velocities is insufficient and provided the overall pitch angle can be changed. In that case a high overall pitch angle is chosen to start the rotor from $\Omega = 0$, where the internal resistance (friction in bearings, gearbox, etc.) is large. When an angular speed Ω of a few degrees per second is reached, the pitch angle is diminished to a value close to the optimum one (otherwise the torque will pass a maximum and soon start to decrease again, as seen from e.g. the $\theta_0 = 60°$ curve in Fig. 4.74). Usually, the internal resistance diminishes as soon as Ω is non-zero. The internal resistance may be represented by an internal torque, $Q_0(\Omega)$, so that the torque available for some load, "external" to the wind power to shaft power converter (e.g. an electric generator), may be written

$$Q^{available} = Q(\Omega) - Q_0(\Omega).$$

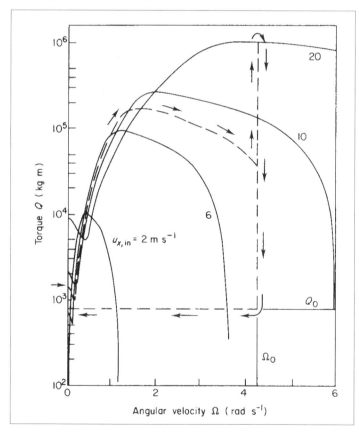

Figure 4.75. Torque variation with angular velocity for a wind energy converter of the type described in the previous figures, with radius $R = 27$ m and overall pitch angle 0°. The dashed, vertical line represents operation at fixed angular velocity (e.g. with a synchronous generator). Q_0 illustrates the internal torque.

If the "load" is characterised by demanding a fixed rotational speed Ω (such as the asynchronous electric generator), it may be represented by a vertical line in the diagram shown in Fig. 4.75. As an example of the operation of a wind energy generator of this type, the dashed lines (with accompanying arrows) in Fig. 4.75 describe a situation with initial wind speed 8 m s^{-1}, assuming that this provides enough torque to start the rotor at the fixed overall pitch angle $\theta_0 = 0°$ (i.e. the torque produced by the wind at 8 m s^{-1} and $\Omega = 0$ is above the internal torque Q_0, here assumed to be constant). The excess torque makes the angular velocity of the rotor, Ω, increase along the specific curve for $u_{x,in} = 8$ m s^{-1}, until the value Ω_0 characterising the electric generator is reached. At this point the load is connected and the wind energy converter begins to deliver power to the load area. If the wind speed later increases, the torque will increase along the vertical line at $\Omega = \Omega_0$ in Fig. 4.75, and if the wind speed decreases, the torque will decline along the same vertical line until it reaches the internal torque value Q_0. Then the angular velocity diminishes and the rotor is brought to a halt.

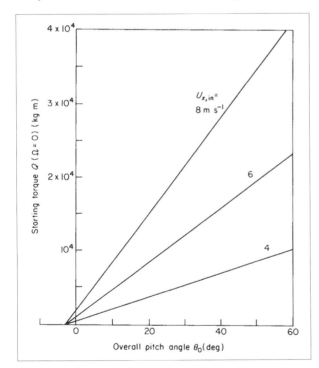

Figure 4.76. Starting torque as a function of overall pitch angle for the wind energy converter considered in the preceding figures.

An alternative type of load may not require a constant angular velocity. Synchronous DC electric generators are of this type, providing an increasing power output with increasing angular velocity (assuming a fixed exchange ratio between rotor angular velocity Ω and generator angular velocity Ω_g). Instead of staying on a vertical line in the torque–versus–Ω diagram, for varying wind speed, the torque now varies along some fixed, monotonically increasing curve characterising the generator. An optimum synchronous generator would be characterised by a $Q(\Omega)$ curve, which for each wind speed $u_{x,in}$ corresponds to the value of $\Omega = \lambda u_{x,in}/R$ which provides the maximum power coefficient C_p (Fig. 4.70). This situation is illustrated in Fig. 4.77, again indicating by a set of dashed curves the torque variation for an initial wind speed of 8 m s^{-1}, followed by an increasing and later again decreasing wind speed. In this case the $Q = Q_0$ limit is finally reached at a very low angular velocity, indicating that power is still delivered during the major part of the slowing-down process.

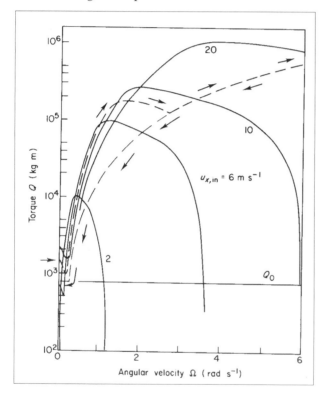

Figure 4.77. Same as Fig. 4.75, but here operation is illustrated for a load (e.g. asynchronous generator) which is optimised, i.e. corresponds to maximum C_p, at each wind speed.

Non-uniform wind velocity

The velocity field of the wind may be non-uniform in time as well as in spatial distribution. The influence of time variations in wind speed on power output of a propeller-type wind energy converter has been touched upon in the previous subsection, although a detailed investigation involving the actual time dependence of the angular velocity Ω was not included. In general, the direction of the wind velocity is also time dependent, and the conversion device should be able to successively align its rotor axis with the long-range trends in wind direction, or suffer a power reduction which is not just the cosine to the angle β between the rotor axis and the wind direction ("yaw angle"), but involves calculating the performance of each blade segment for an apparent wind speed W and angle of attack α different from the ones previously used, and no longer axially symmetric. This means that the quantities W and α (or ϕ) are no longer the same inside a given annular stream-tube, but they depend on the directional position of the segment considered (e.g. characterised by a rotational angle ψ in a plane perpendicular to the rotor axis).

Assuming that both the wind direction and the wind speed are functions of time and of height h (measured from ground level or from the lower boundary of the velocity profile z_0 discussed in section 2.C), but that the direction remains horizontal, then the situation will be as depicted in Fig. 4.78. The coordinate system (x_0, y_0, z_0) is fixed and has its origin in hub height h_0, where the rotor blades are fastened. Consider now a blade element at a distance r from the origin, on the ith blade. The projection of this position on to the (y_0, z_0)-plane is turned the angle ψ_i from vertical, where

$$\psi_i = 2\pi/i + \Omega t, \tag{4.128}$$

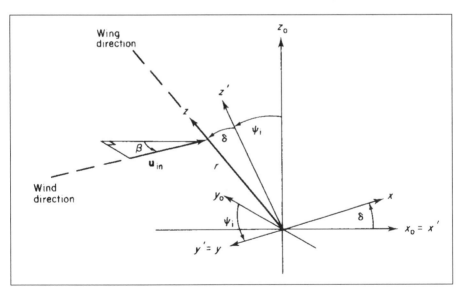

Figure 4.78. Definition of coordinate systems for a coning rotor in a non-uniform wind field.

at the time t. The coning angle δ is the angle between the blade and its projection onto the (y_0, z_0)-plane, and the height of the blade element above the ground is

$$h = h_0 + r \cos \delta \cos \psi_i, \tag{4.129}$$

where h_0 is the hub height. The height h enters as a parameter in the wind speed $u_{in} = |u_{in}(h, t)|$ and the yaw angle $\beta = \beta(h, t)$, in addition to time.

An attempt can now be made to copy the procedure used in the case of a uniform wind velocity along the rotor axis, i.e. to evaluate the force components for an individual stream-tube both by the momentum consideration of section 4.1.4 and by the lift and drag approach of section 4.3.1. The individual stream-tubes can no longer be taken as annuli, but must be of an area A_s (perpendicular to the wind direction) small enough to permit the neglect of variations in wind velocity over the area at a given time. It will still be assumed that the flows inside different stream-tubes do not interact, and also, for simplicity, the stream-tubes will be treated as "straight lines", i.e. not expanding or bending as they pass the region of the conversion device (as before, this situation arises when induced radial velocities are left out).

Consider first the "momentum equation" (4.43), with the suffix s denoting "in the stream-wise direction" or "along the stream-tube",

$$F_s = -J_m\, u_s^{ind}.$$

This force has components along the x-, y-, and z-directions of the local coordinate system of a blade turned the angle ψ from vertical (cf. Fig. 4.69). The y-direction is tangential to the rotation of the blade element, and in addition to the component of F_s there may be an induced tangential velocity u_t^{ind} (and force) of the same type as the one considered in the absence of yaw (in which case F_s is perpendicular to the y-axis). The total force components in the local coordinate system are thus

$$F_x = -J_m\, u_s^{ind} (\cos\beta \cos\delta + \sin\beta \sin\delta \sin\psi),$$
$$F_y = -J_m\, u_s^{ind} \sin\beta \cos\psi + J_m\, u_t^{ind}. \tag{4.130}$$

From the discussion of Fig. 4.5, (4.44) and (4.47),

$$J_m = A_s\, \rho\, u_s = A_s\, \rho\, u_{in}(1-a),$$
$$u_s^{ind} = -2a u_{in},$$

and in analogy with (4.118), a tangential interference factor a' may be defined by

$$u_t^{ind} = 2a' r\, \Omega \cos\delta.$$

However, the other relation contained in (4.118), from which the induced tangential velocity in the rotor plane is half the one in the wake, cannot be derived in the same way without the assumption of axial symmetry. Instead, the variation in the induced tangential velocity as a function of rotational angle ψ is bound to lead to crossing of the helical wake strains and it will be difficult to maintain the assumption of non-interacting stream-tubes. Here, the induced tangential velocity in the rotor plane will still be taken as $\frac{1}{2}u_t^{ind}$, an assumption which at least gives reasonable results in the limiting case of a nearly uniform incident wind field and zero or very small yaw angle.

Secondly, the force components may be evaluated from (4.120)–(4.122) for each blade segment, defining the local coordinate system (x, y, z) as in Fig. 4.78 with the z-axis along the blade and the y-axis in the direction of the blade's rotational motion. The total force, averaged over a rotational period, is obtained by multiplying by the number of blades, B, and by the fraction of time each blade spends in the stream-tube. Defining the stream-tube dimensions by an increment dr in the z-direction and an increment $d\psi$ in the rotational angle ψ, each blade spends the time fraction $d\psi/(2\pi)$ in the stream-tube, at constant angular velocity Ω. The force components are then

$$F_x = B \, (d\psi/2\pi) \, \tfrac{1}{2} \, \rho \, c \, W^2 \, (C_D(\alpha) \sin \phi + C_L(\alpha) \cos \phi) \, dr,$$

$$(4.131)$$

$$F_y = B \, (d\psi/2\pi) \, \tfrac{1}{2} \, \rho \, c \, W^2 \, (-C_D(\alpha) \cos \phi + C_L(\alpha) \sin \phi) \, dr,$$

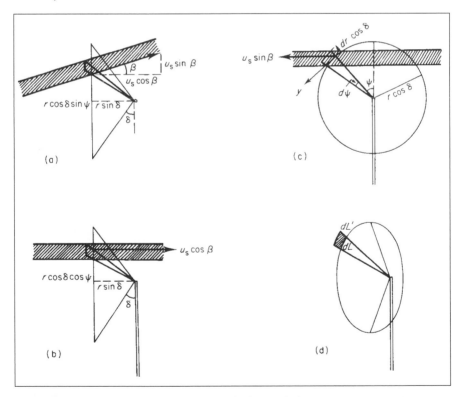

Figure 4.79. Stream–tube definition (hatched areas) for a coning rotor in a non-uniform wind field: (a) view from top; (b) side view; (c) front view; (d) view along stream-tube.

in the notation of (4.122). The angles ϕ and α are given by (4.120) and (4.121), but the apparent velocity W is the vector difference between the stream-wise velocity $u_{in} + \frac{1}{2}\,u_s^{ind}$ and the tangential velocity $r\Omega \cos\delta + \frac{1}{2}\,u_t^{ind}$, both taken in the rotor plane. From Fig. 4.78,

$$W_x = u_{in}\,(1-a)\,(\cos\beta\,\cos\delta + \sin\beta\sin\delta\,\sin\psi),$$
$$W_y = u_{in}\,(1-a)\,\sin\beta\,\cos\psi + r\Omega\,\cos\delta\,(1+a'), \qquad (4.132)$$
$$W_z = u_{in}\,(1-a)\,(-\cos\beta\,\sin\delta + \sin\beta\cos\delta\,\sin\psi).$$

The appropriate W^2 to insert into (4.131) is $W_x^2 + W_y^2$. Finally, the relation between the stream-tube area A_s and the increments dr and $d\psi$ must be established. As indicated in Fig. 4.79, the stream-tube is approximately a rectangle with sides dL and dL', given by

$$dL = r\cos\delta\,(\cos 2\beta\,\cos 2\psi + \sin 2\psi)^{1/2}\,d\psi, \qquad (4.133)$$
$$dL' = (\sin 2\delta\,(1 + \cos 2\beta) + \cos 2\delta\,(\cos 2\beta\,\sin 2\psi + \cos 2\psi))^{1/2}\,dr,$$

and
$$A_s = dL\,dL'.$$

Now, for each stream-tube, a and a' are obtained by equating the x- and y-components of (4.130) and (4.131) and using the auxiliary equations for (W, ϕ) or (W_x, W_y). The total thrust and torque are obtained by integrating F_x and $r\cos\delta\,F_y$, over dr and $d\psi$ (i.e. over all stream-tubes).

Restoration of wind profile in wake, and implications for turbine arrays

For a wind energy converter placed in the planetary boundary layer (i.e. in the lowest part of the Earth's atmosphere), the reduced wake wind speed in the stream-wise direction, $u_{s,out}$, will not remain below the wind speed $u_{s,in}$ of the initial wind field, provided that this is not diminishing with time. The processes responsible for maintaining the general kinetic motion in the atmosphere (cf. section 2.3.1), making up for the kinetic energy lost by surface friction and other dissipative processes, will also act in the direction of restoring the initial wind profile (speed as function of height) in the wake of a power extracting device by transferring energy from higher air layers (or from the "sides") to the partially depleted region (Sørensen, 1996d). The large amounts of energy available at greater altitude (cf. Fig. 3.31) make such processes possible almost everywhere at the Earth's surface and not just at those locations where new kinetic energy is predominantly being created (cf. Fig. 2.56).

In the near wake, the wind field is non-laminar, owing to the induced tangential velocity component, u_t^{ind}, and owing to vorticity shedded from the wing tips and the hub region (as discussed in section 4.3.2). It is then expected that these turbulent components gradually disappear further downstream in the wake, as a result of interactions with the random eddy motion of different scales present in the "unperturbed" wind field. "Disappear" here means "get distributed on a large number of individual degrees of freedom", so that no contributions to the time-averaged quantities considered [cf. (2.13)] remain. For typical operation of a propeller–type wind conversion device, such as the situations illustrated in Figs. 4.72 and 4.73, the tangential interference factor a' is small compared to the axial interference factor a, implying that the most visible effect of the passage of the wind field through the rotor region is the change in stream-wise wind speed. This is a function of r, the distance of the blade segment from the hub centre, as illustrated in Fig. 4.80, based on wind tunnel measurements. The induced r-dependence of the axial velocity is seen to gradually smear out, although it is clearly visible at a distance of two rotor radii in the set-up studied.

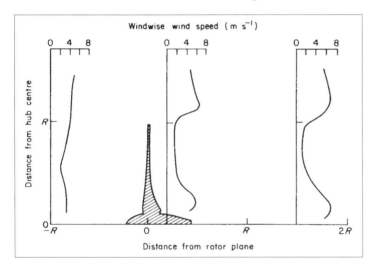

Figure 4.80. Wind tunnel results illustrating the distortion of the wind field along the radial direction (R is the rotor radius) at various distances from the rotor plane (based on Hütter, 1976).

Figure 4.81 suggests that under average atmospheric conditions the axial velocity, $u_{s,in}$, will be restored to better than 90% at about 10 rotor diameters behind the rotor plane and better than 80% at a distance of 5–6 rotor diameters behind the rotor plane, but rather strongly dependent on the amount of turbulence in the "undisturbed" wind field.

A second wind energy converter may be placed behind the first one, in the wind direction, at a distance such that the wind profile and magnitude are reasonably well restored. According to the simplified investigations in wind tunnels (Fig. 4.81), supported by field measurements behind buildings, forests and fences, a suitable distance would seem to be 5–10 rotor diameters (increasing to over 20 rotor diameters if "complete restoration" is required). If there is a prevailing wind direction, the distance between conversion units perpendicular to this direction may be smaller (essentially determined by the induced radial velocities, which were neglected in the preceding subsections, but appear qualitatively in Fig. 4.5). If, on the other hand, several wind directions are important, and the converters are designed to be able to "yaw against the wind", then the distance required by wake considerations should be kept in all directions.

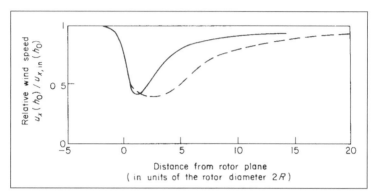

Figure 4.81. Wind tunnel results indicating the restoration at stream-wise wind speed behind a rotor (placed at distance 0, the rotor being simulated by a gauze disc). The approaching wind has a logarithmic profile (solid line) or is uniform with approximately laminar flow (dashed line) (based on Pelser, 1975).

More severe limitations may possibly be encountered with a larger array of converters, say, distributed over an extended area with the average spacing X between units. Even if X is chosen so that the relative loss in stream-wise wind speed is small from one unit to the next, the accumulated effect may be substantial, and the entire boundary layer circulation may become altered in such a way that the power extracted decreases more sharply than expected from the simple wake considerations. Thus, large-scale conversion of wind energy may even be capable of inducing local climatic changes.

A detailed investigation of the mutual effects of an extended array of wind energy converters and the general circulation on each other requires a combination of a model of the atmospheric motion, e.g. along the lines presented in section 2.3.1, with a suitable model of the disturbances induced in the wake of individual converters. In one of the first discussions of this problem, Templin (1976) considered the influence of an infinite two-dimensional array of wind energy converters with fixed average spacing on the boundary layer motion to be restricted to a change of the roughness length z_0 in the logarithmic expression (see section 2.B) for the wind profile, assumed to describe the wind approaching any converter in the array. The change in z_0 can be calculated from the stress τ^{ind} exerted by the converters on the wind, due to the axial force F_x in (4.122) or (4.124), which according to (4.46) can be written

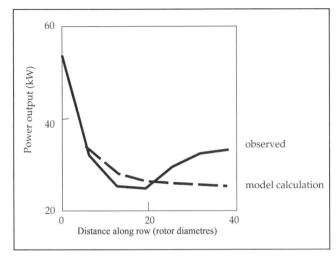

Figure 4.82. Calculated (see text) and measured power outputs from one row of turbines perpendicular to the wind direction, in a wind park located at Nørrekær Enge in Denmark. It consists in total of five rows with a turbine spacing of 6 rotor diameters (Taylor *et al.*, 1993).

$$\tau^{ind} = F_x/S = \tfrac{1}{2}\rho\,(u_{s,in})^2\,2(1-a)^2\,A/S,$$

with S being the average ground surface area available for each converter and A/S being the "density parameter", equal to the ratio of rotor-swept area to ground area. For a quadratic array with regular spacing, S may be taken as X^2. According to section 2.B, $u_{x,in}$ taken at hub height h_0 can, in the case of a neutral atmosphere, be written

$$u_{x,in}(h_0) = \frac{1}{\kappa}\left(\frac{\tau^0 + \tau^{ind}}{\rho}\right)^{1/2}\log\left(\frac{h_0}{z'_0}\right),$$

where τ^0 is the stress in the absence of wind energy converters, and z'_0 is the roughness length in the presence of the converters, which can now be determined from this equation and τ^{ind} from the previous one.

Figure 4.82 shows the results of a calculation for a finite array of converters (Taylor *et al.*, 1993) using a simple model with fixed loss fractions (Jensen, 1994). It is seen that this model is incapable of reproducing the fast restoration of winds through the turbine array, presumably associated with the propagation of the enhanced wind regions created just outside the swept areas (as seen in Fig. 4.80). A three-dimensional fluid dynamics model is required for describing the details of array shadowing effects. Such calculations are in principle possible, but so far no convincing implementation has been presented. The problem is the very accurate description needed, of complex three-dimensional flows around the turbines and for volumes comprising the entire wind farm of maybe hundreds of turbines. This is an intermediate regime between the existing three-dimensional models for gross wind flow over complex terrain and the detailed models of flow around a single turbine used in calculations of aerodynamical stability.

4.3.3 Cross-wind converters

Wind energy converters of the cross-wind type have the rotor axis perpendicular to the wind direction. The rotor axis may be horizontal as in wheel-type converters (in analogy to waterwheels) or vertical as in the panemones used in Iran and China. The blades (ranging from simple "paddles" to optimised airfoil sections) will be moving with and against the wind direction on alternative sides of the rotor axis, necessitating some way of emphasising the forces acting on the blades on one side. Possible ways are simply to shield half of the swept area, as in the Persian panemones (Wulff, 1966); to curve the "paddles" so that the (drag) forces are smaller on the convex than on the concave side, as in the Savonius rotor (Savonius, 1931); or to use aerodynamically shaped wing blades producing high lift forces for wind incident on the "front edge", but small or inadequate forces for wind incident on the "back edge", as in the Darrieus rotor (cf. Fig. 4.83) and related concepts. Another possibility is to allow for changes in the pitch angle of each blade, as, for example, achieved by hinged vertical blades in the Chinese panemone type (Chen Li, 1951). In this case the blades on one side of a vertical axis have favourable pitch angles, while those on the other side have unfavourable settings. Apart for the shielded ones, vertical axis cross-wind converters are omnidirectional, i.e. they accept any horizontal wind direction on equal footing.

Performance of a Darrieus-type converter

The performance of a cross-wind converter, such as the Darrieus rotor shown in Fig. 4.83, may be calculated in a way similar to that used in section 4.3.2, by derivation of the unknown, induced velocities from expressing the forces both in terms of lift and drag forces on the blade segments and in terms of momentum changes between incident wind and wake flow. Also, the flow may be divided into a number of streamtubes (assumed to be non-interacting), according to assumptions on the symmetries of the flow field. Figure 4.74 gives an example of the streamtube definitions, for a two-bladed Darrieus rotor with angular velocity Ω and a rotational angle ψ describing the position of a given blade, according to (4.128). The blade chord, c, has been illustrated as constant, although it may actually be taken as varying to give the optimum performance for any blade segment at the distance r from the rotor axis. As in the propeller rotor case, it is not practical to extend the chord increase to the regions near the axis.

The bending of the blade, characterised by an angle $\delta(h)$ depending on the height h (the height of the rotor centre is denoted h_0), may be taken as a troposkien curve (Blackwell and Reis, 1974), characterised by the absence of bending forces on the blades when they are rotating freely. Since the blade profiles encounter wind directions at both positive and negative forward angles, the profiles are often taken as symmetrical (e.g. NACA 00XX profiles).

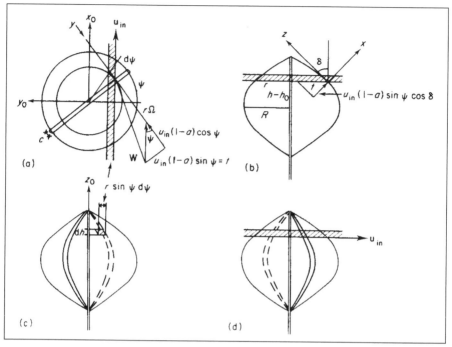

Figure 4.83. Streamtube definition (hatched areas) for two-bladed Darrieus rotor, and determination of apparent wind velocity, in the case of negligible cross-wind induced velocity ($u_{y0}^{ind} = 0$): (a) view from top; (b) view along tangent to blade motion; (c) view along streamtube; (4) view perpendicular to streamtube.

Assuming as in section 4.3.2 that the induced velocities in the rotor region are half of those in the wake, the streamtube expressions for momentum and angular momentum conservation analogous to (4.124) and (4.125) are

$$F_{x0} = -J_m u_s^{ind} = 2\rho A_s (u_{in})^2 (1-a) a,$$
$$F_{y0} = J_m u_{c.w.}^{ind},$$

(4.134)

where the axial interference factor a is defined as in (4.117), and where the streamtube area A_s corresponding to height and angular increments dh and dψ is (cf. Fig. 4.83)

$$A_s = r \sin \psi \, d\psi \, dh.$$

The cross-wind induced velocity $u_{c.w.}^{ind}$ is not of the form (4.118), since the tangent to the blade rotational motion is not along the y_0-axis. The sign of $u_{c.w.}^{ind}$ will be fluctuating with time, and for low chordal ratio c/R (R being the maximum value of r) it may be permitted to put F_{y0} equal to zero (Lissaman, 1976; Strickland, 1975). This approximation will be made in the following. It will also be assumed that the streamtube area does not change by passage through the rotor region and that individual streamtubes do not interact (these assumptions being the same as those made for the propeller-type rotor).

The forces along the instantaneous x- and y-axes due to the passage of the rotor blades at a fixed streamtube location (h, ψ) can be expressed in analogy to (4.131), averaged over one rotational period,

$$F_x = B \, (\mathrm{d}\psi/2\pi) \, \tfrac{1}{2} \, \rho \, c \, W^2 \, (C_D \sin \phi + C_L \cos \phi) \, \mathrm{d}h/\cos \delta,$$

$$F_y = B \, (\mathrm{d}\psi/2\pi) \, \tfrac{1}{2} \, \rho \, c \, W^2 \, (-C_D \cos \phi + C_L \sin \phi) \, \mathrm{d}h/\cos \delta,$$

(4.135)

where

$$\tan \phi = - W_x / W_y; \qquad W^2 = W_x^2 + W_y^2$$

and (cf. Fig. 4.74)

$$W_x = u_{in} \, (1 - a) \sin \psi \cos \delta,$$
$$W_y = - r \, \Omega - u_{in} \, (1 - a) \cos \psi,$$
$$W_z = - u_{in} \, (1 - a) \sin \psi \sin \delta.$$

(4.136)

The angle of attack is still given by (4.121), with the pitch angle θ being the angle between the y-axis and the blade centre chord line (cf. Fig. 4.69c). The force components (4.135) may be transformed to the fixed (x_0, y_0, z_0) coordinate system, yielding

$$F_{x0} = F_x \cos \delta \sin \psi + F_y \cos \psi,$$

(4.137)

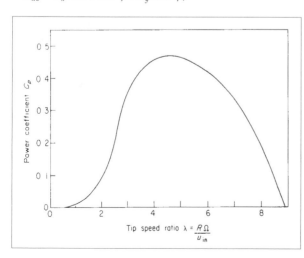

Figure 4.84. Power coefficient as a function of tip-speed ratio ("tip" means point furthest away from axis), for a two-bladed Darrieus rotor, with a chord ratio $c/R = 0.1$ and blade data corresponding to a symmetrical profile (NACA 0012 with Reynolds number Re = 3 × 10^6). A streamtube model of the type described in Fig. 4.83 has been used (based on Strickland, 1975).

with the other components F_{y0} and F_{z0} being neglected due to the assumptions made. Using the auxiliary relations given, a may be determined by equating the two expressions (4.134) and (4.137) for F_{x0}.

After integration over dh and dψ, the total torque Q and power coefficient C_p can be calculated. Figure 4.84 gives an example of the calculated C_p for a low-solidity, two-bladed Darrieus rotor with NACA 0012 blade profiles and a size corresponding to Reynolds number Re = 3×10^6. The curve is rather similar to the ones obtained for propeller-type rotors (e.g. Fig. 4.70), but the maximum C_p is slightly lower. The reasons why this type of cross-wind converter cannot reach the maximum C_p of 16/27 derived from (4.50) (the "Betz limit") are associated with the fact that the blade orientation cannot remain optimal for all rotational angles ψ_i, as discussed (in terms of a simplified solution to the model presented above) by Lissaman (1976).

Figure 4.85 gives the torque as a function of angular velocity Ω for a small three-bladed Darrieus rotor. When compared with the corresponding curves for propeller-type rotors shown in Figs. 4.75 and 4.77 (or generally Fig. 4.74), it is evident that the torque at $\Omega = 0$ is zero for the Darrieus rotor, implying that it is not self-starting. For application in connection with an electric grid or some other back-up system, this is no problem since the auxiliary power needed to start the Darrieus rotor at the appropriate times is very small compared with the wind converter output, on a yearly average basis. However, for application as an isolated source of power (e.g. in rural areas), it is a disadvantage, and it has been suggested that a small Savonius rotor should be placed on the main rotor axis in order to provide the starting torque (Banas and Sullivan, 1975).

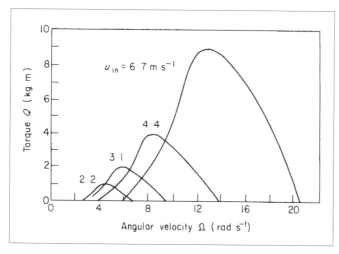

Figure 4.85. Torque as a function of angular velocity for a small three-bladed Darrieus rotor and at selected wind speeds. The experimental rotor had a radius R = 2.25 m, a chord ratio c/R = 0.085 and NACA 0012 blade profiles (based on Banas and Sullivan, 1975).

Another feature of the Darrieus converter (as well as of some propeller-type converters), which is evident from Fig. 4.85, is that for application with a load of constant Ω (as in Fig. 4.75), there will be a self-regulating effect in that the torque will rise with increasing wind speed only up to a certain wind speed. If the wind speed increases further, the torque will begin to decrease. For variable-Ω types of load (as in Fig. 4.77), the behaviour of the curves in Fig. 4.85 implies that cases of irregular increase and decrease of torque, with increasing wind speed, can be expected.

4.3.4 Augmenters and other "advanced" converters

In the preceding sections, it has been assumed that the induced velocities in the converter region were half of those in the wake. This is strictly true for situations where all cross-wind induced velocities (u_t and u_r) can be neglected, as shown in (4.45), but if suitable cross-wind velocities can be induced so that the total stream-wise velocity in the converter region, u_x, exceeds the value of $-\frac{1}{2}(u_{x,in} + u_{x,out})$ by a positive amount δu_x^{ind}, then the Betz limit on the power coefficient, $C_p = 16/27$, may be exceeded,

$$u_x = \frac{1}{2}(u_{x,in} + u_{x,out}) + \delta u_x^{ind}.$$

A condition for this to occur is that the extra induced stream-wise velocity δu_x^{ind} does not contribute to the induced velocity in the distant wake, u_x^{ind}, which is given implicitly by the above form of u_x, since

$$u_{x,out} = u_{x,in} + u_x^{ind} = u_{x,in}(1 - 2a).$$

The stream-tube flow at the converter, (4.44), is then

$$J_m = \rho A_s u_x = \rho A_s u_{x,in}(1 - a + \tilde{a}) \tag{4.138}$$

with $\tilde{a} = \delta u_x^{ind}/u_{x,in}$, and the power (4.48) and power coefficient (4.50) are replaced by

$$E = \rho A_s (u_{x,in})^3 2a(1-a)(1-a+\tilde{a}),$$
$$C_p = 4a(1-a)(1-a+\tilde{a}), \tag{4.139}$$

where the stream-tube area A_s equals the total converter area A, if the single-stream-tube model is used.

Ducted rotor

In order to create a positive increment $\tilde{a}u_{x,in}$ of axial velocity in the converter region, one may try to take advantage of the possibility of inducing a particular type of cross-wind velocity, which causes the stream-tube area to contract in the converter region. If the stream-tube cross section is circular, this may be achieved by an induced radial outward force acting on the air, which again can be caused by the lift force of a wing section placed at the periphery of the circular converter area, as illustrated in Fig. 4.86.

Figure 4.86 compares a free propeller-type rotor (top), for which the stream-tube area is nearly constant (as was actually assumed in sections 4.3.2 and 4.3.3) or expanding as a result of radially induced velocities, with a propeller rotor of the same dimensions, shrouded by a duct-shaped wing-profile. In this case the radial inward lift force F_L on the shroud corresponds (by momentum conservation) to a radial outward force on the air, which causes the stream-tube to expand on both sides of the shrouded propeller; in other words, it causes the streamlines passing through the duct to define a stream-tube, which contracts from an initial cross section to reach a minimum area within the duct and which again expands in the wake of the converter. From (4.116), the magnitude of the lift force is

$$F_L{}^{duct} = \tfrac{1}{2}\, \rho\, W_{duct}{}^2\, c_{duct}\, C_L{}^{duct},$$

with the duct chord c_{duct} defined in Fig. 4.86 and W_{duct} related to the incoming wind speed $u_{x,in}$ by an axial interference factor a_{duct} for the duct, in analogy with the corresponding one for the rotor itself, (4.47),

$$W_{duct} = u_{x,in}\,(1 - a_{duct}).$$

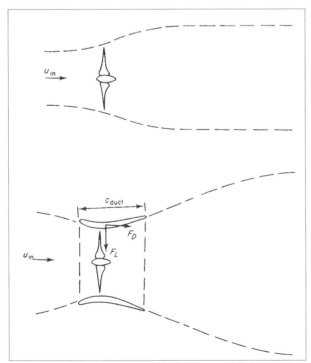

Figure 4.86. Peripheral streamlines for free stream (top) and ducted rotor (below).

$F_L{}^{duct}$ is associated with a circulation Γ around the shroud profile (shown in Fig. 4.86), given by (4.114). The induced velocity $\delta u_x{}^{ind}$ inside the duct appears in the velocity being integrated over in the circulation integral (4.114), and it may to a first approximation be assumed that the average induced velocity is simply proportional to the circulation (which in itself does not depend on the actual choice of the closed path around the profile),

$$\delta u_x^{ind} = k_{duct}\, \Gamma_{duct} / R, \qquad\qquad (4.140)$$

where the radius of the duct, R (assumed to be similar to that of the inside propeller-type rotor), has been introduced, because the path-length in the integral Γ_{duct} is proportional to R and the factor k_{duct} appearing in (4.140) is therefore reasonably independent of R. Writing $\Gamma_{duct} = (\delta\, W_{duct})^{-1}\, F_L^{duct}$ in analogy to (4.114), and introducing the relations found above,

$$\tilde{a}_{duct} = k_{duct}\, c_{duct}\, C_L^{duct}\, (1 - a_{duct}) / 2R. \qquad\qquad (4.141)$$

If the length of the duct, c_{duc}, is made comparable to or larger than R, and the other factors in (4.141) can be kept near to unity, it is seen from (4.139) that a power coefficient about unity or larger is possible. This advantage may, however, be outweighed by the much larger amounts of materials needed to build a ducted converter, relative to a free rotor.

Augmenters taking advantage of the lift forces on a suitably situated aerodynamic profile need not provide a fully surrounding duct around the simple rotor device. A vertical cylindrical tower structure (e.g. with a wing profile similar to that of the shroud in Fig. 4.86) may suffice to produce a reduction in the widths of the stream-tubes relevant to an adjacently located rotor and thus may produce some enhancement of power extraction.

Rotor with tip-vanes

A formally appealing design, shown in Fig. 4.87, places tip-vanes of modest dimensions on the rotor blade tips (van Holten, 1976). The idea is that the tip-vanes act like a duct, without causing much increase in the amount of materials needed in construction. The smaller areas producing lift forces are compensated for by having much larger values of the apparent velocity W_{vane} seen by the vanes than W_{duct} in the shroud case. This possibility occurs because the tip-vanes rotate with the wings of the rotor (in contrast to the duct), and hence see air with an apparent cross-wind velocity given by

$$W_{vane} = R\Omega\, (1 - a_{vane}).$$

The magnitude of the ratio W_{vane}/W_{duct} is thus of the order of the tip-speed ratio) $\lambda = R\Omega/u_{x,in}$, which may be 10 or higher (cf. section 4.3.2).

The lift force on a particular vane is given by $F_L^{vane} = \tfrac{1}{2}\, \rho\, W_{vane}^2\, c_{vane}\, C_L^{vane}$, and the average inward force over the periphery is obtained by multiplying this expression by $(2\pi R)^{-1}\, Bb_{vane}$, where B is the number of blades (vanes), and where b_{vane} is the lengths of the vanes (see Fig. 4.87).

Using a linearised expression analogous to (4.140) for the axial velocity induced by the radial forces,

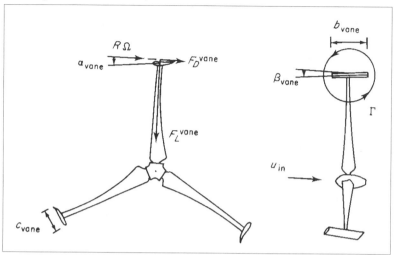

Figure 4.87. Geometry of propeller-type rotor with tip-vanes.

$$\delta u_x^{ind} = \tilde{a}_{vane}\, u_{x,in} = k_{vane}\, \Gamma / R, \qquad (4.142)$$

the additional interference factor $\tilde{a}_{vane}$ to use in (4.139) may be calculated. Here Γ is the total circulation around the length of the tip-vanes (cf. Fig. 4.87), and not the circulation $\Gamma_{vane} = (\rho\, W_{vane})^{-1}\, F_L^{vane}$ in the plane of the vane lift and drag forces (left-hand side of Fig. 4.87). Therefore, Γ may be written

$$\Gamma = F_L^{vane}\, B\, b_{vane} / (2\pi R \rho\, w''),$$

i.e. equal to the average inward force divided by ρ times the average axial velocity at the tip-vane containing peripheral annulus,

$$w'' = u_{x,in}\, (1 + a'').$$

Expressions for a'' have been derived by van Holten (1976). By inserting the above expressions into (4.142), $\tilde{a} = \tilde{a}_{vane}$ is obtained in the form

$$\tilde{a} = k_{vane}\, (1 - a_{vane})^2\, B\, c_{vane}\, b_{vane}\, \lambda^2\, C_L^{vane} / (4\pi R^2 (1 + a'')). \qquad (4.143)$$

The part $(Bc_{vane}b_{vane} / \pi R^2)$ in the above expression is the ratio between the summed tip-vane area and the rotor-swept area. Taking this as 0.05 and [according to van Holten (1976) for $b_{vane}/R = 0.5$] $k_{vane}\, (1 - a_{vane})^2 / (1 + a'')$ as 0.7, the tip-speed ratio λ as 10 and $C_L^{vane} = 1.5$, the resulting $\tilde{a}$ is 1.31 and the power coefficient according to the condition which maximises (4.139) as function of a,

$$a = (2 + \tilde{a} - ((2 + \tilde{a})^2 - 3(1 + \tilde{a}))^{1/2}) = 0.43; \qquad C_p = 1.85.$$

The drag forces F_D^{vane} induced by the tip-vanes (cf. Fig. 4.87) represent a power loss from the converter, which has not been incorporated in the above treatment. According to van Holten (1976), this loss may in the numerical example studied above reduce C_p to about 1.2. This is still twice the Betz limit, for a tip-vane area which would equal the area of the propeller blades, for a three-bladed rotor with average chord ratio $c/R = 0.05$. It is conceivable that the use of such tip-vanes to increase the power output would in some cases be preferable to achieving the same power increase by increasing the rotor dimensions.

Other concepts

A large number of alternative devices for utilisation of wind energy have been studied, in addition to those treated above and in the preceding sections. The lift-producing profile exposed to the wind may be hollow and contain holes through which inside air is driven out by the lift forces. Through a hollow tower structure, replacement air is then drawn to the wing profiles from sets of openings placed in such a way that the air must pass one or more turbine propellers in order to reach the wings. The turbine propellers provide the power output, but the overall efficiency based on the total dimensions of the device is low (Hewson, 1975). Improved efficiency of devices which try to concentrate the wind energy before letting it reach a modest-size propeller may be obtained by first converting the mainly axial flow of the wind into a flow with a large vorticity of a simple structure, such as a circular air motion. Such vortices are, for example, formed over the sharp edges of highly swept delta wings, as illustrated in Fig. 4.88a (Sforza, 1976). There are two positions along the baseline of the delta wing where propeller rotors could be placed in an environment with stream-wise velocities u_s/u_{in} of 2–3 and tangential velocities u_t of the same order of magnitude as u_{in} (see Sforza, 1976).

If the wake streamlines can be diffused over a large region, such that interaction with the wind field not contributing to those stream-tubes passing through the converter can transfer energy to the slipstream motion and thereby increase the mass flow J_m through the converter, then further increase in power extraction can be expected. This is the case for the ducted system shown in Fig. 4.86 (lower part), but it may be combined with the vorticity concept described above to form a device of the general layout shown in Fig. 4.88b (Yen, 1976). Wind enters the vertical cylinder through a vertical slit and is forced to rotate by the inside cylinder walls. The vortex system created in this way (an "artificial tornado") is pushed up through the cylinder by pressure forces and leaves the open cylinder top. Owing to the strong vorticity, the rotating air may retain its identity high up in the atmosphere, where its diffusion extracts energy from the strong winds expected at that height. This energy is supposed to be transferred down the "tornado", strengthening its vorticity and thus providing more power for the turbines placed between the "tornado" bottom and a number of air inlets at the tower foot, which replace air that has left the top of the cylinder (cf. Fig. 4.88b). Neither of these two constructions have found practical applications.

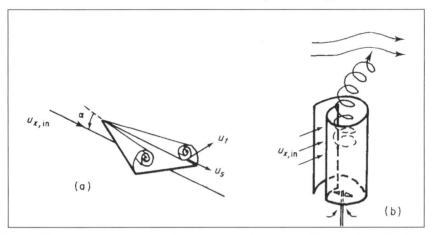

Figure 4.88. Augmenter concepts: (a) delta wing forming trailing vortices; (b) ducted vortex system (based on Sforza, 1976; Yen, 1976).

Conventional (or unconventional) wind energy converters may be placed on floating structures at sea (where the wind speeds are often higher than on land) or may be mounted on balloons (e.g. a pair of counter-rotating propellers beside one other) in order to utilise the increasing wind speed usually found at elevations not accessible to ordinary building structures on the ground. In order to serve as a power source, the balloons must be guyed to a fixed point on the ground, with wires of sufficient strength.

4.3.5 Heat, electrical or mechanical power, and fuel generation

The wind energy converters described in the preceding sections have primarily been converting the power in the wind into rotating shaft power. The conversion system generally includes a further conversion step if the desired energy form is different from that of the rotating shaft.

Examples of this are the electric generators with fixed or variable rotational velocity which were mentioned in connection with Figs 4.75 and 4.77. The other types of energy can, in most cases, be obtained by secondary conversion of electric power. In some such cases the "quality" of electricity need not be as high as that usually maintained by utility grid systems, in respect to voltage fluctuations and variations in frequency in the (most widespread) case of alternating current. For wind energy converters aimed at constant working angular velocity Ω, it is customary to use a gearbox and an induction-type generator. This maintains an a.c. frequency equal to that of the grid and constant to within about 1%. Alternatively, the correct frequency can be prescribed electronically. In both cases, reactive power is created (i.e. power which, like that of a condenser or coil, is phase shifted), which may be an advantage or disadvantage, depending on the loads on the grid.

For variable-frequency wind energy converters, the electric output would be from a synchronous generator and in the form of variable-frequency alternating current. This would have to be subjected to a time-dependent frequency conversion, and for arrays of wind turbines, phase mismatch would have to be avoided. Several schemes exist for achieving this, for example, semiconductor rectifying devices (thyristors) which first convert the variable frequency AC to DC and then in a second step the DC (direct current) to AC (alternating current) of the required fixed frequency.

If the desired energy form is heat, "low quality" electricity may first be produced, and the heat may then be generated by leading the current through a high ohmic resistance. Better efficiency can be achieved if the electricity can be used to drive the compressor of a heat pump (see section 4.6.1), taking the required heat from a reservoir of temperature lower than the desired one. It is also possible to convert the shaft power more directly into heat. For example, the shaft power may drive a pump, pumping a viscous fluid through a nozzle, such that the pressure energy is converted into heat. Alternatively, the shaft rotation may be used to drive a "paddle" through a fluid, in such a way that large drag forces arise and that the fluid is put into turbulent motion, gradually dissipating the kinetic energy into heat. If water is used as the fluid medium, this arrangement is called a "water-brake".

Windmill shaft power has traditionally been used to perform mechanical work of various kinds, including flour milling, threshing, lifting and pumping. Pumping of water, e.g. for irrigation purposes, with a pump connected to the rotating shaft, may be particularly suitable as an application of wind energy, since variable and intermittent power would, in most cases, be acceptable, as long as the average power supply in the form of lifted water over an extended period of time is sufficient.

In other cases an auxiliary source of power may be needed, so that the demand can be met at any time. This can for grid-based systems be achieved by trade of energy (cf. scenarios described in Chapter 6). Demand matching can also be ensured if an energy storage facility of sufficient capacity is attached to the wind energy conversion system. A number of such storage facilities will be mentioned in Chapter 5, and among them will be the storage of energy in the form of fuels, such as hydrogen. Hydrogen may be produced, along with oxygen, by electrolysis of water, using electricity from the wind energy converter. The detailed working of these mechanisms over time and space is simulated in the scenarios outlined in Chapter 6 and related studies (Sørensen *et al.*, 2003; Sørensen, 2004a).

The primary interest may also be oxygen, for example, to be dissolved into the water of lakes which are deficient in oxygen (say, as a result of pollution), or to be used in connection with "ocean farming", where oxygen may be a limiting factor in cases where nutrients are supplied in large quantities, e.g. by artificial upwelling. Such oxygen may be supplied by wind energy converters, while the co-produced hydrogen may be used to produce or move the nutrients. Again, in this type of application large power fluctuations may be acceptable.

4.3.6 Commercial wind power development

Wind turbines have been in substantial use for more than 1000 years, with early text evidence in China and India and the earliest pictures being from Persia around 1300. These were all vertical axis machines, spreading to the Mediterranean region of Europe, while horizontal axis machines were developed in England, Holland, Denmark, the Baltic region and Russia (cf. overview of wind history in Sørensen, 1995a). Cities like Copenhagen had in the 16th century about 100 turbines with a hub height of over 10 m grinding flour just outside the city walls, and in Holland, some wind turbines were used to pump water as part of the drainage system on low land. Around 1900, wind turbines played a significant role in powering new settlements in the North American plains and Australian outposts, and the first hydrogen- and electricity-producing wind turbines were tested in Denmark (la Cour, 1900; cf. Hansen, 1985). The DC electricity-producing wind turbines kept competing with coal-based power during the first decades of the 20th century, gradually being pushed to island sites and eventually loosing them too as undersea cables became extended to most islands. By the end of the 1920s, Denmark had about 30 000 wind turbines or three times as many as today. Experiments with AC generating wind turbines were made before and after World War II, with a comeback in several European countries following the 1957 closure of the Suez Canal (blocking transport of the Middle East oil then used in European power stations). Technically successful was the 200-kW Gedser turbine (Juul, 1961), serving as a model for the subsequent resurrection of wind turbine manufacture following the oil supply cuts in 1973/1974. The development since 1974 has been one of decreasing cost and rapidly increasing installed capacity, as stated in section 1.1.

Figure 4.89 shows the development in turbine size (height and swept diameter) of commercial wind turbines since 1980, and Fig. 4.90 shows corresponding trends in rated capacity and specific power production at the locations of typical installation. During the late 1970s, efforts were directed at several wind technologies: vertical axis machines in the US and Canada, megawatt-size one- and two-bladed horizontal axis machines in the US and Germany, and rotors with tip-vanes in the Netherlands. Yet, the winning technology was the Danish three-bladed, horizontal axis machine, starting at a small unit size of 25-50 kW. The reason for this was primarily that the initial design was a scaled-down version of the 1957 fixed pitch Gedser turbine with blade stall control, gearbox and asynchronous generator, i.e. an already proven design, and that the further development was for a long time taken in very small, safe enlargement steps gradually reaching megawatt size, rather than facing the problems of a totally new design in one jump (cf. Figs. 4.89 and 4.90).

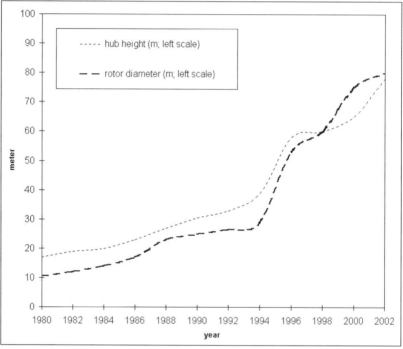

Figure 4.89. Development in average height and rotor diameter of Danish wind turbines. For recent years, turbines typically used in new large wind parks are represented, rather than an average of all turbines sold, which would be less indicative of trends, due to shifting policies for replacement of old wind turbines on land (adapted from Danish wind turbine Price Lists and Overviews, 1981–2002).

Scaling up means controlling the most critical components, such as the rotor blades that must be able to survive the forces on long pieces of suitable material, experiencing wind loads with gust effects that can cause the blades to bend in different directions and exert large forces on the fastening point of the blades. Use of increasingly sophisticated materials and component dimensioning based on both calculation and experience has allowed the scaling up to proceed smoothly, and it is only in recent years that major design principles have been altered, e.g. doing away with the gearbox, using multipole electric generators operating synchronously and employing variable speed AC to DC to constant frequency a.c. electronic conversions. Also pitch change of blade angles has been introduced and thereby new options for operation or shutdown in high winds. The current production of units of a few megawatt machines is spread over a number of different ones among the design principles mentioned. Interesting questions are being asked about the possibility of operating still longer blades. Presumably, this will at some point mean a transition from the current blades made of glass or plant fibre to advanced coal fibres. It should be considered that the operating life of the blades aimed for is equal to the lifetime of the entire turbine, in contrast, e.g., to fast rotating helicopter rotor blades that have to be replaced at short intervals.

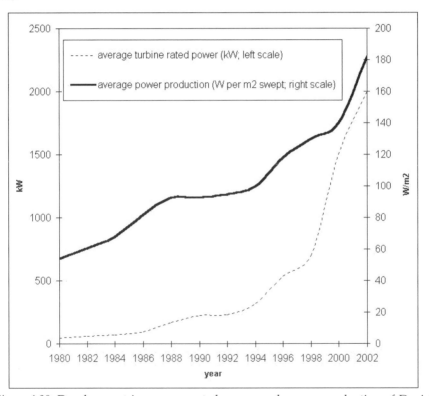

Figure 4.90. Development in average rated power and power production of Danish wind turbines. For recent years, turbines typically installed in new large wind parks are represented, rather than an average based on all sales, which would be less indicative of trends, due to shifting overseas markets and Danish policies for replacement of old wind turbines on land (adapted from Danish Windpower Industry Association, 2003; Danish wind turbine Price Lists and Overviews, 1981–2002).

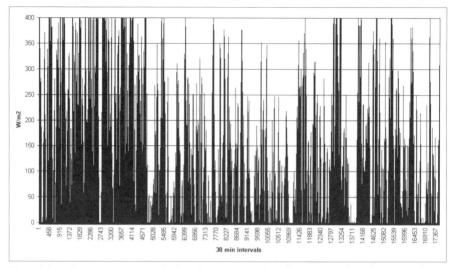

Figure 4.91. Power output calculated for a wind turbine placed at Stigsnæs (near Næstved on inner-Baltic coast of Denmark), based on wind speeds measured during 1995 (Risø, 1999) and a turbine power curve rising from 0 to 400 W m^{-2} between wind speeds of 5–12.27 m s^{-1} and then staying at 400 W m^{-2} until 25 m s^{-1} (typical of the behaviour of actual 2 MW Danish turbines; cf. Petersen, 2001).

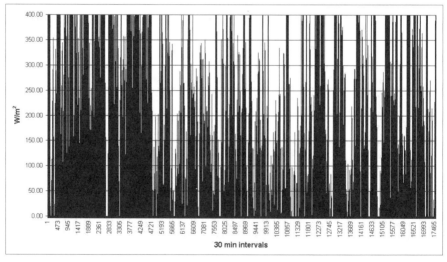

Figure 4.92. Power output from wind turbine placed at Vindeby (off-shore North of Lolland, in the inner-Baltic coast of Denmark slightly South of Stigsnæs), based on wind speeds measured during 1995 and the same turbine power curve as used in Fig. 4.91. Some missing data sequences (of length from few hours to about a week) have been replaced by measured sequences for the same hours of 1994.

Figure 4.89 shows that a major technology shift in terms of increased blade length took place around 1994. In Denmark, this reflects the transition from mainly individually placed wind turbines to utility owned wind "parks" or "farms". Individual owners, mainly farmers, had carried the development of wind power technology to maturity with essentially no help from the established power utility companies. This may be the chief explanation for the early high penetration of wind energy in Denmark, expected to pass 20% in 2003. Fig. 4.90 indicates that a significant increase in turbine rated power has appeared since 1998, but that the average production did not increase nearly as much as the rated power. This is the period of starting to go offshore and may reflect that the distribution of wind speeds over water was not sufficiently known when the machines rated at 1–2 MW were designed. Earlier machines were pretty well optimised to give the maximum energy production over the year, when placed at favourable land sites.

The design optimisation is not a trivial question, as it involves considerations of average energy produced, number of hours operating, and maximum loads exerted upon the structure. Figures 4.91 and 4.92 show the calculated power output at on-shore and off-shore locations in the inner Danish region, indicating that the imposed power ceiling is reached more often offshore. This situation is expected to occur more frequently at the off-shore locations in the North Sea, such as that of the 160-MW Hornsrev wind farm completed in 2002. It suggests that either the choice of blade profiles, the sizing of the power generator, or the regulation strategies used in current megawatt-size machines could be improved to capture more of the potential power in the wind for turbines placed at the reserved Danish off-shore sites, at least during periods of high winds. Deficiencies in aerodynamic behaviour can be determined by repeating the calculations with different power curves, but it is already clear that at most off-shore sites, just a larger generator would increase the annual energy output.

The model calculation assumes a constant C_p at high wind speeds, up to a maximum wind speed of 25 m s^{-1} where the turbine is brought to a stop. It seems that the manufacturers may have let their designs be influenced by complaints from the power companies, that large surpluses were hard to accept, because the auction-based selling schemes used in the newly established liberalised power markets have a tendency to value such surpluses very lowly. On the other hand, the turbine owner can boast of a high power factor, i.e. the ratio of average power production to constant production at full turbine rated power. For the 2-MW Vestas turbines used at Hornsrev, the power factor is 0.54. The actual efficiency of this turbine (power produced over power in the wind) is shown in Fig. 4.93, and again in Fig. 4.94, folded with the frequency of wind speeds. It is seen that the efficiency is high at frequently occurring wind speeds below 9 m s^{-1}, and the weighted efficiency (the integral under the curve) is a decent 0.34. However, the discontinuity in the curve and rapid fall-off for wind speeds over 9 m s^{-1} indicate that improved design could greatly increase annual production by better exploiting wind at speeds 9-15 m s^{-1}, which might be done without serious rotor load problems for blades anyway allowed to operate at up to 25 m s^{-1} winds. It is clear that if the electricity system of a given region can be made to accept larger variations in power production, then the turbine design and regulation may thus be changed in directions that immediately would increase energy production and significantly lower the cost of each kilowatt-hour of energy produced. At present, wind turbine manufacturers insist on offering the same design for use in all wind speed regimes, except for a possible choice of tower height and, for manufacturers interested in arctic markets, some reinforcements for operation in high peak-wind speed, icing-prone environments.

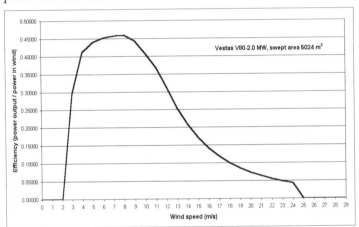

Figure 4.93. Efficiency as a function of wind speed for commercial 2-MW wind turbine (cf. Petersen, 2001; Bonefeld *et al.*, 2002).

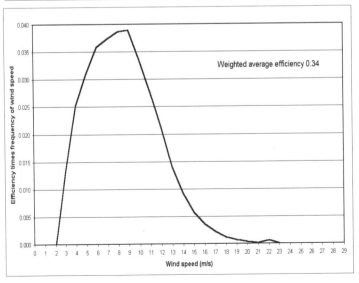

Figure 4.94. The efficiency curve of the commercial wind turbine shown in Fig. 4.93 has here been folded with the wind speed frequency distribution at Hornsrev, site of a 160-MW Danish offshore wind farm (based on data from Bonefeld *et al.*, 2002).

Current wind turbine unit size is around 2 MW, but prototypes of 4–5 MW have been developed. The economic breakdown of current costs are estimated in Table 4.2. There are, of course, variations of installation and grid connection costs depending on location, as there are variations in the cost of the electric energy produced, depending on the actual average power production at the location of the turbines and on financing conditions. Details of the treatment of payments occurring at different times and of interest rates are discussed further in Chapter 7.

Installed cost	On-shore	Off-shore	Unit
Turbines	0.77	0.77	2000-euro/W_{rated}
Foundations	0.06	0.35	2000-euro/W_{rated}
Grid connection[a]	0.15	0.27	2000-euro/W_{rated}
Engineering & administration	0.04	0.04	2000-euro/W_{rated}
Land rights, access roads, facilities	0.09	0.07	2000-euro/W_{rated}
Operation and maintenance (present value of 20 years' cost)	0.25	0.36	2000-euro/W_{rated}
Total capital cost including O&M	**1.36**	**1.86**	**2000-euro/W_{rated}**
Assumed power production	2.6[b]	3.8	kWh/y/W_{rated}
Cost of power produced	**0.035**	**0.033**	**2000-euro/kWh**

Table 4.2. Installation price and cost of energy produced (in year 2000 fixed prices), for wind parks on land or off-shore. Estimates have been made on the basis of current price lists for 2-MW turbines, tender data and trend analysis, with use of Danish wind turbine Price Lists and Overviews (2002), Danish Energy Agency (1999), European Commission (2001). A depreciation time of 20 years and a real interest rate of 3% p.a. (roughly the developed world average taken over the entire 20th century) have been assumed.
[a] The source data on on-shore grid connection costs are relatively low because of fairly short distances to grids of sufficient strength for Danish wind parks used in the studies underlying the value given. Thus, the additional cost of off-shore cables appears more dominant than would be the case in areas less well served with existing grids.
[b] On the very best Danish inland locations, power production may approach 3.5 kWh/y/W_{rated}.

Off-shore foundation and transmission

The possibility of placing wind turbines off-shore, typically in shallow waters of up to 20 m depth, relies on use of low-cost foundation methods developed earlier for harbour and oil-well uses. The best design depends on the material constituting the local water floor, as well as local hydrological conditions, including strength of currents and icing problems. Breakers are currently used to prevent ice from damaging the structure. The presently most common structure in place is a concrete caisson as shown in Fig. 4.95a, but in the interest of cost minimising, the steel solutions shown in Fig. 4.95b,c have received increasing attention and the monopile (Fig. 4.89c) was selected for the recent Samsø wind farm in the middle of Denmark (Birch and Gormsen, 1999; Offshore Windenergy Europe, 2003). Employing the suction effect (Fig. 4.95b) may help cope with short-term gusting forces, while the general stability must be ensured by the properties of the overall structure itself.

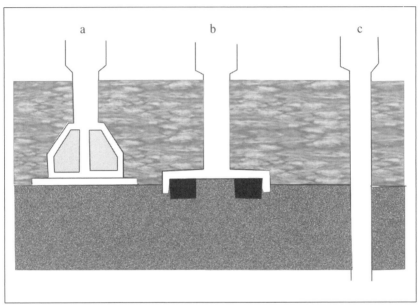

Figure 4.95. Some foundation types for off-shore wind turbine towers placed in shallow waters; a: sand-filled caisson standing on the bottom, b: suction buckets, c: steel monopile. These foundation types are in use at locations with clay-till water floor material.

The power from an off-shore wind farm is transmitted to an on-shore distribution hub by means of one or more undersea cables, the latter providing redundancy that in the case of large farms adds security against cable disruption or similar failures. Current off-shore wind farms use AC cables of up to 150 kV (Eltra, 2003). New installations use cables carrying all three leads plus control wiring. In the interest of loss minimisation for larger installations, it is expected that future systems may accept the higher cost of DC–AC conversion (on shore, the need for AC–DC conversion at sea depends on the generator type used), similar to the technology presently in use for many undersea cable connections between national grids (e.g. between Denmark and Norway or Sweden). Recent development of voltage source-based high voltage direct current control systems to replace the earlier thyristor-based technology promises better means of regulation of the interface of the DC link to the on-shore AC system (Ackermann, 2002).

4.4 Conversion of wave energy

A large number of devices for converting wave energy to shaft power or compression energy have been suggested, and a few of them have been tested on a modest scale. Reviews of wave devices may be found, for example, in Leichman and Scobie (1975), Isaacs *et al.* (1976), Slotta (1976), Clarke (1981), Sørensen (1999a), Thorpe (2001), CRES (2002), and DEA Wave Program (2002). Below, the technical details will be given for two typical examples of actual devices: the oscillating water column device that has been in successful small-scale operation for many years for powering mid-sea buoys, and the Salter duck which theoretically has a very high efficiency, but has not been successful in actual prototyping experiments. However, first some general remarks will be offered.

The resource evaluation in section 3.3.2 indicated that the most promising locations for a wave utilisation apparatus would be in the open ocean rather than in coastal or shallow water regions. Yet all three device types have been researched: (a) shore-fixated devices for concentrating waves into a modestly elevated reservoir, from which the water may drive conventional hydro-turbines; (b) near-shore devices making use of the oscillations of a water column or a float on top of the waves, relative to a structure standing at the sea floor; and (c)floating devices capturing energy by differential movements of different parts of the device.

As a first orientation towards the prospects of developing economically viable wave power devices, a comparison with (e.g. off-shore) wind power may be instructive. One may first look at the weight of the construction relative to its rated power. For an on-shore wind turbine, this number is around 0.1 kg/W_{rated}, while adding the extra foundation weight for off-shore turbines (except caisson in-fill) increases the number to just below 0.2 kg/W_{rated}. For 15 wave power devices studied by the DEA Wave Program (2002), the range of weight to rated power ratios is from 0.4 to 15 kg/W_{rated}. The two numbers below 1.0 are for device concepts not tested and for use on shore or at low water depth, where the power resources are small anyway. So the conclusion is that the weight/power ratio is likely to be at least 2 but likely more than 5 times that of off-shore wind turbines, which to a first approximation is also a statement on the relative cost of the two concepts.

Using the same data, one may instead look at the ratio of actually produced power at a particular location and the weight of the device. For off-shore wind in the North Sea, this is around 20 kWh y^{-1} kg^{-1}. For the 15 wave devices, values of 0.1 to 10 are found, by using for all devices the same wave data estimated for a location some 150 km west of the city Esbjerg in Denmark[*]. Clearly, it is not reasonable to use data for a location 150 km out into the ocean for wave power devices that must stand on the shore or at very shallow water. Omitting these cases, the resulting range reduces to 0.1-1.5 kWh y^{-1} kg^{-1} or over 13 times less that for off-shore wind in the same region. Again, the simplistic translation from weight to cost indicates that wave energy is economically unattractive, because at the same location wind power can be extracted at much lower cost. Also, there are no particular reasons to expect the distribution of weight on less expensive materials (concrete or steel) and more expensive materials (special mechanical and electric equipment) to be substantially different. It is part of this argument, that where wave power would be feasible, off-shore wind power is also available, because it is the wind that creates the waves. Only at large water depths, say, over 20 m depth, where foundation would be problematic, might the wave devices floating on the surface be more attractive than wind. Yet there are not from the above any indications that the cost of such mid-ocean wave power extraction and cable transmission to land will be economically viable, unless all near-shore wind options have already been exploited.

Finally, there is the argument of the time-distribution of the power from wave devices. As seen from Figs. 3.46 and 3.49, wave power exhibits large variations with seasons. Also in the North Sea, variations are large: Rambøll (1999) finds 6 times more average wave power in January than in June, where the corresponding factor for wind power is 2 (Sørensen, 2000a). Acceptance of wave power into grids serving electricity demands is thus going to be considerably more difficult than acceptance of wind, which roughly has the same seasonal variation as demand, at least on the Northern Hemisphere. Thus, in addition to an initial cost likely to be substantially higher than that of wind power, additional costs for energy storage or other supply-demand mismatch management must be considered.

4.4.1 Pneumatic converter

The only wave energy conversion device that has been in practical use, although on a fairly small scale, is the buoy of Masuda (1971), shown in schematic form in Fig. 4.96. Several similar wave power devices exist, based on an oscillating water column driving an air turbine, in some cases shore based and with only one chamber (see e.g. CRES, 2002). The buoy in Fig. 4.96 contains a tube extending downwards, into which water can enter from the

---

[*] The wave data used are calculated from the 10-m wind data shown in Figs. 2.79–2.80 using a fetch model (Rambøll, 1999). As mentioned in Chapters 2 and 6, the 10 m data are not very reliable, and the grid is coarse, but comparison with the actual measurements at Vyl a few kilometres to the west (shown in Fig. 3.45) indicates that the error is well under 50% (however, the calculated data does not, for example, exhibit the two peaks shown in Fig. 3.45, but only one).

bottom, and a "double-action" air turbine, i.e. a turbine that turns in the same way under the influence of pressure and suction (as illustrated by the non-return valves).

The wave motion will cause the whole buoy to move up and down and thereby create an up-and-down motion of the water level in the centre tube, which in turn produces the pressure increases and decreases that make the turbine in the upper air chamber rotate. Special nozzles may be added in order to increase the speed of the air impinging on the turbine blades.

For a simple sinusoidal wave motion, as described in section 2.D, but omitting the viscosity-dependent exponential factor, the variables σ_1 and Z describing the water level in the centre tube and the vertical displacement of the entire buoy, respectively, may also be expected to vary sinusoidally, but with phase delays and different amplitudes,

$$\sigma = a \cos (kx - \omega t),$$
$$\sigma_1 = \sigma_{10} \cos (kx - \omega t - \delta_\sigma),$$
$$Z = Z_0 \cos (kx - \omega t - \delta_Z),$$

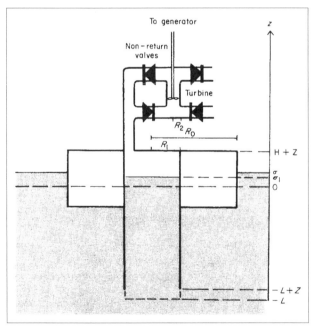

Figure 4.96. Masuda's pneumatic wave energy conversion device.

with $\omega = kU_w$ in terms of the wave number k and phase velocity U_w (cf. section 2.D). The relative air displacement ρ_1 in the upper part of the centre tube (the area of which is $A_1 = \pi R_1^2$) is now (see Fig. 4.96)

$$\rho_1 = \sigma_1 - Z.$$

Assuming the air flow to be incompressible, continuity requires that the relative air velocity $d\rho_2/dt$ in the upper tubes forming the air chamber (of area $A_2 = \pi R_2^2$) be

$$d\rho_2/dt = (A_1/A_2)\, d\rho_1/dt.$$

It is now possible to set up the equations of motion for σ_1 and Z, equating the accelerations $d^2\sigma_1/dt^2$ and d^2Z/dt^2 multiplied by the appropriate masses (of the water column in the centre tube and of the entire device) to the sum of forces acting on the mass in question. These are buoyancy and pressure forces, as well as friction forces. The air displacement variable ρ_2 satisfies a similar equation involving the turbine blade reaction forces. The equations are coupled, but some of the mutual interaction can be incorporated as damping terms in each equation of motion. McCormick (1976) uses linear damping terms $-b(d\sigma_1/dt)$, etc., with empirically chosen damping constants, and drops other coupling terms, so that the determination of σ_1 and Z becomes independent. Yet the $d\rho_2/dt$ values determined from such solutions are in good agreement with measured ones, for small wave amplitudes a (or significant heights $H_S = 2 \times 2^{1/2}a$; cf. section 3.3.1) and wave periods $T = 2\pi/\omega$ which are rather close to the resonant values of the entire buoy, T_0, or to that of the centre tube water column, T_1. Between these resonant periods the agreement is less good, probably indicating that couplings are stronger between the resonances than near them, as might be expected.

The power transferred to the turbine shaft cannot be calculated from the expression (4.12) because the flow is not steady. The mass flows into and out of the converter vary with time, and at a given time they are not equal. However, owing to the assumption that the air is incompressible, there will be no build-up of a compressed air energy storage inside the air chamber, and therefore, the power can still be obtained as a product of a mass flux, i.e. that to the turbine,

$$J_m = \rho_a A_1\, d\rho_1/dt$$

(ρ_a being the density of air), and a specific energy change Δw (neglecting heat dissipation). In addition to the terms (4.34)–(4.36) describing potential energy changes (unimportant for the air chamber), kinetic energy changes and pressure/enthalpy changes, Δw will contain a term depending on the time variation of the air velocity $d\rho_2/dt$ at the turbine entrance,

$$\Delta w \approx \tfrac{1}{2}\left(\left(\frac{d\rho_1}{dt}\right)^2 - \left(\frac{d\rho_2}{dt}\right)^2\right) + \frac{\Delta P}{\rho_a} - \int\left(\frac{d^2\rho_2}{dt^2}\right)d\rho,$$

where ΔP is the difference between the air pressure in the air chamber (and upper centre tube, since these are assumed equal) and in the free outside air, and ρ is the co-ordinate along a streamline through the air chamber

(McCormick, 1976). Internal losses in the turbine have not been taken into account.

Figure 4.97 shows an example of the simple efficiency of conversion, η, given by the ratio of $J_m \Delta w$ and the power in the incident waves, which for sinusoidal waves is given by (3.28). Since the expression (3.28) is the power per unit length (perpendicular to the direction of propagation, i.e. power integrated over depth z), it should be multiplied by a dimension of the device which characterises the width of wave acceptance. As the entire buoy is set into vertical motion by the action of the waves, it is natural to take the overall diameter $2R_0$ (cf. Fig. 4.96) as this parameter, such that

$$\eta = 8\pi J_m \Delta w/(2R_0 \, \rho_w g^2 T a^2) = 32\pi J_m \Delta w/(\rho_w \, g^2 T H_S^2 R_0). \tag{4.144}$$

The efficiency curves as functions of the period T, for selected significant wave heights H_S, which are shown in Fig. 4.97, are based on a calculation by McCormick (1976) for a 3650-kg buoy with overall diameter $2R_0 = 2.44$ m and centre tube diameter $2R_1 = 0.61$ m. For fixed H_S, η has two maxima, one just below the resonant period of the entire buoy in the absence of damping terms, $T_0 \approx 3$ s, and one near the air chamber resonant period, $T_1 \approx 5$ s. This implies that the device may have to be "tuned" to the periods in the wave spectra giving the largest power contributions (cf. Fig. 3.46), and that the efficiency will be poor in certain other intervals of T. According to the calculation based on solving the equations outlined above and with the assumptions made, the efficiency η becomes directly proportional to H_S. This can only be true for small amplitudes, and Fig. 4.97 shows that unit efficiency would be obtained for $H_S \approx 1.2$ m from this model, implying that non-linear terms have to be kept in the model if the power obtained for waves of such heights is to be correctly calculated. McCormick (1976) also refers to experiments, which suggest that the power actually obtained is smaller than that calculated.

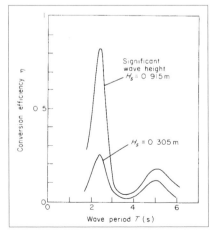

Figure 4.97. Efficiency of a pneumatic wave energy converter as function of wave period, calculated for two wave amplitudes (based on McCormick, 1976).

Masuda has proposed placing a ring of pneumatic converters (floating on water of modest depth) such that the diameter of the ring is larger than the wavelength of the waves of interest. He claims that the transmission of wave power through such a ring barrier is small, indicating that most of the wave energy will be trapped (e.g. multiply reflected on the interior barriers of the ring) and thus eventually be absorbed by the buoy and internal fluid motion in the centre tubes. In other developments of this concept, the converters are moored to the sea floor or leaning towards a cliff on the shore (Clarke, 1981; Vindeløv, 1994; Thorpe, 2001).

4.4.2 Oscillating vane converter

Since water particles in a sinusoidal wave are moving in circular orbits, it may be expected that complete absorption of the wave energy is only possible with a device possessing two degrees of freedom, i.e. one allowing displacement in both vertical and horizontal directions. Indeed, it has been shown by Ogilvie (1963) that an immersed cylinder of a suitable radius, moving in a circular orbit around a fixed axis (parallel to its own axis) with a certain frequency, is capable of completely absorbing the energy of a wave incident perpendicular to the cylinder axis, i.e. with zero transmission and no wave motion "behind" the cylinder.

Complete absorption is equivalent to energy conversion at 100% efficiency, and Evans (1976) has shown that this is also possible for a half-immersed cylinder performing forced oscillations in two dimensions, around an equilibrium position, whereas the maximum efficiency is 50% for a device which is only capable of performing oscillations in one dimension.

The system may be described in terms of coupled equations of motion for the displacement co-ordinates X_i ($i = 1, 2$),

$$m \, \mathrm{d}^2 X_i / \mathrm{d}t^2 = -d_i \, \mathrm{d}X_i / \mathrm{d}t - k_i X_i + \sum_j F_{ij},$$

where m is the mass of the device, d_i is the damping coefficient in the ith direction, k_i correspondingly is the restoring force or "spring constant" (the vertical component of which may include the buoyancy forces), and F_{ij} is a matrix representing the complete hydrodynamic force on the body. It depends on the incident wave field and on the time derivatives of the co-ordinates X_i, the non-diagonal terms representing couplings of the motion in vertical and horizontal directions.

The average power absorbed over a wave period T is

$$E = \sum_i T^{-1} \int_0^T \frac{\mathrm{d}X_i}{\mathrm{d}t} \sum_j F_{ij} \, \mathrm{d}t, \tag{4.145}$$

and the simple efficiency η is obtained by inserting each of the force components in (4.145) as forces per unit length of cylinder, and dividing by the power in the waves (3.28). One may regard the following as design parameters: the damping parameters d_i, which depend on the power extraction system; the spring constants k_i and the mass distribution inside the cylinder, which determine the moments and hence influence the forces F_{ij}. There is one value of each d_i for a given radius R of the cylinder and a given wave period, which must be chosen as a necessary condition for obtaining maximum efficiency. A second condition involves k_i, and if both can be fulfilled the device is said to be "tuned" and it will be able to reach the maximum efficiency (0.5 or 1, for oscillations in one or two dimensions). Generally, the second condition can be fulfilled only in a certain interval of the ratio R/λ between the radius of the device and the wavelength $\lambda = gT^2/(2\pi)$.

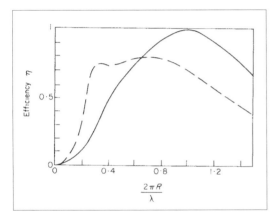

Figure 4.98. Efficiency of wave energy absorption by oscillating motion of a half-immersed cylinder of radius R. The calculation assumes parameters describing damping and coupling terms, such that the device becomes fully tuned at $2\pi R/\lambda = 1$ (solid line) or partially tuned at $2\pi R/\lambda = 0.3$ (dashed line) (based on Evans, 1976).

Figure 4.98 shows the efficiency as a function of $2\pi R/\lambda$ for a half-immersed cylinder tuned at $2\pi R/\lambda_0 = 1$ (full line) and for a partially tuned case, in which the choice of damping coefficient corresponds to tuning at $2\pi\cdot R/\lambda_0 = 0.3$, but where the second requirement is not fulfilled. It is seen that such "partial tuning" offers a possibility of widening the range of wave periods accepted by the device.

A system whose performance may approach the theoretical maximum for the cylinder with two degrees of freedom is the oscillating vane or cam of Salter (1974), illustrated in Fig. 4.99. Several structures of this cross section indicated are supposed to be mounted on a common backbone providing the fixed axis around which the vanes can oscillate, with different vanes not necessarily moving in phase. The backbone has to be long in order to provide an approximate inertial frame, relative to which the oscillations can be utilised for power extraction, for example, by having the rocking motion create pressure pulses in a suitable fluid (contained in compression chambers between the backbone and oscillating structure). The necessity of such a backbone is also the weakness of the system, owing to the large bending forces along the structure, which must be accepted during storms.

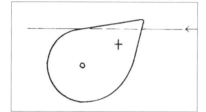

Figure 4.99. Salter's design of an oscillating vane wave energy converter. The cross section shows the backbone and axis of rotation as a small circle and the location of the centre of gravity as a cross. The waves are incident from the right-hand side.

Figure 4.100. Efficiency of Salter's device based on model experiments (solid line) and extrapolations (dashed line) (based on Mollison *et al.*, 1976). See also the discussion topics 6.5.2 and 6.5.3 and Figs. 6.98 and 6.99.

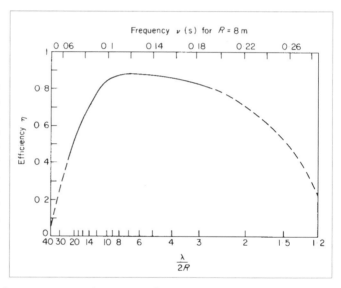

The efficiency of wave power absorption for a Salter "cam" or "duck" of the type shown in Fig. 4.99 is indicated in Fig. 4.100, based on measurements on a model of modest scale ($R = 0.05$ m). Actual devices for use in, say, the good wave energy location in the North Atlantic (cf. Fig. 3.46) would need a radius of 8 m or more if tuned at the same value of $\lambda/(2\pi R)$ as the device used to construct Fig. 4.100 (Mollison *et al.*, 1976). Electricity generators of high efficiency are available for converting the oscillating vane motion (the angular amplitude of which may be of the order of half a radian) to electric power, either directly from the mechanical energy in the rocking motion or via the compression chambers mentioned above, from which the fluid may be let to an electricity-producing turbine. The pulsed compression is not incompatible with a continuous power generation (Korn, 1972). Later experiments have shown that the structural strength needed for the backbone is a problem. For all wave devices contemplated placed in the favourable locations at mid-ocean, transmission to shore is a problem for the economy of the scheme (cf. discussion topics 6.5.2 and 6.5.3).

4.5 Conversion of water flows or elevated water

Electricity generation from water possessing potential, kinetic or pressure energy (4.34)–(4.36) can be achieved by means of a turbine, the general theory of which was outlined in section 4.1.4. The particular design of turbine to be used depends on whether there is a flow J_m through the device, which must be kept constant for continuity reasons, or whether it is possible to obtain zero fluid velocity after the passage through the turbine.

The form of energy at the entrance of the turbine may be kinetic or pressure energy, causing the forces on the turbine blades to be a combination of "impulse" and "reaction" forces, which can be modified at ease. The potential energy of elevated water may be allowed to "fall", i.e. forming kinetic energy, or it may act as a pressure source through a water-filled tube connecting the elevated water source with the turbine placed below. Conversely, pressure energy may be transformed into kinetic energy by passage through a nozzle.

Typical classical turbine designs are illustrated in Fig. 4.101. For high specific energy differences $w_{in}-w_{out}$ (large "heads"), the Pelton turbine, which is a high-speed variant of the simple undershot waterwheel, may be used. It has the inflow through a nozzle, providing purely kinetic energy, and negligible w_{out} (if the reference point for potential energy is taken to correspond to the water level after passing the turbine). Also, the Francis turbine (Fig. 4.101b) is used with large water heads. Here the water is allowed to approach the entire rotor, guided to obtain optimum angles of attack, and the rotor moves owing to the reaction forces resulting from both the excess pressure at entrance and the suction at exit.

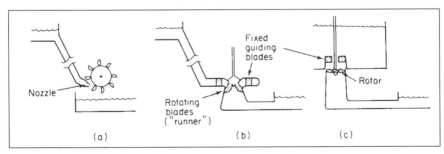

Figure 4.101. Pelton (a), Francis (b) and Kaplan (c) water turbines.

The third type of turbine, illustrated in Fig. 4.101c, can be used for low water heads. Here the rotor is a propeller, designed to obtain high angular speeds. Again, the angle of attack may be optimised by installation of guiding blades at the entrance to the rotor region. If the blade pitch angle is fixed, it is called a Nagler turbine. If it can be varied, it is called a Kaplan turbine.

Figure 4.102 gives examples of actual efficiencies for the types of turbines described above (Fabritz, 1954) as functions of the power level. The design point for these turbines is about 90% of the rated power (which is set to 100% on the figure), so the power levels below this point correspond to situations in which the water head is insufficient to provide the design power level.

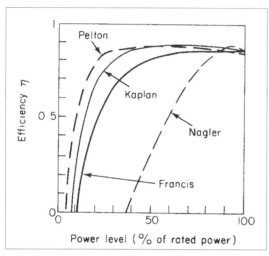

Figure 4.102. Efficiency of water turbines as a function of power level (based on Fabritz, 1954).

Pelton and Francis turbines have been used in connection with river flows with rapid descent, including waterfalls, and in many cases it has been possible by building dams to provide a steady energy source throughout most of the year. This implies storing the water at an elevated level in natural or artificial reservoirs and letting the water descend to the turbines only when needed. An example of natural variations inviting regulation was shown in Fig. 3.59. The operation of systems that include several sources of flow (springs, glacier or snow melt, rainy regions), several reservoirs of given capacity, a given number of turbine power stations and a load of electricity usage which also varies with time, has been studied and optimised by simulation techniques (see e.g. Jamshidi and Mohseni, 1976).

For many years, hydropower has been the most widely used renewable source of electricity and also – among all types of power plants including fossil and nuclear – the technology inviting the largest schemes of power plants rated at several gigawatts and involving often artificial water reservoirs of gigantic size. The development of such schemes with disregards of social and environmental problems has given hydropower a negative reputation. In developing countries thousands of people have been forcefully removed from their homes with no compensation to give way for flooded reservoirs, causing monumental environmental disruption (this not being re-stricted to developing countries, as the examples of Tasmania or Norway show) and in some cases destroying priceless sites of archaeological value (e.g. Turkey). In recent decades, it has become clear that in many cases there exist ways of minimising the environmental damage, albeit at a higher cost. The discussion in section 3.4.2 mentioned Swiss efforts to use cascading sys-tems to do away with any large reservoirs, accepting that the smaller reser-voirs located along the flow of water and designed to minimise local impacts would provide somewhat less regulation latitude. Full consideration of these concerns is today in most societies a prerequisite for considering hy-dropower as a benign energy source.

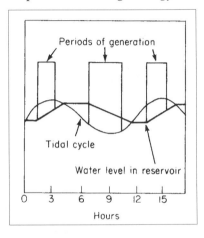

Figure 4.103. Operation of the tidal power plant at la Rance (based on André, 1976).

Kaplan (or Nagler) turbines are used in connection with low water heads (e.g. the local community power plants in China, cf. *China Reconstructs*, 1975) and tidal plants (André, 1976), and they may be used if ocean currents are to be exploited. The 240-MW tidal plant at la Rance in France has turbines placed in a dam structure across an inlet, which serves as a reservoir for two-way operation of the turbines (filling and emptying of the reservoir). Ac-cording to André (1976), the turbine efficiency is over 90%, but the turbines are only generating electricity during part of the day, according to the scheme outlined in Fig. 4.103. The times at which generation is possible are determined by the tidal cycle, according to the simple scheme of operation, but modifications to suit the load variations better are possible, e.g. by using the turbines to pump water into the reservoir at times when this would not occur as a result of the tidal cycle itself. The installation has had several problems due to siltation, causing it to be operated most of the time only for water flows in one direction, despite its design to accept water inflow from both sides.

4.6 Conversion of heat

4.6.1 Application to heating

In some cases, it is possible to produce heat at precisely the temperature needed by primary conversion. However, often the initial temperature is higher than required, even considering losses in transmission and heat drop across heat exchangers of the type shown in Fig. 4.20. In such cases, appropriate temperatures are commonly obtained by mixing (if the heat is stored as sensible heat in a fluid such as water, this water may be mixed with colder water, a procedure often used in connection with fossil fuel burners). This procedure is wasteful in the sense of the second law of thermodynamics, since the energy is, in the first place, produced with a higher quality than subsequently needed. In other words, the second law efficiency of conversion (4.20) is low, because there will be other schemes of conversion by which the primary energy can be made to produce a larger quantity of heat at the temperature needed at load. An extreme case of a "detour" is the conversion of heat to heat by first generating electricity by thermal conversion (cf. section 4.6.2) and then degrading the electricity to heat of low temperature by passing a current through an ohmic resistance ("electric heaters").

Heat pumps

If heat of modest temperature is required, and a high-quality form of energy is available, some device is needed which can derive additional benefits from the high quality of the primary energy source. This can be achieved by using one of the thermodynamic cycles described in section 4.1.2, provided that a large reservoir of approximately constant temperature is available. The cycles (cf. Fig. 4.3) must be traversed "anti-clockwise", such that high-quality energy (electricity, mechanical shaft power, fuel combustion at high temperature, etc.) is added, and heat energy thereby delivered at a temperature T higher than the temperature T_{ref} of the reference reservoir from which it is drawn. Most commonly the Rankine cycle, also described in section 4.1.2 and with a maximum efficiency bounded by (4.22), is used (e.g. in an arrangement of the type shown in Fig. 4.104). The fluid of the closed cycle, which should have a liquid and a gaseous phase in the temperature interval of interest, may be a fluorochloromethane compound (which needs to be recycled owing to climatic effects caused if it is released to the atmosphere). The external circuits may contain an inexpensive fluid (e.g. water), and they may be omitted if it is practical to circulate the primary working fluid directly to the load area or to the reference reservoir.

The heat pump contains a compressor, which performs step 7-5 in the Rankine cycle depicted in Fig. 4.3, and a nozzle, which performs step 2-9. The intermediate steps are performed in the two heat exchangers, giving the working fluid the temperatures T_{up} and T_{low}, respectively. The equations for determining these temperatures are of the form (4.99). There are four such equations, which must be supplemented by equations for the compressor and nozzle performance, in order to allow a determination of all the unknown temperatures indicated in Fig. 4.104, for given T_{ref}, given load and a certain energy expenditure to the compressor. Losses in the compressor are in the form of heat, which in some cases can be credited to the load area.

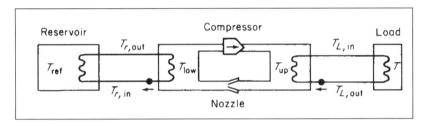

Figure 4.104. Schematic picture of a heat pump.

An indication of the departures from the Carnot limit of the "coefficients of performance", $\varepsilon^{heat\ pump}$, encountered in practice, is given in Fig. 4.105, as a function of the temperature difference $T_{up}-T_{low}$ at the heat pump and for selected values of T_{up}. In the interval of temperature differences covered, the $\varepsilon^{heat\ pump}$ is about 50% of the Carnot limit (4.22), but it falls more and more below the Carnot value as the temperature difference decreases, although the absolute value of the coefficient of performance increases.

Several possibilities exist for the choice of the reference reservoir. Systems in use for space heating or space cooling (achieved by reversing the flow in the compressor and expansion-nozzle circuit, cf. section 4.1.2) have utilised river, lake and sea water, and air, as well as soil as reservoirs. The temperatures of such reservoirs are not entirely constant, and it must therefore be acceptable that the performance of the heat pump systems will vary with time. Such variations are damped if water or soil reservoirs at sufficient depth are used, as seen from Fig. 3.66 and the discussion in section 2.B. Alternative types of reservoirs for use with heat pumps are city waste sites, livestock manure, ventilation air from households or from livestock barns (where the rate of air exchange has to be particularly high), and heat storage tanks connected to solar collectors, etc.

In connection with solar heating systems, the heat pump may be connected between the heat store and the load area (whenever the storage temperature is too low for direct circulation), or it may be connected between the heat store and the collector, such that the fluid let into the solar collector is cooled in order to improve the collector performance. Of course, a heat pump operating on its own reservoir (soil, air, etc.) may also provide the auxiliary heat for a solar heating system of capacity below the demand.

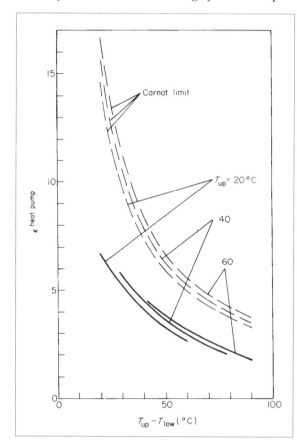

Figure 4.105. Measured coefficient of performance, $\varepsilon^{heat\ pump}$, for a heat pump with a semi-hermetic piston-type compressor (solid lines, based on Trenkowitz, 1969), and corresponding curves for ideal Carnot cycles.

The high-quality energy input to the compressor of a heat pump may also come from a renewable resource, e.g. by wind or solar energy conversion, either directly or via a utility grid carrying electricity generated by renewable energy resources. As for insulation materials, concern has been expressed over the use of CFC gases in the processing or as a working fluid, and substitutes believed to have less negative impacts have been developed.

4.6.2 Conversion of heat into work or electricity

The conversion of heat to shaft power or electricity is achieved by one of the thermodynamic cycles, examples of which are shown in Fig. 4.3. The cycles may be closed as in Fig. 4.3, or they may be "open", in that the working fluid is not recycled through the cooling step (4-1 in most of the cycles shown in Fig. 4.3). Instead, new fluid is added for the heating or compression stage, and "used" fluid is rejected after the expansion stage.

It should be kept in mind that the thermodynamic cycles convert heat of temperature T into work plus some residual heat of temperature above the reference temperature T_{ref} (in the form of heated cooling fluid or rejected working fluid). Emphasis should therefore be placed on utilising both the work and the "waste heat". This is done, for example, by co-generation of electricity and water for district heating.

Conversion of heat from solar collectors

Examples of the use of thermodynamic cycles in the conversion of heat derived from solar collectors into work have been given in Figs. 4.28–4.30 and 4.32. The dependence of the limiting Carnot efficiency on temperature is shown in Fig. 4.34 for selected values of a parameter describing the concentrating ability of the collector and its short-wavelength absorption to long-wavelength emission ratio. The devices shown in Figs. 4.29 and 4.30 aim at converting solar heat into mechanical work for water pumping, while the device in Fig. 4.32 converts heat from a solar concentrator into electricity.

Ericsson hot-air engine

The engines in these examples were based on the Rankine or the Stirling cycle. It is also possible that the Ericsson cycle (which was actually invented for the purpose of solar energy conversion) will prove advantageous in some solar energy applications. It is based on a gas (usually air) as a working fluid and may have the layout shown in Fig. 4.106. In order to describe the cycle depicted in Fig. 4.3, the valves must be closed at definite times and the pistons must be detached from the rotating shaft (in contrast to the situation shown in Fig. 4.106), such that the heat may be supplied at constant volume. In a different mode of operation, the valves open and close as soon as a pressure difference between the air on the two sides begins to develop. In this case, the heat is added at constant pressure, as in the Brayton cycle, and the piston movement is approximately at constant temperature, as in the Sterling cycle (this variant is not shown in Fig. 4.3).

The efficiency can easily be calculated for the latter version of the Ericsson cycle, for which the temperatures T_{up} and T_{low} in the compression and expansion piston cylinders are constant, in a steady situation. This implies that the air enters the heating chamber with temperature T_{low} and leaves it with temperature T_{up}. The heat exchanger equations [corresponding to (4.98), but without the assumption that T_3 is constant] take the form

$$-J_m^f\, C_p^f\, dT^f(x)/dx = h'\, (T^f(x) - T^g(x)),$$

$$J_m^g\, C_p^g\, dT^g(x)/dx = h'\, (T^f(x) - T^g(x)),$$

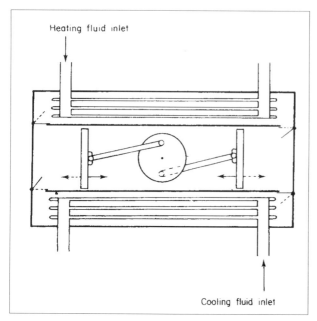

Figure 4.106. Example of an Ericsson hot-air engine.

Cooling fluid inlet

where the superscript g stands for the gas performing the thermodynamic cycle, f stands for the fluid leading heat to the heating chamber heat exchanger, and x increases from zero at the entrance to the heating chamber to a maximum value at the exit. C_p is a constant-pressure heat capacity per unit mass, and J_m is a mass flow rate. Both these and h', the heat exchange rate per unit length dx, are assumed constant, in which case the equations may be explicitly integrated to give

$$T_{c,in} = T_{c,out} - \frac{J_m^g C_p^g}{J_m^f C_p^f}(T_{up} - T_{low}) = \frac{J_m^g C_p^g (1-H)}{J_m^g C_p^g + J_m^f C_p^f} T_{low} + \left(H + \frac{J_m^f C_p^f (1-H)}{J_m^g C_p^g + J_m^f C_p^f}\right) T_{c,out}.$$

(4.146)

Here $T_{c,out} = T$ is the temperature provided by the solar collector or other heat source, and $T_{c,in}$ is the temperature of the collector fluid when it leaves the heat exchanger of the heating chamber, to be re-circulated to the collector or to a heat storage connected to it. H is given by

$$H = \exp(-h((J_m^f C_p^f)^{-1} + (J_m^g C_p^g)^{-1})),$$

where $h = \int h' \, dx$. Two equations analogous to (4.146) may be written for the heat exchange in the cooling chamber of Fig. 4.106, relating the reject temperature $T_{r,in}$ and the temperature of the coolant at inlet, $T_{r,out} = T_{ref}$ to T_{low} and T_{up}. $T_{c,in}$ may then be eliminated from (4.146) and $T_{r,in}$ from the analogous equation, leaving two equations for determination of T_{up} and T_{low} as functions of known quantities, notably the temperature levels T and T_{ref}. The reason for not having to consider equations for the piston motion in order to determine all relevant temperatures is, of course, that the processes associated with the piston motion have been assumed to be isothermal.

The amounts of heat added, Q_{add}, and rejected, Q_{rej}, per cycle are

$$Q_{add} = mC_p^g (T_{up} - T_{low}) + n\mathscr{R}\, T_{up} \log(V_{max}/V_{min}),$$

$$Q_{rej} = mC_p^g (T_{up} - T_{low}) + n\mathscr{R}\, T_{low} \log(V_{max}/V_{min}) + Q'_{rej},$$

(4.147)

where m is the mass of air involved in the cycle and n is the number of moles of air involved. $\mathscr{R}$ is the gas constant, and V_{min} and V_{max} are the minimum and maximum volumes occupied by the gas during the compression or expansion stages (for simplicity the "compression ratio" V_{max}/V_{min} has been assumed to be the same for the two processes, although they take place in different cylinders). The ideal gas law has been assumed in calculating the relation between heat amount and work in (4.147), and Q'_{rej} represents heat losses not contributing to transfer of heat from working gas to coolant flow (piston friction, etc.). The efficiency is

$$\eta = (Q_{add} - Q_{rej})/Q_{add},$$

and the maximum efficiency which can be obtained with this version of the Ericsson engine is obtained for negligible Q'_{rej} and ideal heat exchangers providing $T_{up} = T$ and $T_{low} = T_{ref}$,

$$\max(\eta) = \left(1 - T_{ref}/T\right)\Bigg/\left(1 + \frac{mC_p^g}{n\mathscr{R}\log(V_{max}/V_{min})}\left(1 - T_{ref}/T\right)\right). \quad (4.148)$$

The ideal Carnot efficiency may even be approached, if the second term in the denominator can be made small (however, to make the compression ratio very large implies an increase in the length of time required per cycle, such that the rate of power production may actually go down, as discussed in section 4.1.1.) The power may be calculated by evaluating (4.147) per unit time instead of per cycle.

Conversion of geothermal heat

Geothermal heat sources have been utilised by means of thermodynamic engines (e.g. Brayton cycles), in cases where the geothermal heat has been in the form of steam (water vapour). In some regions, geothermal sources exist which provide a mixture of water and steam, including suspended soil and rock particles, such that conventional turbines cannot be used. Work has been done on a special "brine screw" that can operate under such conditions (McKay and Sprankle, 1974).

However, in most regions the geothermal resources are in the form of heat-containing rock or sediments, with little possibility of direct use. If an aquifer passes through the region, it may collect heat from the surrounding layers and allow a substantial rate of heat extraction, for example, by drilling two holes from the surface to the aquifer, separated from each other, as indicated in Fig. 4.107a. Hot water (not developing much steam unless the aquifer lies very deep or its temperature is exceptionally high, cf. Fig. 3.68) is pumped or rises by its own pressure to the surface at one hole and is re-injected through a second hole, in a closed cycle, in order to avoid pollution from various undesired chemical substances often contained in the aquifer water. The heat extracted from a heat exchanger may be used directly (e.g. as district heating; cf. Clot, 1977) or may generate electricity through one of the "low-temperature" thermodynamic cycles considered above in connection with solar collectors (Mock *et al.*, 1997).

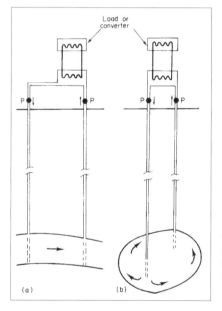

Figure 4.107. Examples of the utilisation of geothermal heat: (a) based on the presence of an aquifer; (b) based on a region of fractured rock.

If no aquifer is present to establish a "heat exchange surface" in the heat-containing rock, it may be feasible to create suitable fractures artificially (by explosives or induced pressure). An arrangement of this type is illustrated in Fig. 4.107b, counting on the fluid which is pumped down through one drilling hole to make its way through the fractured region of rock to the return drilling hole in such a way that continued heat extraction can be sustained. The heat transfer can only be predicted in highly idealised cases (see e.g. Gringarten *et al.*, 1975), which may not be realised as a result of the fairly uncontrolled methods of rock fractionation available.

One important result of the model calculations is that the heat extraction rate deemed necessary for practical applications is often higher than the geothermal flux into the region of extraction, so that the temperature of the extracted heat will be dropping (e.g. by about 1°C per year for one case considered, the number being highly dependent on fracture distribution, rock structure, etc.). This non-sustainable use of geothermal energy is apparent in actual installations in New Zealand and Italy.

Conversion of ocean thermal energy

As discussed in section 3.5.1, downward gradients of temperature exist in most oceans, and they are particularly stable (i.e. without variations with time) in the tropical oceans. The utilisation of such temperature gradients for electricity generation (e.g. by use of a Rankine cycle) has been considered several times since the first suggestions by d'Arsonval (1881).

The temperature differences available over the first 500–1000 m of water depth are only about 25°C, as shown in Fig. 3.63. Considering a closed Rankine cycle, with a working fluid (e.g. ammonia) which evaporates and condenses at convenient temperatures, placed near the ocean surface, it will be required to pump colder water through a pipe from the depth to a heat exchanger for condensation of the working fluid. Further, a warm water heat exchanger is required for evaporating the working fluid. If the heat exchange surface is such that, say, 5°C is "lost" at each heat exchanger, the temperature difference available to the thermodynamic cycle is only 15°C, corresponding to a limiting Carnot efficiency of roughly 0.05. For an actual engine, the efficiency is still lower, and from the power generated should be subtracted the power needed to pump hot and cold water through the heat exchangers and to pump cold water from its original depth to the converter level. It is expected that overall efficiencies around 0.02 may be achieved (cf. e.g. McGowan, 1976).

In order to save energy to pump the hot water through the heat exchanger, it has been suggested that these converters be placed in strong currents such as the Gulf Stream (Heronemus, 1975). The possibility of adverse environmental effects from the power extraction from ocean thermal gradients was touched upon in section 3.5.1. Such dangers may be increased if ocean currents are incorporated into the scheme, because of the possible sensitivity of the itinerary of such currents to small perturbations, and because the dependence of climatic zones on the course of currents such as the Gulf Stream and the Kuro Shio (cf. section 2.3.2).

Open thermodynamic cycles have also been suggested for conversion of ocean thermal energy (Claude, 1930; Beck, 1975; Zener and Fetkovich, 1975), for example, based on creating a rising mixture of water and steam bubbles or "foam", which is separated at a height above sea-level, such that the water can be used to drive a turbine rotor.

If viable systems could be developed for conversion of ocean thermal energy, then there would be a number of other applications of such conversion devices in connection with other heat sources of a temperature little higher than that of the surroundings, especially when such heat sources can be regarded as "free". Examples are the reject or "waste" heat flows from the range of other conversion devices operating at higher initial temperature differences, including fuel-based power plants.

4.7 Conversion of fuels

A number of technologies are available to convert fuels among themselves and into electricity and heat or motive power. Many of them can also be used for fuels derived from renewable energy sources. Those derived from biomass will be dealt with separately in section 4.8. For renewable energy sources such as wind and solar power, one way of dealing with fluctuating production would be to convert surplus electricity to a storable fuel. This could be hydrogen obtained by electrolysis (or by reversible fuel cells, see section 5.2.2), in which case the further conversion of hydrogen to the energy forms in demand can be accomplished by methods already available, for example, for natural gas, such as boilers, engines, gas turbines and fuel cells.

4.7.1 Fuel cell technologies

The idea of converting fuel into electricity by an electrode–electrolyte system originated in the 19th century (Grove, 1839). The basic principle behind a hydrogen–oxygen fuel cell was described in section 4.1.6. The first practical applications were in powering space vehicles, starting during the 1970s.

Developed for stationary power applications, the phosphoric acid cells use porous carbon electrodes with a platinum catalyst and phosphoric acid as electrolyte and feed hydrogen to the negative electrode, with electrode reactions given by (4.72) and (4.73). The operating temperature is in the range 175–200°C, and water is continuously removed.

Alkaline cells use KOH as electrolyte and have electrode reactions of the form

$$H_2 + 2OH^- \rightarrow 2H_2O + 2e^-,$$
$$\tfrac{1}{2} O_2 + H_2O + 2e^- \rightarrow 2OH^-.$$

These cells operate in the temperature range 70–100°C, but specific catalysts require maintenance of fairly narrow temperature regimes. Also, the hydrogen fuel must have a high purity and notably not contain any CO_2. Alkaline fuel cells have been used extensively on spacecraft and recently for road vehicles (Hoffmann, 1998a). Their relative complexity and use of corrosive compounds requiring special care in handling make it unlikely that the cost can be reduced to levels acceptable for general-purpose use.

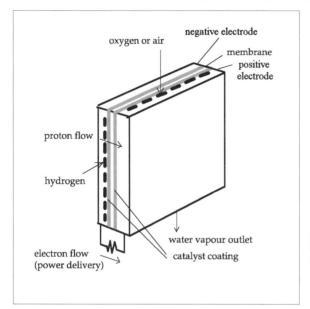

Figure 4.108. Layout of a PEM fuel cell layer, several of which may be stacked.

The third fuel cell type in commercial use is the proton exchange membrane (PEM) cell. It has been developed over a fairly short period of time and is considered to hold the greatest promise for economic application in the transportation sector. It contains a solid polymer membrane sandwiched between two gas diffusion layers and electrodes. The membrane material may be polyperfluorosulphonic acid. A platinum or Pt–Ru alloy catalyst is used to break hydrogen molecules into atoms at the negative electrode, and the hydrogen atoms are then capable of penetrating the membrane and reaching the positive electrode, where they combine with oxygen to form water, again with the help of a platinum catalyst. The electrode reactions are again (4.72) and (4.73), and the operating temperature is 50–100°C (Wurster, 1997). Figure 4.108 shows a typical layout of an individual cell. Several of these are then stacked on top of each other. This modularity implies that PEM fuel cells can be used for applications requiring little power (1 kW or less). PEM cell stacks are dominating the current wealth of demonstration projects in road transportation, portable power and special applications. The efficiency of conversion for the small systems is between 40% and 50%, but a 50 kW system has recently shown an efficiency near 60%. As indicated in Fig. 4.109, an advantage of particular importance for automotive applications is the high efficiency at part loads, which alone gives a factor of two improvement over current internal combustion engines.

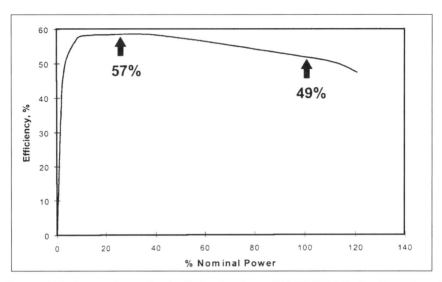

Figure 4.109. Expected part-load efficiencies for a 50-kW PEM fuel cell, projected from measurements involving 10–20 cell test stacks (Patil, 1998).

For use in automobiles, compressed and liquefied hydrogen are limited by low energy density and safety precautions for containers and, first of all, by the requirement of a new infrastructure for fuelling. The hydrogen storage problem, which until recently limited fuel cell projects to large vehicles such as buses, may be solved by use of metal hydride or carbon nanofibre stores (see Chapter 5). In order to avoid having to make large changes to the current gasoline and diesel fuel filling stations, several current schemes use methanol as the fuel distributed to the vehicle fuel tank. The energy density of methanol is 4.4 kWh litre^{-1}, which is half that of gasoline. This is quite acceptable owing to the higher efficiency of conversion. Hydrogen is then formed onboard by a methanol reformer, before being fed to the fuel cell to produce the electric power for an electric motor. The set-up is illustrated in Fig. 4.110. Prototype vehicles with this set-up have recently been tested (cf. Takahashi, 1998; Brown, 1998).

Methanol, CH_3OH, may even be used directly in a fuel cell, without the extra step of reforming to H_2. PEM fuel cells accepting as feedstock a mixture of methanol and water are under development (Bell, 1998). As methanol can be produced from biomass, hydrogen may in this way be eliminated from the energy system. On the other hand, handling of surplus production of wind or photovoltaic power might still conveniently involve hydrogen as an intermediate energy carrier, as it may have direct uses and thus may improve the system efficiency by avoiding the losses in methanol production. The electric power to hydrogen conversion efficiency is about 65% in high-pressure alkaline electrolysis plants (Wagner *et al.*, 1998), and the efficiency of the further hydrogen (plus CO or CO_2 and a catalyst) to methanol conversion is around 70%. Also, the efficiency of producing methanol from biomass is about 45%, whereas higher efficiencies are obtained if the feedstock is methane (natural gas) (Jensen and Sørensen, 1984; Nielsen and Sørensen, 1998).

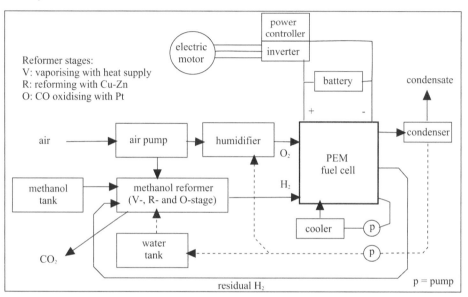

Figure 4.110. Layout of the power system for a methanol-to-hydrogen powered vehicle with fuel cell conversion and an electric motor. The power controller allows shift from direct drive to battery charging.

For stationary applications, fuel cells of higher efficiency may be achieved by processes operating at higher temperatures. One line of research has been molten carbonate fuel cells, with electrode reactions

$$H_2 + CO_3^{2-} \rightarrow CO_2 + H_2O + 2e^-,$$
$$\tfrac{1}{2} O_2 + CO_2 + 2e^- \rightarrow CO_3^{2-}.$$

The electrolyte is a molten alkaline carbonate mixture retained within a porous aluminate matrix. The carbonate ions formed at the positive electrode travel through the matrix and combine with hydrogen at the negative electrode at an operating temperature of about 650°C. The process was originally aimed at hydrogen supplied from coal gasification or natural gas conversion. The first full-scale test (250 kW) is planned to take place at a power utility company in Bielefeld, Germany (Hoffmann, 1998b). The expected conversion efficiency is about 55%, but additional high-temperature heat may be utilised.

Considerable efforts are dedicated to the solid electrolyte cells. Solid oxide fuel cells (SOFC) use zirconia as the electrolyte layer to conduct oxygen ions formed at the positive electrode. Electrode reactions are

$$H_2 + O^{2-} \rightarrow H_2O + 2e^-,$$
$$\tfrac{1}{2} O_2 + 2e^- \rightarrow O^{2-}.$$

The reaction temperature is 700–1000°C. The lower temperatures are desirable, due to lower corrosion problems, and may be achieved by using a thin electrolyte layer (about 10 μm) of yttrium-stabilised zirconia sprayed onto the negative electrode as a ceramic powder (Kahn, 1996). A number of prototype plants (in the 100-kW size range) are in operation. Current conversion efficiency is about 55%, but could reach 70–80% in the future (Hoffmann, 1998b).

Particularly for vehicle applications of hydrogen-based technologies, efforts are needed to ensure a high level of safety in collisions and other accidents. Because of the difference between the physical properties of hydrogen fuel and the hydrocarbon fuels currently in use (higher diffusion speed, flammability and explosivity over wider ranges of mixtures with air), there is a need for new safety-related studies, particularly where hydrogen is stored on board in containers at pressures typically of 20–50 MPa. A few such studies have already been made (Brehm and Mayinger, 1989).

A more detailed discussion of fuel cells can be found in a sequel to this book (Sørensen, 2004a).

4.8 Conversion of biological material

Traditional uses of organic fuels as energy carriers comprise the use of fresh biomass, i.e. storage and combustion of wood fuels, straw and other plant residues, and in recent centuries notably storage and combustion of fossilised biomass, i.e. fuels derived from coal, oil and natural gas. While the energy density of living biomass is in the range of 10–30 MJ per kg of dry weight, the several million years of fossilation processes typically increase the energy density by a factor of two (see Fig. 4.111). The highest energy density is found for oil, where the average density of crude oil is 42 MJ kg^{-1}.

The known reserves of fossil fuels that may be economically extracted today, and estimates of further resources exploitable at higher costs, are indicated in Table 4.3. The finiteness of such resources is, of course, together with the environmental impact issues, the reason for turning to renewable energy sources, including renewable usage of biomass resources.

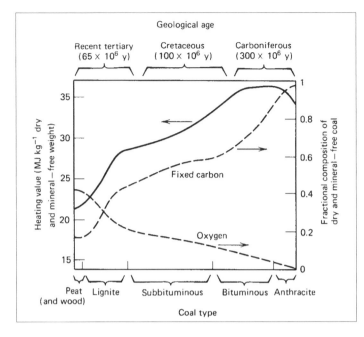

Figure 4.111. Coal to peat classification and selected properties. The dashed lines indicate the fractional content of oxygen and of fixed carbon. The remaining fraction consists of volatile matter (based on US DoE, 1979).

Source	Reserves (EJ)	Resources (> 50% prob.)	Occurrence (speculative)	1990 use (EJ)	Accum. use 1860–1990
Coal				91	5 203
hard coal	15 000	100 000	900 000		
brown coal/lignite	4 000	30 000	90 000		
peat	0	4 000	4 000		
Oil				128	3 343
conventional	5 000	3 000	13 000		
unconventional	7 000	9 000	20 000		
Natural gas				71	1 703
conventional	5 000	5 000	15 000		
unconventional	7 000	20 000	25 000		
in hydrates	0	0	100 000		

Table 4.3. Fossil reserves, resources and consumption in EJ (UN, 1981; Jensen and Sørensen, 1984; Nakicenovic *et al.*, 1996).

However, fresh biomass is a large potential source of renewable energy that in addition to use for combustion may be converted into a number of liquid and gaseous biofuels. This may be achieved by thermochemical or by biochemical methods, as described below.

Biomass is not just stored energy, it is also a store of nutrients and a potential raw material for a number of industries. Therefore, bio-energy is a topic that cannot be separated from food production, timber industries (serving construction purposes, paper and pulp industries, etc.), and organic feedstock-dependent industries (chemical and biochemical industries, notably). Furthermore, biomass is derived from plant growth and animal husbandry, linking the energy aspect to agriculture, livestock, silviculture, aquaculture, and quite generally to the management of the global ecological system. Thus, procuring and utilising organic fuels constitute a profound interference with the natural biosphere, and the understanding of the range of impacts as well as the elimination of unacceptable ones should be an integral part of any scheme for diversion of organic fuels to human society.

4.8.1 Heat production from biomass

Producing heat by burning

Heat may be obtained from biological materials by burning, eventually with the purpose of further conversion. Efficient burning usually requires the reduction of water content, for example, by drying in the Sun. The heat produced by burning cow dung is about 1.5×10^7 J per kg of dry matter, but initially only about 10% is dry matter, so the vaporisation of 9 kg of water implies an energy requirement of 2.2×10^7 J, i.e. the burning process is a net energy producer only if substantial sun-drying is possible. Firewood and other biomass sources constitute stores of energy, since the drying may be performed during summer, such that these biological fuels can be used during winter periods when the heating requirement may be large and the possibility of sun-drying may not exist. The heat produced by burning 1 kg of dry wood or sawmill scrap is about 1.25×10^7 J (1 kg in these cases corresponds to a volume of approximately 1.5×10^{-3} m^3), and the heat from burning 1 kg of straw (assumed water content 15%) is about 1.5×10^7 J (Eckert, 1960). The boilers used for firing with wood or straw have efficiencies that are often considerably lower than those of oil or gas burners. Typical efficiencies are in the range 0.5–0.6 for the best boilers. The rest of the enthalpy is lost, mostly in the form of vapour and heat leaving the chimney (or other smoke exit), and therefore is not available at the load area.

Combustion is the oxidation of carbon-containing material in the presence of sufficient oxygen to complete the process

$$C + O_2 \rightarrow CO_2.$$

Wood and other biomass is burned for cooking, for space heating, and for a number of specialised purposes, such as provision of process steam and electricity generation. In rural areas of many Third World countries a device consisting of three stones for outdoor combustion of twigs is still the most common. In the absence of wind, up to about 5% of the heat energy may reach the contents of the pot resting on top of the stones. In some countries, indoor cooking on simple chulas is common. A chula is a combustion chamber with place for one or more pots or pans on top, resting in such a way that the combustion gases will pass along the outer sides of the cooking pot and leave the room through any opening. The indoor air quality is extremely poor when such chulas are in use, and village women in India using chulas for several hours each day are reported to suffer from severe cases of eye irritation and diseases of the respiratory system (Vohra, 1982).

Earlier, most cooking in Europe and its colonies was done on stoves made of cast iron. These stoves, usually called European stoves, had controlled air intake and both primary and secondary air inlets, chimneys with regulation of gas passage, and several cooking places with ring systems allowing the pots to fit tightly in holes, with a part of the pot indented into the hot gas stream. The efficiency would be up to about 25%, counted as energy delivered to the pots divided by wood energy spent, but such efficiencies would only be reached if all holes were in use and if the different temperatures prevailing at different boiler holes could be made useful, including after-heat. In many cases the average efficiency would hardly have exceeded 10%, but in many of the areas in question the heat lost to the room could also be considered as useful, in which case close to 50% efficiency (useful energy divided by input) could be reached. Today, copies of the European stove are being introduced in several regions of the Third World, with use of local materials such as clay and sand–clay mixtures instead of cast iron.

Wood-burning stoves and furnaces for space heating have conversion efficiencies from below or about 10% (open furnace with vertical chimney) up to 50% (oven with two controlled air inlets and a labyrinth-shaped effluent gas route leading to a tall chimney). Industrial burners and stokers (for burning wood scrap) typically reach efficiencies of about 60%. Higher efficiencies require a very uniform fuel without variations in water content or density.

Most biological material is not uniform, and some pre-treatment can often improve both the transportation and storage processes and the combustion (or other final use). Irregular biomass (e.g. twigs) can be chopped or cut to provide unit sizes fitting the containers and burning spaces provided. Furthermore, compressing and pelletising the fuel can make it considerably more versatile. For some time, straw compressors and pelletisers have been available, so that the bulky straw bundles can be transformed into a fuel with volume densities approaching that of coal. Other biomass material can conceivably be pelletised with advantage, including wood scrap, mixed biomass residues, and even aquatic plant material (Anonymous, 1980). Portable pelletisers are available (e.g. in Denmark) which allow straw to be compressed in the fields so that longer transport becomes economically feasible and so that even long-term storage (seasonal) of straw residues becomes attractive.

A commonly practised conversion step is from wood to charcoal. Charcoal is easier to store and to transport. Furthermore, charcoal combustion – for example, for cooking – produces less visible smoke than direct wood burning and is typically so much more efficient than wood burning that, despite wood-to-charcoal conversion losses, less primary energy is used to cook a given meal with charcoal than with wood.

Particularly in the rich countries, a considerable source of biomass energy is urban refuse, which contains residues from food preparation and discarded consumer goods from households, as well as organic scrap material from commerce and industry. Large-scale incineration of urban refuse has become an important source of heat, particularly in Western Europe, where it is used mainly for district heating (de Renzo, 1978).

For steam generation purposes, combustion is performed in the presence of an abundant water source ("waterwall incineration"). In order to improve pollutant control, fluidised bed combustion techniques may be utilised (Cheremisinoff *et al.*, 1980). The bed consists of fine-grain material, for example, sand, mixed with material to be burned (particularly suited is sawdust, but any other biomass including wood can be accepted if finely chopped). The gaseous effluents from combustion, plus air, fluidise the bed as they pass through it under agitation. The water content of the material in the bed may be high (in which case steam production entails). Combustion temperatures are lower than for flame burning, and this partly makes ash removal easier and partly reduces tar formation and salt vaporisation. As a result, the reactor life is extended and air pollution can better be controlled.

In general, the environmental impacts of biomass utilisation through combustion may be substantial and comparable to, although not entirely of the same nature as, the impacts from coal and oil combustion (see Table 4.4). In addition, ashes will have to be disposed of. For boiler combustion, the sulphur dioxide emissions are typically much smaller than for oil and coal combustion, which would give 15–60 kg t^{-1} in the absence of flue gas cleaning. If ash is re-injected into the wood burner, higher sulphur values appear, but these values are still below the fossil fuel emissions in the absence of sulphur removal efforts.

Substance emitted	Emissions (kg/10^3kg)
Particulates	12.5–15.0
Organic compounds[a]	1.0
Sulphur dioxide	0–1.5[b]
Nitrogen oxides	5.0
Carbon monoxide	1.0

Table 4.4. Uncontrolled emissions from biomass combustion in boilers (kg per tonne of fuel, based on woody biomass; US EPA, 1980).
[a] Hydrocarbons including methane and traces of benzo(a)pyrine.
[b] Upper limit is found for bark combustion. Ten times higher values are reported in cases where combustion ashes are re-injected.

Particulates are not normally regulated in home boilers, but for power plants and industrial boilers, electrostatic filters are employed with particulate removal up to over 99%. Compared to coal burning without particle removal, wood emissions are 5–10 times lower. When starting a wood boiler there is an initial period of very visible smoke emission, consisting of water vapour and high levels of both particulate and gaseous emissions. After reaching operating temperatures, wood burns virtually without visible smoke. When stack temperatures are below 60°C, again during start-up and incorrect burning practice, severe soot problems arise.

Nitrogen oxides are typically 2–3 times lower for biomass burning than for coal burning (per kilogram of fuel), often leading to similar emissions if taken per unit of energy delivered.

Particular concern should be directed at the organic compound emissions from biomass burning. In particular, benzo(a)pyrene emissions are found to be up to 50 times higher than for fossil fuel combustion, and the measured concentrations of benzo(a)pyrene in village houses in Northern India (1.3-9.3 $\times 10^{-9}$ kg m^{-3}), where primitive wood-burning chulas are used for several (6–8) hours every day, exceed the German standards of 10^{-11} kg m^{-3} by 2–3 orders of magnitude (Vohra, 1982). However, boilers with higher combustion temperatures largely avoid this problem, as indicated in Table 4.4.

The lowest emissions are achieved if batches of biomass are burned at optimal conditions, rather than regulating the boiler up and down according to heating load. Therefore, wood heating systems consisting of a boiler and a heat storage unit (gravel, water) with several hours of load capacity will lead to the smallest environmental problems (Hermes and Lew, 1982). This example shows that there can be situations where energy storage would be introduced entirely for environmental reasons.

Finally, it should be mentioned that occupational hazards arise during tree felling and handling. The accident rate is high in many parts of the world, and safer working conditions for forest workers are imperative if wood is to be used sensibly for fuel applications.

Finally, concerning carbon dioxide, which accumulates in the atmosphere as a consequence of rapid combustion of fossil fuels, it should be kept in mind that the carbon dioxide emissions during biomass combustion are balanced in magnitude by the net carbon dioxide assimilation by the plants, so that the atmospheric CO_2 content is not affected, at least by the use of biomass crops in fast rotation. However, the lag time for trees may be decades or centuries, and in such cases, the temporary carbon dioxide imbalance may contribute to climatic alterations.

Composting

Primary organic materials form the basis for a number of energy conversion processes other than burning. Since these produce liquid or gaseous fuels, plus associated heat, they will be dealt with in the following sections on fuel production. However, conversion aiming directly at heat production has also been utilised, with non-combustion processes based on manure from livestock animals and in some cases on primary biomass residues.

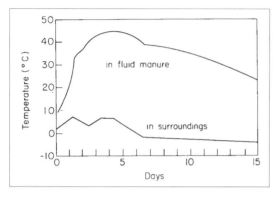

Figure 4.112. Temperature development in composting device based on liquid manure from pigs and poultry and with a blower providing constant air supply. Temperature of surroundings is also indicated (based on Popel, 1970).

Two forms of composting are in use one based on fluid manure (less than 10% dry matter), and the other based on solid manure (50–80% dry matter). The chemical process is in both cases a bacterial decomposition under aerobic conditions, i.e. the compost heap or container has to be ventilated in order to provide a continuous supply of oxygen for the bacterial processes. The bacteria required for the process (of which lactic acid producers constitute the largest fraction; cf. McCoy, 1967) are normally all present in manure, unless antibiotic residues that kill bacteria are retained after some veterinary treatment. The processes involved in composting are very complex, but it is believed that decomposition of carbohydrates [the inverse of the reaction (3.41)) is responsible for most of the heat production (Popel, 1970]. A fraction of the carbon from the organic material is found in new-bred microorganisms.

A device for treating fluid manure may consist of a container with a blower injecting fresh air into the fluid in such a way that it becomes well distributed over the fluid volume. An exit air channel carries the heated airflow to, say, a heat exchanger. Figure 4.112 shows an example of the temperature of the liquid manure, along with the temperature outside the container walls, as a function of time. The amount of energy required for the air blower is typically around 50% of the heat energy output, and is in the form of high-quality mechanical energy (e.g. from an electrically driven rotor). Thus, the simple efficiency may be around 50%, but the second law efficiency (4.20) may be quite low.

Heat production from solid manure follows a similar pattern. The temperature in the middle of a manure heap ("dunghill") may be higher than that of liquid manure, owing to the low heat capacity of the solid manure (see Fig. 4.113). Air may be supplied by blowers placed in the bottom of the heap, and in order to maintain air passage space inside the heap and remove moisture, occasional re-stacking of the heap is required. A certain degree of variation in air supply can be allowed, so that the blower may be driven by a renewable energy converter, for example, a windmill, without storage or back-up power. With frequent re-stacking, air supply by blowing is not required, but the required amount of mechanical energy input per unit of heat extraction is probably as high as for liquid manure. In addition, the heat extraction process is more difficult, demanding, for example, that heat exchangers be built into the heap itself (water pipes are one possibility). If an insufficient amount of air is provided, the composting process will stop before the maximum heat has been obtained.

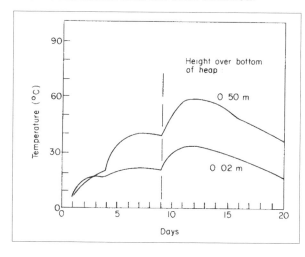

Figure 4.113. Temperature development at different locations within a solid manure composting heap (total height 0.8 m). After nine days, air is added from below (based on Olsen, 1975).

Owing to the bacteriological decomposition of most of the large organic molecules present, the final composting product has considerable value as fertiliser.

Metabolic heat

Metabolic heat from the life processes of animals can also be used by man, in addition to the heating of human habitats by man's own metabolic heat. A livestock shed or barn produces enough heat, under most conditions of winter occupancy, to cover the heating requirements of adjacent farm buildings, in addition to providing adequate temperature levels for the animals. One reason for this is that livestock barns must have a high rate of ventilation in order to remove dust (e.g. from the animal's skin, fur, hair or feathers) and water vapour. For this reason, utilisation of barn heat may not require extra feeding of the animals, but may simply consist of heat recovery from air that for other reasons has to be removed. Utilisation for heating a nearby residence building often requires a heat pump (see section 4.6.1), because the temperature of the ventilation air is usually lower than that required at the load area, so that a simple heat exchanger would not work.

In temperate climates, the average temperature in a livestock shed or barn may be about 15°C during winter. If young animals are present, the required temperature is higher. With an outside temperature of 0°C and no particular insulation of walls, the net heat production of such barns or sheds is positive when the occupants are fully grown animals, but negative if the occupants are young individuals and very much so if newborn animals are present (Olsen, 1975). In chicken or pig farms, the need for heat input may be avoided by having populations of mixed age or heat exchange between compartments for young and adult animals. The best candidates for heat extraction to other applications might then be dairy farms.

A dairy cow transfers about 25% of the chemical energy in the fodder to the milk and a similar amount to the manure (Claesson, 1974). If the weight of the cow remains constant, the rest is converted to heat and is rejected as heat radiation, convection or latent heat of water vaporisation. The distribution of the heat production in sensible and latent forms of heat is indicated in Fig. 4.114. It is seen to be strongly dependent on the air temperature in the barn. At 15°C, about two-thirds of the heat production is in the form of sensible heat. Heat transfer to a heat pump circuit may take place from the ventilation air exit. Water condensation on the heat exchanger surface involved may help to prevent dust particles from accumulating on the heat exchanger surface.

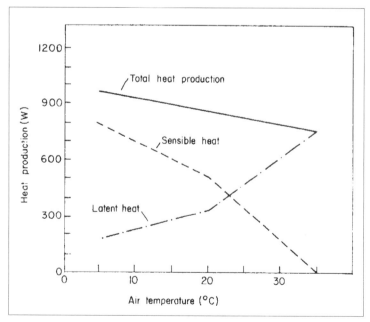

Figure 4.114. Average heat production and form of heat for a "standard" cow (the heat production of typical cows of "red" or "black-spotted" races is about 20% higher, while that of typical "Jersey" cows is about 30% lower) (based on Petersen, 1972).

4.8.2 Fuel production from biomass: overview and generation of gaseous fuels

Fuels are biological material (including fossilised forms), hydrocarbons or just hydrogen, which can be stored and used at convenient times. "Usage" traditionally means burning in air or pure oxygen, but other means of conversion exist, for example, fuel cells (see sections 4.1.6 and 4.7).

A number of conversion processes aim at converting one fuel into another which is considered more versatile. Examples are charcoal production from wood (efficiency of conversion about 50%), liquefaction and gasification (Squires, 1974). Figure 4.115 gives an overview of all the non-food uses of biomass. Before discussing the conversion of fresh biomass, the gasification of coal is briefly discussed because of its importance for possible continued use of fossil biomass (coal being the largest such source) and also because of its similarity to processes relevant for other solid biomass.

Inefficient conversion of coal to oil has historically been used by isolated coal-rich but oil-deficient nations (Germany during World War II, South Africa). Coal is gasified to carbon monoxide and hydrogen, which is then, by the Fischer–Tropsch process (passage through a reactor, e.g. a fluidised bed, with a nickel, cobalt, or iron catalyst), partially converted into hydrocarbons. Sulphur compounds have to be removed as they would impede the function of the catalyst. The reactions involved are of the form

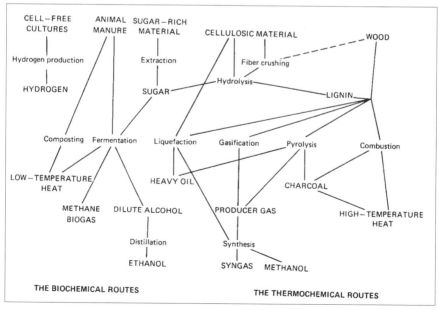

Figure 4.115. Non-food energy uses of biomass.

$$(2n + 1)H_2 + nCO \rightarrow C_nH_{2n+2} + nH_2O,$$
$$(n + 1)H_2 + 2nCO \rightarrow C_nH_{2n+2} + nCO_2$$

and conversion efficiencies range from 21 to 55% (Robinson, 1980). Further separation of the hydrocarbons generated may then be performed, for instance gasoline corresponding to the range $4 \leq n \leq 10$ in the above formulae.

Alternative coal liquefaction processes involve direct hydrogenation of coal under suitable pressures and temperatures. Pilot plants have been operating in the United States, producing up to 600 t a day (slightly different processes are named "solvent refining", "H-coal process", and "donor solvent process"; cf. Hammond, 1976). From an energy storage point of view, either coal or the converted product may be stockpiled.

For use in the transportation sector, production of liquid hydrocarbons such as methanol from natural gas could be advantageous. In the long term, methanol is likely to be produced from renewable biomass sources as described below.

Biogas

Conversion of (fresh) biological material into simple hydrocarbons or hydrogen can be achieved by a number of anaerobic fermentation processes, i.e. processes performed by suitable microorganisms and taking place in the absence of oxygen. Such anaerobic "digestion" processes work on the basis of most fresh biological material, wood excepted, provided that the proper conditions are maintained (temperature, population of microorganisms, stirring, etc.). Thus, biological material in forms inconvenient for storage and use may be converted into liquid or gaseous fuels that can be utilised in a multitude of ways, like oil and natural gas products.

The "raw materials" that may be used are slurry or manure (e.g. from dairy farms or "industrial farming" involving large feedlots), city sewage and refuse, farming crop residues (e.g. straw or parts of cereal or fodder plants not normally harvested), or direct "fuel crops", such as ocean-grown algae or seaweeds, water hyacinths (in tropical climates) or fast-growing bushes or trees. The deep harvesting necessary in order to collect crop residues may not be generally advisable, owing to the role of these residues in preventing soil erosion.

Among the fermentation processes, one set is particularly suited for producing gas from biomass in a wet process (cf. Fig. 4.115). It is called *anaerobic digestion*. It traditionally used animal manure as biomass feedstock, but other biomass sources can be used within limits that are briefly discussed in the following. The set of biochemical reactions making up the digestion process (a word indicating the close analogy to energy extraction from biomass by food digestion) is schematically illustrated in Fig. 4.116.

There are three discernible stages. In the first, complex biomass material is decomposed by a heterogeneous set of microorganisms, not necessarily confined to anaerobic environments. These decompositions comprise hydrolysis of cellulosic material to simple glucose, using enzymes provided by the microorganisms as catalysts. Similarly, proteins are decomposed to amino acids and lipids to long-chain acids. The significant result of this first phase is that most of the biomass is now water soluble and in a simpler chemical form, suited for the next process step.

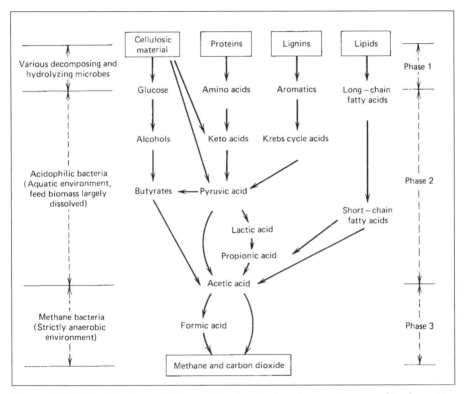

Figure 4.116. Simplified description of biochemical pathways in anaerobic digestion of biomass (based on Stafford *et al.*, 1981). Used with permission: from D. Stafford, D. Hawkes, and R. Horton, *Methane Production from Waste Organic Matter.* Copyright 1981 by The Chemical Rubber Co., CRC Press, Inc., Boza Raton, FL.

The second stage involves dehydrogenation (removing hydrogen atoms from the biomass material), such as changing glucose into acetic acid, carboxylation (removing carboxyl groups) of the amino acids, and breaking down the long-chain fatty acids into short-chain acids, again obtaining acetic acid as the final product. These reactions are fermentation reactions accomplished by a range of acidophilic (acid-forming) bacteria. Their optimum performance requires a pH environment in the range of 6–7 (slightly acid), but because the acids already formed will lower the pH of the solution, it is sometimes necessary to adjust the pH, for example, by adding lime.

Finally, the third phase is the production of biogas (a mixture of methane and carbon dioxide) from acetic acid by a second set of fermentation reactions performed by methanogenic bacteria. These bacteria require a strictly anaerobic (oxygen-free) environment. Often, all processes are made to take place in a single container, but separation of the processes into stages will allow greater efficiencies to be reached. The third phase takes of the order of weeks, the preceding phases on the order of hours or days, depending on the nature of the feedstock.

Starting from cellulose, the overall process may be summarised as

$$(C_6H_{10}O_5)_n + nH_2O \rightarrow 3nCO_2 + 3nCH_4 + 19n \text{ J mol}^{-1}. \tag{4.149}$$

The phase one reactions add up to

$$(C_6H_{10}O_5)_n + nH_2O \rightarrow nC_6H_{12}O_6. \tag{4.150}$$

The net result of the phase two reactions is

$$nC_6H_{12}O_6 \rightarrow 3nCH_3COOH, \tag{4.151}$$

with intermediate steps such as

$$C_6H_{12}O_6 \rightarrow 2C_2H_5OH + 2CO_2 \tag{4.152}$$

followed by dehydrogenation:

$$2C_2H_5OH + CO_2 \rightarrow 2CH_3COOH + CH_4. \tag{4.153}$$

The third phase reactions then combine to

$$3nCH_3COOH \rightarrow 3nCO_2 + 3nCH_4. \tag{4.154}$$

In order for the digestion to proceed, a number of conditions must be fulfilled. The bacterial action is inhibited by the presence of metal salts, penicillin, soluble sulphides, or ammonia in high concentrations. Some source of nitrogen is essential for the growth of the microorganisms. If there is too little nitrogen relative to the amount of carbon-containing material to be transformed, then bacterial growth will be insufficient and biogas production low. With too much nitrogen (a ratio of carbon to nitrogen atoms below around 15), "ammonia poisoning" of the bacterial cultures may occur. When the carbon–nitrogen ratio exceeds about 30, the gas production starts diminishing, but in some systems carbon–nitrogen values as high as 70 have prevailed without problems (Stafford et al., 1981). Table 4.5 gives carbon–nitrogen values for a number of biomass feedstocks. It is seen that mixing feedstocks can often be advantageous. For instance, straw and sawdust would have to be mixed with some low C:N material, such as livestock urine or clover/lucerne (typical secondary crops that may be grown in temperate climates after the main harvest).

If digestion time is not a problem, almost any cellulosic material can be converted to biogas, even pure straw. One initially may have to wait for several months, until the optimum bacterial composition has been reached, but then continued production can take place, and despite low reaction rates an energy recovery similar to that of manure can be achieved with properly prolonged reaction times (Mardon, 1982).

Average manure production for fully bred cows and pigs (in Europe, Australia, and the Americas) is 40 and 2.3 kg wet weight d^{-1}, corresponding to 62 and 6.2 MJ d^{-1}, respectively. The equivalent biogas production may reach 1.2 and 0.18 m^3 d^{-1}. This amounts to 26 and 3.8 MJ d^{-1}, or 42 and 61% conversion efficiency, respectively (Taiganides, 1974). A discussion of the overall efficiency including transportation of biomass to the plant is given in Berglund and Börjesson (2002), finding maximum acceptable transport distances of 100–150 km.

The residue from the anaerobic digestion process has a higher value as a fertiliser than the feedstock. Insoluble organics in the original material are, to a large extent, made soluble, and nitrogen is fixed in the microorganisms.

Pathogen populations in the sludge are reduced. Stafford *et al.* (1981) found a 65–90% removal of *Salmonella* during anaerobic fermentation, and there is a significant reduction in streptococci, coliforms, and viruses, as well as an almost total elimination of disease-transmitting parasites such as *Ascaris*, hookworm, *Entamoeba*, and *Schistosoma*.

Material	Ratio	Material	Ratio
Sewage sludge	13:1	Bagasse	150:1
Cow dung	25:1	Seaweed	80:1
Cow urine	0.8:1	Alfalfa hay	18:1
Pig droppings	20:1	Grass clippings	12:1
Pig urine	6:1	Potato tops	25:1
Chicken manure	25:1	Silage liquor	11:1
Kitchen refuse	6–10:1	Slaughterhouse wastes	3–4:1
Sawdust	200–500:1	Clover	2.7:1
Straw	60–200:1	Lucerne	2:1

Table 4.5. Carbon–nitrogen ratios for various materials (based on Baader *et al.*, 1978; Rubins and Bear, 1942).

For this reason, anaerobic fermentation has been used fairly extensively as a cleaning operation in city sewage treatment plants, either directly on the sludge or after growing algae on the sludge to increase fermentation potential. Most newer plants make use of the biogas produced to fuel other parts of the treatment process, but with proper management, sewage plants may well be net energy producers (Oswald, 1973).

The other long-time experience with biogas and associated fertiliser production is in rural areas of a number of Asian countries, notably China and India. The raw materials here are mostly cow dung, pig slurry, and what in India is referred to as human night soil, plus in some cases grass and straw. The biogas is used for cooking, and the fertiliser residue is returned to the fields. The sanitary aspect of pathogen reduction lends strong support to the economic viability of these schemes.

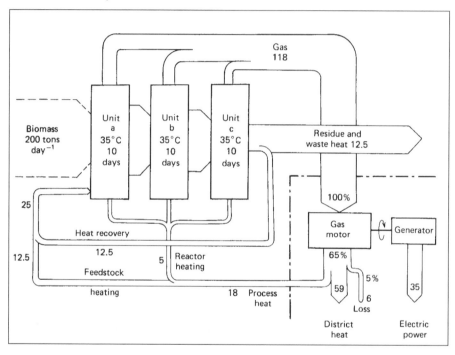

Figure 4.117. Calculated energy flows for a town biogas reactor plant, consisting of three successive units with ten days residence time in each. A biogas motor drives an electric generator, and the associated heat is, in part, recycled to the digestion process, while the rest is fed into the town district heating lines. Flows (all numbers without indicated unit) are in GJ d^{-1} (based on energy planning for Nysted commune, Denmark; Kraemer, 1981).

The rural systems are usually based on simple one-compartment digesters with human labour for filling and emptying of material. Operation is either in batches or with continuous new feed and removal of some 3–7% of the reactor content every day. Semi-industrialised plants have also been built during the last decade, for example, in connection with large pig-raising farms, where mechanised and highly automated collection of manure has been feasible. In some cases, these installations have utilised separate acid and methanation tanks.

Figure 4.117 shows an example of a town-scale digester plant, where the biogas is used in a combined electric power and district heat generating plant (Kraemer, 1981). Expected energy flows are indicated. Storage of biogas for the rural cooking systems is accomplished by variable-volume domes collecting the gas as it is produced (e.g., an inverted, water-locked can placed over the digester core). Biogas contains approximately 23 MJ m^{-3} and is therefore a medium-quality gas. CO_2 removal is necessary in order to achieve pipeline quality. An inexpensive way of achieving over 90% CO_2 removal is by water spraying. This way of producing compressed methane gas from biogas allows for automotive applications, such as the farmer producing tractor fuel on site. In New Zealand such uses have been developed since 1980 (see Fig. 4.118; Stewart and McLeod, 1980). Several demonstration experiences have recently been obtained in Sweden (Losciale, 2002).

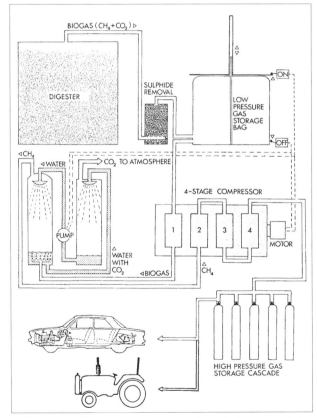

Figure 4.118. Schematic view of New Zealand scheme for methane production and vehicle use (from Stewart, D. and McLeod, R., New Zealand Journal of Agriculture, Sept. 1980, 9-24. With permission).

Storage of a certain amount of methane at ambient pressure requires over a thousand times more volume than the equivalent storage of oil. However, actual methane storage at industrial facilities uses pressures of about 140 times ambient (Biomass Energy Institute, 1978), so the volume penalty relative to oil storage would then be a factor of 9. Storage of methane in zeolitic material for later use in vehicles has been considered.

If residues are recycled, little environmental impact can be expected from anaerobic digestion. The net impact on agriculture may be positive, owing to nutrients being made more accessible and due to parasite depression. Undesirable contaminants, such as heavy metals, are returned to the soil in approximately the same concentrations as they existed before collection, unless urban pollution has access to the feedstock. The very fact that digestion depends on biological organisms may imply that warning signals in terms of poor digester performance may direct early attention to pollution of crop land or urban sewage systems. In any case, pollutant-containing waste, for example, from industry, should never be mixed with the energy-valuable biological material in urban refuse and sewage. The methane-forming bacteria are more sensitive to changes in environment, such as temperature and acidity, than the acid-forming ones.

The digestion process itself does not emit pollutants if it operates correctly, but gas cleaning, such as H_2S removal, may lead to emissions. The methane gas itself shares many of the accident hazards of other gaseous fuels, being asphyxiating and explosive at certain concentrations in air (roughly 7–14% by volume). For rural cooking applications, the impacts may be compared with those of the fuels being replaced by biogas, notably wood burned in simple stoves. In these cases, as follows from the discussion in section 4.8.1, the environment is dramatically improved by introducing biogas digesters.

An example of early biogas plants for use on a village scale in China, India and Pakistan is shown in Fig. 4.119. All the reactions take place in one compartment, which does not necessarily lead to optimum conversion efficiency. The time required for the acid-forming step is less than 24 h, whereas the duration of the second step should be 10–30 days. The heating of the fermentation tank required for application in many climatic regions may be derived from solar collectors, which could form the top cover of the tank. Alternatively, the top cover may be an inflatable dome serving as a store of gas, which can smooth cut a certain degree of load variation. Some installations obtain the highest efficiency by batch operation, i.e. by leaving one batch of biological material in the tank for the entire fermentation period. The one shown in Fig. 4.119 allows continuous operation, i.e. a fraction of the slurry is removed every day and replaced by fresh biological material.

Examples of predicted biogas production rates, for simple plants of the type shown in Fig. 4.119 and based on fluid manure from dairy cows, pigs or poultry, are shown in Table 4.6 and Fig. 4.120. Table 4.6 gives typical biogas production rates, per day and per animal, while Fig. 4.120 gives the conversion efficiencies measured, as functions of fermentation time (tank residence time), in a controlled experiment. The conversion efficiency is the ratio of the energy in the biogas (approximately 23 MJ m^{-3} of gas) and the energy in the manure which would be released as heat by complete burning of the dry matter (some typical absolute values are given in Table 4.6). The highest efficiency is obtained with pigs' slurry, but the high bacteriological activity in this case occasionally has the side-effect of favouring bacteria other than those helping to produce biogas, e.g. ammonia-producing bacteria, the activity of which may destroy the possibility of further biogas production (Olsen, 1975).

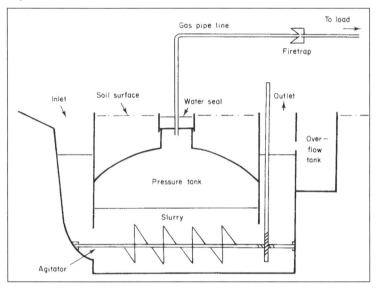

Figure 4.119. Single-chamber biogas plant (based on Chinese installations described by Appropriate Technology Development Organization, 1976).

Source	Manure per day		Biogas per day	
	kg wet weight	*MJ*	*m³*	*MJ*
Cows	40	62	1.2	26
Pigs	2.3	6.2	0.18	3.8
Hens	0.19	0.9	0.011	0.26

Table 4.6. Manure and potential biogas production for a typical animal per day (based on Taiganides, 1974).

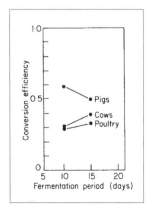

Figure 4.120. Measured conversion efficiencies (ratio between energy in biogas gas produced and energy in the manure) for simple biogas plants (like Fig. 4.119). Some 10–13 kg of fresh manure was added per day and per m³ of tank volume (based on Gramms *et al.*, 1971).

As mentioned, the manure residue from biogas plants has a high value as fertiliser because the decomposition of organic material followed by hydrocarbon conversion leaves plant nutrients (e.g. nitrogen that was originally bound in proteins) in a form suitable for uptake. Malignant bacteria and parasites are not removed to as high a degree as by composting, owing to the lower process temperature.

Some city sewage plants produce biogas (by anaerobic fermentation of the sewage) as one step in the cleaning procedure, using the biogas as fuel for driving mechanical cleaning devices, etc. In this way it is in many cases possible to avoid any need for other energy inputs and in some cases to become a net energy producer (Danish Energy Agency, 1992). Figure 4.121 shows the system layout for a 300 t of biomass per day biogas plant accepting multiple types of feedstock and capable of delivering both power, process and district heat, and fertiliser. Figure 4.122 gives the measured performance data for nine large prototype biogas plants in Denmark.

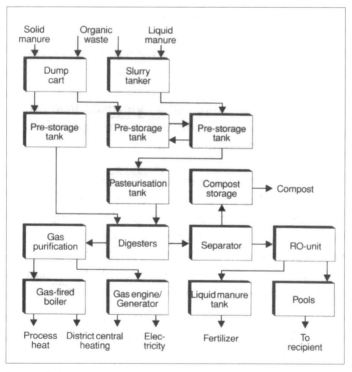

Figure 4.121. Layout of Lintrup biogas plant (Danish Energy Agency, 1992).

Energy balance

The 1992 average production of the large-size Danish biogas plants was 35.1 m^3 per m^3 of biomass (at standard pressure, methane content in the biogas being on average 64%), or 806 MJ m^{-3} (Tafdrup, 1993). In-plant energy use amounted to 90 MJ m^{-3} distributed on 28 MJ electricity and 50 MJ heat, all produced by the biogas plant itself. Fuel used in transporting manure to the plant totalled 35 MJ, and the fertiliser value of the returned residue was estimated at 30 MJ. Thus, the net outside energy requirement is 5 MJ for a production of 716 MJ, or 0.7%, corresponding to an energy payback time of 3 days. If the in-plant biogas use is added, the energy consumption in the process is 13%. To this should be added the energy for construction of the plant, which has not been estimated. However, the best plants roughly break even economically, indicating that the overall energy balance is acceptable.

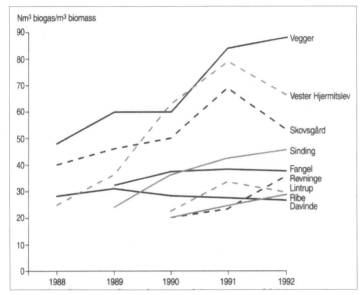

Figure 4.122. Annual average production efficiency (m^3 of biogas at standard pressure, denoted Nm^3, per m^3 of biomass feedstock) for nine community-size biogas plants in Denmark (Danish Energy Agency, 1992).

Greenhouse gas emissions

Using, as in the energy balance section above, the average of large Danish plants as an example, the avoided CO_2 emission from having the combined power and heat production use biomass instead of coal as a fuel is 68 kg per m^3 of biomass converted. Added should be emissions from transportation of biomass, estimated at 3 kg, and the avoided emissions from conventional production of fertiliser replaced by biogas residue, estimated at 3 kg. Reduced methane emissions, relative to the case of spreading manure directly on the fields, is of the order of 61 kg CO_2 equivalent (Tafdrup, 1993). As regards nitrous oxide, there is a possible gain by avoiding denitrification in the soil, but high uncertainty has made an actual estimate fortuitous at the present. The overall CO_2 reduction obtained by summing up the estimates given above is then 129 kg for each m^3 of biomass converted to biogas.

Other environmental effects

Compared with the current mix of coal, oil or natural gas plants, biogas plants have a 2–3 times lower SO_2 emission but a correspondingly higher NO_x emission. Higher ammonia content in the digested residue calls for greater care in using fertiliser from biogas plants, in order to avoid loss of ammonia. This is also true as regards avoiding loss of nutrients from fertiliser to waterways. Compared to spreading manure not refined by the biogas production process, a marked gain in fertiliser quality has been noted, including a much better-defined composition, which will contribute to assisting correct dosage and avoiding losses to the environment. The dissemination of biogas plants removes the need for landfills, which is seen as an environmental improvement. Odour is moved from the fields (where manure and slurry would otherwise be spread) to the biogas plant, where it can be controlled by suitable measures (filters, etc.) (Tafdrup, 1993).

Hydrogen-producing cultures
Biochemical routes to fuel production include a number of schemes not presently developed to a technical or economic level of commercial interest. Hydrogen is a fuel that may be produced directly by biological systems. Hydrogen enters in the process of photosynthesis, as it proceeds in green plants, where the net result of the process is

$$2H_2O + \text{solar radiation} \rightarrow 4e^- + 4H^+ + O_2.$$

However, the electrons and protons do not combine directly to form hydrogen,

$$4e^- + 4H^+ \rightarrow 2H_2,$$

but instead transfer their energy to a more complex molecule ($NADPH_2$; cf. Chapter 3), which is capable of driving the CO_2 assimilation process. By this mechanism, the plants avoid recombination of oxygen and hydrogen from the two processes mentioned above. Membrane systems keep the would-be reactants apart, and thus the energy-rich compound may be transported to other areas of the plant, where it takes part in plant growth and respiration.

Much thought has been given to modifications of plant material (e.g., by genetic engineering), in such a way that free hydrogen is produced on one side of a membrane and free oxygen on the other side (Berezin and Varfolomeev, 1976; Calvin, 1974; Hall *et al.*, 1979).

While dissociation of water (by light) into hydrogen and oxygen (photolysis; cf. Chapter 5) does not require a biological system, it is possible that utilisation of the process on a realistic scale can be more easily achieved if the critical components of the system, notably the membrane and the electron transport system, are of biological origin. Still, a breakthrough is required before any thought can be given to practical application of direct biological hydrogen production cultures.

Thermochemical gasification of biomass

Conversion of fossil biomass such as coal into a gas is considered a way of reducing the negative environmental impacts of coal utilisation. However, in some cases the impacts have only been moved but not eliminated. Consider, for example, a coal-fired power plant with 99% particle removal from flue gases. If it were to be replaced by a synthetic gas-fired power plant with gas produced from coal, then sulphur could be removed at the gasification plant using dolomite-based scrubbers. This would practically eliminate the sulphur oxide emissions, but on the other hand, dust emissions from the dolomite processing would represent particle emissions twice as large as those avoided at the power plant by using gas instead of coal (Pigford, 1974). Of course, the dust is emitted at a different location.

This example, as well as the health impacts associated with coal mining, whether on the surface or in mines (although not identical), has sparked interest in methods of gasifying coal *in situ*. Two or more holes are drilled. Oxygen (or air) is injected through one, and a mixture of gases, including hydrogen and carbon oxides, emerges at the other hole. The establishment of proper communication between holes, and suitable underground contact surfaces, has proved difficult, and recovery rates are modest.

The processes involved would include

$$2C + O_2 \rightarrow 2CO, \tag{4.155}$$

$$CO + H_2O \leftrightarrow CO_2 + H_2. \tag{4.156}$$

The stoichiometric relation between CO and H_2 can then be adjusted using the shift reaction (4.156), which may proceed in both directions, depending on steam temperature and catalysts. This opens the way for methane synthesis through the reaction

$$CO + 3H_2 \rightarrow CH_4 + H_2O. \tag{4.157}$$

At present, the emphasis is on improving gasifiers using coal already extracted. Traditional methods include the Lurgi fixed-bed gasifier (providing gas under pressure from non-caking coal at a conversion efficiency as low as 55%) and the Koppers–Totzek gasifier (oxygen input, the produced gas unpressurised, also of low efficiency).

Improved process schemes include the hy-gas process, requiring a hydrogen input; the bi-gas concept of brute force gasification at extremely high temperatures; and the slagging Lurgi process, capable of handling powdered coal (Hammond, 1976).

Promising, but still at an early stage of development, is catalytic gasification (e.g. potassium catalyst), where all processes take place at a common, relatively low temperature, so that they can be combined in a single reactor (Fig. 4.123). The primary reaction here is

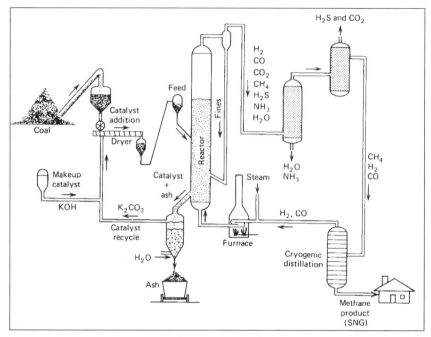

Figure 4.123. Schematic diagram of catalytic gasification process (SNG is synthetic natural gas). (From Hirsch *et al.*, 1982. Reprinted from *Science*, **215**, 121-127, 8, January 1982, with permission. Copyright 1982 American Association for the Advancement of Science.)

$$C + H_2O \rightarrow H_2 + CO \qquad\qquad (4.158)$$

(H_2O being in the form of steam above 550°C), to be followed by (4.156) and (4.157). The catalyst allows all processes to take place at 700°C. Without catalyst, the gasification would have to take place at 925°C and the shift reaction and methanation at 425°C, that is, in a separate reactor where excess hydrogen or carbon monoxide would be lost (Hirsch *et al.*, 1982).

In a coal gasification scheme storage would be (of coal) before conversion. Peat can be gasified in much the same way as coal, as can wood (with optional subsequent methanol conversion as described in section 4.8.3).

Fresh biomass gasification
Gasification of biomass, and particularly wood and other lignin-containing cellulosic material, has a long history. The processes may be viewed as "combustion-like" conversion, but with less oxygen available than needed for burning. The ratio of oxygen available and the amount of oxygen that would allow complete burning is called the "equivalence ratio". For equivalence ratios below 0.1 the process is called "pyrolysis", and only a modest

fraction of the biomass energy is found in the gaseous product – the rest be-ing in char and oily residues. If the equivalence ratio is between 0.2 and 0.4, the process is called a proper "gasification". This is the region of maximum energy transfer to the gas (Desrosiers, 1981).

Chemical reaction	Energy consumed $(kJ\ g^{-1})^a$	Products / process
$C_6H_{10}O_5 \rightarrow 6C + 5H_2 + 2.5\ O_2$	5.94[b]	Elements, dissociation
$C_6H_{10}O_5 \rightarrow 6C + 5H_2O(g)$	–2.86	Charcoal, charring
$C_6H_{10}O_5 \rightarrow 0.8\ C_6H_8O + 1.8\ H_2O(g) + 1.2\ CO_2$	–2.07[c]	oily residues, pyrolysis
$C_6H_{10}O_5 \rightarrow 2C_2H_4 + 2CO_2 + H_2O(g)$	0.16	Ethylene, fast pyrolysis
$C_6H_{10}O_5 + \tfrac{1}{2}O_2 \rightarrow 6CO + 5H_2$	1.85	Synthesis gas, gasification
$C_6H_{10}O_5 + 6H_2 \rightarrow 6"CH_2" + 5\ H_2O(g)$	–4.86[d]	Hydrocarbons, –genera-tion
$C_6H_{10}O_5 + 6O_2 \rightarrow 6CO_2 + 5\ H_2O(g)$	–17.48	heat, combustion

Table 4.7. Energy change for idealised cellulose thermal conversion reactions. (Source: T. Reed (1981), in *Biomass Gasification* (T. Reed, ed.), reproduced with per-mission. Copyright 1981, Noyes Data Corporation, Park Ridge, NJ)
[a] Specific reaction heat.
[b] The negative of the conventional heat of formation calculated for cellulose from the heat of combustion of starch.
[c] Calculated from the data for the idealised pyrolysis oil C_6H_8O ($\Delta H_c = -745.9$ kcal mol^{-1}, $\Delta H_f = 149.6$ kcal g^{-1}, where H_c = heat of combustion and H_f = heat of fusion).
[d] Calculated for an idealised hydrocarbon with ΔH_c as above. H_2 is consumed.

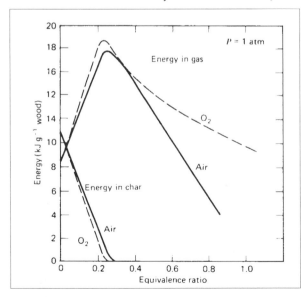

Figure 4.124. Calculated energy content in gas and char produced by equilib-rium processes between air (or oxygen) and bio-mass, as a function of equivalence ratio. (From Reed, 1981. Reprinted from *Biomass Gasification* (T. Reed, ed.), with per-mission. Copyright 1981, Noyes Data Corporation, Park Ridge, NJ).

The chemical processes involved in biomass gasification are similar to the reactions (4.155)–(4.158) for coal gasification. Table 4.7 lists a number of reactions involving polysaccharidic material, including pyrolysis and gasification. In addition to the chemical reaction formulae, the table gives enthalpy changes for idealised reactions (i.e., neglecting the heat required to bring the reactants to the appropriate reaction temperature). Figure 4.124 gives the energy of the final products, gas and char, as a function of the equivalence ratio, still based on an idealised thermodynamical calculation. The specific heat of the material is 3 kJ g^{-1} of wood at the peak of energy in the gas, increasing to 21 kJ g^{-1} of wood for combustion at equivalence ratio equal to unity. Much of this sensible heat can be recovered from the gas, so that process heat inputs for gasification can be kept low.

Figure 4.125 gives the equilibrium composition calculated as a function of the equivalence ratio. By equilibrium composition is understood the composition of reaction products occurring after the reaction rates and reaction temperature have stabilised adiabatically. The actual processes are not necessarily adiabatic; in particular the low-temperature pyrolysis reactions are not. Still, the theoretical evaluations assuming equilibrium conditions serve as a useful guideline for evaluating the performance of actual gasifiers.

The idealised energy change calculation of Table 4.7 assumes a cellulosic composition such as the one given in (4.150). For wood, the average ratios of carbon, hydrogen and oxygen are 1:1.4:0.6 (Reed, 1981).

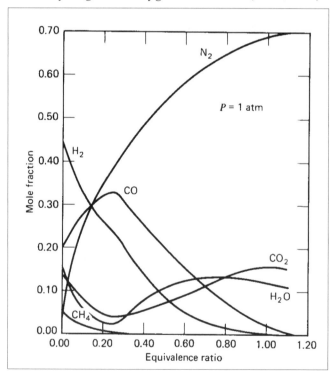

Figure 4.125. Calculated gas composition resulting from equilibrium processes between air and biomass, as a function of equivalence ratio. (From Reed, 1981. Reprinted from *Biomass Gasification* (T. Reed, ed.), with permission. Copyright 1981, Noyes Data Corporation, Park Ridge, NJ).

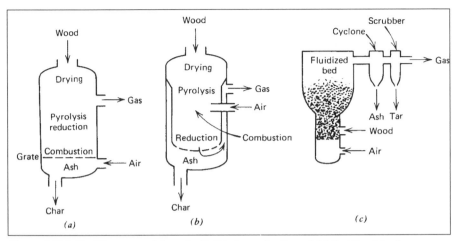

Figure 4.126. Gasifier types: (a) updraft, (b) downdraft, and (c) fluidised bed.

Figure 4.126 shows three examples of wood gasifiers: the updraft, the downdraft, and the fluidised bed types. The drawback of the updraft type is a high rate of oil, tar, and corrosive chemical formation in the pyrolysis zone. This problem is solved by the downdraft version, where such oils and other matter pass through a hot charcoal bed in the lower zone of the reactor and become cracked to simpler gases or char. The fluidised bed reactor may prove superior for large-scale operations, because passage time is smaller. The drawback of this is that ash and tars are carried along with the gas and have to be removed later in cyclones and scrubbers. Several variations on these gasifier types have been suggested (Drift, 2002; Gøbel *et al.*, 2002).

The gas produced by gasification of biomass is a "medium-quality gas"; meaning a gas with burning value in the range 10–18 MJ m^{-3}. This gas may be used directly in Otto or diesel engines, it may be used to drive heat pump compressors, or alternatively, it may by upgraded to pipeline-quality gas (about 30 MJ m^{-3}) or converted to methanol, as discussed in section 4.8.3.

Environmental impacts derive from biomass production, collection (e.g. forestry work) and transport to gasification site, from the gasification and related processes, and finally from the use made of the gas. The gasification residues –ash, char, liquid waste water, and tar– have to be disposed of. Char may be recycled to the gasifier, while ash and tars could conceivably be utilised in the road or building construction industry. The alternative of landfill disposal would represent a recognised impact. Investigations of emissions from combustion of producer gas indicate low emissions of nitrous oxides and hydrocarbons, as compared with emissions from combustion of natural gas. In one case, carbon monoxide emissions were found to be higher than for natural gas burning, but it is believed that this can be rectified as more experience in adjusting air-to-fuel ratios is gained (Wang *et al.*, 1982).

4.8.3. Fuel production from biomass: generation of liquid biofuels

Anaerobic fermentation processes may also be used to produce liquid fuels from biological raw materials. An example is the ethanol production (4.152) from glucose, known as standard yeast fermentation in the beer, wine and liquor industries. It has to take place in steps, such that the ethanol is removed (by distillation or dehydrator application) whenever its concentration approaches a value (around 12%) which would impede reproduction of the yeast culture.

In order to reduce the primary biological material (e.g. molasses, cellulose pulp or citrus fruit wastes) to glucose, the hydrolysis process (4.154) may be used. Some decomposition takes place in any acid solution, but in order to obtain complete hydrolysis, specific enzymes must usually be provided, either directly or by adding microorganisms capable of forming such enzymes. The yeast fungi themselves contain enzymes capable of decomposing polysaccharides into glucose.

The theoretical maximum efficiency of glucose-to-ethanol conversion (described in more detail below) is 97%, and, according to Calvin (1977), the Brazilian alcohol industry in 1974 obtained 14% of the energy in the raw sugar input, in the form of ethanol produced by fermentation of just the molasses residues from sugar refining, i.e. in addition to the crystallised sugar produced. A current figure is 25% (see Fig. 4.128) for an optimised plant design (EC, 1994).

Mechanical energy input, e.g. for stirring, could be covered by the fermentation wastes if they were burned in a steam power plant. In the European example (EC, 1994), these inputs amount to about a third of the energy inputs through the sugar itself.

Alternative fermentation processes based on molasses or other sugar-containing materials produce acetone–butanol, acetone–ethanol or butanol–isopropanol mixtures, when the proper bacteria are added. In addition, carbon dioxide and small amounts of hydrogen are formed (see e.g. Beesch, 1952; Keenan, 1977).

Conversion of fossilised biomass into liquid fuels was briefly mentioned at the beginning of section 4.8.2, in conjunction with the overview Fig. 4.99 showing the conversion routes open for biofuel generation. Among the non-food energy uses of biomass, there are several options leading to liquid fuels which may serve as a substitute for oil products. The survey of conversion processes given in Fig. 4.115 indicates liquid end products as the result of either biochemical conversion using fermentation bacteria or a thermochemical conversion process involving gasification and, for example, further methanol synthesis. The processes, which convert biomass into liquid fuels that are easy to store, are discussed below, but first the possibility of direct production of liquid fuels by photosynthesis is presented.

Direct photosynthetic production of hydrocarbons

Oil from the seeds of many plants, such as rape, olive, groundnut, corn, palm, soy bean, and sunflower, is used as food or in the preparation of food. Many of these oils will burn readily in diesel engines and can be used directly or mixed with diesel oil of fossil origin, as they are indeed in several pilot projects around the world.

However, in most of these cases the oil does not constitute a major fraction of the total harvest yield, and any expansion of these crops to provide an excess of oil for fuel use would interfere with food production. A possible exception is palm oil, because inter-cropping of palm trees with other food crops may provide advantages such as retaining moisture and reducing wind erosion.

Much interest is therefore concentrated on plants that yield hydrocarbons and that, at the same time, are capable of growing on land unsuited for food crops. Calvin (1977) first identified the *Euphorbia* family as an interesting possibility. The rubber tree, *Hevea brasiliensis*, is of this family, and its rubber is a hydrocarbon–water emulsion, the hydrocarbon of which (polyisoprenes) has a large molecular weight, about a million, making it an elastomer. However, other plants of the genus *Euphorbia* yield latex of much lower molecular weight which could be refined in much the same way as crude oil. In the tropical rainforests of Brazil, Calvin found a tree, *Cobaifera langsdorfii*, which annually yields some 30 litres of virtually pure diesel fuel (Maugh, 1979). Still, the interest centres on species that are likely to grow in arid areas such as the deserts of the southern United States, Mexico, Australia, and so on.

Possibilities include *Euphorbia lathyris* (gopher plant*)*, *Simmondsia chinensis* (jojoba), *Cucurdia foetidissima* (buffalo gourd) and *Parthenium argentatum* (guayule). The gopher plant has about 50% sterols (on a dry weight basis) in its latex, 3% polyisoprene (rubber), and a number of terpenes. The sterols are suited as feedstocks for replacing petroleum in chemical applications. Yields of first-generation plantation experiments in California are 15–25 barrels of crude oil equivalent or some 144 GJ ha^{-1} (i.e., per 10^4 m^2). In the case of *Hevea*, genetic and agronomic optimisation has increased yields by a factor of 2000 relative to those of wild plants, so quite high hydrocarbon production rates should be achievable after proper development (Calvin, 1977; Johnson and Hinman, 1980; Tideman and Hawker, 1981). Other researchers are less optimistic (Stewart *et al.*, 1982; Ward, 1982).

Alcohol fermentation

The ability of yeast and bacteria such as *Zymomonas mobilis* to ferment sugar-containing material to form alcohol is well known from beer, wine, and liquor manufacture. If the initial material is cane sugar, the fermentation reaction may be summarised as

$$C_6H_{12}O_6 \rightarrow 2C_2H_5OH + 2CO_2. \tag{4.159}$$

The energy content of ethanol is 30 MJ kg^{-1}, and its octane rating is 89–100. With alternative fermentation bacteria, the sugar may be converted into butanol, $C_2H_5(CH_2)_2OH$. In Brazil, the cost of ethanol is finally coming down to that of gasoline (Johansson, 2002).

In most sugar-containing plant material, the glucose molecules exist in polymerised form such as starch or cellulose, of the general structure $(C_6H_{10}O_5)_n$. Starch or hemicellulose is degraded to glucose by hydrolysis (cf. Fig. 4.99), while lignocellulose resists degradation owing to its lignin content. Lignin glues the cellulosic material together to keep its structure rigid, whether it be crystalline or amorphous. Wood has a high lignin content (about 25%), and straw also has considerable amounts of lignin (13%), while potato or beet starch contain very little lignin.

Some of the lignin seals may be broken by pre-treatment, ranging from mechanical crushing to the introduction of swelling agents causing rupture (Ladisch *et al.*, 1979).

The hydrolysis process is given by (4.150). In earlier times, hydrolysis was always achieved by adding an acid to the cellulosic material. During both world wars, Germany produced ethanol from cellulosic material by acid hydrolysis, but at very high cost. Acid recycling is incomplete; with low acid concentration the lignocelluloses is not degraded, and with high acid concentration the sugar already formed from hemicellulose is destroyed.

Consequently, alternative methods of hydrolysis have been developed, based on enzymatic intervention. Bacterial (e.g., of the genus *Trichoderma*) or fungal (such as *Sporotrichum pulverulentum*) enzymes have proved capable of converting cellulosic material, at near ambient temperatures, to some 80% glucose and a remainder of cellodextrins (which could eventually be fermented, but in a separate step with fermentation microorganisms other than those responsible for the glucose fermentation) (Ladisch *et al.*, 1979).

The residue left behind after the fermentation process (4.159) can be washed and dried to give a solid product suitable as fertiliser or as animal feed. The composition depends on the original material, in particular with respect to lignin content (small for residues of molasses, beets, etc., high for straws and woody material, but with fibre structure broken as a result of the processes described above). If the lignin content is high, direct combustion of the residue is feasible, and it is often used to furnish process heat to the final distillation.

The outcome of the fermentation process is a water–ethanol mixture. When the alcohol fraction exceeds about 10%, the fermentation process slows down and finally halts. Therefore, an essential step in obtaining fuel alcohol is to separate the ethanol from the water. Usually, this is done by distillation, a step that may make the overall energy balance of the ethanol production negative. The sum of agricultural energy inputs (fertiliser, vehicles, machinery) and all process inputs (cutting, crushing, pre-treatment, enzyme recycling, heating for different process steps from hydrolysis to distillation), as well as energy for transport, is, in existing operations such as those of the Brazilian alcohol programme (Trinidade, 1980), around 1.5 times the energy outputs (alcohol and fertiliser if it is utilised). However, if the inputs are domestic fuels, for example, combustion of residues from agriculture, and if the alcohol produced is used to displace imported oil products, the balance might still be quite acceptable from a national economic point of view.

If, further, the lignin-containing materials of the process are recovered and used for process heat generation (e.g. for distillation), then such energy should be counted not only as input but also as output, making the total input and output energy roughly balance. Furthermore, more sophisticated process design, with cascading heat usage and parallel distillation columns operating with a time displacement such that heat can be reused from column to column (Hartline, 1979), could reduce the overall energy inputs to 55–65% of the outputs.

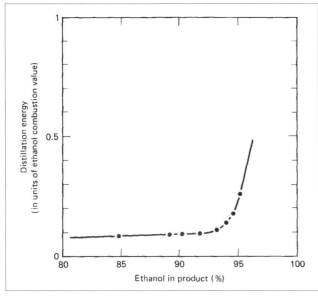

Figure 4.127. Distillation energy for ethanol–water mixture, as a function of ethanol content (assumed initial ethanol fraction 12%) (based on Ladisch and Dyck, 1979).

Radically improved energy balances would emerge if distillation could be replaced by a less energy intensive separation method. Several such methods for separating water and ethanol have been demonstrated on a laboratory scale, including: drying with desiccants such as calcium hydroxide, cellulose, or starch (Ladisch and Dyck, 1979); gas chromatography using rayon to retard water, while organic vapours pass through; solvent extraction using dibutyl phthalate, a water-immiscible solvent of alcohols; and passage through semi-permeable membranes or selective zeolite absorbers (Hartline, 1979) and phase separation (APACE, 1982). The use of dry cellulose or starch appears particularly attractive, because over 99% pure alcohol can be obtained with less than 10% energy input, relative to the combustion energy of the alcohol. Furthermore, the cellulosic material may be cost-free, if it can be taken from the input stream to the fermentation unit and returned to it after having absorbed water (the fermentation reaction being "wet" anyway). The energy input of this scheme is for an initial distillation, bringing the ethanol fraction of the aqueous mixture from the initial 5–12% up to about 75%, at which point the desiccation process is started. As can be seen from Fig. 4.127, the distillation energy is modest up to an alcohol content of 90% and then starts to rise rapidly. The drying process thus substitutes for the most energy-expensive part of the distillation process.

The ethanol fuel can be stored and used in the transportation sector much the same way as gasoline. It can be mixed with gasoline or can fully replace gasoline in spark ignition engines with high compression ratios (around 11). The knock resistance and high octane number of ethanol make this possible, and with pre-heating of the alcohol (using combustion heat that is recycled), the conversion efficiency can be improved. Several countries presently use alcohol–gasoline blends with up to 10% ethanol. This does not require any engine modification. Altering the gasoline Otto engines may be inconvenient in a transition period, but if alcohol distribution networks are implemented and existing gas stations modified, then the car engines could be optimised for alcohol fuels without regard to present requirements. A possible alternative to spark ignition engines is compression ignition engines, where auto-ignition of the fuel under high compression (a ratio of 25) replaces spark or glow plug ignition. With additives or chemical transformation into acetal, alcohol fuels could be used in this way (Bandel, 1981). Ethanol does not blend with diesel oil, so mixtures do require the use of special emulsifiers (Reeves *et al.*, 1982). However, diesel oil can be mixed with other biofuels without problems, e.g. the plant oils (rapeseed oil, etc.) presently in use in Germany.

A number of concerns with regard to the environmental impacts of the ethanol fermentation energy conversion chain must be considered. First of all, the biomass being used may have direct uses as food or may be grown in competition with production of food. The reason is, of course, that the easiest ethanol fermentation is obtained by starting with a raw material with as high a content of elementary sugar as possible, that is, starting with sugar cane or cereal grain. Since sugar cane is likely to occupy prime agricultural land, and cereal production must increase with increasing world population, neither of these biomass resources should be used as fermentation inputs. However, residues from cereal production and from necessary sugar production (present sugar consumption is in many regions of the world too high from a health and nutrition point of view) could be used for ethanol fermentation, together with urban refuse, extra crops on otherwise committed land, perhaps aquatic crops and forest renewable resources. The remarks made in chapter 3 about proper soil management, recycling nutrients, and covering topsoil to prevent erosion are very appropriate in connection with the enhanced tillage utilisation that characterises combined food and ethanol production schemes.

The hydrolysis process involves several potential environmental impacts. If acids are used, corrosion and accidents may occur, and substantial amounts of water would be required to clean the residues for re-use. Most acid would be recovered, but some would follow the sewage stream. Enzymatic hydrolysis would seem less cumbersome. Most of the enzymes would be recycled, but some might escape with waste water or residues. Efforts should be made to ensure that they are made inactive before any release. This is particularly important when, as envisaged, the fermentation residues are to be brought back to the fields or used as animal feed. A positive impact is the reduction of pathogenic organisms in residues after fermentation.

Transport of biomass could involve dust emissions, and transport of ethanol might lead to spills (in insignificant amounts, as far as energy is concerned, but with possible local environmental effects), but overall the impacts from transport would be very small.

Finally, the combustion of ethanol in engines or elsewhere leads to pollutant emissions. Compared with gasoline combustion, emissions of carbon monoxide and hydrocarbons diminish, while those of nitrous oxides, aromatics, and aldehydes increase (Hespanhol, 1979), assuming that modified ignition engines are used. With special ethanol engines and exhaust controls, critical emissions may be controlled. In any case, the lead pollution still associated with gasoline engines in some countries would be eliminated.

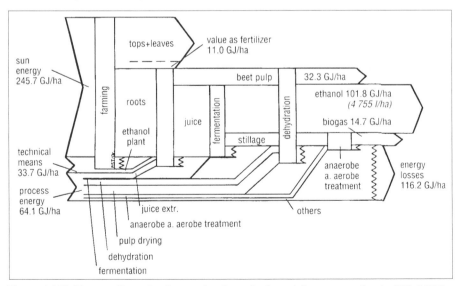

Figure 4.128. Energy flows in the production of ethanol from sugar beets (EC, 1994). Energy inputs to biomass growth, harvesting and transport are not indicated.

The energy balance of current ethanol production from biomass is not very favourable. A European study has estimated the energy flows for a number of feedstocks (EC, 1994). The highest yield of about 100 GJ ha^{-1} is found for sugar beets, shown in Fig. 4.128, but the process energy inputs and allotted life-cycle inputs into technical equipment are as large as the energy of the ethanol produced. A part of this may be supplied from biogas co-produced with the ethanol, but the overall energy efficiency remains low.

In a life-cycle analysis of ethanol production (cf. Chapter 7.4), the fact that such production is currently based upon energy crops rather than on residues (sugar cane or beets rather than straw and husk) means that all energy inputs and environmental impacts from the entire agricultural feed-stock production should be included along with the impacts pertaining to the ethanol plants and downstream impacts. Clearly, it is very difficult in this mode to balance energy outputs and inputs and to reach acceptable impact levels. The interest should therefore be limited to bio-energy processes involving only residues from an otherwise sensible production (food or wood).

Methanol from biomass

There are various ways of producing methanol from biomass sources, as indicated in Fig. 4.115. Starting from wood or isolated lignin, the most direct routes are by liquefaction or by gasification. The pyrolysis alternative gives only a fraction of the energy in the form of a producer gas.

By high-pressure hydrogenation, biomass may be transformed into a mixture of liquid hydrocarbons suitable for further refining or synthesis of methanol (Chartier and Meriaux, 1980), but all methanol production schemes so far have used a synthesis gas, which may be derived from wood gasification or coal gasification. The low-quality "producer gas" resulting directly from the wood gasification (used in cars throughout Europe during World War II) is a mixture of carbon monoxide, hydrogen gas, carbon dioxide, and nitrogen gas (see section 4.8.2). If air is used for gasification, the energy conversion efficiency is about 50%, and if pure oxygen is used instead, some 60% efficiency is possible, and the gas produced has less nitrogen content (Robinson, 1980). Gasification or pyrolysis could conceivably be performed with heat from (concentrating) solar collectors, for example, in a fluidised bed gasifier maintained at 500°C.

The producer gas is cleaned, CO_2 and N_2 as well as impurities are removed (the nitrogen by cryogenic separation), and methanol is generated at elevated pressure by the reaction

$$2H_2 + CO \rightarrow CH_3OH. \tag{4.160}$$

The carbon monoxide and hydrogen gas (possibly with additional CO_2) is called the "synthesis gas", and it is usually necessary to use a catalyst in order to maintain the proper stoichiometric ratio between the reactants of

(4.160) (Cheremisinoff *et al.*, 1980). A schematic process diagram is shown in Fig. 4.129.

An alternative is biogas production from the biomass (section 4.8.2) followed by the methane to methanol reaction,

$$2CH_4 + O_2 \rightarrow 2CH_3OH, \qquad (4.161)$$

also used in current methanol production from natural gas (Wise, 1981). Change of the H_2/CO stoichiometric ratio for (4.160) is obtained by the "shift reaction" (4.156) discussed in section 4.8.2. Steam is added or removed in the presence of a catalyst (iron oxide, chromium oxide).

The conversion efficiency of the synthesis gas to methanol step is about 85%, implying an overall wood to methanol energy efficiency of 40–45%. Improved catalytic gasification techniques raise the overall conversion efficiency to some 55% (Faaij and Hamelinck, 2002). The currently achieved efficiency is about 50%, but all life-cycle estimates of energy inputs have not been included or performed (EC, 1994).

The octane number of methanol is similar to that of ethanol, but the heat of combustion is less, amounting to 18 MJ kg^{-1}. However, the engine efficiency of methanol is higher than that of gasoline, by at least 20% for current motor car engines, so an "effective energy content" of 22.5 MJ kg^{-1} is sometimes quoted (EC, 1994). Methanol can be mixed with gasoline in standard engines, or used in specially designed Otto or diesel engines, such as a spark ignition engine run on vaporised methanol, with the vaporisation energy being recovered from the coolant flow (Perrin, 1981). Uses are similar to those of ethanol, but several differences exist in the assessment of environmental impacts, from production to use (e.g. toxicity of fumes at filling stations).

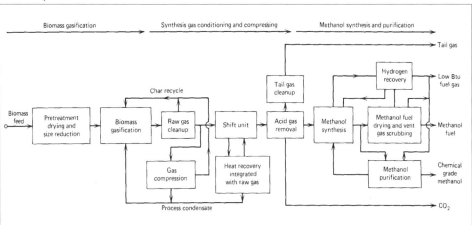

Figure 4.129. Schematic flow diagram for biomass to methanol conversion process (From Wan, Simmins, and Nguyen, 1981. Reprinted from *Biomass Gasification* (T. Reed, ed.), with permission. Copyright 1981, Noyes Data Corp., Park Ridge, NJ).

The future cost of methanol fuel is expected to reach US$ 8/GJ (Faaij and Hamelinck, 2002). The gasification can be made in closed environments, where all emissions are collected, as well as ash and slurry. Cleaning processes in the methanol formation steps will recover most catalysts in re-usable form, but other impurities would have to be disposed of along with the gasification products. Precise schemes for waste disposal have not been formulated, but it seems unlikely that all nutrients could be recycled to agri- or silviculture as in the case of ethanol fermentation (SMAB, 1978). However, the production of ammonia by a process similar to the one yielding methanol is an alternative use of the synthesis gas. Production of methanol from *eucalyptus* rather than from woody biomass has been studied in Brazil (Damen *et al.*, 2002). More fundamental studies aiming to better understand the way in which methanol production relies on degradation of lignin are ongoing (Minami *et al.*, 2002).

4.9 Other conversion processes

This chapter has described a number of general principles, which have been or may become of use in designing actual conversion devices, primarily those aimed at utilising the renewable flows of energy. Not all the existing or proposed devices have been covered.

An attempt has been made to describe at least one example of conversion into a useful form of energy, for each of the renewable resources described in Chapter 3. For this reason, the order in which conversion devices have been described in Chapter 4 roughly follows the order in which the distribution and magnitude of individual renewable flows or energy stores were discussed in Chapter 3. A few digressions from this scheme have been necessary, because some conversion devices (e.g. turbines) are common to several types of resources. This section contains examples of converting more speculative sources of energy, such as the salinity differences identified in section 3.7.2 as a renewable energy resource of possible interest. Conversion of nuclear energy is briefly mentioned, without details of actual devices. It is a non-renewable but potentially important energy source, if the long-held visions of safe breeder fission reactors or fusion devices should ever bear fruit.

4.9.1 Conversion of salinity gradient resources

As discussed in section 3.7.2 (Fig. 3.81), a salinity difference, such as the one existing between fresh (river) and saline (ocean) water, may be used to drive an osmotic pump, and the elevated water may in turn be used to generate electricity by an ordinary turbine (cf. section 4.5).

An alternative method, aiming directly at electricity production, takes advantage of the fact that the salts of saline water are ionised to a certain degree and thus may be used to derive an electrochemical cell of the type discussed in section 4.1.6 (Pattie, 1954; Weinstein and Leitz, 1976).

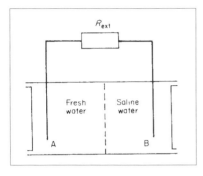

Figure 4.130. Schematic picture of dialytic battery.

This electrochemical cell, which may be called a "dialytic battery" since it is formally the same as the dialysis apparatus producing saline water with electric power input, is shown schematically in Fig. 4.130. In this case, the membrane allows one kind of ion to pass and thereby reach the corresponding electrode (in contrast to the osmotic pump in Fig. 3.81, where water could penetrate the membrane but ions could not). The free energy difference between the state with free Na^+ and Cl^- ions (assuming complete ionisation), and the state with no such ions is precisely the difference between saline and fresh water, calculated in section 3.7.2 and approximately given by (3.51). Assuming further that each Na^+ ion reaching the electrode A neutralises one electron, and that correspondingly each Cl^- ion at the electrode B gives rise to one electron (which may travel in the external circuit), then the electromotoric force, i.e. the electric potential ϕ of the cell, is given in analogy to (4.74), with the number of positive ions per mole equal to n_{Na^+}, and ϕ is related to the change in free energy (3.51) just as (4.74) was related to (4.71),

$$n_e \mathscr{F} \phi = n_{Na^+} \mathscr{F} \phi = \delta G \approx 2 \mathscr{R} T n_{Na^+},$$

or

$$\max (\Delta\phi_{ext}) = \phi \approx 2 \mathscr{R} T / \mathscr{F}. \tag{4.162}$$

Inserting the numerical values given in section 3.7.2, and $T = 300$ K, the cell electromotive force becomes $\phi \approx 0.05$ V. The actual value of the external potential, $\Delta\phi_{ext}$, may be reduced as a result of various losses, as described in section 4.1.6 (e.g. Fig. 4.11). ϕ is also altered if the "fresh" water has a finite amount of ions or if ions other than Na^+ and Cl^- are present.

If the load resistance is R_{ext}, the power delivered by the cell is given by the current $I = \Delta\phi_{ext} R_{ext}^{-1}$,

$$E = I\Delta\phi_{ext},$$

and as usual the maximum power corresponds to an external potential difference $\Delta\phi_{ext}$, which is smaller than the open circuit value (4.162). The internal losses may also be represented by an internal resistance R_{int} defined by

$$\Delta\phi_{ext} = \phi - IR_{int} = \phi - \Delta\phi_{ext}\,(R_{int}/R_{ext}).$$

Thus, the power output may also be written

$$E = \phi^2 R_{ext}/(R_{int} + R_{ext})^2.$$

R_{int} depends on electrode properties, as well as on n_{Na^+} and n_{Cl^-} and their variations across the cell.

Several anode-membrane plus cathode-membrane units may be stacked beside each other in order to obtain a sufficiently large current (necessary because the total internal resistance has a tendency to increase strongly, if the current is below a certain value). Figure 4.131 shows the results of a small-scale experiment (Weinstein and Leitz, 1976) for a stack of about 30 membrane-electrode pairs and an external load, R_{ext}, close to the one giving maximum power output. The power is shown, however, as a function of the salinity of the fresh water, and it is seen that the largest power output is obtained for a fresh water salinity which is not zero but 3–4% of the sea water salinity. The reason is that although the electromotive force (4.162) diminishes with increasing fresh water salinity, this is at first more than compensated for by the improved conductivity of the solution (decrease in R_{int}), when ions are also initially present in the fresh water compartment.

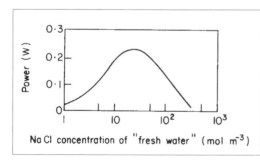

Figure 4.131. Measured performance of dialytic battery, containing about 30 membrane pairs in a stack. The external load was $R_{ext} = 10\ \Omega$ (close to power optimisation), and electrode losses had been compensated for by insertion of a balancing power source (based on Weinstein and Leitz, 1976).

Needless to say, small-scale experiments of the kind described above are a long way from a demonstration of viability of the salinity gradient conversion devices on a large scale, and it is by no means clear whether the dialysis battery concept will be more or less viable than power turbines based on osmotic pumps. The former seems likely to demand a larger membrane area than the latter, but no complete evaluation of design criteria has been made in either case.

4.A Energy bands in semiconductors

The electrons in a solid move in a potential, which for crystalline forms is periodic, corresponding to the lattice structure of the material. The forces on a given electron are electromagnetic, comprising the attractive interaction with the positively charged nuclei (the effective charge of which may be diminished by the screening effect of more tightly bound electrons) as well as repulsive interaction with other electrons. Owing to the small mass ratio between electrons and nuclei, the positions of the nuclei (R_i, $i = 1,2,3...$) may to a first approximation be regarded as fixed. This is the basis for saying that the electrons are moving in a "potential". When an electron is close to a definite nucleus, its wavefunction may be approximated by one of the atomic wavefunctions $\psi_{i,n(i)}$ $(r - R_i)$ for this isolated nucleus. One may therefore attempt to describe an arbitrary electron wavefunction as a linear combination of such atomic wavefunctions ("orbitals"),

$$\psi(r) = \sum_{i=1}^{N} \sum_{n(i)} c_{i,n(i)} \; \psi_{i,n(i)}(r - R_i). \tag{4.163}$$

Here N is the total number of atomic nuclei, and r is the position vector of the electron.

A simple example is that of a diatomic molecule ($N = 2$) with only one important level in each atom,

$$\psi(r) = c_{1,n(1)} \; \psi_{1,n(1)}(r-R_1) + c_{2,n(2)} \; \psi_{2,n(2)}(r-R_2),$$

for which two normalised solutions exist, with opposite relative sign between the two terms. If the overlap

$$S = \int \psi_{1,n(1)}{}^*(r-R_1) \; \psi_{2,n(2)}(r-R_2) \; dr$$

(the asterisk denotes complex conjugation) is non-zero, the energies of the two superposition states will be different from those ($W_{1,n(1)}$ and $W_{2,n(2)}$) of the atomic wavefunctions $\int \psi_{1,n(1)}$ and $\psi_{2,n(2)}$. The most tightly bound solution, $\psi_b(r)$, will correspond to an energy lower than the lowest of the original ones $W_{1,n(1)}$ and $W_{2,n(2)}$, while the other solution, $\psi_a(r)$, will correspond to an energy higher than the highest of the original ones (see e.g. Ballhausen and Gray, 1965).

If the energies W_b and W_a of the bonding solution $\psi_b(r)$ and the antibonding solution $\psi_a(r)$ are calculated for various relative positions $R = |R_1-R_2|$, one may obtain a picture as shown in Fig. 4.116 (approximately describing an ionised hydrogen molecule H_2^+).

If the number of (say, identical) nuclei, N, is larger than two, but still placed in a regular geometrical relationship characterised by a single distance R (e.g. the distance between atomic nuclei in a cubic lattice or a diamond type lattice), the superposition type wavefunction (4.163) still gives an energy picture similar to Fig. 4.132, but now with N curves, the highest and lowest of which may look like the two curves for $N = 2$.

For a solid crystal, N may be of the order of 10^{24}, so the region between the lowest energy state and the highest one will be filled out with energy levels that for all practical purposes can be regarded as continuous. Such a region is called an *energy band*. For each pair of overlapping atomic orbits, $n(1)$ and $n(2)$, there will be a set of linear combinations of wavefunctions in the $N = 2$ case, with bonding and anti-bonding characteristics. In the large-N case, each combination of overlapping atomic orbits will lead to an energy band, such that the overall picture of energy levels as a function of lattice characteristics (in the simple case just R) will be one of a number energy bands inside which all electron energies are "allowed", plus the spaces between energy bands in which no allowed energy is present. The energy band structure may, for example, resemble that shown in Fig. 4.133.

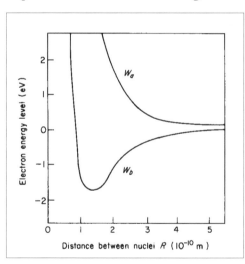

Figure 4.132. Electronic energy levels in a diatomic molecule as a function of inter-atomic distance (e.g. levels formed from $1s$ atomic orbitals in the H_2^+ molecule).

If the distance between the nuclei is large, an electron "feels" only one nucleus (plus a number of tightly bound "inner" electrons, if the nucleus has $Z > 2$) at a time, and the energy spectrum is identical to that of an individual atom (right-hand side of Fig. 4.133). As the assumed distance between nuclei becomes smaller, energy bands develop and become wider. When the energy bands begin to overlap and cross (left-hand side of Fig. 4.133), the basis for the linear assumption (4.163) breaks down, at least in the form discussed so far. If the restriction that the bands are formed from one definite atomic orbital [so that there is no sum over $n(i)$ in (4.163)], is relaxed, it may be possible still to use an expression of the form (4.163) to describe some features of the region of overlapping bands.

As an example, it is possible to obtain the two separate bands in silicon (see Fig. 4.133) by first defining two suitable superpositions of 3s and 3p atomic orbits [corresponding to the summation over $n(i)$, this procedure being called hybridisation in molecular theory], and then by applying the band theory [corresponding to the summation over i in (4.163)]. It is evident that in this region near the band crossing point, it is possible to find energy gaps between adjacent bands, which are much smaller than in the non-overlapping region.

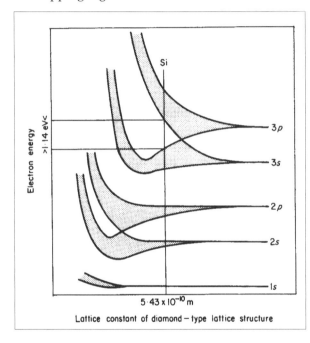

Figure 4.133. Energy band structure of a solid as a function of lattice constant (assumed to completely specify inter-atomic distances, as it does for crystals of diamond structure). Atomic physics labelling of orbits are indicated to the right. The band structure in the left-hand region of overlapping bands should be regarded as highly schematic. It may take different forms for different solids. The figure indicates how the overlapping 3s and 3p bands in silicon form the 1.14-eV energy gap further illustrated in Fig. 4.134.

The electron potentials initially represent a situation with one electron per atom being considered. The simplest approximation of many-electron states in one atom is to regard the electrons as independent of each other, except that they have to obey the Pauli exclusion principle according to which there can be at most one electron in each quantum state. Because the electron has an intrinsic spin of $S = \frac{1}{2}$, there are two spin states ("spin-up" and "spin-down") for each spatial orbit. Thus, the periodic system would be derived by first placing two electrons in the 1s-orbit (cf. Fig. 4.133), then two in the 2s-orbit and then six in the 2p-orbit (comprising three degenerate spatial orbits, because the orbital angular momentum is $L = 1$), etc.

A more accurate solution of the Schrödinger equation of quantum mechanics is necessary in order to include the effects of electron–electron interactions, which make the entire energy spectrum depend on electron number in addition to the smooth dependence on the charge of the nucleus ("atomic number"). In a number of cases, these interactions even influence the succession of orbits being filled in the periodic system.

Similar statements can be made for the energy bands in many-atom lattices. The spectrum of energy bands can be used to predict the filling systematics, as a function of the number of electrons per atom, but again the spectrum is not independent of the degree of filling (for this reason no energy scale is provided in Fig. 4.133). Thus, a diagram such as Fig. 4.133 cannot be used directly to predict the atomic distance (e.g. represented by the lattice constant) that will give the lowest total energy and hence the ground state of the system. This is in contrast to the H_2^+ molecule (Fig. 4.132), for which the minimum of the W_b curve determines the equilibrium distance between the nuclei.

Knowing the lattice structure and the lattice constant (the dimension of an elementary cube in the lattice), a vertical cut in a diagram of the type in Fig. 4.133 will give the allowed and forbidden energy values, with the reservations made above. Filling the proper number of electrons per atom into the energy bands, with reference to the Pauli principle, the energy of the latest added electron may be determined, as well as the lowest available energy level into which an additional electron added to the system may go.

If the electrons available exactly fill a band, and if there is a substantial distance to the next higher level, the material is an electrical insulator (no electrons can change state, i.e. they cannot "move"). If a band is partially filled, the material is a good conductor (e.g. a metal). The continuum of levels in the partially filled band allows the electrons to move throughout the lattice. If the distance between the highest filled band and an empty band is small, the material is called a semiconductor. At zero absolute temperature a semiconductor is an insulator, but because of the thermal energy spread at higher temperatures, given by (4.23) because electrons are Fermi particles, some electrons will be excited into the higher band (the "conduction band"). The conductance of silicon increases by a factor 10^6 between 250 and 450 K.

The bands corresponding to 1s, 2s and 2p atomic orbits are completely filled in Si, but then four electrons remain per atom. According to Fig. 4.133, the 3s and 3p bands are overlapping and allow two mixed bands to be constructed, with an intermediate gap of about 1 eV (1.6×10^{-5} J). The lower band (the "valence band") can hold four electrons per atom, so that this band will be completely full and the other one empty at zero temperature.

If a few of the atoms in a silicon lattice are replaced by an atom with higher Z (atomic number), e.g. phosphorus, the additional electrons associated with the impurity cannot be accommodated in the valence band, but will occupy a discrete level (named the "donor level") just below the conduction band (the energy depression being due to the larger attractive force from the atom of higher Z). A semiconductor material with this type of impurity is called *n*-type. The electrons in the donor-level are very easily excited into the conduction band. Adding *n*-type impurities makes the Fermi level [μ_i in (4.23)] move upwards from the initial position approximately half way between the valence and conduction bands.

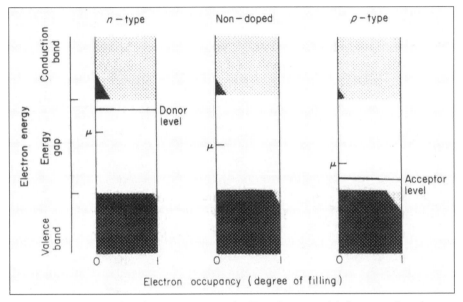

Figure 4.134. Energy band structure near the Fermi energy (μ) for a semiconductor material without impurities (middle column) or with *n*- or *p*-type doping (cf. text). The dark shading indicates the electron occupancy as a function of energy for a finite temperature (occupancy equal to unity corresponds to the maximum number of electrons at a given energy or energy interval which is consistent with the exclusion principle for Fermi-type particles).

Impurities with lower Z (e.g. Al in a Si-lattice) lead to electron vacancies or "holes", which are associated with an energy slightly higher than the top of the valence band, again due to the Z-dependence of the attractive force. These "acceptor levels" easily transfer holes to the valence band, a process which may, of course, alternatively be described as the excitation of electrons from the valence band into the acceptor level. Semiconductor material with this type of impurity is called *p*-type. The holes can formally be treated as particles like the electrons, but with positive charges and provided that the state of the pure semiconductor at zero temperature is taken as reference ("vacuum state" in quantum theory).

The energy diagrams of "doped" semiconductors of *n*- and *p*-types are sketched in Fig. 4.134. For a more complete account of basic semiconductor theory, see, for example, Shockley (1950).

4.10 Suggested topics for discussion

4.10.1

Discuss the production of heat of the temperature T_L from an amount of heat available, of temperature T above T_L, and access to a large reservoir of temperature T_{ref} below T_L.

Quantities of heat with temperatures T and T_{ref} may be simply mixed in order to obtain heat of the desired temperature.

Alternatively, a thermodynamic cycle may be used between the temperatures T and T_{ref} to produce a certain amount of mechanical or electrical energy, which then is used to power the compressor of a heat pump, lifting heat from T_{ref} to the desired temperature T_L.

As a variant, use a thermodynamic cycle between the temperatures T and T_L, producing less work but on the other hand providing reject heat at the desired temperature. Again, a heat pump is used to provide more heat of temperature T_L on the basis of the work output from the thermodynamic cycle.

What are the simple and the second law efficiencies in the individual cases?

4.10.2

Discuss design aspects of a propeller-type wind energy converter for which no possibility of changing the overall pitch angle θ_0 is to be provided.

Consider, for instance, a design similar to the one underlying Figs. 4.70–4.75, but which is required to be self-starting at a wind speed of 2 m s^{-1} and to have its maximum power coefficient at the wind speed of 4 m s^{-1}. Find the tip-speed ratio corresponding to this situation for a suitable overall pitch angle θ_0 and for given assumptions on the internal torque $Q_0(\Omega)$, e.g. that Q_0 is independent of the angular velocity Ω, but proportional to the third power of the blade radius R, with the proportionality factor being given by

$$Q_0 = 10^{-3}\, \pi\rho R^3 u^2_{cut\text{-}in},$$

with $u_{cut\text{-}in} = 2$ m s^{-1}.

Compare the power produced per square metre swept at a wind speed of 4 and 6 m s^{-1} to that which could be derived from a solar cell array of the same area.

For an actual design of wind energy converters, the blade shape and twist would not be given in advance, and one would, for a device with fixed angular velocity, first decide on this angular velocity (e.g. from generator and gearbox considerations) and then choose a design wind speed and optimise the blade shape $c(r)$ and blade twist $\theta(r)$, in order to obtain a large power coefficient at the design speed and possibly to avoid a rapid decrease in the power coefficient away from the design point.

The blade shape $c(r)$, of course, depends on blade number B, but to lowest order, the performance is independent of changes, which leave the product of $c(r)$ for the individual blades and B unchanged. Name some other considerations which could be of importance in deciding on the blade number.

4.10.3

In section 4.3.1, a maximum power coefficient C_p far above unity is suggested for certain cases of sail-driven propulsion on friction-free surfaces. Is this not a contradiction, like taking more power out than there is in the wind?

(Hint: C_p is defined relative to the wind power per unit area, which is of immediate relevance for rotors sweeping a fixed area. But what is the area seen by a sail-ship moving with velocity U?).

4.10.4

What is, e.g. based on local data, the ratio between the daily amount of solar radiation, which on a clear day at different seasons intercepts a fully tracking collector system, and that reaching a flat plate of fixed, south-facing orientation and a certain tilt angle s?

What is the relative importance of east–west tracking alone and solar height tracking alone?

4.10.5

Discuss shore-based wave energy converters, e.g. based on letting the wave trains ascend an inclined (and maybe narrowing) ramp, such that the kinetic energy is converted into potential energy of elevation, which may be used to drive a turbine.

The simple solutions for gravity waves in a deep ocean (section 2.D) cannot be used directly for waves approaching the shallow coastal waters. However, it may be assumed that the wave period remains unchanged, but that the amplitude a of the wave surface increases in such a way that the total power remains constant as the velocity potential becomes altered due to the boundary condition $d\phi/dt = 0$ at the sea bottom. This assumption implies the neglect of frictional dissipation at the sea floor.

There is a maximum ramp height that will permit a wave to reach the top and cross it into an elevated reservoir. If a larger height is required by the turbine, a narrowing ramp construction may be used to give some fraction of the water mass the desired elevation.

4.10.6

Combine productivity data such as those reported in section 3.6.2 with estimated efficiencies of fuel production by anaerobic fermentation processes (section 4.8.2) to give overall conversion efficiencies of bioconversion of solar energy into fuels.

Compare this to other methods of converting solar radiation into mechanical or electrical work, and discuss relative virtues other than conversion efficiency.

4.10.7

Estimate the magnitude of and seasonal variation in stored food energy, separately for standing crops on the field, for livestock to be used for food, and for actually stored food (in grain stores, freezers, etc.). Compare food energy stored to the energy value of emergency stores of oil and natural gas, for example, for your own country.

Hint: Some data can be found in Chapter 6.

4.10.8

Estimate the potential world production of equivalent crude oil from *Euphorbia* plants, assuming only arid land to be utilised and annual harvest yields of 40 MJ m^{-2}.

4.10.9

Consider a continuous operation biogas digester for a one-family farm. The digester feedstock is pig slurry (collected daily) and straw (stored). The biogas is used for cooking, for hot water, and for space heating. Estimate the load curves for different seasons and calculate the volume of uncompressed gas storage required, if load is to be met at any time during the year.

4.10.10

Consider pure methanol- and pure ethanol-driven cars, and for comparison a gasoline-driven car, weighing 800 kg and going on average 18 km per litre of gasoline and 500 km on a full tank. Calculate the mass penalties for methanol and ethanol fuel tanks, if the same operation range is required. Assume that the fuel usage varies linearly with total mass of the car.

CHAPTER

5

ENERGY TRANSMISSION AND STORAGE

5.1 Energy Transmission

Transport of energy may, of course, be in the form of transportation of fuels to the site of conversion. With regard to renewable energy resources, such transport is useful in connection with biomass-derived energy, either by transporting the biological materials themselves or by conversion into bio-fuels (cf. section 4.8) which may be more convenient to move. For most other types of renewable energy, the resource itself cannot be "moved" (possible exceptions may exist, such as diverting a river flow to the place of hydro-power utilisation). Instead, an initial conversion process may be performed, and the emerging energy form may be transmitted to the load areas, where it may be used directly or subjected to a second conversion process before de-livery to the actual users.

Like fuels (which represent chemical energy), heat, mechanical and pos-sibly electrical energy may be stored, and the storage "containers" may be transported. Alternatively, energy may be transmitted through a suitable transmission system, which may involve pipeline transmission (for heat, fu-els and certain types of mechanical energy, e.g. pressure or kinetic energy of a gas or a fluid), electric transmission lines (for electricity) or radiant trans-mission (for heat or electricity).

Energy transmission is used not only to deliver energy from the conven-ient sites of generation (such as where the renewable resource are) to the dominant sites of energy use, but also to deal with mismatch between the time distribution of (renewable) energy generation and time variations in demand. As such, energy transmission and energy storage may supplement each other. Some demands may be displaceable, while others are time urgent. The latter ones often have a systematic variation over the hours of the day and over the seasons. This may be taken advantage of by long-distance transmission of energy across time-zones (east–west).

5.1.1 Transmission of heat

District heating lines

Most heating systems involve the transport of sensible heat in a fluid or gas (such as water or air) through pipes or channels. Examples are the solar heating system illustrated in Fig. 4.21, the heat pump illustrated in Fig. 4.104 and the geothermal heating plants illustrated in Fig. 4.107. Solar heating systems and heat pumps may be used in a decentralised manner, with an individual system providing heat for a single building, but they may also be used on a community scale, with one installation providing heat for a building block, a factory complex, a whole village or a city of a certain size. Many combined heat and power (CHP) systems involve heat transmission through distances of 10–50 km (decentralised and centralised CHP plants). In some regions, pure heating plants are attached to a district heating grid.

Assuming that a hot fluid or gas is produced at the central conversion plant, the transmission may be accomplished by pumping the fluid or gas through a pipeline to the load points. The pipeline may be placed underground and the tubing insulated in order to reduce conduction and convection of heat away from the pipe. If the temperature of the surrounding medium can be regarded as approximately constant, equal to T_{ref}, the variation in fluid (or gas) temperature $T^{fluid}(x)$ along the transmission line (with the path-length co-ordinate denoted x) can be evaluated by the same expression (4.98) used in the simple description of a heat exchanger in section 4.2.1. Temperature variations across the inside area of the pipe perpendicular to the stream-wise direction are not considered.

The rate of heat loss from the pipe section between the distances x and $(x+dx)$ is then of the form

$$dE/dx = J_m\, C_p^{fluid}\, dT^{fluid}(x)/dx = h'\, (T_{ref} - T^{fluid}(x)), \qquad (5.1)$$

where h' for a cylindrical insulated pipe of inner and outer radii r_1 and r_2 is related to the heat transfer coefficient λ^{pipe} [conductivity plus convection terms, cf. (3.34)] by

$$h' = 2\pi\, \lambda^{pipe} / \log(r_2/r_1).$$

Upon integration from the beginning of the transmission pipe, $x = x_1$, (5.1) gives, in analogy to (4.98),

$$T^{fluid}(x) = T_{ref} + (T^{fluid}(x_1) - T_{ref}) \exp(-h'(x-x_1) / J_m C_p^{fluid}), \qquad (5.2)$$

where J_m is the mass flow rate and C_p^{fluid} is the fluid (or gas) heat capacity. The total rate of heat loss due to the temperature gradient between the fluid in the pipe and the surroundings is obtained by integrating (5.1) from x_1 to the end of the line, x_2,

$$\Delta E = \int_{x_1}^{x_2} \frac{dE}{dx} dx = -J_m C_p^{fluid} (T^{fluid}(x_1) - T_{ref}) \left(1 - \exp\left(-\frac{h'(x_2 - x_1)}{J_m C_p^{fluid}} \right) \right).$$

The relative loss of heat supplied to the pipeline, along the entire transmission line, is then

$$\frac{\Delta E}{E} = -\left(1 - \exp\left(-\frac{h'(x_2 - x_1)}{J_m C_p^{fluid}} \right) \right), \qquad (5.3)$$

provided that the heat entering the pipe, E, is measured relative to a reservoir of temperature T_{ref}.

The total transmission loss will be larger than (5.3) because of friction in the pipe. The distribution of flow velocity, v, over the pipe inside cross section depends on the roughness of the inner walls (v being zero at the wall), and the velocity profile generally depends on the magnitude of the velocity (say at the centreline). Extensive literature exists on the flow in pipes (see e.g. Grimson, 1971). The net result is that additional energy must be provided in order to make up for the pipe losses. For an incompressible fluid moving in a horizontal pipe of constant dimensions, flow continuity demands that the velocity field is constant along the pipeline, and the friction can in this case be described by a uniformly decreasing pressure in the stream-wise direction. A pump must provide the energy flux necessary to compensate for the pressure loss,

$$\Delta E^{pump} = - Av \, \Delta P,$$

where A is the area of the inner pipe cross section and ΔP is the total pressure drop along the pipeline.

Transmission lines presently in use for city district heating have average heat losses of 10–15%, depending on insulation thickness (see e.g. WEC, 1991). The pump energy added is in the form of high-quality mechanical work, but with proper dimensioning of the tubes it may be kept to a small fraction of the heat energy transmitted. This energy is converted into heat by the frictional dissipation, and some of this heat may actually be credited to the heat transmission. Maximum heat transmission distances currently considered economic are around 30 km, with the possible exception of some geothermal installations. Figure 5.1 shows an integrated design that allows much easier installation than earlier pipes with separate insulation. In countries with a high penetration of combined heat and power production, such as Denmark, a heat storage facility holding some 10 h of heat load is sometimes added to each plant, in order that the (fuel-based) plant does not have to generate more electric power than needed, say, at night-time when the heating load is large (Danish Energy Agency, 1993).

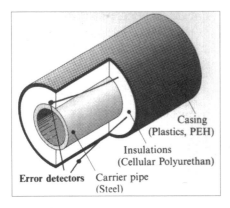

Figure 5.1. Pre-insulated district heating pipe (Danish Energy Agency, 1993).

Heat pipes

A special heat transmission problem exists in cases where very large amounts of heat have to be delivered to or removed from a small region, for example, in connection with highly concentrating solar energy collectors. Such heat transfer may be accomplished by a heat pipe, a device described in section 4.1.3 in relation to thermionic generators.

5.1.2 Transmission of electricity

Normal conducting lines

At present, electric current is transmitted in major utility grids, as well as distributed locally to each load site by means of conducting wires. Electricity use is dominated by alternating current (AC), as far as utility networks are concerned, and most transmission over distances up to a few hundred kilometres is by AC. For transmission over longer distances (e.g. by ocean cables), conversion to direct current (DC) before transmission and back to AC after transmission is common. Cables are either buried in the ground (with appropriate electric insulation) or are overhead lines suspended in the air between masts, without electrical insulation around the wires. Insulating connections are provided at the tower fastening points, but otherwise the low electric conductivity of air is counted on. This implies that the losses will comprise conduction losses depending on the instantaneous state of the air (the "weather situation"), in addition to the ohmic losses connected with the resistance R of the wire itself, $E^{heat} = RI^2$, with I being the current. The leak current between the elevated wire and the ground depends on the potential difference as well as on the integrated resistivity (cf. section 3.7.1), such that the larger the voltage, the further the wires must be placed from the ground.

Averaged over different meteorological conditions, the losses in a standard AC overhead transmission line (138–400 kV, at an elevation of some 15– 40 m) are currently a little under 1% per 100 km of transmission (Hammond *et al.*, 1973), but the overall transmission losses of utility networks, including the finely branched distribution networks in the load areas, may for many older, existing grids amount to some 12–15% of the power production, for a grid extending over a land area of about 10^4 km^2 (Blegaa *et al.*, 1976). Losses are down to 5–6% for the best systems installed at present and are expected to decrease further to the level of 2–3% in the future, when the currently best technologies penetrate further (Kuemmel *et al.*, 1997). This loss is calculated relative to the total production of electricity at the power plants attached to the common grid, and thus includes certain in-plant and transformer losses. The numbers also represent annual averages for a power utility system occasionally exchanging power with other utility systems through interconnecting transmission lines, which may involve transmission distances much longer than the linear extent of the load area being serviced by the single utility system in question.

The trend is to replace overhead lines by underground cables, primarily for visual and environmental reasons. This has already happened for the distribution lines in Europe and is increasingly also being required for transmission lines. In Japan and the USA, overhead lines are still common.

Underground transmission and distribution lines range from simple co-axial cables to more sophisticated constructions insulated by a compressed gas. Several trans-ocean cables (up to 1000 km) have been installed in the Scandinavian region in order to bring the potentially large surpluses of hydropower production to the European continent. The losses through these high-voltage (up to 1000 kV) DC lines are under 0.4% per 100 km, to which should be added the 1–2% transmission loss occurring at the thyristor converters on shore that transform AC into DC and vice versa (Ch. 19 in IPCC, 1996a). The cost of these low-loss lines is currently approaching that of conventional AC underwater cables (about 2 euro kW^{-1} km^{-1}; Meibom *et al.*, 1997, 1999; Wizelius, 1998).

One factor influencing the performance of underground transmission lines is the slowness of heat transport in most soils. In order to maintain the temperature within the limits required by the materials used, active cooling of the cable could be introduced, particularly if large amounts of power have to be transmitted. For example, the cable may be cooled to 77 K (liquid nitrogen temperature) by means of refrigerators spaced at intervals of about 10 km (cf. Hammond *et al.*, 1973). This allows increased amounts of power to be transmitted in a given cable, but the overall losses are hardly reduced, since the reduced resistance in the conductors is probably outweighed by the energy spent on cooling. According to (4.22), the cooling efficiency is limited by a Carnot value of around 0.35, i.e. more than three units of work have to be supplied in order to remove one unit of heat at 77 K.

Superconducting lines

For DC transmission, the ohmic losses may be completely eliminated by use of superconducting lines. A number of elements, alloys and compounds become superconducting when cooled to a sufficiently low temperature. Physically, the onset of superconductivity is associated with the sudden appearance of an energy gap between the "ground state", i.e. the overall state of the electrons, and any excited electron state (similar to the situation illustrated in Fig. 4.134, but for the entire system rather than for individual electrons). A current, i.e. a collective displacement (flow) of electrons, will not be able to excite the system away from the "ground state" unless the interaction is strong enough to overcome the energy gap. This implies that no mechanism is available for the transfer of energy from the current to other degrees of freedom, and thus the current will not lose any energy, which is equivalent to stating that the resistance is zero. In order that the electron system remains in the ground state, the thermal energy spread must be smaller than the energy needed to cross the energy gap. This is the reason why superconductivity occurs only below a certain temperature, which may be quite low (e.g. 9 K for niobium, 18 K for niobium–tin, Nb_3Sn). However, there are other mechanisms that in more complex compounds can prevent instability, thereby explaining the findings in recent years of materials that exhibit superconductivity at temperatures approaching ambient (Pines, 1994; Demler and Zhang, 1998).

For AC transmission, a superconducting line will not be loss-free, owing to excitations caused by the time variations of the electromagnetic field (cf. Hein, 1974), but the losses will be much smaller than for normal lines. It is estimated that the amount of power that can be transmitted through a single cable is in the gigawatt range. This figure is based on suggested designs, including the required refrigeration and thermal insulation components within overall dimensions of about 0.5 m (cable diameter). The power required for cooling, i.e. to compensate for heat flow into the cable, must be considered in order to calculate the total power losses in transmission.

For transmission over longer distances it may, in any case, be an advantage to use direct current, despite the losses in the AC–DC and DC–AC conversions (a few per cent as discussed above). Future intercontinental transmission using superconducting lines has been discussed by Nielsen and Sørensen (1996) and by Sørensen and Meibom (1998) (cf. scenario in section 6.4).

5.1.3 Other types of transmission

Fuels such as natural gas, biogas, hydrogen and other energy-carrier gases may be transmitted through pipelines, at the expense of only a modest amount of pumping energy (at least for horizontal transfer). Pipeline oil transmission is also in abundant use. Alternatives used for sea transportation between continents are containers onboard ships for solid fuels, oil, compressed gases or liquefied gases. Similar containers are in use for shorter distance transport by rail or road. Higher energy densities may be obtained by the energy storage devices discussed in section 5.2.2, such as metal hydrides or carbon nanotubes, e.g. for hydrogen transport. Light container materials are, of course, preferable in order to reduce the cost of moving fuels by vessel, whether on land or at sea.

Current natural gas networks consist of plastic distribution lines operated at pressures of 0.103 to about 0.4 MPa and steel transmission lines operated at pressures of 5-8 MPa. With some upgrading of valves, some modern natural gas pipelines could be used for the transmission of hydrogen (Sørensen *et al.*, 2001). Certain steel types may become brittle with time, as a result of hydrogen penetration into the material, and cracks may develop. It is believed that H_2S impurities in the hydrogen stream increases the problem, but investigations of the steel types currently used for new pipelines indicate little probability of damage by hydrogen (Pöpperling *et al.*, 1982; Kussmaul and Deimel, 1995) .

Mechanical devices have been used to transfer mechanical energy over short distances, but mechanical connections with moving parts are not practical for distances of transfer which may be considered relevant for transmission lines. However, mechanical energy in such forms as hydraulic pulses can be transmitted over longer distances in feasible ways, as, for example, mentioned in connection with wave energy conversion devices placed in open oceans (section 4.4.2).

Finally, radiant transmission of energy may be mentioned. The technique for transmitting radiation and re-collecting the energy (or some part of it) is well developed for wavelengths near or above visible light. Examples are laser beams (stimulated atomic emission) and microwave beams (produced by accelerating charges in suitable antennas), ranging from the infrared to the wavelengths used for radio and other data transmission, e.g. between satellites and ground-based receivers. Large-scale transmission of energy in the form of microwave radiation has been proposed in connection with satellite solar power generation, but is not currently practical. Short distance transmission of signals, e.g. between computers and peripheral units, do involve minute transfers of energy.

5.2 Energy storage

Storage of energy provides a way of adjusting to variations in the energy demand, i.e. a way of meeting a load with a time-dependence different from that of generation. For fuel-type energy, storage can help burn the fuel more efficiently, by avoiding those situations where demand variations would otherwise require regulation of combustion rates beyond what is technically feasible or economic. For renewable energy sources of fluctuating nature, storage can help make energy systems including such sources as dependable as the conventional systems.

Ideal requirements of energy storage include rapid access and versatility of the energy form in which energy from the store is delivered. Conversion of one type of stored energy (initial fuel) into another form of stored energy may be advantageous in order to better suit utilisation. For example, the production of electricity in large, fossil or nuclear power plants may involve long start-up times including additional costs when used for load levelling, while the use of pumped water storage allows delivery upon demand in less than a minute. The economic feasibility of such energy storage depends on the relative fixed and variable costs of the different converter types and on the cost and availability of different fuels. Another example is a fuel-driven automobile, which operates at or near peak power only during short intervals of acceleration. If short-term storage is provided, which can accumulate energy produced off-peak by the automobile engine and deliver the power for starting and acceleration, then the capacity of the primary engine may be greatly reduced.

In connection with renewable energy resources, most of which are intermittent and of a fluctuating power level, supplementing conversion by energy storage is essential if the actual demand is to be met at all times. The only alternative would seem to be fuel-based back-up conversion equipment, but this, of course, is just a kind of energy storage which in the long run may require fuels provided by conversion processes based on renewable primary energy sources.

5.2.1 Storage of heat

Heat capacity storage

Heat capacity, or "sensible heat" storage, is accomplished by changing the temperature of a material without changing its phase or chemical composition. The amount of energy stored by heating a piece of material of mass m from temperature T_0 to temperature T_1 at constant pressure is

$$E = m \int_{T_0}^{T_1} c_P \, dT, \tag{5.4}$$

where c_P is the specific heat capacity at constant pressure.

Energy storage at low temperatures is needed in renewable systems such as solar absorbers delivering space heating, hot water, and eventually heat for cooking (up to 100°C). The actual heat storage devices may be of modest size, aiming at delivering heat during the night after a sunny day, or they may be somewhat larger, capable of meeting the demand during a number of consecutive overcast days. Finally, the storage system may provide seasonal storage of heat, as required at high latitudes where seasonal variations in solar radiation are large, and, furthermore, heat loads are to some extent inversely correlated with the length of the day.

Another aspect of low-temperature heat storage (as well as of some other energy forms) is the amount of decentralisation. Many solar absorption systems are conveniently placed on existing rooftops, that is, in a highly decentralised fashion. A sensible-energy heat store, however, typically loses heat from its container, insulated or not, in proportion to the surface area. The relative heat loss is smaller, the larger the store dimensions, and thus more centralised storage facilities, for example, of communal size, may prove superior to individual installations. This depends on an economic balance between the size advantage and the cost of additional heat transmission lines for connecting individual buildings to a central storage facility. One should also consider other factors, such as the possible advantage in supply security offered by the common storage facility (which would be able to deliver heat, for instance, to a building with malfunctioning solar collectors).

Water storage

Heat energy intended for later use at temperatures below 100°C may conveniently be stored as hot water, owing to the high heat capacity of water (4180 J kg^{-1} K^{-1} or 4.18 × 10^6 J m^{-3} K^{-1} at standard temperature and pressure), combined with the fairly low thermal conductivity of water (0.56 J m^{-1} s^{-1} K^{-1} at 0°C, rising to 0.68 J m^{-1} s^{-1} K^{-1} at 100°C).

Most space heating and hot water systems of individual buildings include a water storage tank, usually in the form of an insulated steel container with a capacity corresponding to less than a day's hot water usage and often only a small fraction of a cold winter day's space heating load. For a one-family dwelling, a 0.1-m^3 tank is typical in Europe and the United States.

A steel hot water container may look like the one sketched in Fig. 5.2. It is cylindrical with a height greater than the diameter in order to make good temperature stratification possible, an important feature if the container is part of a solar heating system. A temperature difference of up to 50°C between the top and bottom water can be maintained, with substantial improvements (over 15%) in the performance of the solar collector heating system, because the conversion efficiency of the collector declines (see Fig. 5.3) with the temperature difference between the water coming into the collector and the ambient outdoor temperature (van Koppen *et al.*, 1979). Thus, the water from the cold lower part of the storage tank would be used as input to the solar collector circuit, and the heated water leaving the solar absorber would be delivered to a higher-temperature region of the storage tank, normally the top layer. The take-out from the storage tank to load (directly or through a heat exchanger) is also from the top of the tank, because the solar system will, in this case, be able to cover load over a longer period of the year (and possibly for the entire year). There is typically a minimum

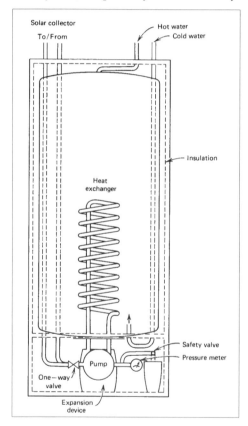

Figure 5.2. Water container for heat storage with possibility of temperature stratification (e.g. for use in connection with solar collectors) (Ellehauge, 1981).

temperature required for the fluid carrying heat to the load areas, and during winter, the solar collector system may not always be able to raise the entire storage tank volume to a temperature above this threshold. Thus, temperature stratification in storage containers is often a helpful feature. The minimum load-input temperatures are around 45–50°C for space heating through water-filled "radiators" and "convectors," but only 25–30°C for water-filled floor heating systems and air-based heating and ventilation duct systems.

For hot water provision for a single family, using solar collector systems with a few square meters of collectors (1 m^2 in sunny climates, 3–5 m^2 at high latitudes), a storage tank of around 0.3 m^3 is sufficient for diurnal storage, while a larger tank is needed if consecutive days of zero solar heat absorption can be expected. For complete hot water and space heating solar systems, storage requirements can be quite substantial if load is to be met at all times and the solar radiation has a pronounced seasonal variation.

Most solar collector systems aiming at provision of both hot water and space heating for a single-family dwelling have a fairly small volume of storage and rely on auxiliary heat sources. This is the result of an economic trade-off due to the rapid reduction in solar collector gains with increasing coverage, that is, the energy supplied by the last square meter of collector added to the system becomes smaller and smaller as the total collector area increases. Of course, the gain is higher with increased storage for fixed collector area over some range of system sizes, but this gain is very modest (see section 6.3.1).

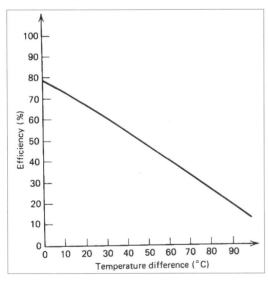

Figure 5.3. Efficiency curve for flat-plate solar collector as a function of the temperature difference between the average temperature of the fluid, which is removing heat from the collector, and the ambient, outside air temperature. The efficiency is the percentage of the absorbed solar radiation, which is transferred to the working fluid. The curve is based on measurements for a selective-surface collector, tilted 45° and receiving 800 W m^{-2} of solar radiation at incident angles below 30°. Wind speed along the front of collector is 5 m s^{-1} (Svendsen, 1980).

For this reason, many solar space heating systems only have diurnal storage, say, a hot water storage. In order to avoid boiling, when the solar radiation level is high for a given day, the circulation from storage tank to collector is disconnected whenever the storage temperature is above some specified value (e.g. 80°C), and the collector becomes stagnant. This is usually no problem for simple collectors with one layer of glazing and black paint absorber, but with multilayered cover or selective-surface coating of the absorber, the stagnation temperatures are often too high and materials would become damaged if these situations were allowed to occur. Instead, the storage size may be increased to such a value that continuous circulation of water from storage through collectors can be maintained during such sunny days, without violating maximum storage temperature requirements at any time during the year. If the solar collector is so efficient that it still has a net gain above 100°C (such as the one shown in Fig. 5.3), this heat gain must be balanced by heat losses from piping and from the storage itself, or the store must be large enough for the accumulated temperature rise during the most sunny periods of the year to be acceptable. In high-latitude climatic regions, this can be achieved with a few square metres of water storage, for collector areas up to about 50 m².

Larger amounts of storage may be useful if the winter is generally sunny, but a few consecutive days of poor solar radiation do occur from time to time. This, for example, is the situation for an experimental house in Regina, Saskatchewan (Besant *et al.*, 1979). The house is superinsulated and is designed to derive 33% of its heating load from passive gains of south-facing windows, 55% from activities in the house (body heat, electric appliances), and the remaining 12% from high-efficiency (evacuated tube) solar collectors. The collector area is 18 m² and there is a 13-m³ water storage tank. On a January day with outdoor temperatures between −25°C (afternoon) and −30°C (early morning), about half the building heat loss is provided by indirect gains, and the other half (about 160 MJ d^{-1}) must be drawn from the store, in order to maintain indoor temperatures of 21°C, allowed to drop to 17°C between midnight and 0700 h. On sunny January days, the amount of energy drawn from the storage reduces to about 70 MJ d^{-1}. Since overcast periods of much over a week occur very rarely, the storage size is such that 100% coverage can be expected, from indirect and direct solar sources, in most years.

The situation is very different in, for instance, Denmark. Although the heating season has only 2700 celsius degree days (due to Gulf Stream warming), as compared with 6000 degree days for Regina, there is very little solar gain (through windows or to a solar collector) during the months of November through February. A modest storage volume, even with a large solar collector area, is therefore unable to maintain full coverage during the winter period, as indicated by the variations in storage temperatures in a concrete case, shown in Fig. 5.4 (the water store is situated in the attic, losing heat to about ambient air temperatures; this explains why the storage temperatures approach freezing in January).

In order to achieve near 100% coverage under conditions such as the Danish ones, very large water tanks would be required, featuring facilities to maintain stable temperature stratification and containing so much insulation (over 1 m) that truly seasonal mismatch between heat production and use can be handled. This is still extremely difficult to achieve for a single house, but for a communal system for a number of reasonably close buildings and with storage facility (and maybe also collectors) placed centrally, 100% coverage should be possible (see section 6.3.1).

Community-size storage facilities
With increasing storage container size, the heat loss through the surface – for a given thickness of insulation – will decrease per unit of heat stored. There are two cases, depending on whether the medium surrounding the container (air, soil, etc.) is rapidly mixing or not. Consider first the case of a storage container surrounded by air.

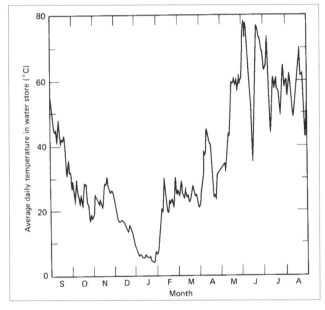

Figure 5.4. Measured average daily temperature in a 5.5-m^3 water storage tank (having 0.2 m of rock wool insulation) fed by 50 m^2 of solar flatplate collector (two layers of glass cover, non-selective absorber, tilt angle 38°). The solar system covers 12% of the building heat load (based on Jørgensen *et al.*, 1980).

The container may be cylindrical such as the one illustrated in Fig. 5.2. The rate of heat loss is assumed proportional to the surface area and to the temperature difference between inside and outside, with the proportionality constant being denoted U. It is sometimes approximated in the form

$$U = 1/(x/\lambda + \mu),$$

where x is the thickness of insulation, λ is the thermal conductivity of the insulating material (about 0.04 W m^{-1} per °C for mineral wool) and μ (around 0.1 m^2 W^{-1} per °C) is a parameter describing the heat transfer in the boundary layer air [may be seen as a crude approximation to (4.89)].

The total heat loss rate is

$$P^{loss} = 2\pi R (R + L) U (T_s - T_a), \tag{5.5}$$

where R and L are radius and height of the cylinder, T_s is the average temperature of the water in the store, and T_a is the outside ambient air temperature. The fraction of the stored heat energy lost per unit time is

$$\frac{P^{loss}}{E^{sens}} = \frac{2U(1+R/L)}{R\,c_P^{water}\,\rho^{water}}, \tag{5.6}$$

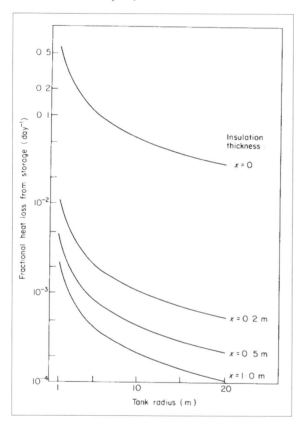

Figure 5.5. Fractional heat loss from a cylindrical storage tank with different degrees of insulation, assuming well-mixed conditions inside as well as outside the tank.

that is, the loss is independent of the temperatures and inversely proportional to a linear system dimension (c_P is heat capacity and ρ is density).

Figure 5.5 shows, for $L = 3R$, the ratio of heat loss rate P^{loss} and stored energy E^{sens}, according to (5.6), as a function of R and for different values of the insulation thickness. This ratio is independent of the temperature difference, owing to the linear assumption for the heat loss rate. If the storage times required imply maximum fractional losses of 10–20%, uninsulated tanks in the size range $5\ \text{m} \le R \le 20\ \text{m}$ are adequate for a few days of storage. If storage times around a month are required, some insulation must be added to tanks in the size range under consideration. If seasonal storage is needed, the smallest tank sizes will not work even with a metre of mineral or glass wool wrapping. Community-size tanks, however, may serve for seasonal storage, with a moderate amount of insulation (0.2 m in the specific example).

A hot water tank with $R = 11.5$ m and $L = 32$ m has been used since 1978 by a utility company in Odense, Denmark, in connection with combined production of electricity and heat for district heating (Jensen, 1981). The hot water store is capable of providing all necessary heating during winter electricity peak hours, during which the utility company wants a maximum electricity production. With a hot water store two or three times larger, the cogenerating power units could be allowed to follow the electricity demand, which is small relative to the heat demand during winter nights.

A hot water store of similar magnitude, around 13 000 m^3, may serve a solar heated community system for 50–100 one-family houses, connected to the common storage facility by district heating lines. The solar collectors may still be on individual rooftops, or they may be placed centrally, for example, in connection with the store. In the first case, more piping and labour is required for installation, but in the second case, land area normally has to be dedicated to the collectors. Performance is also different for the two systems, as long as the coverage by solar energy is substantially less than 100%, and the auxiliary heat source feeds into the district heating lines. The reason is that when storage temperature is below the minimum required, the central solar collector will perform at high efficiency (Fig. 5.3), whereas individual solar collectors will receive input temperatures already raised by the ancillary heat source and thus not perform as well. Alternatively, auxiliary heat should be added by individual installations on the load side of the system, but, unless the auxiliary heat is electrically generated, this is inconvenient if the houses do not already possess a fuel-based heating system.

Most cost estimates speak against storage containers placed in air. If the container is buried underground (possibly with its top facing the atmosphere), the heat escaping the container surface will not be rapidly mixed into the surrounding soil or rock. Instead, the region closest to the container will reach a higher temperature, and a temperature gradient through the soil or rock will slowly be built up. An exception is soil with ground water infiltration. Here the moving water helps to mix the heat from the container into the surroundings. However, if a site can be found with no ground water (or at least no ground water in motion), then the heat loss from the store will be greatly reduced, and the surrounding soil or rock can be said to function as an extension of the storage volume.

As an example, let us consider a spherical water tank embedded in homogeneous soil. The tank radius is denoted R, the water temperature is T_s and the soil temperature far away from the water tank is T_0. If the transport of heat can be described by a diffusion equation, then the temperature distribution as function of distance from the centre of the storage container may be written (Shelton, 1975)

$$T(r) = T_0 + (T_s - T_0) \, R/r, \tag{5.7}$$

where the distance r from the centre must be larger than the tank radius R in order for the expression to be valid. The corresponding heat loss is

$$P^{sens} = \int_{sphere} \lambda \, \partial \, T(r)/\partial \, r \, \mathrm{d}A = -\lambda \, (T_s - T_0) \, 4\pi R, \tag{5.8}$$

where λ is the heat conductivity of the soil and (5.8) gives the heat flux out of any sphere around the store, of radius $r \geq R$. The flux is independent of r. The loss relative to the heat stored in the tank itself is

$$P^{loss} / E^{sens} = -3\lambda / (R^2 \, c_p^{water} \, \rho^{water}). \tag{5.9}$$

Compared to (5.6), it is seen that the relative loss from the earth-buried store is declining more rapidly with increasing storage size than the loss from a water store in air or other well-mixed surroundings. The fractional loss goes as R^{-2} rather than as R^{-1}.

Practical considerations in building an underground or partly underground water store suggest an upside-down obelisk shape and a depth around 10 m for a 20 000-m^2 storage volume. The obelisk is characterised by tilting sides, with a slope as steep as feasible for the soil type encountered. The top of the obelisk (the largest area end) would be at ground level or slightly above it, and the sides and bottom would be lined with plastic foil not penetrable by water. Insulation between lining and ground can be made with mineral wool foundation elements or similar materials. As a top cover, a sailcloth held in bubble shape by slight overpressure is believed to be the least expensive solution. Top insulation of the water in the store can be floating foam material. If the bubble cloth is impermeable to light, algae growth in the water can be avoided (Danish Department of Energy, 1979).

Two community-size seasonal hot water stores placed underground are operating in Sweden. They are both shaped as cut cones. One is in Studsvik. Its volume is 610 m^3, and 120 m^2 of concentrating solar collectors float on the top insulation, which can be turned to face the sun. Heat is provided for an office building of 500-m^2 floor area. The other system is in the Lambohov district of Linköping. It serves 55 semidetached one-family houses having a total of 2600-m^2 flat-plate solar collectors on their roofs. The storage is 10 000 m^3 and situated in solid rock (excavated by blasting). Both installations have operated since 1979, and they furnish a large part of the heat loads of the respective buildings. Figure 5.6 gives the temperature changes by season for the Studsvik project (Andreen and Schedin, 1980; Margen, 1980; Roseen, 1978).

Another possibility is to use existing ponds or lake sections for hot water storage. Top insulation would normally be required, and, in the case of lakes used only in part, an insulating curtain should separate the hot and the cold water.

Keeping the time derivative in (3.35), a numerical integration will add information about the time required for reaching the steady-state situation. According to calculations such as Shelton's (1975), this may be about one year for typical large heat store cases.

The assumption of a constant temperature throughout the storage tank may not be valid, particularly not for large tanks, owing to natural stratification that leaves the colder water at the bottom and the warmer water at the top. Artificial mixing by mechanical stirring is, of course, possible, but for many applications temperature stratification would be preferable. This could be used to advantage, for example, with storage tanks in a solar heating system (see Fig. 4.21), by feeding the collector with the colder water from the lower part of the storage and thus improving the efficiency (4.102) of the solar collector.

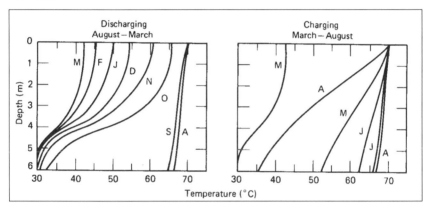

Figure 5.6. Temperature profiles in the Studsvik hot water store, monthly during (right) charging period March (M) to August (A) and during (left) discharging period from August (A) to March (M) (based on Roseen, 1978).

An alternative would be to divide the tank into physically separated sub-units (cf. e.g. Duffie and Beckman, 1974), but this requires a more elaborate control system to introduce and remove heat from the different units in an optimised way, and if the storage has to be insulated the sub-units should at least be placed with common boundaries (tolerating some heat transfer among the units) in order to keep the total insulation requirement down.

Multi-unit storage systems may be most appropriate for uninsulated storage in the ground, owing to the participation of the soil between the storage units. Figure 5.7 shows a possible arrangement (Brüel *et al.*, 1976) based on cylindrical heat exchangers placed in one or more circles around a central heat exchanger cylinder in the soil. If no active aquifers are traversing the soil region, a steady-state temperature distribution with a high central temperature can be built up. Used in connection with, for example, a flat-plate solar collector, it is possible to choose to take the fluid going to the collector circuit from a low temperature and to deliver the return fluid from the solar collector to successive heat exchangers of diminishing temperatures in order to maximise the collector efficiency as well as the amount of heat transferred to the storage. Dynamic simulations of storage systems of this type, using a solution of (3.33) with definite boundary conditions and source terms, have been performed by Zlatev and Thomsen (1976).

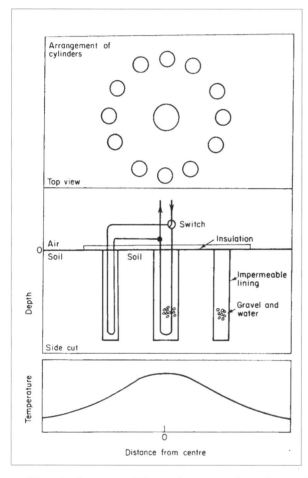

Figure 5.7. Uninsulated heat storage in soil based on arrangements of cylindrical holes (e.g. drilled with use of water flushing-type drills) (based on Brüel *et al.*, 1976).

Use of other materials, such as gravel, rock or soil, has been considered in connection with heating systems at temperatures similar to those relevant for water. Despite volume penalties of a factor 2–3 (depending on void volume), these materials may be more convenient than water for some applications. The main problem is to establish a suitable heat transfer surface for the transfer of heat to and from the storage. For this reason, gravel and rock stores have been mostly used with air, large volumes of which can be blown through the porous material, as a transfer fluid.

For applications in other temperature ranges, other materials may be preferred. Iron (e.g. scrap material) is suitable for temperatures extending to several hundred degrees Celcius. In addition, for rock-based materials the temperatures would not be limited to the boiling point of water.

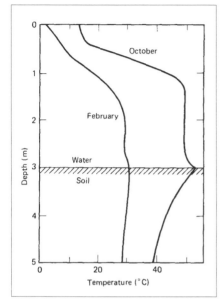

Figure 5.8. Temperature profiles for solar salt gradient pond in Miamisburg, Ohio, for the months of October and February (maximum and minimum temperature in convective layer). The continuation of the profiles in the soil beneath the pond is also included. (From Wittenberg and Harris, 1979. Reprinted with permission from *Proceedings of 14th Intersociety Energy Conversion Engineering Conference.* Copyright 1979 American Chemical Society.)

Solar ponds and aquifer storage

A solar pond is a natural or artificial hot water storage system much like the ones described above, but with the top water surface exposed to solar radiation and operating like a solar collector. In order to achieve both collection and storage in the same medium, layers from top to bottom have to be "inversely stratified", that is, stratified with the hottest zone at the bottom and the coldest one at the top. This implies that thermal lift must be opposed, either by physical means such as placing horizontal plastic barriers to separate the layers or by creating a density gradient in the pond, which provides gravitational forces to overcome the buoyancy forces. This can be done by adding certain salts to the pond, taking advantage of the higher density of the more salty water (Rabl and Nielsen, 1975).

An example of a solar pond of obelisk shape is the 5200-m^3 pond installed at Miamisburg, Ohio. Its depth is 3 m, and the upper half is a salt gradient layer of NaCl, varying from 0% at the top to 18.5% at a 1.5-m depth. This gradient layer opposes upward heat transport and thus functions as a top insulation without impeding the penetration of solar radiation. The bottom layer has a fixed salt concentration (18.5%) and contains heat exchangers for withdrawing energy. In this layer, convection may take place without problems. Most of the absorption of solar radiation takes place at the bottom surface (this is why the pond should be shallow), and the heat is subsequently released to the convective layer. At the very top, however, some absorption of infrared solar radiation may destroy the gradient of temperature.

Figure 5.8 shows temperature gradients for the Miamisburg pond during its initial loading period (no load connected). From the start of operation in late August the first temperature maximum occurred in October, and the subsequent minimum occurred in February. The two situations are shown in Fig. 5.8. The temperature in the ground just below the pond was also measured. In October, the top layer disturbance can be seen, but in February, it is absent, due to ice cover on the top of the pond.

Numerical treatment of seasonal storage in large top-insulated or solar ponds may be by time simulation, or by a simple approximation, in which solar radiation and pond temperature are taken as sine functions of time, with only the amplitude and phase as parameters to be determined. This is a fairly good approximation because of the slow response of a large seasonal store, which tends to be insensitive to rapid fluctuations in radiation or air temperature. However, when heat is extracted from the store, it must be checked that disturbance of the pond's temperature gradient will not occur, say, on a particularly cold winter day, where the heat extraction is large. Still, heat extraction can, in many cases, also be modelled by sine functions, and if the gradient structure of the pond remains stable, such a calculation gives largely realistic results.

In Israel, solar ponds are being operated for electricity generation by use of Rankine cycle engines with organic working fluids in order to be able to accept the small temperature difference available. A correspondingly low thermodynamic efficiency must be accepted (Winsberg, 1981).

Truly underground storage of heat may also take place in geological formations capable of accepting and storing water, such as rock caverns and aquifers. In the aquifer case, it is important that water transport be modest, that is, that hot water injected at a given location stay approximately there and exchange heat with the surroundings only by conduction and diffusion processes. In such cases, it is estimated that high cycle efficiencies (85% at a temperature of the hot water some 200°C above the undisturbed aquifer temperature – the water being under high pressure) can be attained after breaking the system in, that is, after having established stable temperature gradients in the surroundings of the main storage region (Tsang *et al.*, 1979).

Medium- and high-temperature storage
In relation to industrial processes, temperatures regimes are often defined as medium in the interval from 100 to 500°C and high above 500°C. These definitions may also be used in relation to thermal storage of energy, but it may be useful to single out also the lower medium temperature range from 100°C to about 300°C, as the discussion below indicates.

Materials suitable for heat storage should have a large heat capacity, they must be stable in the temperature interval of interest, and it should be convenient to add heat to or withdraw heat from them.

The last of these requirements can be fulfilled in different ways. Either the material itself should possess good heat conductivity, as do metals, for example, or it should be easy to establish heat transfer surfaces between the material and some other suitable medium. If the transfer medium is a liquid or a gas, it could be passed along the transfer surface at a velocity sufficient for the desired heat transfer, even if the conductivities of the transfer fluid and the receiving or delivering material are small. If the storage material is arranged in a finite geometry, such that insufficient transfer is obtained by a single pass, then the transfer fluid may be passed along the surface several times. This is particularly relevant for transfer media such as air, which has a very low heat conductivity, and when air is used as a transfer fluid, it is important that the effective transfer surface be large. This may be achieved for storage materials of granular form such as pebble or rock beds, where the nodule size and packing arrangement can be such that air can be forced through and reach most of the internal surfaces with as small an expenditure of compression energy as possible.

Material	Temperature interval (°C)	Mass spec. heat (kJ kg^{-1} °C^{-1})	Volume spec. heat (MJ m^{-3} °C^{-1})	Heat conductivity (W m^{-1} °C^{-1})
Solids				
Sodium chloride	< 800	0.92	2.0	9[a, b]
Iron (cast)	< 1500	0.46	3.6	70[b]–34[c]
Rock (granite)	< 1700	0.79	2.2	2.7[b]
Bricks		0.84	1.4	0.6
Earth (dry)		0.79	1.0	1.0
Liquids				
Water	0–100	4.2	4.2	0.6
Oil ("thermal")	–50 to 330	2.4	1.9	0.1
Sodium	98 to 880	1.3	1.3	85[b]–60[c]
Diethylene glycol	–10 to 240	2.8	2.9	

Table 5.1 Heat capacities of various materials (Kaye and Laby, 1959; Kreider, 1979; Meinel and Meinel, 1976). All quantities have some temperature dependence. Standard atmospheric pressure has been assumed, that is, all heat capacities are c_p.
[a] Less for granulates with air-filled voids.
[b] At 1000°C.
[c] At 700°C.

These considerations lie behind the approaches to sensible heat storage, which are exemplified by the range of potential storage materials listed in

Table 5.1. Some are solid metals, where transfer has to be by conduction through the material. Others are solids that may exist in granular form for blowing by air or another gas through the packed material. They exhibit more modest heat conductivity. The third group comprises liquids, which may serve both as heat storage materials and also as transfer fluids. The dominating path of heat transfer may be conduction, advection (moving the entire fluid), or convection (turbulent transport). For highly conducting materials such as liquid sodium, little transfer surface is required, but for the other materials listed substantial heat exchanger surfaces may be necessary.

Solid metals, such as cast iron, have been used for high-temperature storage in industry. Heat delivery and extraction may be by passing a fluid through channels drilled into the metal. For the medium- to high-temperature interval the properties of liquid sodium (cf. Table 5.1) make this a widely used material for heat storage and transport, despite the serious safety problems (sodium reacts explosively with water). It is used in nuclear breeder reactors and in concentrating solar collector systems, for storage at temperatures between 275 and 530°C in connection with generation of steam for industrial processes or electricity generation. The physics of heat transfer to and from metal blocks and of fluid behaviour in pipes is a standard subject covered in several textbooks (see, e.g. Grimson, 1971).

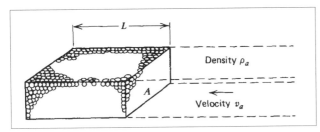

Figure 5.9. Rock bed sensible heat store. Air of density ρ_a and velocity v_a may be blown through the bed cross-section A, travelling the length L of the bed.

Fixed beds of rock or granulate can be used for energy storage both at low and at high temperatures, normally using air blown through the bed to transfer heat to and from the store. The pressure drop ΔP across a rock bed of length L, such as the one illustrated in Fig. 5.9 where air is blown through the entire cross-sectional area A, may be estimated as (Handley and Heggs, 1968)

$$\Delta P \approx \rho_a v_a^2 \, L \, d_s^{-1} \, m_s^2 \, (368 + 1.24 \, Re/m_s)/(Re(1 - m_s)^3), \tag{5.10}$$

where ρ_a and v_a are density and velocity of the air passing through the bed in a steady-state situation, d_s is the equivalent spherical diameter of the rock particles, and m, their mixing ratio, is one minus the air fraction in the volume $L \times A$. Re is the Reynolds number describing the ratio between "inertial" and "viscous" forces on the air passing between the rock particles. Re may be estimated as $\rho_a v_a d_a/\mu$, where μ is the dynamic viscosity of air. If the rock particles are not spherical, the equivalent diameter may be taken as

$$d_s = (6ALm_s/n)^{1/3},$$

where n is the number of particles in the entire bed volume. The estimate (5.10) assumes the bed to be uniform and the situation stationary. The total surface area of the particles in the bed is given by

$$A_s = 6m_s AL/d_s = n\pi d_s^2.$$

Optimal storage requires that the temperature gradient between the particle surfaces and their interior be small and that the pressure drop (5.10) also be small, leading to optimum particle diameters of a few centimetres and a void fraction of about 0.5 (implying that m_s is also about 0.5).

Organic materials such as diethylene glycol or special oil products (Table 5.1) are suitable for heat storage between 200 and 300°C and have been used in concentrating solar collector test facilities (Grasse, 1981). Above 300°C, the oil decomposes.

Despite low volume heat capacities, gaseous heat storage materials could also be considered, such as steam (water vapour), which is often stored under pressure, in cases where steam is the form of heat energy to be used later (in industrial processes, power plants, etc.).

Latent heat storage associated with structural or phase change

The energy associated with a change of phase for a given material can be used to store energy. The phase change may be one of melting or evaporating, or it may be associated with a structural change, e.g. in lattice form, content of crystal-bound water, etc. When heat is added or removed from a given material, a number of changes may take place successively or in some cases simultaneously, involving phase changes as well as energy storage in thermal motion of molecules, i.e. both latent and sensible heat. The total energy change, which can serve as energy storage, is given by the change in enthalpy.

Solid–solid phase transitions are observed in one-component, binary, and ternary systems, as well as in single elements. An example of the latter is solid sulphur, which occurs in two different crystalline forms, a low-temperature orthorhombic form and a high-temperature monoclinic form (cf. Moore, 1972). However, the elementary sulphur system has been studied merely out of academic interest in contrast to the one-component systems listed in Table 5.2. Of these systems, which have been studied in relation to practical energy storage, Li_2SO_4 exhibits both the highest transition temperature T_t and the highest latent heat for the solid–solid phase change ΔH_{ss}. Pure Li_2SO_4 has a transition from a monoclinic to a face-centred cubic structure with a latent heat of 214 KJ kg^{-1} at 578°C. This is much higher than the heat of melting (-67 KJ kg^{-1} at $860°C$). Another one-component material listed in Table 5.2 is Na_2SO_4, which has two transitions at 201 and 247°C, with the total latent heat of both transitions being ~ 80 KJ kg^{-1}.

Material	Transition temperature T_t (°C)	Latent heat ΔH_{ss} (kJ kg^{-1})
V_2O_2	72	50
FeS	138	50
KHF_2	196	135
Na_2SO_4	210, 247	80
Li_2SO_4	578	214

Table 5.2. Solid–solid transition enthalpies ΔH_{ss} (Fittipaldi, 1981).

Recently, mixtures of Li_2SO_4 with Na_2SO_4, K_2SO_4 and ZnSO have been studied. Also, some ternary mixtures containing these and other sulphates were included in a Swedish investigation (cf. Sjoblom, 1981). Two binary systems (Li_2SO_4–Na_2SO_4, 50 mol% each, T_t = 518°C; and 60% Li_2SO_4–40% $ZnSO_4$, T_t = 459°C) have high values of latent heat, ~190 KJ kg^{-1}, but they exhibit a strong tendency for deformation during thermal cycling. A number of ternary salt mixtures based on the most successful binary compositions have been studied experimentally, but there is a lack of knowledge of both phase diagrams, structures, and re-crystallisation processes that lead to deformation in these systems.

Salt hydrates

The possibility of energy storage by use of incongruently melting salt hydrates has been intensely investigated, starting with the work of Telkes (Telkes, 1952, 1976). The molten salt consists of a saturated solution and additionally some undissolved anhydrous salt because of its insufficient solubility at the melting temperature, considering the amount of released crystal water available. Sedimentation will develop, and a solid crust may form at the interface between layers. In response to this, stirring is applied, for example, by keeping the material in rotating cylinders (Herrick, 1982), and additives are administered in order to control agglomeration and crystal size (Marks, 1983).

An alternative is to add extra water to prevent phase separation. This has led to a number of stable heat of fusion storage systems (Biswas, 1977; Furbo, 1982). Some melting points and latent heats of salt hydrates are listed in Table 5.3. Here we use as an example Glauber salt ($Na_2SO_4 \cdot 10H_2O$), the storage capacity of which is illustrated in Fig. 5.10, both for the pure hydrate and for a 33% water mixture used in actual experiments. Long-term verification of this and other systems in connection with solar collector installations have been carried out by the European Economic Community. For hot water systems the advantage over sensible heat water stores is minimal, but this may change when space heating is included, because of the seasonal storage need (Furbo, 1982).

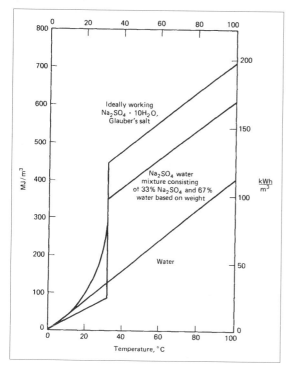

Figure 5.10. Heat storage capacity as a function of temperature for ideally melting Glauber salt, for Glauber salt plus extra water, and for pure water (Furbo, 1982).

Salt hydrates release water when heated and release heat when they are formed. The temperatures at which the reaction occurs vary for different compounds, ranging from 30 to 80°C; this makes possible a choice of storage systems for a variety of water-based heating systems such as solar, central, and district heating. Table 5.3 shows the temperatures T_m of incongruent melting and the associated quasi-latent heat Q (or ΔH) for some hydrates that have been studied extensively in heat storage system operation. The practical use of salt hydrates faces physicochemical and thermal problems such as supercooling, non-congruent melting, and heat transfer difficulties imposed by locally low heat conductivities (cf., e.g. Achard *et al.*, 1981).

Generally, storage in a chemical system with two separated components of which one draws low-temperature heat from the environment and the other absorbs or delivers heat at fairly low (30–100°C) or medium (100–200°C) temperature is referred to as a *chemical heat pump* (McBride, 1981). Simply stated, the chemical heat pump principle consists in keeping a substance in one of two containers, although it would prefer to be in the other one. In a classical example, the substances are water vapour and sulphuric acid. Because the pressure over sulphuric acid is much lower than over liquid water (see Fig. 5.11), the water vapour will prefer to move from the water surface to the H_2SO_4 surface and become absorbed there, with a heat gain deriving in part from the mixing process and in part from the heat of evaporation. The temperature of the mixture is then above what is needed at the water source. Heat is stored when the temperature of the sulphuric acid/water container is made still higher, so that the equilibrium pressure of vapour above the acid surface at this temperature becomes higher than that above the water surface at its temperature. The pressure gradient will in this situation move water vapour back to the water surface for condensation.

Hydrate	Incongruent melting point, T_m (°C)	Specific latent heat ΔH (MJ m^{-3})
$CaCl_2 \cdot 6H_2O$	29	281
$Na_2SO_4 \cdot 10H_2O$	32	342
$Na_2CO_3 \cdot 10H_2O$	33	360
$Na_2HPO_4 \cdot 12H_2O$	35	205
$Na_2HPO_4 \cdot 7H_2O$	48	302
$Na_2S_2O_3 \cdot 5H_2O$	48	346
$Ba(OH)_2 \cdot 8H_2O$	78	655

Table 5.3. Characteristics of salt hydrates.

A similar system, but with the water being attached as crystal water in a salt (i.e., salt hydration), is the Na_2S/water system being developed in Sweden ("System Tepidus"; Bakken, 1981). Figure 5.12 shows this chemical heat pump, which is charged by the reaction

$$Na_2S + 5H_2O \text{ (vapour)} \rightarrow Na_2S \cdot 5H_2O + 312 \text{ kJ mol}^{-1}. \qquad (5.11)$$

The heat for the evaporation is taken from a reservoir of about 5°C, that is, a pipe extending through the soil at a depth of a few metres (as in commercial electric heat pumps with the evaporator buried in the lawn), corresponding to roughly 10°C in the water container (B in Fig. 5.12) owing to heat exchanger losses. The water vapour flows to the Na_2S container (A in Fig. 5.12) through a connecting pipe that has been evacuated of all gases other than water vapour and where the water vapour pressure is of the order of 1% of atmospheric pressure. During charging, the temperature in the sodium sulphide rises to 65–70°C, owing to the heat formed in the process (5.11). When the temperatures in containers A and B and the equilibrium pressures of the water vapour are such that they correspond to each other by a horizontal line in the pressure–temperature diagram shown in Fig. 5.13, the flow stops and the container A has been charged.

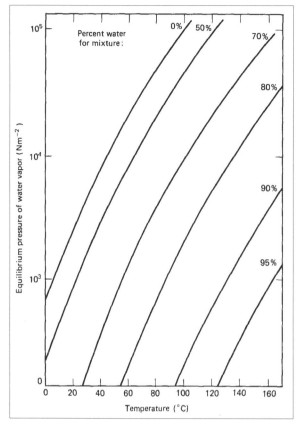

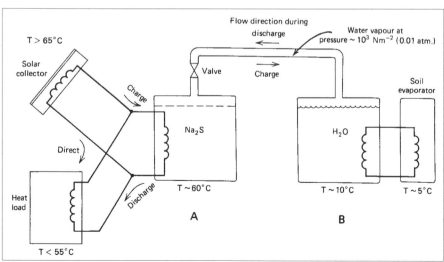

Figure 5.11 (left). Equilibrium pressure of water vapour over sulphuric acid/water mixtures as a function of temperature and percentage of water in the mixture (Christensen, 1981).

Figure 5.12 (below). Schematic picture of a chemical heat pump operating between a sodium sulphide and a water container and based on the formation of the salt hydrate $Na_2S \cdot 5H_2O$. Typical temperatures are indicated. There is a low-temperature heat source connected to the water container and a high-temperature heat source (a solar collector) connected to the salt container, along with the heat demand (load). A switch allows either load or solar collector to be connected.

To release the energy, a load area of temperature lower than the container A is connected to it, and heat is transferred through a heat exchanger. Lowering the temperature in A causes a pressure gradient to form in the connecting pipe, and new energy is drawn from B to A. In order to prevent the heat reservoir (the soil) from cooling significantly, new heat must be added to compensate for the heat withdrawn. This takes place continuously by transfer processes in the soil (solar radiation absorbed at the soil surface is conducted to the subsurface location of the evaporator pipes). However, in the long term a lower temperature would develop in the soil environment if no active makeup heat were supplied. This is done by leading surplus heat from a solar collector to the sodium sulphide container when the solar heat is not directly required in the load area. When the temperature of container A is raised in this way, the pressure gradient above the salt will be in the direction of driving water vapour to container B, thereby removing some of the crystallisation water from the salt.

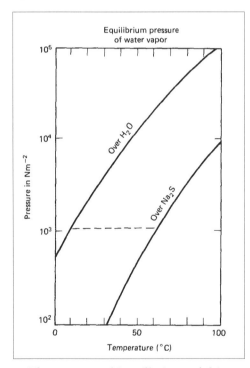

Figure 5.13. Equilibrium pressure of water vapour over water and over sodium sulphide as a function of temperature. For a given pressure (dashed line), the equilibrium temperatures in the water and the salt containers will differ by roughly 55°C.

The two actual installations of this type of chemical heat pump are a one-family dwelling with a storage capacity of 7000 kWh (started in 1979) and an industrial building (Swedish Telecommunications Administration) with 30 000 kWh worth of storage, started in 1980. Future applications may comprise transportable heat stores, since container A may be detached (after closing the valve indicated in Fig. 5.12) and carried to another site. Once the container is detached, the sensible heat is lost, as the container cools from its 60°C to ambient temperatures, but this only amounts to 3–4% of the energy stored in the $Na_2S \cdot 5H_2O$ (Bakken, 1981).

It should be mentioned that a similar loss will occur during use of the storage unit in connection with a solar heating system. This is because of the intermittent call upon the store. Every time it is needed, its temperature increases to 60°C (i.e. every time the valve is opened), using energy to supply the sensible heat, and every time the need starts to decrease, there is a heat loss associated either with making up for the heat transfer to the surroundings (depending on the insulation of the container) to keep the container temperature at 60°C or, if the valve has been closed, with the heat required to re-heat the container to its working temperature. These losses could be minimised by using a modular system, where only one module at a time is kept at operating temperature, ready to supply heat if the solar collector is not providing enough. The other modules would then be at ambient temperatures except when they are called upon to become re-charged or to replace the unit at standby.

The prototype systems are not cost effective, but the estimates for system cost in regular production is 4–5 euro per kWh of heat supplied for a 15-m³ storage system for a detached house with half the cost taken up by the solar collector system and the other half taken up by the store. For transport, container sizes equivalent to 4500 kWh of 60°C heat are envisaged. However, the charge-rate capacity of roughly 1 W kg⁻¹ may be insufficient for most applications.

Although the application of the chemical heat pump considered above is for heating, the concept is equally useful for cooling applications. Here the load would simply be connected to the cold container. Several projects are underway to study various chemical reactions of the gas/liquid or gas/solid type based on pressure differences, either for cooling alone or for both heating and cooling. The materials are chosen on the basis of temperature requirements and long-term stability while allowing many storage cycles. For example, $NaI–NH_3$ systems have been considered for air conditioning purposes (Fujiwara *et al.*, 1981). A number of ammoniated salts that react on heat exchange could be contemplated.

Chemical reactions
The use of high-temperature chemical heat reactions in thermal storage systems is fairly new and to some extent related to attempts to utilise high-temperature waste heat and to improve the performance of steam power plants (cf. Golibersuch *et al.*, 1976). The chemical reactions that are used to store the heat allow, in addition, upgrading of heat from a lower temperature level to a higher temperature level, a property that is not associated with phase transition or heat capacity methods.

Conventional combustion of fuel is a chemical reaction in which the fuel is combined with an oxidant to form reaction products and surplus heat. This type of chemical reaction is normally irreversible, and there is no easy way that the reverse reaction can be used to store thermal energy. The process of burning fuel is a chemical reaction whereby energy in one form (chemical energy) is transformed into another form (heat) accompanied by an increase in entropy. In order to use such a chemical reaction for storage of heat, it would, for example, in the case of hydrocarbon, require a reverse process whereby the fuel (hydrocarbon) could be obtained by adding heat to the reaction products carbon dioxide and water. So, use of chemical heat reactions for thermal energy storage requires suitable reversible reactions.

The change in bond energy for a reversible chemical reaction may be used to store heat, but although a great variety of reversible reactions are known, only a few have so far been identified as being technically and economically acceptable candidates. The technical constraints include temperature, pressure, energy densities, power densities, and thermal efficiency. In general, a chemical heat reaction is a process whereby a chemical compound is dissociated by heat absorption, and later, when the reaction products are recombined, the absorbed heat is again released. Reversible chemical heat reactions can be divided into two groups: thermal dissociation reactions and catalytic reactions. The thermal dissociation reaction may be described as

$$AB\ (+\Delta H \text{ at } T_1 \text{ and } p_1) \leftrightarrow A + B\ (-\Delta H \text{ at } T_2 \text{ and } p_2), \tag{5.12}$$

indicating that the dissociation takes place by addition of heat ΔH to AB at temperature T_1 and pressure p_1, whereas the heat is released ($-\Delta H$) at the reverse reaction at temperature T_2 and pressure p_2. The reciprocal reaction (from right to left) occurs spontaneously if the equilibrium is disturbed, that is, if $T_2 < T_1$ and $p_2 > p_1$. To avoid uncontrolled reverse reaction, the reaction products must therefore be separated and stored in different containers. This separation of the reaction products is not necessary in catalytic reaction systems where both reactions (left to right and right to left) require a catalyst in order to obtain acceptable high reaction velocities. If the catalyst is removed, neither of the reactions will take place even when considerable changes in temperature and pressure occur. This fact leads to an important advantage, namely, that the intrinsic storage time is, in practice, very large and, in principle, infinite. Another advantage of closed-loop heat storage systems employing chemical reactions is that the compounds involved are not consumed, and because of the high energy densities (in the order of magnitude: 1 MWh m^{-3} compared to that of the sensible heat of water at $\Delta T = 50$ K, which is 0.06 MWh m^{-3}), a variety of chemical compounds are economically acceptable.

The interest in high-temperature chemical reactions is derived from the work of German investigators on the methane reaction

$$Q + CH_4 + H_2O \leftrightarrow CO + 3H_2, \tag{5.13}$$

(Q being heat added) which was studied in relation to long-distance transmission of high-temperature heat from nuclear gas reactors (cf. Schulten *et al.*, 1974). The transmission system called "EVA-ADAM", an abbreviation of the German "Einzelrohrversuchsanlage und Anlage zur dreistufigen adiabatischen Methanisierung", is being further developed at the nuclear research centre at Jülich, West Germany. It consists of steam reforming at the nuclear reactor site, transport over long distances of the reformed gas (CO + $3H_2$), and methanation at the consumer site where heat for electricity and district heating is provided (cf. e.g. Harth *et al.*, 1981).

Closed loop system	Enthalpy[a] ΔH^0 (kJ mol^{-1})	Temperature range (K)
$CH_4 + H_2O \leftrightarrow CO + 3H_2$	206 (250)[b]	700–1200
$CH_4 + CO_2 \leftrightarrow 2CO + 2H_2$	247	700–1200
$CH_4 + 2H_2O \leftrightarrow CO_2 + 4H_2$	165	500–700
$C_6H_{12} \leftrightarrow C_6H_6 + 3H_2$	207	500–750
$C_7H_{14} \leftrightarrow C_7H_8 + 3H_2$	213	450–700
$C_{10}H_{18} \leftrightarrow C_{10}H_8 + 5H_2$	314	450–700

Table 5.4. High-temperature closed-loop chemical C-H-O reactions. (Hanneman *et al.*, 1974; Harth *et al.*, 1981).
[a] Standard enthalpy for complete reaction.
[b] Including heat of evaporation of water.

The reaction in (5.13) is a suitable candidate for energy storage that can be accomplished as follows: heat is absorbed in the endothermic reformer where the previously stored low-enthalpy reactants (methane and water) are converted into high-enthalpy products (carbon monoxide and hydrogen). After heat exchange with the incoming reactants, the products are then stored in a separate vessel at ambient temperature conditions, and although the reverse reaction is thermodynamically favoured, it will not occur at these low temperatures and in the absence of a catalyst. When the heat is needed, the products are recovered from storage and the reverse, exothermic reaction (methanation) is run (cf. Golibersuch *et al.*, 1976). Enthalpies and temperature ranges for some high-temperature closed-loop C-H-O systems, including the reaction (5.13), are given in Table 5.4. The performance of the cyclohexane to benzene and hydrogen system (listed fourth in Table 5.4) has been studied in detail by Italian workers, and an assessment of a design storage plant has been made (cf. Cacciola *et al.*, 1981). The complete design storage plant consists of hydrogenation and dehydrogenation multistage adiabatic reactors, storage tanks, separators, heat exchangers, and multistage compressors. Thermodynamic requirements are assured by independent closed-loop systems circulating nitrogen in the dehydrogenation and hydrogen in the hydrogenation units.

A number of ammoniated salts are known to dissociate and release ammonia at different temperatures, including some in the high-temperature range (cf. e.g. Yoneda *et al.*, 1980). The advantages of solid–gas reactions in general are high heats of reaction and short reaction times. This implies, in principle, high energy and power densities. However, poor heat and mass transfer characteristics in many practical systems, together with problems of sagging and swelling of the solid materials, lead to reduced densities of the total storage system.

Metal hydride systems are primarily considered as stores of hydrogen, as will be discussed later. However, they have also been contemplated for heat storage, and in any case the heat-related process is integral in getting the hydrogen into and out of the hydride. The formation of a hydride MeH_x (metal plus hydrogen) is usually a spontaneous exothermic reaction:

$$Me + \tfrac{1}{2}\, x\, H_2 \rightarrow MeH_x + Q, \tag{5.14}$$

which can be reversed easily by applying the amount of heat Q,

$$MeH_x + Q \rightarrow Me + \tfrac{1}{2}\, x\, H_2. \tag{5.15}$$

Thus, a closed-loop system, where hydrogen is not consumed but is pumped between separate hydride units, may be used as a heat store. High-temperature hydrides such as MgH_2, Mg_2NiH_2 and TiH_2 have, owing to their high formation enthalpies (e.g. for MgH_2, $\Delta H \geq 80$ kJ per mol of H_2; for TiH_2, $\Delta H > 160$ KJ per mol of H_2), heat densities of up to 3 MJ kg^{-1} or 6 GJ m^{-3} in a temperature range extending from 100 to 600°C (cf. Buchner, 1980).

5.2.2 Storage of high-quality energy forms

A number of storage systems may be particularly suitable for the storage of "high-quality" energy, such as mechanical energy or electric energy. If the energy to be stored is derived from a primary conversion delivering electricity, for example, then one needs an energy storage system which will allow the regeneration of electricity with a high cycle efficiency, i.e. with a large fraction of the electricity input recovered in the form of electricity. Thermal stores, such as the options mentioned in the previous section, may even at T = 800–1500°C (metals, etc.) not achieve this, because of the Carnot limit to the efficiency of electricity regeneration, as well as storage losses through insulation, etc. Thermal stores fed via a heat pump drawing from a suitable reference reservoir may reach tolerable cycle efficiencies more easily.

Storage form	Energy density		Cycle efficiency
	kJ kg^{-1}	MJ m^{-3}	
Conventional fuels			
Crude oil	42 000	37 000	
Coal	32 000	42 000	
Dry wood	12 500^a	10 000	
Synthetic fuels			
Hydrogen, gas	120 000	10	0.4–0.6
Hydrogen, liquid	120 000	8 700	
Hydrogen, metal hydride	2 000–9 000	5 000–15 000	
Methanol	21 000	17 000	
Ethanol	28 000	22 000	
Thermal – low quality			
Water, 100°C → 40°C	250	250	
Rocks, 100°C → 40°C	40–50	100–140	
Iron, 100°C → 40°C	~30	~230	
Thermal – high quality			
Rocks, e.g. 400°C → 200°C	~160	~430	
Iron, e.g. 400°C → 200°C	~100	~800	
Inorganic salts, heat of fusion > 300°C	> 300	> 300	
Mechanical			
Pumped hydro, 100 m head	1	1	0.65–0.8
Compressed air		~15	0.4–0.5
Flywheels, steel	30–120	240–950	
Flywheels, advanced	> 200	> 100	~0.95
Electrochemical			
Lead–acid	40–140	100–900	0.7–0.8
Nickel–cadmium	~350	~350	varying
Lithium ion (other advanced batteries)	700 (> 400)	1400 (> 300)	0.7 (> 0.8)
Superconducting		~100	~0.85

Table 5.5. Energy density by weight and volume for various storage forms, based on measured data or expectations for practical applications. For the storage forms aimed at storing and regenerating high-quality energy (electricity), cycle efficiencies are also indicated. Hydrogen gas density is quoted at ambient pressure and temperature. For compressed air energy storage, both electricity and heat inputs are included on equal terms in estimating the cycle efficiency (with use of Jensen and Sørensen, 1984).

^{a} Oven-dry wood may reach values up to 20 000 kJ kg^{-1}.

Table 5.5 gives an indication of the actual or estimated energy densities and cycle efficiencies of various storage systems. The theoretical maximum of energy density is, in some cases, considerably higher than the values quoted. For comparison, energy densities of thermal stores and a number of fuels are also given. Some of the fuels (methanol, wood and hydrogen) may be produced by conversion based on renewable energy sources (without having to wait for fossilisation processes to occur). The cycle efficiency is defined with the assumption that the initial energy form is electricity or another high-quality energy form, and the value quoted for hydrogen is based on electrolysis of water as the first step in the storage cycle. Methanol may also be reversibly transformed into hydrogen and carbon oxide to play the role of a closed storage cycle (Prengle and Sun, 1976). The most striking feature is the low volume energy density of nearly all the reversible storage concepts considered, relative to that of solid or liquid fossil fuels.

The magnitudes of stores that might be associated with renewable energy use (such as indicated in Fig. 2.87) are discussed in more detail in sections 6.2.4–6.2.7, with a summary in Tables 6.9 and 6.10 discussed in connection with future energy scenario construction. For comparison, reservoirs of fossil fuels may be found in Table 4.2.

Pumped hydro storage

The total exploitable hydro potential is of the order of 10^{12} W on average over the year (section 3.4.2), and only the fraction of this associated with reservoirs can be considered relevant for energy storage. Those river flows that have to be tapped as they come may be interesting as energy sources, but not as energy storage options.

The hydro reservoirs feeding into turbine power plants may be utilised for storage of electric energy generated by non-hydropower plants (e.g. wind or photovoltaic energy converters), provided that all the power plants are connected by a common grid, and provided that transmission capacity is sufficient to accommodate the extra burden of load-levelling storage type operation of the system. The storage function in a system of this kind is primarily obtained by displacement of load. This means that the hydropower units are serving as backup for the non-hydro generators by providing power when non-hydropower production falls short of load. The small start-up time for hydro turbines (½–3 minutes) makes this mode of operation convenient. When there is a surplus power generation from the non-hydro units, then the hydro generation is decreased, and non-hydro produced power is transmitted to the load areas otherwise served by hydropower (Sørensen, 1981a; Meibom et al., 1999). In this way, there is no need to pump water up into the hydro reservoirs, as long as the non-hydropower generation stays below the combined load of hydro and non-hydro load areas. To fulfil this condition, the relative sizes of the different types of generating units must be chosen carefully.

When the surplus energy to be stored exceeds the amounts that can be handled in the displacement mode described above, then upward pumping of water into the hydro reservoirs may be considered by use of two-way turbines, so that the energy can be stored and recovered by the same installation. Alternatively, pumped storage may utilise natural or artificially constructed reservoirs not associated with any exploitable hydropower.

Figure 5.14 shows an example of the layout of a pumped storage facility. Installations where reservoirs are not part of a hydro flow system are typically intended for short-term storage. They may be used for load-levelling purposes, providing a few hours of peak load electric power per day, based on night-time pumping. In terms of average load covered, the storage capacities of these installations are below 24 h. On the other hand, some of the natural reservoirs associated with hydro schemes have storage capacities corresponding to one or more years of average load (e.g. the Norwegian hydro system; cf. Sørensen, 1981a; Meibom *et al.*, 1999). Pumping schemes for such reservoirs could serve for long-term storage of energy.

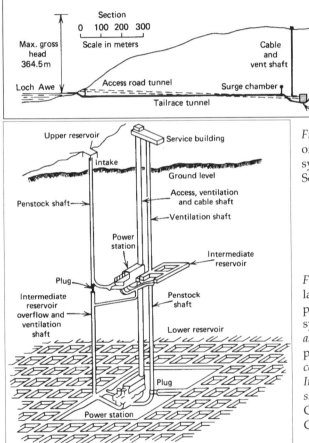

Figure 5.14 (above). Layout of pumped hydro storage system at Cruachan in Scotland.

Figure 5.15 (left). Schematic layout of underground pumped hydro storage system. From Blomquist *et al.*, 1979. Reprinted with permission from *the Proceedings of the 14th Intersociety Energy Conversion Engineering Conference.* Copyright 1979 American Chemical Society.

If no natural elevated reservoirs are present, pumped storage schemes may be based on underground lower reservoirs and surface level upper reservoirs. The upper reservoirs may be lakes or oceans. The lower ones should be excavated or should make use of natural cavities in the underground. If excavation is necessary, a network of horizontal mine shafts, such as the one illustrated in Fig. 5.15, may be employed in order to maintain structural stability against collapse (Blomquist *et al.*, 1979; Hambraeus, 1975).

The choice of equipment is determined primarily by the size of head, that is, the vertical drop available between the upper and the lower reservoir. Figure 4.101 shows in a schematic form the three most common types of hydro turbines. The Kaplan turbine (called a Nagler turbine if the position of the rotor blades cannot be varied) is most suited for low heads, down to a few metres. Its rotor has the shape of a propeller, and the efficiency of converting gravitational energy into shaft power is high (over 0.9) for the design water velocity, but lower for other water speeds. The efficiency drop away from the design water velocity is rapid for the Nagler turbine, less so for the Kaplan version. These turbines are inefficient for upward pumping, although they can be made to accept water flow from either side (André, 1976). A displacement pump may be used in a "tandem arrangement" (i.e. separate turbine and pump). The electric generator is easily made reversible, so that it may serve either as generator or as motor.

For larger heads, the Francis and Pelton turbines may be used. Francis turbines have a set of fixed guiding blades leading the water onto the rotating blades (the "runner") at optimum incident angle. It can be used with water heads up to about 600 m in the simple version illustrated in Fig. 4.101b, but multistage versions have been considered, guiding the water through a number of runners (five for an actual French installation; cf. Blomquist *et al.*, 1979). In this way, heads above 1000 m can be accepted, and the arrangement may be completely reversible, with very modest losses. For either pumping or generating, the turbine efficiency at design water flow may be over 0.95, but for decreasing flow the efficiency drops. Typical overall efficiencies of the storage cycle (pumping water up by use of surplus electric power, regenerating electric power based on downward flow through turbines) are around 0.8 for existing one-stage Francis turbine installations. Shifting from pumping to generating takes about 1 min (Hambraeus, 1975). The total cycle efficiency of the multistage Francis turbines for heads of 1000–1500 m is about 0.7 (Blomquist *et al.*, 1979).

If the head is larger than the limit for single-stage, reversible Francis turbines, an alternative to the multistage Francis turbines is offered by the tandem units consisting of separate impulse turbines and pumps. The pump units for pumping upward over height differences exceeding 1000 m are usually multistage pumps (six stages for an actual installation in Italy), with efficiency over 0.9 being achieved. The impulse turbine part is of Pelton type (see Fig. 4.101a), consisting of a bucket-wheel being driven by the impulse of one or more water jets created by passing the water through nozzles. The power for this process is the pressure force created by the column of water, from the turbine placed at the lower reservoir level to the upper reservoir level. The pressure energy can be converted partially or fully into linear kinetic energy according to the requirements of the different turbine types,

$$mg \, \Delta z = W^{pot}_{initial} = m' \, \tfrac{1}{2}u^2 + (m - m') \, P \, \rho^{-1} = W^{kin} + H.$$

Here the initial potential energy associated with the head Δz is transformed into a kinetic energy part associated with partial mass m' moving with velocity u and a pressure energy part with the enthalpy H given by the pressure P over the density of water ρ, times the remaining mass $m-m'$. The conversion efficiency of Pelton turbines is about 0.9 over a wide range of power levels, and the tandem arrangement of separate turbine and pump (but generator/motor, turbine, and pump all mounted on a common shaft) allows quick shifts between generating and pumping or vice versa.

The losses in conversion are associated in part with "leakage", that is, with water that passes round the turbine without contributing to power, and in part with energy dissipation in the form of heat, for example, due to friction (cf. Angrist, 1976). Further losses are associated with evaporation of water, especially from solar-exposed upper reservoirs.

Excavation for underground storage limits the application to short-term storage (up to about 24 h of average load) because the cost scales approximately linearly with storage capacity. For large natural reservoirs, seasonal energy storage can be considered, since the cost has a large component determined by the maximum load requirement and therefore becomes fairly independent of storage capacity beyond a certain point, as long as the reservoir is available.

Flywheels

Mechanical energy may be stored in the form of rotational motion under conditions of low frictional losses. A flywheel is such a rotating structure, capable of receiving and delivering power through its shaft of rotation. Friction is diminished by use of high-performance bearings, and the entire rotating structure may be enclosed in a space with partial vacuum or filled with an inert gas.

The amount of energy stored in a body of mass distribution $\rho(x)$ rotating about a fixed axis with angular velocity ω is

$$W = \tfrac{1}{2} I \, \omega^2, \tag{5.16}$$

with the moment of inertia I given by

$$I = \int \rho(x) \, r^2 \, dx.$$

It would appear from these expressions that high angular velocity and a majority of the mass situated at large distance r from the axis of rotation would lead to high amounts of energy stored. The relevant question to ask, however, is how to obtain the highest energy density, given material of a certain strength.

The strength of materials is quantified as the tensile strength, defined as the highest stress not leading to a permanent deformation or breaking of the material. If the material is inhomogeneous, the breaking stresses, and hence the tensile strengths, are different in different directions. For fibrous materials, there are characteristic tensile strengths in the direction of the fibres and perpendicular to the direction of the fibres, the former in some cases being orders of magnitude larger than the latter ones.

The components in x-, y-, and z-directions of the force per unit volume, f, are related to the stress tensor τ_{ij} by

$$f_i = \Sigma_j \, \partial \, \tau_{ij} / \partial x_j, \tag{5.17}$$

and the tensile strength σ_l in a direction specified by i is

$$\sigma_l = \max \left(\Sigma_j \, \tau_{ij} \, n_j \right), \tag{5.18}$$

where n is a unit vector normal to the "cut" in the material (see Fig. 5.16) and the maximum sustainable stress in the direction i is to be found by varying the direction of n, that is, varying the angle of slicing. In other words, the angle of the cut is varied until the stress in the direction i is maximum, and the highest value of this maximum stress not leading to irreversible damage defines the tensile strength. If the material is isotropic, the tensile strength is independent of direction and may be denoted σ.

Figure 5.16. Definition of internal stress force (Jensen and Sørensen, 1984).

Consider now a flywheel such as the one illustrated in Fig. 5.17, rotating with angular velocity ω about a fixed axis. The mass distribution is symmetric around the axis of rotation, that is, invariant with respect to rotations through any angle θ about the rotational axis. It is further assumed that the material is homogeneous, so that the mass distribution is fully determined by the mass density ρ and the variation of disk width $b(r)$ as a function of radial distance r from the axis, for a shape symmetric about the midway plane normal to the axis of rotation.

The internal stress forces (5.17) plus any external forces f_{ext} determine the acceleration of a small volume of the flywheel situated at the position x:

$$\rho \frac{d^2 x_i}{dt^2} = f_{ext,i} + \sum_j \frac{\partial \tau_{ij}}{\tau x_j}, \tag{5.19}$$

which is the Newtonian equation of motion. Integrated over some volume V, the force becomes

$$F_i = \int_V f_{ext,i}\, dx + \int_V \sum_j \frac{\partial \tau_{ij}}{\partial x_j}\, dx = F_{ext,i} + \int_V \sum_j \tau_{ij}\, n_j\, da, \qquad (5.20)$$

where in the last line the volume integral over V has been transformed into a surface integral over the surface A enclosing the volume V, with n being a unit vector normal to the surface.

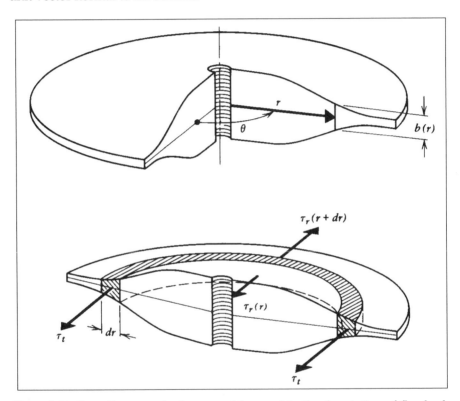

Figure 5.17. Co-ordinates and other quantities used in the description of flywheels. The lower part of the figure illustrates the half-torus shape, confined between radii r and $r + dr$, used in the evaluation of stresses in the direction perpendicular to the cut (Jensen and Sørensen, 1984).

The Constant stress disk

For uniform rotation with constant angular velocity ω, the acceleration on the left-hand side of (5.19) is radial and given by $r\omega^2$ at the distance r from the axis. Disregarding gravitational forces, the centrifugal force alone must be balanced by the internal stresses, and one may proceed to find the conditions under which all parts of the material experience the same stress τ. If τ equals the tensile strength σ or a specified fraction of it (in order to have a safety margin, as one always would in practice), then the material is utilised optimally, and the energy stored is the maximum that can be achieved using the given material properties.

Taking the volume V as that enclosed between the radius r and $r + dr$ and between the centre angle $\theta = -x/2$ and $\theta = x/2$, with the full widths being $b(r)$ and $b(r+dr)$, the balance between the centrifugal force and the internal stresses is obtained from (5.19) and (5.20),

$$2\rho r\omega^2 b(r)\, r\, dr = 2\tau\left((r + dr)\, b(r + dr) - r\, b(r) - b(r)\, dr\right). \tag{5.21}$$

The factors two come from the angular integrals over $\cos\theta$. The first two terms on the right-hand side of (5.21) are derived from the radial stresses, while the last term represents the tangential stresses on the cuts halving the torus shape considered (cf. Fig. 5.16). To first order in dr, (5.21) may be re-written as

$$\rho r^2\omega^2 b(r) = \tau r\, db(r)\, /\, dr \tag{5.22}$$

from which the disc thickness leading to constant stress is found as

$$b(r) = b_0 \exp\left(-\tfrac{1}{2}\,\rho\, r^2\omega^2/\tau\right) \tag{5.23}$$

The optimum shape is seen to be an infinitely extending disc of exponentially declining thickness.

Other flywheel shapes

The approach used above may be generalised. Instead of assuming constant stress, the shape of the flywheel is assumed to be known [i.e., $b(r)$ is known, the material still being homogeneous and the shape symmetrical around the axis of rotation as well as upon reflection in the midway plane perpendicular to the axis]. Then the stresses will have to be calculated as a function of rotational speed ω. Owing to the assumptions made, there are only two radially varying stress functions to consider, the radial stress $\tau_r(r)$ and the tangential stress $\tau_t(r)$, both depending only on the distance r from the axis of rotation. Stress components parallel to the axis of rotations are considered absent. Considering again a half-torus shape (see Fig. 5.17), the forces perpendicular to the cut plane may be written in a generalisation of (5.21):

$$2\rho r^2\omega^2 b(r)\, dr = 2(\tau_r\left((r + dr)\, b(r + dr)\, (r + dr) - r\, \tau_r(r)\, b(r)\right) - \tau_t(r)\, b(r)\, dr. \tag{5.24}$$

To first order in dr, this gives, after rearrangement,

$$\tau_t(r) = \rho r^2\, \omega^2 + \tau_r(r) + d\,\tau_r(r)/dr + r\tau_r(r)\, db(r)\, /\, (b(r)\, dr). \tag{5.25}$$

This is one equation relating radial and tangential stresses. In order to determine the stresses, a second relation must be established. This is the relation between stresses and strains, corresponding to Hooke's law in the theory of elasticity. Introduce deformation parameters ε_t and ε_r for tangential and radial stretching by the relations

$$2\pi\Delta r = 2\pi\varepsilon_t, \tag{5.26}$$

$$d(\Delta r)/dr = \varepsilon_r = \varepsilon_t + r\, d\varepsilon_t/dr, \tag{5.27}$$

where the first equation gives the tangential elongation of the half-torus confined between r and $r + dr$ (see Fig. 5.17), and (5.27) gives the implied radial stretching. Then the stress–strain relations may be written

$$\varepsilon_t(r) = Y^{-1}(\tau_t(r) - \mu\,\tau_r(r)), \tag{5.28}$$

$$\varepsilon_r(r) = Y^{-1}(\tau_r(r) - \mu\,\tau_t(r)), \tag{5.29}$$

where the strength of compression Y is called Young's module and μ is Poisson's ratio, being the ratio between Y and the corresponding quantity Z measuring the strength of expansion in directions perpendicular to the compression (Shigley, 1972). Eliminating the deformations from (5.27)–(5.29), a new relation between the stresses is obtained:

$$(1 + \mu)\,(\tau_r(r) - \tau_t(r)) = r\, d\tau_t(r)\,/\,dr - r\mu\, d\tau_r(r)\,/\,dr. \tag{5.30}$$

Inserting (5.25) into (5.30), a second-order differential equation for the determination of $\tau_r(r)$ results. The solution depends on materials properties through ρ, Y and μ and on the state of rotation through ω. Once the radial stress is determined, the tangential one can be evaluated from (5.25).

As an example, consider a plane disc of radius r_{max}, with a centre hole of radius r_{min}. In this case, the derivatives of $b(r)$ vanish, and the solution to (5.30) and (5.25) is

$$\tau_r(r) = (3 + \mu)\rho\,\omega^2\,(r^2_{min} + r^2_{max} - r^2_{min}\, r^2_{max}\,/\,r^2 - r^2)\,/8$$

$$\tag{5.31}$$

$$\tau_t(r) = (3 + \mu)\rho\,\omega^2\,(r^2_{min} + r^2_{max} + r^2_{min}\, r^2_{max}\,/\,r^2 - (1 + 3\mu)\,r^2\,/\,(3 + \mu))\,/8.$$

The radial stress rises from zero at the inner rim, reaches a maximum at $r = (r_{min}r_{max})^{1/2}$, and then declines to zero again at the outer rim. The tangential stress is maximum at the inner rim and declines outwards. Its maximum value exceeds the maximum value of the radial stress for most relevant values of the parameters (μ is typically around 0.3).

Comparing (5.25) with (5.16) and the expression for I, it is seen that the energy density W in (5.16) can be obtained by isolating the term proportional to ω^2 in (5.25), multiplying it by $\frac{1}{2}r$, and integrating over r. The integral of the remaining terms is over a stress component times a shape-dependent expression, and it is customary to use an expression of the form

$$W/M = \sigma\, K_m\,/\,\rho, \tag{5.32}$$

where $M = \int \rho b(r) r\, d\theta\, dr$ is the total flywheel mass and σ is the maximum stress [cf. (5.18)]. K_m is called the shape factor. It depends only on geometry, if all stresses are equal as in the "constant stress disc", but as the example of a flat disc has indicated [see (5.31)], the material properties and the geometry can not generally be factorised. Still, the maximum stress occurring in the flywheel may be taken out as in (5.32), in order to leave a dimensionless quantity K_m to describe details of the flywheel construction (also, the factor ρ has to be there to balance ρ in the mass M, in order to make K_m dimensionless). The expression (5.32) may now be read in the following way: given a maximum acceptable stress σ, there is a maximum energy storage density given by (5.32). It does not depend on ω, and it is largest for light materials and for large design stresses σ. The design stress is typically chosen as a given fraction ("safety factor") of the tensile strength. If the tensile strength itself is used in (5.32), the physical upper limit for energy storage is obtained, and using (5.16), the expression gives the maximum value of ω for which the flywheel will not fail by deforming permanently or by disintegrating.

Flywheel performance
Some examples of flywheel shapes and the corresponding calculated shape factors K_m are given in Table 5.6. The infinitely extending disc of constant stress has a theoretical shape factor of unity, but for a practical version with finite truncation, K_m of about 0.8 can be expected. A flat, solid disc has a shape factor of 0.6, but if a hole is pierced in the middle, the value reduces to about 0.3. An infinitely thin rim has a shape factor of 0.5 and a finite one of about 0.4, and a radial rod or a circular brush (cf. Fig. 5.18) has K_m equal to one–third.

Shape	K_m
Constant stress disc	1
Flat, solid disc ($\mu = 0.3$)	0.606
Flat disc with centre hole	~0.3
Thin rim	0.5
Radial rod	1/3
Circular brush	1/3

Table 5.6. Flywheel shape factors.

According to (5.32), the other factors determining the maximum energy density are the maximum stress and the inverse density, in the case of a homogeneous material. Table 5.7 gives tensile strengths and/or design stresses with a safety factor included and gives densities for some materials contemplated for flywheel design.

For automotive purposes, the materials with the highest σ/ρ values may be contemplated, although they are also generally the most expensive. For stationary applications, weight and volume are less decisive, and low material cost becomes a key factor. This is the reason for considering cellulosic materials (Hagen *et al.*, 1979). One example is plywood discs, where the disc is assembled from layers of unidirectional plies, each with different orientation. Using (5.32) with the unidirectional strengths, the shape factor should be reduced by almost a factor of 3. Another example in this category is paper roll flywheels, that is, hollow, cylindrically wound shapes, for which the shape factor is $K_m = (1 + (r_{min}/r_{max})^2)/4$ (Hagen *et al.*, 1979). The specific energy density would be about 15 kJ kg^{-1} for the plywood construction and 27 kJ kg^{-1} for "super-paper" hollow torus shapes.

Material	Density (kg m^{-3})	Tensile strength (10^6 N m^{-2})	Design stress (10^6 N m^{-2})
Birch plywood	700	125	30
"Super-paper"	1100	335	
Aluminium alloy	2700	500	
Mild steel	7800		300
Maraging steel	8000	2700	900
Titanium alloy	4500		650
Carbon fibre (40% epoxy)	1550	1500	750
E-glass fibre (40% epoxy)	1900	1100	250
S-glass fibre (40% epoxy)	1900	1750	350
Kevlar fibre (40% epoxy)	1400	1800	1000

Table 5.7. Properties of materials considered for flywheels (Davidson *et al.*, 1980; Hagen *et al.*, 1979).

Unidirectional materials may be used in configurations such as the flywheel illustrated in Fig. 5.18b, where tangential (or "hoop") stresses are absent. Volume efficiency is low (Rabenhorst, 1976). Generally, flywheels made from filament have an advantage in terms of high safety, because disintegration into a large number of individual threads makes a failure easily contained. Solid flywheels may fail by expelling large fragments, and for safety such flywheels are not proper in vehicles, but may be placed underground for stationary uses.

Approximately constant stress shapes (cf. Fig. 5.18a) are not as volume efficient as flat discs. Therefore, composite flywheels of the kind shown in Fig. 5.18c have been contemplated (Post and Post, 1973). Concentric flat rings (e.g. made of Kevlar) are separated by elastomers that can eliminate breaking stresses when the rotation creates differential expansion of adjacent rings. Each ring must be made of a different material in order to keep the variations in stress within a small interval. The stress distribution inside each ring can be derived from the expressions in (5.31), assuming that the elastomers fully take care of any interaction between rings. Alternatively, the elastomers can be treated as additional rings, and the proper boundary conditions can be applied (see e.g. Toland, 1975).

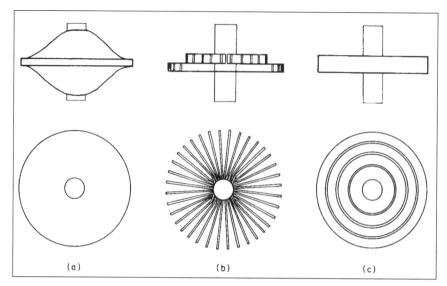

Figure 5.18. Different flywheel concepts. The upper line gives side views, and the lower line gives top views.

Flywheels of the types described above may attain energy densities of up to 200 kJ kg^{-1}. The problem is to protect this energy against frictional losses. Rotational speeds would typically be 3–5 revolutions per second. The commonly chosen solution is to operate the flywheel in near vacuum and to avoid any kind of mechanical bearings. Magnetic suspension has recently become feasible for units of up to about 200 t, using permanent magnets made from rare-earth cobalt compounds and electromagnetic stabilisers (Millner, 1979). In order to achieve power input and output, a motor generator is inserted between the magnetic bearing suspension and the flywheel rotor. If the motor is placed inside the vacuum, a brushless type is preferable.

For stationary applications, the weight limitations may be circumvented. The flywheel could consist of a horizontally rotating rim wheel of large dimensions and weight, supported by rollers along the rim or by magnetic suspension (Russell and Chew, 1981; Schlieben, 1975). Energy densities of about 3000 kJ kg^{-1} could, in principle, be achieved by using fused silica composites (cf. fibres of Table 5.7); if the installations were placed underground in order to allow reduced safety margins. Unit sizes would be up to 10^5 kg.

Compressed gas storage

Gases tend to be much more compressible than solids or fluids, and investigations of energy storage applications of elastic energy on a larger scale have therefore concentrated on the use of gaseous storage media.

Storage on a smaller scale may make use of steel containers, such as the ones common for compressed air used in mobile construction work. In this case the volume is fixed and the amount of energy stored in the volume is determined by the temperature and the pressure. If air is treated as an ideal gas, the (thermodynamic) pressure P and temperature T are related by the equation of state

$$PV = \nu \mathcal{R} T, \tag{5.33}$$

where V is the volume occupied by the air, ν is the number of moles in the volume, and $\mathcal{R} = 8.315 \ \text{J K}^{-1} \ \text{mol}^{-1}$. The pressure P corresponds to the stress in the direction of compression for an elastic cube, except that the sign is reversed (in general the stress equals $-P$ plus viscosity-dependent terms). The container may be thought of as a cylinder with a piston, enclosing a given number of moles of gas, say, air, and the compressed air is formed by compressing the enclosed air from standard pressure at the temperature of the surroundings, that is, increasing the force f_x applied to the piston, while the volume decreases from V_0 to V. The amount of energy stored is

$$W = A \int_{x_0}^{x} f_x \, dx = - \int_{V_0}^{V} P \, dV, \tag{5.34}$$

where A is the cylinder cross-sectional area, x and x_0 are the piston positions corresponding to V and V_0, and P is the pressure of the enclosed air.

For large-scale storage applications, underground cavities have been considered. The three possibilities investigated until now are salt domes, cavities in solid rock formations, and aquifers.

Cavities in salt deposits may be formed by flushing water through the salt. The process has, in practical cases, been extended over a few years, in which case the energy spent (and cost) has been very low (Weber, 1975). Salt domes are salt deposits extruding upwards toward the surface, therefore allowing cavities to be formed at modest depths.

Rock cavities may be either natural or excavated, and the walls are properly sealed to ensure air-tightness. If excavated, they are much more expensive to make than salt caverns.

Aquifers are layers of high permeability, permitting underground water flows along the layer. In order to confine the water stream to the aquifer, there have to be encapsulating layers of little or no permeability above and below the water-carrying layer. The aquifers usually do not stay at a fixed depth, and thus, there will be slightly elevated regions where a certain amount of air can become trapped without impeding the flow of water. This possibility of air storage (under the elevated pressure corresponding to the depth involved) is illustrated in Fig. 5.19c.

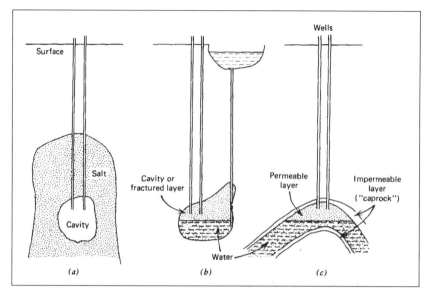

Figure 5.19. Types of underground compressed air storage: (a) storage in salt cavity, (b) rock storage with compensating surface reservoir, and (c) aquifer storage.

Figure 5.19 illustrates the forms of underground air storage mentioned: salt, rock, and aquifer storage. In all cases, the site selection and preparation is a fairly delicate process. Although the general geology of the area considered is known, the detailed properties of the cavity will not become fully disclosed until the installation is complete. The ability of the salt cavern to keep an elevated pressure may not live up to expectations based on sample analysis and pressure tests at partial excavation. The stability of a natural rock cave, or of a fractured zone created by explosion or hydraulic methods, is also uncertain until actual full-scale pressure tests have been conducted. For the aquifers, the decisive measurements of permeability can only be made at a finite number of places, so that surprises are possible due to rapid permeability change over small distances of displacement (cf. Adolfson *et al.*, 1979).

The stability of a given cavern is influenced by two design features that the operation of the compressed air storage system will entail, notably the temperature variations and the pressure variations. It is possible to keep the cavern wall temperature nearly constant, either by cooling the compressed air before letting it down into the cavern or by performing the compression so slowly that the temperature only rises to the level prevailing on the cavern walls. The latter possibility (isothermal compression) is impractical for most applications, because excess power must be converted at the rate at which it comes. Most systems therefore include one or more cooling steps. With respect to the problem of pressure variations, when different amounts of energy are stored, the solution may be to store the compressed air at constant pressure but variable volume. In this case either the storage volume itself should be variable, as it is by aquifer storage (when variable amounts of water are displaced), or the underground cavern should be connected to an open reservoir (Fig. 5.19b), so that a variable water column may take care of the variable amounts of air stored at the constant equilibrium pressure prevailing at the depth of the cavern. This kind of compressed energy storage system may alternatively be viewed as a pumped hydro storage system, with extraction taking place through air-driven turbines rather than through water-driven turbines.

Adiabatic storage
Consider now the operation of a variable-pressure type of system. The compression of ambient air takes place approximately as an adiabatic process, that is, without heat exchange with the surroundings. Denoting by γ the ratio between the partial derivatives of pressure with respect to volume at constant entropy and at constant temperature,

$$(\partial P / \partial V)_S = \gamma (\partial P / \partial V)_T, \tag{5.35}$$

the ideal gas law (5.33) gives $(\partial P / \partial V)_T = - P/V$, so that for constant γ,

$$PV^\gamma = P_0 V_0^\gamma. \tag{5.36}$$

The constant on the right-hand side is here expressed in terms of the pressure P_0 and volume V_0 at a given time. For air at ambient pressure and temperature, $\gamma = 1.40$. The value decreases with increasing temperature and increases with increasing pressure, so (5.36) is not entirely valid for air. However, in the temperature and pressure intervals relevant for practical application of compressed air storage, the value of γ varies less than $\pm 10\%$ from its average value.

Inserting (5.36) into (5.34) we get the amount of energy stored,

$$W = - \int_{V_0}^{V} P_0 \left(\frac{V_0}{V} \right)^\gamma dV = \frac{P_0 V_0}{\gamma - 1} \left(\left(\frac{V_0}{V} \right)^{\gamma - 1} - 1 \right), \tag{5.37}$$

or, alternatively,

$$W = \frac{P_0 V_0}{\gamma - 1} \left(\left(\frac{P}{P_0} \right)^{(\gamma-1)/\gamma} - 1 \right). \tag{5.38}$$

More precisely, this is the work required for the adiabatic compression of the initial volume of air. This process heats the air from its initial temperature T_0 to a temperature T, which can be found by rewriting (5.33) in the form

$$T / T_0 = PV / (P_0 V_0)$$

and combining it with the adiabatic condition (5.36),

$$T = T_0 \, (P / P_0)^{\,(\gamma-1)/\gamma}. \tag{5.39}$$

Since desirable pressure ratios in practical applications may be up to about $P/P_0 = 70$, maximum temperatures exceeding 1000 K can be expected. Such temperature changes would be unacceptable for most types of cavities considered, and the air is therefore cooled before transmission to the cavity. Surrounding temperatures for underground storage are typically about 300 K for salt domes and somewhat higher for storage in deeper geological formations. Denoting this temperature T_s, the heat removed if the air is cooled to T_s at constant pressure amounts to

$$H = c_P(T - T_s), \tag{5.40}$$

where c_P is the heat capacity at constant pressure. Ideally, the heat removed would be kept in a well-insulated thermal energy store, so that it can be used to re-heat the air when it is taken up from the cavity to perform work by expansion in a turbine, with the associated pressure drop back to ambient pressure P_0. Viewed as a thermodynamic process in a temperature–entropy (T, S)-diagram, the storage and retrieval processes in the ideal case look as indicated in Fig. 5.20. The process leads back to its point of departure, indicating that the storage cycle is loss-free under the idealised conditions assumed so far.

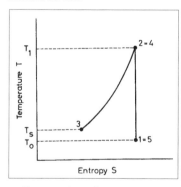

Figure 5.20. Idealised operation of adiabatic compressed air storage system. The charging steps are 1-2 adiabatic compression and 2-3 isobaric cooling to cavern temperature. The unloading steps are 3-4 isobaric heating and 4-5 adiabatic expansion through turbine. The diagram follows a given amount of air, whereas an eventual thermal energy store is external to the "system" considered. T_0 is surface ambient temperature, T_1 is temperature after compression, and T_s is the cavern temperature.

In practice, the compressor has a loss (of maybe 5–10%), meaning that not all the energy input (electricity, mechanical energy) is used to perform compression work on the air. Some is lost as friction heat and so on. Further, not all the heat removed by the cooling process can be delivered to re-heat the air. Heat exchangers have finite temperature gradients, and there may be losses from the thermal energy store during the time interval between cooling and re-heating. Finally, the exhaust air from actual turbines has temperatures and pressures above ambient. Typical loss fractions in the turbine may be around 20% of the energy input, at the pressure considered above (70 times ambient) (Davidson et al., 1980). If under 10% thermal losses could be achieved, the overall storage cycle efficiency would be about 65%.

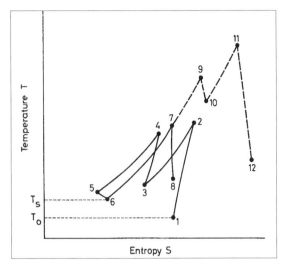

Figure 5.21. Operation of compressed air storage systems with finite losses. The solid path corresponds to a scheme with two compression and two cooling stages, temperature adjustment of the stored air, re-heating, and a single turbine stage with reject air injected in the atmosphere (open cycle 1-8). As an alternative, the path from step 7 to step 12 exhibits two heating and expansion steps, corresponding to the operating plant at Huntorf, Germany. See text for further details.

The real process may look as indicated in Fig. 5.21 in terms of temperature and entropy changes. The compressor loss in the initial process 1-2 modifies the vertical line to include an entropy increase. Further, the compression has been divided into two steps (1-2 and 3-4) in order to reduce the maximum temperatures. Correspondingly, there are two cooling steps (2-3 and 4-5), followed by a slight final cooling performed by the cavity surroundings (5-6). The work retrieval process involves in this case a single step 6-7 of re-heating by use of heat stored from the cooling processes (in some cases more than one re-heating step is employed). Finally, 7-8 is the turbine stage, which leaves the cycle open by not having the air reach the initial temperature (and pressure) before it leaves the turbine and mixes into the ambient atmosphere. Also, this expansion step shows deviations from adiabaticity, seen in Fig. 5.21 as an entropy increase.

There are currently only a few utility integrated installations. The earliest full-scale compressed storage facility has operated since 1978 at Huntorf in Germany. It is rated at 290 MW and has about 3×10^5 m^3 storage volume (Lehmann, 1981). It does not have any heat recuperation, but it has two fuel-based turbine stages, implying that the final expansion takes place from a temperature higher than any of those involved in the compression stages (and also at higher pressure). This is indicated in Fig. 5.21 as 7-9-10-11-12, where steps 7-9 and 10-11 represent additional heating based on fuel, while steps 9-10 and 11-12 indicate expansion through turbines. If heat recuperation is added, as it is in the 110-MW plant operated by the Alabama Electric Corp. (USA) since 1991, this will move point 7 upwards towards point 9, and point 8 will move in the direction of 12, altogether representing an increased turbine output (Linden, 2003).

The efficiency calculation is changed in the case of additional fuel combustion. The additional heat input may be described by (5.40) with appropriate temperatures substituted, and the primary enthalpy input H_0 is obtained by dividing H by the fuel to heat conversion efficiency. The input work W_{in} to the compressor changes in the case of a finite compressor efficiency η_c from (5.38) to

$$W_{in} = \frac{P_0 V_0}{\gamma - 1}\left(\left(\frac{P}{P_0}\right)^{(\gamma-1)/(\gamma\eta_c)} - 1\right). \tag{5.41}$$

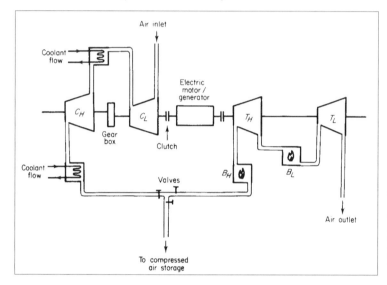

Figure 5.22. Layout of the Huntorf compressed air storage facility. Compressors are denoted C, turbines are T, and burners are B. The subscripts H/L stand for high/low pressure.

The work delivered by the turbine receiving air of pressure P_1 and volume V_1, and exhausting it at P_2 and V_2, with a finite turbine efficiency η_t, is

$$W_{out} = \frac{P_1 V_1}{\gamma - 1}\left(1 - \left(\frac{P_2}{P_1}\right)^{\eta_t(\gamma-1)/\gamma}\right), \tag{5.42}$$

which except for the appearance of η_t is just (5.38) rewritten for the appropriate pressures and volume.

Now, in case there is only a single compressor and a single turbine stage, the overall cycle efficiency is given by

$$\eta = W_{out}/(W_{in} + H_0). \tag{5.43}$$

For the German compressed air storage installation mentioned above, η is 0.41. Of course, if the work input to the compressor is derived from fuel (directly or through electricity), W_{in} may be replaced by the fuel input W_0 and a fuel efficiency defined as

$$\eta_{fuel} = W_{out} / (W_0 + H_0) \tag{5.44}$$

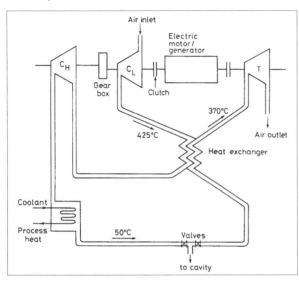

Figure 5.23. Layout of an "advanced" compressed air storage facility, with heat recuperation and no fuel input (symbols are explained in legend to Fig. 5.22).

If W_{in}/W_0 is taken as 0.36, η_{fuel} for the example becomes 0.25, which is 71% of the conversion efficiency for electric power production without going through the store. The German installation is used for providing peak power on weekdays, charging during nights and weekends.

Figure 5.22 shows the physical layout of the German plant, and Fig. 5.23 shows the layout of a more advanced installation with no fuel input, corresponding to the two paths illustrated in Fig. 5.21.

Aquifer Storage.
The aquifer storage system shown in Fig. 5.19c would have an approximately constant working pressure, corresponding to the average hydraulic pressure at the depth of the air-filled part of the aquifer. According to (5.34) the stored energy in this case simply equals the pressure P times the volume of air displacing water in the aquifer. This volume equals the physical volume V times the porosity p, that is, the fractional void volume accessible to intruding air (there may be additional voids that the incoming air cannot reach), so the energy stored may be written

$$W = pVP. \tag{5.45}$$

Typical values are $p = 0.2$ and P around 6×10^6 N m^{-2} at depths of some 600 m, with useful volumes of 10^9 to 10^{10} m^3 for each site. Several such sites have been investigated with the idea of storing natural gas.

An important feature of an energy storage aquifer is the time required for charging and emptying. This time is determined by the permeability of the aquifer. The permeability is basically the proportionality factor between the flow velocity of a fluid or gas through the sediment and the pressure gradient causing the flow. The linear relationship assumed may be written

$$v = - K (\eta\rho)^{-1} \partial P/\partial s, \tag{5.46}$$

where v is the flow velocity, η is the viscosity of the fluid or gas, ρ is its density, P is the pressure, and s is the path length in the downward direction. K is the permeability, being defined by (5.46). In metric (SI) units, the permeability has the dimension of m^2. The unit of viscosity is $m^2\,s^{-1}$. Another commonly used unit of permeability is the *darcy*. One darcy equals 1.013×10^{12} m^2. If filling and emptying of the aquifer storage are to take place in a matter of hours rather than days, the permeability has to exceed 10^{11} m^2. Sediments such as sandstone are found with permeabilities ranging from 10^{10} to 3×10^{12} m^2, often with considerable variation over short distances.

In actual installations, losses occur. The cap-rock bordering the aquifer region may not have negligible permeability, implying a possible leakage loss. Friction in the pipes leading to and from the aquifer may cause a loss of pressure, as may losses in the compressor and turbine. Typically, losses of about 15% are expected in addition to those of the power machinery. Large aquifer stores for natural gas are in operation, e.g. at Stenlille, Denmark: 10^9 m^3 total volume, from which 3.5×10^8 m^3 of gas can be extracted at the employed pressure of 17 MPa (DONG, 2003). The same gas utility company operates an excavated salt dome gas storage at Lille Thorup of slightly smaller volume, but allowing 4.2×10^8 m^3 of gas to be extracted at 23 MPa.

Hydrogen storage

Hydrogen can be stored like other gases, compressed in suitable containers capable of taking care of the high diffusivity of hydrogen, as well as sustaining the pressures required to bring the energy density up to useful levels. However, the low energy density of gaseous hydrogen has brought alternative storage forms into the focus of investigation, such as liquid hydrogen and hydrogen trapped inside metal hydride structures or inside carbon-based or other types of nanotubes (cf. Sørensen, 2004a).

Hydrogen is an energy carrier, not a primary energy form. The storage cycle therefore involves both the production of hydrogen from primary energy sources and the retrieval of the energy form demanded by a second conversion process.

Hydrogen production

Conventional hydrogen production is by catalytic steam reforming of methane (natural gas) or gasoline with water vapour. The process, which typically takes place at 850°C and 2.5×10^6 Pa, is

$$C_nH_m + n\ H_2O \rightarrow n\ CO + (n+m/2)\ H_2, \tag{5.47}$$

followed by the catalytic shift reaction

$$CO + H_2O \rightarrow CO_2 + H_2. \tag{5.48}$$

Finally, CO_2 is removed by absorption or membrane separation. The heat produced by (5.48) often cannot be directly used for (5.47). For heavy hydrocarbons, including coal dust, a partial oxidation process is currently in use (Zittel and Wurster, 1996). An emerging technology is high-temperature plasma-arc gasification, based on which a pilot plant operates on natural gas at 1600°C in Norway (at Kvaerner Engineering). The advantage of this process is the pureness of the resulting products (in energy terms, 48% hydrogen, 40% carbon and 10% water vapour) and therefore low environmental impacts. Since all the three main products are useful energy carriers, the conversion efficiency may be said to be 98% minus the energy needed for the process. However, conversion of natural gas to carbon is not normally desirable, and the steam can be used only locally, so a 48% efficiency is more meaningful.

Production of hydrogen from biomass may be achieved by biological fermentation or by high-temperature gasification similar to that of coal. These processes were described in more detail in section 4.8.2. Production of hydrogen from (wind- or solar-produced) electricity may be achieved by conventional electrolysis (demonstrated by Faraday in 1820 and widely used since about 1890) or by reversible fuel cells, with current efficiencies of about 70% and over 90%, respectively.

Electrolysis conventionally uses an aqueous alkaline electrolyte, with the anode and cathode areas separated by a microporous diaphragm (replacing earlier asbestos diaphragms),

$$H_2O \rightarrow H_2 + \tfrac{1}{2}\ O_2, \tag{5.49}$$

$$\Delta H = \Delta G + T\ \Delta S. \tag{5.50}$$

At 25°C, the change in free energy, ΔG, is 236 kJ mol^{-1}, and the electrolysis would require a minimum amount of electric energy of 236 kJ mol^{-1}, while the difference between enthalpy and free energy changes, $\Delta H - \Delta G$, in theory could be heat from the surroundings. The energy content of hydrogen (equal to ΔH) is 242 kJ mol^{-1} (lower heating value), so the $T\ \Delta S$ could exceed 100%. However, if heat at 25°C is used, the process is exceedingly slow. Temperatures used in actual installations are so high that the heat balance is positive and cooling has to be applied. This is largely a consequence of electrode overvoltage mainly stemming from polarisation effects. The cell potential V for water electrolysis may be expressed by

$$V = V_r + V_a + V_c + Rj, \tag{5.51}$$

where V_r is the reversible cell potential. The overvoltage has been divided into the anodic and cathodic parts V_a and V_c. The current is j, and R is the internal resistance of the cell. The three last terms in (5.51) represent the electrical losses, and the voltage efficiency η_V of an electrolyser operating at a current j is given by

$$\eta_V = V_r / V, \tag{5.52}$$

while the thermal efficiency is

$$\eta_t = \Delta H / \Delta G = |\Delta H / (n \mathscr{F} V)|, \tag{5.53}$$

with the Faraday constant being $\mathscr{F} = 96493$ coulombs mol^{-1} and n being the number of moles transferred in the course of the overall electrochemical reaction to which ΔG relates.

Efforts are being made to increase the efficiency above the current 50 to 80% (for small to large electrolysers) by increasing operating temperature and optimising electrode materials and cell design; in that case the additional costs should be less than for the emerging solid-state electrolysers, which are essentially fuel cells operated in reverse fashion, i.e. using electric power to produce hydrogen and oxygen from water in an arrangement and with reaction schemes formally the same as those of fuel cells described in section 4.7. If the same fuel cell allows operation in both directions, it is called a *reversible fuel cell*.

A third route contemplated for hydrogen production from water is thermal decomposition of water. As the direct thermal decomposition of the water molecule requires temperatures exceeding 3000 K, which is not possible with presently available materials, attempts have been made to achieve decomposition below 800°C by an indirect route using cyclic chemical processes. Such thermochemical or water-splitting cycles were originally designed to reduce the required temperature to the low values attained in nuclear reactors, but could, of course, be used with other technologies generating heat at around 400°C. An example of the processes studied (Marchetti, 1973) is the three-stage reaction

$6FeCl_2 + 8H_2O \rightarrow 2Fe_3O_4 + 12HCl + 2H_2$ (850°C)
$2Fe_3O_4 + 3Cl_2 + 12HCl \rightarrow 6FeCl_3 + 6H_2O + O_2$ (200°C)
$6FeCl_3 \rightarrow 6FeCl_2 + 3Cl_2$ (420°C).

The first of these reactions still requires a high temperature, implying a need for energy to be supplied, in addition to the problem of the corrosive substances involved. The research is still a long way from having created a practical technology.

The process of water photodissociation was described in section 3.6.1. There have been several attempts to imitate the natural photosynthetic process, using semiconductor materials and membranes to separate the hydrogen and oxygen formed by application of light (Calvin, 1974; Wrighton *et al.*, 1977). So far no viable reaction scheme has been found, either for artificial photodissociation or for hybrid processes using heat and chemical reactions in side processes (Hagenmuller, 1977).

Processing of the hydrogen produced involves removal of dust and sulphur, plus other impurities depending on the source material (e.g. CO_2 if biogas is the source).

Hydrogen storage forms
The storage forms relevant for hydrogen are connected with the physical properties of hydrogen, such as the energy density shown in Table 5.5 for hydrogen in various forms. Combustion and safety-related properties are shown in Table 5.8 and compared with those of methane, propane and gasoline. The high diffusivity has implications for container structure, together with the large range of flammability/explosivity for all applications.

Property	*Unit*	*Hydrogen*	*Methane*	*Propane*	*Gasoline*
Minimum energy for ignition	10^{-3} J	0.02	0.29		0.24
Flame temperature	°C	2045	1875		2200
Auto-ignition temperature in air	°C	585	540	510	230–500
Maximum flame velocity	m s^{-1}	3.46	0.43	0.47	
Range of flammability in air	vol.%	4–75	5–15	2.5–9.3	1.0–7.6
Range of explosivity in air	vol.%	13–65	6.3–13.5		1.1–3.3
Diffusion coefficient in air	10^{-4} m^2 s^{-1}	0.61	0.16		0.05

Table 5.8. Safety-related properties of hydrogen and other fuels (with use of Zittel and Wurster, 1996).

Compressed storage in gaseous form. The low volume density of hydrogen at ambient pressure (Table 5.5) makes compression necessary for energy storage applications. Commercial hydrogen containers presently use pressures of 20–30×10^6 Pa, with corresponding energy densities of 1900–2700×10^6 J m^{-3}, which is still less than 10% of that of oil. Research is in progress for increasing pressures to about 70 MPa, using high-strength composite materials such as Kevlar fibres. Inside liners of carbon fibres (earlier glass/aluminium) are required to reduce permeability. Compression energy requirements influence storage cycle efficiencies and involve long transfer times. The work required for isothermal compression from pressure p_1 to p_2 is of the form

$$W = A\,T \log (p_2/p_1),$$

where A is the hydrogen gas constant 4124 J K^{-1} kg^{-1} times an empirical, pressure-dependent correction (Zittel and Wurster, 1996). To achieve the compression, a motor rated at Bm must be used, where m is the maximum power throughput and B, depending on engine efficiency, is around 2.

Liquid hydrogen stores. Because the liquefaction temperature of hydrogen is 20 K (−253°C), the infrastructure and liquefaction energy requirements are substantial (containers and transfer pipes must be super-insulated). On the other hand, transfer times are low (currently 3 min to charge a passenger car). The energy density is still 4–5 times lower than for conventional fuels (see Table 5.5). The liquefaction process requires very clean hydrogen, as well as several cycles of compression, liquid nitrogen cooling, and expansion.

Metal hydride storage. Hydrogen diffused into appropriate metal alloys can achieve storage at volume densities over two times that of liquid hydrogen. However, the mass storage densities are still less than 10% of those of conventional fuels (Table 5.5), making this concept doubtful for mobile applications, despite the positive aspects of near loss-free storage at ambient pressures (0–6 MPa) and transfer accomplished by adding or withdrawing modest amounts of heat (plus high safety in operation), according to

$$Me + \tfrac{1}{2}\,xH_2 \leftrightarrow MeH_x, \qquad\qquad (5.54)$$

where the hydride may be body-centred cubic lattice structures with about 6 × 10^{28} atoms per m^3 (such as $LaNi_5H_6$, $FeTiH_2$). The currently highest density achieved are for metal alloys absorbing two hydrogen atoms per metal atom (Toyota, 1996). The lattice absorption cycle also performs a cleaning of the gas, because impurities in the hydrogen gas are too large to enter the lattice.

Methanol storage. One current prototype hydrogen-fuelled vehicle uses methanol as storage, even if the desired form is hydrogen (because the car uses a hydrogen fuel cell to generate electricity for its electric motor; Daimler-Chrysler-Ballard, 1998). This is due to the simplicity of methanol storage and filling infrastructure. In the long run, transformation of hydrogen to methanol and back seems too inefficient, and it is likely that the methanol concept will be combined with methanol fuel cells (cf. section 4.7), while hydrogen-fuelled vehicles must find simpler storage alternatives.

Graphite nanofibre stores. Current development of nanofibres has suggested wide engineering possibilities, regarding both electric and structural adaptation, including the storage of foreign atoms inside "balls" or "tubes" of large carbon structures (Zhang *et al.*, 1998). Indications are that hydrogen may be stored in nanotubes in quantities exceeding that of metal hydrides, and at a lower weight penalty, but no designs exists yet (Service, 1998).

Regeneration of power from hydrogen
Retrieval of energy from stored hydrogen may be by conventional low-efficiency combustion in Otto engines or gas turbines, or it may be through fuel cells at a considerably higher efficiency, as described in section 4.7 and in Sørensen (2004a).

Batteries

Batteries may be described as fuel cells where the fuels are stored inside the cell rather than outside it. Historically, batteries were the first controlled source of electricity, with important designs being developed in the early 19th century by Galvani, Volta and Daniell, before Grove's discovery of the fuel cell and Planté's construction of the lead–acid battery. Today, available batteries use a wide range of electrode materials and electrolytes, but despite considerable development efforts aimed at the electric utility sector, battery storage is still in practice restricted to small-scale use (consumer electronics, motor cars, etc.).

Type	Electrolyte	Energy efficiency (%)	Energy density (Wh kg^{-1})	Power-densities Peak (W kg^{-1})	Power-densities Sustained (W kg^{-1})	Cycle life (cycles)	Operating temperatures (°C)
Commercial:							
Lead–acid	H_2SO_4	75	20–35	120	25	200–2000	−20 to 60
Nickel–cadmium	KOH	60	40–60	300	140	500-2000	−40 to 60
Ni-metal-hydride	KOH	50	60–80	440	220	<3000	10 to 50
Lithium-ion	$LiPF_6$	70	100–200	720	360	500–2000	−20 to 60
Under development:							
Sodium–sulphur	β-Al_2O_3	70	120	240	120	2000	300 to 400
Lithium-sulphide	AlN	75	130	200	140	200	430 to 500
Zinc–chlorine	$ZnCl_2$	65	120	100			0
Lithium–polymer	Li-β-Alu	70	200			>1200	−20 to 60
1977 goal cells:							
High energy		65	265		55-100	2500	
High power		70	60	280	140	1000	

Table 5.9. Characteristics of selected batteries (Jensen and Sørensen, 1984; Cultu, 1989a; Scrosati, 1995; Buchmann, 1998) and comparison with 1977 development goals (Weiner, 1977).

Efforts are being made to find systems of better performance than the long-employed lead–acid batteries, which are restricted by a low energy density (see Table 5.5) and a limited life. Alkaline batteries such as nickel–cadmium cells, proposed around 1900 but first commercialised during the 1970s, are the second largest market for use in consumer electronics equipment and recently for electric vehicles. Despite high cost, a rapid impact has been made by lithium-ion batteries since their introduction in 1991 (see below). They allow charge topping (i.e. charging before complete discharge) and have a high energy density, suitable for small-scale portable electronic equipment.

Rechargeable batteries are called accumulators or *secondary batteries*, whereas use-once-only piles are termed *primary batteries*. Table 5.9 gives some important characteristics of various battery types. It is seen that the research goals set 15 years ago for high-power batteries have been reached in commercially available products, but not quite the goals for high-energy density cells. One reason for the continued high market share of lead–acid batteries is the perfection of the technology, which has taken place over the last decades.

An electrochemical storage battery has properties determined by cell voltage, current and time constants. The two electrodes delivering or receiving power to or from the outside are called e_n and e_p (*negative* and *positive electrode*). The conventional names *anode* and *cathode* are confusing in the case of rechargeable batteries. Within the battery, ions are transported between the negative and positive electrodes through an *electrolyte*. This as well as the electrodes may be solid, liquid or, in theory, gaseous. The electromotive force E_0 is the difference between the electrode potentials for an open external circuit,

$$E_o = E_{ep} - E_{en},\tag{5.55}$$

where it is customary to measure all potentials relative to some reference state. The description of the open cell behaviour uses standard steady-state chemical reaction kinetics. However, when current flows to or from the cell, equilibrium thermodynamics is no longer applicable, and the cell voltage V_c is often parametrised in the form

$$V_c = E_0 - \eta\, IR,\tag{5.56}$$

where I is the current at a given time, R is the internal resistance of the cell, and η is a "polarisation factor" receiving contributions from the possibly very complex processes taking place in the transition layer separating each electrode from the electrolyte. Figure 5.24 illustrates in a highly schematic form the different potential levels across the cell for open and closed external circuit (cf. e.g. Bockris and Reddy, 1973).

The lead–acid battery
In the electrolyte (aqueous solution of sulphuric acid) of a lead–acid battery, three reactions are at work,

$$\begin{aligned}
&H_2O \leftrightarrow H^+ + OH^-,\\
&H_2SO_4 \leftrightarrow 2H^+ + SO_4^{2-},\\
&H_2SO_4 \leftrightarrow H^+ + HSO_4^-,
\end{aligned}\tag{5.57}$$

and at the (lead and lead oxide) electrodes, the reactions are

negative electrode: $Pb + SO_4^{2-} \leftrightarrow PbSO_4 + 2e^-,$

$$positive\ electrode:\ PbO_2 + SO_4^{2-} + 4H^+ + 2e^- \leftrightarrow PbSO_4 + 2H_2O. \qquad (5.58)$$

The electrolyte reactions involve ionisation of water and either single or double ionisation of sulphuric acid. At both electrodes, lead sulphate is formed, from lead oxide at the positive electrode and from lead itself at the negative electrode. Developments have included sealed casing, thin-tube electrode structure and electrolyte circulation. As a result, the internal resistance has been reduced, and the battery has become practically maintenance free throughout its life. The energy density of the lead–acid battery increases with temperature and decreases with discharge rate (by about 25% when going from 10 to 1 h discharge, and by about 75% when going from 1 h to 5 min discharge, cf. Jensen and Sørensen, 1984). The figures given in Table 5.9 correspond to an average discharge rate and an 80% depth of discharge.

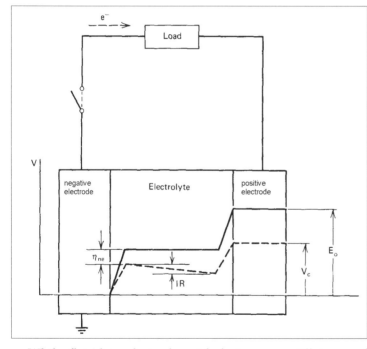

Figure 5.24. Potential distribution through an electrochemical cell: solid line, open external circuit; dashed line, load connected (Jensen and Sørensen, 1984).

While flat-plate electrode grid designs are still in use for automobile starter batteries, tubular-plate designs have a highly increased cycle life and are used for electric vehicles and increasingly for other purposes. The claimed life is about 30 years, according to enhanced test cycling. Charging procedures for lead–acid batteries influence battery life.

Alkaline electrolyte batteries

Among the alkaline electrolyte batteries, nickel–cadmium batteries have been used since about 1910, based upon the investigations during the 1890s by Jungner. Their advantage is a long lifetime (up to about 2000 cycles) and with careful use a nearly constant performance, independent of discharge rate and age (Jensen and Sørensen, 1984). However, they do not allow drip charging and easily drop to low capacity if proper procedures are not followed. During the period 1970–90, they experienced an expanding penetration in applications for powering consumer products, such as video cameras, cellular phones, portable players and portable computers, but have now lost most of these markets to the more expensive lithium-ion batteries.

Iron–nickel oxide batteries, which were used extensively in the early part of the 20th century, in electric motorcars, are inferior as regards cell efficiency and peaking capability, owing to low cell voltage and high internal resistance, which also increases the tendency for self-discharge. Alternatives such as nickel–zinc batteries are hampered by low cycle life.

The overall reaction may be summarised as

$$2NiOOH + 2H_2O + Cd \leftrightarrow 2Ni(OH)_2 + Cd(OH)_2. \tag{5.59}$$

The range of cycle lives indicated in Table 5.9 reflects the sensitivity of NiCd batteries to proper maintenance, including frequent deep discharge. For some applications, it is not practical to have to run the battery down to zero output before recharging.

An alternative considered is nickel–metal hydride batteries, which exhibit a slightly higher energy density but so far with a lower cycle life.

High-temperature batteries

Research on high-temperature batteries for electric utility use (cf. Table 5.9) has been ongoing during several decades, without decisive breakthroughs. Their advantage would be fast, reversible chemistry, allowing for high current density without excess heat generation. Drawbacks include serious corrosion problems, which have persisted and curtailed the otherwise promising development of, for example, the sodium–sulphur battery. This battery has molten electrodes and a solid electrolyte, usually of tubular shape and made from ceramic beta-alumina materials. Similar containment problems have faced zinc–chlorine and lithium–sulphur batteries.

Lithium-ion batteries

Lithium metal electrode batteries attracted attention several decades ago (Murphy and Christian, 1979), owing to the potentially very high energy density. However, explosion risks stood in the way of commercial applications until the current lithium-ion concept was developed by Sony in Japan (Nagaura, 1990). Its electrode materials are $LiCoO_2$ and Li_xC_6 (carbon or graphite), respectively, with an electrolyte of lithium hexafluorophosphate dissolved in a mixture of ethylene carbonate and dimenthyl carbonate (Scrosati, 1995). The battery is built from alternating layers of electrode materials, between which the Li-ions oscillate cyclically (Fig. 5.25). The cell potential is

high, 3.6 or 7.2 V. Li-ion batteries do not accept overcharge, and a safety device is usually integrated in the design to provide automatic venting in case
of emergency. Owing to its high power density (by weight or volume) and
easy charging (topping up is possible without loss of performance), the resulting concept has rapidly penetrated to the high-end portable equipment
sector. The safety record is excellent, justifying abandoning the lithium metal
electrode despite some loss of power. The remaining environmental concern
is mainly due to use of the very toxic cobalt, which requires an extremely
high degree of recycling to be acceptable.

Ongoing research aims at bringing the price down and avoiding the use
of toxic cobalt, while maintaining the high level of safety and possibly improving performance. The preferred concept uses an all solid-state structure
with lithium–beta–alumina forming a layered electrolyte and $LiMn_2O_4$ or
$LiMnO_2$ as the positive electrode material (Armstrong and Bruce, 1996), selected from the family of intercalation reactions (Ceder *et al.*, 1998),

$$Li_{x1}MO_2 + (x_2-x_1)Li \rightarrow Li_{x2}MO_2.$$

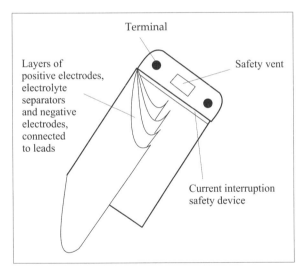

Terminal

Layers of
positive electrodes,
electrolyte
separators
and negative
electrodes,
connected
to leads

Safety vent

Current interruption
safety device

Figure 5.25. Layout of
commercial lithium-ion rechargeable battery.

The aim is to reduce the cost of these lithium–polymer batteries to about
20% of the Li-ion battery costs, opening up use not just in small-scale portable equipment but also in electric vehicles. Currently, the ionic conductivity
of solid polymer materials is too low to allow ambient temperature operation, but operating temperatures of around 80°C should be acceptable for
non-portable applications (Tarascon and Armand, 2001).

Mandatory recycling of batteries has already been introduced in several
countries and is clearly a must for some of the recently emerged concepts.

If batteries ever get to the point of being suitable for use in the grid-based
electricity supply sector, it is likely to be for short-term and medium-term

storage. Batteries have already been extensively incorporated into off-grid systems, such as rural power systems (whether based upon renewable energy or diesel sets) and stand-alone systems such as security back-up facilities. Utility battery systems could serve as load-levelling devices and emergency units, as well as back-up units in connection with variable primary supplies from variable renewable sources.

The small-scale battery options currently extremely popular in the consumer sector carry one very promising possibility, namely, that of creating a de-coupling between production and load in the sector of electricity supply, which traditionally has been strongly focused upon a rigid expectation that any demand will be instantaneously met. If many of the household appliances become battery operated, the stringent requirements placed upon the electricity system to meet peak loads may become less of a burden. A considerable off-peak demand for charging rechargeable batteries could provide precisely the load (as seen from the utility end) control desired, whether it be nearly a constant load pattern or a load following the variations in renewable energy supplies. The same goes for batteries serving the transportation sector (electric vehicles). Freedom from having to always be near an electric plug to use power equipment is seen as a substantial consumer benefit, which many people are willing to pay a premium price for.

Reversible fuel cells (flow batteries)

As mentioned above, fuel cells in reverse operation may replace alkaline electrolysis for hydrogen production, and if fuel cells are operated in both directions, they constitute an energy storage facility. Obvious candidates are further developments of current PEM fuel cells with hydrogen or methanol as the main storage medium. Presently, prototypes of hydrogen/oxygen or hydrogen/air reversible PEM fuel cells suffer from the inability of currently used membranes to provide high efficiency both ways. Whereas a 50% efficiency is acceptable for electricity production, only the same 50% efficiency is achieved for hydrogen production, i.e. lower than conventional electrolysis (Proton Energy Systems, 2003). Fuel cells operated in reverse mode has reached near 100% hydrogen production efficiency, using a few, very large membranes (0.25 m^2), but such cells are not effective for power production (Yamaguchi *et al.*, 2000).

Early development of reversible fuel cells, sometimes called flow batteries and then referred to as redox flow batteries, used the several ionisation stages of vanadium as a basis for stored chemical energy. The reactions at the two electrodes would typically involve

$$V^{4+} \rightarrow V^{5+} \text{ (positive terminal) } and \text{ } V^{3+} \rightarrow V^{2+} \text{ (negative terminal)}.$$

The electrode material would be carbon fibre, and an ion exchange membrane placed between the electrodes would ensure the charge separation.

Specific energy storage would be about 100 kJ kg^{-1}. It has recently been proposed (Skyllas-Kazacos, 2003) to replace the positive electrode process by a halide ion reaction such as

$$Br^- + 2Cl^- \leftarrow BrCl^- + 2e^-$$

in order to increase the storage density. Other developments go in the direction of replacing expensive vanadium by cheaper sodium compounds, using reactions such as

$$3NaBr \rightarrow NaBr_3 + 2Na^+ + 2e^-$$
$$2e^- + 2Na^+ + Na_2S_4 \rightarrow 2Na_2S_2$$

and the reverse reactions. A 120-MWh store with a maximum discharge rate of 15 MW, based upon these reactions, has just been completed at Little Barford in the UK. A similar plant is under construction by the Tennessee Valley Authority in the USA (Anonymous, 2002). Physically, the fuel cell hall and the cylindrical reactant stores take up about equal spaces. The cost is estimated at 190 euro per kWh stored or 1500 euro per kW rating (Marsh, 2002).

In the longer range, reversible fuel cell stores are likely to be based on hydrogen, as this storage medium offers the further advantage of being useful as a fuel throughout the energy sector. A scenario exploring this possibility in detail is presented elsewhere (Sørensen, 2004a; and in condensed form in Sørensen et al., 2003; Sørensen, 2002a, 2003c, 2003d), but the same idea penetrates the energy scenarios presented in section 6.4.

Other storage concepts

Direct storage of light
Energy storage by photochemical means is essential for green plants, but attempts to copy their technique have proved difficult, because of the nature of the basic ingredient: the membrane formed by suitably engineered double layers, which prevents recombination of the storable reactants formed by the solar-induced, photochemical process. Artificial membranes with similar properties are difficult to produce at scales beyond that of the laboratory. Also, there are significant losses in the processes, which are acceptable in the natural systems, but which would negatively affect the economy of man-made systems.

The interesting chemical reactions for energy storage applications accepting direct solar radiation as input use a storing reaction and an energy retrieval reaction of the form

$$A + h\nu \rightarrow B,$$
$$B \rightarrow A + useful\ energy.$$

An example would be the absorption of a solar radiation quanta for the purpose of fixing atmospheric carbon dioxide to a metal complex, for example, a complex ruthenium compound. By adding water and heating, methanol can be produced,

$$[M]CO_2 + 2H_2O + 713 \text{ kJ mol}^{-1} \rightarrow [M] + CH_3OH + 1.5O_2.$$

The metal complex has been denoted [M] in this schematic reaction equation. The solar radiation is used to recycle the metal compound by reforming the CO_2-containing complex. The reaction products formed by a photochemical reaction are likely to back-react if they are not prevented from doing so. This is because they are necessarily formed at close spatial separation distances, and the reverse reaction is energetically favoured as it is always of a nature similar to the reactions by which it is contemplated to regain the stored energy.

One solution to this problem is to copy the processes in green plants by having the reactants with a preference for recombination form on opposite sides of a membrane. The membranes could be formed by use of surface-active molecules. The artificial systems consist of a carbohydrate chain containing some 5–20 atoms and in one end a molecule that associates easily with water ("hydrophilic group"). A double layer of such cells, with the hydrophilic groups facing in opposite directions, makes a membrane. If it is closed, for example, forming a shell (Fig. 5.26), it is called a micelle.

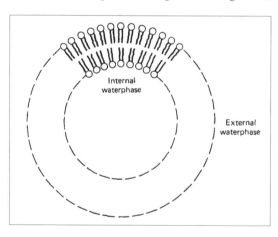

Internal waterphase

External waterphase

Figure 5.26. Double-shell type of micelle. The water-accepting compounds are illustrated by small circles, and the hydrocarbon chain molecules are illustrated by short lines.

Consider now the photochemical reaction bringing A to an excited state A^*, followed by ionisation,

$$A + h\nu \rightarrow A^*$$
$$A^* + B \rightarrow A^+ + B^-.$$

Under normal circumstances, the two ions would have a high probability of recombining, and the storage process would not be very efficient. But if A^+ can be made to form in a negatively charged micelle, the expelled electron would react with B to form B⁻ outside the micelle, and B^- will not be able to react with A^+. The hope is, in this way, to be able to separate macroscopic quantities of the reactants, which would be equivalent to storing meaningful amounts of energy for later use (see e.g. Calvin, 1974), but efficiencies of artificial photosynthetic processes have remained low (Steinberg-Yfrach *et al.*, 1998). Recently, a number of matrix structures have been identified which delay the back-reaction by hours or more, but as yet the retrieval of the stored energy is made correspondingly more difficult. The materials are layered viologen compounds [such as *N,N'*-dimethyl-4,4'-bipyridinium chloride, methyl viologen (Slama-Schwok *et al.*, 1992), or zirconium phosphate–viologen compounds with Cl and Br inserts (Vermeulen and Thompson, 1992)].

The research on photochemical storage of energy is at a very early stage and certainly not close to commercialisation. If the research is successful, a new set of storage options will become available. However, it should be stressed that storage cycle efficiencies will not be very high. For photo-induced processes the same limitations exist as for photovoltaic cells, for example, only part of the solar spectrum being useful and losses associated with having to go through definite atomic and molecular energy states. Further losses are involved in transforming the stored chemicals to the energy form demanded.

Superconducting storage

A magnetic field represents energy stored. When a magnet is charged by a superconducting coil (e.g. a solenoid), the heat losses in the coil may become practically zero, and the system constitutes an electromagnetic energy store with rapid access time for discharging. The maintenance of the coil materials (type II superconductors such as NbTi) near the absolute zero temperature requires cryogenic refrigeration by e.g. liquid helium. Owing to the cost of structural support as well as protection against high magnetic flux densities around the plant, underground installation is envisaged. A storage level of 1 GJ is found in the scientific installation from the mid-1970s at the European physics laboratory CERN at Geneva (aimed at preserving magnetic energy across intervals between the accelerator experiments performed at the laboratory). A 100-GJ superconducting store has been constructed in the USA by the Department of Defence, who want to exploit the fact that superconducting stores can accumulate energy at modest rates but release it during very short times, as required in certain anti-missile defence concepts (Cultu, 1989b). This is still a prototype development. Economic viability is believed to require storage ratings of 10–100 TJ, and one may hope to be able to employ high-temperature superconductors in order to reduce the cooling requirements (however, limited success has been achieved in raising the critical temperature; cf. Locquet *et al.*, 1998).

Only "type II" superconductors are useful for energy storage purposes, since the superconducting property at a given temperature T (below the critical temperature which determines the transition between normal and superconducting phase) disappears if the magnetic field exceeds a critical value, $B_c(T)$ ("magnetic field" will here be taken to mean "magnetic flux density", a quantity which may be measured in units of tesla = V s m^{-2}). Type II superconductors are characterised by high $B_c(T)$, in contrast to type I superconductors.

A magnetic field represents a storage of energy, with the energy density w related to the magnetic flux density (sometimes called "magnetic induction") B by

$$w = B^2 / (2\mu_0),$$

where μ_0 (= 1 26 × 10^{-6} henry m^{-1}) is the vacuum permeability. On connecting the superconducting coil to a power source, the induced magnetic field represents a practically loss-free transfer of energy to the magnetic storage. By a suitable switch, the power source may be disconnected, and energy may later be retrieved by connecting the coil to a load circuit. Cycle efficiencies will not be 100%, owing to the energy required for refrigerating the storage. Since the magnetic field must be kept below the critical value, increased capacity involves building larger magnets.

Conventional circuit components such as capacitors also represent an energy store, as do ordinary magnets. Such devices may be very useful to smooth out short-term variations in electrical energy flows from source to user, but are unlikely to find practical uses as bulk storage devices.

5.3 Suggested topics for discussion

5.3.1

Consider a thermal storage of temperature T (say, 200°C) aimed at receiving electric energy through a heat pump (with access to a reservoir of temperature T_{ref}) and delivering electrical energy by means of a thermodynamic engine cycle (available coolant also of temperature T_{ref}). What is the theoretical maximum cycle efficiency, and what can be expected in practice?

5.3.2

Discuss energy and power levels for the following "natural hydro energy" concept. Behind a seashore mountain chain there are a number of elevated water reservoirs (e.g. 200 m above sea-level). The sea is characterised by a high evaporation rate (using solar energy), and the prevailing winds carry the evaporated water towards the coast, where a high percentage of the water returns as rain over the mountain region reservoirs (or imagine that precipitation is stimulated by seeding the clouds with some kind of condensation nuclei). Sample evaporation and precipitation rates may be found in meteorological tables or may be deduced from Fig. 2.109.

The water is returned from reservoirs to the sea through hydropower plants. In a refined model, storage losses from the reservoirs due to evaporation should be included.

5.3.3

On the basis of time sequences of data for an automobile (e.g. your own), try to construct a load-duration curve for the entire period. Based on this, estimate a suitable level of rated power, if the engine should only cover average load, and specify the amount of storage needed for operation with this engine (which may be taken as an ideal, loss-free electric motor).

5.3.4

Construct the load-duration curve for space heating of a dwelling at your geographical location. Assume that this load is to be covered by an energy converter providing constant power year round and that loss-free heat storage is available. Determine the magnitude of the constant converter output that, through use of the storage, will suffice to cover the load at all times. Further, determine the minimum storage size needed. Compare the constant converter output to the range of power provided by currently used heating systems. Use estimated losses for actual storage systems that may be relevant to assess what the required storage capacity would be in case the storage involved realistic losses.

5.3.5

A steam turbine power plant with a steam temperature of 700 K and a condenser temperature of 350 K has an efficiency of 0.36 (electric output energy divided by input steam energy) and the turbine efficiency is 20% less than the Carnot efficiency (ε_{max}). Calculate the efficiency of the electric generator.

5.3.6

An electric vehicle of mass 1000 kg excluding the battery is designed for a 50-km range in a city driving cycle. The vehicle is equipped with a new 500-kg battery, and the average energy consumption is measured to be 0.3 kWh km^{-1} (from the battery). At a speed of 36 km h^{-1} the required acceleration is 2.0 m s^{-2}, and the power required to overcome the frictional losses is one-third of the total power required. Do the same calculation for an acceleration $a = 1.5$ m s^{-2}.

- Calculate the required average energy density of the new battery.
- Calculate the required power density of the new battery.
- What type of battery could fulfil these requirements (alone or in combination as a hybrid system)?

5.3.7

Estimate the total hydro resource for storage application for a country or region. Characterise the natural reservoirs in terms of water volume and feasible turbine head, and identify the geological formations that could be considered for underground reservoir construction. In this way arrive at rough storage capacities for short- and long-term storage separately.

5.3.8

Use (5.39) to express the compressed storage energy (5.38) in terms of the temperature difference $T - T_0$ rather than in terms of the pressure P.

5.3.9

Compare the energy needed to compress a certain amount of air by an adiabatic process with that required for an isothermal and an isobaric process.

5.3.10

Calculate the shape factor for a thin-rim-type flywheel, as well as mass and volume specific energy density, assuming the material to be steel. Do the same numerically for some of the flywheel types shown in Fig. 5.18 (using data from Tables 5.6 and 5.7) and compare the properties. Discuss priorities for various types of application.

ENERGY SUPPLY SYSTEMS

CHAPTER 6

6.1 Energy systems

In Chapter 4, the energy conversion devices were largely viewed as disjoint entities. Here, they will be regarded as parts of larger systems, which may comprise several converter units, storage facilities and transmission networks. The decisive feature of such systems is, of course, the purpose they are meant to serve, i.e. the end-point of the energy conversion processes. Therefore, a systems study must comprise analysis of both the end-uses of energy (called "loads" or "demands") and the chains of conversion and transport steps connecting the primary energy source extraction with the loads, including considerations of spatial and temporal variations of supply and demand.

It is important to agree on the definition of "end-use energy". There is the energy delivered to the end-user, and after subtraction of the losses in the final conversion taking place at the end-user, there is a net amount of energy being made useful. However, the real demand is always a product or a service, not the energy in itself. This implies that the true energy demand is often difficult to pinpoint, as one may find novel ways of satisfying the non-energy final need that could radically change the end-use energy that would seem "required" or "demanded". A typical definition used in practice is to consider the lowest amount of energy associated with delivering the product or service demanded, as measured after the final conversion at the end-user and selected among all conversion schemes that can be realised with present knowledge. For some technologies, there will be a thermodynamic limit to energy conversion efficiency, but whether it is fundamental or may be circumvented by going to a different technology depends upon the nature of the final product or service. For many services, the theoretical minimum energy input is zero. Future advances in technologies or new ideas for satisfying a particular need may change the minimum amount of end-use energy, and one should decide on which of the following to use in the definition:

- only technology available in the marketplace today, or
- only technology proven today and with a potential for market penetration, or
- any technology that can be imagined at present.

The outcome will be different, and when comparing system solutions, it is important to use the same definition of demand throughout. My preference is to use the best technology available today in defining end-use, but any definition will do if it is used consistently.

6.2 Simulation of system performance

This chapter gives a number of examples of energy system simulations. Some of them simulate the function of existing systems, while others deal with hypothetical systems that may be created in the future. These will be analysed using the scenario method, which is a way of checking the consistency of proposed energy futures as parts of visions of the general future development of a society.

Scenario studies are meant to assist a decision-making process, by describing a given energy system in its social context, in a way suitable for assessing systems not yet implemented. Simple forecasts of demand and supply based on economic modelling cannot directly achieve this, as economic theory deals only with the past and occasionally the present structure of society (cf. Chapter 7). In order to deal with the future, economic modellers may attempt to invoke the established quantitative relations between parts of the economic system and assume that they stay valid in the future. This produces a "business-as-usual" forecast. Because the relations between the ingredients of the economy, e.g. expressed through an input–output matrix, vary with time, one can improve the business-as-usual forecast by taking into account trends already present in the past development. However, even such trend forecasts cannot be expected to retain their validity for very long periods, and it is not just the period of forecasting time that matters, but also changes in the rules governing society. These may change due to abrupt changes in technology used (in contrast to the predictable, smooth improvements of technological capability or average rate of occurrence of novel technologies), or they may be changed by deliberate policy choices. It is sometimes argued that econometric methods could include such non-linear behaviour, e.g. by replacing the input–output coefficients by more complex functions. However, to predict what these should be cannot be based on studies of past or existing societies, because the whole point in human choice is that options are available that are different from past trends. The non-linear, non-predictable relations that may prevail in the future, given certain policy interventions at appropriate times, must therefore be postulated on normative grounds. This is what the scenario method does, and any attempt to mend economic theory also implies invoking a scenario construction and analysis, so in any case this is what has to be done (Sørensen, 2001b).

It is important to stress that scenarios are not predictions of the future. They should be presented as policy options that may come true only if a pre-scribed number of political actions are indeed carried out. In democratic so-cieties, this can only happen if preceded by corresponding value changes af-fecting a sufficiently large fraction of the society. Generally, the more radical the scenario differs from the present society, the larger must the support of a democratically participating population be. For non-democratic societies im-plementing a scenario future by decree, negative implications will likely en-tail.

The actual development may comprise a combination of some reference scenarios selected for analysis, with each reference scenario being a clear and perhaps extreme example of pursuing a concrete line of political preference. It is important that the scenarios selected for political consideration be based on values and preferences that are important in the society in question. The value basis should be made explicit in the scenario construction. Although all analysis of long-term policy alternatives is effectively scenario analysis, particular studies may differ in the comprehensiveness of the treatment of future society. A simple analysis may make normative scenario assumptions only for the sector of society of direct interest for the study (e.g. the energy sector), assuming the rest to be governed by trend rules similar to those of the past. A more comprehensive scenario analysis will make a gross scenario for the development of society as a whole, as a reference framework for a deeper investigation of the sectors of particular interest. One may say that the simple scenario is one that uses trend extrapolation for all sectors of the economy except the one focused upon, whereas the more radical scenario will make normative, non-linear assumptions regarding the development of society as a whole. The full, normative construction of future societies will come into play for scenarios describing an ecologically sustainable global so-ciety, whereas scenarios aiming only at avoiding or coping with one par-ticular problem, such as the climate change induced by greenhouse warm-ing, are often of a simpler kind. Both types are exemplified in the scenario work described below.

6.2.1 Treatment of the time variable

In order to predict the performance of systems consisting of one or several energy converters, stores and transmission devices, a mathematical model of the energy flow may be constructed. Such a model is composed of a number of energy conversion and transport equations, including source and sink terms corresponding to the renewable energy input and the output to load areas, both of which are varying with time. The conversion processes depend on the nature of the individual devices, and the description of such devices (cf. Chapter 4) aims at providing the necessary formulae for a sufficiently complete description of the processes involved. In a number of cases (e.g. among those considered in Chapter 4) only a steady-state situation is studied, and the energy outputs are calculated for a given level of energy input. In a time-dependent situation, this type of calculation is insufficient, and a dynamic description must be introduced in order to evaluate the response time and energy flow delay across the converter (cf. e.g. section 4.4.1). Similar remarks apply to the description of the storage systems, and, finally, the transmission network introduces a further time dependence and a certain delay in the energy flow reaching the load areas. The transmission network is often in the form of pipelines carrying a flow of some fluid (e.g. natural gas, hydrogen or hot water) or an electrical conductor carrying a flow of electric current. Additional transport of energy may take place in containers (e.g. oil products or methanol carried as ship, rail or road cargo).

In order to arrive at manageable problems, it is, in most cases, necessary to simplify the time dependence for some parts of the system. First of all, short-term fluctuations in the source energy flow may, in some circumstances, be left out. This is certainly possible if the conversion device is itself insensitive to fluctuations of sufficiently high frequency. It could be so due to the inertia of the rotating mass for a wind energy converter or the time constant for temperature changes in the absorber plate (and also in the circulating fluid) of a solar heat collector. It may also be a valid approximation if the short-term variations in the energy flux from the source can be regarded as random and if the collection system consists of a large number of separate units placed in such a way that no coherence in the fluctuating inputs can be expected.

Secondly, the performance of the conversion devices may often be adequately described in terms of a quasi steady-state approximation. This consists of calculating an instantaneous energy output from the converter on the basis of an instantaneous energy input as if this input flux were permanent, i.e. doing a steady-state calculation for each moment of time. This excludes an assessment of the possible time delay between the input flux and the output flux. If a solid mechanical connection transfers the energy through the converter (e.g. the rotor–shaft–gearbox–electric generator connections in a horizontal-axis wind energy converter), the neglect of time delays is a meaningful approximation. It may also be applicable for many cases of non-rigid transfer (e.g. by a fluid) if short-term correlations between the source flux and the load variations are not essential (which they seldom are in connection with renewable energy sources). For the same reason, time delays in transmission can often be neglected. The flow received at the load points may be delayed by seconds or even minutes, relative to the source flow, without affecting any of the relevant performance criteria of the system.

On the other hand, the delays introduced by the presence of energy storage facilities in the system are essential features which cannot and should not be neglected. Thus, the storage devices will have to be characterised by a time-dependent level of stored energy, and the input and output fluxes will, in general, not be identical. The amount of energy $W(S_i)$ accumulated in the storage S_i can be determined from a differential equation of the form

$$\frac{dW(S_i)}{dt} = \sum_j E_{ji}^+ - \sum_k E_{ik}^- - E_i^{loss},$$ (6.1)

or from the corresponding integral equation. The individual terms in the two expressions involving summation on the right-hand side of (6.1) represent energy fluxes from the converters to and from the storage devices. The loss term E_i^{loss} may depend on the ingoing and outgoing fluxes and on the absolute amount of energy stored in the storage in question, $W(S_i)$.

In practice, the simulation is performed by calculating all relevant quantities for discrete values of the time variable and determining the storage energy contents by replacing the time integral of (6.1) by a summation over the discrete time points considered. This procedure fits well with the "quasi-steady-state approximation", which at each integration step allows the calculation of the converter outputs (some of which are serving as storage inputs E_{ji}^+) for given renewable energy inputs, and similarly allows the calculation of conversion processes in connection with the storage facilities, and the energy fluxes E_{ij}^- to be extracted from the storage devices in order to satisfy the demands at the load areas. If the time required for conversion and transmission is neglected, a closed calculation can be made for each time integration step. Interdependence of storage inputs and outputs, and of the primary conversion on the system variables in general (e.g. the dependence of collector performance on storage temperature for a flat-plate solar collector), may lead to quite complex calculations at each time step, such as the solution of non-linear equations by iteration procedures (section 4.2.1).

If the finite transmission times cannot be neglected, they may to a first approximation be included by introducing simple, constant delays, such that the evaluations at the mth time step depend on the values of certain system variables at earlier time steps, $m - d$, where d is the delay in units of time steps. The time steps need not be of equal length, but may be successively optimised to obtain the desired accuracy with a minimum number of time steps, by standard mathematical methods (see e.g. Patten, 1971, 1972).

The aim of modelling may be to optimise either performance or system layout. In the first case, the system components are assumed to be fixed, and the optimisation aims at finding the best control strategy, i.e. determining how best to use the system at hand ("dispatch optimisation"). In a multi-input–multi-output conversion system, this involves choosing which of several converters to use to satisfy each load and adjusting inputs to converters in those cases where this is possible (e.g. biofuels and reservoir-based hydro as opposed to wind and solar radiation). For system optimisation, the structure of the conversion system may also be changed, with recognition of time delays in implementing changes, and the performance over some extended period would be the subject of optimisation. For simple systems (without multiple inputs or outputs from devices), linear programming can furnish a guaranteed optimum dispatch of existing units, but in the general case, it is not possible to prove the existence of an optimum. Still, there are systematic ways to approach the optimisation problem, e.g. by using the method of steepest descent for finding the lowest minimum of a complex function, combined with some scheme for avoiding shallow, secondary minima of the function to minimise (Sørensen, 1996a).

6.2.2 Load structure

An energy system may serve a load characterised by demands of one or more forms of energy to be delivered at one or several different locations. In other words, loads may range from very specific and spatially confined energy demands (e.g. for performing one specific task such as food refrigeration) over several stages of intermediate-size systems (such as individual buildings with requirements of heating, hot water and electricity for appliances, building blocks, industrial facilities, entire cities or agricultural districts) to the total demands of regional areas, nations and further aggregated communities, with a complex combination of energy needs.

Considering the loads on a regional or national level, it is customary to divide the loads into a number of sectors, such as energy demands of industry, agriculture, commerce, service and residential sectors, and transportation. Each sector may require one or more types of energy, e.g. heat, electricity or a "portable source" capable of producing mechanical work (typically a fuel). The distribution of the loads on individual sectors depends on the organisation of the society and on climatic conditions. The latter items influence the need for space heating or cooling, while the former includes settlement patterns (which influence the need for transportation), building practices, health care and social service levels, types of industry, etc. A systematic way of treating energy demands is described in the following, with a concrete example of its application to be used in the example of system simulation following in section 6.3.3.

The development of energy demands is sometimes discussed in terms of marginal changes relative to current patterns. For changes over extended periods of time, this is not likely to capture the important issues. Another approach, called the "bottom-up model", is offered by looking at human needs, desires and goals and building up first the material demands required for satisfying these and then the energy required under certain technology assumptions (Kuemmel *et al.*, 1997, appendix A). This approach is based on the view that certain human needs are basic needs, i.e. non-negotiable, while others are secondary needs that depends on cultural factors and stages of development and knowledge and could turn out differently for different societies, subgroups or individuals within a society. The basic needs include those of adequate food, shelter, security and human relations, and there is a continuous transition to more negotiable needs that incorporate material possessions, art, culture and human interactions and leisure. Energy demand is associated with satisfying several of these needs, with manufacture and construction of the equipment and products entering into the fulfilment of the needs, and with procuring the materials needed along the chain of activities and products.

In normative models such as the scenarios for the future, the natural approach to energy demand is to translate needs and goal satisfaction into energy requirements consistent with environmental sustainability. For market-driven scenarios, basic needs and human goals play an equally important role, but secondary goals are more likely to be influenced by commercial interest rather than by personal motives. It is interesting that the basic needs approach is always taken in discussions of the development of societies with low economic activity, but rarely in discussions of industrialised countries.

The methodology suggested here is to first identify needs and demands, commonly denoted human goals, and then to discuss the energy required to satisfy them in a chain of steps backwards from the goal-satisfying activity or product to any required manufacture and then further back to materials. This will be done on a per capita basis (involving averaging over differences within a population), but separately for different geographical and social settings, as required for any local, regional or global scenarios.

The primary analysis assumes a 100% goal satisfaction, from which energy demands in societies that have not reached this can later be determined. The underlying assumption is that it is meaningful to specify the energy expenditure at the end-use level without caring about the system responsible for delivering the energy. This is only approximately true. In reality, there may be couplings between the supply system and the final energy use, and the end-use energy demand therefore in some cases becomes dependent on the overall system choice. For example, a society rich in resources may take upon it to produce large quantities of resource-intensive products for export, while a society with less resources may instead focus on knowledge-based production, both doing this in the interest of balancing an economy to provide satisfaction of the goals of their populations, but possibly with quite different implications for energy demand. The end-use energy demands will be distributed on energy qualities, which may be categorised as follows:

1. Cooling and refrigeration 0–50°C below ambient temperature
2. Space heating and hot water 0–50°C above ambient
3. Process heat below 100°C
4. Process heat in the range 100–500°C
5. Process heat above 500°C
6. Stationary mechanical energy
7. Electrical energy (no simple substitution possible)
8. Energy for transportation (mobile mechanical energy)
9. Food energy

The goal categories used to describe the basic and derived needs can then be chosen as follows:

A: Biologically acceptable surroundings
B: Food and water
C: Security
D: Health
E: Relations and leisure
F: Activities
       f1: Agriculture
       f2: Construction
       f3: Manufacturing industry
       f4: Raw materials and energy industry
       f5: Trade, service and distribution
       f6: Education
       f7: Commuting

Here categories A–E refer to direct goal satisfaction, f1–f4 to primary derived requirements for fulfilling the needs, and finally, f5–f7 to indirect requirements for carrying out the various manipulations stipulated. Ranges of estimated energy requirements for satisfying needs identified by present societies are summarised in Table 6.1 (Kuemmel *et al.*, 1997), with a more detailed, regional distribution given in Table 6.2 (Sørensen and Meibom, 1998). For comparison an analysis of current energy end-uses according to the principles presented above is given in Table 6.3, indicating the present low average efficiencies of the final conversion steps. The assumptions behind the "full goal satisfaction" energy estimates are given below.

	1. Cooling & refrigeration	2. Space heating	3. Process heat under 100°C	4. Process heat 100-500°C	5. Process heat over 500°C	6. Stationary mechanical energy	7. Electric appliances	8. Transportation work	9. Food energy	TOTAL
A. Biologically acceptable surroundings	0-200	0-650	0	0	0	0	0	0	0	0-650
B. Food and water	21	0	5	7	0	0	0	0	120	153
C. Security	(0)	(0)	0	0	0	(0)	0	1	0	1
D. Health	0	0	115	17	0	0	(0)	(0)	0	132
E. Relations, leisure	(0)	(0)	0	0	0	0	60-82	25-133	0	85-215
F. Activities: Construction	0	0	0	0	0	30-60	0	7-15	0	37-75
Trade, service and distribution	1-80	0-150	0-10	0	0	5	20	30-100	0	56-325
Agriculture	0	0	0	0-6	0	1-3	1	3-6	0	5-16
Manufacturing industry	1-40	0-150	10-100	20-70	12-30	20-40	20-40	7-15	0	90-445
Raw Materials and energy industry	0	0	0-30	0-30	0-250	0-170	0-30	0-20	0	0-530
Education	0-20	0-65	0	0	0	0	1-2	0	0	1-67
Commuting	0	0	0	0	0	0	0	0-30	0	0-30
TOTAL	23-361	0-1015	130-260	44-130	12-280	56-278	102-175	73-320	120	560-2639

Table 6.1. Global end-use energy demand based upon bottom-up analysis of needs and goal satisfaction in different parts of the world, using best available currently available technologies (average energy flow in W/cap.) (From Kuemmel *et al.*, 1997).

For use in the global scenario for year 2050 described in section 6.4 below, the end-use energy components for each category are estimated on the basis of actual assumed goal fulfilment by the year 2050, given in Table 6.4 on a regional basis. This analysis assumes a population development based on the United Nations population studies (UN, 1996), taking the alternative corresponding to high economic growth (in the absence of which population is estimated to grow more). Figures 6.1 and 6.2 show the present and assumed 2050 population density, including the effect of increasing urbanisation, particularly in developing regions, leading to 74% of the world population living in urban conglomerates by the year 2050 (UN, 1997b).

Biologically acceptable surroundings

Suitable breathing air and shelter against wind and cold temperatures, or hot ones, may require energy services indirectly to manufacture clothes and structures and directly to provide active supply or removal of heat. Insulation by clothing makes it possible to stay in cold surroundings with a modest increase in food supply (which serves to heat the layer between the body and the clothing). The main heating and cooling demands occur in extended spaces (buildings and sheltered walkways, etc.) intended for human occupation without the inconvenience of heavy clothing that would impede e.g. manual activities.

Regions &/ Energy quality:	1. USA, Canada	2. Western Europe, Japan, Australia	3. Eastern Europe, Ex-Soviet, Middle East	4. Latin America, SE Asian "tigers"	5. China, India, rest of Asia	6. Africa	Average & total
Space heating*	205	212	164	19	96	10	
Other low-temp. heat	150	150	150	150	150	150	
Medium-temp. heat	50	50	50	50	50	50	
High-temp. heat	40	40	40	40	40	40	
Space cooling*	9	2	43	37	71	44	
Other refrigeration	35	35	35	35	35	35	
Stationary mechanical	150	150	150	150	150	150	
Electric applicances	150	150	150	150	150	150	
Transportation	200	150	200	150	150	150	
Food energy	120	120	120	120	120	120	
Total 2050	1109	1059	1102	901	1012	899	986 W/cap.
End-use energy	420	559	1146	1243	4008	1834	9210 GW
Population 2050, mill.	379	528	1040	1380	3960	2040	9340
Total area million km^2	20.1	15.4	28.3	26.3	20.1	30.9	141.1

Table 6.2. Per capita energy use for "full goal satisfaction" in W/cap. and total in GW for the assumed 2050 population stated. Rows marked * are based on temperature data for each cell of geographical area (0.5° longitude-latitude grid used). Manufacturing and raw materials industries are assumed to be distributed in proportion to population between regions (this and following tables and geographically based figures are reprinted from Sørensen and Meibom, 1998, with permission). The full list of the countries included in each region is given in the reference.

Region: 1994 end-use energy	1. USA, Canada	2. Western Europe, Japan, Australia	3. Eastern Europe, Ex-Soviet, Middle East	4. Latin America, SE Asian "tigers"	5. China, India, rest of Asia	6. Africa	Average & total	
Space heating	186	207	61	2	19	1	46	W/cap.
	52	116	41	1	48	1	260	GW
Other low-temp. heat	120	130	40	15	18	10	36	W/cap.
	34	73	27	12	47	7	199	GW
Medium-Temp. heat	40	50	30	10	10	5	17	W/cap.
	11	28	20	8	26	3	97	GW
High-temp. heat	35	40	30	10	10	3	16	W/cap.
	10	22	20	8	26	2	88	GW
Space cooling	9	1	13	2	3	0	4	W/cap.
	2	1	9	2	8	0	22	GW
Other re-frigeration	29	23	14	2	2	0	7	W/cap.
	8	13	9	1	5	0	37	GW
Stationary mechanical	100	130	80	25	5	4	34	W/cap.
	28	73	53	21	13	3	191	GW
Electric appliance	110	120	50	20	5	4	29	W/cap.
	31	67	33	16	13	3	164	GW
Transpor-tation	200	140	40	20	5	3	34	W/cap.
	56	79	27	16	13	2	193	GW
Food energy	120	120	90	90	90	90	95	W/cap.
	34	67	60	74	233	61	530	GW
Total end-use	948	962	448	195	167	121	318	W/cap.
energy	268	540	298	160	432	83	1781	GW
Population 1994	282	561	666	820	2594	682	5605	million
Region area	20	15	28	26	20	31	141	million km²

Table 6.3. Estimated end-use energy 1994 (Sørensen and Meibom, 1998). Due to the nature of available statistical data, the division into categories is not identical to that used in the scenarios. Furthermore, some end-use energies are extrapolated from case studies. These procedures aim to provide a more realistic scenario starting point. However, as the scenario assumptions are based upon basic principles of goal satisfaction, the inaccuracy of current data and thus scenario starting points do not influence scenario reliability, but only stated differences between now and the future.

Regions: / Energy quality:	1. USA, Canada	2. Western Europe, Japan, Australia	3. Eastern Europe, Ex-Soviet, Middle East	4. Latin America, SE Asian "tigers"	5. China, India, rest of Asia	6. Africa
Space heating	0.9 (0.9)	1.0 (0.96)	0.75 (0.25)	0.67 (0.08)	0.63 (0.16)	0.15 (0.10)
Other low-temp. heat	0.87 (0.8)	1.0 (0.87)	0.53 (0.27)	0.6 (0.10)	0.67 (0.12)	0.13 (0.07)
Medium-temp. heat	0.9 (0.8)	1.0 (1.0)	0.8 (0.6)	0.8 (0.2)	0.8 (0.2)	0.1 (0.1)
High-temp. heat	0.88 (0.88)	1.0 (1.0)	0.75 (0.75)	0.75 (0.25)	0.75 (0.25)	0.13 (0.08)
Refrigeration	1.0 (0.83)	1.0 (0.67)	0.4 (0.4)	0.5 (0.05)	0.33 (0.05)	0.05 (0.01)
Stationary mechanical	0.93 (0.67)	1.0 (0.87)	0.53 (0.53)	0.67 (0.17)	0.67 (0.03)	0.13 (0.03)
Electric appliances	1.0 (0.73)	1.0 (0.8)	0.53 (0.33)	0.67 (0.13)	0.67 (0.03)	0.1 (0.03)
Transportation	0.9 (1.0)	1.0 (0.93)	0.35 (0.20)	0.67 (0.13)	0.33 (0.03)	0.1 (0.02)
Food energy	1.0 (1.0)	1.0 (1.0)	1.0 (0.75)	1.0 (0.75)	1.0 (0.75)	0.83 (0.75)
Averages of the above	**0.93 (0.85)**	**1.0 (0.90)**	**0.62 (0.45)**	**0.68 (0.21)**	**0.64 (0.18)**	**0.20 (0.13)**

Table 6.4. The fraction of "full goal satisfaction" assumed in the year 2050 scenario, with estimated values for 1994 given in parenthesis (Sørensen and Meibom, 1998). These estimates involve assumptions on conditions for development in different parts of the world: on one hand positive conditions such as previous emphasis on education being a prerequisite for economic development, and on the other hand negative conditions such as social instability, frequent wars, corrupt regimes, lack of tradition for democracy, for honouring human rights, and so on. It is recognised that making assumptions on such issues does constitute a considerable amount of subjectivity. As an example, UN projections have a tradition for disregarding non-economic factors and, for instance, assume a much higher rate of development for African countries.

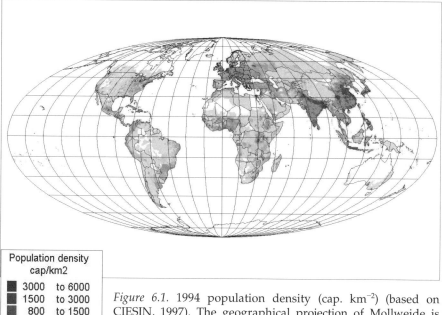

Population density cap/km2	
3000	to 6000
1500	to 3000
800	to 1500
550	to 800
380	to 550
250	to 380
150	to 250
80	to 150
40	to 80
20	to 40
10	to 20
0.1 to	10
all others	

Figure 6.1. 1994 population density (cap. km^{-2}) (based on CIESIN, 1997). The geographical projection of Mollweide is used in order to obtain a faithful reproduction of equal areas (but at the expense of angular relations).

Figure 6.2. Population density assumed in 2050 scenario (cap. km^{-2}) (consistent with UN, 1996, 1997, using an urbanisation model described in Sørensen and Meibom, 1998). Note that the higher population densities in large cities are inconspicuous, while the lower densities in rural areas, relative to 1994, do show up clearly.

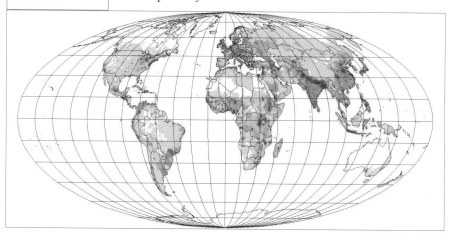

Rather arbitrarily, but within realistic limits, it is assumed that a fulfilment of the goals related to shelter on average requires a space of 40 m^2 times a height of 2.3 m to be at the disposal of each individual in society and that this space should be maintained at a temperature of 18–22°C, independent of outside temperatures and other relevant conditions. As a "practical" standard of housing technology, I shall further use a rate of heat loss P from this space of the approximate form $P = C \times \Delta T$, where ΔT is the temperature difference between the desired indoor temperature and the outside temperature (cf. section 6.2). The constant C consists of a contribution from heat losses through the external surfaces of the space, plus a contribution from exchanging indoor air with outside air at a minimum rate of about once every 2 h. Half of the surfaces of the "person space" are considered as external, with the other half being assumed to face another heated or cooled space. Best current technology solutions would suggest that the heat loss and ventilation values of $C = 0.35$ (heat loss) + 0.25 (air exchange) = 0.6 W °C^{-1} per m^2 of floor area can be attained. The precise value, of course, depends on building design and particularly on window area. The air exchange part assumed above is about 40% lower than it would be without use of heat exchangers in a fraction of the buildings. The 40 m^2/cap. assumed dwelling space will below be augmented with 20 m^2/cap. for other activities (work, leisure).

Now the energy needs[*] for heating and cooling, at a given location and averaged over the year, can be calculated with the use of climate tables giving the ambient temperature, e.g. hour by hour, for a typical year. If there are periods where temperatures are such that constant heating or cooling is required, the corresponding energy needs can be determined from the average temperatures alone. Heat capacity of the building will smooth out short-term variations, such that it is often a good approximation to determine heating and cooling demands from daily or even monthly average temperatures. An example of such estimations is shown in Figs. 6.3 and 6.4, for the annual heating and cooling requirements separately, as a function of geographical location. The assumption is that space cooling is required for outdoor temperatures above 24°C and heating is required for temperatures below 16°C. In combination with heat from indoor activities and the thermal

---

[*] Several studies use a thermodynamic concept called *free energy* (F). It is related to the conventional, conserved energy P by the relation $F = P \, \Delta T/T$, where T is the absolute temperature of the space to be heated or cooled and the temperature of the surroundings is $T_0 = T - \Delta T$ (all temperatures measured in Kelvin). The free energy concept incorporates the quality of energy, being less than the conserved energy for heat and equal to it only for mechanical or electrical energy. The rate of free energy supply needed to provide the heat energy flux P into a space of temperature T embedded in surroundings of temperature T_0 is $F = P \, \Delta T / T = C \, (\Delta T)^2/T$. This expression is valid both for heating and cooling. In the latter case both ΔT and P are negative.

properties of suitable building techniques and materials, this can provide indoor temperatures within the desired range of 18–22°C. The local space heating and cooling requirements are then obtained by folding the climate-related needs with the population density (taken from Fig. 6.2 for the year 2050 and used in the scenarios for the future described in section 6.4).

Here are a few examples: For Irkutsk in Siberia, the annual average temperature of –3°C gives an average dwelling energy requirement for heating of 651 W (per capita and neglecting the possibility of heat gain by heat exchangers included in Figs. 6.3 and 6.4 below). For Darwin in Australia no heating is needed. These two values are taken as approximate extremes for human habitats in the summary table. A very few people world–wide live in harsher climates, such as that of Verkhoyansk (also in Siberia, average temperature –17°C, heating need 1085 W/cap.). Other examples are $P = 225$ W/cap. (New York City), $P = 298$ W/cap. (Copenhagen) and P very nearly zero for Hong Kong. Cooling needs are zero for Irkutsk and Copenhagen, while for Darwin, with an annual average temperature of 29°C, there is a cooling energy requirement of $-P = 209$ W/cap., assuming that temperatures above 22°C are not acceptable. The range of cooling energy demands is assumed to be roughly given by these extremes, as $-P = 0$ to 200 W/cap. For New York City, the annual average cooling requirement is 10 W/cap. (typically concentrated within a few months), and for Hong Kong it is 78 W/cap.

Food and water

Energy-wise, the food intake corresponding to full satisfaction of food needs is about 120 W/cap. Today, 28% of this is in the form of meat and other animal products (Alexandratos, 1995), but in the 2050 scenario presented below it is assumed that this fraction declines to 23%, partly owing to better balance in the diet of the industrialised regions and partly owing to an increased meat fraction in those regions presently having the lowest (e.g. East Asia at 9%). The distribution of food energy is shown in Figs. 6.5–6.6.

To store the food adequately, short– and long–term refrigeration are assumed to be used. The weight of the average per capita food intake is of the order of 2×10^{-5} kg s^{-1}, of which 0.8×10^{-5} kg s^{-1} is assumed to have spent 5 days in a refrigerator at a temperature $\Delta T = 15$°C below the surrounding room temperature, and 0.4×10^{-5} kg s^{-1} is assumed to have spent 2 months in a freezer at $\Delta T = 40$°C below room temperature. The heat loss rate through the insulated walls of the refrigerator or freezer is taken as 2×10^{-2} W °C^{-1} per kg of stored food. The energy requirement then becomes

$P \sim 0.8 \times 10^{-5} \times 5 \times 24 \times 3600 \times 2 \times 10^{-2} \times 15 = 1.04$ W/cap. (refrigerator)
$P \sim 0.4 \times 10^{-5} \times 2 \times 720 \times 3600 \times 2 \times 10^{-2} \times 40 = 16.6$ W/cap. (freezer),

plus the energy needed to bring the food down to the storage temperatures,

$P \sim$ 0.72+2.12 =2.84 W/cap. (assuming a heat capacity of 6000 J kg^{-1} °C^{-1} above 0°C and half that value below the freezing point, and a phase change energy of 350 kJ kg^{-1}). The energy is assumed to be delivered at the storage temperatures. Some energy could be regained when melting frozen food.

Cooking the food requires further energy. Assuming that 40% of the food intake is boiled at ΔT = 70°C above room temperature, and that 20% of the food intake is fried at ΔT = 200°C above room temperature, the energy needed to bring the food up to the cooking temperatures is $P \sim$ 3.36 + 4.80 = 8.16 W/cap., and the energy required for keeping the food cooking is $P \sim$ 1.45 + 2.08 = 3.53 W/cap., assuming daily cooking times of 30 minutes for boiling and 15 minutes for frying (some food cultures certainly use more), and heat losses from the pot/pan/oven averaging 1 W °C^{-1} for the quantities of food cooked per person per day.

Provision of water involves pumping and cleaning/purification. The pumping and treatment energy needs are negligible on a per capita basis, but both are included in the industry sector considered below.

Security

Heating and cooling of buildings used by courts, police, military and other security-related institutions are included as part of the 40-m^2 floor area accorded each person. The remaining energy use for personal and national security would be for transportation and energy depreciation of materials and would hardly amount to more than 1 W/cap., except for very belligerent or crime-ridden nations or nations with badly disliked regimes.

Health

Hot water for personal hygiene is taken as 50 litres/day/cap. at T = 40°C above the waterworks' supply temperature, implying a rate of energy use averaging roughly P = 97 W/cap. Some of this could be recycled. Clothes washing and drying may amount to treating about 1 kg of clothes per day per capita. Washing is assumed to handle 5 kg of water/kg of clothes, at T = 60°C (in practice often more water, at different temperatures some of which are closer to inlet temperature), or an average energy of P = 15 W/cap. For drying it is assumed that 1 kg of water has to be evaporated (heat of evaporation about 2.3 × 10^6 J kg^{-1}) per day per capita, at an effective temperature elevation of 80°C (the actual temperature is usually lower, but mechanical energy is then used to enhance evaporation by blowing air through rotating clothes containers). Local air humidity plays a considerable role in determining the precise figure. Condensing dryers recover part of the evaporation heat, say, 50%. The energy use for the case considered is then 17 W/cap. Hospitals and other buildings in the health sector use energy for space conditioning and equipment. These are included in the household energy use (where they contribute 1–2%).

Relations

Full goal satisfaction in the area of human relations involves a number of activities which are not independent on cultural traditions, habitats and individual preferences. One possible combination of energy services for this sector will be used to quantify energy demands.

The need for lighting depends on climate and habits regarding the temporal placement of light-requiring activities. Taking 40 W of present "state-of-the-art" commercial light sources (about 50 lumen per watt) per capita for 6 h/day, an average energy demand of 10 W/cap. results. Still, the radiant energy from the light sources represents some ten times less energy, and more efficient light sources are likely to become available in the future.

Radio, television, telecommunication, music, video, computing, etc. take some 65–130 W/cap. (say 3 h a day on average for each appliance), or an energy flux of 8–16 W/cap. The range of appliances used in this sector is rapidly changing, and it is hardly realistic to predict quantities other than the total. New devices (scanners, printers, DVD burners plus those not known today) are rapidly added, but energy savings made possible by the new technology may well balance the increase in the number of devices. Examples are flat displays using 10-50 times lower energy than the CRT screens still dominating at least television equipment. Stand-by power for much of the equipment, as well as the power to computer screens and peripherals is currently not yet energy optimised (cf. Sørensen, 1991). This is assumed to change in the future. Still, as the hoped-for balance between the increase in new equipment and their energy use is not proven, an additional energy expenditure of some 30 W/cap. has been included in the estimates used. Things like permanent Internet access and automatisation of a number of maintenance jobs in buildings could be among the reasons for increased energy use, although it would be coupled to reductions elsewhere in society.

Social and cultural activities taking place in public buildings are assumed to be included above (electricity use) or under space conditioning (heating and cooling).

Recreation and social visits entail a need for transportation, by surface or by sea or air. A range of 25–133 W/cap. is taken to be indicative of full goal satisfaction: the upper figure corresponds to travelling 11 000 km y^{-1} in a road-based vehicle occupied by two persons and using for this purpose 100 litres of gasoline equivalent per year per person. This amount of travel could be composed of 100 km weekly spent on short trips, plus two 500-km trips and one 5000 km trip a year. Depending or habitat and where friends and relatives live, the shorter trips could be reduced or made on bicycle, and there would be variations between cultures and individuals. Road and air congestion increases the likelihood of flattening out the current increase in transportation activity (as would teleconferencing and videophones), and transport is increasingly seen as a nuisance, in contrast to the excitement associated with early motoring and air travel. Road and air guidance systems would reduce the energy spending of stop–go driving and aircraft circling on hold. Hence, the lower limit is some 5–6 times less than the upper limit.

Activities

Education (understood as current activities plus lifelong continued education required in a changing world) is assumed to entail building energy needs corresponding to 10% of the residential value, i.e. an energy flux of 0–20 W/cap. for cooling and 0–65 W/cap. for heating.

Construction is evaluated on the basis of 1% of structures being replaced per year. This would be higher in periods of population increase. Measuring structures in units of the one-person space as defined above under *Biologically acceptable surroundings*, it is assumed that there are 1.5 such structures per person (including residential, cultural, service and work spaces). This leads to an estimated rate of energy spending for construction amounting to 30–60 W/cap. of stationary mechanical energy plus 7–15 W/cap. for transportation of materials to building site. The energy hidden in materials is deferred to industrial manufacture and raw materials industry.

Agriculture, including fishing, the lumber industry and food processing, in some climates requires energy for food crop drying (0–6 W/cap.), for water pumping, irrigation and other mechanical work (about 3 W/cap.), for electrical appliances (about 1 W/cap.) and for transport (tractors and mobile farm machinery, about 6 W/cap.).

The distribution and service (e.g. repair or retail) sector is assumed, depending on location, to use 0–80 W/cap. of energy for refrigeration, 0–150 W/cap. for heating of commerce or business related buildings, about 20 W/cap. of electric energy for telecommunications and other electrical appliances, and about 5 W/cap. of stationary mechanical energy for repair and maintenance service. Transportation energy needs in the distribution and service sectors, as well as energy for commuting between home and working places outside home, depend strongly on the physical location of activities and on the amount of planning that has been made to optimise such travel, which is not in itself of any benefit. Estimated energy spending is in the range of 30–100 W/cap. depending on these factors. All the energy estimates here are based on actual energy use in present societies, supplemented with reduction factors pertaining to the replacement of existing equipment by technically more efficient types, according to the "best available and practical technology" criterion, accompanied by an evaluation of the required energy quality for each application.

In the same way, the energy use of the manufacturing industry can be deduced from present data, once the volume of production is known. Assuming the possession of material goods to correspond to the present level in the

USA or Scandinavia, and a replacement rate of 5% per year, one is led to a rate of energy use of about 300 W/cap. Less materialistically minded societies would use less. Spelled out in terms of energy qualities there would be 0–40 W/cap. for cooling and 0–150 W/cap. for heating and maintaining comfort in factory buildings, 7–15 W/cap. for internal transportation and 20–40 W/cap. for electrical appliances. Most of the electrical energy would be used in the production processes, for computers and for lighting, along with another 20–40 W/cap. used for stationary mechanical energy. Finally, the process heat requirement would include 10–100 W/cap. below 100°C , 20–70 W/cap. from 100–500°C and 12–30 W/cap. above 500°C, all measured as average rates of energy supply, over industries very different in regard to energy intensity. Some consideration is given to heat cascading and reuse at lower temperatures, in that the energy requirements at lower temperatures have been reduced by what corresponds to about 70% of the reject heat from the processes in the next higher temperature interval.

Most difficult to estimate are the future energy needs of the resource industry. This is for two reasons: one is that the resource industry includes the energy industry and thus will be very different depending on what the supply option or supply mix is. The second factor is the future need for primary materials: will it be based on new resource extraction as it is largely the case today, or will recycling increase to near 100% for environmental and economic reasons connected with depletion of mineral resources?

As a concrete example, let us assume that renewable energy sources are used, as in the scenario considered below in section 6.4. The extraction of energy by a mining and an oil and gas industry as we know it today will disappear, and the energy needs for providing energy will take a quite different form, related to renewable energy conversion equipment which in most cases is more comparable to present utility services (power plants, etc.) than to a resource industry. This means that energy equipment manufacturing becomes the dominant energy-requiring activity.

For other materials, the ratios of process heat, stationary mechanical energy and electricity use depend on whether mining or recycling is the dominant mode of furnishing new raw materials. In the ranges given, not all maxima are supposed to become realised simultaneously, and neither are all minima. The numbers are assumed to comprise both the energy and the material provision industries. The basis assumption is high recycling, but for the upper limits not quite 100% and adding new materials for a growing world population. The assumed ranges are 0–30 W/cap. for process heat below 100°C and the same for the interval 100–500°C, 0–250 W/cap. above 500°C, 0–170 W/cap. of stationary mechanical energy, 0–30 W/cap. of electrical energy and 0–20 W/cap. of transportation energy.

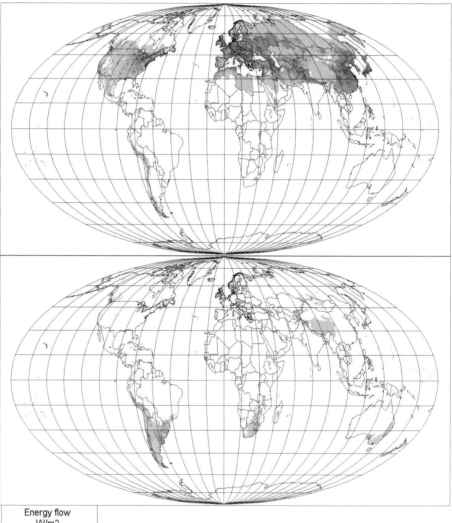

Energy flow
W/m2

5	to 10
2	to 5
1	to 2
0.5	to 1
0.2	to 0.5
0.1	to 0.2
0.05	to 0.1
0.02	to 0.05
0.01	to 0.02
0.005	to 0.01
0.002	to 0.005
0.0001 to	0.002
all others	

Figure 6.3. Energy requirement for January (above) and July (below) space heating (W m^{-2}), assuming the year 2050 population model (UN, 1996, 1997) and full satisfaction of space heating needs in an average space of 60 m^2 (floor area) by 2.3 m (ceiling height) per capita, including domestic as well as other indoor spaces in use (Sørensen and Meibom, 1998).

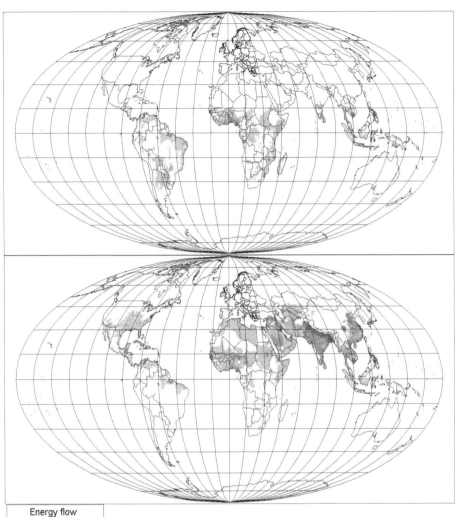

Figure 6.4. Energy requirement for January (above) and July (below) space cooling (W m^{-2}), assuming the year 2050 population model (UN, 1996, 1997) and full satisfaction of space cooling needs in an average space of 60 m^2 (floor area) by 2.3 m (ceiling height) per capita, including domestic as well as other indoor spaces in use (Sørensen and Meibom, 1998).

Energy flow
W/m2

5	to	10
2	to	5
1	to	2
0.5	to	1
0.2	to	0.5
0.1	to	0.2
0.05	to	0.1
0.02	to	0.05
0.01	to	0.02
0.005	to	0.01
0.002	to	0.005
0.0001	to	0.002

all others

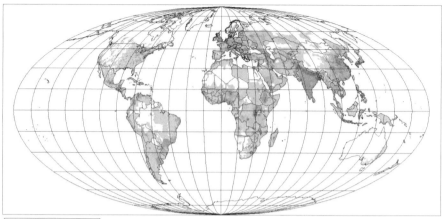

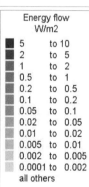

Energy flow
W/m2

5	to	10
2	to	5
1	to	2
0.5	to	1
0.2	to	0.5
0.1	to	0.2
0.05	to	0.1
0.02	to	0.05
0.01	to	0.02
0.005	to	0.01
0.002	to	0.005
0.0001	to	0.002

all others

Figure 6.5. Vegetable food required in the 2050 scenario, expressed in W m^{-2} and including 30% losses in preparation at the end-user.

Figure 6.6. Animal-based food required in the 2050 scenario, expressed in W m^{-2} and including 20% losses in preparation at the end-user.

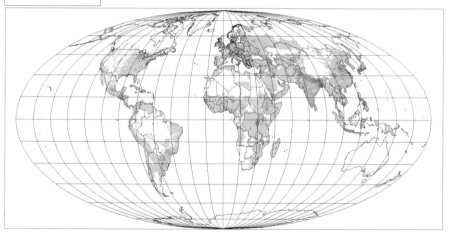

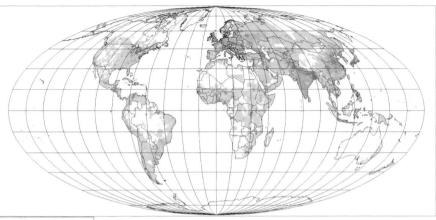

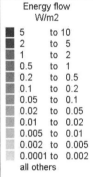

Figure 6.7. Average energy input to heat pumps used for space conditioning and low-temperature heat production, required in 2050 scenario and expressed in W m⁻².

Figure 6.8. Electric and mechanical energy plus medium- to high-temperature heat required in the 2050 scenario, expressed in W m⁻² (the heat pump inputs of Fig. 6.7 are excluded).

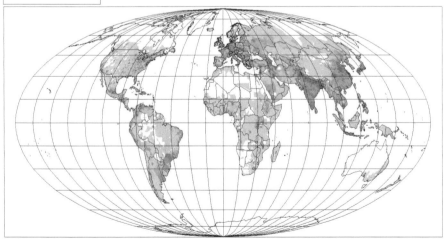

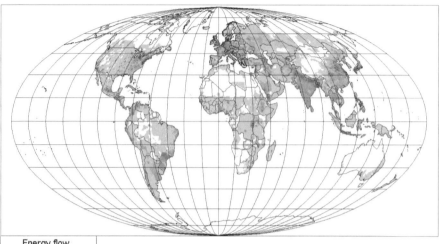

Energy flow W/m2	
5	to 10
2	to 5
1	to 2
0.5	to 1
0.2	to 0.5
0.1	to 0.2
0.05	to 0.1
0.02	to 0.05
0.01	to 0.02
0.005	to 0.01
0.002	to 0.005
0.0001	to 0.002
all others	

Figure 6.9. Energy for all forms of transportation required in the 2050 scenario and expressed in W m^{-2} (including losses in vehicle energy conversion and storage equipment).

Figure 6.10. Total energy required in the 2050 scenario for delivery to the end-user, expressed in W m^{-2} (food energy and environmental heat inputs to heat pumps included).

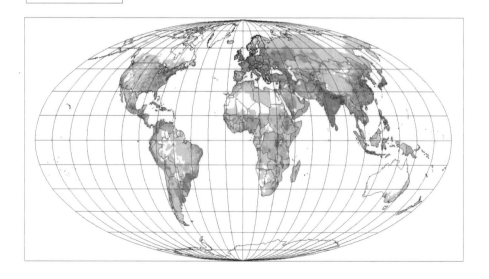

Summary of end-use energy requirements

Table 6.5 summarises the 2050 scenario estimates of energy that has to be delivered to the end-users, distributed on main energy qualities. The per capita energy services are more than twice those of today, and yet the energy that on average has to be delivered to each end-user is only half that of to-day, owing to higher efficiency in each conversion step. In section 6.4 the primary energy required for the scenario will be determined.

Regions: / Energy quality:	1. USA, Canada	2. Western Europe, Japan, Australia	3. Eastern Europe, Ex-soviet, Middle East	4. Latin America, SE Asian "tigers"	5. China, rest of Asia	6. Africa	Average & total
Food based on animals	30	30	30	25	25	20	23%
	45	45	45	37	37	25	36 W/cap.
	17	24	47	52	148	51	339 GW
Food based on grain & vegetables	70	70	70	75	75	80	77%
	119	119	119	128	128	114	123 W/cap.
	45	63	124	177	506	232	1148 GW
Gross transportation energy	359	299	140	201	99	30	125 W/cap.
	136	158	146	277	392	61	1170 GW
Heat pump input for low-temp. heat and cooling	103	110	87	43	80	22	65 W/cap.
	39	58	90	60	318	45	610 GW
Environmental heat	240	256	203	100	186	51	151 W/cap.
	91	135	210	140	741	105	1421 GW
Direct electric & all other energy	420	424	245	288	283	47	240 W/cap.
	153	224	255	398	1116	96	2242 GW
Total delivered energy	1272	1252	838	800	814	290	742 W/cap.
	482	661	871	1104	3225	591	6934 GW
Population 2050	379	528	1040	1380	3960	2040	9340 millions

Table 6.5. Energy delivered to end-user in 2050 scenario, including environmental heat (Sørensen and Meibom, 1998). This differs from end-use energy by losses taking place at the end-users location.

Table 6.5 gives the gross energy input to transportation needs, accumulating the individual demands for personal transportation, for work-related transport of persons and goods, and for transport between home and work ("commuting"). The net energy required to overcome frictional resistance and the parts of potential energy (for uphill climbing) and acceleration that are not reclaimed is multiplied by a factor of 2 to arrive at the gross energy delivery. This factor reflects an assumed 2050 energy conversion efficiency for fuel cell-driven (with electric motor) road vehicles of 50%, as contrasted with about 20% for present-day combustion engines. For the fraction of vehicles (presumably an urban fleet) using batteries and electric drives, the 50% efficiency is meant to reflect storage cycle losses (creep current discharge during parking and battery cycle efficiency). The geographical distribution of transportation energy demands is shown in Fig. 6.9.

The row "direct electric and all other energy" in Table 6.5 comprises medium- and high-temperature heat, refrigeration other than space cooling (which was included in the space heating and cooling energy), stationary mechanical energy and dedicated electrical energy for appliances and motors outside the transportation sector. For these forms of energy, no end-use efficiency is estimated, as the final service efficiency depends on factors specific to each application (reuse and cascading in the case of industrial process heat, sound- and light-creating technologies, computing, display and printing technologies, all of which are characterised by particular sets of efficiency considerations). The geographical distribution of these energy requirements in the 2050 scenario is shown in Fig. 6.8. That of the electric energy input to heat pumps (COP = 3), which in the scenario covers space heating, cooling and other low-temperature heat demands, is shown in Fig. 6.7, and Fig. 6.10 adds all the scenario energy demands, including environmental heat drawn by the heat pump systems. The low COP assumed for the heat pumps reflects the location of the corresponding loads at high latitude, where suitable low-temperature reservoirs are difficult to establish.

There is obviously a lack of accuracy, due to both technical and normative factors, in estimating such future energy demands: on one hand new activities involving energy use may emerge, and on the other hand the efficiency of energy use may further increase by introduction of novel technology. Yet it is reassuring that this gross estimate of energy demands associated with full goal satisfaction (for a choice of goals not in any way restrictive) is much lower than the present energy use in industrialised countries. It is demonstrating that bringing the entire world population, including little-developed and growing regions, up to a level of full goal satisfaction is not precluded for any technical reasons. The scenarios in section 6.4 below are consistent with this in assuming energy efficiency gains of about a factor of 4.

Figure 6.11 illustrates the scenario assumptions in different regions, by giving factors between 2050 and 1994 population and end-use energy with and without food energy. Compared to many other scenarios (see e.g. WEA,

2000), the present one has more emphasis on using the most efficient technology. The reason for this is an assumption of economic rationality that strangely contrasts with observed behaviour: Current consumers often choose technology, which entails a much higher cost for energy inputs during operation, than the increase in cost of the most energy-efficient brand of the technology in question. This economic irrationality pervades our societies in several other ways, including the emphasis on RD&D programmes on supply solutions much more expensive than solutions directed at diminishing energy use. Psychologically, the "growth paradigm" thus seems deeply rooted in current economic thinking, so much that many people find it "wrong" to satisfy their needs with less energy and "wrong" to invest in technology that may turn their energy consumption on a declining path. The problem is rarely made explicit in economic theory, but is buried under headings such as "free consumer choice", coupled with encouraging the "freedom" of advertisement, allowing substandard equipment to be marketed without mention of the diseconomies of its lifetime energy consumption. When will consumers be educated to rather respond with the thought that "if a product needs advertising, there must be something wrong with it"? Either the product does not fulfil any need, or it is sub-quality relative to other products filling the same need, e.g. with far lower energy consumption. New, expensive energy production technology, such as solar or fuel cells, get much more media attention than freezers with a four times lower-than-average energy consumption. Given that the world development at present is not governed by economic rationality, the supply–demand matching exercises presented in the following subsections, based on the scenarios defined above, will take up the discussion of robustness against underestimates of future energy demand.

The specification of the energy form required at the end-user at first amounts to distributing the sector loads on electrical, mechanical, heat, radiant energy, etc., but the heat energy demand must be further specified, as described above, in order to provide the type of information necessary in planning and selecting the energy systems capable of serving the loads. As an example of the investigations needed, Fig. 6.12 shows the temperature distribution of heat demand for the Australian food processing industry.

In a complex system with decisions being taken by a large number of individual consumers, the time pattern of the load cannot be precisely predicted. Yet the average composition and distribution of the load is expected to change smoothly, and only modest ripples appear relative to the average. There may be extreme situations, with a correspondingly low probability of occurrence, which should be taken into account in designing energy system solutions.

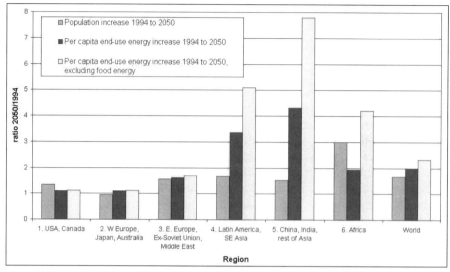

Figure 6.11. Ratio of the 2050 scenario end-use energy to estimated 1994 value.

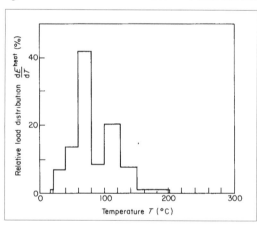

Figure 6.12. Temperature distribution of heat demand in the Australian food processing industry (based on Morse, 1977).

Fluctuations in electricity demand have routinely been surveyed by utility companies, many of which are using simulation models to optimise the use of their supply system. Figure 6.13 gives two examples of seasonal variations in electricity loads, one being for a high-latitude area, exhibiting a correlation with the number of dark hours in the day, and the other being for a metropolitan area in the south of the USA, reflecting the use of electricity for space cooling appliances.

Superimposed on the seasonal variations is a daily cycle, often exhibiting a different pattern on weekdays and weekends. The example given in Fig. 6.14 shows a further irregularity for Mondays (called a "peak working day"), probably reflecting a higher energy requirement for the "cold starts" of industrial processes, relative to the continuance of operation during the following days. Not all regions exhibit a pronounced "peak working day" load such as the one in Fig. 6.14. Figure 6.15 shows the daily load variations at Danish utilities for a summer and a winter weekday. It is seen that the ratio between the maximum and the minimum load is considerably higher than in the example of Fig. 6.14, presumably reflecting a higher proportion of industries requiring continuous operation in the USA.

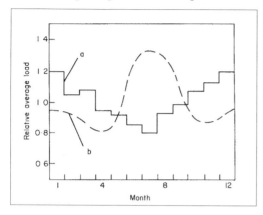

Figure 6.13. Seasonal variations in relative electricity load in areas with dominating components of lighting and air conditioning, respectively (full line: Denmark, latitude 56°N; Danish Electricity Supply Undertakings, 1971; dashed line: metropolitan area in southern USA, based on Brown and Cronin, 1974).

In connection with renewable energy sources, a particularly attractive type of load is one which need not be performed at a definite time but can be placed as suited within a certain time period. Examples of such loads (well known from historical energy use) are flour grinding, irrigation of fields and perhaps a substantial number of household tasks, such as bathing and washing. The patterns associated with the energy demand structure for many such tasks rest on habits, which in many cases have been developed under the influence of definite types of energy supply, and which may change as a result of a planned or unplanned transition to different energy sources. The distinction between casual habits and actual goals requiring energy conversion is not always an easy one, and such distinctions are worthy subjects for social debate.

6.2.3 Source data

Data on the renewable energy flows (or reservoirs) that may serve as inputs to the energy conversion system have been a major theme of Chapter 2 and particularly of Chapter 3.

As mentioned in section 6.1.2, the performance of most conversion systems can be calculated with sufficient accuracy by a "quasi-steady-state approximation", and most of these systems possess "buffering mechanisms" which make them insensitive to input fluctuations of short duration (such as high-frequency components in a spectral decomposition). For this reason, the input data describing the renewable energy fluxes may be chosen as time averages over suitable time intervals. What constitutes a "suitable" time interval depends on the response time ("time constant") of the conversion system, and it is likely that the description of some systems will require input data of a very detailed nature (e.g. minute averages), whereas other systems may be adequately described using data averaged over months.

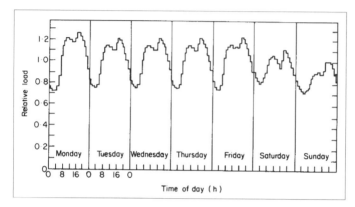

Figure 6.14. Relative electricity load (hourly averages) during a typical week for a utility system in the northeastern USA (based on Fernandes, 1974).

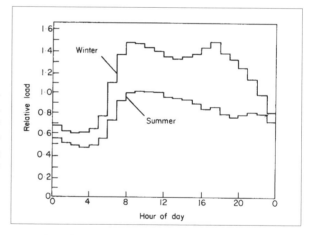

Figure 6.15. Relative electricity load (hourly averages) on a typical summer and winter weekday for the Danish utility systems (Danish Electricity Supply Undertakings, 1971).

A situation often encountered is that the data available are not sufficient for a complete system simulation on the proper time scale (i.e. with properly sized time steps). Furthermore, interesting simulations usually concern the performance of the system some time in the future, whereas available data by necessity pertain to the past. For demand data, a model must be established, while for resource data, it may be possible to use "typical" data. By "typical" is meant not just average data, but data containing typical variations in the time scales that are of importance for the contemplated energy conversion system.

Examples of such data are hourly values of wind velocities selected from past data, which, incorporated into a simulation model, may serve to estimate system performance for future years. Clearly, this type of "reference year" should not contain features too different from long-term statistics.

For some renewable energy systems, e.g. hydroelectricity, the variations between years are often important system design parameters.

Synthetic data bases (covering e.g. a year) have been constructed by combining sequences of actual data picked from several years in such a way that averages of important variables take values close to their long-range averages, but such that "typical" fluctuations are preserved in the data. The "Reference Year" data of solar radiation and other meteorological data that have for some decades been available for engineering design purposes for selected locations (notably certain cities in Europe and North America, cf. section 3.1.3) is an example of this approach.

For use in computer simulations, standard procedures exist which will generate random numbers with a specified distribution (e.g. Gaussian) once the fluctuation (or "half-width") and average value of the distribution are specified. This approach may be described as a *stochastically generated data base*. It has found widespread application, e.g. in prediction of wind turbine production characteristics (Sørensen, 1986; Troen and Petersen, 1989).

In order to illustrate the selection of data for system simulations such as those to be presented in section 6.4, we here describe the data used to describe some major renewable energy flows that will be used in the scenarios for the next century.

6.2.4 Photovoltaic power production

A solar radiation model is constructed on the basis of data for solar radiation incident on a horizontal plane. One set of data is based on an analysis (Pinker and Laszlo, 1992) of satellite measurements of radiation, albedo (reflectance), cloud cover and attenuation of radiation in the air (NASA, 1997). The other study considered collects many types of weather and climate data (NCEP-NCAR, 1998) and uses balance equations and a global circulation model (horizontal resolution about 210 km and 28 vertical levels) to improve the consistency of data (Kalnay *et al.*, 1996). However, for solar radiation no ground-based observations are used, so this is again based on top-of-the-atmosphere fluxes and various absorption, reflection and attenuation processes plus the balance requirements at the Earth's surface. The two types of analysis give similar results for solar radiation. The balance equations (difference between upward and downward short- and long-wavelength radiation and heat fluxes) do suggest too high an albedo over oceans, but the land data that we are using appear to be reliable. Figure 2.21 showed the radiation on a horizontal plane at the Earth's surface for selected months of 1997 (NCEP-NCAR, 1998).

In order to calculate the solar radiation incident on inclined surfaces, such as the ones characterising most solar installations, one would ideally need hourly data for direct and scattered radiation (or equivalently for direct and total global radiation). One would then assume that the scattered radiation is uniform to calculate its value for differently inclined surfaces. For the direct part, ray calculations with the appropriate angles between the direction to the sun and the normal to the collector surface have to be performed hourly. Such calculations have been performed and compared with measurements on inclined surfaces at several major solar installations and some weather stations (see Chapter 3). Such data are not globally available, but can be approximately inferred from monthly average radiation data.

The task is to estimate the radiation on a surface tilted towards either north or south by an angle approximately equal to the latitude (as this gives the optimum performance) on the basis of the horizontal surface solar radiation data available. No great accuracy is aimed at because the actual solar panel installations in the 2050 scenario will be of two types:

1. Building integrated panels that will anyway be influenced by the structures at hand: some solar panels will be mounted on vertical facades, with others on roofs being either flat or tilted, typically by angles 30°, 45° or 60°. In all cases, the orientation may not be precisely towards south or north, although we estimate the resource as comprising only those buildings facing approximately correctly and not being exposed to strong shadowing effects from other structures. The penalty of incorrect orientation and tilting angle is often modest and mostly influences the distribution of power across the seasons (cf. Chapter 3).

2. Centralised solar farms will generally be oriented in an optimum way, using total production maximisation for panels not tracking the sun. However, the majority of locations suited for central photovoltaic (PV) installations will be the desert areas of Sahara, the Arabian Peninsula, the Gobi desert and Australian inland locations. As these are all fairly close to the Equator, there is modest seasonal variations in solar radiation, and the horizontal surface data are often quite representative.

In consequence of the many unknown factors regarding the precise inclination of actual installations, the following simple approximation is used, originating from the analysis of the Danish (latitude 56°N) data given in Fig. 3.15. Here the radiation on a latitude-inclined surface in January and July very nearly equals the horizontal radiation in October and April, whereas the horizontal surface data for January and July are lower and higher than the inclined surface measurements, respectively. The October and April horizontal data are thus used as a proxy for January and July inclined surface radiation, and the April and October inclined surface values are constructed as simple averages of the January and July adopted values. This procedure, which works well for the Danish latitude, will also be less inaccurate for low latitudes, because of the relative independence of seasons mentioned above, and we simply use it for all locations.

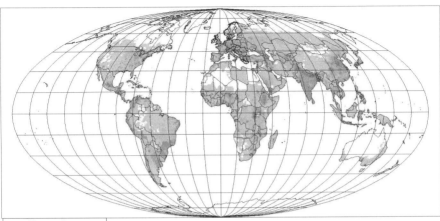

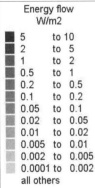

Energy flow
W/m2

5	to	10
2	to	5
1	to	2
0.5	to	1
0.2	to	0.5
0.1	to	0.2
0.05	to	0.1
0.02	to	0.05
0.01	to	0.02
0.005	to	0.01
0.002	to	0.005
0.0001	to	0.002
all others		

Figure 6.16. Potential power production from building-integrated solar cells in the 2050 scenario after subtraction of transmission and storage cycle losses. The January estimates are shown above and the July estimates are shown below.

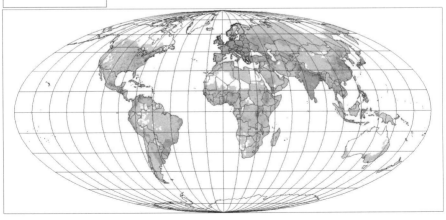

Figure 2.21 showed the NCEP-NCAR (1998) data for a horizontal surface. An alternative calculation using the European general circulation model HADCM2-SUL (Mitchell and Johns, 1997) gave very similar monthly 1997 solar radiation values.

In order to derive the actual energy extracted from PV solar panels, a fixed conversion efficiency of 15% is assumed. This is a fairly conservative estimate for the year 2050 technology, considering that the current efficiency of the best monocrystalline cells is above 20% and that of amorphous cells is near 10%, with multicrystalline cells falling in between. The 2050 choice is likely to be thin-film technology, but not necessarily amorphous or element-group III-V based, as new techniques allow crystalline or multicrystalline silicon material to be deposited on substrates without the complicated process of ingot growth and cutting. The efficiency of such cells is lower than that of the best current crystalline cells, but increasing, and multicrystalline cells have since 2002 surpassed monocrystalline cells in world-wide shipments.

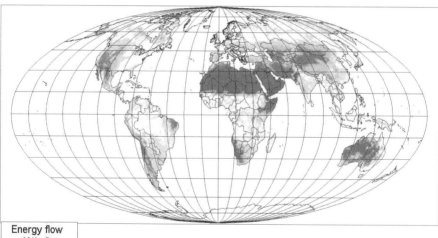

Energy flow W/m2	
■	1.8 to 2
■	1.6 to 1.8
■	1.4 to 1.6
■	1.2 to 1.4
■	1 to 1.2
■	0.8 to 1
■	0.6 to 0.8
■	0.4 to 0.6
■	0.2 to 0.4
■	0.01 to 0.2
	all others

Figure 6.17. Potential annual average power production for the 2050 scenario from centralised photovoltaic plants, placed on marginal land, after subtraction of transmission and storage cycle losses. Seasonal variations are small.

Finally, the availability of sites for mounting PV panels either in a decentralised fashion or centrally in major energy parks is estimated: the decentralised potential is based on availability of suitably inclined, shadow-free surfaces. The area of suitably oriented surfaces that may be used for building-integrated PV energy collection is assumed to be 1% of the urban horizontal land area plus 0.01% of the cropland area. The latter reflects the density of farmhouses in relation to agricultural plot sizes and is based on European estimates, roughly assuming that 25% of rural buildings may install solar panels. The potential PV production is thus taken as 15% of radiation times the fraction of the above two area types and the stated percentage for each. A final factor of 0.75 is applied in order to account for transmission and storage cycle losses, assuming about 5% transmission losses and that roughly half the energy goes through a storage facility of 60% round trip efficiency (e.g. reversible fuel cells). The flow of energy reduced by all these factors represents the energy delivered to the end-use customers in the form of electricity. It is shown in Fig. 6.16 for two months, with regional sums being discussed in section 6.4.

For centralised PV, the potential is taken as that derived from 1% of all rangeland plus 5% of all marginal land (deserts and scrubland), again times 15% of the incoming radiation and times 0.75 to account for transmission and storage losses. As shown in Fig. 6.17, this is a huge amount of energy, and it shows that such centralised PV installations can theoretically cover many times the demand of our 2050 scenario. Setting aside 1% of rangeland would imply an insignificant reduction in other uses of the land, even if this would rather be a small number of large plants. The same is true for desert and marginal land, where only the 5% most suited needs to be invoked, which in reality might be 10% for the entire installation (including frames, access roads), of which the solar panels are only part. Because of the huge area of e.g. the Sahara desert, this would suffice to supply more than the entire world need for energy. Of course, this implies the availability of intercontinental transmission, which may be quite realistic by the year 2050, e.g. through superconducting trunk lines (Nielsen and Sørensen, 1996).

6.2.5 Wind power production

The data used to assess potential wind power production are 1997 values from a re-analysis of meteorological station data and elevated data from balloons and aircraft, analysed according to the method of Kalnay *et al.* (1996), using a general circulation model to improve consistency (NCEP-NCAR, 1998). For the scenario, a simple average of pressure level 1000 mb and 925 mb data is used to represent the wind speeds at typical turbine hub heights of around 70 m. However, as the data are monthly mean wind speeds $<v>$ constructed on the basis of zonal and meridional winds, we need a model of the relationship between $<v>^3$ and $<v^3>$ to go from wind speeds to power in the wind. Simple models imply a rough proportionality between these two quantities, as used e.g. in the US wind atlas prepared by the Pacific NW Laboratory (Swisher, 1995) and also apparent from the Weibull distribution approach of the European wind atlas (Troen and Petersen, 1989). Inspired by these sources, one may estimate the relation between the power in wind P_w and the wind speed v (m s^{-1}) from the relation

$$P_w = 0.625 <v^3> \approx 1.3 <v>^3 \ \text{Wm}^{-2}.$$

The power in the wind obtained in this way is illustrated in Fig. 3.27 for the four seasons.

Going from the power in the wind to the power that may be produced by a wind turbine, the non-linear response of wind turbines has to be taken into consideration. Turbines of modern designs aim at a high annual production and typically start producing only at around 5 m s^{-1} and reach a fixed maximum production at around 12 m s^{-1}. Thus, there would not be any production for a monthly average wind speed below 5 m s^{-1}, if this were made up of nearly constant values throughout the time range. However, actual time series of wind speeds reflect the passage of weather fronts and typically oscillates with periods of some two weeks and entail some power production from the above type of wind turbine at practically all monthly average wind speeds, down to zero. The approach is then to parametrise the average power production from wind turbines as

$$W_p = \text{Minimum}(0.33 \ P_w, 500) \ \text{W m}^{-2},$$

representing the 21st century wind turbine constructions through a power factor slightly above the best of the current ones. A survey of power curves from existing wind turbines (Petersen and Lundsager, 1998) has been used as a basis for this estimate. This simple model is in reasonable agreement with data in regions of present wind exploitation. Owing to the coarseness of the GIS* grids used (for the wind speed data about 300-km cells at the Equator, diminishing by the cosine of the latitude towards the poles), the method does not at present compete with wind atlas methods available in certain parts of the world. It is, however, suited for planning purposes and can be refined as needed. The use of general circulation models in recalculating the data ensures that both surface roughness and general circulation determinants are included.

The wind power estimates calculated from the US data have been checked by performing a similar calculation using a recent model of the European Hadley Center (Mitchell and Johns, 1997; IPCC Data Distribution Centre, 1998), giving substantial agreement, within the variability from year to year, where the Hadley model of course differs from the NCAR model, which is based on actual measurements. The year 1997 is about 10% under the long-term average for Danish locations (according to the data collected by *Naturlig Energi*, 1998), but the monthly variations are somewhat atypical. For in-

---

* GIS = geographical information system.

stance, the average of the four months included in our model is 20% under the long-term average. The previous year (1996) was the poorest wind year of the 20-year survey, and we wanted to include realistic seasonal variations in our data set and thus avoid further averaging.

It should be said that the data from ground-based stations (which has a substantial impact on the NCEP-NCAR data set owing to the large number of such measurements relative to elevated data) may be a poor representation of actual wind conditions in a given area because of the sheltered location of many meteorological measuring stations. For this reason, the European wind atlas used only selected station data believed to be more relevant for wind turbine exploitation. However, the comparison with the Hadley data, which have only a weak dependence on station data, because they are generated from a general circulation model using topological features of wind roughness (friction), indicates that the method is quite reliable. The reason for this is presumably that we use pressure level data for both of the two lowest model pressure levels, rather than the 10-m data also available in the NCEP-NCAR data base. Repeating the calculation for the 10-m data indeed gives a considerably stronger dependence on the dubious ground level measurements.

The final step in estimating potential production from wind turbines is to appraise the locations suited for erection of wind turbines. They are divided into three categories: decentralised wind power production on farmland and centralised production in wind parks, either on-shore or off-shore. For the decentralised production, we assume that a (vertical) turbine swept area amounting to 0.1% of the cropland and rangeland areas may be used for wind production. Only the part of cropland designated as "open cropland" (cf. section 6.2.6) is used. This excludes mixed forest and cropland, where the trees are likely to impede wind flow. The assumed density of turbines corresponds roughly to the current density of wind turbines in many parts of Denmark, although the current average size is smaller than the turbines envisaged for the mid-21st century.

Current new turbines are typically of unit size 1.5 MW and have hubheights of 50–60 m, while the average standing wind turbine is around 500 kW with a hubheight of around 40 m. Owing to the combination of swept area increasing as the square of the linear dimension, and wind speeds increasing with height, the 2- to 4-MW turbines envisaged in the 2050 scenario will not be conspicuously larger than those of the current generation. It is therefore assumed that visual intrusion aspects will allow a turbine density similar to the present one. Other environmental concerns such as noise are very minor. The noise emitted by the present generation of wind turbines in most cases falls off to levels not significantly exceeding the background noise (wind passing trees and structures such as power lines, highway signs and buildings) only a few (5–10) rotor diameters away from the turbines (Søren-

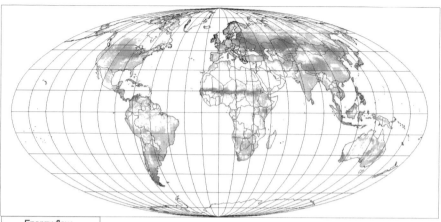

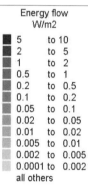

Energy flow
W/m2

5	to	10
2	to	5
1	to	2
0.5	to	1
0.2	to	0.5
0.1	to	0.2
0.05	to	0.1
0.02	to	0.05
0.01	to	0.02
0.005	to	0.01
0.002	to	0.005
0.0001	to	0.002
all others		

Figure 6.18a,b. Potential wind power production from decentralised wind turbines placed near farms for January 1997 (*a*, above) and for April 1997 (*b*, below). Transmission and storage cycle losses at an estimated 25% have been subtracted.

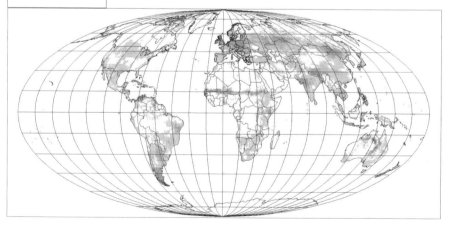

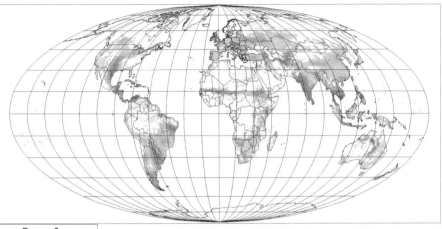

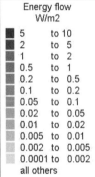

Energy flow
W/m2

5	to 10
2	to 5
1	to 2
0.5	to 1
0.2	to 0.5
0.1	to 0.2
0.05	to 0.1
0.02	to 0.05
0.01	to 0.02
0.005	to 0.01
0.002	to 0.005
0.0001	to 0.002
all others	

Figure 6.18c,d. Potential wind power production from decentralised wind turbines placed near farms for July 1997 (*c*, above) and for October 1997 (*d*, below). Transmission and storage cycle losses at an estimated 25% have been subtracted.

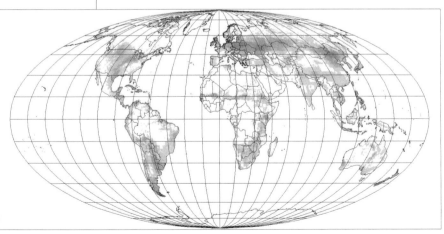

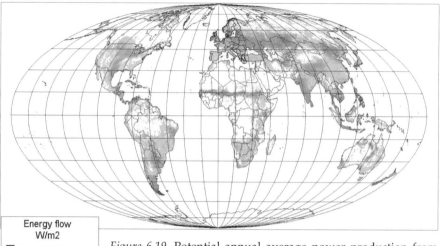

Energy flow W/m2		
5	to	10
2	to	5
1	to	2
0.5	to	1
0.2	to	0.5
0.1	to	0.2
0.05	to	0.1
0.02	to	0.05
0.01	to	0.02
0.005	to	0.01
0.002	to	0.005
0.0001 to		0.002
all others		

Figure 6.19. Potential annual average power production from decentralised wind turbines placed near farms on cropland or rangeland. An assumed 25% transmission and storage cycle loss has been subtracted.

Figure 6.20. Potential annual average power production from centralised wind turbine clusters placed on marginal land. An assumed 25% transmission and storage cycle loss has been subtracted.

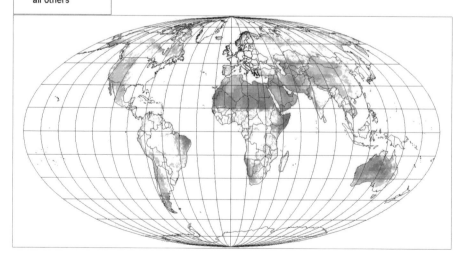

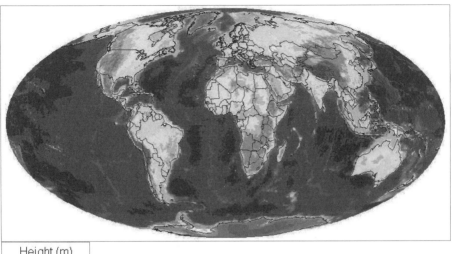

Height (m)

5.000 to 9.000
3.000 to 5.000
2.000 to 3.000
1.000 to 2.000
500 to 1.000
200 to 500
20 to 200
0 to 20
-20 to 0
-200 to -20
-500 to -200
-1.000 to -500
-2.000 to -1.000
-3.000 to -2.000
-5.000 to -3.000
-9.000 to -5.000

Figure 6.21. Land and ocean topography (based on data from Sandwell *et al.*, 1998).

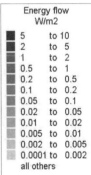

Energy flow
W/m2

5   to 10
2   to 5
1   to 2
0.5   to 1
0.2   to 0.5
0.1   to 0.2
0.05   to 0.1
0.02   to 0.05
0.01   to 0.02
0.005 to 0.01
0.002 to 0.005
0.0001 to 0.002
all others

Fig. 6.22. Potential net annual average power production from off-shore wind turbine parks (at water depths under 20 m).

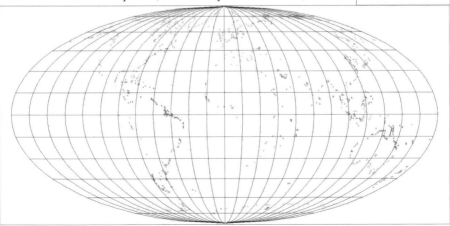

sen, 1995a, 1997a). In terms of ownership, the density of decentralised turbines would correspond to 1 out of 5–10 farmers owning a wind turbine (with large variations due to farm size variations between regions). Current wind power utilisation in the country with highest per capita wind power production, Denmark, is still dominated by farm-attached wind turbines, while in many other countries the emphasis has been on centralised production from turbine clusters. The amount of decentralised wind power potentially available is shown in Figs. 6.18 and 6.19. A strong seasonality is evident, and repeating the evaluation using data for other years than 1997 indicates that variations between years is also large.

Centralised wind power is here to be understood as power not produced in connection with individual farms. For land-based wind parks, only placement on marginal land is accepted in the scenario. In contrast to photovoltaic power, there is a fundamental limit to the fraction of land that can be used for wind power extraction, owing to shadowing and interference between wake flows that create turbulence and reduce the power output. This limit would be a vertical swept area of about 1% of the horizontal land area (Sørensen, 1995a). The scenario assumes that only 10% of the marginal land areas are used for placement of wind turbines at such spacing, implying a maximum wind swept area constituting 0.1% of the land area. This is small enough not to interfere with other uses of the marginal land, such as the solar cell parks considered in section 6.2.4. Figure 6.20 gives the annual production that potentially may be derived from such wind parks.

Finally, there are off-shore wind parks, meaning turbines placed in shallow waters at a distance from the shore that minimises visual impact. This mode of wind utilisation has become popular in recent years, as innovative low-cost solutions to the off-shore foundation problem have been found (Sørensen, 1995a). In order to obtain an estimate for the potential, off-shore water bodies with a depth under 20 m are first identified (Fig. 6.21). Of these, it is assumed that only 10% can be used for wind power generation, owing to competing uses (fishing, military, ship routes, etc.). Again taking the 1% maximum utilisation rate imposed by turbine interference into account, one arrives at a 0.1% maximum use of off-shore locations for wind power extraction. The average amount of power that can be derived from these locations is shown in Fig. 6.22. There is an additional number of inland water bodies that could also be used for wind power production, but which is here disregarded due to their possible recreational value.

6.2.6 Food production

The general model used in the scenario model to describe the biomass sector is shown in Fig. 6.23. It is a refinement of a model developed earlier (Sørensen *et al.*, 1994; Sørensen, 1994a). Each part of the model is discussed below, and numerical assumptions are stated.

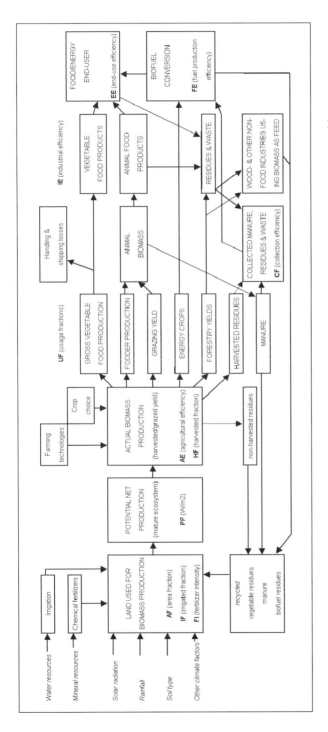

Figure 6.23. Overview of model used for describing the agricultural and silvicultural sectors in a scenario of future integrated food and energy production.

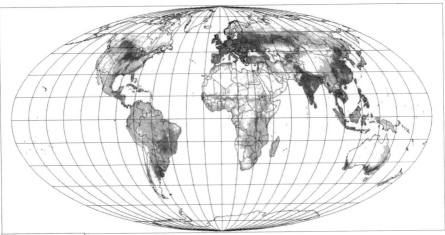

Area fraction	
■	0.9 to 1
■	0.8 to 0.9
■	0.7 to 0.8
■	0.6 to 0.7
■	0.5 to 0.6
■	0.4 to 0.5
■	0.3 to 0.4
■	0.2 to 0.3
■	0.1 to 0.2
■	0.001 to 0.1
	all others

Figure 6.24. The fraction of area taken up by cropland and cropland in rotation (taken as the fraction of each $0.5° \times 0.5°$ grid cell used for these purposes, on the basis of the classification in US Geological Survey, 1997).

Figure 6.25. The fraction of area used as rangeland, cf. Fig. 6.24.

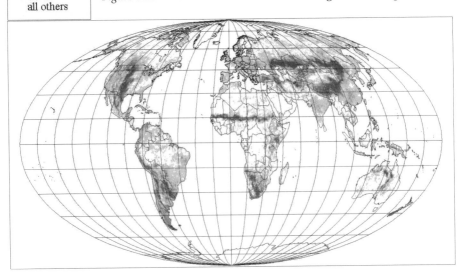

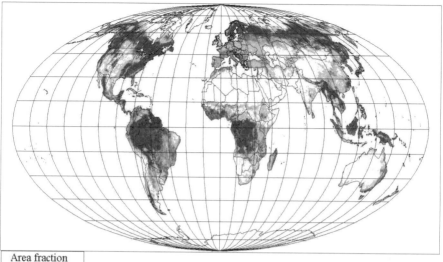

Area fraction	
■	0.9 to 1
■	0.8 to 0.9
■	0.7 to 0.8
■	0.6 to 0.7
■	0.5 to 0.6
■	0.4 to 0.5
■	0.3 to 0.4
■	0.2 to 0.3
■	0.1 to 0.2
■	0.001 to 0.1
	all others

Figure 6.26. The fraction of area occupied by forest, cf. Fig. 6.24.

Figure 6.27. The fraction of area constituted by marginal land, cf. Fig. 6.24.

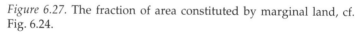

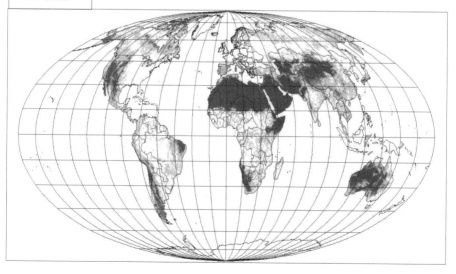

The land area used for food crops is considered to be the same in 2050 as now. This includes the cropland area fraction (Fig. 6.25; *AF* in Fig. 6.24) and for grazing also the rangeland (Fig. 6.26). Marginal land (Fig. 6.27) is not used for food crops. Some of the latter is today used for grazing in a little intensive way, in contrast to the use of cropland in rotation for occasional grazing. Crop cultivation on the cropland fraction is in some areas (e.g. Africa) little intensive, and present yields strongly reflect the agricultural practices of each region. As an indication of the potential biomass production on these areas, the model shown in Fig. 6.24 uses calculated net primary production data from the "Terrestrial Ecosystem Model (TEM)" of the Woods Hole group (Melillo and Helfrich, 1998; model evolution and description in Raich *et al.*, 1991; Melillo *et al.*, 1993; McGuire *et al.*, 1997; Sellers *et al.*, 1997), shown in Fig. 3.79. Global warming may induce increased primary production in a fairly complex pattern and the borders of natural vegetation zones will change, sometimes by several hundred kilometres (IPCC, 1996a).

Greenhouse warming-induced changes in area fractions are not included in the model, assuming that farming practice is able to gradually change and replace the crops cultivated in response to such altered conditions, which are anyway long-term compared to the lives of annual crops. The model does not specify which crops will be cultivated at a given location, but simply assumes productivity consistent with growing crops suited for the conditions. The TEM data are for a mature ecosystem, taking into account natural water, humidity and nutrient constraints along with solar radiation and temperature conditions. Annual crops are likely to have smaller yields because of only partial ground cover during part of the year and the corresponding smaller capture of radiation. On the other hand, the crops selected for cultivation may be favourably adapted to the conditions and therefore give higher yields than the natural vegetation at the location. Furthermore, irrigation may prevent yield losses in dry periods, and application of chemical fertilisers may improve overall yields.

It is conceivable that ecological concern will dictate a restrictive use of these techniques and will lead to a move towards increased use of the organic agricultural principles currently showing at the 10% level, area-wise, in Europe and a few other places. The 2050 scenario assumes what is called "integrated agriculture" (Sørensen *et al.*, 1994), a concept for which use of pesticides is banned and recycled vegetable residues and animal manure are the main sources of nutrient restoration, but where biological pest control and limited use of chemical fertilisers are not excluded. The yield losses implied by this method of farming are under 10% relative to current chemical agriculture, according to the experience in Northern Europe.

On cultivated land (including grazing land and managed forests) in regions such as Denmark (modest radiation and good soil and water access), the average annual biomass production is 0.62 W m^{-2} (of which 0.3 W m^{-2} are cereal crops; Sørensen *et al.*, 1994). This equals the value for a grid cell in

Denmark given in the TEM data base for mature natural productivity. In Southern Europe, the current production is about half (Nielsen and Sørensen, 1998), while the TEM data base gives a slightly higher value than for Denmark. The reasons for this are less intensive agricultural practice and water limitations for the growth pattern of the crops cultivated (limitations that would be less severe for a mature ecosystem). It thus seems reasonable in the scenario to use the TEM as a proxy for cultivation yields, provided than one assumes better farming techniques used by the year 2050 and assumes that irrigation and chemical fertilisers are used when necessary. This is precisely the assumption stated above as the basis for our scenario. The model then uses the net natural primary production data of the TEM globally, but without adding further increases on the basis of irrigation (which in dry regions can double agricultural output) and use of chemical fertilisers (which can provide a further doubling, if the soil is poor in nutrients or nutrients are not returned to the fields). In other words, the disadvantage in going from mature vegetation to annual crops is offset against the advantage of reducing limiting factors related to water and nutrients. In Fig. 6.23, this means disregarding the irrigation and fertiliser parameters IF and FI and proceeding with the potential production PP taken from the TEM data base.

The TEM global biomass production estimates for PP shown in Fig. 3.79 are expressed in energy units (1 g carbon per year is equal to a rate of energy production of 0.00133 W).

Currently, in Denmark only about 10% of this energy is contained in the food consumed domestically. The indication from this is that there is room for altered management of the system, by diverting residues to energy extraction and later returning the nutrients to the fields. One may also note that the current system is based on high meat consumption and the associated emphasis on animal raising and, in the Danish case export. By even the modest change in vegetable to animal food consumption ratio that is assumed in the demand scenario shown in Figs. 6.5 and 6.6, it is possible globally to divert substantial amounts of biomass to energy purposes, without jeopardising the need to provide food for a growing world population.

It is not assumed that the intensive agricultural practices of Northern Europe will have been adopted globally by year the 2050. The agricultural efficiency factor AE in Fig. 6.23 is taken as unity only for regions 1 and 2 (cf. Table 6.3). For Africa (region 6) it is taken as 0.4, and for the remaining regions it is taken as 0.7. The fraction of the biomass production actually harvested is taken globally as $HF = 0.4$. The remaining fraction consists of roots and residues ploughed down in order to provide natural fertilisation for the following growth season.

Regarding the land areas classified as cropland, the assumed distribution on uses is given in Table 6.6, based on considerations of cropland scarcity and traditions for animal raising. The low animal fodder value for Africa reflects the fact that the African tradition for animal raising is based on range-land, not cropland providing fodder.

Region	AE (cropland)	HF	UF (vegetable food)	UF (fodder)	UF (energy crops)
1	1	0.4	0.4	0.5	0.1
2	1	0.4	0.4	0.5	0.1
3	0.7	0.4	0.5	0.5	0
4	0.7	0.4	0.4	0.5	0.1
5	0.7	0.4	0.7	0.3	0
6	0.4	0.4	0.8	0.2	0

Table 6.6. Parameters used for cropland biomass production in the 2050 scenario (regions were defined in Table 6.2, cf. also Fig. 6.23 for explanation of abbreviations).

The amounts of vegetable-type food that can potentially be produced and delivered to the end-users in this scenario can now be calculated on an area basis, assuming the losses in going from vegetable food produced to vegetable food delivered as 25% [*IE(veg. products)* = 0.75 for vegetable food products in Fig. 6.23],

Delivered vegetable food =
  $AF(cropl.) \times PP[Wm^{-2}] \times AE \times HF \times UF(veg. food) \times IE(veg. prod.),$

where *AF* and *PP* depend on the precise geographical location and the others only on region. The calculated distribution of vegetable food delivered to the consumers is shown in Fig. 6.28.

For food from animals, such as meat, milk and eggs, the average efficiency in transforming biomass to delivered animal products is assumed to be *IE(animal products)* = 0.15, a value reflecting a typical average of a smaller efficiency for meat production and a higher one for milk and eggs (Sørensen *et al.*, 1994). The amounts of animal-based food using cropland-derived fodder and delivered to the consumer is thus

Delivered animal food(1) =
  $AF(cropl.) \times PP[Wm^{-2}] \times AE \times HF \times UF(fodder) \times IE(anim. prod.).$

The distribution of potential animal food deliveries based on the route where livestock is fed fodder produced on cropland is shown in Fig. 6.29.

The other part of animal food is from animals grazing on rangeland, where we shall assume that livestock grazes *HF* = 0.4 of the biomass production per unit of area, and put *AE* = 1. The use of rangeland is assumed to be 50% for grazing and 50% for other purposes (such as energy crops or no commercial use). Thus, the utilisation factor is taken as *UF(grazing)* = 0.5:

Delivered animal food (2) =
  AF(rangeland)× PP[Wm^{-2}]× HF× UF(grazing)× IE(anim. prod.).

The distribution of potential animal foodstuff delivered to the end-users through the route of rangeland grazing is shown in Fig. 6.30. The ratio of the two contributions (crop feeding and grazing routes) is determined by the area set aside for each. The resulting fraction of animal food derived from rangeland grazing is 37%, in terms of energy content.

The efficiency in the end-user's making use of the delivered food, denoted *EE* in Fig. 6.23, has for all the bio-energy routes been included in the definition of gross demand in section 6.2.2.

6.2.7 Biofuel production

A number of fuels may be produced from biomass and residues derived from vegetable and animal production or from forestry and dedicated energy crops, ranging from fuels for direct combustion over biogas (mainly methane mixed with carbon dioxide) to liquid biofuels such as ethanol or methanol, or gaseous fuels such as synthesis gas (a mixture of mainly carbon monoxide and hydrogen, also being an intermediate step in producing methanol) or pure hydrogen. The production of biofuels by thermochemical processes is based on high-temperature gasification and various cleaning and transformation processes (Jensen and Sørensen, 1984; Nielsen and Sørensen, 1998).

Whether the biofuel production is by thermal or biological processes, the expected conversion efficiency is of the order of *FE* = 50% (cf. Fig. 6.23). This is to be compounded with a factor describing the ability of the biofuel production industry to collect the necessary feedstock. This collection efficiency factor, which we call *CF*, describes the efficiency in collecting biomass for such industrial uses. For vegetable foods, it is assumed that *CF (veg. waste)* = 25% of the gross production is available for energy production [some of this would come from the food industry losses of *(1 − IE(veg. prod.))* = 25%, some from the subsequent household losses of 30%, cf. sections 6.2.2 and 6.2.6]. The overall yield of biofuels from vegetable crops is then

Biofuels from vegetable foodcrops =
  AF(cropl.)× PP[Wm^{-2}]× AE× HF× UF(veg. food)× CF(veg. waste)× FE.

Considering manure, this would be available only when livestock are in stables or otherwise allows easy collection. The model assumes that grazing animals leave manure in the fields and that this is not collected (although it could be in some cases), but that animals being fed fodder from crops will be in situations where collection of manure is feasible. Furthermore, although the 85% of animal biomass not ending up in food products will be used both to maintain the metabolism of livestock animals and the process of producing manure, it will also contain a fraction that may be used directly for fuel production (e.g. slaughterhouse wastes). Combined with manure, this is considered to amounts to $CF(anim.) = 0.6$, giving for the fodder to animal route to biofuels:

Biofuels from manure and other animal residues =
$AF(cropl.) \times PP[Wm^{-2}] \times AE \times H \times UF(fodder) \times (1-IE(anim.\ prod.)) \times CF(anim.) \times FE$

The further possibility of producing biofuels from forestry residues (either scrap derived from the wood industry or residues collected as part of forest management) may be described by a factor $CF(forestry) = 0.3$, defined as a percentage of the total forest biomass production. This is the fraction collected, expressed in energy units. For managed forests it depends on the fraction of wood suitable for the wood manufacturing industry (furniture, etc.), which again depend on the tree type. Adding wood scrap from industry and discarded wooden items, as well as from forest management, would in many regions exceed 30%. However, starting from an enumeration of all forests, including the rainforests and other preservation-worthy forest areas that are not proposed to be touched, and considering only managed forests that deliver to wood industries, 30% is probably a maximum for the year 2050. The forest residue to biofuel route is then

Biofuels from forest management =
$AF(forestland) \times PP[Wm^{-2}] \times HF \times CF(forestry) \times FE.$

The potential amounts of biofuels that could be derived from forestry are shown in Fig. 6.31, and the sum of the three routes to biofuels described above are given on an area bases in Fig. 6.32. This may be denoted "decentralised fuel production", although forestry may not be seen as entirely decentral in nature. However, it is an ongoing activity and distinct from producing biofuels from land used exclusively for energy crops. Two energy crop routes will be considered.

While the biomass production on rangeland does not give rise to biofuel production because manure from grazing livestock is not collected, some of the rangeland may be suited for cultivation of dedicated energy crops. As only 50% is assumed to be used for grazing, one may consider the remaining 50% as potentially exploitable for energy purposes, in the scenario versions where centralised energy schemes are considered acceptable [i.e. $UF(rangeland\ energy\ crops) = 0.5$]. On cropland there is additionally assumed a maximum of 10% set aside for energy crops in areas of generous resources, such as Western Europe and the Americas [see the $UF(cropland\ energy\ crops)$ values of 0.1 or 0 in Table 6.6]. The potential biofuel production from these areas is

Biofuels from energy crops on cropland =
$AF(cropl.) \times PP[Wm^{-2}] \times AE \times HF \times UF(cropland\ energy\ crops) \times FE,$

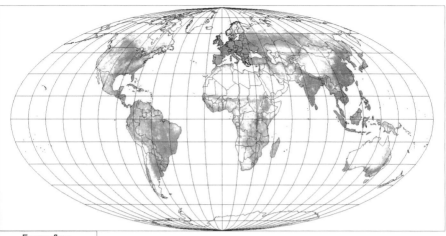

Energy flow W/m2	
5	to 10
2	to 5
1	to 2
0.5	to 1
0.2	to 0.5
0.1	to 0.2
0.05	to 0.1
0.02	to 0.05
0.01	to 0.02
0.005	to 0.01
0.002	to 0.005
0.0001 to	0.002
all others	

Figure 6.28. Potential delivery to final consumer of vegetable-based food from cropland production, expressed in units of energy flow (W m^{-2}).

Figure 6.29. Potential delivery to final consumer of animal-based food from livestock being fed with fodder grown on cropland, in energy units.

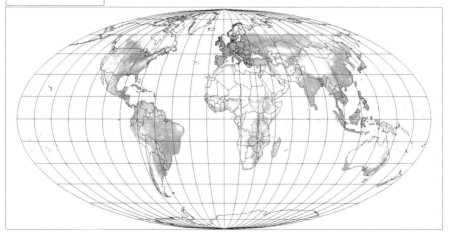

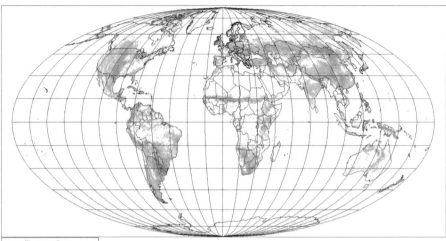

Energy flow
W/m2

5	to 10
2	to 5
1	to 2
0.5	to 1
0.2	to 0.5
0.1	to 0.2
0.05	to 0.1
0.02	to 0.05
0.01	to 0.02
0.005	to 0.01
0.002	to 0.005
0.0001	to 0.002
all others	

Figure 6.30. Potential delivery to final consumer of animal-based food from livestock grazing on rangeland, expressed in units of energy flow (W m^{-2}).

Figure 6.31. Potential delivery to final consumer of biofuels based on forestry residues and wood waste, expressed in energy units.

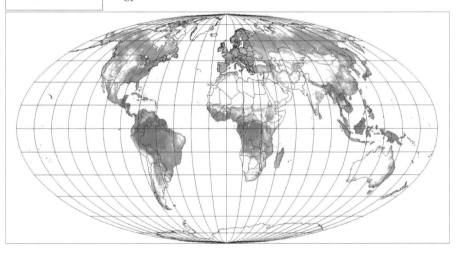

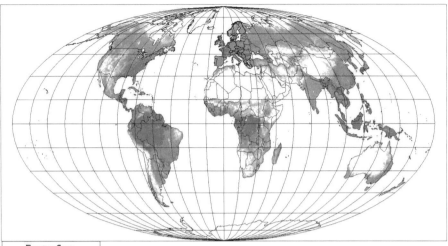

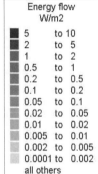

Figure 6.32. Potential delivery to final consumer of all "decentralised" biofuels based on forestry, agricultural residues, manure and household waste, in units of energy flow (W m^{-2}).

Figure 6.33. Potential delivery to final consumer of "centralised" biofuels based on dedicated energy crops grown on part of rangeland and minor parts of cropland, in energy units.

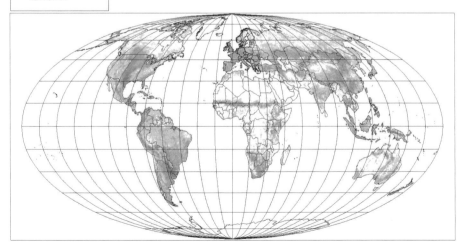

Biofuels from energy crops on rangeland =
AF(rangeland)× PP[Wm^{-2}]× HF× UF(rangeland energy crops)× FE.

The value of *FE* is 0.5. The area distribution for the two potential sources of biofuels from energy crops is shown in Fig. 6.33.

6.2.8 System choice

Given the available energy sources and the magnitude and distribution of energy demands, a viable conversion, storage and transmission system must be found. If the demands can be met by the sources at hand, it is likely that there will be a number of different systems that will be able to serve the purpose. Thus a question of optimising the choice of energy system arises, involving a range of considerations including economy, environmental impact and supply security.

Some of these considerations will be discussed in Chapter 7. The physical constraints on the system are given in part by the availability of different renewable energy sources in different parts of the world (which was surveyed in Chapter 3), and, in part, by the characteristics of different system components (discussed in Chapters 4 and 5). Assuming the load to have components of the type discussed in section 6.2.2, but without entering into the details in the relative demand of different energy forms, as distributed in the heat demand spectrum and in the time dependence of the load for different geographical regions and societies, Fig. 6.34 provides a sketch of some possible components of energy systems based on renewable resources. More detailed scenarios are illustrated in section 6.4.

The total delivery capacity of the energy supply system will generally be determined by the maximum load, with some adjustments made according to the expected time distribution of source fluxes and loads. This implies that situations may arise in which not all incident energy can be converted, because the demand is already met and the stores (or those for a given energy form) are full. In this case, as well as in situations where different parts of the system would each be capable of satisfying the load, system operational problems arise. Regulation procedures would then have to be defined which could determine which converter to tune down or stop in cases of system overflow and which stores should be depleted first when more than one is available to serve a given load at a given time. Also strategies for exchange of energy between regions with different load structure (or time zones) offer possibilities for avoiding over- or underflow (Meibom *et al.*, 1999).

Once a renewable energy-based system has been established and optimised in various respects, its implementation may be discussed. Supposing that a society has found such a system acceptable and desirable, the immediate problem will be to plan a smooth transition from an existing (e.g. fuel-based) energy supply system to the renewable energy system, which would also have to be open to modifications caused by future changes in demand structure and technology. Replacement of the present fuel-based energy systems with renewable energy-based systems may require quite an extended transitional period (Sørensen, 1975b; Sørensen *et al.*, 1994; Sørensen and Meibom, 1998).

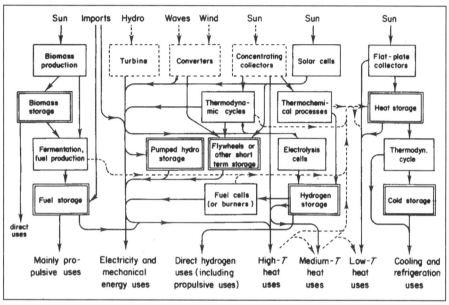

Figure 6.34. Example of the components of a supply–demand matching energy system based on renewable sources, with primary and secondary conversion processes, storage and transmission.

6.3 Examples of local system simulations

The renewable energy conversion systems or combinations of systems for use in different regions and in differently structured societies will very probably have to be different. The contribution of a single type of conversion in the system may be determined on the basis of climate and social structure, and it may change in time, particularly the period of introducing renewable energy technologies in the presence of an existing system based on depletable resources. In the first phase, a given type of renewable energy conversion, such as flat-plate solar collectors for heating of buildings or electricity-producing wind energy converters, may be integrated with a fuel-based auxiliary system in such a way that the part of the total load which would at present be least economic to replace by renewable energy is still covered by fuels. While it may be possible to cover most of the low-temperature heat load by solar collectors in some regions of the world, there are many regions where the extra cost of a solar energy system that would be able to increase the coverage from, say, 70% to 90% is much higher per energy unit supplied than a system aimed only at covering less than 70% of the load on a yearly average basis. The examples given in this section will try to deal with some of these problems, which are of immediate interest for current implementations of energy supply systems, by investigating the relation between fractional load coverage and various system parameters characterising the size and complexity of the system.

6.3.1 Solar heat or heat-and-electricity producing systems

The examples described in this subsection will be based on flat-plate solar collector devices (cf. section 4.2.1) in various configurations, and the system simulations will use as a data base the Danish Reference Year, which was summarised in section 3.1.3 (cf. Figs. 3.11–3.15). It is clear that use of the same systems at latitudes with higher levels of solar radiation will improve the performance and make it possible to cover loads with smaller and less expensive systems. On the other hand, the simulation of solar energy systems at fairly high latitudes (56°N for the Danish reference data) leads to consideration of a number of design problems, which may show up more clearly than for latitudes with higher and more stable solar input and smaller heating loads, but which may still be of general interest. In any case, the need for heating is higher at higher latitudes, making the optimisation of systems for such use particularly relevant.

Systems involving solar thermal collection into a heat store (cf. Fig. 4.21) have a complex dynamical behaviour, due to the direct influence on absorber performance of heat storage temperature, collector fluid inlet temperature and speed of circulation. Indirectly, the collector circuit is influenced by heat storage temperature of load patterns governing the heat exchange between store and load circuits. If a photovoltaic part is present (thus forming a hybrid photovoltaic-thermal or PVT system), its performance depends on the complex temperature behaviour only through the temperature dependence of power production efficiency (cf. Fig. 4.38), which in any case is much weaker than that of the thermal collection part.

Model description

The system modelled (using the software package NSES, 2001) is a building with solar collectors, heat stores (that could be communal, i.e. common for several buildings or even part of a district heating system) and load patterns consisting of time profiles of the use of electricity, space heating and hot water. The design of the building has to be specified in order to calculate heat losses and energy gains through passive features (such as windows), as does the activities in the building relevant for energy flows [people occupancy, use of electric appliances and lighting, opening and closing of shutters (if any) in front of windows, etc.]. The external parameters include ambient temperature, wind speed and solar radiation.

In case both thermal and electric energy are produced, the thermal absorber could be the back plate of a photovoltaic collector sheet, or the thermal collection could be from the air flow over the PV panel but below a transparent cover. In the first case, the heat transfer fluid would typically be water and in the second case it would be air. Heat collection in front of the PV panel could have negative effects such as more hours with dew formation on the cover glass. Heat collection behind the PV panel could be lowered, if reflective back layers are used in the PV cell design to increase light capture probability by prolonging the radiation path-length within the cell.

The simulations reported below use the model described in section 4.2.1 for solar thermal collection and heat exchange in the collector–storage circuit. Details of the radiation data used in the numerical examples and the load description entering the modelling of the storage-load circuit are given in the following subsections.

Solar radiation data

Radiation data are taken from a stream of "reference year data" for a position 56°N and 10°E (cf. section 3.1.3). The data are collections of hourly global and diffuse radiation covering a year and exhibiting variations typical of what is contained in 30-year meteorological time series. A reference year (of which several have been constructed, covering cities in Europe and the USA) is put together by pieces of actual data from the 30-year data source (30 years is the selected time span for defining a "climate"). Typical segment length is 14 days, in order to preserve the short-term correlations in the data, arising from continuity and the typical passage time of weather front systems. Thus, the data are supposed to have both the typical short-term correlations and also the long-term variability that can ensure a realistic modelling of solar system performances over the lifetime of the system in question.

The most recent[*] Danish reference year data set employed is illustrated in Figs. 6.35 and 6.36 and shows the hourly Danish 1996 reference year profiles

---

[*] The Danish reference year data set was first established in 1974 (Andersen et al., 1974). This set was not based on a full 30-year underlying data base and had some unrealistic features, including too much solar radiation in spring. International projects launched by the International Energy Agency and the European Commission's

of ambient temperature and solar radiation incident on a south-facing plane tilted 45°.

The collection of short-wavelength radiation by the solar panel takes place in a system comprising one or several cover layers followed by one or more absorbing layers. If present, the photovoltaic modules constitute one such layer. Part of the absorbed radiation is transformed into an electric current, while the rest is transformed into long-wavelength radiation, which is either converted into useful heat or lost to the surroundings.

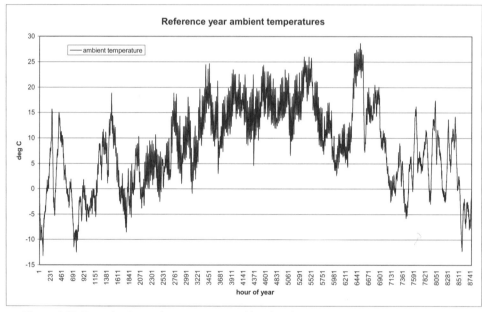

Figure 6.35. Danish 1996 reference year profile of ambient temperatures.

R&D programme tried to establish a common prescription for the reference year sets of participating countries. A Danish reference year data set from this effort is included in the European publication (Lund *et al.*, 1985). However, not all problems were resolved, including the mentioned unrealistic feature of the data set, which reappeared in the 1985 data. In recent years, new efforts have been made to obtain a more consistent tool for building calculations, and at least the 1996 Danish part of this effort is now free from obvious inconsistencies (DTI, 2000). There are other objections which could be made, as the new data sets are now primarily directed at small solar collector systems aimed only at providing hot water during summer periods. For example, the snow cover present in early data sets is now omitted, making it difficult to predict reflected light received by collectors with tilt angles near 90°, which in Denmark would be very relevant for winter collection.

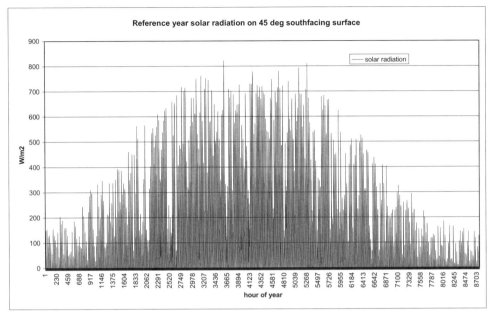

Figure 6.36. Danish 1996 reference year profile of 45°S solar radiation over the year.

In current thermal solar collectors, a heat transfer fluid is passed over or under the usually building-integrated solar collector. The connection with the rest of the system (heat store and building loads), illustrated in Fig. 4.21, usually contains two circuits called the collector and the load circuit, both connected to the heat store by heat exchangers.

The heat transfer fluid is typically air or water. Use of water will for PVT systems entail a requirement for sealed paths and protection against corrosion, just as for thermal solar collectors. Although the use of air as a transfer fluid is easier, the low heat capacity of air and low energy transfer rates to and from air place limits on its applicability.

Thermal collection in front of the solar cell (but below a cover glass) may alter the efficiency of the PV conversion in a negative way, whereas collection behind the solar cell should pose few problems once it has been made sure that the PV and thermal plate layers are well connected, so that heat transfer (conduction) is high. Incoming solar radiation is either made useful for electricity production, is transformed into heat taken up by components of the collector, or leaves the cell through front or rear walls. A reflective rear layer may be used at the back-side of the PV module. It causes the light to pass through the solar cell twice, with a greater overall chance of capture. On the other hand, this excludes the possibility of light passing through the PV layers and only being absorbed in an underlying thermal collector layer.

Heat load of an individual house

The heating load of the building is fixed by specifying the indoor temperature of the building and the hot water need as a function of time. The indoor temperature may need to be fixed, or it may be allowed to decrease during the night and when the house is unoccupied. For the purpose of simulation, the hot water needs may be considered to be identical on every day of the year and concentrated on certain hours, such as a morning peak plus a period of moderate use from late afternoon and throughout the evening (including clothes washing, dish washing, baths, etc.). The amount of heat required to keep the indoor temperature at the prescribed value does not, on the other hand, remain constant throughout the year, but equals the heat losses through doors, windows, walls, floor and roof, as well as losses associated with air exchange, which may be regulated by a ventilation system or may be entirely "natural", i.e. associated with crevices, etc. in building materials and assembly sites.

The heat losses through building surfaces depend on the difference between inside and outside temperature and may be approximated by a linear expression

$$h = (T_L - T_a) \sum_i A_i U_i , \qquad (6.2)$$

where T_L is the indoor ("load-") temperature, T_a is the outside temperature and A_i and U_i are corresponding areas and specific heat loss rates for different parts of the external building surface (walls, windows, roof, floor, etc.). The floor against the ground may be poorly described by the expression with T_a, and the heat loss expression should perhaps contain a particular term describing the floor loss, being proportional to the difference between the load and soil temperatures, rather than the difference between the load and air temperatures. However, the heat loss rates U_i are in any case only approximately constant, and they may have a further dependence on temperature (e.g. if radiative transfer is involved) and on outside wind speed, moisture, etc., as discussed above in connection with the collecting absorber.

An average contemporary one-family detached dwelling located in the temperate climatic zone may be characterised by wall and ceiling losses amounting to around 57 W °C^{-1} (U_{wall} in W m^{-2} °C^{-1} times wall and ceiling total area) and ground floor losses of 18 W °C^{-1} (U_{floor} often slightly lower than U_{wall}). With 20-m^2 window area and $U_{window} = 2$ W m^{-2} °C^{-1} (two layers of glazing; with intermediate space filled with low-conduction gas the value may be reduced to half), the total "conduction" loss becomes 115 W °C^{-1}. If the house has 1½ storeys (cf. sketch in Fig. 6.37), the ground floor area may be about 100 m^2 and the combined floor area above 150 m^2 (depending on how low ceiling heights are accepted as contributing to floor area for the upper floor). This corresponds to an inside volume of about 330 m^3, and if the rate of air exchange is 0.75 per hour, then $V_{air} = 250$ m^3 of air per hour will have to be heated from T_a to T_L (reduction is possible by letting the outgoing air pre-heat the incoming air in a heat exchanger),

$$(A_i U_i)_{air\,exchange} = V_{air}\, C_v^{air} \approx 87\ \text{W}\ ^\circ\text{C}^{-1},$$

since $C_v^{air} = 0.347$ Wh m^{-3} $^\circ$C^{-1}. For the reference calculation, V_{air} is taken as 250 m^3 h^{-1} only during the period 16:00 to 08:00 and as 125 m^3 h^{-1} between 08:00 and 16:00 (considering lower occupancy during these hours, and assuming a programmable ventilation system). Measurements have suggested that the air exchange rate in naturally ventilated Danish dwellings (individual houses as well as apartments) may be lower than required for reasons of health (Collet *et al.*, 1976). In a sample of 81 dwellings a range of air exchange rates between 0.15 and 1.3 times per hour were found, with an average value of 0.62 times per hour.

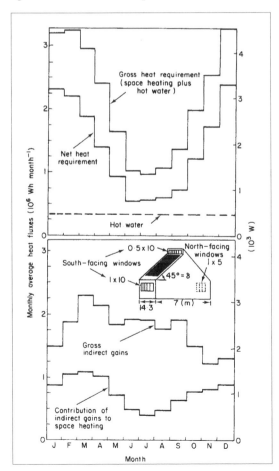

Figure 6.37. Heat load data for a one-family dwelling, in terms of monthly averages of heat fluxes. The layout and dimensions of the house are indicated as an inset.

1. Building heat losses

Loss rate through surface (except windows) 75 W °C^{-1}.

Loss rate through windows (no shutters) 40 W °C^{-1}.

Ventilation air 250 m^3 h^{-1} from 16:00 to 08:00, 125 m^3 h^{-1} from 08:00 to 16:00.

Minimum inlet temperature required for space heating load, 28°C.

Indoor temperature 20°C from 07:00 to 09:00 and from 16:00 to 24:00, 19°C from 09:00 to 16:00 and 16°C from 0:00 to 07:00.

Hot water usage: 50 kg from 07:00 to 08:00, 25 kg h^{-1} from 16:00 to 18:00.

Required hot water temperature: 42°C above an assumed input temperature of 8°C.

2. Indirect heat gains

Persons: 100 W from 22:00 to 07:00, 200 W from 07:00 to 09:00 and from 16:00 to 22:00, and 0 W from 09:00 to 16:00.

Equipment: 60 W from 00:00 to 07:00, 600 W from 07:00 to 08:00 and from 23:00 to 24:00, 120 W from 08:00 to 16:00, 1200 W from 16:00 to 17:00 and from 18:00 to 21:00, 1800 W from 17:00 to 18:00, and 900 W from 21:00 to 24:00.

Lighting: 100 W from 07:30 to 08:30 and from 15:30 to 00:30, unless global solar radiation exceeds 50 W m^{-2}, zero during other hours.

Gains from 13 m^2 vertically mounted windows, of which 8 m^2 are facing south, the rest north. Glazing: three layers of glass with refractive index 1.526, extinction coefficient times glass thickness (the product xL in $P^{t.a.}$) is 0.05. An absorptance $\alpha^{sw} = 0.85$ is used when working with $P^{t.a}$ to specify the rooms behind the windows.

3. Flat-plate solar collector

Area 2, 6 or 40 m^2, tilt angle 45°, azimuth 0° and geographical latitude 55.69°N. Danish Reference Year 1996 version used.

Albedo of ground in front of collectors, 0.2.

Absorbing plates: short-wavelength absorptance $\alpha^{sw} = 0.9$, reflection 0.1, long-wavelength emittance $\varepsilon^{lw} = 0.11$, full transfer between PV and heat absorbing plates.

Efficiency of electricity production by PV layers: 0.15–0.0005 (T_c–20°C).

Cover system: one layer of glass (refractive index 1.526, xL in (4.81) equal to 0.05, emittance $\varepsilon^{lw} = 0.88$).

Heat capacity of collector: 1.4 + 0.8 times number of glass layers (Wh m^{-2} °C^{-1}).

Collector back heat loss coefficient: 0.2 W °C^{-1} m^{-2}.

Collector circuit flow rate times fluid heat capacity, $J_mC_p = 41.8$ W °C^{-1} per m^2 of collector area.

4. Storage facility

Capacity of heat storage given in terms of sensible heat storage in a container holding 0.1, 0.2 or 40 m^3 of water, the latter usually divided into 2 sections of 20 m^3.

Storage loss coefficient, 0.2 or 1.2 W °C^{-1} (surroundings at 8°C for larger stores).

Heat exchange rate between collector circuit and storage (100 W °C^{-1} per m^2 of collector).

Table 6.7. Reference data for simulation of solar system for a Danish one-family house.

The minimum ventilation requirement has been quoted as 8 m^3 h^{-1}/cap. (Ross and Williams, 1975), but may be more if the ventilation air is polluted by e.g. automobile exhaust or if tobacco smoking, open air cooking, fireplace combustion or similar practices take place on the premises. The value quoted would roughly correspond to the smallest values encountered in the Danish measurements, assuming that a family of 4–5 persons occupy the dwelling. Although this minimum of air exchange generously provides the necessary oxygen and removes breathing air CO_2, it may not be sufficient to ensure health and wellbeing. Indeed, when "dust" or other particulate matter is abundantly injected into the air (e.g. from carpets or from particular activities taking place in the building), then increased ventilation is definitely required. Analyses of particulate matter from the air of private dwellings show the largest component is often pieces of human skin tissue. It should also be kept in mind that many building materials release gases (some of which have adverse effects on health) in substantial quantities, for at least the first few years after installation.

The heat required to compensate for the building heat losses may be distributed by a device (radiative and/or convective) with a heat transfer area chosen in such a way that any expected load can be met. In earlier individual or district heating, fuel-based systems, it was customary to work with high inlet temperatures (steam at 100°C, water at 80°C). This was lowered to about 60°C after the first oil crisis in 1973. The minimum working inlet temperature would be about 45°C for a water radiator system, whereas distribution based on air circulation may employ heat supply of temperature around 28°C. The latter choice is made in the present calculations, partly because it is suitable for solar systems aiming at coverage during as large a fraction of the year as possible and partly because efforts to decrease building heat losses through better insulation will make the capital costs of a water distribution system less appealing. For each of the heat exchangers, a temperature loss of 8°C is assumed.

The hot water is assumed to be required at a temperature 42°C above the inlet (waterworks) temperature, which in Denmark is taken as 8°C.

The metabolism of persons present within the house releases heat, corresponding to roughly 125 W per person as an average net heat transfer to the surroundings. The changing occupancy of the house, as well as the lower rate of metabolism during sleep and the higher one during periods of high activity, are modelled by a variation in the gains from persons, assumed to be of an identical pattern every day of the year, as specified in Table 6.7. Also the use of appliances which convert electricity or another form of energy (cookers, washing machines, dishwashers, lights, amplifiers and other electronic equipment, etc.) contributes to indirect heat gains, since the various conversions in most cases result in the release of the entire energy in the form of heat inside the house. The contribution from cooking is reflected in the distribution of the heat releases from appliances on different hours of the day, the assumed pattern being also given in Table 6.7. The gain from lighting, assumed to be dominated by highly efficient bulbs, is assumed to be 100 W during one hour in the morning and eight hours in the evening, except when the outside total solar radiation exceeds 50 W m^{-2}. Again, the use of more efficient equipment including low electricity-consuming light bulbs does have the effect of diminishing indirect contribution to space heating with time.

However, a potentially very large contribution to the gross indirect gains of the house considered is due to solar radiation transmitted through the windows of the house and absorbed in the interior. The magnitude depends on the design of the building, which of course is often different in cold climates from that in regions where shielding against solar heat is desired. Most of the window area has here been assumed to be on the southern side of the building, and, on average, the gain through the windows facing south far exceeds the heat losses due to the temperature difference between inside and outside air. The gains are calculated as if the window glazing were the cover system of a solar collector and as if the inside of the room were acting as an absorber plate of short-wavelength absorptance $\alpha_{sw} = 0.85$ (due to the small fraction of transmitted light which will be multiply reflected on indoor objects and become re-transmitted to the outside through window areas). Indirect gains occurring at times when they are not needed are assumed lost. The surplus gains (or parts of them) would in many be stored as sensible heat in the wall materials, etc., thus potentially providing storage from day to night, but if it is desirable to avoid temperatures higher than the stipulated ones, surpluses may be disposed of by venting. Active storage by air heat exchangers working with a heat store is not considered.

The solar system modelling passes through the steps described above, with the temperature increase in the collector circuit and store being determined iteratively as described, and assuming that the fluid in this circuit is pumped through the collector only when the temperature of the collector has reached that of the energy storage (the collected heat is used to achieve this during a "morning start-up" period) and only if a positive amount of energy can be gained by operating the collector.

A number of simplifying assumptions have been made in formulating the model used, which include the neglect of reduced transparency of the collector cover in case of dirt or dust on the cover surface, the neglect of edge losses from the collector and heat losses associated with special geometry (e.g. where the piping enters and leaves the collector sections) and also of the detailed temperature distribution across the collector, from bottom to top, which may not be adequately described by the averaging procedure used. Another set of complications not included in the simulation model are those associated with possible condensation of water on the outer or inner surfaces of the collector cover system, icing conditions and snow layers accumulating on the top surface of the collector. It is assumed that water condensation conditions, which are expected to occur, especially on some mornings, simply prolong the morning start-up time somewhat, while the collected solar energy provides the evaporation heat as well as the heat required for raising the temperature of the collector. Sample runs of the model with different assumptions regarding the amount of heat required for start-up have revealed that these amounts of heat are anyway very small compared with the total gain, when averaged over the year.

For the purpose of the simulation, the year is divided into 12 identical months, each 730 h long. The monthly averages given below are all based on these artificial months rather than on calendar months. The parameters describing the solar heat system have been summarised in Table 6.7. Most of them are kept fixed, and the variations performed primarily relate to collector size, storage volume and types of operation.

Heat pump systems

The model described above calculates the heat loads not covered by the solar system and evaluates the auxiliary heat required, relative to heat of insufficient temperature from the solar heat stores. For a system with both electricity and heat outputs, it is natural to consider adding a heat pump using as its cold side the solar heat store.

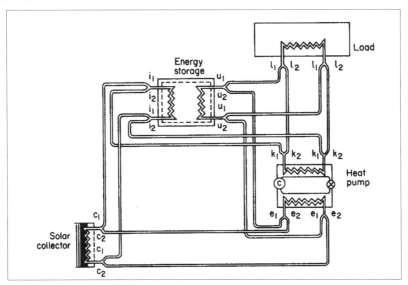

Figure 6.38. Schematic solar heat system with heat pump that may be switched between collector and load side. The main operational modes correspond to opening the following switches indicated on the figure: mode 1: c_1, i_1, u_1 and l_1 (simple operation); mode 2: c_2, e_2, k_1, i_2, u_1 and l_1 (collector cooling); mode 3: c_1, i_1, u_2, e_1, k_2 and l_2 (heat pumping to load) (Sørensen, 1976).

Theoretically, a heat pump may be added to the solar energy system in different ways. Two of these are illustrated in Fig. 6.38. In one, the heat pump is situated between the storage and the collector, in order to increase the temperature difference between inlet and outlet fluid, $T_{c,out}-T_{c,in}$, and thereby improve the efficiency of the collector. In the other mode, the heat pump is inserted between the storage and the load areas, so that it can be used to "boost" the load circuit temperature difference, $T_{L,in}-T_{L,out}$, whenever the storage temperature is below the required minimum value of $T_{L,in}$. The heat pump could use as its low-temperature reservoir the soil or air of the surroundings, rather than the solar storage facility, and thus be an independent energy source only related to the solar heating system by the controls, which determine the times of providing auxiliary heat. The merits of such a system, comparing solar thermal collection with solar electric plus heat pump, are discussed in Sørensen (2000b). The option of using the heat pump to cool the collector fluid has been tested and found unsuitable, because it improves efficiency of collection only when the storage temperature (which in the absence of the heat pump would determine the collector inlet temperature) is high, and in this case the collector output temperature is mostly lower than the store temperature, leading to no heat transfer (although $T_{c,out}-T_{c,in}$ is increased by lowering $T_{c,in}$, $T_{c,out}$ may not be high enough, i.e. above T_s).

For the mode 3 operation considered here as an option, the heat pump compressor may be set into action on the condition that the heat store temperature is below a certain value (e.g. 50°C for hot water, 28°C for space heating). Additional heat exchange losses occur between heat pump and storage (again assumed to entail a temperature difference of 8°C, i.e. the heat pump effectively operates between $T_{up} = T_{load} + 8°C$ and $T_{low} = T_s - 8°C$). The coefficient of performance (COP = heat output over electricity input) is taken by the empirical relation (cf. Fig. 4.105, all temperatures in K)

$$COP^{-1} = (1.8+0.00025(T_{up}-T_{low}-50)^2 + 1000/(T_{up}-T_{low})^3)(T_{up}-T_{low})/T_{up}.$$

Typical COP's for the temperature lifts considered are in the range of 2-5.

Software validation and results of simulations

As examples of hybrid PVT system simulations, three are presented in the following, with 2, 6 and 40 m² of solar collectors, and heat stores ranging from a very small 0.1-m³ water tank to 40 m³. For the largest system, the variant of communal storage common for a number of buildings is considered. Finally, the possibility of supplying auxiliary heat from a heat pump is discussed. However, first the software must be validated.

The NSES software has been tested against an experimentally investigated system similar to the one described in section 6.3.1 (Table 6.7), but making the store so large (10000 m^3) in the calculation, that the collector inlet temperature could be regarded as a constant 30°C. A special set of meteorological input data was used, with ambient temperature constantly at 20°C and wind speed at 5 m s^{-1}, and with solar radiation stepping through a range from zero to maximum magnitude (formally, this was taken as diffuse radiation, because in that way the program would not attempt to calculate incident angle variations for the sample hours). Parameters deviating from the values of Table 6.7 are given in Table 6.8. The resulting solar thermal efficiencies, parametrised in terms of $T^* = (T_{c,in} - T_e)/E_{c+}^{sw}$, are shown in Fig. 6.39.

Solar collector area:	1 m^2
Short-wavelength absorption of PV panel:	0.93
Reflectance of PV panel:	0.07
Long-wavelength emittance of PV panel:	1.00
Plate temperature:	30°C
PV efficiency at plate temperature 20°C:	0.08
Circuit flow times fluid heat capacity:	83.6 W/m^2/K
Back heat loss coefficient:	1 W/m^2/K
Albedo in front of collector:	0
Capacity of heat store:	10 000 m^3

Table 6.8. Parameters changed from the values given in Table 6.7 for the test run of a Racell PVT panel tested empirically.

The most important parameters influencing the normalised efficiency given in Fig. 6.39 are the plate absorptance, emittance, and the back thermal loss. These parameters quoted in Table 6.8 are based on the manufacturer's specifications rather than measured (Bosanac *et al.*, 2001). In cases where they can be compared, such as the test shown in Fig. 6.39, NSES gives results quite similar but not identical to those of the program TRNSYS, which has been in extensive use for many years (Duffie and Beckman, 1991). The water flow parameter indicated in Table 6.8 is fairly high, and the resulting heat exchange is less perfect than that assumed in the simulations below, in keeping with the experimental set-up and the temperature rise found. The back thermal loss is also taken somewhat higher (a factor of 2) than what is indicated by the type of and thickness of insulation. This is because U_{back} is supposed to describe all thermal losses from the collector, including the edge losses that may be substantial for a small collector such as the one tested (actual area 2.2 m^2). Finally, the plate emittance would be expected to be lower than the absorptance (as they would be equal for a black body and the emittance would be lower for a selectively coated surface). The measurements also leave some doubts of the influence of electricity coproduction, as the panel's quoted 8% electric efficiency does not match the under 5% difference between measurements with and without electricity collection. In the calculation shown in Fig. 6.39, the emittance is left at unity, but it should be said that the slope of the curve diminishes quite rapidly with falling ε_p^{lw}. Although the experimental errors are not known, the agreement appears quite good, both between theory and experiments and also between the two software packages. The methodology used in both software packages is highly dependent on *ad hoc* parametrisations derived from other experiments, and they do not give exactly the same results for identical parameters.

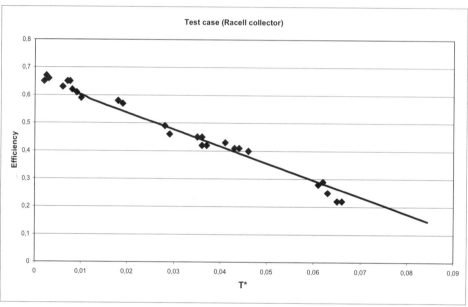

Figure 6.39. NSES efficiency results for test case (solid line) compared with measured data (♦). The effective temperature used is $T^* = (T_{c,in} - T_e)/E_{c+}^{sw}$.

Small PVT system simulation

The smallest system simulated is similar to the Racell prototype used for program verification: the PV collecting panel is 2 m², and the thermal collection takes place under the PV panel by tubes attached to or part of a thermally conducting plate. The thermal storage tank a mere 100-litre water container, with heat exchangers to collector and load circuits. The photovoltaic collection efficiency is taken as 15%, representing typical current commercial silicon modules, and only a linear term is included in the thermal performance degradation of the PV efficiency (cf. Fig. 4.38). The only simulation parameter not discussed above is the initial temperature, which for consistency has to be taken in such a way that it resembles the temperature at the end of the simulation period, when this as here is an entire year. This is not important for the small size stores, as they have no long-term memory.

HEAT PUMP BASICS ARE IN 4.6

The resulting monthly average behaviour is shown in Fig. 6.40, and Figs. 6.41–6.42 show the hour-by-hour behaviour of storage temperature and the requirement for auxiliary heat.

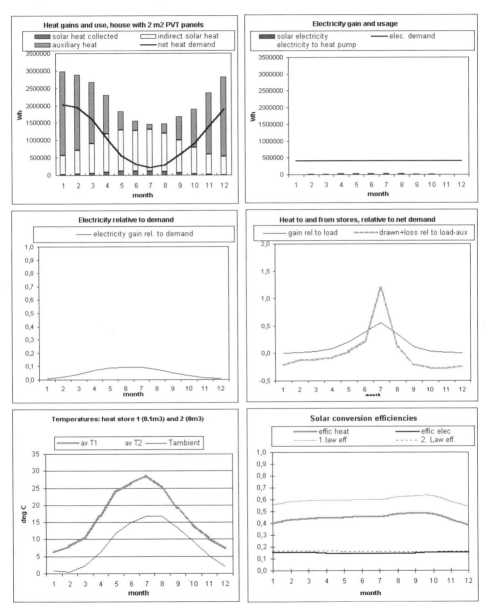

Figure 6.40. Monthly simulation results for a small PVT system. See discussion in text.

The top two panels in Fig. 6.40 show the monthly collection of heat and electricity, and compare it with demands. The heat diagram also shows the indirect heat gains from persons, activities and windows and the auxiliary heat needed in order to cover the total hot water and heating needs. Evidently, the solar system contributes little heat or electricity during winter months. During summer, solar electricity production is still far below building demands, whereas solar heat production covers up to two-thirds of demands. Percentages relative to loads are shown in the middle line of panels, and for heat also relative to the load remaining after application of auxiliary heat input.

The bottom line of diagrams shows the average temperatures of the water tank and of the ambient air and the efficiencies of electric and thermal conversion, plus first and second law thermodynamic efficiencies. The electric efficiency deviates a little from the 15% in downward direction, due to the thermal effect during summer months. The thermal efficiency remains over 40% year-round, the first law efficiency is the sum of electric and thermal efficiencies, and the second low efficiency is only slightly over the electric efficiency, due to the modest temperature rises accomplished by the solar system.

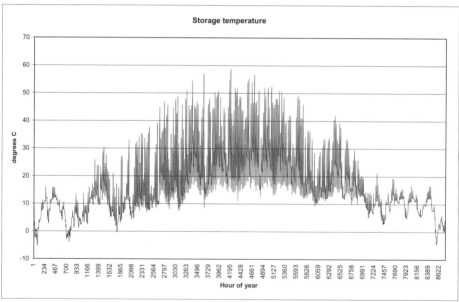

Figure 6.41. Hourly storage temperatures from NSES simulation of a small PVT system.

The simulated hourly course of storage temperatures given in Fig. 6.41 shows that the solar system contributes to space heating during the period April through September (storage temperature above 28°C). The store temperature rarely exceeds 50°C, so some auxiliary energy is needed for hot water year-round, as it was seen in Fig. 6.40. This could be partially avoided by lowering the flow rate of the water passed through the collector circuit, with some associated decrease in thermal efficiency. Figure 6.42 gives the temporal distribution of auxiliary heat added in order to cover the demand at all times. It is seen that there are some days even in winter where the system is able to cover the space heating need.

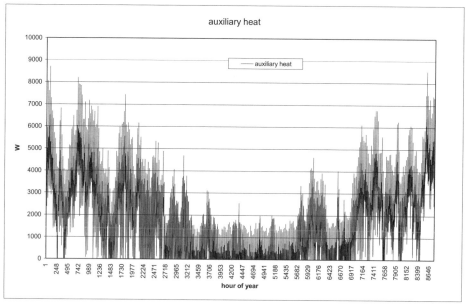

Figure 6.42. Hourly auxiliary heat requirements from NSES simulation of a small PVT system.

Medium-size PVT system simulation

The medium system considered has 6 m² of PVT collectors, but still a very modest heat store (200 litres), because long-term storage is not intended, given the collector size, solar radiation conditions at the location (Denmark), and load. As the middle left entry in Fig. 6.44 indicates, it is now possible to cover over 30% of the electricity load during the months of May, June and July. Also for heat, the coverage in summer is high (roughly 100% in July), due to the higher summer storage temperature (see Fig. 6.46), relative to the small system considered in the previous subsection. From May to October 1, the storage temperature oscillates between 15 and 80°C.

The 15% electricity production makes the heat collection efficiency (bottom line, Fig. 6.44) almost constantly 15% below that of a pure thermal system of the same size, the monthly simulation results of which are shown in Fig. 6.43. During summer, the system without electricity production has a higher average store temperature (in July 60°C as compared to 50°C), whereas during winter months there is no difference. The thermal efficiency of the PVT system is, on average, 35%, with small variations: during spring and autumn, 40% is reached, whereas during the summer, a minimum of 30% is coinciding with the peak store temperature.

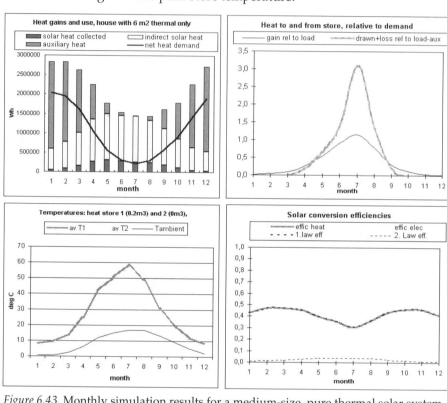

Figure 6.43. Monthly simulation results for a medium-size, pure thermal solar system.

Figure 6.45 shows the solar thermal gain for each hour of the year and also the energy gain by the heat store, which is a little less, due to incomplete transfer through the heat exchangers (solar absorber to working fluid and working fluid to store). Figure 6.46 gives the hourly store temperatures, which are more appropriate for the heating applications than those of the small system considered in Fig. 6.41. As a consequence, the need for auxiliary heat, shown in Fig. 6.48, is reduced, particularly during the summer period. Figure 6.47 shows the net heat drawn from the store to the load circuits (hot water and air duct heating), including heat losses through the store insulation. During winter, the net heat drawn is seen to be often negative. This is when auxiliary energy is used to satisfy the load. The auxiliary energy brings the amount of hot water needed up to the required temperature and likewise the amount of air circulated between the heat store and load areas. The return air and water (after the load heat exchange) may in winter have temperatures above that of the store and thus actually help to increase the store temperature. The implied "net solar gain to store" is therefore negative, as non-solar heat is used to heat the store.

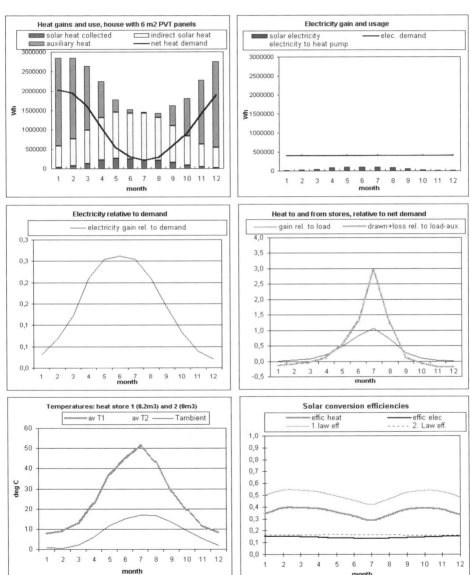

Figure 6.44. Monthly simulation results for a medium-size PVT system. See discussion in text.

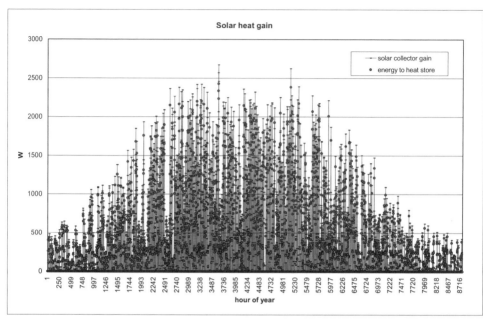

Figure 6.45. Hourly solar heat gains from simulation of a medium-size PVT system.

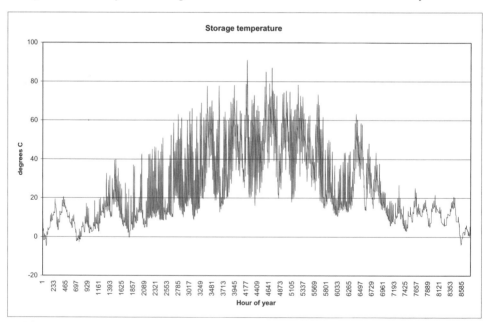

Figure 6.46. Hourly storage temperatures from simulation of a medium-size PVT system.

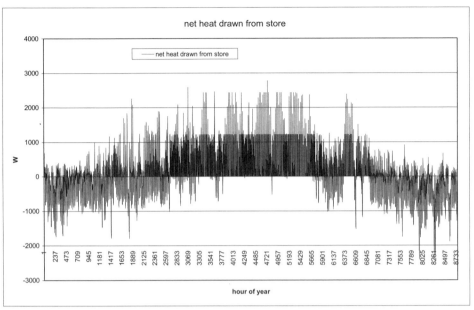

Figure 6.47. Net heat drawn hourly from store from simulation of a medium-size PVT system.

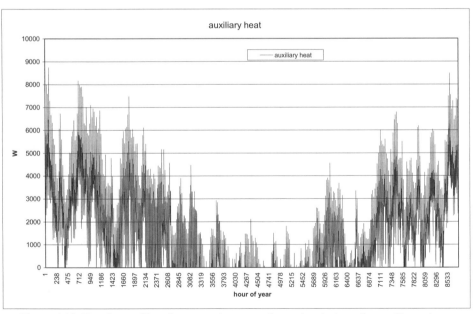

Figure 6.48. Hourly auxiliary heat requirements from simulation of a medium-size PVT system.

Large PVT system simulation

The large system considered has 40 m² of PVT collectors and 40 m³ of water storage. This may be divided into two, e.g. hemispherical parts with a common interface, to simulate stratification. The input to the collector is from the lower store, and the heat to load is extracted from the upper part, provided it has a temperature higher than the lower one. The water passing through the solar collector is delivered at the top of the upper store, where a heat exchanger spiral is continuing down to the lower store. This means that if the temperature of the water heated by the solar collector exceeds that of the upper store, it delivers energy to that store and reaches the lower store at a diminished temperature, possibly allowing it to pass further heat to the lower part. This is not meant as a substitute for a more detailed simulation of stratification, which is not done here.

The input parameters are as in Table 6.7, but with two storage compartments of 20 m³ of water and heat loss coefficients (*U*-values) of 0.55. The heat loss from the stores is diminished to about a tenth of the value that would be proper for a single store with 20–30 cm of insulation, because it is supposed that some 100 buildings share a common store, in which case the surface-to-volume advantage allows this kind of reduction (cf. the simulations made in Sørensen 1979; 2000a). The program has been instructed to perform a 1% mixing of the upper and lower store for each iteration step (here 1 h), as a crude model of the stratification of a single storage volume. The resulting storage average store temperatures are indicated in the lower line of Fig. 6.49 and in more detail in Fig. 6.51. It is seen that the mixing does not allow temperature differences between top and bottom to persist for very long.

The top diagrams of Fig. 6.49 show that electricity needs are covered from late March to early September, but that solar heat is only able to satisfy demand without assistance from late March to the end of August. There is a strong asymmetry between the behaviour in spring and autumn, which is directly caused by the temperature variations of the store and thus the collector circuit inlet temperature. The store temperature remains high until November, which causes the thermal solar collection to perform poorly, while on the other hand the store is capable of satisfying heat demands from heat collected earlier, as shown in the middle, right-hand diagram: there is no need for auxiliary heat from May to November. This proves that the store is large enough to act as a seasonal store, but on the other hand, there is a very negative effect on the solar thermal collection during the later months of the year. This is borne out very clearly by the efficiency curves on the lower right.

The hourly solar gain shown in Fig. 6.50 supports this interpretation, showing a clear asymmetry between spring and autumn, with a very poor performance towards the end of the year. The hourly course of storage temperature (1 is lower store, 2 is upper) is shown in Fig. 6.51. It peaks in

August, and has its minimum value in February, i.e. displaced by 1–2 months relative to the solar radiation. Due to the size of the store, there are no rapid variations as for the store temperature of the smaller systems shown in Fig. 6.41 and 6.46.

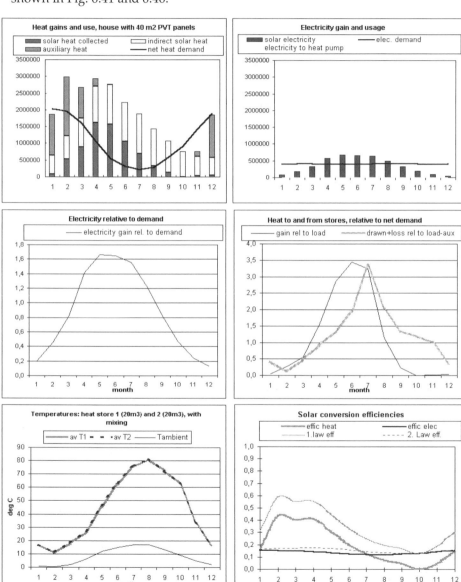

Figure 6.49. Monthly NSES simulation results for a large PVT system. See text for details.

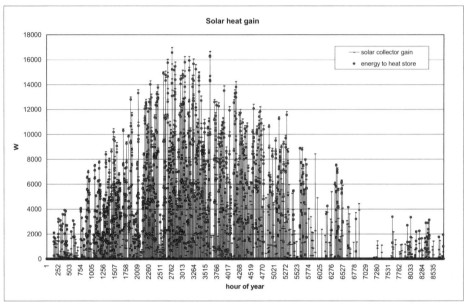

Figure 6.50. Hourly solar gains from simulation of a large PVT system.

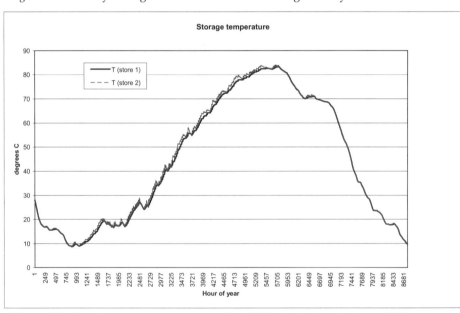

Figure 6.51. Hourly storage temperatures for NSES simulation of a large PVT system.

The net heat drawn from the (two) stores is given in Fig. 6.52. The negative values explained above are now confined to a few weeks in January and

February. Finally, Fig. 6.53 gives the necessary auxiliary heat inputs, which are still substantial during winter, although the solar heating system now satisfies all loads from May to November.

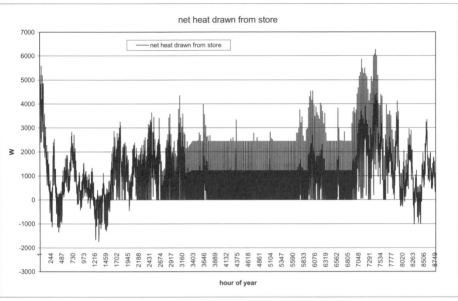

Figure 6.52. Net heat drawn hourly from the two stores of the large PVT system.

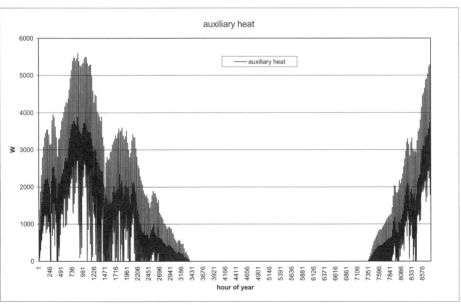

Figure 6.53. Hourly auxiliary heat requirements for the large PVT system.

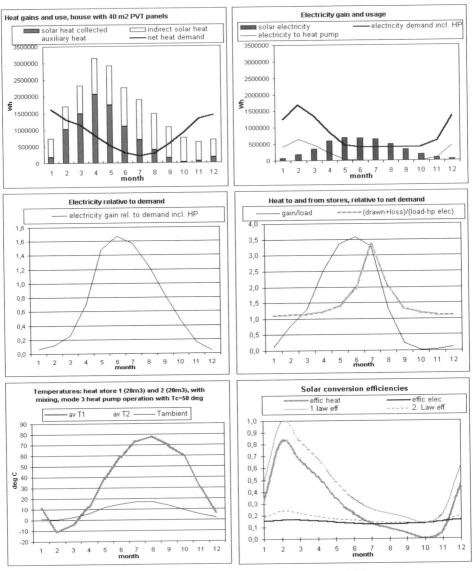

Figure 6.54. Monthly simulation results for large PVT system with heat pump operating in mode 3 with control temperature 50° C (see text for details).

PVT systems with heat pump

The final set of calculations adds a heat pump operating in mode 3 (see the discussion associated with Fig. 6.38) to the large PVT system. In mode 3, the heat pump operates on the heat store as its lower temperature reservoir and delivers heat at appropriate temperatures for covering space heating and hot water needs. Setting the control temperature of the heat pump to 50°C, it is possible to deliver all auxiliary heat from the heat pump system. The monthly system performance curves are shown in Fig. 6.54. In the electricity diagram (upper right), the electric energy inputs to the heat pump are indicated. At the peak they increase the electricity demand by nearly 70%.

Looking at the store temperature curves (lower left), it is seen that the heat pump during February and March cools the store down to a temperature below the ambient one. As a consequence, the apparent total efficiency (lower right) exceeds 100%, because heat from the surroundings is drawn into the heat store. Figure 6.55 shows how the net energy drawn from the store now always is positive, and in order to make the distinction between hot water and space heating coverage easier, the space heating demand is shown separately in Fig. 6.56.

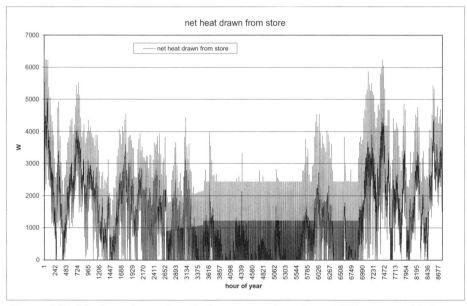

Figure 6.55. Net heat drawn hourly from stores for a large PVT system with heat pump operating in mode 3 with control temperature 50°C. No other auxiliary heat is needed.

The store temperatures below 0°C are not quite realistic, partly because a water store has been assumed, so that large quantities of anti-freeze additives are required (in addition to the small amounts used for the fluid in the collector circuit, due to occasional operation of the collector in situations of solar radiation and low temperatures), and partly because having a large cold store may increase the need for electric energy to the heat pump (if the store is within the building) or make district heating transmission more diffi-cult (frost in pumps, etc.).

For these reasons, an alternative heat pump run is made, where the con-trol temperature is set at 29°C. This means that the heat pump will take care of all auxiliary energy for space heating, but not for hot water. The summary results are shown in Fig. 6.57. The store temperature now stays above 7°C (also in the detailed hourly output), and there is no longer any anomaly in the efficiency curves. It is seen on the upper right diagram that the electric energy to the heat pump now is modest, but (in the upper left diagram) that some direct auxiliary heat has to be provided during winter months.

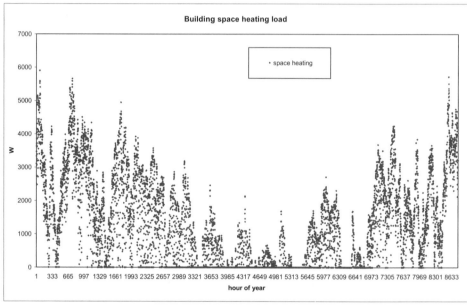

Figure 6.56. Hourly space heating demand (used for all simulation cases).

6.3.2 Wind electricity systems

Figures 6.58 and 6.59 gives the variation in wind power production for a large portion of the currently installed wind turbines in Denmark. Such data can be used for system simulation and inquiry into questions of surplus and deficit management through import/export or through storage. It is seen that for turbines on land, maximum output peaks over four times the aver-age are found during winter months, whereas for off-shore plants, the maximum is only 2.3 times the average. This may indicate a greater resource stability over water, but compounded by the likelihood that the power curve of the turbines used off-shore is not approximating that leading to maximum annual production as well as the turbines used on land. This design choice is illustrated in Fig. 6.61. Figure 6.60 shows the power duration curves for wind turbines placed on land and off-shore in Denmark, as well as a combined curve for the on-shore/off-shore mix of a particular scenario for future wind coverage (Sørensen *et al.*, 2001). The latter is seen to produce power during every hour of the year.

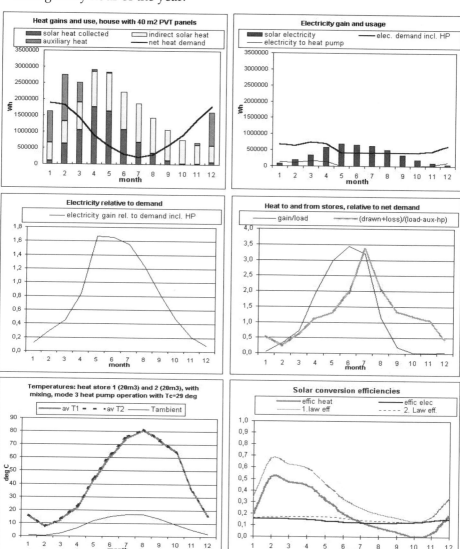

Figure 6.57. Monthly NSES simulation results for a large PVT system with heat pump operating in mode 3 with control temperature 29°C (see text for details).

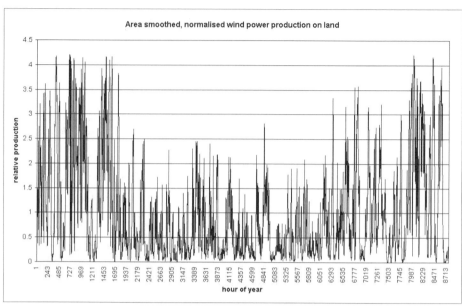

Figure 6.58. Hourly variation in wind power production during the year 2000 for some 3000 Danish wind turbines placed on land. The overall annual average power output is normalised to one (Ravn, 2001; Eltra/Elkraft, 2001).

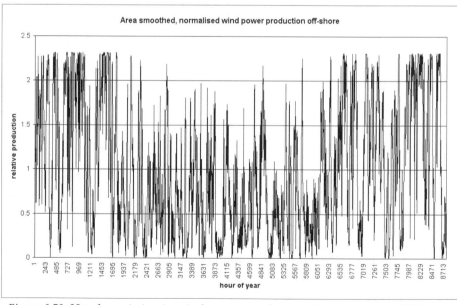

Figure 6.59. Hourly variation in wind power production during the year 2000, normalised to annual average, for about 100 Danish wind turbines placed off-shore (Ravn, 2001; Eltra/Elkraft, 2001).

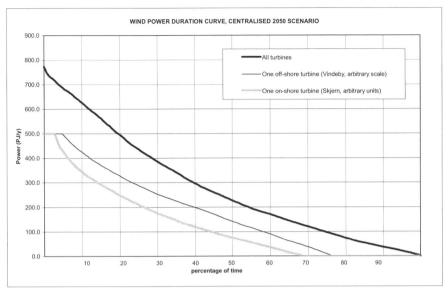

Figure 6.60. Power duration curves for wind power produced at a single site on land or off-shore in Denmark and for a combined scenario for future generation on many sites, with substantial expansion of the off-shore wind contribution, but only use of on-shore sites already hosting a wind turbine today (Sørensen *et al.*, 2001).

An example of how the power distribution of a propeller-type wind energy converter depends on design is shown in Fig. 6.61. The highest annual power production is obtained with the design (a) and not design (c), which peaks at the average wind speed of 10 m s^{-1}, due to the asymmetry of the curves. The assumed hub height is 56 m, and 1961 data from Risø are used (cf. Fig. 3.34). Several of the calculations described in this section (taken from Sørensen 1976c, 1977, 1978a, b) use these data and remarkably agree with newer results, notably because hub heights of 50–60 m are the actual standard for present commercial wind turbines.

System without energy storage

Power duration curves (or "time duration curves for power") can be constructed using detailed hour-by-hour simulation, or even without it, provided that the wind speed data exist in the form of a frequency distribution (such as the dashed curves in Fig. 3.38). Once a storage facility or an auxiliary conversion system has to be operated together with the wind energy converters, the system simulation requires the retention of the time sequences of the meteorological data.

For a system without energy storage, there may be an advantage in choosing a power coefficient curve that peaks well below the peak of the frequency distribution of the power in the wind, rather than just maximising annual production.

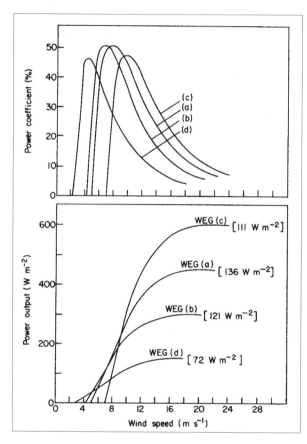

Figure 6.61. Power coefficients (per cent of power in the wind) are given as a function of wind speed for four typical designs of wind energy generators (cf. Fig. 4.70), with the corresponding power output (below). The 1961 average power production at Risø, Denmark (height 56 m) is given in square brackets for each type of design. (Sørensen, 1978a).

Figures 6.60 and 6.62 give examples of power duration curves that illustrate the advantage of dispersing the locations of turbines. In Fig. 6.60, the substantial smoothing obtained for the Danish sites currently in use, supplemented with continued build-up at off-shore locations, is seen to significantly stretch the power duration curve. A similar conclusion can be drawn from Fig. 6.62, in this case for 18 locations in Denmark, Ireland and Germany. If the wind speed frequency distribution function is denoted $h(u) = df(u)/du$, where u is the scalar wind speed and $f(u)$ is the accumulated frequency of wind speeds less than or equal to u, then the average power production is given by

$$E = \int \tfrac{1}{2} \rho\, C_p(u) u^3 h(u) du \qquad (6.3)$$

WIND ENERGY BASICS ARE IN 3.2, 4.3

for a wind energy converter with sufficient yawing capability to make the $\cos \theta$ factor in (3.24) insignificantly different from unity during all (or nearly all) the time covered by the integral in (6.3). The power duration curve is then

$$F(E) = \int \delta \left(\tfrac{1}{2} \rho \, C_p(u) u^3 \geq E\right) h(u) \, du, \tag{6.4}$$

where the δ function is unity if the inequality in its argument is fulfilled and zero otherwise. If the data are in the form of a time series, the integration may be made directly over dt (t is then time, in units of the total interval being integrated over) rather than over $h(u)du$. $F(E)$ gives the fraction of time [i.e. of the time period being integrated over in (6.4)], during which the power generation from the wind energy converter exceeds the value E.

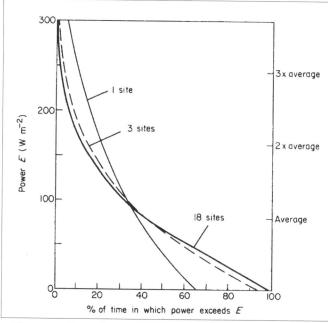

Figure 6.62. Annual power duration curves for wind energy systems with converters at 1, 3 or 18 different sites, based on 1973 data from West German weather stations, taken at the height 10 m but extrapolated to 72 m. The assumed power coefficient curve is similar to type (b) in Fig. 6.61 (based on Molly, 1977).

The coupling of just three units, at different locations in Germany (typical distances about 240 km), is seen in Fig. 6.62 to imply practically as much improvement as the coupling of 18 units. The reason is that the maximum distances within Germany are still not sufficient to involve the frequent occurrence of entirely different wind regimes within the area, and thus the general meteorological conditions are often the same for all the converters except for minor differences which can already be taken advantage of by selecting three different locations. In order to take advantage of more different meteorological front system patterns, distances of 1000 km or more are required. A study of the synergy of 60 European sites made by Giebel (1999) still finds variations in wind power production substantially at variance with demand, indicating a fundamental need for back-up or storage.

Regulation of a fuel-based back-up system

Consider now an existing fuel-based electricity-producing system, to which a number of wind energy converters is added with the purpose of saving fuel. It is conceivable that this will increase the requirements for regulation of the fuel-based units and lead to an increased number of starts and stops for the fuel-based units. Since most of these units will not be able to start delivering power to the grid at short notice, the decision to initiate the start-up procedure for a given unit must be taken some time in advance, based on expectations concerning the load and the fraction of load which the wind energy converters may be able to cover.

The time required for start-up of a conventional fuel-based unit is associated with pre-heating the boiler and turbine materials and thus depends on the time elapsed since the last usage. Traditionally, for start-up times less than 7 h, the start is called a "warm start". Start-up times are generally diminishing due to new technology, so variations between plants of different age is often large, with quoted start-up times being 20–60 min for coal-fired base and intermediate load units, 1–7 h for pressurised water fission reactors and 5–6 h for boiling water fission reactors (Nordel, 1975). For fission reactors, "warm starts" assume start from an under-critical reactor with control rods in, and operating pressure. If the fuel-based units have cooled completely since the last usage, start-up times are 0.6–6 h for large coal plants, 18 h for a pressurised water reactor and 14 h (to reach 90% of the rated power) to 48 h (full power) for a boiling water reactor. These start-up times are representative, and actual times may be smaller for some newer plants.

Now the question is to what extent the power production of the wind energy converters can be predicted in advance by time intervals corresponding to the different start-up times for the relevant fuel-based units. Two types of prediction may be considered, one being actual forecasts based on (mental or mathematical) models of the dynamics of general circulation and the other being associated with analysis of the statistical behaviour of wind patterns, possibly including correlations that make simple methods of extrapolation permissible.

Standard meteorological forecasting, as practised routinely by meteorological services throughout the world, may be very useful in predicting storms and thus identifying periods of very high wind power production, with advance notice usually the same as that required by the quoted start-up times for fuel-based units. However, forecasts at present are usually unable to specify very precisely the expected wind speed, if this is in the critical range of 5–10 m s^{-1} covering the very steep part of the converter performance curve from zero to quite high production (cf. Fig. 6.61).

Therefore, it may be worth taking advantage of the correlations present in sequences of wind production data and in wind speed data, such as those implied by the spectral decomposition (see e.g. Fig. 3.37). The continuity of the general circulation and the nature of the equations describing the atmospheric motion (section 2.3.1) combine to produce sequences of wind energy production figures, which exhibit regularities of seasonal and diurnal scale as well as regularities of 3–5 days duration, associated with the passage of mesoscale front systems over a given geographical location. In the following, models based on trend extrapolation of power production sequences, on models adding elements of stochastic fluctuations, and on climate modelling based physical forecasting will be described.

Simple extrapolation model

Figure 6.63 shows a sequence of hourly, sequential production data for a wind energy converter [power coefficient (a) in Fig. 6.61] placed at the Risø turbine test site 56 m above the ground. After a period of insufficient wind (insufficient to start the wind energy converter), the average production rises, goes through a peak and decreases again in a matter of about 4 days. Then follows a period of small and fluctuating production. Also during the 4 days of large production, hourly fluctuations are superimposed on the general trend. However, it is evident that a power prediction simply saying that the production will be high, modest or near zero on a given day, if the power production was high, modest or low, respectively, on the previous day, will be correct quite often, and similarly the production during a given hour is strongly correlated with the production for the previous hour.

An extrapolation model may then be tried which bases the decision to initiate the start-up procedures for fuel-based units on some (possibly weighted) average of the known wind energy production data for the time period immediately before.

For simplicity, the fuel-based supply system will be assumed to consist of a number of identical units, each with a maximum power generation which is a tenth of the average load, E_{av}, for the entire grid of the utility system considered. It is assumed that the units can be regulated down to a third of their rated power, and that some are operated as base load plants (i.e. at maximum power whenever possible), while others are operated as intermediate load plants (i.e. at most times allowing for regulation in both upward and downward direction). The start-up time for the base load units is assumed to be 6 h, and once such a unit has been started it will not be stopped for at least 24 h. The start-up time assumed for the intermediate load units is 1 h. For simplicity, the load is factorised in the form

$$E_{load} = E_{av}C_1C_2, \tag{6.5}$$

where E_{av} is the annual average load, C_1 is the seasonally varying monthly averages (such as those shown in Fig. 6.13), and C_2 is the average diurnal variations (such as those shown in Fig. 6.15). Both C_1 and C_2 are normalised so that the annual average implied by (6.5) is E_{av}. The approximation involved in (6.5) implies neglect of seasonal variations in diurnal load profiles, as well as weekday/holiday differences, and is used only for the purpose of illustration. The actual hourly data is used in the forecasting subsection below.

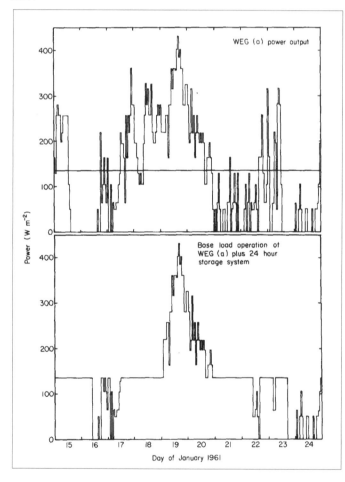

Figure 6.63. Hourly output from wind energy system without (top) and with short-term storage (bottom) for a 10-day period in January 1961. The wind energy converters are assumed to be of type (a) (Fig. 6.61), 56 m Risø data are used, and the 24 h energy storage is assumed to be loss-free and operated in a base load mode (Sørensen, 1978a).

In the first set of calculations to be discussed, the extrapolation of wind power production will be based on two averages of past production. One is the average production of the preceding 24 h, A_{24}, which is used once every 24 h in conjunction with the minimum value of the expected load (6.5) for the following day to decide on the number n_{base} of base load fuel-fired units to start. If this number is different from the number already in use, the change will become effective 6 h after the decision is made, according to the assumed base load start-up time. Thus,

$$n_{base} = \left[\frac{E_{av} C_1 \min(C_2) - A_{24}}{0.1 E_{av}} \right],$$

where the square bracket denotes integer truncation.

The other wind power average used in the regulation of the utility system is the average wind energy production in the previous 4 h, A_4, which is used every hour in conjunction with the expected load for the following hour to make a trial estimate of the required number of intermediate load units, n_{int}^{trial},

$$n_{int}^{trial} = \text{integer} \left(\frac{E_{av} C_1 C_2 (t + 1h) - A_4}{0.1 E_{av}} \right) - n_{base} + 2,$$

where the operation "integer" rounds to the closest integer number. This trial value is two units larger than the minimum value indicated by the predicted wind power production, in order to make room for regulation of the intermediate load units in both directions if the wind power prediction should turn out to be inaccurate. The value of additional units is chosen as 2, because this is the number that for the Danish power system gives the smallest number of start-ups per year for the fuel-based units, in the absence of wind energy contributions. If the trial value, n_{int}^{trial}, of intermediate load units for the following hour exceeds the number already in use, a number of start-ups are initiated, so that the number of units actually operating 1 h later will be $n_{int} = n_{int}^{trial}$. If the number indicated by n_{int}^{trial} is smaller than the one already operating, it is not necessarily implemented, but rather the units already in operation are regulated downwards. If it can be calculated that the regulation of all operating units down to the technical minimum level (assumed to be a third of the rated power) is insufficient, then one unit is stopped. It has been found advantageous on an annual basis not to stop more than one unit each consecutive hour, even if the difference between the minimum power production of the operating units and the expected demand will be lost. This is because of the uncertainty in the wind power predictions, which in any case leads to losses of energy from the fuel-based units every time the wind power production exceeds expectation by an amount which is too large to be counterbalanced by regulating the fuel-based units down to their technical minimum level (in this situation it is also permitted to regulate the base load units). Not stopping too many of the intermediate load units makes it easier to cope with an unexpected occurrence of a wind power production below the predicted one during the following hour.

There will be situations when the wind power deficit is so large that for a given hour it cannot be covered even by regulating all fuel-based units in operation to maximum power level. In this case, peak load units must be called in. They can be fuel-based units with very short start-up times (some gas turbines and diesel units), hydropower units (with reservoirs) or "imports", i.e. transfers from other utility systems through (eventually international) grid connections. In a sufficiently large utility pool, there will always be units somewhere in the system which can be regulated upwards at short notice. The role of international exchange in regulating a wind power system has been investigated by Sørensen (1981a) and by Meibom et al. (1997, 1999). A further option is, of course, dedicated energy storage (including again hydro reservoirs).

All of these peak load possibilities are in use within currently operated utility systems, and they are used to cover peaks in the expected load variation, as well as unexpected load increases, and to make up for power production from suddenly disabled units ("outage").

Summary of simulation results

Based on the one-year simulation of the fuel-based system described above, with an installed capacity I_{wind} of wind energy converters added, a number of quantities are calculated. The installed wind energy capacity I_{wind} is defined as the ratio between annual wind energy production from the number of converters erected, E_{wind}^{prod} given by (6.3), and the average load,

$$I_{wind} = E_{wind}^{prod} / E_{av} . \tag{6.6}$$

The quantities recorded include the number of starts of the base and intermediate load units, the average and maximum number of such units in operation, and the average power delivered by each type of unit. For the peak load facilities, a record is kept of the number of hours per year in which peak units are called and the average and the maximum power furnished by such units. Finally, the average power covered by wind is calculated. It may differ from the average production I_{wind} by the wind energy converters because wind-produced power is lost if it exceeds the load minus the minimum power delivered by the fuel-based units. If the annual average power lost is denoted E_{lost}, the fraction of the load actually covered by wind energy, C_{wind}, may be written as

$$C_{wind} = \frac{E_{wind}^{prod} - E^{lost}}{E_{av}} . \tag{6.7}$$

Figure 6.64 shows the average number of base and intermediate load unit start-up operations per unit and per year, as a function of I_{wind}. Also shown is the average number of base and intermediate load units in operation. The maximum number of base load units operating is six, and the maximum number of base plus intermediate load units is 15.

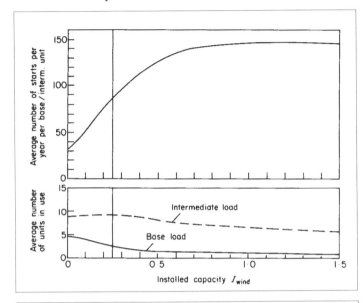

Figure 6.64. Start-up requirements and average number of fuel-based units in operation, as a function of installed wind capacity in a combined system without storage possibilities (Sørensen, 1978b).

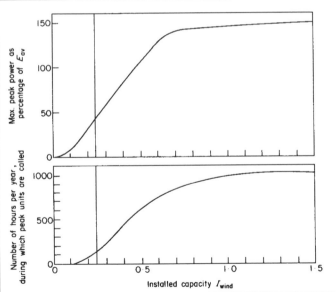

Figure 6.65. The number of hours requiring utilisation of peak units (or imports) and the maximum amount of peak power required during the year for a combined fuel- plus wind-based electricity-producing system without storage possibilities (Sørensen, 1978b).

Up to about $I_{wind} = 0.3$, the wind energy converters are seen to replace mainly base load units. The number of starts in the absence of wind energy converters is, on average, 35 per unit, rising to 98 for $I_{wind} = 0.3$ and saturating at about 144.

In Fig. 6.65, the number of hours during which peak units are called is given, together with the maximum power drawn from such units in any hour of the year.

In the simulation, both of these quantities are zero in the absence of wind energy converters. This is an unrealistic feature of the model, since actual fuel-based utility systems do use peak load facilities for the purposes described in the previous subsection. However, it is believed that the additional peak load usage excluded in this way will be independent of whether the system comprises wind energy converters or not.

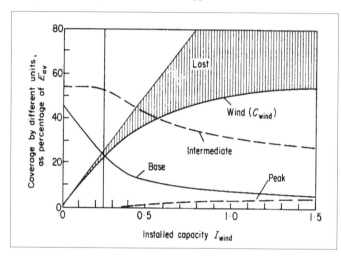

Figure 6.66. Fractional coverage of the annual average power required (total annual load) by wind and three types of fuel-based units for a combined system without storage. The wind energy production not used to cover loads is indicated by the shaded region "lost" (from Sørensen, 1978b).

The number of peak unit calls associated with inappropriate wind predictions rises with installed wind capacity. For $I_{wind} = 0.25$, it amounts to about 140 h per year or 1.6% of the hours, with a maximum peak power requirement which is about 40% of E_{av}. The asymptotic values for large I_{wind} are 1000 h per year (11.4%) and 155% of E_{av}.

The total amount of energy delivered by the peak units is not very large, as seen in Fig. 6.66. For I_{wind} up to about 0.4, it is completely negligible, and it never exceeds a few per cent of the total annual load for any I_{wind}. This is considered important because the cost of fuels for peak load units is typically much higher than for base or intermediate load units.

In Fig. 6.66, the fractional coverage by wind, base, intermediate and peak units is shown as a function of I_{wind}, and the fraction of produced wind energy which is lost, $I_{wind}-C_{wind}$, is indicated as a shaded area. Up to an installed wind energy capacity of about $I_{wind} = 0.1$, the lost fraction is negligible, and up to I_{wind} around 0.25, it constitutes only a few per cent. For larger I_{wind}, the lost fraction increases rapidly, and the wind coverage, C_{wind}, actually approaches a maximum value of about 60%. Figure 6.66 also indicates how the wind energy converters first replace base load units and only for larger I_{wind} affect the coverage by intermediate load units.

Stochastic model

The regulation procedures described have been quite rigorous, and one may ask how much the system performance will be altered if the prescriptions are changed. By modifying the prescription for choosing n_{base}, the relative usage of base and intermediate load units may be altered, and by modifying the prescription for choosing n_{int}, the relative usage of intermediate and peak load units can be changed. The prescriptions chosen for n_{base} and n_{int} give the smallest total number of starts of base plus intermediate units for zero wind contribution ($I_{wind} = 0$).

Also amenable to modification is the extrapolation procedure used to predict the future wind energy production. Rather than using plain averages over the preceding 4 and 24 h (A_4 and A_{24}), one may try using a weighted average, which would emphasise the most recent hours more strongly. A number of such weighted averages were tried, of which one is illustrated in Fig. 6.67. Here A_{24} was replaced by ($0.8A_{24} + 0.2A_4$) in the expression for n_{base}, and A_4 was replaced by ($0.8A_4 + 0.2A_1$), A_1 being the wind energy production for the preceding hour, in the expression for n_{int}^{trial}. The resulting changes are very modest, for this as well as for other combinations tried, and no clear improvements over the simple averages are obtained in any of the cases.

One may then ask if knowing the recent history of wind power production is at all important for regulating the fuel-based back-up system. To demonstrate the non-random nature of the hour-to-hour variation in wind power production, the predictive model based on previous average production was replaced by a stochastic model, which used the power production at a randomly selected time of the year, each time a prediction was required.

The basis for this procedure is a power duration curve [such as the ones in Fig. 6.60; however, for the present case the Risø curve obtained by combining Fig. 3.40 and Fig. 6.61(a) was used]. A random number in the interval from zero to one (0–100%) was generated, and the power level prediction was taken as the power E from the power duration curve which corresponds to the time fraction given by the random number. In this way, the stochastic model is consistent with the correct power duration curve for the location considered and with the correct annual average power. The procedure outlined was used to predict the power generation from the wind, to replace the averages A_{24} and A_4 in the expressions for n_{base} and n_{int}^{trial}.

The results of the stochastic model (c) are compared to those of the extrapolation type model (a) in Fig. 6.67. The most dramatic difference is in the number of start-ups required for the fuel-based units, indicating that a substantial advantage is achieved with respect to regulation of the fuel-based system by using a good predictive model for the wind-power contribution and that the extrapolation model serves this purpose much better than the model based on random predictions.

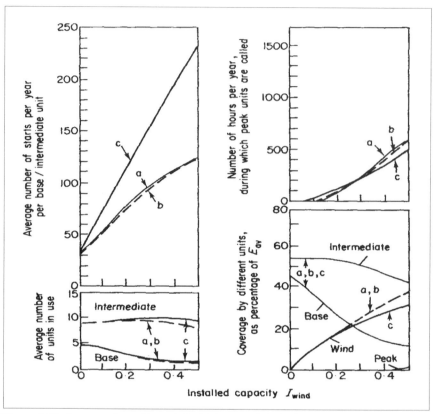

Figure 6.67. Excerpts from Figs. 6.64–66 (curves labelled a), compared to two other calculations of the performance of a combined fuel/wind energy system. Dashed curves (b) represent a calculation where the extrapolation type wind energy predictions are based on previous average production data with a higher weighting of the most recent hours; heavy lines (c) represent a calculation where the wind power predictions are not extrapolated, but randomly chosen within a stochastic model described in the text (Sørensen, 1978b).

Some differences are also seen in the requirements for peak units, which are smaller for the extrapolation type model up to $I_{wind} = 0.3$ and then larger. The latter behaviour is probably connected with the lowering of the total wind coverage (bottom right in the figure) when using the stochastic model. This means that the number of fuel-based units started will more often be wrong, and when it is too high, more of the wind-produced power will be lost, the wind coverage hence will be lower, and the need for auxiliary peak power will be less frequent. The reason why these situations outnumber those of too few fuel-based units started, and hence an increased need for peak power, is associated with the form chosen for the n_{int} regulating procedure. If this is the explanation for the lowering of the number of peak unit calls in the stochastic model for I_{wind} above 0.3 (upper right in Fig. 6.67), then the fact that the number of peak unit calls is lower for the extrapolation-type model than for the stochastic model in the region of I_{wind} below 0.3 implies that the extrapolation model yields incorrect predictions less often.

The economic evaluation of the requirements for additional start-ups, increased utilisation of peak load facilities, etc., of course, depends on the relative cost of such operations in the total cost calculation. For a particular choice of all such parameters, as well as fuel costs and the capital cost of the wind energy converters, a maximum installed wind capacity I_{wind} can be derived from the point of view of economic viability (cf. section 7.5.1).

Wind forecasting model

For the reasons explained in Chapter 2, general weather forecasts become inaccurate for periods longer than a few days, with the exact period depending on the strength of coupling between large-scale and chaotic atmospheric motion. For short periods, say, under 12 h, the extrapolations methods described above in connection with time delays in starting up back-up power plants are expected to perform as well as meteorological forecasts based on general circulation models. Thus, the role of wind forecasting for power systems with wind power is in the intermediate time range of 1–2 days. The interest in this interval has recently increased, due to establishment of power pools, where surpluses and deficits in power production are auctioned at a pool comprising several international power producers interconnected by transmission lines of suitable capacity. The Nordic Power Pool currently functions with a 36-h delay between bidding and power delivery. Therefore, the possibility of profitably selling surplus wind power or buying power in case of insufficient wind depends on the ability to predict wind power production 36 h into the future.

The quality of wind power predictions, even for as little as 12 h, using the circulation models currently used in 1–5 day forecasts, is illustrated in Fig. 6.68. It clearly reflects the problem caused by non-linearity, as explained above. Yet, for the 36-h forecasts needed for bidding at the power pool, the average error of the forecasts of wind power production (at Danish wind farms) is only 60% of that of extrapolation values (Meibom *et al.*, 1997; 1999). The effect that this has on pool bidding is illustrated in Figs. 6.69 and 6.70 for a model in which the Danish power system is 100% based on wind power, while the other Nordic countries primarily use hydropower, as today. Figure 6.60 shows the amount of wind power buying and selling, after covering own needs, for Eastern and Western Denmark. Fig. 6.70 similarly gives the power duration curve for deviations of actual wind power delivery from the bid submitted 36 h earlier. In terms of cost, using current trading prices at the Nordic Pool, the incorrect bids will imply a monetary penalty amounting to 20% of the revenue from pool trade (this depends on a facility called the adjustment market, where additional bids for short-notice "regulation power" are used to avoid any mismatch between supply and demand). No other energy storage systems could presently provide back-up for wind power at such a low cost.

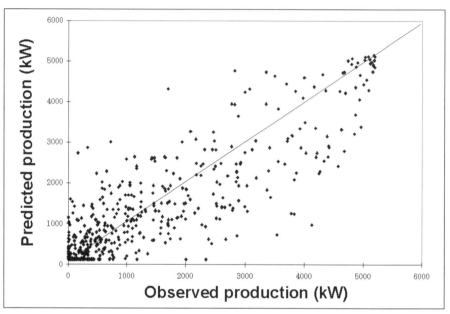

Figure 6.68. Correlation between 12 h meteorological forecasts of and observed wind power production at Danish wind turbine park (based on Landberg *et al.*, 1999).

Systems with short-term storage

All of the storage types mentioned in section 5.2.2 may be utilised as short-term stores in connection with wind energy systems. The purpose of short-term stores may be to smooth out the hourly fluctuations (the fluctuations on a minute or sub-minute scale being smoothed simply by having the system consist of several units of modest spacing). A back-up system must still be available if the extended periods of insufficient wind (typical duration one or a few weeks) are not to give rise to uncovered demands.

If the inclusion of short-term stores is to be effective in reducing the use of fuel-based peak load facilities, it is important to choose storage types with short access times. It is also preferable for economic reasons to have a high storage cycle efficiency. This may be achieved for flywheel storage, pumped hydro storage and superconducting storage, according to Table 5.5.

A wind energy system including storage may be operated in different ways, such as:

(i) *Base load operation.* The system attempts to produce a constant power E_0. If the production is above E_0 and the storage is not full, the surplus is delivered to the storage. If the production is below E_0 and the storage is not empty, the complementary power is drawn from the storage. If the production is above E_0 and the storage is full, energy will be lost. If the production is below E_0 and the storage is empty, a deficit occurs which has to be covered by auxiliary (e.g. fuel-based) generating systems.

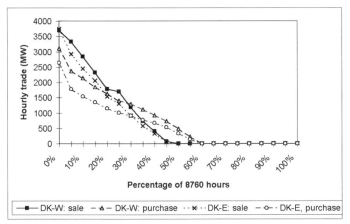

Figure 6.69. Annual duration curves for trade of Danish wind power at the Nordic Power Pool for East (DK-E) and West Denmark (DK-W) (Meibom *et al.*, 1997, 1999).

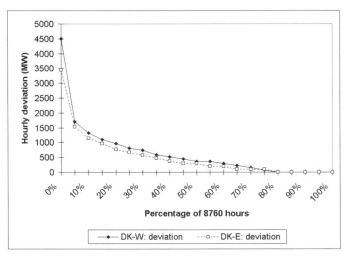

Figure 6.70. Annual duration curve for adjustment in Danish wind power delivery to Nordic Pool, relative to bid made 36 h in advance. Notation as in Fig. 6.69.

(ii) *Load-following operation.* The wind energy system attempts to follow the load (6.3), i.e. to produce power at a level E proportional to E_{load},

$$E = E_0 C_1 C_2.$$

If the production by the wind energy converters is above the actual value of E and the storage is not full, the surplus is delivered to the storage. If the production is below E and the storage is not empty, the difference is drawn from the storage. The possibilities of energy loss or need for auxiliary energy may arise as in (i).

(iii) *Peak load operation.* Outside the (somehow defined) time periods of peak load, all wind-produced energy is transferred to the storage facility. During the peak load periods, a prescribed (e.g. constant) amount of power is drawn from the wind energy converters and their storage. Surplus production (if the storage is full) during off-peak hours may be transferred to the grid to be used to replace base or intermediate load energy, if the regulation of the corresponding (fuel-based) units allows. A deficit during peak hours may, of course, occur. This system has two advantages. One is that the power to be used in a given peak load period (say, 2 h) can be collected during a longer period (say, 10 h), so that one might think that the probability of having adequate wind during part of this period is higher than the probability of having adequate wind during the peak hour, when the energy is to be used. The other advantage is that the wind energy in this mode of operation will be replacing fuel for peak load units, i.e. usually the most expensive type of fuel. With respect to the first claimed advantage, the weight of the argument depends on the rating of the wind energy converters relative to their planned contribution to covering the peak load. If the wind energy converters chosen are very small, they would have to operate during most of the off-peak hours to provide the storage with sufficient energy. If they are very large, the advantage mentioned will probably arise, but, on the other hand, surplus energy which cannot be contained in the storage will be produced more often. It all depends then on the relative costs of the wind energy converters and the storage facilities. If the latter are low, a storage large enough to hold the produced energy at any time may be added, but if the storage costs are high relative to the costs of installing more converters, the particular advantage of an extended accumulation period may be weakened.

(iv) *Other modes of operation.* Depending, among other things, on the installed wind energy capacity relative to the total load and on the load distribution, other schemes of operating a wind energy system including a short-term storage may be preferable. For example, some fraction of the wind energy production may always be transferred to the storage, with the rest going directly to the grid. This may increase the probability that the storage will not be empty when needed. Alternatively, wind energy production above a certain fixed level may be transferred to the storage, but the system will still attempt to follow the actual load. There is also the possibility of having a certain fraction of the wind energy production load following, but not all of it. Clearly, a number of relevant optimisation problems of this type can be assessed by performing computer simulations of various kinds.

Results of simulation studies

If a short-term storage is added to a wind energy system in the base load operation mode, the hourly sequence of power deliveries from the system may change in the way indicated in Fig. 6.63. On a yearly basis, the modifications may be summarised in the form of a modified power duration curve (compare Fig. 6.71 with e.g. the single turbine curves in Fig. 6.60), where some of the "excess" power production occupying levels above average has been moved over to the region below the average level to fill up part of what would otherwise represent a power "deficit".

The results of Fig. 6.71 have been calculated on the basis of the 1961 Risø data (height 56 m), using the power coefficient curve (a) in Fig. 6.61, in an hour-by-hour simulation of the system performance. The storage size is kept fixed in Fig. 6.71 at the value 10 h, defined as the number of hours of average production (from the wind energy converters), to which the maximum energy content of the storage facility is equivalent. Thus, the storage capacity W_s may be written as

$$W_s = E_0 t_s, \qquad\qquad (6.8)$$

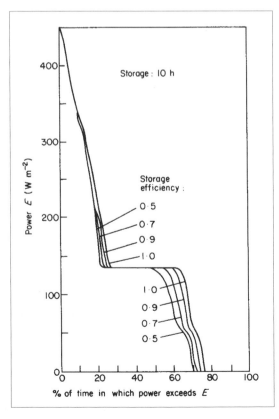

Figure 6.71. Annual power duration curves for a wind energy system with 10-h storage capacity $[t_A(E_0)]$, as a function of the storage cycle efficiency. The average wind power production is E_0. The calculation is for a wind energy converter of type (a) (Fig. 6.61), operated in base-load mode, and using 1961 Risø data at the height 56 m (Sørensen, 1978a).

where W_s is in energy units and t_s is the storage capacity in hours. E_0 is the power level which the wind energy system in base load operation mode aims at satisfying (for other modes of operation, E_0 is the average power aimed at).

Storage cycle efficiencies lower than unity are considered in Fig. 6.71, assuming for simplicity that the entire energy loss takes place when drawing energy from the storage. Thus, an energy amount W/η_s must be removed from the storage in order to deliver the energy amount W, where η_s is the storage cycle efficiency.

The effect of the storage is to make the fraction of time, during which the power level E_0 (or power levels $0 < E \leq E_0$) is available, larger, causing the fractions of time during which the power level exceeds E_0 to become smaller. The availability of a given power level E is the time fraction $t_A(E)$, during which the power level exceeds E, and Fig. 6.72 shows the availability of the base load power level, $t_A(E_0)$, for the case of a loss-free storage (full line), as a function of storage capacity.

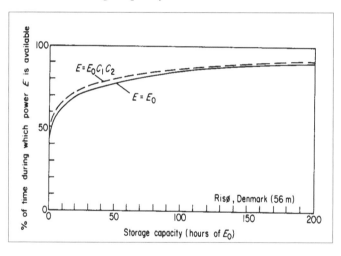

Figure 6.72. Power availability as a function of storage capacity $t_A(E_0)$ for a loss-free storage, wind energy converters of type (a) (Fig. 6.61) with average power production E_0 and 1961 Risø data. The system is operated in a baseload (solid line) or a load-following mode (dashed line) (Sørensen, 1978a).

It is shown that a modest storage leads to a substantial improvement in availability, but that improvements above availabilities of 75–80% come much more slowly and require storage sizes which can no longer be characterised as short-term stores. However, the availabilities that can be achieved with short-term stores would be considered acceptable by many utility systems presently operating, owing to the system structure and grid connections.

Another way to characterise the effect of adding a storage facility is to calculate the decrease in the average fluctuation σ of the system output E away from E_0,

WIND ENERGY BASICS ARE IN 3.2, 4.3

$$\sigma = \left(N^{-1} \sum_{i=1}^{N} (E(t_i) - E_0)^2 \right)^{1/2},$$ (6.9)

where t_i is the time associated with the ith time interval in the data sequence of N intervals. The annual fluctuation is given in Fig. 6.73 as a function of storage capacity t_s, again for a loss-free storage.

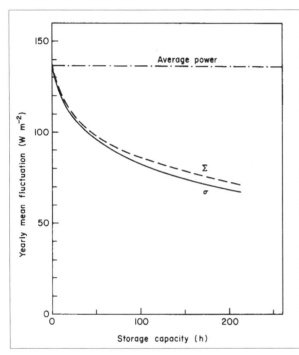

Figure 6.73. Annual average power fluctuations as a function of storage capacity $t_A(E_0)$ for a loss-free storage, wind energy converters of type (a) (Fig. 6.61) and 1961 Risø 56-m data. The system is operated in baseload mode (solid line) or in a load-following mode (dashed curve) (Sørensen, 1978a).

In the hourly simulation method, it is easy to keep track of the number of times when the operation of the storage switches from receiving to delivering power, or vice versa. This number is of particular relevance for some types of storage facilities, for which the lifetime may be approximately proportional to the number of charge/discharge cycles. Figure 6.74 gives the annual number of charge/discharge cycles, as a function of storage capacity, for the same assumptions as those made in Fig. 6.73.

All these ways of illustrating the performance of a store indicate that the improvement per unit of storage capacity is high at first, but that each additional storage unit above $t_s \approx 70$ h gives a very small return.

If the wind energy system including short-term storage is operated in a load-following mode, the storage size may still be characterised by t_s derived from (6.8), but in calculating fluctuations (called Σ for convenience of distinction) E_0 in (6.9) should be replaced by $E_0 C_1 C_2$ and in constructing a time duration curve the relevant question is whether the power delivered by the system exceeds $E_0C_1C_2$. Therefore, the power duration curve may be replaced by a curve plotting the percentage of time during which the difference $E - E_0C_1C_2$ between the power produced at a given time, E, and the load-following power level aimed at, exceeds a given value ΔE. Such curves are given in Fig. 6.75 for an "average load" E_0 equal to the annual average power production from the wind energy converters and assuming a loss-free store. The effect of augmenting the storage capacity t_s is seen to be to prolong the fraction of time during which the power difference defined above exceeds the value zero, i.e. the fraction of time during which the load aimed at can indeed be covered.

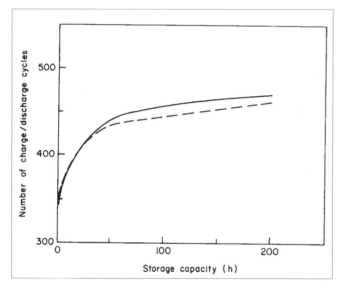

Figure 6.74. Number of annual charge/ discharge cycles of the energy storage for the systems considered in Fig. 6.73. Again the solid line corresponds to a baseload operation, and the dashed line corresponds to a load-following operation (Sørensen, 1978a).

The results of this load-following operation are also indicated in Figs. 6.72–6.74 (dashed lines), which show that the wind energy system actually performs better in the load-following mode than in the base-load mode, when a non-zero storage capacity is added.

Finally, Fig. 6.76 gives some results of operating a wind energy system with short-term storage in the peak load mode, based on a simulation calculation performed by Coste (1976). The storage facility is imagined as an advanced battery type of cycle efficiency 70%. The system is aimed at delivering a constant peak load power level, E_{peak}, but only during certain hours of the day and only on weekdays. The remaining time is used to accumulate energy in the storage whenever the wind energy converters are productive. If the storage units become filled, surplus generation may be delivered to the grid as off-peak power (at reduced credits). Figure 6.76 gives the storage capacity needed in this mode of operation, in order to achieve an availability of 80%, as a function of the peak load time fraction, i.e. the ratio between the accumulated time designed as peak load periods and the total time period considered (one year). The lack of continuity of the calculated points is due to the dependence on the actual position of the peak load hours during the day. It is determined by the load variation (cf. Fig. 6.14) and represents an increasing amount of "peak shaving" as the peak load time fraction increases. This influences the length of accumulation times between peak load periods (of which there are generally two, unevenly spaced over the day), which again is responsible for some of the discontinuity. The rest is due to the correlation between wind power production and time of the day [cf. the spectrum shown in Fig. 3.37 or the average hourly variation, of which an example is given in Sørensen (1978a)], again in conjunction with the uneven distribution of the peak load periods over the day.

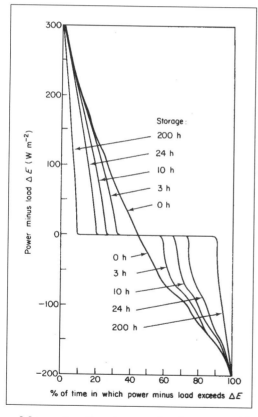

Figure 6.75. Annual time duration curve of the difference between the power produced by a wind energy system operated in the load-following mode and the actual load (or the fixed fraction E_0/E_{av} of it which it is aimed at covering) for a wind energy system with average power production E_0 and a loss-free storage of the capacity $t_A(E_0)$ indicated. Wind energy data from Risø (1961, 56 m) are used together with the power coefficient of type (a) (Fig. 6.61) (Sørensen, 1978a).

Measuring the storage size in terms of hours of peak load power E_{peak}, the storage size required to reach an 80% availability increases with the peak load time fraction, but relative to the capacity of the wind energy converters the trend is decreasing or constant. For small peak load time fractions, the converter capacity is "small" compared to the storage size, because the accumulation time is much longer than the "delivery" time. In the opposite extreme, for a peak load time fraction near unity, the system performs as in the base load operation mode, and the storage size given as hours of E_{peak} coincides with those given in the previous figures in terms of E_0.

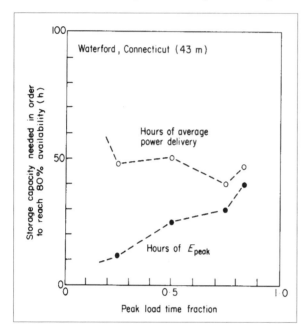

Figure 6.76. Storage capacity needed in order to reach 80% availability for a wind energy system in peak load operational mode: $t_A(E_{peak})$ and $t_A(E_0)$ as a function of peak load time fraction. The assumed wind converter design has an average power production of 57 W m^{-2} at Waterford, Connecticut (cf. Fig. 3.39), and the storage facility has a cycle efficiency of 70%. The average power production (related to the number of installed converters) is slightly above the average power delivery E_0 (which simply equals E_{peak} times the peak load time fraction) (based on Coste, 1976).

Systems with long-term storage

Long-term storage will be of interest in connection with wind energy systems aiming at covering a high fraction of the actual load. In this case, it is not possible to rely entirely on fuel-based back-up units to provide auxiliary power, and the wind energy system itself has to approach autonomy (to a degree determined by grid connections and the nature of the utility systems to which such connections exist).

The mode of operation used in connection with wind energy systems including long-term storage would therefore usually be the load-following mode. The most simple system of this sort may be one using pumped hydro storage (eventually sharing this storage system with a hydropower system, as mentioned earlier), in which case a high storage cycle efficiency can be obtained. The performance of such a system would be an extrapolation of the trends shown in Fig. 6.72.

In many cases, storage options using pumped hydro would not be available and other schemes would have to be considered (e.g. among those discussed in section 5.2.2). The best candidate for long-term storage and versatility in use is presently thought to be hydrogen storage (Gregory and Pang-born, 1976; Lotker, 1974; Bockris, 1975; Sørensen et al., 2003). The cycle efficiency is fairly low (cf. Table 5.5) for electricity re-generation by means of fuel cells (and still lower for combustion), but the fuel cell units could be fairly small and placed in a decentralised fashion, so that the "waste heat" could be utilised conveniently (Sørensen, 1978a; Sørensen et al., 2003). The production of hydrogen would also provide a fuel which could become important in the transportation sector. The use of hydrogen for short-term storage has also been considered (Fernandes, 1974), but, provided that other short-term stores of higher cycle efficiency are available, a more viable system could be constructed by using storage types with high cycle efficiency for short-term storage and hydrogen for long-term storage only.

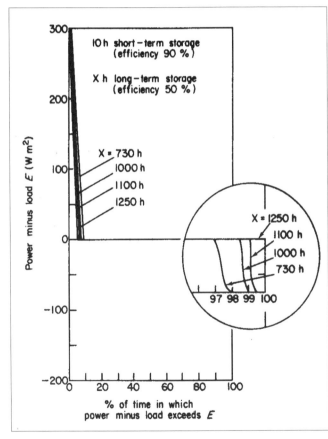

Fig. 6.77. Annual power duration curve for a system comprising both short- and long-term storage facilities (with cycle efficiencies 0.9 and 0.5, respectively). Type (a) converters (Fig. 6.50) and 1961 Risø 56-m data are used, and the average wind energy production (i.e. the number of converters installed) is chosen as 36% above the average load E_{av} (Sørensen, 1978a).

An example of the performance of a wind energy system of this kind is given in Fig. 6.77, using the simulation model described in the previous parts of this section. A 10-h (of average production E_0, which has to exceed average load E_{av} owing to the storage losses) storage with 90% cycle efficiency is supplemented by a long-term storage of assumed efficiency 50% (e.g. electrolysis–hydrogen–fuel cells), and the capacity of the long-term storage is varied, until the minimum value which will enable the system to cover the total load at any time of the year is found. This solution, characterised by a power-minus-load duration curve which is positive 100% of the time, is seen to require a long-term storage of capacity $t_A(E_{av})$ = 1250 h. If the availability requirement is reduced to 97% of the time, the long-term storage size may be reduced to 730 h (one month). Hydrogen storage in underground caverns or aquifers, that satisfies these requirements, is available in many locations (Sørensen *et al.*, 2001, 2003; Sørensen, 2004a).

6.4 A global energy scenario

In section 6.2.2, a demand scenario was developed for the year 2050, and in sections 6.2.4–6.2.7 estimates were made of potential energy utilisation from the most important renewable energy sources for the same period of time. Resources were divided into those suited for decentralised and centralised supply, both on a geographical basis. Estimates of additional renewable inputs, such as hydropower and geothermal energy, exist and can be added (Sørensen and Meibom, 1998, 2000).

The next task is to match supply and demand, adding new system components as necessary, e.g. if the energy form supplied is not that demanded, and pointing out requirements for energy imports and exports between regions when the location of supply and of demand is not the same. In practice, the form of transport will depend on the type of energy (electricity transmission, gas piping, heat in district lines, or fuels to be moved, e.g. by vehicle or ship). Also temporal mismatch can be identified, and energy storage requirements can be determined. In most of the cases considered, some storage cycle losses were already incorporated into the supply estimates, notably for those renewable energy sources where the source variability already indicated a storage need. The supply–demand matching is now made, first by using only those amounts of renewable energy that are estimated to be available locally, in a decentralised form, and subsequently, the possible advantages of also including centralised production are investigated.

6.4.1 Decentralised renewable energy 2050 scenario

In going from Table 6.2 to Table 6.5, the demand categories were already simplified under the assumption of an abundant fraction of the supply being in the form of electric energy. What remains now is to determine the sources of supply for each demand type.

For the vegetable food fraction, the results of comparing local supply and demand are shown in Fig. 6.78, where Fig. 6.78a shows the amount of surplus for those geographical grid cells where supply exceeds demand, and Fig. 6.78b shows the amount of deficit for those local cells where demand exceeds supply. Regional sums are given in Table 6.9. It follows that, on average, world-wide supply exceeds demand by 35%. This must be considered reasonable, as there has to be room for variations in crop harvests and thus food production from year to year, and further the transportation required for evening out supply and demand will entail some additional losses.

Like today, there is surplus vegetable food production in the Americas and Western Europe (regions 1, 2 and 4, cf. Table 6.5) and by the year 2050 also in region 3 (including Russia), owing to substantial improvements in agricultural practices assumed for this region. Region 5 (including China and India) will be just self-sufficient by the year 2050, whereas Africa (region 6) will have a deficit that must be covered by imports. In the scenario, Africa is the only region that by 2050 is in a development situation where it may offer labour at lower expense tha the other regions, and thus, there will be the possibility of paying for food imports by industrial revenues, provided that an education policy is pursued that will give the working force the necessary skills. In addition to inter-regional exchange, upon closer inspection, Fig. 6.78 indicates scenario requirements for transport of vegetable food within regions, especially from farming areas into cities.

The scenario assumptions for inter-regional trade in food are summarised in Fig. 6.83, where the regional exports have been selected from the surpluses available. The substantial needs for both vegetable and animal foods in Africa are uniformly imported from regions 1–4.

For animal-based food from either rangeland or fodder-fed animals, the surpluses and deficits are shown in Fig. 6.79. The picture is similar to that of vegetable foods, with surpluses in the region 1–4, but here with deficits in regions 5 and 6. This is due to the increase in the meat and milk fractions of diets assumed for Asia (Table 6.5), but the amounts are easily covered by imports from other regions, as indicated in Fig. 6.82. Overall, the animal food supply exceeds demand by 27%, which again is considered adequate in view of additional losses. Variations between years are smaller than for primary crops (because of the storage functions performed by livestock), but fairly frequent epidemics of animal disease are known to require a reserve.

Figure 6.80 shows the surplus and deficit of potential liquid biofuels derived from agriculture and silviculture, relative to the energy demand for transportation. The assumed fraction of biofuels used in this way is 48.5% (chosen such that the global average demand is covered), and the remaining is considered going into industrial uses such as medium-temperature process heat, where it is assumed used with 90% efficiency.

When constructing the demand scenario, we left it open to what extent electric vehicles would be used, but the availability of liquid biofuels is such

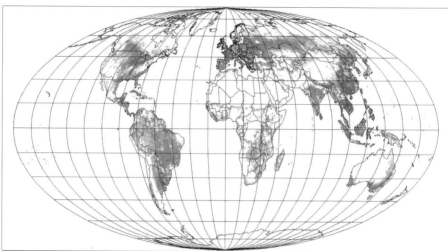

Energy flow
W/m2

5	to 10
2	to 5
1	to 2
0.5	to 1
0.2	to 0.5
0.1	to 0.2
0.05	to 0.1
0.02	to 0.05
0.01	to 0.02
0.005	to 0.01
0.002	to 0.005
0.0001 to	0.002
all others	

Figure 6.78. Local surplus (*a*: above, scale left, annual average supply minus demand in W m^{-2} is shown if positive) and deficit (*b*: below, scale right, annual average demand minus supply in W m^{-2} is shown if positive) of vegetable-based food supply relative to demand, on an area basis, valid for both decentralised and centralised 2050 scenarios (this and the following figures are from Sørensen and Meibom, 1998).

Energy flow
W/m2

-0.002 to -0.0001	
-0.005 to -0.002	
-0.01 to -0.005	
-0.02 to -0.01	
-0.05 to -0.02	
-0.1 to -0.05	
-0.2 to -0.1	
-0.5 to -0.2	
-1 to -0.5	
-2 to -1	
-5 to -2	
all others	

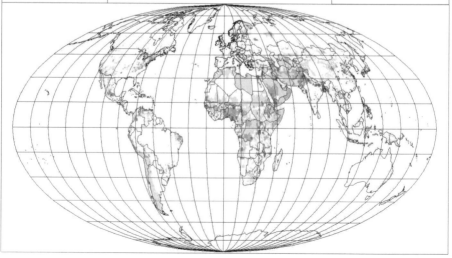

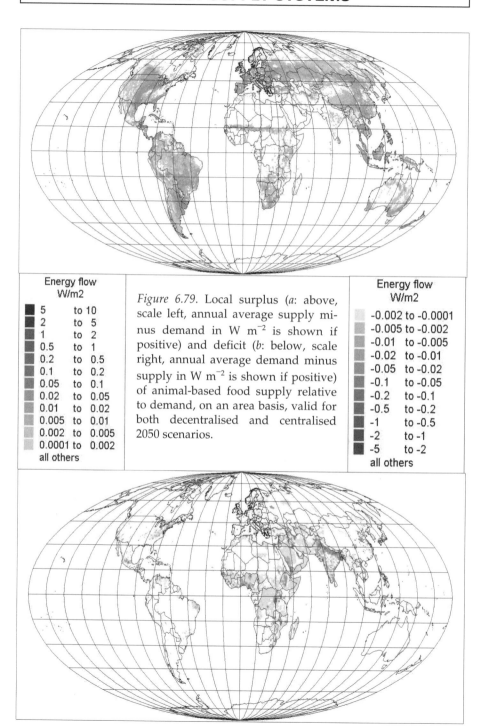

Energy flow W/m2	
5	to 10
2	to 5
1	to 2
0.5	to 1
0.2	to 0.5
0.1	to 0.2
0.05	to 0.1
0.02	to 0.05
0.01	to 0.02
0.005	to 0.01
0.002	to 0.005
0.0001 to	0.002
all others	

Figure 6.79. Local surplus (*a*: above, scale left, annual average supply minus demand in W m^{-2} is shown if positive) and deficit (*b*: below, scale right, annual average demand minus supply in W m^{-2} is shown if positive) of animal-based food supply relative to demand, on an area basis, valid for both decentralised and centralised 2050 scenarios.

Energy flow W/m2	
-0.002 to -0.0001	
-0.005 to -0.002	
-0.01 to -0.005	
-0.02 to -0.01	
-0.05 to -0.02	
-0.1 to -0.05	
-0.2 to -0.1	
-0.5 to -0.2	
-1 to -0.5	
-2 to -1	
-5 to -2	
all others	

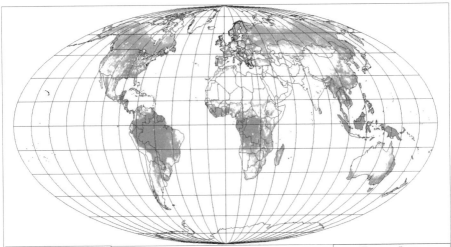

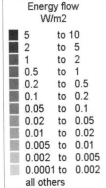

Energy flow W/m2

5	to	10
2	to	5
1	to	2
0.5	to	1
0.2	to	0.5
0.1	to	0.2
0.05	to	0.1
0.02	to	0.05
0.01	to	0.02
0.005	to	0.01
0.002	to	0.005
0.0001	to	0.002
all others		

Figure 6.80. Local surplus (*a*: above, scale left, annual average supply minus demand in W m^{-2} is shown if positive) and deficit (*b*: below, scale right, annual average demand minus supply in W m^{-2} is shown if positive) of transportation biofuel supply (such as methanol) relative to transportation demand, on an area basis, according to the decentralised 2050 scenario.

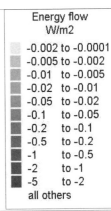

Energy flow W/m2

-0.002	to	-0.0001
-0.005	to	-0.002
-0.01	to	-0.005
-0.02	to	-0.01
-0.05	to	-0.02
-0.1	to	-0.05
-0.2	to	-0.1
-0.5	to	-0.2
-1	to	-0.5
-2	to	-1
-5	to	-2
all others		

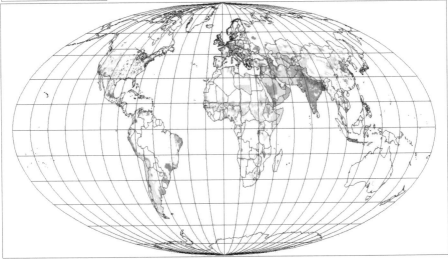

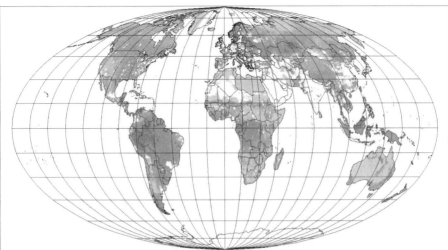

Energy flow W/m2		
5	to	10
2	to	5
1	to	2
0.5	to	1
0.2	to	0.5
0.1	to	0.2
0.05	to	0.1
0.02	to	0.05
0.01	to	0.02
0.005	to	0.01
0.002	to	0.005
0.0001 to		0.002
all others		

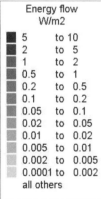

Figure 6.81. Annual average local surplus (*a*: above, scale left, annual average supply minus demand in W m^{-2} is shown if positive) and deficit (*b*: below, scale right, annual average demand minus supply in W m^{-2} is shown if positive) of electricity and biofuels for stationary use (such as biogas or hydrogen) relative to demand, on an area basis, according to the decentralised 2050 scenario.

Energy flow W/m2	
-0.002 to -0.0001	
-0.005 to -0.002	
-0.01 to -0.005	
-0.02 to -0.01	
-0.05 to -0.02	
-0.1 to -0.05	
-0.2 to -0.1	
-0.5 to -0.2	
-1 to -0.5	
-2 to -1	
-5 to -2	
all others	

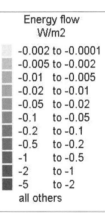

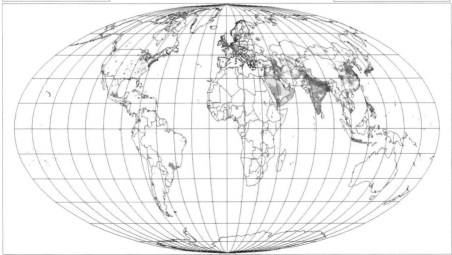

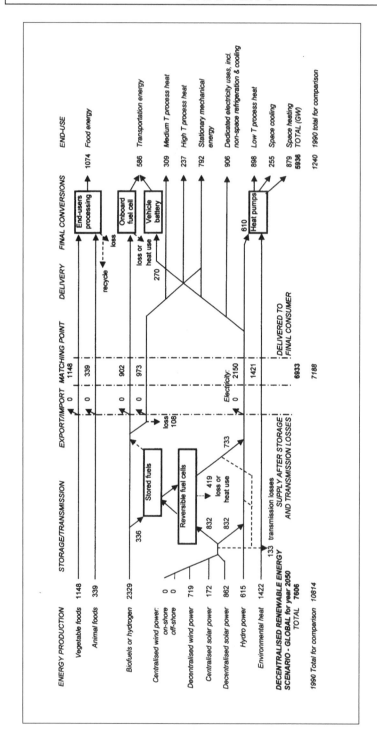

Figure 6.82. Overview of the decentralised 2050 scenario (all energy flows in GW or GWy/y).

that they could cover all transportation needs (with methanol use in fuel cells determining the assumed conversion efficiency). However, the regional supply–demand situation is such that substantial energy imports into region 5 are necessary. As we shall see below, this requires that as much energy trade as possible be based on biofuels and hence dictates a large electricity use in the transportation sector of the exporting regions.

The deficits in Fig. 6.80b are in urban areas and areas without biomass growth (e.g. Andes and Saudi Arabia) and in large parts of India. On average, global surplus and deficit balance, as seen from Table 6.9. There is surplus only in South America and Africa. The numbers for North America balance and those for region 3 almost balance (and are remedied in the scenario by increasing the fraction going into the transportation sector), whereas region 5 (including China and India) has a deficit of some 65%. The high production value for South America is related to the high forest productivity, and one may consider part of this as difficult to realise without imposing on (what should be) preservation areas. However, as we shall see, exports from South America of biofuels to the lacking regions in Asia are essential. Other areas of the model, such as regions 2 and 3 which also have average deficits, would have to import from regions 1 and 4, albeit on a smaller scale. As explained below, the actual scenario will not use the 48.5/51.5% split of the biofuel use, notably because electric vehicles take over some of the transportation demand.

The remaining energy produced is in the form of electricity. Figure 6.81 shows annual average surplus and deficit in produced electricity plus the 51.5% of biofuels that in the initial assessment is not used by the transportation sector, relative to demand for electricity and all other energy not covered above. The regional summaries are given in Table 6.9. Although annual totals are in balance, there are very large mismatches on both a spatial and a temporal scale. Consistent surpluses of electricity production occur in the Americas and Africa, and large, consistent deficits occur in region 5 (notably India and China). Western Europe exhibits small deficits, and region 3 exhibits strong seasonal variations, adding up to a substantial annual surplus. The reason is the variation in wind power production, which is ten times higher in January than in April. This is true for the year 1997, from which wind data were taken, but variations between years are large, implying considerable instability of power supply in region 3. A reason is that continental winds in the Eurasian continent are less persistent than West European coastal winds. The implication of the low wind power production in April in region 3 is that there is a substantial global deficit of power, affecting mostly region 5, which might have depended on imports from region 3.

Both the Figs. 6.78–6.81 and Table 6.9 show, that the large exchange of decentralised power that would solve the global mismatch problem would have to be from North and South America and Africa to southeast Asia. This is not a very feasible proposition, unless a global, superconducting transmission grid could be established before the year 2050. It would also appear that this level of inter-continental exchange is contrary to the idea of mainly local supply underlying the decentralised scenario. In the following subsection, we shall see if the uneven supply situation can be changed if a certain amount of centralised renewable energy production is also permitted.

In constructing the actual scenario, we first try to make as much biofuel available for export to region 5 as possible, because fuels can easily be transported over long distances (by ship), in contrast to electricity, where we assume that even by 2050 superconducting transmission will not be feasible, say, from South America to India or China. The regional scenario details shown in Fig. 6.82 indicate the assumed regional shift in biofuel and electricity use that allows larger amounts of biofuels to be exported to region 5. Basically, electricity is allowed to enter the transportation sector through electric vehicles used for all urban transport as well as track-based regional transport, except for region 5, which has to use the biofuels imported. As even the necessary electricity demand in region 5 is larger than possible local production, the decentralised scenario can only be realised if there is enough biomass (after imports) in region 5 to allow the missing electricity to be generated from biofuels. Since there is a loss in converting fuels to electricity, and since there was just enough biofuel available globally before redistribution, according to the assumptions of the decentralised scenario, this simply cannot be done. A solution would be to relax the assumptions restricting the decentralised energy production or, alternatively, to introduce a little centralised energy.

We have chosen the latter option, adding some photovoltaic power production in region 5 not based on building-integrated installations. This is a highly available resource, and there are more than enough areas available to make up for the electricity deficit. Still, it does negate the philosophy of all energy being produced locally, but this is probably less unacceptable than the huge imports and exports of fuels which are necessary in this scenario and which introduce a dependency on non-local resources difficult to reconcile with the value basis of the decentralised scenario.

Figure 6.82 summarises the decentralised scenario, and Fig. 6.83 shows the main flows of inter-regional energy trade.

6.4.2 Centralised renewable energy 2050 scenario

By the "centralised" scenario I understand a scenario exploiting central supply installations in addition to the decentralised ones, with a new attempt to optimise the amounts obtained from each category and the trade between regions. Sections 6.2.4–6.2.7 have provided estimates of the total additionally exploitable resources available in a centralised mode. Their combined magnitude is of the order of ten times the 2050 average demand, and the regional distribution, particularly of the potential photovoltaic power production, is fairly even. The regional energy sources employed in the centralised scenario are listed in Table 6.10, with an indication of those additional to the ones used in the decentralised scenario.

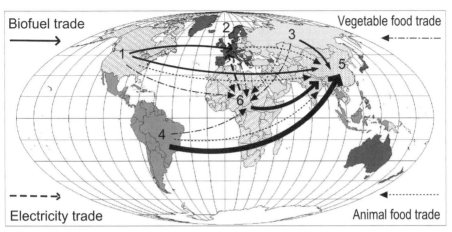

Decentralised scenario for regions 1-6

Figure 6.83. Regional patterns of energy trade in the decentralised 2050 scenario.

Region (cf. Table 6.5):	1	2	3	4	5	6	Total	Unit
Vegetable food	108	113	91	192	40	−140	**402**	GW
Animal food	43	31	16	91	−63	−27	**91**	GW
Transportation (relative to 48.5% of biofuels)	9	−69	−19	202	−230	112	**2**	GW
All other energy, annual average	123	−14	103	528	−879	201	**72**	GW
All other energy, January	106	−71	297	559	−1000	234	**100**	GW
All other energy, April	101	−3	−16	488	−900	165	**−165**	GW
All other energy, July	155	26	37	577	−750	213	**253**	GW
All other energy, October	129	−8	93	489	−800	191	**98**	GW

Table 6.9. Balance of potential regional supply, in decentralised mode, and demand. "All other" energy supply includes electricity from hydro, solar cells and wind turbines, plus 51.5% of biofuels. "All other" demand comprises electricity for heat pumps supplying heating, cooling and low-temperature heat, medium- and high-temperature energy, stationary mechanical energy, and electricity for appliances and other apparatus, including refrigeration.

For hydropower, the existing plants and those already under construction were already used in the decentralised scenario, but no additional plants were allowed to be built, in consideration of the environmental impacts that might be caused. The same choice is made for the centralised scenario. The bulk of the remaining resources (shown in Fig. 3.58) are large-reservoir type hydro sites, notably in South America.

Region (cf. Table 6.2)	1	2	3	4	5	6	Total	Unit
Total food balance (as in Table 6.9)	151	144	107	283	−23	−167	**493**	GW
Total biofuels used	250	190	300	640	327	192	**1899**	GW
Of which decentralised biofuels	250	166	236	640	295	192	**1779**	GW
Balance: total biofuels minus use for transportation	136	158	146	277	392	61	**1170**	GW
Hydropower	80	70	50	120	110	10	**440**	GW
Decentralised solar power	27	50	67	109	189	23	**465**	GW
Centralised solar power	40	20	114	19	400	100	**693**	GW
Decentralised wind power	20	37	80	100	100	10	**347**	GW
On-shore wind parks	12	10	20	38	46	0	**126**	GW
Off-shore wind power	0	40	0	0	13	0	**53**	GW
Balance: other energy, annual average	100	−63	139	288	−660	217	**22**	GW
Balance: other energy, Jan.	74	−120	169	306	−890	214	**−250**	GW
Balance: other energy, Apr.	87	−51	72	243	−710	200	**−160**	GW
Balance: other energy, July	129	−17	180	352	−410	233	**461**	GW
Balance: other energy, Oct.	110	−64	129	252	−610	220	**35**	GW

Table 6.10. Assumed energy supply (after storage conversion cycles and transmission but before inter-regional imports/exports) in the centralised 2050 scenario, and corresponding supply–demand balances.

Wind turbines not attached to farm buildings are introduced in the centralised scenario. In Western Europe, part of the substantial off-shore wind potential should first be put into operation, because this is already economical and installation is in progress. In region 3, there are also additional centralised wind resources, but since the problem for the decentralised scenario was the seasonal variation of continental winds, it will give a more stable system to exploit some of the centralised photovoltaic potential. For region 5, the only realistic proposal is to exploit substantial amounts of photovoltaic power, plus the smaller additional centralised biomass potential available. A little off-shore wind is exploitable in the south.

The amounts of centralised sources proposed in Table 6.10 to be exploited in the centralised scenario are shown in Table 6.11 as fractions of the potential resources that they constitute (as estimated in sections 6.2.4–6.2.7). For region 4, we have reduced the use of biomass for fuels, in line with the arguments of forest preservation given in section 6.4.1. Still, region 4 has the option of a very large energy export.

Region	1	2	3	4	5	6	Total
Decentralised biofuels	0.91	0.98	0.96	0.93	1.00	0.92	**0.94**
Centralised biofuels	0	0.12	0.64	0	0	0	**0.21**
Hydropower	0.88	0.82	0.95	0.48	0.90	0.67	**0.71**
Decentralised solar power	0.54	0.93	0.71	0.76	0.99	0.20	**0.72**
Centralised solar power	0.013	0.003	0.014	0.004	0.041	0.005	**0.013**
Decentralised wind power	0.56	0.85	0.45	0.75	1.00	0.21	**0.64**
On-shore wind parks	0.75	0.068	0.27	0.66	1.00	0	**0.21**
Off-shore wind power	0	0.52	0	0	0.70	0	**0.16**

Table 6.11. Fraction of potential resources used in the centralised 2050 scenario.

The resulting energy system is much more robust than the decentralised one, and none of the renewable energy sources is utilised to the ceiling. This modesty should actually be considered realistic, owing to the variations between seasons and years of renewable energy production (as already discussed for food production). Table 6.10 indicates substantial variations in supply/demand balances between seasons. We believe that we have taken this into account by the storage cycle introduced (see Figs. 6.90–6.96). However, a detailed simulation needs to be performed to ensure this, as was done for Western Europe in a similar study (Nielsen and Sørensen, 1998).

In an economic evaluation, the cost of establishing a surplus generation capacity should be weighed against the cost of providing more energy storage than envisaged in our reference case (50% of wind and solar cell power passing through storage facilities, as assumed in sections 6.2.4 and 6.2.5). It is in the spirit of the centralised scenario that imports and exports are used in those cases where they ease the strain on ecosystems. It is therefore a conclusion that the exploitation of centralised renewable resources allows more environmental concerns to be considered than insisting on producing as much as possible of the energy needed locally.

Table 6.10 gives the regional balances of supply and demand for the resources exploited in the centralised scenario (before import and export), and Figs. 6.90–6.96 show the overall scenario and its regional implications. The food balances are identical to those of Figs. 6.78 and 6.79, and the local surpluses and deficits for biofuels and electricity (before trade of energy) are

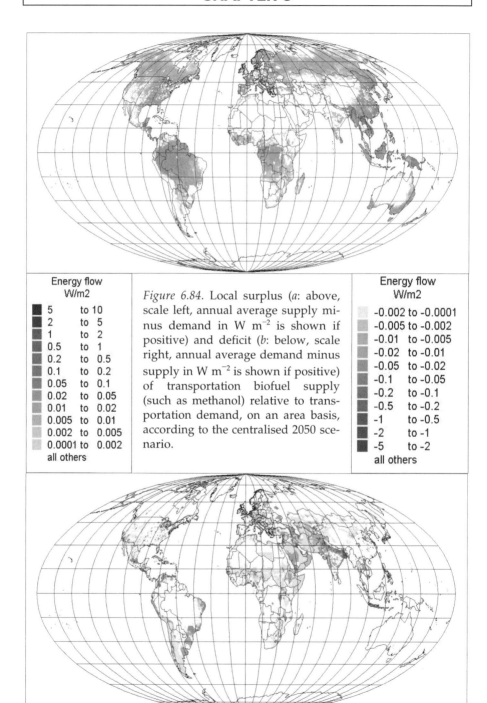

Energy flow
W/m2

5	to	10
2	to	5
1	to	2
0.5	to	1
0.2	to	0.5
0.1	to	0.2
0.05	to	0.1
0.02	to	0.05
0.01	to	0.02
0.005	to	0.01
0.002	to	0.005
0.0001	to	0.002
all others		

Figure 6.84. Local surplus (*a*: above, scale left, annual average supply minus demand in W m^{-2} is shown if positive) and deficit (*b*: below, scale right, annual average demand minus supply in W m^{-2} is shown if positive) of transportation biofuel supply (such as methanol) relative to transportation demand, on an area basis, according to the centralised 2050 scenario.

Energy flow
W/m2

-0.002	to	-0.0001
-0.005	to	-0.002
-0.01	to	-0.005
-0.02	to	-0.01
-0.05	to	-0.02
-0.1	to	-0.05
-0.2	to	-0.1
-0.5	to	-0.2
-1	to	-0.5
-2	to	-1
-5	to	-2
all others		

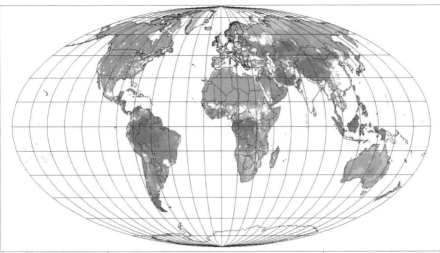

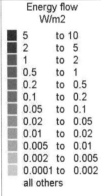

Energy flow W/m2	
5	to 10
2	to 5
1	to 2
0.5	to 1
0.2	to 0.5
0.1	to 0.2
0.05	to 0.1
0.02	to 0.05
0.01	to 0.02
0.005	to 0.01
0.002	to 0.005
0.0001 to	0.002
all others	

Figure 6.85. January local surplus (*a*: above, scale left, average supply minus demand in W m^{-2} is shown if positive) and deficit (*b*: below, scale right, average demand minus supply in W m^{-2} is shown if positive) of electricity and biofuels for stationary use (such as biogas or hydrogen) relative to demand, on an area basis, according to the centralised 2050 scenario.

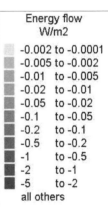

Energy flow W/m2	
-0.002 to -0.0001	
-0.005 to -0.002	
-0.01 to -0.005	
-0.02 to -0.01	
-0.05 to -0.02	
-0.1 to -0.05	
-0.2 to -0.1	
-0.5 to -0.2	
-1 to -0.5	
-2 to -1	
-5 to -2	
all others	

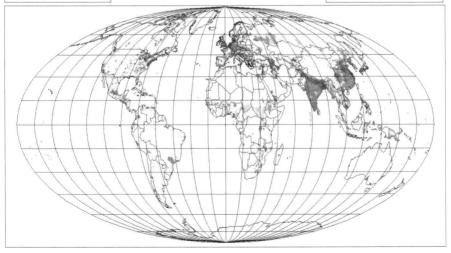

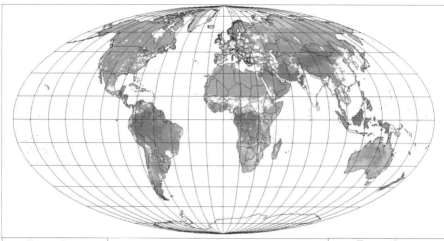

Energy flow
W/m2

5	to 10
2	to 5
1	to 2
0.5	to 1
0.2	to 0.5
0.1	to 0.2
0.05	to 0.1
0.02	to 0.05
0.01	to 0.02
0.005	to 0.01
0.002	to 0.005
0.0001	to 0.002
all others	

Figure 6.86. April local surplus (*a*: above, scale left, average supply minus demand in W m^{-2} is shown if positive) and deficit (*b*: below, scale right, average demand minus supply in W m^{-2} is shown if positive) of electricity and biofuels for stationary use (such as biogas or hydrogen) relative to demand, on an area basis, according to the centralised 2050 scenario.

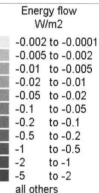

Energy flow
W/m2

-0.002 to	-0.0001
-0.005 to	-0.002
-0.01 to	-0.005
-0.02 to	-0.01
-0.05 to	-0.02
-0.1 to	-0.05
-0.2 to	-0.1
-0.5 to	-0.2
-1 to	-0.5
-2 to	-1
-5 to	-2
all others	

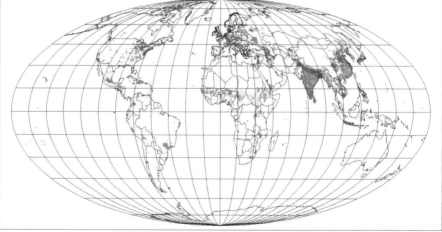

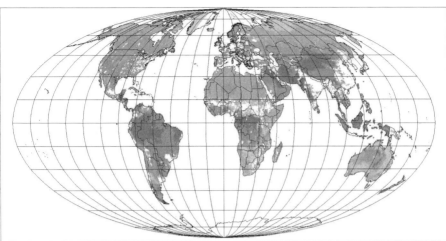

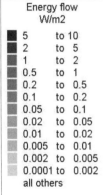

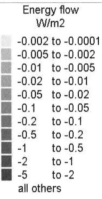

Energy flow
W/m2

5	to 10
2	to 5
1	to 2
0.5	to 1
0.2	to 0.5
0.1	to 0.2
0.05	to 0.1
0.02	to 0.05
0.01	to 0.02
0.005	to 0.01
0.002	to 0.005
0.0001 to	0.002
all others	

Energy flow
W/m2

-0.002 to -0.0001	
-0.005 to -0.002	
-0.01 to -0.005	
-0.02 to -0.01	
-0.05 to -0.02	
-0.1 to -0.05	
-0.2 to -0.1	
-0.5 to -0.2	
-1 to -0.5	
-2 to -1	
-5 to -2	
all others	

Figure 6.87. July local surplus (*a*: above, scale left, average supply minus demand in W m^{-2} is shown if positive) and deficit (*b*: below, scale right, average demand minus supply in W m^{-2} is shown if positive) of electricity and biofuels for stationary use (such as biogas or hydrogen) relative to demand, on an area basis, according to the centralised 2050 scenario.

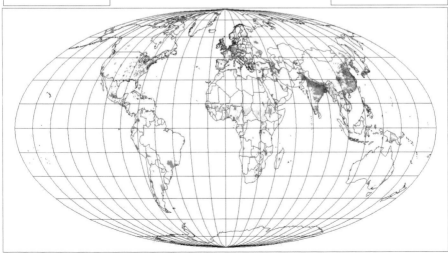

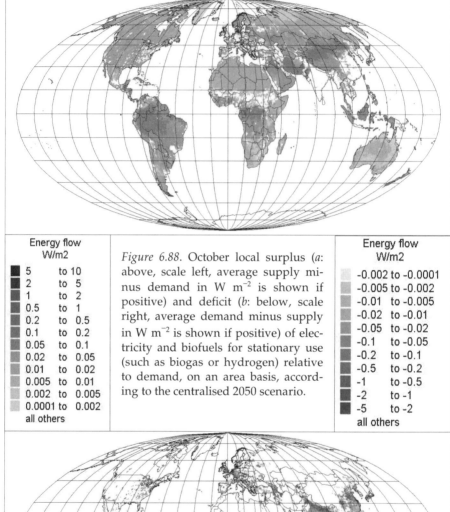

Energy flow W/m2		
■	5	to 10
■	2	to 5
■	1	to 2
■	0.5	to 1
■	0.2	to 0.5
■	0.1	to 0.2
■	0.05	to 0.1
■	0.02	to 0.05
■	0.01	to 0.02
■	0.005	to 0.01
■	0.002	to 0.005
■	0.0001 to	0.002
	all others	

Energy flow W/m2	
-0.002 to	-0.0001
-0.005 to	-0.002
-0.01 to	-0.005
-0.02 to	-0.01
-0.05 to	-0.02
-0.1 to	-0.05
-0.2 to	-0.1
-0.5 to	-0.2
-1	to -0.5
-2	to -1
-5	to -2
all others	

Figure 6.88. October local surplus (*a*: above, scale left, average supply minus demand in W m^{-2} is shown if positive) and deficit (*b*: below, scale right, average demand minus supply in W m^{-2} is shown if positive) of electricity and biofuels for stationary use (such as biogas or hydrogen) relative to demand, on an area basis, according to the centralised 2050 scenario.

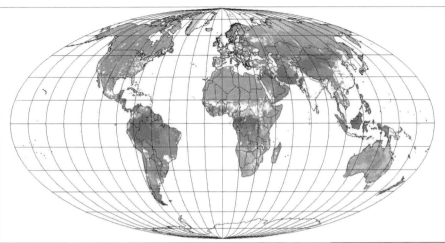

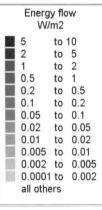

Energy flow
W/m2

5	to	10
2	to	5
1	to	2
0.5	to	1
0.2	to	0.5
0.1	to	0.2
0.05	to	0.1
0.02	to	0.05
0.01	to	0.02
0.005	to	0.01
0.002	to	0.005
0.0001	to	0.002
all others		

Figure 6.89. Annual average local surplus (*a*: above, scale left, average supply minus demand in W m^{-2} is shown if positive) and deficit (*b*: below, scale right, average demand minus supply in W m^{-2} is shown if positive) of electricity and biofuels for stationary use (such as biogas or hydrogen) relative to demand, on an area basis, according to the centralised 2050 scenario.

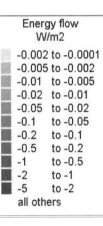

Energy flow
W/m2

-0.002 to -0.0001
-0.005 to -0.002
-0.01 to -0.005
-0.02 to -0.01
-0.05 to -0.02
-0.1 to -0.05
-0.2 to -0.1
-0.5 to -0.2
-1 to -0.5
-2 to -1
-5 to -2
all others

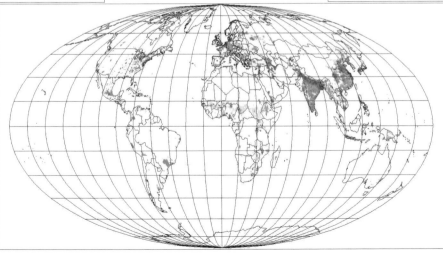

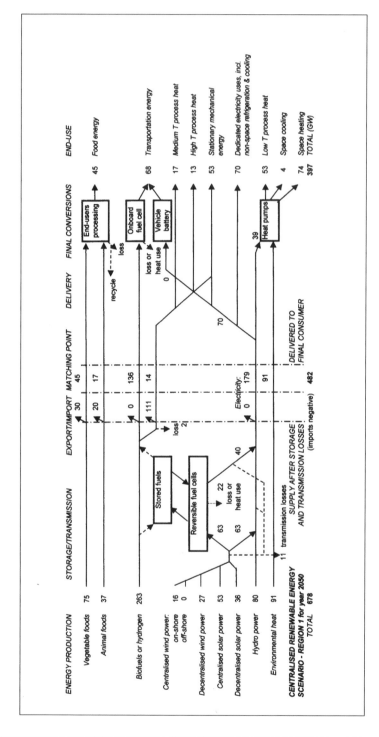

Figure 6.90. Region 1 centralised 2050 scenario with indication of import and export of energy (the unit used for energy flows is GW or GWy y⁻¹).

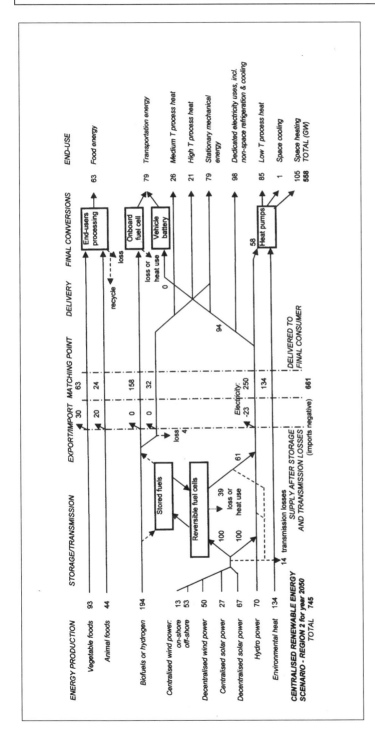

Figure 6.91. Region 2 centralised 2050 scenario with indication of import and export of energy (the unit used for energy flows is GW or GWy y^{-1}).

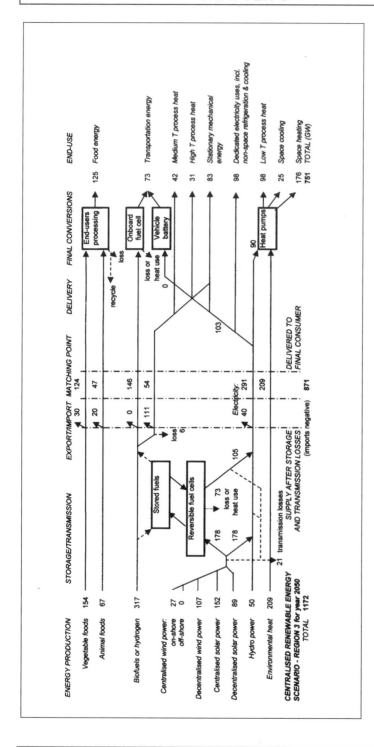

Figure 6.92. Region 3 centralised 2050 scenario with indication of import and export of energy (the unit used for energy flows is GW or GWy y^{-1}).

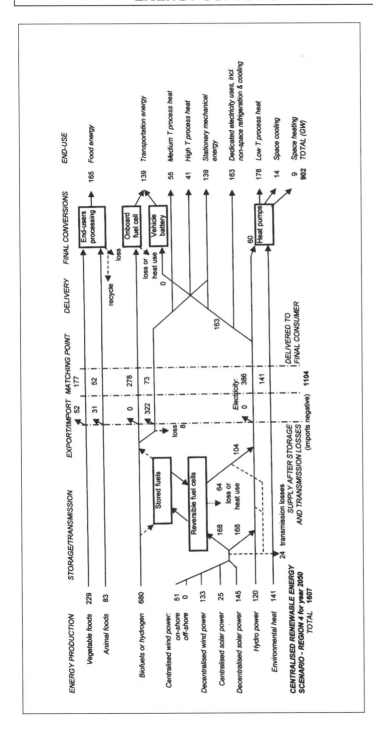

Figure 6.93. Region 4 centralised 2050 scenario with indication of import and export of energy (the unit used for energy flows is GW or GWy y^{-1}).

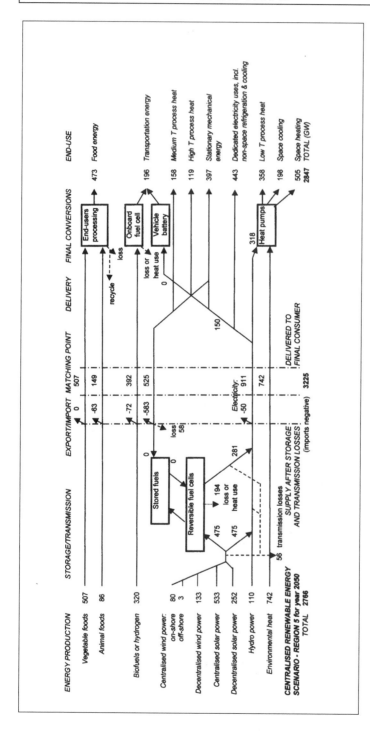

Figure 6.94. Region 5 centralised 2050 scenario with indication of import and export of energy (the unit used for energy flows is GW or GWy y⁻¹).

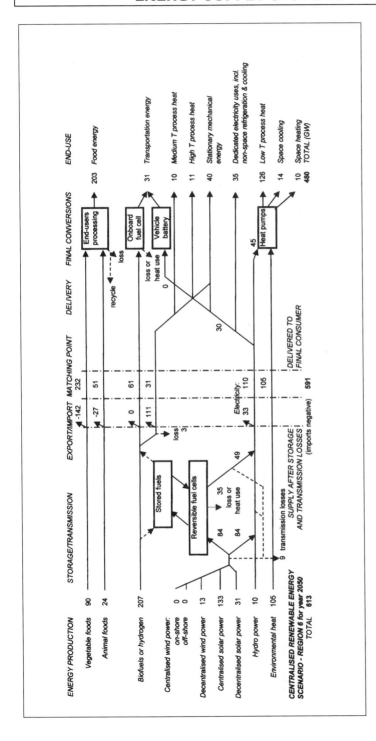

Figure 6.95. Region 6 centralised 2050 scenario with indication of import and export of energy (the unit used for energy flows is GW or GWy y^{-1}).

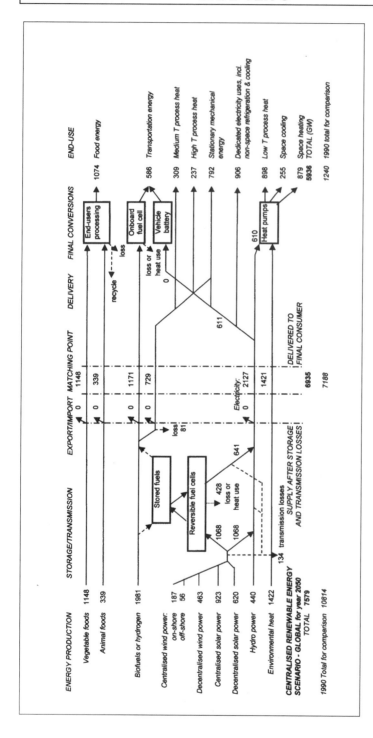

Figure 6.96. Overview of the centralised 2050 scenario (the unit used for energy flows is GW or GWy y^{-1}).

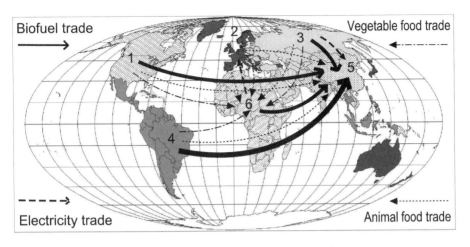

Centralised scenario for regions 1-6

Figure 6.97. Regional patterns of energy trade in the centralised 2050 scenario.

shown in Figs. 6.84–6.89, the latter both on a seasonal and on an annual basis. The main routes of import/export are depicted in Fig. 6.97. The overall picture is similar to that of the decentralised scenario, except for the higher resilience obtained from the centralised production in wind farm, photovoltaic parks and energy forests or crops. The amount of centralised energy used to cover the scenario demand is still very modest compared to the potential listed in sections 6.2.4–6.2.7, even when it is used to ensure no abuse of new large hydro installations or rainforest conversion. The PV and wind energy export options arising e.g. for desert land give such areas new opportunities for development and, in some cases, make up for current oil export revenues lost. Also the large biofuel exports considered for South America give credibility to the assumed substantial economic development of this continent (see section 6.2.2)

While it is not surprising that the decentralised scenario would have difficulties for the most densely populated regions, it may cause a little surprise to see how large a surplus of energy the centralised scenario offers. As discussed in the following section, this has implications for other demand scenarios than the one discussed here. The surpluses are illustrated in Table 6.11, which gives the percentages of estimated available resources actually used in the centralised scenario.

6.4.3 Implementation and resilience of scenario

In discussing the robustness of the two scenarios described above against changes in assumptions, it is useful to separate the demand and the supply scenarios. The demand scenario of section 6.2.2 assumes that efficiency measures are implemented that would entail smaller costs than any of the options for additional energy supply. This may appear a very reasonable assumption, but it is a feature of the present system that in several demand categories there are efficiency improvements not made which would be highly economic even with current prices. The reason for this may be sought in the lower prestige seemingly accorded to demand-side measures as compared with new supply. Whether or not this element of irrationality can be eliminated in the future is uncertain, and many scenarios of future energy demand estimate substantially higher consumption levels than the one assumed in the present study. In the centralised 2050 scenario of section 6.4.2, there is a low utilisation of available renewable energy sources, and substantially higher demands could be satisfied, notably with photovoltaic plants not attached to buildings. The centralised scenario is thus highly resilient to changes in demand, whereas the tight matching of demand and supply in the decentralised scenario makes this more vulnerable.

With respect to the supply scenarios, it is found that the decentralised option, defined as using only building-integrated PV and less than one detached wind turbine per farmhouse, plus existing hydro, leaves little room for increase of supply. For buildings, a higher solar utilisation would require urban planning with orientation of streets and buildings optimised for solar capture. This is unlikely to be feasible, as half the buildings in 2050 already exist, and the other half will be built gradually over the 50-year period, implying the necessity of very rapid action if urban planning aspects should be significantly altered. For farm-attached wind turbines, it is assumed that not every farm possesses a turbine by the year 2050. This is justified both because of resource variations (even on extended lands belonging to a farm, no site with suitable wind conditions may be found, owing to nearby forests or other obstructing landscape features) and because not all farmers may choose to install a turbine. The model adopted corresponds to the most dense dispersed wind turbine density found today in Denmark, and the assumption is that current smaller turbines will have been replaced by larger ones by the year 2050. This is beginning to happen in Denmark, where the regulation after some years with increasing planning restrictions on replacement of existing wind turbines is now generally encouraging replacement with larger units. However, new zoning restrictions still prevent a number of the existing turbines to be allowed replacement once their service life runs out. Thus, for wind turbines it is also difficult to imagine a substantial increase within the decentralised mode relative to the scenario proposed.

One area where the proposed decentralised scenario may be conservative is in coverage of heat demands. There are a number of possibilities for deriving additional heat from the scenario, provided that district heating lines are available for transporting the heat to the sites of demand: the storage cycle, transforming wind and solar power to fuels and later regaining electricity through reversible fuel cells, gives rise to waste heat that could be led to demand sites. The same is true of industrial high-temperature heat usage, where waste heat could be cascaded down to users of lower temperature heat. If these methods are adopted, the generation of low-temperature heat through heat pumps could be more or less eliminated, leaving more electricity to high-quality uses. In practical terms, these options may allow a demand increase of some 25%.

There are a few features of the regional scenario building that may be unrealistic. For instance, in region 2 there is a high production of solar power in Australia, which could not practically be transmitted to the demand sites in Japan and Europe. The same is true for off-shore wind power produced along the coasts of Greenland and northern Norway. This affects only the centralised scenario, and since it has many unused, centralised renewable energy resources nearer to the demand centres, it is no serious problem.

The centralised renewable energy scenario can easily remedy such local problems and can sustain much higher levels of demand, while still offering seasonal smoothing of supply through the storage conversion cycles devised. As mentioned, it also makes it much easier to solve the geographical mismatch between energy generation and demand. As a result of the use of wind parks and central solar plants, it has less long-distance transport and transmission of energy than the decentralised scenario, but at the same time it readily accepts such energy trade, which on the other hand may be considered incompatible with the value basis of a decentralised scenario.

Implementation of either of the 2050 scenarios involves sketching a path of moving from the current system primarily based on fossil fuels to a very different system, and identifying the conditions that have to be fulfilled to make the transition happen. These include economic milestones for the new technologies involved, but they also involve political decisions that need to be taken. It is assumed that the social climate in which the transitions happen is governed by the mixture of free market competition and regulations by society, as is the case in most regions of the world today. The regulatory part would impose requirements, such as building codes and minimum technical standards for reasons of safety and consumer protection, and minimum energy use for appliances. Another available public handle is environmental taxes that incorporate indirect costs which otherwise would distort the competition between different solutions in the marketplace. A consistent method of estimating the environmental taxes that will make the market a fair one is life-cycle analysis of the entire energy supply chains and the technologies involved. The methodology for doing this is described in Chapter 7, and further examples for many of the renewable energy systems considered may be found in Kuemmel *et al.* (1997).

In a fair market, the price that new technologies have to measure up against is the price of currently used coal, oil, natural gas, hydro and nuclear technologies, all supplemented with the externalities not included in the present market prices. Renewable technologies such as wind power, the cost of which today is only slightly above current fossil fuel-based systems, will clearly be economic if externalities are included (as these are very small for wind power and more than twice the actual price for fossil fuels). Also technologies such as biofuel production, which today involve a cost about twice that of fossil fuels, would be able in the fair market to enter by standard competitive forces. For photovoltaic power and the new conversion and storage technologies (e.g. fuel cells), present costs are higher than can be remedied by introducing externalities in the comparison with fossil fuels. Therefore, these have to be assumed to pass through a technical development over the 50-year transition period that brings the price down to below the threshold value. Subsidies may be contemplated in order to speed up this process in the initial phase, but already the political readiness to include externalities in prices would constitute a strong motivation for the development of alternative solutions.

The assumptions that the future transition will be driven by fair market rules are at variance with the present situation. On one side, there are hidden subsidies in some regions (e.g. to nuclear power), and on the other side, monopolies and generally differences in size and power of the industries involved in different technologies make the price setting likely not to follow those prescribed by the life-cycle analysis in a fair market philosophy. An obvious solution is to regulate the market not by taxation but by legislation, requiring, for example, power providers to use specific technologies. This makes it unnecessary to accumulate tax money at the state level (which by some nations is seen as a positive feature), but makes the system rather stiff, as each technical change possibly has to be followed up by altering the legislative regime. The taxation method is more flexible, and, once the level of environmental tax is decided by governments, the market functions exactly as before. However, the taxation method should give the manufacturers of the new technologies with smaller environmental impacts a good chance to compete even if they are initially smaller than the established market players. It is also important that externalities are set politically, thereby doing away with the uncertainties of scientific assessment, once the legislation is in place. A problem is the possible differences in tax levels that different nations may see as fair. International synchronisation is highly desirable, as in all policy aimed at reducing global threats. Depending on the degree of planning tradition in different societies, the energy transition might also benefit from "setting goals" and continually monitoring whether they are fulfilled. If, for example, the market does not respond well enough to the price signals set by the environmental taxes, it is then possible to adjust the size of the imposed externality (which would often not violate the scientific basis for it) or to introduce specific legislation to remove the obstacles to a free and fair market.

A natural extension of the scenario work described above would be to look at scenarios with a higher energy demand, despite the economic reasons given for doing energy efficiency improvements before adding supply, whenever the former are less expensive. In addition to political opposition to improving efficiencies (due to the political/economic misconception that "growth" is always good and "stagnation" or "decline" is bad, which may hold only for non-material qualities), there are specific assumptions in the demand scenario used above that could be altered. This would define new demand scenarios with higher demand, which along with the discussion above suggests combination only with the scenario called "centralised". For instance, the present scenario increase in end-use services has been assumed to take place mainly in regions currently less developed, whereas Europe and the USA will not increase end-use demand, but only economic and social welfare, through activities of low energy intensity (information technology, arts and leisure).

A possible alternative future would assume that e.g. physical fitness and leisure time enjoyment will not be achieved by bicycling or walking in the neighbourhood forest or visits to theatres and concerts, but rather by attending fitness institutions with energy-intensive work-out appliances and excursions by private plane, snowmobile, motorboat and the like. In this case, end-use energy may increase also in the already rich countries, and only centralised renewable energy supply can satisfy the additional energy demand. Exactly how may be explored in future scenario work.

A final remark may be appropriate regarding the use of the expression "centralised" to characterise the second supply scenario presented above. To some, this expression has a negative connotation, reminiscent of centralised economic planning or lack of power to decide by the individual citizen. Here, the word is used solely to signal the use of technologies that are not integrated into homes or controlled by individuals, as are the rooftop solar panels, individually owned wind turbines and fuel cell plants located in each building in the decentralised scenario. Such community-size installations are here found beneficial for a robust energy system, regardless of the type of ownership structure that may be associated with them.

6.5 Suggested topics for discussion

6.5.1

Discuss the possibility that solar electricity could play the roles considered for wind electricity in section 6.3.2.

What is the diurnal and seasonal match between variations in solar energy generation and electricity load? For example, construct power duration curves or "power-minus-load" curves and discuss the need for short- and

long-term storage of energy, in comparison with the corresponding information regarding wind energy systems in geographical regions of good wind
access.

Are there climatic regimes which from a system structure point of view
(as distinct from a purely economic point of view) would make solar electricity the preferable solution?

Does it make any difference to the answers to the above questions
whether the solar system in mind is based on solar cells or on concentrating
collectors followed by thermal conversion?

6.5.2

On the basis of Fig. 6.98, discuss the potential of wave power in the North
Atlantic Ocean for electricity production. Figure 6.98 is the annual power
duration curve for one particular wave device (a Salter duck, Fig. 4.99) with
the efficiency curve given in Fig. 4.100. The diameter of the device is 16 m,
and it is assumed to be omnidirectional, in the absence of directional data
(Mollison *et al.*, 1976). The corresponding power duration curve for the
waves themselves is given in Fig. 3.49.

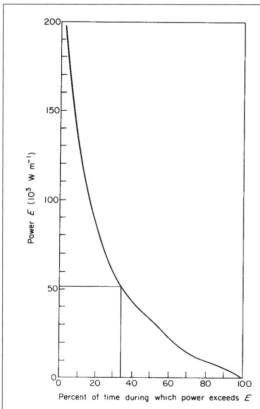

Figure 6.98. Annual power duration curve for omnidirectional oscillating vane wave
power converter of diameter 16
m, using the efficiency curve
given in Fig. 4.100. The wave
data are for Station India in the
northeast Atlantic (cf. Fig. 3.49),
having an average power level
of 91 kW m^{-1}. The thin straight
lines indicate the availability of
the average power production
(average efficiency 56%) (based
on Mollison *et al.*, 1976).

Compare the availability of the average power with that of the wind power systems considered in section 6.3.2. Why does the wave power go right to the 100% time fraction in Fig. 6.98, in contrast to the single wind turbine power duration curves (e.g. Fig. 6.60).

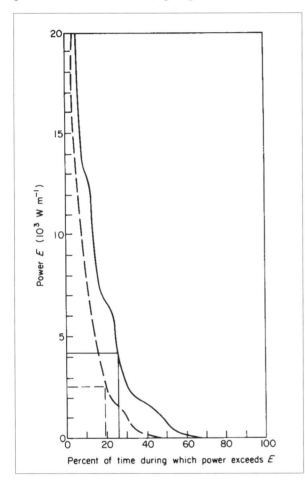

Figure 6.99. Annual power duration curves for omni-directional (solid, heavy line) or one-directional (west, dashed line) oscil-lating vane wave power converter of diameter 6 m, using the efficiency curve given in Fig. 4.84. The wave data are for Station Vyl in the North Sea (for 1971, courtesy of Nielsen, 1977), having an average power level of 5.35 kW m^{-1}. The omnidirectional device has an average efficiency of 79%. If the device were as-sumed to accept two oppo-site directions (east and west), the power duration curve would lie about midway between the two curves drawn. Diminishing the diameter of the device does not increase the maximum availability.

6.5.3

Try to answer the same questions as in 6.5.2 on the basis of Fig. 6.99 rather than Fig. 6.98. Figure 6.99 is based on data from the North Sea, and the size of the Salter device (6 m) is chosen to give the maximum annual power pro-duction.

Consider the statement that "waves accumulate wind energy over time and over fetch distance, hence wave power is more stable and dependable than wind power".

6.5.4

Wind energy converters might be placed on platforms at sea. What advantages could be expected from such an arrangement? The structures associated with wave power devices might conceivably be able to support wind energy converters on top. Would there be advantages in such a combination?

6.5.5

What size solar collector would be able to provide power for a small automobile (requiring, say, 25 kW peak power)?

Such a collector is not for integration in the roof of the car, but consider a car with a suitable storage facility (e.g. flywheel) which could be charged by electricity from solar cells covering e.g. a garage rooftop.

The (annual) average power requirement of the car is much lower than the peak power. Assume e.g. an average net power requirement of 300 W (roughly corresponding to 1300 W of fuel for the more efficient types of automobiles produced at present), and estimate the solar cell area which would be required in order to produce power at about this level for most of the year (depending on latitude).

The flywheel storage, which has also been considered for "peak shaving" in present generations of cars, has the advantage of being capable of delivering very high peak power levels for start and acceleration (if this is considered a feature of petrol-driven car engines worth preserving).

6.5.6

Try to build a renewable energy demand and supply scenario for your country or region, based on the maps and data given in this chapter and possibly supplemented with additional local data (considering that the geographical resolution of about 50 × 50 km may not be sufficient for the study of a particular region).

Based on this work, determine whether your region is likely to be an energy-importing or energy-exporting one.

SOCIO-ECONOMIC ASSESSMENT OF ENERGY SUPPLY SYSTEMS

CHAPTER 7

7.1 Framework of socio-economic analysis

The choice of energy systems, like any other feature of major social organisation, is a product of a historical development, of prevailing social value systems and often of the role of influential individuals from various layers of society: political decision-makers, industrialists or intellectual figures.

In this chapter, the tools available for analysis and comparative assessment of energy systems are presented. As they are not independent of social preferences, one should expect to find different tools available, catering to different positions in the social debate. However, when presented in a systematic fashion, the effects of differences in assumptions become transparent, and the outcome is to exhibit different results as due not just to uncertainty but to specific differences in underlying choices. This would also force the users of these methods to specify their normative positions.

In the following, both simple methods of including direct economic impacts of an energy system and more advanced methods trying to include or at least keep track of indirect and non-monetary kinds of impacts will be presented.

7.1.1 Social values and the introduction of monetary economies

Social value systems are by no means laws of nature. They differ from one culture to another, and they change with time. They can be influenced to a substantial degree by conscious or unconscious propaganda (from teachers, advertisements, spiritual or political indoctrinators, etc.) and by willingness to imitate (thus making it possible for small groups within society to affect alterations in values, e.g. the rich, the eloquent, the pious, the well-educated).

However, there are a number of more fundamental social values, associated with basic needs, which are determined by the biology of human beings and thus less subject to modification. Food and (at least in some climates) shelter, an acceptable biological environment (an atmosphere with oxygen, temperatures within certain limits, etc.) and human relations (comprising at least sexual interaction) are in this category. The list may be continued with a number of additional values, which become increasingly linked to the type of society in question. Some such values may be quite basic, but owing to the substantial adaptability of human beings, conditions which violate these values or "needs" may persist without apparent dissatisfaction, at least not on a conscious level. Examples of social values of this type are human interaction in a more broad sense, meaningful activities ("work") and stimulating physical surroundings.

This view of basic values was utilised in building the scenarios for energy demand described in Chapter 6. The present world is, to a large extent, characterised by a widespread deprivation of even the basic social values: people suffer from hunger, inadequate shelter and bad health; they are offered unsatisfying work instead of "meaningful activities" and they are surrounded by a polluted environment, ugly varieties of urbanisation, etc.

Many current societies place more emphasis on pursuing the single goal of maximising the gross national income, which is then effectively taken to represent all the desired needs in a single number (see e.g. Mishan, 1969; the definition of gross national product is the added value of all activities or transactions involving money). It is then assumed that the "economic growth" represented by an increasing gross national product positively influences most of the social values, such as access to food, housing and a range of consumer goods, the production and sale of which forms the basis for the economic growth. In recent decades, measures (mostly legislative) which aim at reducing a number of the side-effects of production and material consumption have been implemented in many nations, causing definite improvements with regard to environmental quality, health and working conditions.

The emphasis on economic figures is associated with the use of monetary units to discuss these questions, implying a risk of disregarding those values that cannot be assigned a monetary value.

In principle, any measure of value could be used as a foundation for economic relations between people and nations. The important thing is the way in which monetary values are assigned. Consider, for example, oil. The present extraction price at some oil well in the Near East, followed by transport to Europe and refining may amount to a total of a few US cents or European euros per kg. However, the different refining products are sold to the consumers in Europe at prices (valid for the period 1979–1998) above 1 US\$/euro per kg (for gasoline). Included in these figures are government taxation and royalties, with the taxes in some countries being earmarked as reflecting environmental costs, including that of CO_2 emissions. To a large extent these are substitution prices, determined by the cost of alternative energy sources (e.g. wind power), or the probable cost of developing new types of liquid fuels. In other parts of the world, such as the USA, no environmental taxes are added to oil extraction costs, but only the profits to producers and distributors.

It may then be asked whether any of the oil prices quoted above reflect the social value of oil for present societies. Faced with the threat of a complete cut of oil supplies, most industrial nations would currently attach such a high value to a number of their activities based on oil consumption that they would probably be willing to pay several times the present oil price for some amount of oil. Thus, the social value of oil may be much higher than its present cost in terms of money. One may also consider the value which oil could have to future generations, not just as a handy fuel with high energy density, but also as a raw material for chemical industries. Perhaps a reflection on these aspects would lead to assigning a value to oil not just a few times the present price, but maybe several orders of magnitude higher.

The purpose of the oil example is to illustrate some of the arbitrariness in the ways in which monetary values are assigned by present societies. Oil is a physical commodity, and it is clear that the assignment of a monetary value to social "commodities" of a more abstract kind is no less arbitrary.

Historically, the price of products has also been influenced by the bargaining strength of sellers and buyers, directly or through currency exchange rates. Some see this as a way to pave the way for development of less developed economies. Still, emphasising the substantial arbitrariness of any evaluation in terms of money, it should be said that certain cost comparisons, for example, of alternative technologies for satisfying the same purpose, can be made in a fairly consistent manner by focusing on the expenditure of essential ingredients such as non-renewable raw materials and energy input into the production process. These costs may be quoted in terms of a common monetary unit and may be meaningfully compared, once a suitable framework has been established for assigning monetary values to environmental impacts and depletion of resources in a manner consistent with their long-term social value. These challenges will be taken up in sections 7.1.3 and 7.1.4 on indirect economy and life-cycle analysis.

7.1.2 Economic theory

Economic theories propose to identify quantitative relations between variables describing economic behaviour in society, based on observations of the past. The part of society that may be described in this way depends on the outcome of debates over the possibility of monetising (i.e. assigning a monetary value to) various social variables, ranging from perceived life quality to impacts of climate change on village life in Papua New Guinea. The normative positions of economists range from believing that everything can be monetised to considerably more modest claims. As external conditions, technology, knowledge and preferences can all change with time, the economic rules extracted from past experience are not necessarily valid for the future, and economic theory therefore cannot be used for guiding national or international planning and policy, as these activities depend on perceptions of the future conditions. However, economic theory is often used to make short-term forecasts that are adequately termed "business-as-usual" forecasts, precisely because they assume that the future is based on the same "laws" or relationships as those prevailing in the past.

Economic theories usually deal with a homogeneous society, often in the form of a nation and containing a private business sector as well as a public service sector. There is clearly a need to consider interactions between such individual "economies", through foreign trade and other mutual contacts, which is another subject for economic studies. Simple economic theories often neglect the effect of the size of individual enterprises and the size of each production unit. The latter may have an influence on the time required for changing the level of production, and the former has a considerable influence on the prices.

Economic theories may be regarded as tools for answering "if–then" questions on the basis of past experience. The parameters entering into economic theories are necessarily obtained from shorter or longer time series of past societies. The parameters entering could reflect past changes by incorporating trends in a functional form, and the theory would then predict a dynamic change constituting a non-linear extrapolation of history.

Future conditions are necessarily uncertain for two reasons. One is the possibility of changing attitudes (e.g. to environmental quality) and political prescriptions (which are the result of "free will"); the other is forced changes in conditions (resource depletion, appearance of new diseases, planetary variations causing e.g. new ice ages, etc.). The political task of planning therefore essentially requires techniques for "planning under uncertainty".

A number of features of an economy are highly dependent on the organisation of society. The presence of an influential public sector is characteristic of a substantial number of societies today, but not long ago the industrial sectors in some countries were largely unaffected by governments, and the rapid growth of industrialisation in the Western world about 200 years ago (and somewhat later in a number of other regions) was influenced by the transition from feudal societies, with a large number of local concentrations of power, to nations with increasing amounts of power concentrated in the hands of a central government. In the industrial sector, an increasing division has taken place between the economy of the enterprise itself and that of the owners (stockholders or the state) and the employees. On the other hand, such distinctions are largely absent in many agricultural and small-scale commercial enterprises (farms, family shops, etc.).

There are important formal differences between the ways in which current economic theories view the production process. In a socialist economy, the power to produce goods is entirely attributed to the workers, i.e. only labour can produce a "surplus value" or profit, and machinery and raw materials are expenses on the same footing. In a capitalist economy, machinery and other equipment are considered capable of producing a profit, implying that if one machine operated by one worker is producing the same output as 100 workers, then the machine is attributed 99% of the profit, in contrast to the socialist economy, in which the single worker operating the machine would be attributed all the profit – formally increasing his productivity hundred-fold as compared to before the machines were introduced. Alternatively formulated, in a capitalist economy, only capital is capable of producing profit, and labour is on the same footing as raw materials and machinery: commodities bought at the lowest possible prices by an optimal allocation of capital to these types of expenses.

It is, of course, only of theoretical interest to discuss which of the production factors are assigned the power to produce profit, but it is of decisive practical interest to determine how the profits are actually shared between labour and capital, i.e. how the profits of a given enterprise are distributed among workers and employees on one side and owners (capital providers) on the other side. The state may, as in socialist economies, be considered as collectively representing the workers and administering some of (or all of) their profit shares, but in many non-socialist economies, public spending includes a number of direct or indirect supports to the capitalists, as well as public goods to be shared by all members of society, and a certain amount of recycling of capital back to the enterprises. In mixed economies, which have developed from the basic capitalistic ones, the net distribution of profits may thus be said to involve one portion being pre-distributed to benefit society as a whole and only the residual profit being placed at the disposal of the capitalists.

The *long-term prices* of goods produced in the industrial sector are, in most cases today, determined by the full-cost rule or mark-up procedure, counting the total production costs and adding a standard profit (percentage). In order to change prices, changes in production method, wages or costs of raw materials or duties must take place, or it may temporarily be decided to leave the *fixprice* method. In a fixprice economy, equalisation of output (supply) and demand is accomplished by regulating the output up or down, if supply falls short of or exceeds demand.

In other sectors, such as the agricultural one, there is often a declining rate of profit with increasing production, and pricing is not done by the mark-up procedure, but instead according to a *flexprice* economy, in which an equalisation between supply and demand is achieved by putting the price up if demand exceeds supply and down if supply exceeds demand. The flexprice system is also in evident use (in capitalist economies) for determining *short-term* price variations of industrial products.

In a socialist economy, it is possible to maintain fixed prices on essential goods, such as food, despite fluctuations in the cost of the raw materials due to resource scarcity or abundance, by shifting funds (e.g. to import food, if local production fails) from other sectors, but such procedures are, of course, only applicable if the scarce resources are physically available at some higher cost.

Production planning

The organisation of the production processes in a number of inter-related industries, which are buying and supplying materials and goods between themselves, may be described in terms of a *production function* Y (see e.g. Morishima, 1976), giving a measure of net total output,

$$Y = Y(\{M_i, L_i, R_i\}).$$

Here M_i, L_i and R_i stand for the machinery, labour and raw materials required by the ith process, and the net output z_j of one of the goods entering somewhere in the hierarchy of production processes may be assumed to be given by a linear expression,

$$z_j = \sum_i b_{ji} x_i, \tag{7.1}$$

where x_i is the rate of the ith process and $\{b_{ji}\}$ is a set of coefficients. The entire collection of process rates, $\{x_i\}$, constitutes a *production plan*. It is seen that the conglomerate production function Y involves a weighing of the individual goods (index j), which will normally be described in different units. Equation (7.1) describes both inputs and outputs of goods, prescribing that those z_j which are inputs are given a negative sign and those which are outputs are given a positive sign.

A production plan aiming at maximising the total output, as specified by Y, for given inputs, makes little sense, unless the weighting of different goods is properly chosen. The choice of this weighting may be taken to transform the measure of output into a measure of profit. If the monetary value (price) assigned to each good z_j is denoted p_j (in monetary units per unit quantity), then the net profit, y, may be written

$$y = \sum_j p_j z_j. \tag{7.2}$$

The production plan $\{x_i\}$ may now through (7.1) and (7.2) be chosen so as to maximise the total profit y. The method may be generalised to include all sectors of society, i.e. also the public one.

The allocation of the net total profit to the individual processes, which may be performed by different enterprises (with different owners in a capitalist economy), presents a further problem. A "rational" (but not necessarily socially acceptable) distribution might be to return the total profit to the individual processes, in the same ratios as the contributions of the processes to the total profit. The prices of the individual goods defined in this way are called *"shadow prices"*. Thus, if all the prices p_j appearing in (7.2) are taken to be the shadow prices, then the net total profit y becomes zero. In practice, this would rarely be the case, and even if individual prices within the production hierarchy are at one time identical to the shadow prices, then changes in, for example, raw material costs or improved methods in some of the processes will tend to make y non-zero and will lead to frequent adjustments of shadow prices.

This behaviour emphasises the static nature of the above methodology. No dynamic development of the production and demand is included, and the changes over time can only be dealt with in a "quasi-static" manner by evaluating the optimum production plan, the pricing and the profit distribution, each time the external conditions change and by neglecting the time delays in implementing the planning modifications. Dynamic descriptions of the time development of non-static economies have been formulated in qualitative terms, e.g. in the historical analysis of Marx (1859) and in the approach of Schumpeter (1961).

The quasi-static theory is particularly unsuited for describing economies with a long planning horizon and considerations of resource depletion or environmental quality. By assigning a monetary value to such *"externalities"* or *"external diseconomies"*, a dynamic simulation calculation can be carried out, defining the basic economic rules and including time delays in implementing changes, delayed health effects from pollution, and the time constants for significant depletion of non-renewable resources, either physical depletion or depletion of raw materials in a given price range. Rather than basing the planning on instantaneous values of profit, it would then be based on suitably weighted sums of profits during the entire length of the planning horizon (see e.g. Sørensen, 1976a). The determination of values to be ascribed to externality costs is discussed below under life-cycle analysis.

Furthermore, the applicability of the simulation approach is limited because of the uncertainty inherent in time integration over very long time intervals, using economic parameters which are certainly going to be modified during the course of time or replaced by completely different concepts in response to changes in external conditions as well as major changes in the organisation and value systems of societies. Examples of problems which defy description in terms of conventional economic theories with use of simulation techniques may be the socio-economic costs of possible climatic impact from human (profit-seeking) activities, the global contamination resulting from nuclear warfare and the more subtle aspects of passing radioactive waste to future generations. Such problems, it would seem, should be discussed in a general framework of social values, including the values associated with not restricting the possibilities for future generations to make their own choices by restricting their access to natural resources or by for-cing them to spend a considerable part of their time on our leftover problems rather than on their own.

The quantification of, for example, environmental values and their inclusion in economic planning still suffers from the lack of a suitable theoretical framework. It has been suggested to employ contingent evaluation, e.g. implying that funds are allocated to environmental purposes (reducing pollution of given production processes, setting aside land for recreational purposes, etc.) to an extent determined by the willingness of people (salary earners) to accept the reduction in salaries which corresponds to the different allocation of a given amount of profit (see e.g. Hjalte *et al.*, 1977; Freeman *et al.*, 1973). It is clearly difficult to include long-term considerations in this type of decision process. In many cases, the adverse environmental effects of present activities will not show for many years, or our knowledge is insufficient to justify statements that adverse effects will indeed occur. Only recently have public debates over such issues started to take place before the activity in question has begun. Thus, the discussion has often been one of halting an ongoing activity for environmental purposes, and the argument against doing so has been the loss of the investments already made. It is therefore important to refer the evaluation of environmental problems of production or of consumption to the planning phase, which implies that if environmental costs can be evaluated, they can be directly included in the planning prescription [such as (7.1) and (7.2)] as a cost of production on the same level as the raw materials and machinery (Sørensen, 1976a).

Another method that is in fairly widespread use in present economies is to implement government regulations, often in the form of threshold values of pollution levels (emissions, concentrations, etc.) which should not be exceeded. If then, for a given production process, the emission of pollutants is proportional to the output, this implies a government-prescribed ceiling on the output z_j of a given product in (7.2). This is similar to the ceilings which could be prescribed in cases of a scarce resource (e.g. cereal after a year of poor harvest) or other rationing requirement (e.g. during wartime). Such an environmental policy will stimulate the development of new production processes that have a diminished level of pollution per unit of output, so that in this way the output of goods may still be increased without violating the pollution limits.

The problem is, however, that the government-enforced limits on pollution usually decrease with time, because they were initially chosen too leniently and because additional long-term adverse effects keep appearing (an example is radiation exposure standards). Thus, the socio-economic impacts of environmental offences are constantly underestimated in this approach, and industries and consumers are indirectly stimulated to pollute up to the maximum permitted at any given time. By contrast, the approach whereby environmental impact must be considered earlier, in the production-planning phase, and whereby governments may determine the cost of polluting to be at such a high value that long-term effects are also included, has the effect of encouraging firms and individuals to keep pollution at the lowest possible level.

Distribution problems

In classical market theory, a completely free competition between a large number of small firms is assumed. There are no monopolies and no government intervention in the economy. Extension of this theory to include, for example, a public sector and other features of present societies, without changing the basic assumption, is called *"neoclassical theory"*. According to this, the distribution of production on the firms, and the distribution of demand on the goods produced, may attain an equilibrium, with balanced prices and wages determined from the flexprice mechanism (a *Walras equilibrium*).

In the neoclassical theory, it was assumed that unemployment was a temporary phenomenon, which occurred when the economy was not in equilibrium. Based on observations in times of actual economic crises occurring in capitalist countries, Keynes (1936) suggested that in reality this was not so, for a number of reasons. Many sectors of the "late capitalistic" economies were of the fixprice category rather than the flexprice one, wages had a tendency to be rigid against downward adjustment, the price mechanism was not functioning effectively (e.g. owing to the presence of monopolies), and the relation between prices and effective demand was not following the Walras equilibrium law. Since prices are rigid in the fixprice economy, full employment can be obtained only when demand and investments are sufficiently high, and Keynes suggested that governments may help stimulate demand by lowering taxes and may help increase the economic activity by increasing public spending (exhibiting a "multiplier effect" due to its in-creased demand on consumer goods) or by making it more attractive (for capitalists) to invest their money in means of production. The distinction between consumer goods and capital goods is important, since unemploy-ment may exist together with a high, unsatisfied demand for consumer goods if capital investments are insufficient (which again is connected with the expected rates of interest and inflation).

In a socialist economy, allocation of resources and investments in means of production is made according to an overall plan (formulated collectively or by representatives), whereas in a capitalistic society the members of the capitalist class make individual decisions on the size of investment and the types of production in which to invest. It may then well be that the highest profit is in the production of goods, which are undesirable to society, but still in demand by customers, if sales efforts are backed by aggressive advertising campaigns. It is in any case not contradictory that there may be a demand for such goods, if more desirable goods are not offered or are too expensive for most consumers. Yet relatively little is done, even in present mixed-economy societies, to influence the quality of the use of capital, although governments do try to influence the level of investments.

Unemployment is by definition absent in a socialist society. In a capitalist society, the effect of introducing new technology that can replace labour is often to create unemployment. For this reason, such technology is often fiercely opposed by workers and unions, even if the benefits of the machin-ery include relieving from workers the hardest work, avoiding direct human contact with dangerous or toxic substances, etc. In such cases, the rules of capitalistic economy clearly oppose improvements in social values.

Actual pricing policies

The mechanism of price adjustments actually used in a given economy is ba-sically a matter of convention, i.e. it cannot be predicted by a single theory, but theories can be formulated which reflect some features of the price mechanism actually found in a given society at a given time. A mechanism characteristic of present growth-oriented economies (both socialist and capitalist) is to adjust prices in such a way that excess profit per unit of good is reduced, whenever an excess demand exists or can be created, in order to be able to increase output (Keynes–Leontief mechanism).

Important parameters in an empirical description of these relations are the *price flexibility*,

$$\eta_i = \frac{\Delta p_i / p_i}{\delta d_i / d_i}, \tag{7.3}$$

and the *demand elasticity*,

$$\varepsilon_i = -\frac{\Delta d_i / d_i}{\Delta p_i / p_i}, \tag{7.4}$$

of the ith good or sector. Here p_i and d_i represent price and demand, and δd_i is an exogenous change in demand, whereas Δd_i is the change in demand resulting from a change in price from p_i to $p_i + \Delta p_i$. The change in output, Δz_i, being indicated by an exogenously produced change in demand, Δd_i, is

$$\Delta z_i = \Delta d_i + \delta d_i = (1 - \eta_i \varepsilon_i)\, \delta d_i. \tag{7.5}$$

The use of the indicators η_i and d_i for predicting the results of changing prices or demand is equivalent to a quasi-static theory, whereas a dynamic theory would require η_i and ε_i to be dynamic variables coupled to all other parameters describing the economy through a set of non-linear, coupled differential equations (i.e. including feedback mechanisms left out in simple extrapolative approaches).

The large number of assumptions, mostly of historical origin, underlying the present theoretical description of pricing policy (actual or "ideal") does suggest that comparative cost evaluations may constitute an inadequate or at least incomplete basis for making major decisions with social implications. Indeed, many important political decisions are made in bold disregard of economic theory input. This is relevant for the discussion of energy systems. As a result, a statement that, for example, one system is 10% cheaper than another one would be seen in a new light in cases where the combination of ingredients (raw materials, labour, environmental impact, availability of capital) determining the prices of the two systems are different. There might be a choice between changing the production process to make a desired system competitive or, alternatively, making society change the rules by which it chooses to weight the cost of different production factors.

7.2 Direct cost evaluation

For the purpose of comparing the direct costs of alternative energy systems, actually incurred costs at a given time may not be suitable. The reason may be that the costs of some alternatives are on a rapidly changing course, while those of others are stable. In particular, the cost of some alternatives may be misleading if full-scale production is not in progress. Price fluctuations derived from market changes and temporary campaigns may be avoided by using mark-up prices with identical profit rates directly for comparing the alternatives. However, many renewable energy technologies, such as photovoltaics, are presently in a phase of rapid development and are expected to lead to a lowering of prices, at least in proportion to accumulated volume of production. On the other hand, fuel-based energy systems are influenced decisively by future fuel costs, which are uncertain both because of uncertainties in predicting the rate of resource depletion and because they are artificial prices which will be modified as the technical and political expectations regarding alternative energy systems change.

This section is restricted to a discussion of methods of meaningfully comparing known direct costs of fuels, equipment and operation, in cases where these different types of expenses are differently distributed over time.

7.2.1 Treatment of inflation

In many economies, changes in prices have taken place, which are neither associated with changes in the value of the goods in question nor reflect increased difficulties in obtaining raw materials, etc. Such changes may be called inflation (negative inflation is also called deflation). If they affect all goods, they merely amount to changing the unit of monetary measure and thus would have no effect on the economy, were it not for the fact that the rate of inflation cannot be precisely predicted. In negotiated labour markets, inflation is a way to make up for labour cost increases considered too high by industry. However, people and institutions dealing with money cannot simply correct their interest rate for inflation, but have to estimate the likely inflation over an entire depreciation time in order, for example, to offer a loan on fixed interest terms.

In order to reduce the chance of losing money due to an incorrect inflation estimate, it is likely that the interest rate demanded will be higher than necessary, which again has a profound influence on the economy. It makes the relative cost of capital as a production factor too high, and for a (private or public) customer considering a choice between alternative purchases (e.g. of energy systems performing the same task), a bias in the direction of favouring the least capital-intensive system results. Governments in many capitalistic countries try to compensate (and sometimes overcompensate) for the adverse effects of unreasonably high interest rates by offering tax credits for interest paid.

Often inflation is taken to have a somewhat extended meaning. Individual nations publish indices of prices, giving the relative change in the cost of labour, raw materials, etc. with time. Each of the indices is derived as a weighted average over different sectors or commodities. To the extent that the composition of these averages represents the same quantity, the price indices thus include not only the inflation as defined above, but also real changes (e.g. due to increased difficulties of extracting non-renewable resources, increased living standard, etc.). If the rate of inflation is taken to include all price changes, then the use of the price indices offers one way of comparing the cost of goods produced at different times by correcting for inflation in the extended sense, i.e. by transforming all costs to monetary values pertaining to a fixed point in time.

If the annual inflation of prices relevant to a given problem is denoted i (fraction per year), and the market interest is r_m (fraction per year), then the effective interest rate r is given by

$$(1+r) = (1+r_m)/(1+i). \tag{7.6}$$

Other production costs, such as those of raw materials, machinery and labour, may be similarly corrected, and if (7.6) and the analogous relations are applied repeatedly to transform all costs to the levels of a given year, then the calculation is said to be "in fixed prices".

The difficulty in carrying out a fixed price calculation is again that the inflation rate i may differ from year to year, and it has to be estimated for future years if the calculation involves interest payments, fuels or working expenses, such as operation and maintenance of equipment.

If all changes in prices from year to year are small, (7.6) may be approximated by $r \approx r_m - i$.

7.2.2 Present value calculations

The costs of energy conversion may be considered a sum of the capital costs of the conversion devices, operation and maintenance costs, plus fuel costs if there are any (e.g. conversion of non-renewable fuels or commercially available renewable fuels such as wood and other biomass products). Some of these expenses have to be paid throughout the operating lifetime of the equipment, and some have to be paid during the construction period, or the money for the capital costs (or part of them) has to be borrowed and paid back in instalments.

Thus, different energy systems would be characterised by different distributions of payments throughout the operating lifetime or at least during a conventionally chosen depreciation time. In order to compare such costs, they must be referred to a common point in time or otherwise be made commensurable. A series of yearly payments may be referred to a given point in time by means of the *"present value"* concept. The present value is defined by

$$P = \sum_{i=1}^{n} p_i (1+r)^{-1}, \tag{7.7}$$

where $\{p_i\}$ are the annual payments and n is the number of years considered. The interest rate r is assumed to be constant and ascribed only once every year (rather than continuously), according to present practice. The interest is paid *"post-numerando"*, i.e. the first payment of interest would be after the first year, $i = 1$. Thus, P is simply the amount of money that one must have at year zero in order to be able to make the required payments for each of the following years, from the original sum plus the accumulated interest earned, if all remaining capital is earning interest at the annual rate r.

If the payments p_i are equal during all the years, (7.7) reduces to

$$P = p_{(i)} (1+r)^{-1} ((1+r)^{-n}-1) / ((1+r)^{-1}-1),$$

and if the payments increase from year to year, by the same percentage, $p_i = (1+e)p_{i-1}$ for $i = 2,..., n$, then the present value is

$$P = p_0 (1+r)^{-1} \frac{((1+e)/(1+r))^n -1}{(1+e)/(1+r) -1}. \tag{7.8}$$

Here e is the rate of price escalation (above inflation rate if the average inflation is corrected for in the choice of interest r, etc.) and p_0 is the (fictive) payment at year zero, corresponding to a first payment at year one equal to $p_0(1+e)$.

The present value of a capital cost C paid at year zero is $P = C$, whether the sum is paid in cash or as n annual instalments based on the same interest rate as used in the definition of the present value. For an annuity-type loan the total annual instalment A (interest plus back-payment) is constant and given by

$$A = C(1+r) \frac{(1+r)^{-1} -1}{(1+r)^{-n} -1}, \tag{7.9}$$

which combined with the above expression for the present value proves that $P = C$. When r approaches zero, the annuity A approaches C/n.

The present value of running expenses, such as fuels or operation/maintenance, is given by (7.7), or by (7.8), if cost increases yearly by a constant fraction. It is usually necessary to assign different escalation rates e for different fuels and other raw materials; for labour of local origin, skilled or unskilled; and for imported parts or other special services. Only rarely will the mix of ingredients behave like the mix of goods and services defining the average rate of inflation or price index (in this case prices will follow inflation and the escalation will be zero). Thus, for a given energy system, the present value will normally consist of a capital cost C plus a number of terms, which are of the form (7.7) or (7.8) and which generally depend on future prices that can at best be estimated on the basis of existing economic planning and expectations.

7.2.3 Cost profiles and break-even prices

If successive present values of all expenditures in connection with, say, an energy system are added from year zero to n, an *accumulated present value* is obtained. As a function of n it is monotonically increasing. It is illustrated in

Fig. 7.1 by two different systems. One is a system in which the only expenses are fuels and the related handling. Its accumulated present value starts at zero and increases along a curve which bends downwards owing to the effect of the finite interest level in diminishing the present value of future payments. If fuel prices were to increase, the curve would show an overall tendency to bend upwards. The other system, which may be thought of as a renewable energy conversion system, has a high initial present value reflecting the capital cost of the conversion equipment, but no fuel costs and therefore only a slightly increasing accumulated present value due to working costs.

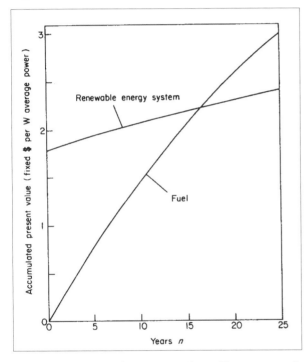

Figure 7.1. Accumulated present value to the year n of total costs for a renewable energy system and for converting fuel in an existing system, as a function of n. The $-values indicated at the ordinate would correspond to typical wind turbine costs and typical coal prices as experienced during the late 20th century in Europe.

The accumulated present value offers a way of comparing the costs of different energy supply systems over a period of time. Curves like those of Fig. 7.1 give a measure of total costs in commensurable units, as a function of the time horizon of the comparison. If the planning horizon is taken as 25 years (e.g. the life expected for the conversion equipment), then the total cost of the renewable energy system in the figure is lower than that of buying fuel to produce the same amount of energy, whereas if the planning horizon is set at 15 years (the standard depreciation period adopted by some utility companies), then there is roughly a break-even.

The accumulated present value may also be used to define the *break-even* capital cost of a new energy system, relative to an existing system. In this case, the depreciation time n used in expressions such as (7.8) is kept fixed, but the capital cost of the new system, C, is considered a variable. The break-even cost, $C(b.e.)$, is then defined as the lowest capital cost C, for which the total present value cost (accumulated to the year n) of the new system is no greater than that of the existing system, including all relevant costs for both systems, such as the (known) capital cost of the existing system (if the new system is aimed at replacing its capacity), the fuel costs and costs of operation and maintenance. If the depreciation time selected for the comparison is taken as the physical lifetime of the equipment, situations may also arise in which different periods n should be used for the systems being compared.

If the construction period is extended, and systems with different construction times are to be compared, the distribution of capital costs over the construction period is simply treated as the running payments and, referred to the common year (year 0) by means of (7.7).

The example shown in Fig. 7.1 assumes a given capital cost C of the renewable energy system. For a depreciation period of 25 years, it is seen that the break-even cost relative to the fuel (that would be saved by installing the renewable energy system) is higher than the assumed cost, $C(b.e.) > C$, implying that the renewable system could be more expensive and still be competitive according to the break-even criterion. For n below 15 years, the cost C is above the break-even value, and the system's cost will have to be reduced in order to make the system competitive according to given assumptions regarding fuel cost, etc.

The production cost of energy depends on the distribution of the capital costs over the depreciation time. Whether the method of financing is by loan-taking, by use of equity or a combination of the two, the cost can (but does not have to) be distributed over different years according to the scheme of annuity payments or according to the return demanded on capital assets.

Assume first that the capital is raised on the basis of an annuity loan or an equivalent scheme with annual instalments given by an expression of the form (7.9), where the interest r is to be taken as the market interest rate r. This implies that no correction for inflation is made and that the annual instalments A are of equal denomination in current monetary values. The left-hand side of Fig. 7.2 illustrates the energy production costs according to this scheme for the two alternatives considered in Fig. 7.1. The rate of inflation has been assumed to be 10% per year, the rate of interest corrected for inflation is 3% per year, and thus according to (7.6) the market interest $r \approx 13.3\%$ per year.

If these same energy costs are corrected for inflation, so that they correspond to values in fixed year-0 monetary units, the results are as shown in the right-hand side of Fig. 7.2 (solid lines). The fuel cost is constant in fixed prices (assumption), but the cost of the renewable energy declines rapidly, especially during the first few years (although it is increasing in current prices, as shown to the left).

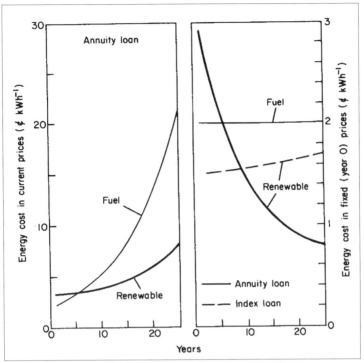

Figure 7.2. Energy costs for a renewable energy or fuel-based system (cf. Fig. 7.1), excluding cost of transmission. Solid lines correspond to financing through annuity loans, and the dashed line (for the renewable energy system) corresponds to financing by index loan, required to give the same 25-year accumulated present value of payments as the annuity loan. The left-hand part of the figure is in current prices, whereas the right-hand side is in fixed prices, assuming an annual rate of inflation equal to 10%. Note the difference in scale.

This behaviour illustrates a barrier that the annuity loan system poses against any capital-intensive system. Although the renewable energy system is competitive according to the accumulated present value criterion (when considered over 25 years as in Fig. 7.1), the energy cost during the first year is about 50% higher than that of the fuel it replaces, and only after the sixth year does it become lower. From a private economy point of view, it may not be possible to afford the expenses of the first five years, although in the long term the system is the cheaper one.

The example also illustrates the fallacy of comparing costs of different (energy) systems by just quoting the payments (running plus instalments on loans) for the first year of operation, as is sometimes done.

If cost comparisons are to be made in terms of energy costs, an improved prescription is to quote the average cost during the entire depreciation period (or lifetime) considered. This is meaningful only if fixed prices are used, but then it is equivalent to the present value method, in the sense that a system that is competitive according to one criterion will also be according to the other.

The creation of loan conditions which would eliminate the adverse effect of inflation on the back-payment profile has been considered (Hvelplund, 1975). The idea is to abandon the constancy (in current prices) of the annual instalments and instead modify each payment according to the current price index. If this procedure is followed in such a way that the accumulated present value of all the payments during the depreciation period equals that of the annuity loan, then both types of instalment are given by the expression (7.9), but for the index loan, the interest r is to be corrected for inflation, and A is given in fixed year-0 prices. The index-regulated instalments in current prices are then

$$A_m (j) = (1+i)^j C (1+r) ((1+r)^{-1} - 1) / (((1+r)^{-n} - 1),$$

for the ith year, assuming the inflation to be constant and equal to the fraction i per year. For the annuity loan, the instalments in current prices are

$$A_m (j) = C (1+r_m) ((1+r_m)^{-1} - 1) / (((1+r_m)^{-n} - 1).$$

In fixed prices, both expressions should be divided by $(1+ i)^j$. For the example considered in Fig. 7.2, both payment schemes are illustrated on the right-hand side, with the index type loan given by a dashed line. It shows that the system with the lowest cost in terms of present value is also lowest in annual energy cost, for each of the years, and in fixed as well as in current prices. While loans based on price indices have not been used in practice, the use of variable interest has been tried, in order to remove the premium paid for uncertainty in future interest rates.

The energy prices discussed so far are actual production prices, for different types of financing. The price actually charged for energy is, of course, a matter of policy by the enterprise or state selling the energy. They could choose to distribute the prices over the time interval in a manner different from the distribution of actual production costs.

The "free parameter" in the index loan type is the effective interest rate r (i.e. the interest corrected for inflation), which determines the initial instalment that is to be regulated annually according to the price index. Historically, there have been substantial fluctuations in the value of r (approximately $r - i$), but decade averages have been in the range 0–5% for most nations, with short periods of higher values, e.g. following the 1974 Arab oil embargo. The 20th century average is 3% per year in most industrialised countries. Banks presently offering "index loans" demand quite a high interest r, implying that the present value of their returns on such loans is higher than for annuity loans. It will be argued in section 7.3.1 that society may prescribe a low (or even zero) value of r for projects considered appropriate. It is clear that index regulation of instalments could be used in this connection.

Cost comparisons in industry are often made by use of a simple "rule of thumb" for the capital investments, assuming that the annual "cost of capital" is a fixed percentage of the investment. This may be a reasonable first approximation, if the fixed percentage chosen includes the effect of inflation over the depreciation period, rather than just reflecting the annuity or return on the investment during the first year after the investment is made. Another concept often used in capitalist industry is "internal interest", which is the return on capital investments, and it thus endorses the capitalistic assumption that all profits are assigned to capital and that, for example, workers are "production costs" which are not assigned further shares of the surplus value produced (cf. section 7.1.2).

7.3 Indirect economics

The term "indirect economics" may be taken to relate to those social values (costs or benefits) which are not or cannot be evaluated in monetary units. To some extent, the discussion below is simply a list of considerations which at present are not included in most direct economic evaluations, but which may be considered if agreement is reached on their social importance. The specific interest in the indirect economic factors in connection with energy supply systems using renewable energy sources is derived from the hypothesis that many of the factors presently regarded as "indirect economics" will place the renewable energy systems in a relatively more favourable position, relative both to the short-term fuel-based alternatives (fossil and fissile sources) and to the long-term alternative (fusion sources).

7.3.1 Resource and environmental management

The effort required in order to extract non-renewable resources is generally expected to increase as easily accessible deposits become depleted and less and less accessible sources have to be exploited. This tendency is counteracted by the development of more ingenious extraction methods, but if effort is measured in terms of the energy required for extraction of a given raw material (e.g. bringing a certain amount of composite rock to the surface and breaking the physical or chemical bonds by which the desired material is attached to other rock materials), then there is a well-defined minimum effort which has to be provided.

For a number of raw materials, methods exist or could be developed for recycling, i.e. using discarded production goods as a resource basis rather than materials extracted from the natural surroundings. This would seem of obvious advantage if the concentration of the desired material is higher in the scrap material than in available natural deposits and if the collection and extraction costs for the recycled material are lower than the extraction and transportation costs for the new natural resource. A question to be raised in this connection is whether it might be advantageous to base the decisions on recycling not on instantaneous costs of recycling versus new resource extraction, but instead on the present value cost of using either method for an extended period of time (e.g. the planning horizon discussed earlier).

The recycling decision would not have to be taken in advance if the scrap resources are kept for later use, waiting for the process to be profitable. However, in many cases a choice is being made between recycling and further dispersion or dilution of the waste materials to a form and concentration that would be virtually impossible to use later as a resource base. Thus, the consideration of resource depletion is one that contributes to the indirect economy of a given production process and that should be considered if the planning horizon is extended.

Of all the non-renewable resources, energy resources play a special role, partly because they cannot be recycled after the energy conversion has taken place (combustion, nuclear transformation) and partly because of their part in the extraction of other raw materials and manufacturing processes.

Energy analysis

For energy conversion systems, the energy accounting is of vital interest because the *net energy* delivered from the system during its physical lifetime is equal to its energy production minus the energy inputs into the materials forming the equipment, its construction and maintenance. Different energy supply systems producing the same gross energy output may differ in regard to energy inputs, so that a cost comparison based on net energy outputs may not concur with a comparison based on gross output.

The energy accounting may be done by assigning an *"energy value"* to a given commodity, which is equal to the energy used in its production and in preparing the raw materials and intermediate components involved in the commodity, all the way back to the resource extraction, but no further. Thus, nuclear, chemical or other energy already present in the resource in its "natural" surroundings are not counted. It is clear that the energy values obtained in this way are highly dependent on the techniques used for resource extraction and manufacture.

For goods which may be used for energy purposes (e.g. fuels), the "natural" energy content is sometimes included in the energy value. There is also some ambiguity in deciding on the accounting procedure for some energy inputs from "natural" sources. Solar energy input into, for example, agricultural products is usually counted along with energy inputs through fertilisers, soil conditioning and harvesting machinery, and (if applicable) active heating of greenhouses, but it is debatable what fraction of long-wavelength or latent heat inputs should be counted (e.g. for the greenhouse cultures), which may imply modifications of several of the net energy exchanges with the environment. Clearly, the usefulness of applying energy analysis to different problems depends heavily on agreeing on the definitions and prescriptions that will be useful in the context in which the energy value concept is to be applied.

Early energy analyses were attempted even before sufficient data had become available (Chapman, 1974; Slesser, 1975; von Hippel et al., 1975). Still, they were able to make specific suggestions, such as not building wave energy converters entirely of steel (Musgrove, 1976). More recently, a large number of energy systems have been analysed for their net energy production (see e.g. the collection in Nieuwlaar and Alsema, 1997).

It should be said, in anticipation of the life-cycle approach discussed in section 7.4, that if all impacts of energy systems are included in a comparative assessment, then there is no need to perform a specific net energy analysis. Whether a solar-based energy device uses 10 or 50% of the originally captured solar radiation internally is of no consequence as long as the overall impacts are acceptable (including those caused by having to use a larger collection area to cover a given demand).

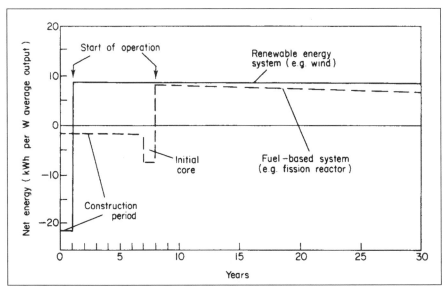

Figure 7.3. Sketch of net energy profiles for renewable energy and fuel-based energy systems.

In order to illustrate that net energy is also characterised by time profiles that may differ for different systems, Fig. 7.3 sketches the expected time profiles for the net energy production of two different types of energy conversion systems. They are derived from cost estimates in Sørensen (1976d) by simply assuming a constant energy content of 30 MJ per (1978)US$ spent as capital investment or on operation and maintenance, apart from the uranium fuel. The energy value of the fission reactor fuel elements is estimated as 75 MJ per (1978)US$ initially, but rising linearly to 150 MJ per (1978)US$ at the end of the 30-year period considered. For the fuel-based system, the energy used in extraction will constitute an increasing fraction of the total effort (cost), as the grade of resource (concentration in ore, etc.) becomes lower, as the non-renewable resource is increasingly depleted. For both systems, the initial energy "investment" is recovered over 2–3 years of energy production, according to the estimate of Fig. 7.3.

The prolonged construction period of some energy systems does present a problem if such systems are planned to be introduced on a large-scale over a short period of time. The extended delay between the commencement of energy expenditures and energy production makes the construction period a potentially serious drain of energy (Chapman, 1974). This problem is less serious for modular systems with short construction periods for each individual unit.

Social interest rate

The concept of interest is associated with placing a higher value on present possession of assets than on future ones. Using the monetary value system, a positive interest rate thus reflects the assumption that it is better to possess a certain sum of money today than it is to be assured of having it in 1 year or in 20 years from today. The numerical value of the interest rate indicates how much better it is to have the money today, and if the interest has been corrected for the expected inflation, it is a pure measure of the advantage given to the present.

For the individual with a limited life in front of her or him, this attitude seems a very reasonable one. But for a society, it is difficult to see a justification for rating the present higher than the future. New generations of people will constitute the society in the future, and placing a positive interest on the measures of value implies that the "present value" (7.7) of an expense to be paid in the future is lower than that of one to be paid now. This has a number of implications.

Non-renewable resources left for use by future generations are ascribed a value relatively smaller than those used immediately. This places an energy system requiring an amount of non-renewable fuels to be converted every year in a more favourable position than a renewable energy system demanding a large capital investment now.

Part of the potential pollution created by fuel-based energy conversion, as well as by other industrial processes, is in the form of wastes, which may either be treated (changed into more acceptable substances), diluted and spread in the environment, or concentrated and kept, i.e. stored, either for good or for later treatment, assuming that future generations will develop more advanced and appropriate methods of treatment or final disposal. A positive interest rate tends to make it more attractive, when possible, to leave the wastes to future concern, because the present value of even a realistic evaluation of the costs of future treatment is low.

For instance, any finite cost of "cleaning-up" some environmental side-effect of a technology (e.g. nuclear waste) can be accommodated as a negligible increase in the capital cost of the technology, if the cleaning-up can be postponed sufficiently far into the future. This is because the cleaning-up costs will be reduced by the factor $(1+r)^n$ if the payment can wait n years (cf. the discussion in section 7.1.2 of proposed economic theories for dealing with environmental side-effects, in connection with "production planning").

It may then be suggested that zero interest rate be used in matters of long-term planning seen in the interest of society as a whole. The social values of future generations, whatever they may be, would, in this way, be given a weight equal to those of the present society. In any case, changing from the market interest rate to one corrected for inflation represents a considerable step away from the interest (in both senses) of private capitalistic enterprises and towards that of society as a whole. As mentioned above, the interest rate corrected for inflation according to historical values of market interest and inflation, and averaged over several decades, amounts to a few per cent per year. However, when considering present or presently planned activities which could have an impact several generations into the future (e.g. resource depletion, environmental impact, including possible climatic changes), then even an interest level of a few per cent would appear to be too high to take into proper consideration the welfare of future inhabitants of the Earth. For such activities, the use of a social interest rate equal to zero in present value calculations does provide one way of giving the future a fair weight in the discussions underlying present planning decisions.

One might argue that two facts speak in favour of a positive social interest rate: (i) the capital stock in terms of buildings and other physical structures, that society passes to future generations, is generally increasing with time, and (ii) technological progress may lead to future solutions that are better and cheaper than those we would use to deal with problems of e.g. waste today. The latter point is the reason why decommissioned nuclear reactors and high-level waste are left to future generations to deal with. However, there are also reasons in favour of a negative social interest rate: environmental standards have become increasingly more strict with time, and regulatory limits for injection of pollutants into both humans (through food) and the environment have been lowered as function of time, often as a result of new knowledge revealing negative effects of substances previously considered safe. The problems left to the future may thus appear larger then, giving rise to a negative contribution to social interest, as it will be more expensive to solve the problems according to future standards. The use of a zero social interest rate may well be the best choice, given the uncertainties.

However, it is one thing to use a zero interest rate in the present value calculations employed to compare alternative future strategies, but quite another thing to obtain capital for the planned investments. If this capital is not available and has to be procured by loans, it is unlikely that such loans will be offered at zero interest rate. This applies to private individuals in a capitalistic society and to nations accustomed to financing certain demanding projects by foreign loans. Such borrowing is feasible only if surpluses of capital are available in some economies and if excess demand for capital is available in others. Assuming that this is not the case, or that a certain region or nation is determined to avoid dependence on outside economies, then the question of financing is simply one of allocation of the capital assets created within the economy.

If a limited amount of capital is available to such a society, it does not matter if the actual interest level is zero or not. In any case the choices between projects involving capital investments, which cannot all be carried out owing to the limited available capital, can then be made on the basis of zero interest comparisons. This makes it possible to maximise long-term social values, at the expense of internal interest for individual enterprises. Social or "state" intervention is necessary in order to ensure a capital distribution according to social interest comparisons, rather than according to classical economic thinking, in which scarcity of money resources is directly reflected in a higher "cost of money", i.e. positive interest rate.

7.3.2 Regional economy

The economy of most energy systems is highly dependent on the load structure. Two aspects of load structure are the time and spatial distributions of load, which may influence system choice and mode of operation. Examples of time distributions of load were given in section 6.2.2, and the implications for particular systems were described (e.g. in section 6.3.2). One example of the importance of the spatial distributions of load was given in the section on community-size solar heat systems (section 6.3.1). The important parameter here was identified as a kind of density or *"energy use density"*, for example, specified in terms of the average spacing between load "points" (i.e. individual dwelling units in the example of section 6.3.1). This parameter is decisive in determining the economic feasibility of, for example, district heating lines, gas distribution lines, electricity-carrying grids or the distribution of portable fuels (or other stored energy) by vehicles.

The load structure also has an impact on indirect economic aspects through its possible influence on physical planning and regional infrastructure in general. However, this is not a one-way process. Although certain types of infrastructure could not be imagined without access to concrete energy systems (and other specific technology), the causal relationship will often be from the choice of a given type of infrastructure to an implied energy supply requirement, specifying both quantity and form of energy. Examples are dense city settlements, allowing for transmission line energy supply from central units, in contrast to dispersed individual housing settlements, favouring individual energy systems (at least for heating) and creating a large energy demand for individual transportation, for example using automobiles (because distances are too large for walking or bicycling, and population densities are too small for collective transportation). The increasing commuting distances between home and working place which have developed in most industrialised regions may be said to have been made possible only by access to cheap oil products, as has the transition from small shops within the residential areas to large shopping centres which are often much further away from living areas and thus require substantial amounts of travel.

If energy conservation is considered of sufficient importance, the individual houses with large heating requirements (due to large surface area and better wind access) might be valued less than more densely placed dwelling units. The latter, along with the associated service facilities and perhaps also production industry (the pollution of which should be suppressed, irrespective of siting), would be placed in groups along the routes of public transportation. The ideal structure for such a society would still be characterised as fairly decentralised, because the minimisation of transportation distances would still require many small centres, each possessing the frequently used facilities for service and recreation.

It is sometimes said that the introduction of large energy conversion units and extensive transmission systems will promote centralisation of activities as well as of control, while small conversion units (e.g. based on renewable energy) would promote societies of decentralised structure, with emphasis on regional development. This view obviously cannot be entirely substantiated in regard to the size of conversion plants. An array of 100 wind energy converters or a solar farm may have rated capacities of magnitude similar to those of large fuel-based power plants, and in the case of wind energy, for example, it would not make much sense to disperse the array of converters from a wind-optimal site to the actual load areas, which would typically be characterised by poor wind conditions. Furthermore, many renewable energy systems greatly benefit from the existence of power grids and other energy transmission systems, allowing alternative supply during periods of insufficient local renewable energy, instead of requiring dedicated local energy stores. However, the centralisation–decentralisation aspect may play an important role with regard to the distribution of employment and penetration of technology, based on the requirements in the construction phase and for operation and service of equipment.

Some of the renewable energy technologies (e.g. those considered in the examples of sections 6.3–6.4) depend mainly on materials and skills which are found or may be developed locally in most regions. They are then well placed to encourage regional development in a balanced way, in contrast to technologies which depend heavily on very specialised technology that is available at only a few places. Generally, the types of and organisation of research and development programmes related to individual energy conversion and transmission technologies play an important indirect economic role. In some cases society will subsidise a particular technology by paying its research and development costs; in other situations, private enterprises pay these costs and include them in the final selling price of the energy conversion equipment. Still more indirect is the influence of R&D efforts, as well as the types of skills required to produce and run specific (energy) technological systems, on preferences and directions in education and choice of profession. Clearly, there is a feedback to political decisions on, for example, energy options or military strategy from factors such as a strong emphasis on either nuclear engineering or ecology in the educational system.

7.3.3 National economy

In mixed or socialist economies, certain productions may be maintained even if they yield a negative profit. One purpose of this may be to maintain the capability of supplying important products when external conditions change (e.g. maintaining a food production capacity even in periods when imported foods are cheaper) or to maintain a regional dispersion of activity which might be lost if the economic activities were selected strictly according to direct profit considerations. This may apply to regions within a nation or to nations within larger entities with a certain level of coordination.

As energy supply is important to societies, they are likely to be willing to transfer considerable funds from other sectors to ensure security of energy supplies and national control over the energy supply sources and systems.

International relations comprise trade of raw materials and intermediate and final products. International trade, not only of the unevenly distributed raw materials, has now reached very substantial dimensions. Often quite similar goods are exchanged between nations in different parts of the world, with the only visible result being increased demand for transportation. Of course, all such activities contribute to the gross national product, which has been increasing rapidly (even after correction for inflation) over recent decades in most countries in the world.

A measure of a nation's success in the international trade is its *balance of foreign payments,* i.e. the net sum of money in comparable units, which is received by the nation from all other nations. Since the world-wide sum of foreign payment balances is by definition zero, not all nations can have a positive balance of payments, but they could of course all be zero.

An important factor in foreign trade is each dealer's willingness to accept payments in a given currency. To some extent, currencies are simply being traded in the same way as other commodities, using basically the flexprice method for fixing exchange rates. It is also customary in international trade of goods that governments are called upon to intervene with some form of guarantee for the payments to be made. In questions of financing, the international "trade of money" has led to an increasing amount of foreign loans, given both to individual enterprises and to local and national governments. A positive balance of current foreign payments may thus indicate not that a surplus of goods has been exported, but that a large number of loans have been obtained, which will later have to be returned in instalments including interest. In addition to the balance of current payments, the *balance of foreign debts* must then be considered in order to assess the status of an individual nation. The origin of foreign financing schemes is, of course, the accumulation of capital in some countries, a capital which has to be invested somewhere or other in order to yield a positive interest.

Assuming that all nations try to have a zero or positive balance of foreign payments, they would be influenced in their choice of, for example, energy systems in such a way as to prefer systems most in accordance with this objective. For this reason, the direct economic evaluation should be complemented with a listing of the *import values* of all expenses (equipment, maintenance and fuels, if any) for the energy supply systems considered. The import value is defined as the fraction of the expense which is related to imported goods or services (and which usually has to be paid in foreign currency). The import value of different energy technologies depends on methods of construction and on the choice of firms responsible for the individual enterprises. It is therefore heavily dependent on local and time-dependent factors, but is nevertheless a quantity of considerable importance for national planning. In many cases, the immediate flexibility of the import value of a given system is less than that implied by the above discussion, because of the time lags involved in forming a national industry in a new field or, depending on the number of restrictions placed on foreign trade, because of the existence of well-established foreign enterprises with which competition from new firms would be hard to establish. Examples of calculations of import values for different energy technologies, valid for the small nation of Denmark, may be found in Blegaa *et al.* (1976).

Employment

In a comparison of alternative investments, there may be an evaluation of the influence of each investment on the employment situation. If the social benefits from the different investments are comparable, the one with the highest employment would be chosen in a situation of unemployment, and the one with lowest employment would be chosen in a situation of labour scarcity. This type of prescription may be used directly in capitalistic economies, but the employment impact of different technologies may also be considered in socialistic or "welfare" economies, even if the concepts of unemployment or over-employment are not officially used. One may talk of "hidden unemployment" if there is not enough meaningful work available for distribution among the "work force", and of course, if the work that is available is shared, the work load on the individual may become higher than desired in times of rapid expansion.

The employment situation for a given (energy) system may be described in terms of an *employment factor,* defined as the direct employment (as measured in numbers of man-years or man-hours) associated with a given sum of money spent on the system. In addition, there may be an indirect employment factor associated with the activities initiated on the basis of the profits and payments made to workers (directly and in the supply industries), assuming that they would otherwise be unemployed. This "multiplier effect" is in a socialistic economy directly associated with the surplus value created, and in a (mixed) capitalistic economy it is divided into one part being at the disposal of the workers (salary minus unemployment compensation, after taxes), one part at the disposal of the capitalists (net profit after taxation), and finally, one part administered by the public (the taxes).

Not all the employment is necessarily local, and it may be relevant in a national evaluation to disregard employment created outside the country. If this is done, the employment factor becomes closely related to the import value, in fact proportional to one minus the import fraction, if employment factors and import values are assumed to be the same for all sectors of the economy, to a first approximation.

If the purpose is to satisfy an expected energy demand, it may be more relevant to compare employment for different systems supplying the same average amount of energy, rather than comparing employment for systems with the same cost (the employment factor per unit investment defined above). It should also be kept in mind that if the real end-use can be obtained with a reduced energy expenditure, then this has the same value (both for the individual and for society) as satisfying the demand by addition of new conversion units. Thus, as long as the final service rendered is unaltered, an energy unit saved by conservation is worth as much as an additional energy unit produced. This has the implication for the direct cost comparison that investments in energy conservation efforts should be considered together with investments in new supply systems and that conservation should be given priority, as long as the cost of saving one energy unit is lower than that of producing one. As mentioned in the previous sections, such a priority system should not exclude long-term considerations, including the time delays in the implementation of conservation measures as well as new systems, but it would appear that many conservation efforts not undertaken at present are less costly than any proposed alternative of additional supply systems.

For the employment considerations, the employment factor of schemes for improving conversion efficiency, as well as that of other energy conservation measures which do not change the end-use social values, should be compared with those of energy supply systems. Examples of both kinds may be found (e.g. in Blegaa *et al.*, 1976; Hvelplund *et al.*, 1983). They derive total employment factors, which are fairly independent of the technology considered (all of which are highly industrialised, but on different scales), and are roughly equal to 35 man-years per million (1978)US$ spent. The national employment is obtained by multiplying this number by the fraction not imported.

As mentioned earlier, employment or "work" is in itself not a goal of society, and unemployment can always be eliminated by a distribution procedure which simply shares the available work. However, meaningful work for each individual may well be a social goal which can be pursued by proper incorporation of employment implications of alternative systems in the planning considerations.

Use of subsidies for introducing "appropriate technology"

In capitalistic economies with an influential public sector concerned with social priorities (i.e. a mixed economy), the allocation policy of the government may be used to support individuals or firms that wish to introduce the technological solutions judged most appropriate from the point of view of society, even though other solutions are more attractive in a private economic assessment.

A number of subsidy methods are available, including direct subsidy either to manufacturers or to customers, tax credits, loan guarantees and offers of special loans with favourable conditions.

Subsidies can be used for a limited period to speed up the initial introduction of appropriate technology, assuming that in a mature phase the technology can survive on its own. However, the discussion in the previous sections regarding the arbitrariness of costs included in a direct economic assessment makes it likely that some solutions may be very attractive to society and yet unable to compete in a market ruled by direct costs alone. Here redistribution subsidies can be used, or governments can decide to change, the rules of price fixation (using environmental taxation or regulatory commands regarding choice of technology, as is done e.g. in building regulations for reasons of safety). In both cases, democratic support for these actions must be present, assuming that we do not want to introduce a command economy.

Government allocation of funds generally offers a way of changing the distribution of profits in a given direction. The funds available to a government, in part, derive from taxes which are collected according to a certain distribution key (rates of taxation usually differ for different income sizes, and are different for individual taxpayers and enterprises). In recent years, taxation has been used to deal with externality costs, i.e. taxes directly related to environmental costs picked up by society (e.g. carbon taxes in the case of global warming).

Since renewable energy technologies help relieve the burden of continued exploitation of non-renewable resources, with the associated environmental problems, they may be particularly eligible as candidates for "appropriate technologies", which should be supported and if necessary given government subsidies. Obvious conditions for the renewable energy technologies to fulfil are that they do not themselves create environmental problems comparable to those of the replaced technologies and that the construction and operation of the renewable energy systems do not involve increased exploitation of non-renewable resources. Remarks on the need to consider the environmental impact of large-scale renewable energy utilisation have been made earlier (e.g. in Chapter 3), but clearly, renewable energy extraction offers the possibility of limiting the climatic impact to introducing delays and diversions in naturally occurring cycles, in a way that can be planned and controlled and that the operation of renewable energy systems does not involve pollution of the kind associated with fuel-based energy conversion.

With regard to the resource demand of the renewable energy systems, it may be divided into a part comprising materials that can be recycled and another part comprising non-renewable inputs including fuels. These must stay at acceptable levels, as otherwise the system is not sustainable.

From the national government point of view, it would thus seem that indirect economic considerations make use of renewable energy sources eligible as an optimal long-term solution, worthy of support and possibly subsidy, if short-term economic considerations give a less favourable result than the long-term considerations. In situations of unemployment, governments of nations with mixed "welfare" economies which are providing unemployment benefit could then consider replacing the unemployment benefit by subsidies to build renewable energy systems, without serious resource implications (perhaps even with a positive influence on the balance of foreign payments, if the renewable energy systems are replacing imported fuels) and without having to increase taxes. Historically, unemployment periods have offered opportunities for positive restructuring of infrastructure in society.

In those cases where the energy supply is in the hands of the government (e.g. decided by a department of energy), it can simply choose the appropriate solutions. If, on the other hand, the investments and choices of energy supply system are made by individual citizens (as is often the case for residential heating systems) or by various enterprises (industries with their own energy supplies, electricity-selling utility companies, which in some countries are private companies), then the government has to use either compulsory regulations or taxation/subsidies to influence the choice of systems.

If technological improvements depending on scale and hence demanding large sales are the limiting factors, direct government subsidy to customers for a limited period of time may increase the market sufficiently. Direct subsidies to producers are often refused by industrial enterprises operating in competitive markets. On the other hand, placement of large orders by the government (e.g. energy systems for public buildings, etc.) may have the desired effect. If no obvious mass production benefits in terms of reduced prices are discernible, but the technology is still socially desirable despite needing a long depreciation period to obtain break-even with competing systems, then direct subsidies are not likely to be of much help. However, the government may offer or arrange access to index loans, which have the same present value of all payments as the loan types commercially available. This means that a private investor can borrow money for the renewable energy system in question and can pay back according to the index loan curve (illustrated by dashed line on the right-hand side of Fig. 7.2), rather than according to the annuity type loans common at present (heavy, solid line in Fig. 7.2). The instalments on the index loan, which are constant in fixed (inflation corrected) prices, may be determined each year according to the inflation for the past year and a fixed interest rate in fixed prices, which as mentioned should be just a few per cent per year or even zero, if the social value of the system is considered sufficiently high. A government-induced change of the loan types from annuity to index loans, for the buyers of the subsidised technologies, is not expected to influence the prices of product, because the present value of total payments is unaltered. Direct subsidies to customers, on the other hand, may cause prices to rise, which of course is not the intention.

A widely accepted form of government subsidy is to provide research and basic development efforts. A problem revealed by looking at the history of government involvement in science and technology R&D is that of dual-purpose support, for example, providing the energy industry with highly developed technology transferred from other sectors, such as technology developed for military or space applications. It is not clear that such transferred technology is optimal or even appropriate for use in the energy sector.

One important planning parameter is the amount of uncertainty associated with the evaluation of a given technology which is considered for introduction into the energy supply system. This uncertainty may be economic or physical. Systems based on future fuel supplies have important economic uncertainties associated with the future development of fuel prices. Renewable energy systems, on the other hand, are usually characterised by having most of the lifetime costs concentrated in the initial investment, implying that once the construction is over there is no uncertainty in energy cost. On the other hand, such systems may have physical uncertainties, for example, to do with length of life and maintenance costs, particularly if the technology is new and totally unlike technologies that have been used long enough in other fields for the relevant information to have been gathered.

As an example of the sensitivity of choosing the right subsidy policy, one may look at the introduction of wind power in various countries since 1975. Following the oil supply crisis in 1973-1974, many countries offered subsidies for introduction of renewable energy, including wind turbines, but without reaching the desired effect. In several countries, subsidies were given directly to manufacturers, while in other countries, the subsidies were given to those purchasing wind turbines. However, in most cases, wind turbine prices did not decline as a result of the subsidy. One exception was Denmark, where the subsidy (initially 30% of the cost, declining with time to zero) was given to the customers, but only for approved technologies, where the approval was based on the manufacturer being able to make a plausibility argument (not prove!) that his technology had or would soon have a payback time under 8 years. This scheme was implemented by the Danish government following recommendations from an independent working group (ATV, 1975, 1976). At the time, neither wind power nor the other new renewable energy technologies could achieve an 8-year pay-back time, but the manufacturers provided calculations suggesting the plausibility of getting there in the near future, and once starting selling units with the subsidy, had a strong incentive to lower prices to the promised level. There was no way the subsidy could just be added to the sales price, and the following years saw an effort to improve the technology and streamline the production process that made Danish wind turbines increasingly viable and helped create a strong export market, in addition to the backbone of a healthy home market. Other countries with a more conventional subsidy policy did not succeed in creating a viable wind industry, and it is only in recent years that the Danish wind industry has seen serious competitors.

7.3.4 World economy

At present, there are vast discrepancies between different parts of the world, and sometimes between different regions within a nation, with respect to degree of industrialisation, richness of natural resources, as well as the extent to which basic human needs are satisfied, and the amount and distribution of material goods. The goals of different societies should all comprise the coverage of basic needs (cf. sections 6.2.2 and 7.1.1), but otherwise may differ greatly.

It is important to recognise that some societies may set themselves goals which appear reasonable when seen in isolation, but which actually exclude the fulfilment of (perhaps even basic) goals in other societies. Examples of this may be found in the exploitation, by wealthy industrialised countries, of natural resources found in countries where basic needs are unsatisfied. In the same way, if all the reserves of arable land were in some countries while other countries were unable to provide basic food for their populations, one might ask whether it would be reasonable for the privileged countries to choose to set aside all the surplus arable land for recreational purposes.

In principle, countries with many unsatisfied needs may sell their natural resources, if they possess any that are valued by other nations (e.g. fuels), and in this way obtain capital to use in trying to satisfy their own needs. However, not all countries in need do possess such resources, or the prices they obtain for their resources may not be high enough to cover their needs. It may also be that the resource exportation is in immediate conflict with the use of the same resources within the country, or that the resources exported would have been vital in pursuing developments within the country over a longer time period.

Despite the intense international trade in raw materials and produced goods (industrial or agricultural), it does appear that the inequities in the availability of development options are increasing, implying that the mechanisms of (more or less free) competitive trade are inadequate in the pursuance of global goals. Furthermore, the conditions of trade are traditionally fixed in a fairly unilateral way by the "strong" industrial nations. The countries with basic unsatisfied needs, but without excess natural resources to sell, are forced to offer primarily cheap labour and goods made particularly competitive by low salary levels (e.g. textiles). However, the strong nations have repeatedly enforced regulations on the quantities of such goods to be imported, and as a result, the rate of improving conditions in the "weak" countries has been too slow. The lack of fulfilment of basic needs is still widespread; it has increased as a result of a population growth offsetting the positive effect of trade and international aid (which often returns only a fraction of the profits made, in much the same way as re-allocation of some profits by the state in a mixed national economy).

There is an increasingly urgent need to find new types of distribution keys and national development plans with a global perspective in order to counteract the inequities of the present. As mentioned above, the more serious problems are inequities in development options rather than inequities in absolute levels (of GNP or whatever). Indeed, the national goals in terms of social values may be very different for different nations and regions (e.g. urban, rural), and it would not be right to attempt to push all nations through the same kind of development (and mistakes) as the one undergone by the most "advanced" nations.

This is particularly relevant in discussions of "appropriate" technology, such as energy technology. The energy technologies most likely to provide early benefits in rural areas are necessarily decentralised, as are the renewable energy sources. Furthermore, at least some of the renewable energy technologies are consistent with using the local skills and raw materials present almost everywhere.

In the highly industrialised nations, the most urgent priorities from a global point of view would seem to be stopping those activities that constitute a drain on the development possibilities in other parts of the world. This would probably involve devoting more skills to prudent use of non-renewable resources, increased recycling and higher efficiency with respect to end-use. Use of renewable resources (e.g. for energy production) and securing increased efficiency in conversion to end-use are both examples of policies that are consistent with a global perspective. A transitional period of mixed fuel/renewable energy-based supply systems is, of course, required.

Without doubt, the most important reason for adopting a course with a strong emphasis on producing equitable development options for every nation or region is the desirability of reducing any factors likely to lead to conflict. At present, both the affluent and some of the poor nations are spending large sums on preparing themselves for military confrontations. As weapon arsenals of incredible sizes have already transformed the Earth into a global minefield, no planning considerations should ignore their possible effect on enhancing or damping the tensions in the world.

7.3.5 Privatisation of the energy industry and related problems

In several countries, there has in recent years been a debate over the structure of the energy industry, starting with the electric power utilities. These were in many countries state enterprises or semi-public concessioned companies, i.e. companies with selling prices determined by the government (typically through an "electricity commission") and with local city or province administrations as majority share holders. Privatisation has implied state or local governments selling off these assets at a substantial one-time profit, at the expense of losing detailed control over prices and supply security. The term "deregulation" has been used, although as detailed below it is not quite appropriate. The first countries/states to carry through such a privatisation of the public electricity utilities were the UK and California, followed by Norway and then several other countries.

The outcome of the early privatisations has been a general deterioration of the electric utility sector, leading to poorer service, sometimes higher prices, and in any case a sharp decline in development efforts and new investments, making the power plant stock increasingly outdated and prone to failure causing serious power blackouts. More recent privatisations have attempted to avoid some of these problems by additional legislation, e.g. imposing on the utilities a "public service obligation" (PSO), usually in the form of a fixed kWh-tax earmarked to research and development of new technology relevant to the sector, and by requiring a fixed percentage of power to be produced by non-polluting renewable energy sources. Alternatively, in some countries renewable energy is being offered to environmentally concerned customers at a price above the price for polluting power. Since one initial claim has been that privatisation would lower consumer prices, efforts have been made to invite this outcome. Key among these is the auction or "pool" concept, where power producers bid for the contract to supply power for periods ranging from an hour to a couple of days, and where long-term bids are handled in analogy to "futures options" at a stock market.

This has led to a division of the power industry into power producers, transmitters and distributors, where typically the producers are privatised and the net-providers are still semi-public concessioned companies (regarding it as improper to create competition in power networking, since it would not be rational to have several independent grids in the same region – a problem known from privatised fixed or mobile telephone grids). One company is then given the concession to be the grid balancing administrator, i.e. to reconcile the possibly lowest bids received from producers into a pattern compatible with the grid transmission capabilities available, and to correct mismatch or fall-out situations by drawing on producers having offered bids of emergency power (to be provided with no forward notice). This means that the balancing administrator can use any available transmission facilities regardless of ownership and that the grid owners will receive economic compensation according to set tariffs. This also opens up for customers being able to enter into contractual agreements with specific producers (bypassing the pool system) and yet be assured of transmission through third-party owned grid sections. Given the limited capacity of any grid, this clearly entails some intricate work to be done by the grid responsible agent, making the best use of available transmission capacity by dispatching power from the producers on contract to the end-users [who of course will receive power from nearby producers through their grid, while the far-away contractual producer delivers a corresponding amount of power into the grid for delivery to clients closer to them (cf. system modelling by Meibom et al., 1999)].

The result is the creation of huge power systems, comprising e.g. all of North America or continental Europe, but still limited by the capacity of transmission between countries or regions. Had the transmission capacity been infinite and the grid failure rate zero, this interconnection would, of course, lead to a highly stable and reliable electricity supply system, but given the existing finite ceilings on transfer, the result may well in combination with the free choice of suppliers be the opposite. The issue is very complex, and although simulations of adverse situations have been attempted, no complete answer can be said to exist at the present time, e.g. answering the question of whether the very massive blackouts experienced in various parts of the world are connected to the liberalisation of trading options or not.

The liberalised electricity market invites producers to continue to use old and inefficient power plants longer than would be dictated by national economy concerns. Only when tangible disadvantages emerge, such as inconveniently long start-up times or not being able to meet environmental requirements, will replacement be contemplated. In the case of environmental impacts, some countries have supplemented privatisation with legislation aimed at reducing pollution. This could be in the form of total ceilings on emissions of SO_2, NO_x, particles and CO_2, which may be chosen as decreasing with time. It is then up to the power industry to adjust its generation mix, so that the ceilings are never exceeded. In some cases, these schemes are supplemented with issuing tradable permits for a given amount of pollution. The idea is that the polluting power plants can trade a given amount of pollution permits between them, leading ideally to spending money on pollution abatement where it is most effective. In practice it rather implies that the half-decent power plants of more developed regions can continue to pollute at a constant level, while the power plants of less developed regions, which ought to have been phased out in any case, get an added incentive to do so (this situation is clearly operating in Western and Eastern Europe). The argument against tradable permits is that one could economically assist less developed regions in replacing their most polluting equipment without, at the same time, halting the progress of pollution reduction in the regions most capable of carrying through a more advanced reduction.

These remarks substantiate the inappropriateness of the term "deregulation". In fact, the privatised energy sector needs a substantial increase in legislation to handle all the problem areas seen from a national or world economic point of view.

It is interesting that the wave of privatisation suggestions occurs in countries where the electricity supply has been fully established. Think for a moment what would have happened if these ideological schemes had been implemented 100 years ago when electricity systems were initially being established. The cost of providing electricity was obviously lowest in densely populated areas such as city centres, and much higher in rural areas. On one hand, this would have made it uneconomical to supply electricity outside cities through grids. On the other hand, it would have offered better conditions for local generation by renewable energy sources. The struggle between wind power and coal-based power, which went on in certain European countries up to the 1920s, where coal power finally "won", could have had the opposite outcome. However, we would have lost the benefits offered by the power grids, in terms of stable supply despite failure of some power units and of making optimum use of the different peak power demands in different regions to reduce the required rated capacity. The actual historical development of power supply can be characterised as one of solidarity, where city customers subsidised rural dwellers through the policy of maintaining the same price for all electricity customers, in contrast to current deregulated systems with different prices for different customers according to actual delivery costs and preferences for polluting or non-polluting production methods.

The final concern is about the implication of societies' loosing influence on essential areas. Energy supply is clearly an area of high strategic importance. Having left this sector to private companies, it may be sold to foreign investors with agendas incompatible with the interests of a particular country. The foreign owner may elect supply options which are at variance with the preferences of local customers, as regards accident risks, supply security, environmental impacts, etc. Again, the only possible solution to such concerns is to surround the energy supply sector with such a massive range of regulations and legislation that the owners of the production and grid system do not have any room left to manoeuvre - in which case it would seem simpler to retain the public ownership.

7.4 Life-cycle analysis

The abbreviation LCA if used for both life-cycle analysis and life-cycle assessment. However, they are two different concepts: life-cycle analysis is the scientific and technical analysis of impacts associated with a product or a system, while life-cycle assessment is the political evaluation based upon the analysis.

The need for incorporating study of environmental impacts in all assessment work performed in our societies, from consumer product evaluation to long-term planning decisions, is increasingly being accepted. Energy systems were among the first to be subjected to LCA, trying to identify environmental impacts and social impacts related e.g. to health, or in other words to include in the analysis impacts that have not traditionally been reflected in prices paid in the marketplace. This focuses on the sometimes huge difference between direct cost and full cost, including what are termed externalities: those social costs that are not incorporated in market prices. It is seen as the role of societies (read governments) to make sure that the indirect costs are not neglected in consumer choices or decision-making processes related to planning in a society. The way externalities are included will depend on the political preferences. Possible avenues range from taxation to legislative regulation.

Life-cycle analysis is a tool suited for assisting planners and decision-makers in performing the necessary assessments related to external costs. The LCA method aims at assessing all direct and indirect impacts of a technology, whether a product, an industrial plant, a system or an entire sector of society. LCA incorporates impacts over time, including impacts derived from materials or facilities used to manufacture tools and equipment for the process under study, and it includes final disposal of equipment and materials, whether involving re-use, recycling or waste disposal. The two important characteristics of LCA are

- Inclusion of "cradle-to-grave" impacts
- Inclusion of indirect impacts imbedded in materials and equipment

The ideas behind LCA were developed during the 1970s and went under different names such as "total assessment", "including externalities", or "least cost planning". Some of the first applications of LCA were in the energy field, including both individual energy technologies and entire energy supply systems. It was soon realised that the procurement of all required data was a difficult problem. As a result, the emphasis went towards LCA applied to individual products, where the data handling seemed more manageable. However, it is still a very open-ended process, because manufacture of, say, a milk container requires both materials and energy and to assess the impacts associated with the energy input anyway calls for an LCA of the energy supply system. Only as the gathering of relevant data has been ongoing for a considerable time, has it become possible to perform credible LCAs.

Product LCA has in recent years been promoted by organisations such as SETAC (Consoli *et al.*, 1993), and several applications have appeared over recent years (e.g. Mekel and Huppes, 1990; Pommer *et al.*, 1991; Johnson *et al.*, 1994; DATV, 1995). Site- and technology-specific LCA of energy systems has been addressed by the European Commission (1995f) and by other recent projects (Petersen, 1991; Inaba *et al.*, 1993; Kato *et al.*, 1993; Meyer *et al.*, 1994; Sørensen and Watt, 1993; Yasukawa *et al.* 1996; Sørensen, 1994b, 1995a, 1996c; Kuemmel *et al.*, 1997). Methodological issues have been addressed by Baumgartner (1993), Sørensen (1993b, 1995b, 1996b, 1997b) and Engelenburg and Nieuwlaar (1993), and energy system-wide considerations have been addressed by Knöepfel (1993), Kuemmel *et al.* (1997) and Sørensen (1997c), the latter with emphasis on greenhouse gas emission impacts.

7.4.1 Defining purpose and scope of LCA

The first consideration in formulating a life-cycle assessment strategy is to formulate the purpose of the analysis. Several uses may be contemplated:

(*a*) To determine impacts from different ways of producing the same product
(*b*) To determine impacts from different products serving the same purpose
(*c*) To determine all impacts from a sector of the economy, e.g. the energy sector
(*d*) To determine all impacts from the entire social system and its activities

If the purpose is either (*a*) or (*b*), the analysis is called a product LCA, whereas the purposes (*c*) or (*d*) define a systems LCA. The present section will concentrate on studies pertaining to item (*c*), but there are borderline cases, such as the analysis of power produced by a particular power plant, with its upstream and downstream processes, or the analysis of building insulation, with its inputs of insulation materials and installation work. In such cases, we shall talk about a single chain of energy conversions based on site- and technology-specific components. The course of the actual investigation may thus employ different types of analysis:

(A) Chain analysis (with side-chains)
(B) System level analysis (each device treated separately)
(C) Partial system analysis (e.g. confined to energy sector).

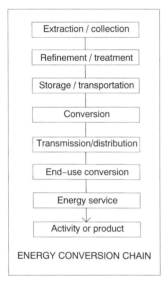

Figure 7.4. Generic energy chain (Sørensen, 1995c).

In a chain analysis (A), impacts include those of the main chain (Fig. 7.4) as well as impacts caused by provision of sideline inputs (Fig. 7.5), which are allocated to the chain investigated by the fraction of production entering the chain. For example, if the equipment used in a chain step such as an oil refinery is provided by a manufacturer who sells 20% of his production to the oil refinery, then 20% of each of his impacts (environmental, social) are allocated to the oil refinery. Each physical component of the chain is going through a number of life-cycle phases, from construction activities through the period of operation and concurrent maintenance, evolving into stages of major repairs or dismantling as part of final decommissioning. Each stage has inputs of materials, energy and labour, and outputs of pollutants and useful components. Impacts are thus positive or negative: the positive impacts are generally the benefits of the activity, notably the products or services associated with energy use, while the negative impacts are a range of environmental and social impacts. Skill of operating the components of a system is a determining factor for the magnitude of impacts, as is of course the technology used and the structure of the society receiving the impacts.

An energy system is a complete system for generating, distributing and supplying energy to a range of end-users, defined by some domain of demand, e.g. specified by type, by geographical coverage or by recipient profile. Examples were shown in section 6.4. Physically, the system components occupying the centre of Fig. 7.5 would be facilities for extracting or collecting energy, for importing or exporting energy, for converting energy from one form to another, for transporting and distributing energy, and finally, for transforming the energy into a useful product or service, as indicated in Fig. 7.4. Products and services are the demanded quantities, as seen from the human society. They obviously change with time and according to development stage of a given society.

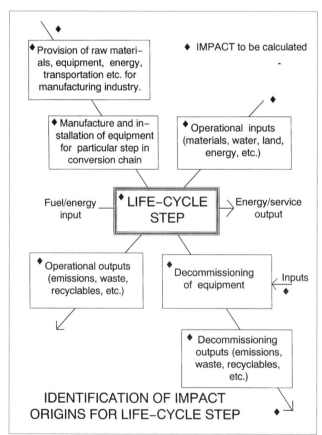

Figure 7.5. Input and output streams for a particular life-cycle step (based on Sørensen, 1997a).

In the system level analysis (*B*), the impacts from each device in the system are calculated separately and summed up in the end. For example, the direct impacts from running the oil refinery are calculated, and the direct impacts of the equipment manufacturer are calculated, as well as any inputs he may receive from other actors. At the end, summing up all the calculated impacts will provide a true total without double counting.

Analyses made so far using the LCA method have mostly been chain calculations (A) and partial system analyses (C), in which the energy sector is treated directly, while other sectors of the economy are treated indirectly. In the latter case, it is necessary to calculate impacts individually according to the system level scheme (B) for the components of the energy system itself, whereas input of equipment from other sectors are to be evaluated as products, with their imbedded impacts from other actors included. A double-counting problem arises in this type of analysis when the imbedded impacts include energy use by one of the manufacturers or other actors not treated explicitly. If all steps of the energy conversion chain are treated directly, the impacts from such inputs should be excluded from the indirect side-chains outside the energy sector. In many cases, this distinction is easy to make, because the impacts found in the literature are normally divided into direct and indirect, and the indirect ones are distributed over their origin, such as from energy use or other materials use. There are, however, cases in which the data do not allow a precise splitting of the energy and non-energy inputs. In such cases, one has to estimate this split judgementally.

Figure 7.6 illustrates the double-counting problem: if a chain LCA is made for each chain identified in Fig. 7.6b, there will be double-counting of both some direct and some indirect impacts. The solution for simple double counting of the major chains is to calculate impacts for each compartment in Fig. 7.6a and then sum them up, but as regards the indirect impacts, one has to make sure that there is no hidden double counting. This may, in principle, be accomplished by including only the fraction of impacts corresponding to the input reaching the compartment focused upon (as in Fig. 7.4), but some of these inputs may involve energy that emerges as output from the energy system (i.e. Fig. 7.6a) itself. In other words, if the entire energy system is included in the analysis, one should simply omit energy-related impacts from the indirect side-chains. If only a partial energy system is being analysed, one would still have to include impacts from other parts not explicitly included within the system.

7.4.2 Treatment of import and export

Many of the available data sources include indirect impacts based on an assumed mix of materials and labour input, taking into account the specific location of production units, mines, refineries, etc. This means that an attempt has been made to trace back the origin of all ingredients in a specific product (such as electricity) in a bottom-up approach, which is suited for comparison of different ways of furnishing the product in question (e.g. comparing wind, coal and nuclear electricity).

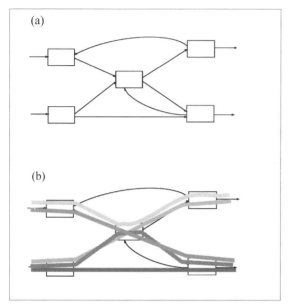

Figure 7.6. Energy system (a) with forward chains indicated (b) (Kuemmel *et al.*, 1997).

In the case of the analysis of an entire energy system, the data for specific sites and technology, as well as specific countries from which to import, are not the best suited. Especially for the future scenarios, it seems improper to use data based on the current location of mines, refineries and other installations. One may, of course, average over many different sets of data, in order to obtain average or "generic" data, but the selection of future energy systems should not depend sensitively on where, for example, our utilities choose to purchase their coal this particular year, and therefore a different approach has to be found.

One consistent methodology is to consider the energy system of each country being studied a part of the national economy, such that if that country chooses to produce more energy than needed domestically in order to export it, this constitutes an economic activity no different from, say, Denmark producing more Lego blocks than can be used by Danish children. The exports are an integral part of the economy of a small country such as Denmark, because it cannot and does not produce every item needed in its society and thus must export some goods in order to be able to import other ones that it needs. Seen in this way, it is "their own fault" if manufacturing the goods they export turns out to have more environmental side-effects there than our imports have in their countries of origin, and the total evaluation of impacts should simply include those generated as part of the economy of the country investigated, whether for domestic consumption or export. Likewise, the impacts of goods imported should be excluded from the evaluation for the importing country, except, of course, for impacts arising during the operation or use of the imported items in the country (for example, burning of imported coal).

A consistent methodology is thus to include all impacts of energy production and use within the country considered, but to exclude impacts inherent in imported materials and energy. Doing this for a Danish scenario (Kuemmel *et al.*, 1997), the impact calculation related to energy for the present system turns out not to differ greatly from one based on the impacts at the place of production, because Denmark is about neutral with respect to energy imports and exports (this varies from year to year, but currently, oil and gas exports roughly balance coal imports). However, for other countries, or for future systems, this could be very different, because the impacts of renewable systems are chiefly from manufacture of conversion equipment, which consists of concrete, metals and other materials, of which 30–50% may be imported.

The argument against confining externality calculations to impacts originating within the country is that this could lead to purchase of the most environmentally problematic parts of the system from countries paying less attention to the environment than we claim to do. Countries in the early stages of industrial development have a tendency to pay less attention to environmental degradation, making room for what is negatively termed "export of polluting activities to the third world". The counter argument is that this may be better for the countries involved than no development and that, when they reach a certain level of industrialisation, they will start to concern themselves with the environmental impacts (examples are Singapore and Hong Kong). Unfortunately, this does not seem universally valid, and the implication of environmental neglect in countries like China, India and South American nations extends outside their borders, e.g. in the case of global warming, which is not confined to the country of origin or its close regional neighbours. Still, from a methodological point of view, the confinement of LCA impacts to those associated with activities in one country does provide a fair picture of the cost of any chosen path of development, for industrialising as well as for highly industrialised countries.

The problem is that a large body of existing data is based on the other methodology, where each energy form is viewed as a product, and impacts are included for the actual pathway from mining through refining and transformation to conversion, transmission and use, with the indirect impacts calculated where they occur, that is in different countries. In actuality, the difficulty in obtaining data from some of the countries involved in the early stages of the energy pathway has forced many externality studies to use data "as if" the mining and refining stages had occurred in the country of use. For example, the European Commission study (1995c) uses coal mined in Germany or England, based on the impacts of mining in these countries, rather than the less known impacts associated with coal mining in major coal ex-porting countries. It has been pointed out that this approach to energy exter-nalities can make the LCA too uncertain to be used meaningfully in decision processes (Schmidt *et al.*, 1994). Recent product LCAs looking at soft-drink bottles and cans clearly suffer from this deficiency, making very specific as-sumptions regarding the place of production of aluminium and glass and of the type of energy inputs to these processes (UMIP, 1996). If the can manu-facturer chooses to import aluminium from Tasmania or the USA instead of, say from Norway, the balance between the two types of packing may tip the other way.

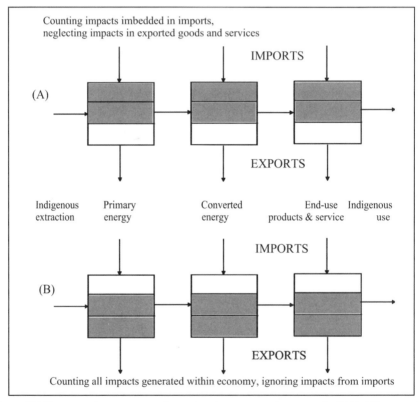

Figure 7.7. Two consistent ways of treating imports and exports in LCA (Kuemmel *et al.*, 1997).

The two types of analysis are illustrated in Fig. 7.7. The method in com-mon use for product LCA, where imbedded impacts are traced through the economies of different countries, not only seems problematic when used in any practical example, but is also unsuited for the system analysis that deci-sion-makers demand. In contrast, one may use a conventional economic approach, where import and export are specific features of a given economy, deliberately chosen in consideration of the assets of the country, and calculate LCA impacts only for the economy in question in order to be consistent.

7.4.3 What to include in an LCA

The history of LCA has taken two distinct paths. One is associated with the energy LCAs, being developed from chain analysis without imbedded impacts to analysis including such impacts, as regards both environmental and social impacts. The history and state-of-the-art of energy LCA is described in Sørensen (1993b, 1996b). The main ingredients suggested for inclusion in an energy LCA are listed below. The other path is associated with product LCA and has been pushed by the food and chemical industries, particularly through the association SETAC (Consoli *et al.*, 1993). SETAC tends to consider only environmental impacts, neglecting most social impacts. The same is true for the prescriptions of standardisation institutions (ISO, 1997, 1999), mainly focusing on establishing a common inventory list of harmful substances to consider in LCA work. Both types of LCA were discussed methodologically around 1970, but only recently have they been transferred to credible calculations, because of the deficiencies in available data and incompleteness. Many absurd LCA conclusions have been published from the late 1960s to the late 1980s. The present stance is much more careful, realising that LCA is not and cannot be made a routine screening method for products or energy systems, but has to remain an attempt to furnish more information to the political decision-maker than has previously been available. The decision process will be of a higher quality if these broader impacts are considered, but the technique is never going to be a computerised decision tool capable of replacing political debate before decisions are made. This is also evident from the incommensurability of different impacts, which cannot always be meaningfully brought to a common scale of units: to emphasise this view on the scope of LCA, I give in this section a list of impacts to consider, without claiming it to be inclusive or complete.

The types of impacts that may be contemplated for assessment reflect to some extent the issues that at a given moment in time have been identified as important in a given society. It is therefore possible that the list will be modified with time and that some societies will add new concerns to the list. However, the following groups of impacts, a summary of which is listed in Table 7.1, constitute a fairly comprehensive list of impacts considered in most studies made to date (Sørensen, 1993b):

- *Economic impacts such as impacts on owners' economy and on national economy, including questions of foreign payments balance and employment.*

This group of impacts aim at the direct economy reflected in market prices and costs. All other impacts can be said to constitute indirect costs or externalities, the latter if they are not included in prices through e.g. environmental taxes. Economy is basically a way of allocating scarce resources. Applying economic assessment to an energy system, the different payment times of different expenses have to be taken into account, e.g. by discounting individual costs to present values. This again gives rise to different economic evaluations for an individual, an enterprise, a nation, and some imaginary global stakeholder. One possible way of dealing with these issues is to apply different sets of interest rates for the above types of actors and, in some cases, even a different interest rate for short-term costs and for long-term, inter-generational costs, for the same actor. Ingredients in these kinds of economic evaluation are the separate private economy and national economy accounts often made in the past. The national economy evaluation includes such factors as import fraction (balance of foreign payments), employment impact (i.e. distribution between labour and non-labour costs), and more subtle components such as regional economic impacts. Impact evaluations must pay particular attention to imports and exports, as many of the indirect impacts will often not be included in trade prices or their presence or absence will be unknown.

a. Economic impacts such as impacts on owners' economy and on national economy, including questions of foreign payments balance and employment.

b. Environmental impacts, e.g. land use, noise, visual impact, local, regional and global pollution of soil, water, air and biota, impacts on climate.

c. Social impacts, related to satisfaction of needs, impacts on health and work environment, risk, impact of large accidents.

d. Security and resilience, including supply security, safety against misuse, terrorist actions as well as sensitivity to system failures, planning uncertainties and changes in future criteria for impact assessment.

e. Development and political impacts, such as degree of consistency with goals of a given society, impacts of control requirements and institutions, openness to decentralisation and democratic participation.

Table 7.1. Impacts to be considered in life-cycle analysis of energy systems (based on Sørensen, 1996f).

- *Environmental impacts, e.g. land use, noise, visual impact, local pollution of soil, water, air and biota, regional and global pollution and other impacts on the Earth–atmosphere system, such as climatic change.*

Environmental impacts include a very wide range of impacts on the natural environment, including atmosphere, hydrosphere, lithosphere and biosphere, but usually with the human society left out (but to be included under the heading social impacts below). Impacts may be classified as local, regional and global. At the resource extraction stage, in addition to the impacts associated with extraction, there is the impact of resource depletion.

In many evaluations, the resource efficiency issue of energy use in resource extraction is treated in conjunction with energy use further along the energy conversion chain, including energy used to manufacture and operate production equipment. The resulting figure is often expressed as an energy pay-back time, which is reasonable because the sole purpose of the system is to produce energy, and thus it would be unacceptable if energy inputs exceeded outputs. In practice, the level of energy input over output that is acceptable depends on the overall cost and should be adequately represented by the other impacts, which presumably would become large compared with the benefits if energy inputs approached outputs. In other words, energy pay-back time is a secondary indicator, which should not itself be included in the assessment, when the primary indicators of positive and negative impacts are sufficiently well estimated. Also issues of the quality of the environment, as seen from an anthropogenic point of view, should be included here. They include noise, smell and visual impacts associated with the cycles in the energy activity. Other concerns could be the preservation of natural flora and fauna.

It is necessary to distinguish between impacts on the natural ecosystems and those affecting human wellbeing or health. Although human societies are, of course, part of the natural ecosystem, it is convenient and often necessary to treat some impacts on human societies separately, which will be done in the following group. However, the situation is often that a pollutant is first injected into the natural environment and later finds its way to humans, e.g. by inhalation or through food and water. In such cases, the evaluation of health impacts involves a number of calculation steps (dispersal, dose–response relation) that naturally have to be carried out in order.

- *Social impacts, related to satisfaction of needs, impacts on health and work environment, risks, impact of large accidents.*

Social impacts include the impacts from using the energy provided, which means the positive impacts that derives from services and products associated with the energy use (usually with other inputs as well) and the negative impacts associated with the energy end-use conversion. Furthermore, social impacts derive from each step in the energy production, conversion and transmission chain. Examples are health impacts, work environment, job satisfaction, and risk, including the risk of large accidents. It is often useful to distinguish between occupational impacts and impacts to the general public. Many of these impacts involve transfer of pollutants first to the general environment and then to human society, where each transfer requires separate investigation as stated above. This is true both for releases during normal operation of the facilities in question and for accidents. Clearly, the accident part is a basic risk problem that involves estimating probabilities of accidental events of increasing magnitude.

- *Security impacts, including both supply security and also safety against misuse, terrorist actions, etc.*

Security can be understood in different ways. One is supply security, and another is the security of energy installations and materials against theft, sabotage and hostage situations. Both are relevant in a life-cycle analysis of an energy system. Supply security is a very important issue, e.g. for energy systems depending on fuels unevenly spread over the planet. Indeed, some of the most threatening crises in energy supply have been related to supply security (1973-1974 oil supply withdrawal, 1991 Gulf War).

- *Resilience, i.e. sensitivity to system failures, planning uncertainties and future changes in criteria for impact assessment.*

Resilience is also a concept with two interpretations. One is the technical resilience, including fault resistance and parallelism, e.g. in providing more than one transmission route between important energy supply and use locations. Another is a more broadly defined resilience against planning errors (e.g. resulting from a misjudgement of resources, fuel price developments, or future demand development). A more tricky, self-referencing issue is resilience against errors in impact assessment, assuming that the impact assessment is used to make energy policy choices. All the resilience issues are connected to certain features of the system choice and layout, including modularity, unit size and transmission strategy. The resilience questions may well be formulated in terms of risk.

- *Development impacts (e.g. consistency of a product or a technology with the goals of a given society).*

Energy systems may exert an influence on the direction of development a society will take or rather may be compatible with one development goal and not with another goal. These could be goals of decentralisation, goals of concentration on knowledge-based business rather than heavy industry, etc. For so-called developing countries, clear goals usually include satisfying basic needs, furthering education and raising standards. Goals of industrialised nations are often more difficult to identify.

- *Political impacts include e.g. impacts of control requirements, and openness to decentralisation in both physical and decision-making terms.*

There is a geopolitical dimension to the above issues: development or political goals calling for import of fuels for energy may imply increased competition for scarce resources, an impact which may be evaluated in terms of increasing cost expectations or in terms of increasing political unrest (more "energy wars"). The political issue also has a local component, pertaining to the amount of freedom of local societies to choose their own solutions, possibly different from the one selected by the neighbouring local areas.

Qualitative or quantitative estimates of impacts

There is a consensus that one should try to quantify as much as possible in any impact assessment. However, items for discussion arise in the handling of those impacts that cannot be quantified (and later for those quantifiable impacts that prove to be hard to monetise). One historically common approach is to ignore impacts that cannot be quantified. Alternatively, one may clearly mark the presence of such impacts and, in any quantitative summary, add a warning that the numbers given for impacts cannot be summed up to a total, as some impacts are missing. As, for example, Ottinger (1991) points out, the danger is that policy-makers will still ignore the warnings and use the partial sums as if they were totals. Hopefully, this is an underestimation of the capacities of decision-makers, as their task is precisely to make decisions in areas, where only part of the consequences are known at any given time and where most quantitative data are uncertain. If this were not the case, there would be no need for decision-makers, as the calculated total impact values would directly point to the best alternative. We shall return to some of these issues in section 7.4.9 below, where we discuss ways of presenting the results of LCAs.

Treatment of risk-related impacts and accidents in LCA

Of the impacts listed above, some involve an element of risk. Risk may be defined as a possible impact occurring or causing damage only with a finite probability (for example, being hit by a meteor or developing lung diseases as a result of air pollution). The insurance industry uses a more narrow definition requiring the event to be *"sudden"*, i.e. excluding the damage caused by general air pollution. In LCA, all types of impacts have to be considered, but the treatment may depend on whether they are *"certain"* or *"stochastic"* in nature and, in the latter case, whether they are insurance type risks or not.

As regards the health impacts associated with dispersal of air pollutants followed by ingestion and an application of a dose–response function describing processes in the human body (possibly leading to illness or death), one can use established probabilistic relationships, provided population densities and distributions are such that the use of statistical data is appropriate. A similar approach can be taken for the risk of accidents occurring fairly frequently, such as motorcar accidents, but a different situation occurs for low-probability accidents with high levels of associated damage (e.g. nuclear accidents).

The treatment of high-damage, low-probability accident risks needs additional considerations. The standard risk assessment used e.g. in the aircraft industry consists of applying fault tree analysis or event tree analysis to trace accident probabilities forward from initiating events or backward from final outcomes. The idea is that each step in the evaluation is a known failure type associated with a concrete piece of equipment and that the probability for failure of such a component should be known from experience. The combined probability is thus the sum of products of partial event probabilities for each step along a series of identified pathways. It is important to realise that the purpose of this evaluation is to improve design by pointing out the areas where improved design is likely to pay off.

Clearly, unanticipated event chains cannot be included. In areas such as aircraft safety, one is aware that the total accident probability consists of one part made up by anticipated event trees and one made up by unanticipated events. The purpose of the design efforts is clearly to make those accidents that can be predicted by the fault tree analysis (and thus may be said to constitute "built-in" weaknesses of design) small compared with the unanticipated accidents, for which no defence is possible, except to learn from actual experiences and hopefully move event chains including e.g. common mode failures from the "unanticipated" category into the "anticipated", where engineering design efforts may be addressed. This procedure has led to overall declining aircraft accident rates, while the ratio between unanticipated and anticipated events has stayed at approximately the value 10.

The term "probability" is often used in a loose manner, with no proof of a common, underlying statistical distribution. For example, technological change makes the empirical data different from the outcome of a large number of identical experiments. This is equally true in cases of oil spills or nuclear accidents, for which the empirical data are necessarily weak, owing to the low frequency of catastrophic events (albeit compounded with potentially large consequences). Here the term "probability" is really out of place and if used should be interpreted as just "a frequency indicator".

7.4.4 Choosing the context

When the purpose of the LCA is to obtain generic energy technology and systems evaluations (e.g. as inputs into planning and policy debates), one should try to avoid using data depending too strongly on the specific site selected for the installation. Still, available studies may depend strongly on location (e.g. special dispersal features, e.g. in mountainous terrain) or a specific population distribution (presence of high-density settlements downstream relative to emissions from the energy installation studied). For policy uses, these special situations should if possible be avoided and be left for the later, detailed plant siting phase to eliminate unsuited locations. This is not normally a problem if the total planning area is sufficiently diverse.

Pure emission data often depend only on the physical characteristics of a given facility (power plant stack heights, quality of electrostatic filters, sulphate scrubbers, nitrogen oxide treatment facilities, etc.) and not on the site. However, the dispersion models are, of course, site dependent, but general concentration versus distance relations can usually be derived in model calculations avoiding any special features of sites. As regards the dose commitment, it will necessarily depend on population distribution, while the dose–response relationship should not depend on this. As a result, a generic assessment can in many cases be performed with only a few adjustable parameters left in the calculation, such as the population density distribution, which may be replaced with average densities for an extended region.

The approach outlined above will only serve as a template for assessing new energy systems, as the technology must be specified and usually would involve a comparison between different new state-of-the-art technologies. If the impact of the existing energy system in a given nation or region has to be evaluated, the diversity of the technologies in place must be included in the analysis, which would ideally have to proceed as a site- and technology-specific analysis for each piece of equipment.

In generic assessments, not only do technology and population distributions have to be fixed, but also a number of features characterising the surrounding society will have to be assumed, in as much as they may influence the valuation of the calculated impacts (and in some cases also the physical evaluation, e.g. as regards society's preparedness for handling major accidents, which may influence the impact assessment in essential ways).

7.4.5 Aggregation issues

Because of the importance of aggregation issues, both for data definition and for calculation of impacts, we shall discuss this topic in a little more detail. There are at least four dimensions of aggregation that play a role in impact assessments:

- Aggregation over technologies
- Aggregation over sites
- Aggregation over time
- Aggregation over social settings

The most disaggregated studies done today are termed "bottom-up" studies of a specific technology located at a specific site. Since the impacts will continue over the lifetime of the installation, and possibly longer (radioactive contamination), there is certainly an aggregation over time involved in stating the impacts in compact form. The better studies attempt to display impacts as a function of time, e.g. as short-, medium- and long-term effects. However, even this approach may not catch important concerns, as it will typically aggregate over social settings, assuming them to be inert as a function of time. This is, of course, never the case in reality, and in recent centuries, the development with time of societies has been very rapid, entailing also rapid changes in social perceptions of a given impact. For example, the importance presently accorded to environmental damage was absent just a few decades ago, and there are bound to be issues that society over the next decades will be concerned about, but which currently are considered as marginal by wide sections of society.

The item of aggregation over social settings also has a precise meaning for a given instance. For example, the impacts of a nuclear accident will greatly depend on the response of the society. Will there be heroic firemen as in Chernobyl, who will sacrifice their own lives in order to diminish the consequences of the accident? Has the population been properly informed about what to do in case of an accident (going indoors, closing and opening windows at appropriate times, etc.)? Have there been drills of evacuation procedures? For the Chernobyl accident in 1986 the answer was no; in Sweden today it would be yes. A study making assumptions on accident mitigation effects must be in accordance with the make-up of the society for which the analysis is being performed. Again, the uncertainty in estimating changes in social context over time stands out.

Aggregation over sites implies that peculiarities in topography (leading perhaps to irregular dispersal of airborne pollutants) are not considered, and that variations in population density around the energy installation studied will be disregarded. This may be a sensible approach in a planning phase, where the actual location of the installation may not have been selected. It also gives more weight to the technologies themselves, making this approach suitable for generic planning choices between classes of technology (e.g. nuclear, fossil, renewable). Of course, once actual installations are to be built, new site-specific analyses will have to be invoked in order to determine the best location.

Aggregation over technologies would in most cases not make sense. However, in the particular case, where the existing stock of e.g. power plants in a region is to be assessed, something like technology aggregation may play a role. For example, one might use average technology for the impact analysis, rather than performing multiple calculations for specific installations involving both the most advanced and the most outdated technology. Thus, in order to asses the total impact of an existing energy system, one might aggregate over coal-fired power stations built at different times, with differences in efficiency and cleaning technologies being averaged. On the other hand, if the purpose is to make cost-benefit analyses of various sulphur and nitrogen cleaning technologies, each plant has to be treated separately.

In a strict sense, aggregation is clearly not allowed in any case, because the impacts that play a role never depend linearly or in simple ways on assumptions of technology, topography, population distribution, and so on. One should, in principle, treat all installations individually and make the desired averages on the basis of the actual data. This may sound obvious, but in most cases it is also unachievable, because available data are always incomplete and so is the characterisation of social settings over the time periods needed for a complete assessment. As regards the preferences and concerns of future societies, or the impacts of current releases in the future (such as climatic impacts), one will always have to do some indirect analysis, involving aggregation and assumptions on future societies (using e.g. the scenario method described in Chapter 4).

One may conclude that some aggregation is always required, but that the level of aggregation must depend on the purpose of the assessment. In line with the general characterisation given in section 7.4.1, one may discern the following specific purposes for conducting an LCA:

- Licensing of a particular installation
- Energy system assessment
- Assistance to energy planning and policy efforts.

For licensing of a particular installation along a fuel chain or for a renewable energy system, clearly a site- and technology-specific analysis has to be performed, making use of actual data for physical pathways and populations at risk (as well as corresponding data for impacts on ecosystems, etc.). For the assessment of a particular energy system, the elements of full chain from mining or extraction through refining, treatment plants and transportation to power plants, transmission and final use must be considered separately, as they would typically involve different locations. A complication in this respect is that, e.g. for a fuel-based system, it is highly probable that over the lifetime of the installation fuel would be purchased from different vendors, and the fuel would often come from many geographical areas with widely different extraction methods and impacts (e.g. Middle East versus North Sea oil or gas, German or Bolivian coal mines, open-pit coal extraction in Australia, and so on). Future prices and environmental regulations will determine the change in fuel mix over the lifetime of the installation, and any specific assumptions may turn out to be invalid.

For the planning type of assessment, it would in most industrialised nations be normal to consider only state-of-the-art technology, although even in some advanced countries there is a reluctance to apply known and available environmental cleaning options (currently for particle, SO_2 and NO_x emissions, in the future probably also for CO_2 sequestering or other removal of greenhouse gases). In developing countries, there is even more of the tendency to ignore available but costly environmental impact mitigation options. In some cases, the level of sophistication selected for a given technology may depend on the intended site (e.g. near to or away from population centres). Another issue is maintenance policies. The lifetime of a given installation depends sensitively on the willingness to spend money on maintenance, and the level of spending opted for is a matter to be considered in the planning decisions.

The following list enumerates some of the issues involved (Sørensen, 1993b):

Technology and organisation
- Type and scale of technology
- Age of technology
- Maintenance state and policy
- Matching technology with the level of skills available
- Management and control set-up

Natural setting
- Topography, vegetation, location of waterways, ground water tables, etc.
- Climatic regime: temperature, solar radiation, wind conditions, water currents (if applicable), cloud cover, precipitation patterns, air stability, atmospheric particle content

Social setting
- Scale and diversity of society
- Development stage and goals
- Types of government, institutions and infrastructure

Human setting
- Values and attitudes, goals of individuals
- Level of participation, level of decentralisation of decision-making

Impact assessments suitable for addressing these issues involve the construction of scenarios for future society in order to have a reference frame for discussing social impacts. Because the scenario method has normative components, it would in most cases be best to consider more than one scenario, spanning important positions in the social debate of the society in question.

Another issue is the emergence of new technologies that may play a role over the planning period considered. Most scenarios of future societies do involve some assumption of new technologies coming into place, based on current research and development. However, the actual development is likely to involve new technologies that were not anticipated at the time of making the assessment. It is possible to some extent to analyse scenarios for sensitivity to such new technologies, as well as to possible errors in other scenario assumptions. This makes it possible to distinguish between those future scenarios that are resilient, i.e. do not become totally invalidated by changes in assumptions, as distinct from those that depend strongly on the assumptions made.

In the case of energy technologies, it is equally important to consider the uncertainty of demand assumptions and assumptions on supply technologies. The demand may vary according to social preferences, as well as owing to the emergence of new end-use technologies that may provide same or better services with less energy input. It is therefore essential that the entire energy chain be looked at, down not to the energy delivered but to the non-energy service derived. No one demands energy, but we demand transportation, air conditioning, computing, entertainment and so on.

The discussion of aggregation issues clearly points to the dilemma of impact analyses: those answers that would be most useful in the political context often are answers that can be given only with large uncertainty. This places the final responsibility in the hands of the political decision-maker, who has to weigh the impacts associated with different solutions and in that process to take the uncertainties into account (e.g. choosing a more expensive solution because it has less uncertainty). But this is, of course, what decision-making is about.

Social context

The social context in which a given energy system is placed may have profound consequences for a life-cycle assessment of the energy system. The social context influences both the nature of and the magnitude of impacts to be considered. Key aspects of describing the social context of an energy system are the natural setting, the social setting and the human setting.

The natural setting has to do with geography and demography. Geography may force people to settle in definite patterns, which again may influence the impact of e.g. pollution of air, waterways and soils. In other words, these impacts will not be the same for the same type of energy equipment if this is placed in different geographical settings. The patterns of releases and dispersal are different, and the chance of affecting the populations is also different, say, for a city placed in a narrow valley as compared with one situated on an extended plain.

The social setting includes many features of a society: its stage of development, its scale and diversity, its institutional infrastructure and its type of government. Many of the social factors are important determinants for the selection of an energy system for a particular society, and they are equally important for determining the way that operation of the system is conducted, as well as the way in which society deals with various impacts. This may pertain to the distribution of positive implications of the energy system, but it may also relate to the actions taken in the case of negative impacts (e.g. the way society deals with a major accident in the energy sector).

The human setting involves the values and attitudes of individual members of society. They are important for the choices made by citizens, e.g. in relation to choices between different types of end-use technology, and of course also to the opinion of people regarding the energy planning and energy future towards which they would like their society to move. In democratic societies, the role of attitudes is to influence the political debate, either by making certain technological choices attractive to decision-makers or by protesting against choices about to be made by governments or political assemblies, thereby expressing the lack of public support for such decisions. Examples of both kinds of political influence are numerous.

The processes are further complicated by feedback mechanisms, such as the one formed by media attention and interest groups' attempts to influence attitudes in the general population.

Data related to social setting should be used in the impact calculation. Health impacts of energy systems depend on the age and health distribution in the society in question, social impacts depend on the social structure, and environmental impacts may depend on the settlement type, geography and climate of the region in question.

Most countries have statistics pertaining to these kinds of issues, but it is rare to see them used in connection with energy impact analyses. It is therefore likely that an effort is required to juxtapose all the relevant types of data, but in principle it can be done with available tools.

More difficult is the question of incorporating values and attitudes of the members of a given society in the assessment. Available studies are often made differently in different societies, and in any case it is unlikely that the impacts associated with such traits of a society can be expressed in terms of numbers that can be compared to economic figures and the other data characterising the energy system.

In other words, one would have to accept that impacts must be described in different phrases or units and that not all of them can be numerically compared. This should not imply that some impacts should *a priori* be given a smaller weight. In fact, what the social evaluation is all about is to discuss in political terms those issues that do not lend themselves to a straightforward numerical evaluation.

The influence of media coverage, which in many societies plays an important role in shaping political perceptions and their feedback on values and attitudes, has been studied e.g. by Stolwijk and Canny (1991), and the influence of protest movements and public hearings have been studied by Gerlach (1987) and Gale (1987) (cf. also the general introductory chapter in Shubik, 1991). The role of institutions has been studied by Lau (1987), by Hooker and van Hulst (1980), and by Wynne (1984).

7.4.6 Monetising issues

The desire to use common units for as many impacts in an LCA as possible is, of course, aimed at facilitating the job of a decision-maker wanting to make a comparison between different solutions. However, it is important that this procedure does not further marginalise those impacts that cannot be quantified or which seem to resist monetising efforts. The basic question is really whether or not the further uncertainty introduced by monetising offsets the benefit of being able to use common units.

Monetising may be accomplished by expressing damage in monetary terms or by substituting the cost of reducing the emissions to some threshold value (avoidance cost). Damage costs may be obtained from health impacts by counting hospitalisation and workday salaries lost, replanting cost of dead forests, restoration of historic buildings damaged by acid rain, and so on. Accidental human death may, for example, be replaced by the life insurance cost.

Unavailability of data on monetisation has led to the alternative philosophy of interviewing cross sections of the affected population on the amount of money they would be willing to pay to avoid a specific impact or to monitor their actual investments (contingency evaluations such as hedonic pricing, revealed preferences, or willingness to pay). Such measures may change from day to day, depending on exposure to random bits of information (whether true or false), and also depend strongly on the income at the respondents' disposal, as well as competing expenses of perhaps more tangible nature. Should the monetised value of losing a human life (the "statistical value of life", SVL, discussed below) be reduced by the fraction of people actually taking out life insurances and should it be allowed to take different values in societies of different affluence?

All of the monetising methods mentioned are clearly deficient: the damage cost by not including a (political) weighing of different issues (e.g. weighing immediate impacts against impacts occurring in the future), and the contingency evaluation by doing so on a wrong basis (influenced by people's knowledge of the issues, by their accessible assets, etc.). The best alternative may be to avoid monetising entirely by using a multivariate analysis, e.g. by presenting an entire impact profile to decision-makers, in the original units and with a time sequence indicating when each impact is believed to occur, and then inviting a true political debate on the proper weighing of the different issues. However, the use of monetary values to discuss alternative policies is so common in current societies that it may seem a pity not to use this established framework wherever possible. It is also a fact that many impacts can meaningfully be described in monetary terms, so the challenge is to make sure that the remaining ones are treated adequately and do not "drop out" of the decision process.

The translation of impacts from physical terms (number of health effects, amount of building damage, number of people affected by noise, etc.) to monetary terms (US$/PJ, DKr/kWh, etc.) is proposed in studies such as the European Commission project (1995a,b) to be carried out by a study of the affected population's willingness to pay (WTP) for avoiding the impacts. This means that the study does not aim at estimating the cost to society, but rather the sum of costs afflicted on individual citizens. The concept of WTP was introduced by Starr (1969). The WTP concept has a number of inherent problems, some of which are:

- Interview studies may lead people to quote higher amounts than they would pay in an actual case.
- The resulting WTPs will depend on disposable income.
- The resulting WTPs will depend on the level of knowledge of the mechanism by which the impacts in question work.

The outcome of an actual development governed by the WTP principle may be inconsistent with agreed social goals of equity and fairness, as it may lead to polluting installations being built in the socially poorest areas.

The accidental deaths associated with energy provision turn out in most studies of individual energy supply chains, such as the European Commission study, to be the most significant impact, fairly independently of details in the monetising procedure selected. We shall therefore deal with choice of the monetised value of an additional death caused by the energy system in a little more detail below. At this point, it should only be said that our preference is to work with a monetised damage reflecting the full LCA cost of energy to society, rather than the cost to selected individual citizens.

Statistical value of life

In calculating externalities, a European Commission (EC) project uses the value of 2.6 Meuro (about 3 million US$) to monetise the loss of a life for all the European energy chains considered (European Commission, 1995a-f). This value is based on a survey of three types of data:

- Willingness to accept a higher risk of death, as revealed by salary increases in risky jobs as compared with similar jobs with small risk
- Contingency valuation studies, i.e. interviews aimed at getting statements of WTP for risks of death
- Actual expenditures paid to reduce risk of loss of life (e.g. purchase of automobile air bags, anti-smoking medication, etc.)

Compensations paid by European governments to families of civil servants dying in connection with their job were also considered in the European Commission study. The scatter in data reviewed ranged from 0.3 to 16 Meuro per death. For use outside Western Europe, the study recommends to use the purchase parity translation of the statistical value of life (SVL) used in the European case studies (i.e. same purchasing power).

A feeling for the statistical value of life can be obtained by considering the salary lost by accidental death. If one assumes that the accidental death on average occurs at the middle of working life and then calculate the total salary that would have been earned during the remaining time to retirement, one would in Denmark get a little over 20 years multiplied by the average salary for the high seniority part of a work career, amounting to between 300 000 and 400 000 DKr per year or around 8 MDKr (some 1.25 MUS$ or 1.1 Meuro). If this was paid to an individual, it should be corrected for interest earning by giving the corresponding present value of annual payments of about 60 keuro/y over 20 years. However, as a cost to society, it may be argued that no discounting should take place because society does not set money aside for future salary payments.

Two other arguments might be considered. One is that in times of unemployment the social value of a person fit to work may be less than the potential salary. Accepting this kind of argument implies that the outcome of technology choices in a society would depend on the ability of that society to distribute the available amount of work fairly (the total amount of salaries involved is not fixed, because salaries are influenced by the level of unemployment). The other argument is that the members of a society have a value to that society above their ability to work. If this were not the case, a society would not provide health services that prolong people's lives beyond retirement age. A judgement on the merits of these arguments would lead to the conclusion that the SVL for society is more than the 1.1 Meuro estimated above, but not how much more. One could say that the European Commission study's value of 2.6 Meuro represents a fairly generous estimate of nontangible values to society of its members and that a lower value may be easier to defend. However, as stated above, the EC estimate has an entirely different basis, representing an individual SVL rather than one seen from the point of view of society. The conclusion may rather be that it is reassuring that two so different approaches do not lead to more different values.

One further consideration is that not all deaths associated with, say, Danish use of energy take place in Denmark. If coal is imported from Bolivia, coal mining deaths would occur there, and the question arises whether a smaller value of life should be used in that case, reflecting the lower salary earnings in Bolivia (and perhaps a smaller concern by society). This would easily be stamped as a colonial view, and the EC study effectively opted to use the same SVL no matter where in the world the death occurs (this is achieved by assuming that e.g. all coal comes from mines in Germany or the UK, whereas in reality Europe imports coal from many different parts of the world).

The global equity problem is one reason why the concept of SVL has been attacked. Another is the ethical problem of putting a value on a human life. The reply to the latter may be that SVL is just a poorly chosen name selected in the attempt to give the political decision-process a clear signal (read: in monetary units) regarding the importance of including consideration of accidental death in decisions on energy system choice and siting. This debate over the use of SVL was taken up by Grubb in connection with the greenhouse warming issue (Grubb, 1996), using arguments similar to those given above.

For the examples of externality and LCA results to be presented below, a monetised SVL value of 2.6 Meuro/death has been used. The discussion above suggests that if this SVL is on the high side, it is so by at most a factor of 2.

Depreciation

Since impacts from energy devices occur throughout the lifetime of the equipment and possibly after decommissioning, one point to discuss is whether expenses occurring in the future should be discounted. This was discussed in section 7.3 in connection with the concept of a social interest rate.

7.4.7 Chain calculations

This section outlines the use of LCA to perform what in section 7.4.1 was termed a chain calculation. The procedure consists in following the chain of conversion steps leading to a given energy output, as illustrated in Fig. 7.4, but considering input from and outputs to side-chains, as exemplified in Fig. 7.5. Such chain LCAs are the equivalent of product LCAs and usually involve specific assumptions on the technology used in each step along the chain. The immediate LCA concerns may, for instance, be emissions and waste from particular devices in the chain. However, before these can be translated into actual damage figures, one has to follow their trajectories through the environment, their uptake by human beings, as well as the processes in the human body possibly leading to health damage. The method generally applied to this problem is called the pathway method.

The pathway method consists of calculating, for each step in the life-cycle, the emissions and other impacts directly released or caused by that life-cycle step, then tracing the fate of the direct impact through the natural and human ecosystems, e.g. by applying a dispersion model to the emissions in order to determine the concentration of pollutants in space and time. The final step is to determine the impacts on humans, on society or on the ecosystem, using for instance dose–response relationships between intake of harmful substances and health effects that have been established separately. The structure of a pathway is indicated in Fig. 7.8.

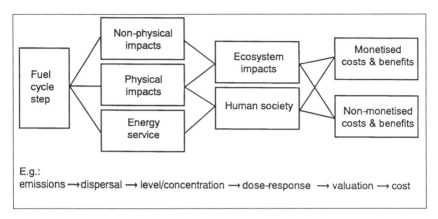

Figure 7.8. Illustration of the pathway method (Sørensen, 1995c).

Consider as an example electricity produced by a fossil fuel power station using coal (Fig. 7.9). The first step would be the mining of coal, which may emit dust and cause health problems for miners, then follows cleaning, drying and transportation of the coal, spending energy such as oil for transportation by ship (here the impacts from using oil have to be incorporated, e.g. taken from a separate study). The next step is storage and combustion of the coal in the boiler of the power station, leading to partial releases of particulate matter, sulphur dioxide and nitrogen oxides through the stack. These emissions would then have to be traced by a dispersion model, calculating air concentrations in different distances and directions away from the stack. Based upon these concentrations, inhalation amounts and the health effects caused by these substances in the human body are obtained by using the relation between dose (exposure) and effect, taken from some other study or e.g. World Health Organisation data bases. Other pathways should also be considered, for instance, pollutants washed out and deposited by rain and subsequently taken up by plants such as vegetables and cereals. They may later find their way to humans and cause health problems.

For each life-cycle step, the indirect impacts associated with the chain of equipment used to produce any necessary installation, the equipment used to produce the factories producing the primary equipment, and so on have to be assessed, together with the stream of impacts occurring during operation of the equipment both for the life-cycle step itself and its predecessors (cf. Fig. 7.5). The same is true for the technology used to handle the equipment employed in the life-cycle step, after it has been decommissioned, in another chain of discarding, recycling and reusing the materials involved.

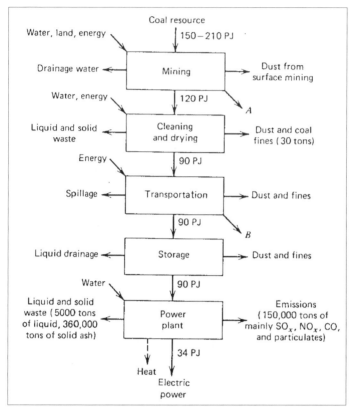

Figure 7.9. Coal-based electricity chain (Sørensen, 1993b). Modern plants reduce the power plant emissions indicated, by use of filters.

In the coal power example, the costs of respiratory diseases associated with particulates inhaled may be established from hospitalisation and lost workday statistics, and the cancers induced by sulphur and nitrogen oxides may be similarly monetised, using e.g. insurance payments as a proxy for deaths caused by these agents.

Generally, the initiating step in calculating chain impacts may be in the form of emissions (e.g. of chemical or radioactive substances) from the installation to the atmosphere, releases of similar substances to other environmental reservoirs, visual impacts or noise. Other impacts would be from inputs to the fuel cycle step (water, energy, materials such as chalk for scrubbers). As regards basic emission data, these are at present routinely being collected for many power plants, whereas the data for other conversion steps are often more difficult to obtain. Of course, emission data, e.g. from road vehicles, may be available in some form, but rarely distributed over driving modes and location (at release) as one would need in most assessment work.

Based on the releases, the next step of calculating the dispersal in the ecosphere may exploit available atmospheric or aquatic dispersion models. In the case of radioactivity, also decay and transformation have to be considered. For airborne pollutants, the concentration in the atmosphere is used to calculate deposition (using dry deposition models, deposition by precipitation scavenging or after adsorption or absorption of the pollutants by water droplets). As a result, the distribution of pollutants (possibly transformed from their original form, e.g. from sulphur dioxide to sulphate aerosols) in air and on ground or in water bodies will result, normally given as a function of time, because further physical processes may move the pollutants down through the soil (eventually reaching ground water or aquifers) or again into the atmosphere (e.g. as dust).

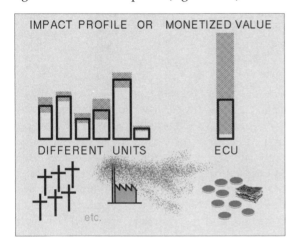

Figure 7.10. Multivariate versus monetised presentation of LCA results ("ECU" was a forerunner for the euro currency; Sørensen, 1996b).

Given the concentration of dispersed pollutants as function of place and time, the next step along the pathway is to establish the impact on human society, such as by human ingestion of the pollutant. Quite extended areas may have to be considered, both for normal releases from fossil fuel power plants and for nuclear plant accidents (typically a distance from the energy installation of a thousand kilometres or more, according to the European Commission study, 1995c). Along with the negative impacts there is, of course, the positive impact derived from the energy delivered. In the end, these are the ones that will have to be weighed against each other. Finally, one may attempt to assist the comparison by translating the dose–responses (primarily given as number of cancers, deaths, workdays lost, and so on) into monetary values. This translation of many units into one should only be done if the additional uncertainty introduced by monetising is not so large that the comparison is weakened (see Fig. 7.10). In any case, some impacts are likely to remain which cannot meaningfully be expressed in monetary terms. The impacts pertaining to a given step in the chain of energy conversions or transport may be divided into those characterising normal operation and those arising in case of accidents. In reality, the borderlines between often occurring problems during routine operation, mishaps of varying degree of seriousness, and accidents of different size are fairly hazy and may be described in terms of declining frequency for various magnitudes of problems. The pathways of impact development are to a considerable extent similar for routine and accidental situations involving injuries and other local effects, e.g. connected with ingestion or inhalation of pollutants, and as regards public impacts of the release and dispersal of substances causing nuisance where they reach inhabited areas, croplands or recreational areas. The analysis of these transfers involves identifying all important pathways from the responsible component of the energy system to the final recipient of the impact, such as a person developing illness or dying, possibly with delays of considerable lapses of time in cases such as late cancers.

7.4.8 Matrix calculations

A full systemic LCA calculation proceeds by the same steps as the chain calculations, but without summing over the indirect impact contributions from side-chains. All devices have to be treated independently, and only at the end may one sum impacts over the totality of devices in the system in order to obtain system-wide results. In practice, the difference between chain and system LCAs in work effort is not great, because in chain LCA calculations also each device is usually treated separately before the chain totals are evaluated. There are exceptions, however, where previous results for energy devices in chains that already include indirect impacts are used. In such cases, the chain results cannot immediately be used as part of a system-wide evaluation.

Particular issues arise when LCA evaluations are made for systems considered for future implementation: there may be possible substitutions between human labour and machinery, linking the analysis to models of employment and reproductive activities. In order to find all the impacts, vital parts of the economic transactions of society have to be studied. The total energy system comprises conversion equipment, transmission lines or transportation, as well as end-use devices converting energy to the desired services or products. The demand modelling involves consideration of the development of society beyond the energy sector.

More factual is the relation between inputs and outputs of a given device, which may be highly non-linear but in most cases are given by a deterministic relationship. Exceptions are, for example, combined heat-and-power plants, where the same fuel input may be producing a range of different proportions between heat and electricity. This gives rise to an optimisation problem for the operator of the plant (who will have to consider fluctuating demands along with different dispatch options involving different costs). A strategy for operation is required in this case before the system LCA can proceed. But once the actual mode of operation is identified, the determination of inputs and outputs is, of course, unique. The impact assessment then has to trace where the inputs came from and keep track of where the outputs are going in order to determine which devices need to be included in the analysis. For each device, total impacts have to be determined, and the cases where successive transfers may lead back to devices elsewhere in the system can be dealt with by setting up a matrix of all transfers between devices belonging to the energy system. In what corresponds to an economic input–output model, the items associated with positive or negative impacts must be kept track of, such that all impacts belonging to the system under consideration in the life-cycle analysis will be covered.

Once this is done, the impact assessment itself involves summation of impacts over all devices in the system, as well as integration over time and space, or just a determination of the distribution of impacts over time and space. As in the chain studies, the spatial part involves use of dispersal models or compartment transfer models (Sørensen, 1992b), while the temporal part involves charting the presence of offensive agents (pollutants, global warming inducers, etc.) as a function of time for each located device and further a determination of the impacts (health impacts, global warming trends, and so on) with their associated further time delays. This can be a substantial effort, as it has to be done for each device in the system, or at least for each category of devices (an example of strong aggregation is shown in Fig. 6.83).

The first step is similar to what is done in the chain analysis, i.e. running a dispersion model that uses emissions from point or area sources as input (for each device in the system) and thus calculating air concentration and land deposition as function of place and time. For example, the RAINS model is used to calculate SO_2 dispersal on the basis of long-term average meteorological data, aggregated with the help of a large number of trajectory calculations (Alcamo et al., 1990; Hordijk, 1991; Amann and Dhoondia, 1994).

Based on the output from the dispersion model, ingestion rates and other uptake routes are again used to calculate human intake of pollutants through breathing air, skin, etc., and a model of disposition in the human body, with emphasis of accumulating organs and rates of excretion, is used. The resulting concentration for each substance and its relevant depository organs is via a dose–response function used to calculate the morbidity and mortality arising from the human uptake of pollution. It is customary to use a linear dose-response function extending down to (0,0) in cases where measurements only give information on effects for high doses. This is done in the study of the European Commission, based on a precautionary point of view as well as theoretical considerations supporting the linear relation (European Commission, 1995a-f). The alternative of assuming a threshold, below which there is no effect, is often used in regulatory schemes, usually as a result of industry pressure rather than scientific evidence.

System-wide calculations are often interpreted as comprising only those components that are directly related to energy conversion. Sometimes this borderline is difficult to determine, e.g. for vehicles of transportation, which cause traffic and thus links to all the problems of transportation infrastructure. A general approach would be to treat all components of the energy system proper according to the system approach, but to treat links into the larger societal institutions and transactions as in the chain LCA. In this way, the overwhelming prospect of a detailed modelling all of society is avoided, and yet the double-counting problem is minimised because energy loops do not occur (although loops of other materials may exist in the chains extending outside the energy sector).

Marginal versus systemic change

Studies such as those described above make the assumption that the energy installations considered are marginal additions to the existing system. For instance, one particular coal-fired power plant is added to a system, which is otherwise unchanged. This implies that in calculating indirect impact, the whole set-up of society and its industrial capacity is the current one. Such an approach would not be valid in a systemic study of some scenario for the future system. Such a scenario will be imbedded into a society that may be very different from the present, as regards energy flows, industry structure and social habits. In studying the future impacts from manufacturing e.g. photovoltaic panels, the process energy input will not come from the present mix of mainly fossil power plants, but will have to reflect the actual structure of the future energy system assumed.

Evidently, this systemic approach gives results very different from those emerging from treating the same future energy system as the result of a series of marginal changes from the present and may thus affect the determination of an optimum system solution (e.g. if the energy input to a future renewable energy installation is higher than that of a competing fossil installation, then the marginal evaluation based on current fossil power supply will be less favourable than one based on a future scenario of non-fossil energy supply).

One workable alternative to the marginal assumption, in case different forms of energy supply have to be compared to each other without being part of an overall scenario, is to consider each system as autonomous, i.e. for the photovoltaic power plant to assume that the energy for manufacture comes from similar photovoltaic plants. This makes the impact evaluation self-contained, and the assumption is generally fair, e.g. if the power for site-specific work mostly comes from nearby installations, rather than from the national average system. As renewable systems like wind turbines and solar plants are of small unit size, the gradual establishment of a sizeable capacity could indeed be seen as involving energy use based on the previously installed plants of the same kind. This suggestion may not always apply, as energy inputs in some cases have to be of forms different from the one associated with the installation studied.

7.4.9 Communicating with decision-makers

As the purpose of LCA is to facilitate decision-making, some thought should be given to the way the results of an analysis are presented to the target group. This underlies, of course, the quest for monetising all impacts: it is believed that decision-makers understand monetary impacts better than physical ones and that qualitative descriptions have little say in policy. From a scientific point of view, the dividing line goes between qualitative and quantitative impact statements. That the quantifiable impacts cannot all be expressed in the same unit is intuitively clear: numbers of cancer deaths, loss of agricultural crops, acid rain damage to Greek temples and traffic noise are fundamentally impacts expressed in "different units". The translation into monetary values, however it is done, is losing part of the message. This is why we said in connection with Fig. 7.10 that monetising should be used only if it does not significantly increase uncertainty, which means that the decision-makers should not be exposed to the monetising simplification unless it preserves their possibility of making a fair assessment.

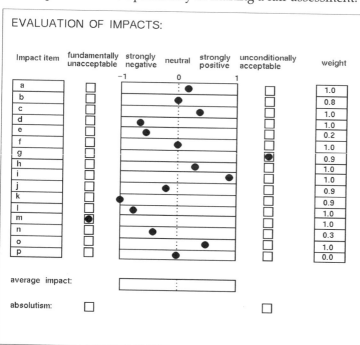

Figure 7.11. Layout of multivariate impact assessment scheme (Sørensen, 1982, 1993b).

If those impacts that can be quantified are kept in different units, the question arises of how they can be presented to the decision-maker in a form facilitating their use. The common answer is to use a multivariate approach, where, as indicated in the left side of Fig. 7.10, each category of impact is presented in its own units. Figure 7.11 expands on one methodology for multivariate presentation (Sørensen, 1993b), suggesting the use of what is called an impact profile. The idea of the profile is that each particular type of impact is evaluated in the same way for different systems. Thus, the magnitudes indicated by the profile are no more subjective than the monetised values, although they cannot be summed across different impact categories. Clearly those impacts that can be meaningfully monetised should be so, but the impact profile actually gives much more information, as it tells the decision-maker whether two energy solutions have the same type (i.e. the same profile) of impacts or whether the profile is different and thus makes it necessary for the decision-maker to assign weights to different kinds of impacts (e.g. comparing greenhouse warming impacts of fossil system with noise impacts of wind turbines). The assignment of weights to different impact categories is the central political input into the decision process.

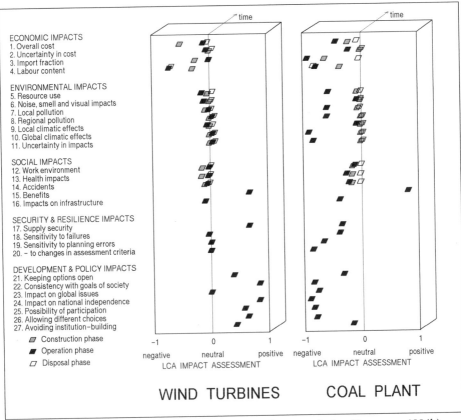

Figure 7.12. LCA impact profiles for coal and wind power chains (Sørensen, 1994b).

The impact profile approach further makes it a little easier to handle qualitative impacts that may only allow a description in words, because such impacts can often be characterised vaguely as "small", "medium" or "large", a classification that can be rendered in the profiles and compared for different energy systems. Hence, the advantage of the profile method is that the decision-maker sees both the bars representing monetised values and the adjacent bars describing the outcome of a qualitative assessment. Thus, the chance of overlooking important impacts is diminished. In any case, the multivariate profile approach does as mentioned give the decision-maker more information than a single monetary value. A further point that may play a role in decision-making is the presence of value systems making certain impacts "unconditionally acceptable" or "unconditionally unacceptable". Such absolute constraints can be accommodated in the assignment of weights (zero or infinite) as indicated in Fig. 7.11. Figure 7.12 shows an example of a profile of the impacts from two energy installations, including both positive and negative impacts (details are given in Sørensen, 1994b).

Section 7.4.10 below gives several examples of monetised impacts based on recent LCA studies, for both individual installations (chain calculations) and systems (present and future scenarios).

It is evident that although the derivation of each single impact figure requires a large effort, the result still may involve substantial uncertainty. The analyses to be presented in the following section will also show that the largest uncertainties are often found for the most important impacts, such as nuclear accidents and greenhouse warming. Clearly, there is a general need to improve data by collecting information pertinent to these types of analysis. This need is most clearly met by doing site- and technology-specific studies. As regards indirect inputs, national input–output data are often based upon statistical aggregation choices failing to align with the needs for characterising transactions relevant for the energy sector. Also there are usually gaps in data availability. It is one conclusion from these observations that there is a need to be able to present qualitative and quantitative impacts to a decision-maker in such a way that the weight and importance of each item become clear, despite uncertainties and possibly different units used. The multivariate presentation tools invite the decision-maker to employ multicriteria assessment.

The difficulties encountered in presenting the results of externality studies and life-cycle analyses in a form suited for the political decision-making process may be partly offset by the advantages of bringing into the debate the many impacts often disregarded (which is, of course, the core definition of "externalities", meaning issues not included in the market prices). It may be fair to say that LCAs and the imbedded risk assessments will hardly ever become a routine method of computerised assessment, but they may continue to serve a useful purpose by focusing and sharpening the debate involved in any decision-making process and hopefully help increase the quality of the basic information upon which a final decision is taken, whether on starting to manufacture a given new product or to arrange a sector of society (such as the energy sector) in one or another way.

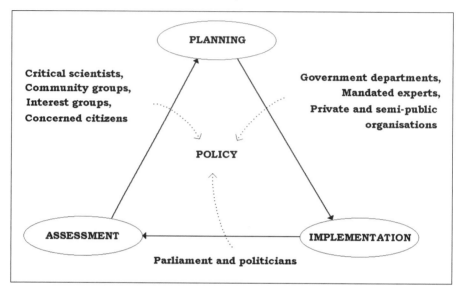

Figure 7.13. The actor triangle, a model of democratic planning, decision-making and continued assessment (Sørensen, 1993a).

Finally, Fig. 7.13 indicates how decision-making is a continuous process, involving planning, implementation and assessment in a cyclic fashion, with the assessment of actual experiences leading to adjustments of plans or in some cases to entirely new planning.

7.4.10 Application of life-cycle analysis

Examples of LCA will be presented in the following sections. Each is important in its own right, but they also illustrate the span of issues that one may encounter in different applications. The first is greenhouse warming impacts associated with emissions of substances such as carbon dioxide. It involves global dispersal of the emitted substance, followed by a subtle range of impacts stretching far into the future and depending on a number of poorly known factors, for instance, the ability of nations in central Africa to cope with vector-borne diseases such as malaria. The uncertainties in arriving at any quantitative impact measure will be discussed.

The second example is that of chain calculations for various power plants, using fuels or renewable energy. It illustrates the state of the art of such calculations, which are indeed most developed in the case of electricity provision chains. Life-cycle impact estimates for the emerging hydrogen and fuel cell technologies may be found in an accompanying book (Sørensen, 2004a).

The third case presents the impacts of road traffic, included to show an example of a much more complex LCA chain, with interactions between the energy-related impacts and impacts that would normally be ascribed to other sectors in the economy of a society.

In a practical approach to life-cycle analysis and assessment, one typically goes through the following steps:

- Establishing an inventory list, i.e. identifying and categorising materials and processes inherent in the production, use and final disposal of the product or system.
- Performing an impact analysis, i.e. describing the impacts on environment, health, etc. for each item in the inventory list.
- Damage assessment, translating the physical impacts into damage figures (in monetary units where possible).
- Overall assessment, identifying the most crucial impact pathways.
- Mitigation proposals, suggesting ways of avoiding damage, e.g. by use of alternative materials or processes.
- Regulatory proposals, describing how norms and regulative legislation or prescriptions for manufacture and use can decrease damage.

One may say that the first two items constitute the life-cycle analysis, the two next items the life-cycle assessment, and the two final ones optional political undertakings.

7.4.11 LCA of greenhouse gas emissions

The injection of carbon dioxide and other greenhouse gases such as methane, nitrous oxide and chlorofluorocarbons into the atmosphere changes the disposition of incoming solar radiation and outgoing heat radiation, leading to an enhancement of the natural greenhouse effect. The modelling of the Earth's climate system in order to determine the long-term effect of greenhouse gas emissions has taken place over the last 40 years with models of increasing sophistication and detail. Still, there are many mechanisms in the interaction between clouds and minor constituents of the atmosphere, as well as coupling to oceanic motion, that are only modelled in a crude form. For instance, the effect of sulphur dioxide, which is emitted from fossil fuel burning and in the atmosphere becomes transformed into small particles (aerosols) affecting the radiation balance, has only become realistically modelled over the last couple of years. Because of the direct health and acid rain impacts of SO_2, and because SO_2 is much easier to remove than CO_2, the emissions of SO_2 are being curbed in many countries. As the residence time of SO_2 in the atmosphere is about a week, in contrast to the 80–120 years for

CO_2, the climate models have to be performed dynamically over periods of 50–100 years (cf. Chapter 2).

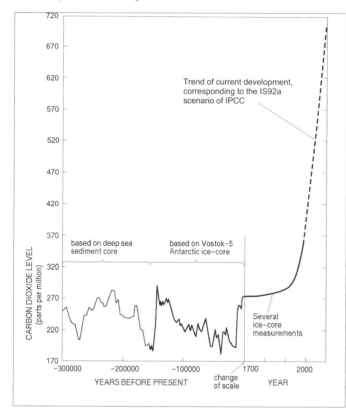

Figure 7.14. History of atmospheric CO_2 levels based on ice-core data (based upon Sørensen, 1991; Emiliani, 1978; Berger, 1982; Barnola *et al.*, 1987; Neftel *et al.*, 1985; Friedli *et al.*, 1986; Siegenthaler and Oeschger, 1987; IPCC 1996a).

Figure 7.14 shows measured values of CO_2 concentrations in the lower atmosphere over the last 300 000 years. During the ice age cycles, systematic variations between 190 and 280 ppm took place, but the unprecedented increase that has taken place since about 1800 is primarily due to combustion of fossil fuels, with additional contributions e.g. from changing land use, including felling of tropical forests. If current emission trends continue, the atmospheric CO_2 concentration will have doubled around the mid-21st century, relative to the pre-industrial value. The excess CO_2 corresponds to slightly over half the anthropogenic emissions, which is in accordance with the models of the overall carbon cycle (IPCC, 1996b). The ice core data upon which the historical part of Fig. 7.14 is based also allow the trends for other greenhouse gases to be established. The behaviour of methane is similar to that of CO_2, whereas there is too much scatter for N_2O to allow strong conclusions. For the CFC gases, which are being phased out in certain sectors, there is less than 40 years of data available. Both CO_2 and methane concentrations show regular seasonal variations, as well as a distinct asymmetry between the Northern and the Southern hemispheres (Sørensen, 1992b).

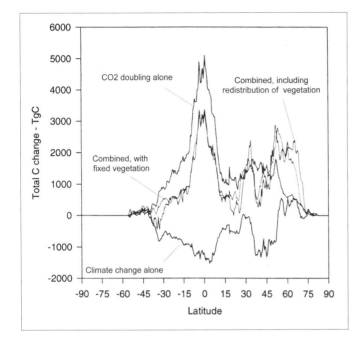

Figure 7.15. Simulated changes in equilibrium soil-stored carbon as a function of latitude (units: 1012 g carbon per half degree latitude), with individual contributions indicated (based upon Melillo *et al.,* 1993 and IPCC, 1996b).

Despite the incompleteness of climate models, they are deemed realistic enough to permit the further calculation of impacts due to the climate changes predicted for various scenarios of future greenhouse gas emissions.

It should be mentioned that the discussion above pertains only to the influence of greenhouse gases on the physical earth–ocean–atmosphere system. There is a subtle interaction with the biological activity in each compartment, with and without anthropogenic interference. Some of the mechanisms identified by the IPCC (1996b, Chapter 9), in addition to the temperature changes discussed in Chapter 2, are changed vegetation cover and agricultural practices would influence carbon balance and amounts of carbon stored in soils; wetlands may dry out owing to temperature increases, thereby reducing methane release; and albedo changes may influence the radiation balance. Figure 7.15 shows how carbon storage is increased owing to the combined effects of higher levels of atmospheric carbon (leading to enhanced plant growth) and the opposite effects arising from changed vegetation zones as predicted by the climate models (affecting moisture, precipitation and temperature). Comprehensive modelling of the combined effect of all the identified effects has not yet been performed.

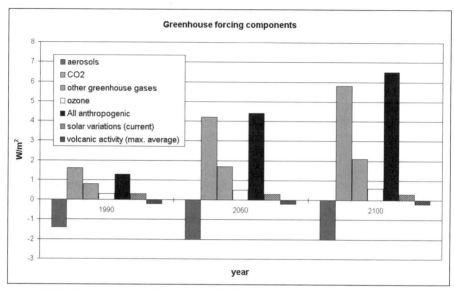

Figure 7.16. Components of present and future (geographically averaged) greenhouse forcing, according to the IS92a scenario of the IPCC (based on IPCC, 1996b, p. 81 and p. 320 for the year 1990, p. 321 for 2060 – the year where CO_2 concentrations have doubled – and finally pp. 320–321 for the year 2100, with future natural forcing components taken at 1990 values) (Kuemmel *et al.*, 1997).

The current anthropogenic greenhouse forcing, i.e. the net excess flow of energy into the atmosphere taking into account anthropogenic effects since the late 18th century, is estimated at 1.3 W m^{-2}, as a balance between a twice as big greenhouse gas contribution and a negative contribution from sulphate aerosols (IPCC, 1996b, p. 320). The further increase in the concentration of CO_2 and other greenhouse gases, together with at least a stabilisation in current emissions leading to the creation of sulphate aerosols, as assumed in the IPCC business as usual scenario (called "IS92a"), is estimated to raise the radiative forcing to some 6–7 W m^{-2} by year 2100 (IPCC, 1996b, p. 320). For the IS92a scenario, the doubling of CO_2 occurs around the year 2060, where the forcing is estimated to be 4.4 W m^{-2}. The estimates are summarised in Fig. 7.16, which also gives the non-anthropogenic contributions from presently observed variations in solar radiation (somewhat uncertain) and from volcanic activity (very irregular variations with time).

It should be noted that it is not strictly allowed to make a "sum" of all anthropogenic forcing components. For example, aerosols currently have a quite different geographical distribution from that of carbon dioxide, and the combined effect is far more complex than the simple difference between the two absolute values.

Greenhouse warming impacts

Several studies have tried to estimate the damage that may occur as a consequence of the enhanced greenhouse effect. Early studies have concentrated on the damage that may occur in industrialised countries or specifically in North America (Cline, 1992; Frankhauser, 1990; Tol, 1995; Nordhaus 1994; summaries may be found in IPCC, 1996c). Exceptions are the top-down studies of Hohmeyer (1988) and Ottinger (1991), using rather uninformed guesses regarding starvation deaths from climate change.

The IPCC Second Assessment Report (IPCC, 1996b) summarises information on the expected physical damage caused by global warming published up to 1994. Based on this work we shall make a monetary evaluation using assumptions similar to those of the European Commission study (cf. Sørensen, 1997c). It should be said that the energy use scenarios upon which the impact studies were based are typically the reference scenario of the 1992 IPCC report (business as usual increase in energy consumption and doubling of CO_2; IPCC, 1992) or earlier work, and the physical warming assumptions are those of either the IPCC 1994 study (IPCC, 1995) or, in most cases, earlier studies. The teams of the 1995 IPCC impact assessments were provided with the IPCC climate model results (seasonal temperatures, precipitation and soil moisture) for three global models (IPCC, 1994), which, however, were all calculations made without inclusion of the sulphur aerosol contributions. Most teams were not even able to use these data, as the impact studies at their disposal were all made earlier than 1994, but in any case the temperature increases assumed reflected the status of models not incorporating the cooling effects and altered geographical distribution of temperatures found in the newer model results presented in the first volume of the IPCC 1995 assessment (IPCC, 1996b). Therefore, impacts must be considered as possibly overestimated, although in many cases they involve uncertainties in precisely defining the warming that will lead to the effect discussed, which probably overshadows the effects of improvements achieved in the basic climate modelling.

The estimation of impacts on agriculture and forestry particularly depends on the change in natural vegetation zones and local changes in soil moisture and temperatures. Although precise location of the zones within a few hundred kilometres may not be essential, the relative magnitudes of them would directly affect the impact assessment; as a result, the confidence we can attach to the impacts predicted is not particularly high. In addition, there is, as mentioned in section 7.4.6, a factor 2 uncertainty in the monetary valuation used for accidental loss of life. One reason for making these crude impact estimations is to find out whether the mitigation or avoidance costs are likely to be lower than the damage costs for not taking any action.

Impact description:	Ref. Below	Valuation H (T$)	Valuation L (T$)
Additional heatwave deaths (doubling, additional 0.1M y^{-1}, valued at 3 M$ each)	a	30	20
Fires due to additional dry spells (doubling)	b	1	0.5
Loss of hardwood and fuelwood yields (20% relative to 1980 production)	c,d,e	4	1
Increase in skin cancer due to UV radiation	b	3	3
Additional asthma and allergy cases (increase in pollen and air pollution due to warming)	b	3	1.5
Financial impact of doubling extreme events	b	3	0.8
Increase in insect attacks on livestock and humans	b	?	?
Food production (failure to adopt new crops, increased pests and insect attacks, production loss 10–30%), (population at risk increasing from present 640M to 700–1000M), additional deaths from starvation due to crop loss (100M additional deaths, chiefly in developing countries)	f,g	300	0
Deaths connected with migration caused by additional droughts or floods (50M deaths, the affected population being over 300M)	b	150	0
Increased mortality and morbidity due to malaria (presently 2400M at risk and 400M affected, increase due to warming 100M cases, in tropical and partly in subtropical regions, a potential 7-fold increase in temperate regions assumed to be curbed)	b,h	300	75
Increased mortality and morbidity due to onchocercosis (presently 120M at risk and 17.5M affected, increase and spread due to warming 10M additional cases, primarily in developing countries)	b,i	30	5
Increased mortality and morbidity due to schistosomosis (presently 600M at risk and 200M cases, increase due to warming 25%, in developing countries)	b	150	20
Increased mortality and morbidity due to dengue (presently 1800M at risk and 20M cases, increase due to warming 25%, in developing countries)	b	15	0
Other effects of sanitation and fresh water problems connected with salt water intrusion, droughts, floods and migration (developing countries)	b	50	0
Loss of tourism, socio-economic problems, loss of species, ecosystem damage, etc.	b	?	?
Total of valuated impacts (order of magnitude)		**1000**	**100**

Table 7.2. Estimated global warming impacts during the entire 21st century, for IPCC reference case (CO_2 doubling) (unit: 10^{12} US$). H: high SVL of 3 M$ used throughout. L: SVL put at zero for Third World (Sørensen, 1997c; see reference keys on next page)

Key to references for Table 7.2:
Base reference IPCC (1996a), specific references in the table are as follows:
a. Kalkstein and Smoyer (1993)
b. McMichael *et al.*, in IPCC (1996a), Chapter 18
c. Zuidema *et al.* (1994)
d. Kirschbaum *et al.*, in IPCC (1996a), Chapter 1
e. Solomon *et al.*, in IPCC (1996a), Chapter 15
f. Reilly *et al.*, in IPCC (1996a), Chapter 13 and Summary for Policymakers
g. Rosenzweig *et al.* (1993); Parry and Rosenzweig (1993)
h. Martens *et al.* (1994)
i. Walsh *et al.* (1981); Mills (1995)
The valuations involve further estimates and should be regarded as very uncertain.

In our externality estimation, we shall use the 2060 situation as an average representing the entire 21st century. It is clear that valuation studies must be made on the basis of impact studies already made and that impact studies must be made on the basis of already available climate model results, which again are based on identified scenarios. The next IPCC assessment of impacts and externality costs should, of course, be based on the newer data, but first of all it should be made more clear that such assessment is by necessity one step behind the impact studies, two steps behind the state-of-the-art climate models, and three steps behind energy use scenarios.

Table 7.2 gives a list of impacts identified by the IPCC Second Assessment Report (IPCC, 1996a), in terms of the additional number of people exposed to a range of risks due to the average warming, sea-level rise and additional extreme events predicted by the climate studies reviewed by the (earlier) IPCC studies. As many impacts involve human deaths, the results scale practically linearly with the value of a statistical life (SVL) used. This means that a choice of a value different from the chosen 3 million US$ (which is close to the 2.6 Meuro used by the European Commission study discussed above) can be accommodated as an overall factor.

The individual entries in Table 7.2 are based on the sources indicated, with monetisation of impacts evaluated in two different ways. The first valuation column uses the high SVL value of 3 million US$ globally, while the second takes the SVL to be zero for developing countries, in order to display the geographical differences of impacts. Here follows a brief explanation of the impacts included:

Heatwave deaths occur in major cities due to the heat island effect, possibly combined with urban pollution. The doubling estimated by Kalkstein and Smoyer (1993) is mostly due to increased occurrence at mid-latitudes (city temperature rise 2.5°C assumed), and thus two-thirds are assumed to happen in industrialised countries. A case study in New York City (Kalkstein, 1993) finds an increased mortality of 4×10^{-5} over a 5- to 10-day period. The present annual New York rate of heatwave deaths is 320, and Kalkstein has collected similar numbers for a number of large cities around the world, experiencing days with temperatures above 33°C. The estimated doubling of incidence will thus imply an additional order of magnitude of 10^5 heatwave deaths, annually and globally, valued at 30 TUS$ over the 100-year period. Uncertainties come from possible acclimatisation effects and from increased populations in large cities expected over the 21st century.

The doubling of fires causes mainly economic loss, assumed to be evenly distributed between developed and developing countries, whereas the 20% loss of hardwood and fuelwood yields is predicted to follow a complex geographical pattern (Kirschbaum et al., IPCC, 1996a, Chapter 1), but with the highest losses occurring in tropical regions (Solomon et al., IPCC, 1996a, Chapter 15). We assume that 75% of the economic losses occur in developing countries. The increased number of skin cancer cases due to an assumed 10% loss of ozone occurs mainly at higher latitudes (IPCC, 1996a, Chapter 18).

Additional allergy cases would be associated with increased levels of pollen and air pollution due to heating and would occur predominantly at lower latitudes, whereas asthma incidence is highest in humid climates, expected to be enhanced at higher latitudes (IPCC, 1996a, Chapter 18). These impacts are assumed to be equally divided between developed and developing regions. Owing to an expected shortfall of hardwood supply relative to demand during the 21st century, the actual economic value may be considerably higher than the estimate given in Table 7.2. The financial loss associated with a predicted doubling of extreme events (draughts and floods, IPCC, 1996a, Chapter 18) is assumed to occur with 75% of the incidences in developing countries. The predicted increased frequence of insect bites is not valued, but could have an economic impact, e.g. on livestock production.

One major issue is the impact of climate change on agricultural production. Earlier evaluations (e.g. Hohmeyer, 1988) found food shortage to be the greenhouse impact of highest importance. However, the 1995 IPCC assessment suggests that in developed countries, farmers will be able to adapt crop choices to the slowly changing climate, such that the impacts will be entirely from Third World farmers lacking the skills to adapt. The estimated production loss amounts to 10–30% of current global production (Reilly et al., IPCC, 1996a, Chapter 13), increasing the number of people exposed to risk of hunger from the present 640 million to somewhere between 700 and 1000 million (Parry and Rosenzweig, 1993). There are also unexploited possibilities for increasing crop yields in developing countries, so the outcome will depend on many factors, including speed of technology transfer and development. Assuming a lower estimate of 100 million additional starvation deaths during the 21st century, one arrives at the 300 T$ figure given in Table 7.2, all occurring in the Third World. Similarly, deaths associated with migration induced by extreme events are estimated at 50 million.

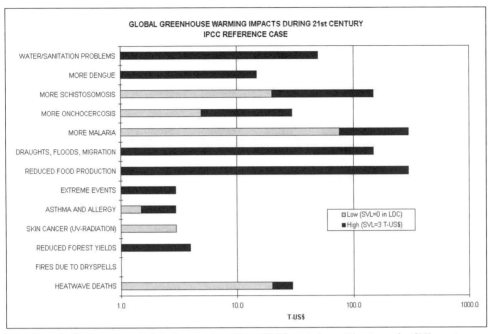

Figure 7.17. Impacts of global warming (from Table 7.2, logarithmic scale; SVL = statistical value of life, LDC = Third World countries) (Kuemmel *et al.*, 1997).

The other major impact area is death from diseases transmitted by insect vectors influenced by climatic factors, such as mosquitoes. For Malaria, there are presently 2400 million people at risk (McMichael *et al.*, IPCC, 1996a, Chapter 18), and an additional 400 million are expected due to greenhouse warming and its implied expansion of the geographical area suited as habitat for the malaria-carrying mosquitoes (Martens *et al.*, 1994). This will involve subtropical and even some temperate regions, but also in tropical regions the incidence of malaria is predicted to increase. Assuming 100 million additional deaths from malaria during the 21st century, of which 75% are in the tropics, one arrives at the figures given in Table 7.2. Large uncertainties are associated with the possibilities of mitigating actions in the subtropical and temperate regions.

Considering the large impacts of diseases such as malaria, it is natural to compare damage costs with possible avoidance costs. Relatively little money has been spent on control of vector-borne diseases, and it is recognised that just as mosquitoes have developed resistance to pesticides, the malaria vectors have grown resistant to both the conventional medication chlorochine and newer substitutes such as meflochine. Many scientists have doubts whether new genetically engineered agents can be introduced effectively into the mosquito populations, and prospects for new, effective drugs are also unclear. On the other hand, the pharmaceutical industry may not have made a wholehearted effort, in consideration of the low purchasing power of e.g. rural African people (Butler *et al.*, 1997). Recent studies of maternal transmission of antibodies raise hope for development of vaccines (Fried *et al.*, 1998).

Also for other tropical diseases such as onchocercosis, vector populations are expected both to increase by 25% at current sites (Mills, 1995) and to spread to new sites (Walsh *et al.*, 1981), leading to 10 million additional cases. Schistosomosis may spread into the subtropical regions, whereas dengue and yellow fever are expected to remain in the tropics. Table 7.2 reflects these expectations through its distribution of impacts between developed and developing regions, and it also gives an estimate of deaths occurring due to sanitation problems connected with salt water intrusion into drinking water (due to sea-level rise) and larger migrating populations, mainly in the developing countries (assuming that immigration into industrialised countries continues to be controlled). The economic consequences of other identified impacts, such as loss of species, effect on tourism, etc., have not been estimated.

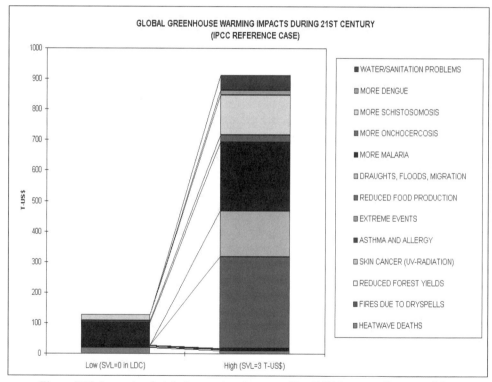

Figure 7.18. Impacts of global warming (same as Fig. 7.17, but on a linear scale).

The overall impacts are of the order of magnitude 10^{15} US\$, when using the SVL of industrialised countries, and one order of magnitude lower, if Third World impacts are valued at or near zero. This spells out the greenhouse impact dilemma that 90% of the damage is in the Third World, much higher than their share in causing the problem. The IPCC Working Group II identification of a number of impacts specifically occurring in the low-latitude regions explains why the impact estimates are so much higher than those of early work based upon and extrapolated from industrialised regions (Cline, 1992; Frankhauser, 1990; Nordhaus, 1990). Other factors contributing to this result include the high SVL and the assumptions of less ability to mitigate impacts in the Third World (by switching food crops, building dikes to avoid floods, and so on). If the impacts of the present study, evaluated with the uniformly high SVL, are distributed equally over the 100-year period, the annual cost of greenhouse warming is found to be roughly 40% of the current global GNP. In Figs. 7.17 and 7.18, the total estimated greenhouse warming impacts, and the part occurring outside the Third World (SVL = 0 in less developed countries, LDC), are shown on both a logarithmic and a linear scale.

Estimating greenhouse warming externalities for fuel combustion

The translation of the 10^{15} US\$ impact from greenhouse emissions into externalities for specific energy activities may be done in the following way. It is assumed that the impacts identified correspond to the IPCC business-as-usual scenario IS92a, and that the situation in the year 2060, when CO_2 has doubled relative to pre-industrial levels, is a representative average of the entire 21st century. The average CO_2 concentration and forcing components were shown in Figs. 7.14 and 7.16. It is seen from Fig. 7.16 that the CO_2 forcing in the year 2060 is 4.2 W m^{-2} or 95.4% of the total anthropogenic forcing. One may therefore make the simplifying assumption that 95.4% of the impacts determined above are to be associated with the CO_2 emissions alone. The non-CO_2 emissions are dominated by emissions arising from other activities than burning of fossil fuels (notably land use changes such as felling of tropical forests), and it does not appear right to give an externality credit to the sulphate aerosols, although they may partially cancel some of the warming trends caused by CO_2. This is a question open for debate, but the view here is that human interference with climate in either direction should be given a negative externality tag. The possible cost of sulphate aerosol formation is not estimated, because in the energy use scenarios considered adequate this source of atmospheric contamination disappears within a few decades, owing to fuel shifts or to stronger SO_2 emission control. Similarly, the impacts of disappearing tropical forests are disregarded, in the hope that this trend is halted but also because it is not an externality of the energy system. It should be mentioned also that some of the CO_2 emissions do not arise from the energy sector, but in considering only CO_2 we assume that these are balanced by non-CO_2 emissions from the energy sector. The IPCC IS92a scenario emission prediction is shown in Fig. 7.19, along with averaged historical data.

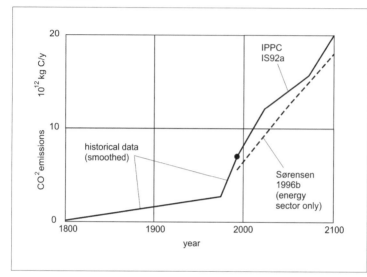

Figure 7.19. Historical and future global carbon dioxide emissions for the scenarios used in estimating greenhouse warming externalities (Kuemmel et al., 1997; based on IPCC, 1996b; Sørensen, 1997a).

Integrating the CO_2 emissions from 1990 to 2060 one obtains a total of 814 $\times 10^{12}$ kg C. If the integration in made from 1765 to 2060 one gets 1151 $\times 10^{12}$ kg C. There are now two ways of assigning the externality cost. One is the straightforward division of 95% of 1000 $\times 10^{12}$ US$ by the total emissions from pre-industrial times to the point of CO_2 doubling, giving 0.83 US$ per kg of C or 0.23 US$ per kg of CO_2. This means distributing the impacts felt through the entire 21st century over the emissions leading to doubling of CO_2 in the atmosphere, which apart from the uncertainty in assigning the impacts to a specific time period represents impacts per unit of the causing offence. However, in the climate debate, it is often argued that as one cannot do anything about the historical emissions but only regulate future emissions, one should distribute the impacts only over future emissions, i.e. from 1990 to 2060. This is called the *grandfathering principle*. It gives a specific externality of 1.17 US$ per kg of carbon or 0.32 US$ per kg of CO_2.

To further illustrate the width of choice, we mention the method used in several earlier publications (Sørensen, 1997a,c). Here only the energy part of CO_2 emissions were considered, and it was assumed that doubling of CO_2 would occur 50 years from now, in 2040 (cf. dashed line in Fig. 7.19). This leads to total energy related CO_2 emissions for the period 1990–2060 amounting to 433 $\times 10^{12}$ kg of C. Assuming that the part of the impact externality caused by the energy sector is 75% (or 750 $\times 10^{12}$ US$), one obtains, using the grandfathering principle, a specific externality of 1.72 US$ per kg of C or 0.47 US$ per kg of CO_2. The range of estimates can then be summarised as in Table 7.3.

Greenhouse warning impacts	Estimate I (IPCC, IS92a) grandfathering	Estimate II (IPCC, IS92a) no grandfathering	Estimate III (Sørensen, 1997a,c) grandfathering
Cause (see text for further explanation)	All CO_2 emitted 1990–2060: 814 × 10¹² kg of C	All CO_2 emitted 1765–2060: 1151 × 10¹² kg of C	Energy-related CO_2 emitted 1990–2060: 433 × 10¹² kg of C
Effect (throughout 21st century)	95% of total impact: 954 × 10¹² US$	95% of total impact: 954 × 10¹² US$	75% of total impact: 750 × 10¹² US$
Specific externality	1.17 US$/kg of C 0.32 US$/kg of CO_2 0.26 ECU/kg of CO_2	0.83 US$/kg of C 0.23 US$/kg of CO_2 0.18 ECU/kg of CO_2	1.72 US$/kg of C 0.47 US$/kg of CO_2 0.38 ECU/kg of CO_2

Table 7.3. Summary of greenhouse warming impact estimates (Kuemmel et al., 1997).

In the following examples, estimate I is used as the central estimate, and the other two estimates are used as defining an uncertainty range, without claiming that values outside this range could not be defended (e.g. owing to different views taken with respect to monetising).

The externalities for specific types of fossil fuel combustion can now be evaluated. Among different types of fossil power production, present coal-fired power stations typically emit 0.27 kgC kWh$_{elec}^{-1}$, gas-fired ones emit 0.16 kgC kWh$_{elec}^{-1}$, and oil-fired plants emit 0.21 kgC kWh$_{elec}^{-1}$, The transportation sector mainly uses oil products, and for gasoline-driven automobiles typical emissions would be 660 gC litre^{-1} or about 49 gC per vehicle-km at an average of 13.5 km per litre of gasoline (approximately the current value).

For biofuels like biogas or methanol, one should not count the CO_2 emissions, owing to the short time interval between the carbon assimilation by plants and the subsequent release (assuming that even woody biomass for methanol production comes from relatively short-rotation crops), but emissions of other greenhouse gases like N_2O or CH_4 from burning of biomass should be included in the calculation. In those cases where heat and electricity are co-generated, one may assign the externality to the fuel input in order to avoid confusion over the distribution between multiple outputs.

7.4.12 LCA of power production chains

This section gives examples of LCA calculations for various power production chains, with the purpose of illustrating the current state of the art.

Fossil fuel chains

Most present production of power takes place at fuel-fired power stations, implying that the chains defining the life-cycles start with fuel extraction and go through conversion steps until the final disposal of residues and release of pollutants to the air. The practice is currently changing as a result of more efficient use of the fuels combined production of power and heat, sometimes in combined cycle plants where energy is extracted from the fuel in more than one step. Also changing is the level of effluent control (currently capturing most SO_2 and NO_x from flue gases, in the future possibly also CO_2 for subsequent controlled storage). A traditional fuel chain for coal-based electricity production was shown in Fig. 7.9.

Recent studies have attempted to evaluate the impacts of each step in the fossil fuel-based power chains, using a bottom-up approach based on particular installations at specific places. The results are summarised in Tables 7.4–7.6 and 7.8. A common feature is the dominating influence on total life-cycle externalities coming from global warming. The estimates made in section 7.4.11 will be used, but one should keep in mind that they pertain to the specific scenario of doubling greenhouse gas concentrations in the atmosphere by the mid-21st century. It is a methodological problem to select the correct externality to include in scenarios involving a reduction of fossil fuel usage. Many studies use the same values for different scenarios, which would lead to an underestimation of the benefits of substituting other forms of energy for those emitting greenhouse gases. The coal-fired power plant considered in Table 7.4 is a proposed plant located at West Burton England, with 99.7% particle removal, fairly decent SO_2 scrubbing but poor NO_x emission control. The quantified impacts are mainly from a study performed under the EC Joule Programme (European Commission, 1995c), except for the greenhouse warming impacts, which are estimated on the basis of Tables 7.2 and 7.3. As mentioned, one of the central assumptions in the EC study is to monetise deaths using a statistical value of life (SVL) amounting to 2.6 Meuro or 3.25 million US$. Other monetising assumptions used in the study are shown in Table 7.7 below. The impacts of the coal fuel cycle are dominated by the global warming effects, where an SVL of 3 million US$ was used (slightly lower than the EC value at current exchange rates). However, the impacts associated with other emissions may be a factor of 10 lower than for an average British coal-fired plant, as inferred from comparing the emission standards assumed here with typical emissions from current installations.

Environmental & public impacts	Type of impact: emissions (g/kWh):	Uncer- tainty*	Monetised value US cents/kWh	Uncertainty & ranges*
1. Plant construction/decommissioning	NA		NA	
2. Plant operation				
CO_2	880	L		
SO_2 (may form aerosols)	1.1	M		
NO_x (may form aerosols)	2.2	M		
particulates	0.16	M		
CH_4	3	M		
N_2O	0.5	H		
Greenhouse warming (cf. Table 7.3)	From CO_2,CH_4,...		37.6	27–55
Degradation of building materials	From acid rain		0.10	H,r,n
Reduced crop yields	From acid rain		0	
Forest and ecosystem impacts			0	
Ozone impacts			NQ	
Domestic impacts only:	Cases:			
Mortality from primary particles (PM10)	0.2 per TWh	H	0.06	H,r,n
from secondary aerosols	1.0 per TWh		0.35	H,r,n
from chronic effects	3 per Twh		NQ	
Morbidity from dust and aerosols,				
major acute	4.6 per Twh		0	M,r,n
minor acute	50 000 workdays lost per TWh		0.07	M,r,n
chronic cases	200 per TWh		0	M,r,m
Noise (from power plant)			0.01	M,l,n
Occupational health and injury				
1. Mining diseases	3 per TWh		0.01	M,l,m
Mining accidents, death	0.1 per TWh		0.03	L,l,n
major injury	3.1 per TWh		0.05	L,l,n
minor injury	27 per TWh		0.01	H,l,n
2. Transport, death	0.06 per Twh		0.03	L,l,n
major injury	0.33 per Twh		0	M,l,n
minor injury	3.17 per TWh		0	H,l,n
3. Construction/decommissioning (injury)	1.1 per TWh		0.01	M,l,n
4. Operation (injury)	0.9 per TWh		0	L,l,n
Economic impacts				
Direct economy			3.1–5.6	
Resource use	Low but finite		NQ	
Labour requirements			NQ	
Import fraction (UK plant)	Local coal assumed		NQ	
Benefits from power			6.5-19	
Other impacts				
Supply security	Many import options		NQ	
Robustness (against technical error,	Fairly low for large		NQ	
planning errors, assessment changes)	Plants			
Global issues	Competition		NQ	
Decentralisation and consumer choice	Not possible		NQ	
Institution building	Modest		NQ	

Table 7.4. Impacts from site-specific British coal fuel chain (Sørensen, 1995c; occupational and environmental impacts except greenhouse warming are based on European Commission, 1995c). See notes below Table 7.6.

Environmental impacts	Type of impact: emissions (g/kWh):	Uncer- tainty*	Monetised value US cents/kWh	Uncertainty & ranges*
1. Plant construction/decommissioning	NA		NA	
2. Plant operation				
CO_2	880	L		
SO_2 (may form aerosols)	0.8	M		
NO_x (may form aerosols)	0.8	M		
particulates	0.2	M		
CH_4	3	M		
N_2O	0.5	H		
Greenhouse warming (cf. Table 7.3)	From CO2,CH4,...		35.8	26–53
Degradation of building materials	From acid rain		0.03	H,r,n
Reduced crop yields	From acid rain, ozone		0	
Forest and ecosystem impacts	Cases:		0	
Ozone impacts	0.1 per Twh	H	0.03	
European-wide impacts:				
Mortality from primary particles (PM_{10})	0.8 per TWh	H	0.25	H,r,n
from secondary aerosols	3.0 per TWh		1.0	H,r,n
from chronic effects	14 per Twh		NQ	
Morbidity from dust and aerosols,				
major acute	17.4 per Twh		0	M,r,n
minor acute	187 000 workdays lost per TWh		0.25	M,r,n
chronic cases	760 per TWh		0.01	M,r,m
Noise (from power plant)			0.01	M,l,n
Occupational health and injury				
1. Mining diseases & death due to radon exp.	0.8 per TWh		0	M,l,m
Mining accidents, death	0.2 per TWh		0.06	L,l,n
major injury	6.8 per TWh		0.15	L,l,n
minor injury	70.5 per TWh		0.01	H,l,n
2. Transport, death	0.03 per TWh		0.01	L,l,n
major injury	0.31 per TWh		0	M,l,n
minor injury	9.8 per TWh		0	H,l,n
3. Construction/decommissioning	0 per TWh		0	M,l,n
4. Operation (injury)	0.08 per TWh		0	L,l,n
Economic impacts				
Direct economy			3.1–5.6	
Resource use	Low but finite		NQ	
Labour requirements			NQ	
Import fraction (German plant)	Local coal assumed		NQ	
Benefits from power			6.5-19	
Other impacts				
Supply security	Many import options		NQ	
Robustness (against technical error,	Fairly low for large		NQ	
planning errors, assessment changes)	Plants			
Global issues	Competition		NQ	
Decentralisation and consumer choice	Not possible		NQ	
Institution building	Modest		NQ	

Table 7.5. Impacts from site-specific German coal fuel chain (Kuemmel *et al.*, 1997; occupational and environmental impacts except greenhouse warming are based on European Commission, 1995c). See notes below Table 7.6.

Environmental impacts	Type of impact: emissions (g/kWh):	Uncer-tainty*	Monetised value US cents/kWh	Uncertainty & ranges*
1. Plant construction/decommissioning	NA		NA	
2. Plant operation				
CO$_2$ equivalent greenhouse gases	1183	L		
SO$_2$ (may form aerosols)	1.2	M		
NO$_x$ (may form aerosols)	1.3	M		
particulates	0.3	M		
CH$_4$	NA	M		
N$_2$O	NA	H		
Greenhouse warming (cf. Table 7.3)	From CO$_2$,CH$_4$,...		39.5	28–58
Degradation of building materials	From acid rain		0.03	H,r,n
Reduced crop yields	From acid rain		0	
Forest and ecosystem impacts			0	
Ozone impacts			NQ	
	Cases:			
Mortality fromprimary particles (PM$_{10}$)	2.6 per TWh	H	0.9	H,r,n
from secondary aerosols	included above		Incl. above	H,r,n
from chronic effects	9 per TWh		NQ	
Morbidity from dust and aerosols			0.19	M,r,n
Noise (from power plant)			0.01	M,l,n
Occupational health and injury				
1. Mining diseases	NA		NQ	M,l,m
Mining accidents	NA		0.25	L,l,n
2. Transport	3.1 per TWh		0	L,l,n
Road damage			0	L,l,n
3. Construction/decommissioning			NQ	H,l,n
4. Operation			0	L,l,n
Economic impacts				
Direct economy			3.1–5.6	
Resource use	Low but finite		NQ	
Labour requirements			NQ	
Import fraction	All coal imported		NQ	
Benefits from power			6.5-19	
Other impacts				
Supply security	Many import options		NQ	
Robustness (against technical error, planning errors, assessment changes)	Fairly low for large Plants		NQ	
Global issues	Competition		NQ	
Decentralisation and consumer choice	Not possible		NQ	
Institution building	Modest		NQ	

Table 7.6. Impacts from site-specific Danish coal fuel chain, distributed according to electricity production (Warming, 1996; greenhouse warming damage and economic impacts from Kuemmel *et al.*, 1997).

NA = not analysed, NQ = not quantified. Values are aggregated and rounded (to zero if below 0.01 USc kWh^{-1}).

* (L,M,H): low, medium and high uncertainty. (l,r,g): local, regional and global impact. (n,m,d): near, medium and distant time frame. Monetary impacts originally given in euro are translated using 1 euro = 1.25 US$ (Tables 7.4–7.6 and 7.8).

Several of the impacts mentioned in Tables 7.4–7.6 are not quantified, so an assessment must deal with the qualitative impacts as well. In those cases where it has been feasible, uncertainties are indicated (L: low, within about a factor of 2; M: medium, within an order of magnitude; and H: high, more than an order of magnitude), and the impacts are labelled according to whether they are local, regional or global (l, r or g), as well as whether they appear in the near term (n: under 1 year), medium term (m: 1–100 years) or distant term (d: over 100 years into the future).

In Table 7.5, the results of a similar calculation for a coal-fired power plant at Lauffen, Germany, are shown. This plant has better NO_x cleaning than the British plant, but is situated in a high population density area, and the calculation does include impacts from pollution across national borders, in contrast to the example shown in Table 7.4, which deals only with pollution impacts inside the UK. Finally, Table 7.6 shows the results of applying the same techniques to a power plant on the island of Funen, Denmark. Lower population densities and more rapid transport of stack effluents by wind make the impact from this plant smaller than that of the German counterpart, despite higher CO_2 emissions. However, like most Danish power plants, this is coproducing power and heat, so the appropriation of impacts on only the kWh of electricity produced is not fair. The energy productions of power and heat are, on average, nearly equal, so one might attribute half to each (Warming, 1996), without considering the difference in energy quality. Per unit of energy, the impacts of Table 7.6 would thus be roughly halved. The pollution impacts do not scale from the German values of Table 7.5 by a simple factor, owing to different travelling times in the atmosphere of different substances. For each kg of emissions, the health impacts of SO_2 in the Danish case are two-thirds of those of the German case, but in the case of NO_x and particles they are only about a third.

Health effect:	Euro	US$
Mortality (SVL)	2 600 000	3 250 000
Respiratory hospital admission	6 600	8 250
Emergency room visits	186	233
Bronchitis	138	173
One day activity restricted	62	78
Asthma attack	31	39
One day with symptoms	6	8

Table 7.7. Valuation assumption of the European Commission study (European Commission, 1995a-f).

Table 7.8 shows the results of the European Commission study for the natural gas power cycle (amended with greenhouse warming and economic impacts, the latter because the study itself considers only externalities and thus not the costs already paid in the marketplace). The assumed combined cycle gas turbine plant located at West Burton, UK, has not been built (European Commission, 1995d). It is seen that the greenhouse warming impacts are considerably smaller than those of the coal fuel cycle. The distribution of the global warming effects on emission sources was discussed in section 7.4.11. Conventional pollution is less for the advanced gas cycle, partly owing to its high efficiency (over 51%).

Nuclear fuel chain

For nuclear power plants, the assessment in principle proceeds as in the fossil case, but one expects more important impacts to come from plant construction, fuel treatment and final storage of nuclear waste. Because of the very long time horizons associated with radiation-induced health problems and deposition of nuclear waste, the impact assessment period by far exceeds the plant operation period, and questions of discounting and of changes in technological skills over periods of several centuries come to play a decisive role. Such monetary issues were discussed in section 7.3.1. Another issue particularly (but not exclusively) important for nuclear power is the accident-related impacts, the treatment of which may call for a risk analysis (section 7.4.3).

The observed number of large nuclear accidents involving core meltdown in commercial reactors is currently two (Three Mile Island and Chernobyl). The implied "probability" in the nuclear case is illustrated in Table 7.9. Counting two accidents over the accumulated power production to the end of 1994 one gets a 10^{-4} per TWh frequency, or a 5×10^{-5} per TWh frequency for an accident with severe external consequences. At the time of the Chernobyl accident, the estimate would have been 3–4 times higher owing to the much lower accumulated power production of reactors world-wide by 1986. For comparison, the built-in probability for an accident with Chernobyl-type consequences for a new, state-of-the-art light-water nuclear reactor is by the fault-tree analysis calculated to be about 1.25×10^{-6} per TWh (ST2-accident; cf. European Commission, 1995e; Dreicer, 1996). The factor 40 difference between the two numbers comes partly from the difference between a state-of-the-art reactor and the average stock and partly from the difference between the probability of anticipated accidents and the actual frequency estimate that includes unanticipated events.

The outcome of an LCA chain analysis (but excluding side-chains) of the French state-of-the-art nuclear power plant considered in the European Commission study (European Commission, 1995e) is shown in Table 7.10, except that the impacts of major accidents are taken from the estimate of Table 7.9. The emphasis is on impacts from release of radioisotopes, and again monetising is involving the assumptions of Table 7.7. The largest normal operation impacts are from the re-processing step. The magnitude is based on estimates made by the operator of the plant at La Hague, COGEMA. However, their appraisal has recently been challenged as constituting a gross underestimate, compared with values derived from actual samples obtained at sea near the plant (Butler, 1997). As regards the deposition of high-level waste, no quantitative estimate was made of incidents that may occur over the required long deposition periods. Also the impacts of proliferation of nuclear materials and know-how have not been quantified.

Environmental impacts	Type of impact: emissions (g/kWh)	Uuncer-tainty*	Monetised value US cents/kWh	Uncertainty & ranges*
Fuel extraction and power plant operation:				
Main emissions:				
CO_2	401	L		
NO_x (may form aerosols)	0.71	M		
CH_4	0.28	M		
N_2O	0.014	M		
Greenhouse warming (cf. Table 7.3)	From CO_2,CH_4,...		14	10–21
Degradation of steel, painted surfaces	From acid rain	M	0.01	H,r,n
	Cases:			
Mortality from acid aerosols	0.16 per TWh	M	0.05	M,r,n
Morbidity from acid aerosols	6200 symptom days,			
	520 with problems,			
	per TWh	M	0.01	H,r,n
Noise (from power station)	Regulatory maximum		0	M,l,n
Occupational health and injury				
Accidents:				
Major off-shore platform accidents	0.016 per TWh		0.01	H,l,n
Other off-shore platform accidents	0.005 per TWh		0	H,l,n
Injury:				
Off-shore platform construction	0.07 per Twh		0	H,l,n
Economic impacts				
Direct economy			5.0–6.3	
Resource use	Low but finite		NQ	
Labour requirements			NQ	
Import fraction (UK plant)	British gas assumed		NQ	
Benefits from power			6.5–19.0	
Other impacts				
Supply security	Depends on pipeline		NQ	
	integrity			
Robustness (against technical error,	Fairly low for large		NQ	
planning errors, assessment changes)	Plants			
Global issues	Competition		NQ	
Decentralisation and consumer choice	Not possible		NQ	
Institution building	Modest		NQ	

Table 7.8. Impacts from CCGT natural gas-based power production chain (European Commission, 1995d; greenhouse warming impacts and economic impacts from Kuemmel *et al.*, 1997). See notes below Table 7.6.

Historical evidence and interpretation:		
Accumulated experience when Three Mile Island accident happened (1979)	3 000 TWh	
Accumulated experience when Chernobyl accident happened (1986)	5 800 TWh	
Accumulated experience to end of 1994	20 000 TWh	
Implied order of magnitude for frequency of core-melt accident	1×10^{-4} TWh^{-1}	
A. Implied order of magnitude for accident with Chernobyl-type releases	5×10^{-5} TWh^{-1}	
Chernobyl dose commitment (UNSCEAR, 1993)	560 000 person sievert	
Valuation:	*million US $*	*million euro*
Induced cancers (SVL = 2.6 M euro, no discounting)	250 000	200 000
Birth defects	25 000	20 000
Emergency teams, clean-up teams, security teams	63	50
Early radiation deaths (SVL = 2.6 M euro)	125	100
Evacuation and relocation	125	100
Food bans and restrictions	125	100
Unplanned power purchases	1 000	1 000
Capacity loss and reduced supply security	12 000	10 000
Cost of encapsulation and clean-up (at plant and elsewhere)	212 000	170 000
Increased decommissioning costs	125 000	100 000
Impact on nuclear industry (reputation, reduced new orders)	125 000	100 000
Monitoring, experts' and regulators' time	12	10
Concerns in general public (psychosomatic impacts)	125	100
B. Total estimate of Chernobyl accident costs	750 000 M US$	600 000 M euro
Average large-accident cost of generating nuclear power (A times B)	3.8 US cents/kWh	30 m euro/kWh

Exchange rate assumptions: 1 m euro = 0.125 US cents = 0.125 ¥

Table 7.9. Frequency of and damage by large nuclear accidents (Sørensen, 1995c).

The use of data for the nuclear accident analysis from historical cases such as Chernobyl[*] may be criticised for not taking into account technological progress (cf. the discussion above on average and state-of-the-art technology). However, the treatment does depend on the type of use to be made of the assessment. For the few countries where nuclear power plants are currently built, there are questions of the standards of operational safety (e.g. absence of provisions for information of the population on the need for staying indoors with controlled closure and opening of windows, evacuation, food bans, etc.). Preparations for such events with associated drills have not even been carried out in all present large nuclear user countries.

The impacts in Table 7.10 are for a nuclear power station located at Tricastin in France, with fuel manufacture and re-processing also taking place at specific French sites. They include those from use of fossil fuels, e.g. in construction of nuclear power plants, because the evaluation performed is based on a marginal addition of one nuclear power station to the presently existing system. The impacts from side-chains yielding inputs to the main

---

[*] Recent disclosure of a possible earthquake occurring at a critical time of the Chernobyl accident sequence (DR, 1997) casts doubt on the causes of the accident, but the reactor design peculiarities certainly contributed to the temporal distribution and high temperatures of radioactivity releases (Sørensen, 1987a).

fuel-cycle are considered small compared to the largest impacts included, just as in the fossil case, although this may not hold as well in the nuclear case because of its higher proportion of construction costs in the total.

Environmental emissions	Dose commitment person-sievert/TWh	Uncer-tainty*	Monetised value US cents/kWh	Uncertainty & ranges*
Plant construction/decommissioning	NA		NA	
CO_2, SO_2, NO_x, particles	NA		NA	
Noise, smell, visual impact	NA		NA	
Radioactivity(according to distance):				
1. Fuel extraction and refinement, local	0.1	L	0.01	M,l,n
regional	0.1	L	0	M,r
global	0	L	0	M,g
2. Normal power plant operation, local	0.4	M	0.01	M,l,m
regional	0	M	0	M,r
global	1.9	M	0.04	M,g,d
3. Power plant accidents (cf. Table 8), local	5	H	0.25	H,l,m
regional	10	H	0.4	H,r,m
global	15	H	0.6	H,g,d
4. Reprocessing and waste handling, local	0.2	H	0	H,l,d
regional	0	H	0	H,r,d
global	10.2	H	0.24	H,g,d
Social impacts				
Occupational injuries	0		NQ	
Occupational radioactivity				
1. Fuel extraction and refinement	0.1	L	0	M
2. Construction and decommissioning	Over 0.02	M	0	M
3. Transport	0	L	0	L
4. Normal power plant operation	0	M	0	M
5. Power plant accidents	0	M	0	H,l,n
6. Re-processing and waste handling	0	H	0	H
Accident handling (evacuation, food ban, clean-up, back-up power, cf. Table 7.9)			1.9	H,r,m
Indirect accident impacts (expert time, mis-trust, popular concern, cf. Table 7.9)			0.63	H,g,m
Economic impacts				
Direct costs			3.75–7.50	L
Resource use	Not sustainable with-out breeders		NQ	
Labour requirements	Low		NQ	
Import fraction (for France)	Low		NQ	
Benefits from power (consumer price)			5.6–16.9	L
Other impacts				
Supply security	Medium		NQ	
Robustness (technical, planning)	Important		NQ	
Global issues (proliferation and weapons)	Important		NQ	
Decentralisation and choice	Not possible		NQ	
Institutions building (safety and control)	Fairly high		NQ	

Table 7.10. Impacts of nuclear fuel cycle (European Commission, 1995e; for accident evaluation Sørensen, 1995c). See notes below Table 7.6.

Renewable energy chains

This section gives examples of LCA chain calculations for renewable energy converters playing a role, e.g. in the scenarios discussed in Chapter 6.

Wind power

Life-cycle analysis has been made for wind turbines of unit size around 500 kW installed in Denmark and elsewhere. These units typically feature three-bladed glass-fibre rotors mounted on steel-tube or concrete towers, transmitting power via a gearbox to an induction generator, which again is attached to the standard utility grid lines via an electronic control box. Several of the impacts exhibit a dependence on the natural, social and human setting. This should be considered when transferring data between settings.

The highest penetration of wind power presently (1999) is in the Danish electricity system, where it is about 12%. If penetrations above some 20% are reached, the question of energy storage may have to be addressed and corresponding impacts included in the LCA. However, for grid systems such as the European one, characterised by strong international power transmission links, e.g. from Denmark to the Norwegian hydro system based on annual storage cycles, the cost of dealing with the variability of wind energy production could be very low (Sørensen, 1981a; Meibom *et al.*, 1997).

The overall cost of producing wind energy at the best locations in Denmark is currently about 0.25 DKr kWh^{-1} (4 US cents kWh^{-1}, evaluated by Sørensen, 1995a; Aastrand *et al.*, 1996), with O&M constituting an average of 0.07 DKr kWh^{-1} (1 US cents kWh^{-1}) out of this, averaged over an assumed 20-year lifetime. Because the capital cost is dominating, there is much less uncertainty in cost after the turbine is installed than there would be for a fuel-based system. The import fraction of the capital cost is 28%, that of the running cost 15% (Sørensen, 1986). The employment factor (full-time equivalent per million Danish kroner, spent) is about 3. Whether labour is considered a positive or negative impact within an economy depends on whether there is unemployment or not, while for the individual it depends on how tightly social assets are bound to work in a given society and on that individual's attitude to work (work as a satisfying activity versus work as just the means of funding life activities). In any case, creating jobs within a society is often viewed as preferable to creating jobs abroad.

The land use associated with wind turbines may be assessed in the following way. In order not to experience reduced wind access, wind turbines in a wind park have to be placed several rotor diameters apart. Danish wind turbines are placed both in park configurations and individually. The land between them and right up to the towers may be used for agriculture, so that the largest cost in terms of land use is often the access roads needed for service purposes. Typical values are 10 m^2 per kW of rated power (Sørensen, 1986). Land resources may be returned after decommissioning.

The visual impact of wind turbines also depends on whether individual turbines or arrays of turbines are occupying a given location. Aesthetically, slender towers and white blades have been found to produce a positive reception. A factor in the assessment is also the long history of wind power in Denmark, which causes wind turbines to be more easily accepted as parts of the visual landscape, along with farmhouses and churches often having towers as conspicuous as those of wind turbines. For wind parks, architects are usually employed to ensure a visually acceptable integration into the particular traits of a given landscape (Sørensen, 1981b). The further expansion of Danish wind power supply is believed to increasingly involve off-shore siting, owing to the dense population of the country, and several suitable locations have been identified.

The mechanical noise from Danish wind turbines erected in 1993 is 97 dB(A) at the point of emission[*]. The noise mostly originates from the gearbox and can be reduced by sound insulation. Aerodynamic noise from the blades depends on wind speed and is similar to that of other structures or vegetation (Sørensen, 1981b). The noise from a single wind turbine becomes inaudible just a few rotor diameters from the site, and even for wind farms the increment over background noise is less than 2 dB(A) at a distance of 1.5 km (European Commission, 1995f). Average background noise is typically 35–37 dB(A) in quiet rural locations, and legislation in Denmark requires that this level is not noticeably increased by human activities.

Telecommunication interference has been studied and found similar to that of static structures (e.g. buildings), with the exception of frequency modulations propagating in particularly shaped mountain locations (Sørensen, 1986).

The extraction of power from the wind has a slight influence on the micro-climate below and behind the turbines, but otherwise there is no pollution associated with the operation of wind turbines, assuming lubricants to be contained. The chief potential source of pollution is in the manufacture and maintenance operations, which in Table 7.11 have been assumed to employ the current mix of fuels for the contribution from energy inputs (to both manufacture and transportation). The impacts will decrease in proportion to the possible future transition to renewable energy sources. Available accident statistics show negligible risk to members of the general public being hit by expelled rotor blades in cases of rotor failure (Sørensen, 1981b).

The work environment at manufacturers of windmill components is similar to that of other equipment manufacturers, while the work environment for tower building, assembly and maintenance is similar to that of work in the building industry, with open air activities and scaffold work at a height. With proper safety precautions, such work is generally offering a

---

[*] dB(A) is an average logarithmic sound scale weighted so as to reflect the human perception of noise.

varied and challenging environment. The accident rates assumed in Table 7.11 are taken from industrial statistical data for the relevant sectors. One significant component has historically been electrocution accidents during improper maintenance attempts, usually by the owner rather than by professional teams.

Health problems are primarily associated with the industrial part of manufacture, including in particular the use of epoxy resins in blade manufacture. Modern production technology should confine this step of production to closed spaces with fully automated facilities, implying that employees only do control-room work, so that the risk of exposure to harmful chemicals would basically be via accidents. Numerical estimates of health and accident risks are uncertain because of the aggregate nature of available statistical data and because of the different production techniques used by different wind turbine manufacturers.

The social benefit of wind turbines comes from using the electric power produced. As discussed above, this benefit might decrease if, at higher penetration of wind energy in the power supply system, the variations of wind-produced power led to occasional dumping of excess power. As explained, this is unlikely to become an issue for Denmark, even for large wind energy penetrations.

Wind energy can have impacts on the infrastructure of the electricity supply system, particularly if wind turbines are sited in a dispersed fashion, such that the demands on the power transmission network might be reduced owing to shorter average distances to consumers or, on the other hand, might be increased as a result of connecting distant wind farms, e.g. at offshore locations.

Security of supply is generally high, with the qualifications related to the variability of wind conditions made in the preceding subsection. Failures occur in all parts of the system, but not according to any particular pattern. Around 10% of Danish wind turbines experience a component failure in any given month. About 4% of the failures result in replacement of a component (e.g. a blade, the generator, the control unit), according to the newsletter *Windpower Monthly/Windstats Quarterly*. These failures are represented in the overall O&M costs, which include an insurance premium. A wind power system has little sensitivity to individual turbine failures because of the modular nature of the system. Because of the short time lag between deciding to build a turbine and its operation, wind technology is much less sensitive to planning errors (e.g. wrong forecast of future loads) and changes in criteria used for selecting technology than are systems with several years lag between decision and operation.

Wind power is consistent with keeping options open for the future. It might be abandoned without repercussions (in contrast to e.g. nuclear installations and waste repositories, which have to be looked after long after decommissioning). In Denmark, wind power is consistent with present goals of creating an energy system with as little environmental impact as possible. It is a national solution, and it avoids global issues such as the overuse of resources by the rich countries, while at the same time it ensures national independence of energy supply, in proportion to its penetration.

Environmental impacts	Impact type: emissions (g/kWh)	Uncer-tainty*	Monetised value US cents/kWh	Uncertainty & ranges*
Releases from fossil energy currently used:				
1. Turbine manufacture (6.6 GJ/kW rated)				
CO_2 (leading to greenhouse effect)	12.1	L	0.39	0.28–0.57
SO_2 (leading to acid rain and aerosols)	0.05	L	0.01	H,r,n
NO_x (possibly aerosols and health impacts)	0.04	L	0	H,r,n
particulates (lung diseases)	0.002	L	0.01	H,r,n
2. Operation (2.2 GJ/kW over 20 year lifetime)				
CO_2 (leading to greenhouse effect)	3.8	L	0.12	0.09–0.18
SO_2 (leading to acid rain and aerosols)	0.01	L	0	
NO_x (possibly aerosols and health impacts)	0.02	L	0	
particulates (lung diseases)	0	L	0	
	Other:			
Noise increase from gearbox at about 1 km distance from wind-blade interaction	<1 dB(A) <3 dB(A)		0.01 total	H.l.n
Land use	10m²/kW		NQ	
Visual intrusion	+/- like buildings, church towers, etc.		NQ	
Social impacts				
Occupational injuries (manuf. & materials):				
1. Turbine manufacture, death	0.03/Twh	L	0	L,l,n
major injury	0.9/TWh	L	0.01	L,l,n
minor injury	5/TWh	M	0	M,l,n
2. Operation (same categories combined)			0	M,l,n
Economic impacts				
Direct costs			5–11	
Resource use (energy pay-back time given)	1.1 y	L	NQ	
Labour requirements (manufacture)	9 person y/MW	L	NQ	
Import fraction (for Denmark)	0.28	L	NQ	
Benefits from power sold (penetration < 30%)			5–15	
Other impacts				
Supply security (variability in wind is high, entry based on plant availability)	High		NQ	
Robustness (up-front investment binds, entry based on technical reliability)	High		NQ	
Global issues (non-exploiting)	Compatible		NQ	
Decentralisation & choice (less with large size)	Good		NQ	
Institution building (grid required)	Modest		NQ	

Table 7.11. Impacts from current Danish wind energy systems (Sørensen, 1996e; with use of data from Sørensen, 1981b, 1993b; Meyer *et al.*, 1994; European Commission, 1995f). See notes below Table 7.6.

Wind power allows for broad participation in the decision-making in the energy field, and as one of several possible decentralised solutions, it would permit different local communities to chose different solutions. Few institutional constructions are needed if ownership is dispersed, but one possible set-up would be for present utility companies to own the wind turbines, in which case the regulatory implications are no different from current ones. With dispersed ownership, essential questions are free access to the power distribution grid and price fixation for buying from and selling to the national grid. Access is at present in Denmark ensured by legislation, and price negotiations are conducted between the utility companies and the wind turbine owner's organisation. However, they are overseen by the Department of Energy, which is supposed to intervene if fair pricing is not obtained between the unevenly sized institutions. Some of the social and systemic impacts discussed above cannot be quantified; they are flagged in Table 7.11.

Photovoltaic power

Even more than for wind power, the PV power system LCA is dominated by the manufacturing process. The manufacture of crystalline silicon panels involves the main steps depicted in Figure 7.20, whereas for amorphous silicon panels the wafer formation, cutting and surface doping steps are replaced by direct vapour deposition of doped gases onto a substrate. Different types of solar panels differ with respect to use of substrates, glass cover material and/or films. The cells may have surface textures and elaborate grooving patterns for conductors, and modules may incorporate reflectors and, in some cases, bypass diodes and even inverters (which otherwise would be system components). Finally, mounting of panels in arrays may involve dedicated support structures or may be building integrated.

On the system side, further transformer equipment may appear, as well as battery storage or back-up devices in the case of stand-alone systems. Decommissioning and dismantling of the solar equipment is expected to follow recycling and reuse patterns emerging for the building industry in general, probably as a front-runner industry.

Assessing the impacts is helped by the experience with similar production process steps found throughout the microelectronics industry. The basic raw material for silicon cells is silicon dioxide (sand, quartzite). It is reduced to metallurgical-grade silicon in arc furnaces. Both mining and reduction may produce dust (and hence risk of silicoses). The furnaces additionally produce carbon monoxide and a range of silicon-containing compounds that appear as dust that might be inhaled e.g. during cleaning operations (Boeniger and Briggs, 1980). At current low penetration, the photovoltaic industry has used scrap material obtained inexpensively from the microelectronics industry, but in the future dedicated solar-grade material (much less expensive to produce than microelectronics grade) will be used.

In the case of amorphous cells, the next step is production of silane (SiH_4). For crystalline cells it is usually production of trichlorosilane in a fluidised bed and then purification to semiconductor grade multicrystalline silicon. Subsequent processes are doping and growth of monocrystalline ingots. These are then ground to cylindrical shape and sliced into wafers, which are then cleaned. Multicrystalline cells may be produced by slicing ingots made of cast multicrystal silicon in a process similar to that for crystalline, or they may be formed by vapour deposition like for amorphous cells, albeit at considerably higher deposition temperatures. The material used for monocrystalline cells is currently expensive, also in input energy, but future production is expected to use thin-film technology for monocrystalline cells.

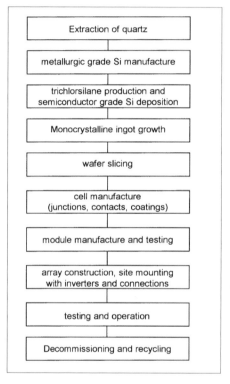

Figure 7.20. Main steps in LCA chain for monocrystalline silicon photovoltaic cells (Sørensen, 1995c).

The chlorosilane production involves hydrochloric acid, and the chlorosilanes themselves are corrosive and an irritant to skin and lung as well as toxic. Workers are required to use protective clothing and face masks with filters. Further risks are posed by hydrogen–air mixtures, which could ignite and explode. One such case has been reported (Moskowitz et al., 1994). For amorphous silicon, special precautions are needed for handling silane gas, as it ignites spontaneously. One solution is never to store large quantities of silane gas and to use special containers designed to avoid leakage even in case of strong pressure increases. Also, fabrication sites are equipped with automatically releasing fire-extinguishing devices.

Vacuum growth of crystalline material may involve dispersal of oily aerosols that have to be controlled by wet scrubbers and electrostatic filters (CECSMUD, 1982). Doping of p-type material may involve boron trichloride, which reacts with water vapour to form acids easily absorbed through the skin, or diborane, which is a strong irritant and flammable as well. The n-type doping at the top layer of a crystalline cell uses phosphorus diffusion of $POCl_3$ or P_2O_5 in sealed environments, whereas the n-type doping of amorphous cells may involve phosphine (PH_3), a highly toxic substance widely used in the semiconductor industry (Watt, 1993).

Grinding and cleaning of wafers produce silicon-containing slurry with remains of detergents used. An alternative is ribbon-growth, which avoids these problems (CECSMUD, 1982). Amorphous cell manufacture also involves a number of cleaning agents. Etching of surface textures may employ a variety of techniques, selected on the basis of concern for recycling of chemicals and reduction of the use of toxic substances (Watt, 1993). Workers have to wear protective clothing, and high levels of ventilation are required. Drying uses liquid nitrogen and may be fairly energy intensive.

Screen printing of electric circuits involves possible work environment problems familiar to those in the microelectronics industry (caused by metal particles and organic solvents). Laser grooving involves the laser safety precautions for radiation and fires, and the application of coatings such as titanium oxide or silicon dioxide is considered relatively harmless. Cell testing and light soaking of amorphous cells (in order to avoid restructuring degradation) should be done in special rooms because of the risk of exposure to ultraviolet radiation. Personnel replacing bulbs should wear safety masks and gloves if pressurised krypton lamps are used. Polymer coatings such as ethylene vinyl acetate (EVA) or polyvinyl fluoride (Tedlar) may have some health impacts during their manufacture. If soldering is used in module assembly, fumes should be controlled. The tendency is for increasing use of robots in the manufacturing process lines, leading normally to reductions in health impacts for remaining workers. Cells based on cadmium telluride and gallium arsenide involve different types of potential impacts (Moskowitz *et al.*, 1995), which could lead to very high life-cycle impacts (Alsema and Engelenburg, 1992). However, these cells are based on expensive materials, the natural abundance of which is much smaller than that of silicon, and therefore total recycling is required. A large fraction of the calculated impacts is involved with releases of toxic materials in fires involving solar panels. The photovoltaic industry is aware of the problems of current pilot production lines and aims at controlling or replacing the chemicals identified as troublesome (Patterson, 1997).

There are few impacts during the operation of photovoltaic installations. Land use may be an issue for central plants, but not for building-integrated systems. Albedo changes caused by the presence of the panels are not significantly different from those of alternative roof surfaces, in the case of roof-integrated cells, but could have climatic impacts in case of large, centralised solar plants located, for example, in desert areas. Reflections from panels located in cities could be annoying, and considerations of visual impacts will generally require careful architectural integration of panels. In some areas, cleaning of solar panel surfaces for dust may be required, and electronic control equipment such as inverters may cause radiofrequency disturbances if not properly shielded. As mentioned for non-silicon cells, the behaviour of panels during fires is an important consideration. Recycling of solar cell equipment has also been mentioned as a requirement (Sørensen, 1993b).

Environmental impacts	Impact type: emissions (g/kWh)	Uncer-tainty*	Monetised value US cents/kWh	Uncertainty & ranges*
Releases from fossil energy used in the steps of the PV conversion cycle:				
CO₂ (m-Si now and around 2010)	75, 30	L		H,g,m
(a-Si now and around 2010)	44, **11**	L		H,r,n
SO₂ and NOₓ (m-Si now and around 2010)	0.3, 0.1	L		H,r,n
(a-Si now and around 2010)	0.2, **0.04**	L		H,r,n
Greenhouse effect from fossil emissions (m-Si)			2.4, 1.0	0.7–3.5
(a-Si, both either now or in 2010)			1.4, **0.4**	0.3–2.1
(if non-fossil energy is used in PV production)			or **0**	0
Mortality and morbidity from fossil air pollution				
described above (m-Si, now and 2010)			0.05, 0	H,r,n
(a-Si, now and 2010)			0.03, **0**	H,r,n
Land use	0		0	L,l,n
Visual intrusion			NQ	
Social impacts				
Occupational injuries:				
1. From fossil fuel use (m-Si now and 2010)			0.01, 0	L,l,n
(a-Si now and 2010)			0, **0**	L,l,n
2. From panel manufacture			NA	
3. From construction and decommissioning (dif-				
ferential from using other building materials)			0	L,l,n
4. From operation			0	L,l,n
Economic impacts				
Direct costs (at present)			38–76	
(around 2010)			**3.8–11.3**	H
Energy pay-back time (now and 2010, a-Si)	3y, 0.5y		NQ	
Labour requirements (now and 2010)	40, 4 man-y/MW		NQ	
Benefits from power sold (penetration < 20%)			**5–15**	H
Other impacts				
Supply security (plant availability)	High		NQ	
Robustness (technical reliability)	High		NQ	
Global issues (non-exploiting)	Compatible		NQ	
Decentralisation & choice	Good		NQ	
Institution building (grid required)	Modest		NQ	

Table 7.12. Estimated impacts from present and future rooftop photovoltaic energy systems based on multicrystalline (m-Si) or amorphous (a-Si) silicon cells, placed at average European locations (Sørensen and Watt, 1993; Yamada *et al.*, 1995; Sørensen, 1994b, 1995c, 1997a). The boldface numbered alternatives pertain to what are considered the most likely technologies for the future. See notes below Table 7.6.

For a silicon-based photovoltaic system integrated into a building, the LCA impact evaluation presented in Table 7.12 shows modest negative impacts, most of which occur during the manufacturing phase, and substantial positive impacts in the area of impacts on the local and global society. The impacts during manufacturing to a large degree result from the use of fossil fuels for mining, manufacture and transport, according to the marginal approach taken in the references used. A comprehensive analysis of a renewable energy scenario would instead use the new energy system to determine indirect energy inputs, with substantially altered results as a consequence.

The source used for energy pay-back times and carbon dioxide emissions is Yamada *et al.* (1995). Other estimates give current energy pay-back times for a-Si and c-Si (monocrystalline silicon) systems as 2–3 years and future ones below 1 year (Alsema, 1997; Frankl *et al.*, 1997), while current greenhouse gas emissions have been estimated at 100–200 g of CO_2 equivalent per kWh of power produced, declining to some 40 g of CO_2-eq. kWh^{-1} after the year 2010 (Dones and Frischknecht, 1997).

The photovoltaic case is special, because the direct cost at present is far higher than that of the alternatives. If the cost goes down in the future, many of the impacts will also go down, because they are associated with material use or processes that will have to be eliminated or optimised in order to reach the cost goals. The lower cost estimate for 2010 is based on the stacked-cell concept of Wenham *et al.* (1995a). Yamada *et al.* (1995) quote about 6.7 US cents kWh^{-1} for amorphous cell systems integrated into roofs. The cost per kWh produced obviously also depends on the location of the building. The different spans of economic benefits from the power sold, exhibited in Tables 7.4–7.12, are meant to reflect the differences in load-following capability of the three different types of plants. For a variable resource such as solar energy, there will be additional costs if the penetration becomes large compared with the size of the grid system (say, over 30%), because in that case additional equipment must be introduced to deal with the fluctuating power production.

Biogas plants
As described in Chapter 4, biogas can be produced on the basis of several biomass resources, e.g. manure, plant residues or household waste. Table 7.13 shows a typical composition of Danish waste. The biogas plant at Ribe in Denmark is a typical example of current biogas technology accepting such waste. It annually converts about 110 million kg of manure and 30 million kg of other organic waste into 100 TJ of biogas. This is at the lower end of conversion efficiencies for the ten large biogas plants operated in Denmark, ranging in energy terms from 30 to 60%, depending on feedstock (Danish Energy Agency, 1996; Nielsen and Holm-Nielsen, 1996). A future average conversion efficiency of 50% can be expected.

Biogas consists of methane (CH_4) plus varying amounts of carbon dioxide (CO_2). Methane in particular has a high greenhouse warming potential

(GWP), so that leakages during production and transport, e.g. via pipelines or in containers, have to be avoided or kept at a low value. Feedstock for the Ribe biogas plant is mainly slurry from farms transported by road to the plant, but there is a small solid component.

Substance	Weight per cent	Range
Carbon	25	15–35
Oxygen	18	12–24
Hydrogen	3	2–5
Nitrogen	0.6	0.2–1.0
Sulphur	0.003	0.002–0.6
Chlorine	0.7	0.5–1.0
Water	20	15–35
Ash	25	15–40
Lower heating value, municipal solid waste	8.8 GJ t^{-1}	8.4–9.2 GJ t^{-1}
Lower heating value, industrial waste	13.7 GJ t^{-1}	8.7–19.0 GJ t^{-1}

Table 7.13 Elementary analysis and energy content of Danish waste. Source: Formidlingsrådet (1989).

Table 7.14 gives some LCA-relevant technical data for the Ribe plant. One issue of concern is the presence of small amounts of hydrogen sulphide (H_2S), which could contribute to acid rain formation. For one other Danish biogas plant (at Fangel), a cleaning process aimed at reducing the H_2S content to between 700 and 1500 ppm is being tested (Danish Energy Agency, 1995). This is equivalent to a weight percentage of 0.09 to 0.2, the upper value being equal to the one assumed in Table 7.14. More stringent sulphur emission limits are likely to be implemented if biogas production reaches an important penetration in future energy system.

Other important emissions are those of methane liberated from the slurry tanks before or during slurry collection and while being transported to the biogas plant. It is estimated that collecting the slurry and using it in a biogas plant will reduce methane emissions compared to emissions from current practice, where the manure is stored for several months before being spread onto fields. Nielsen and Holm-Nielsen (1996) calculate the reduction for the Ribe plant to be 160 t of CH_4 per year, minus 40 t due to leakage from storage at the biogas plant. The greenhouse warming potential of these amounts of methane should be credited the biogas energy system. This is because the agricultural sector life-cycle impacts are not part of the investigation and thus any change must be included as an indirect cost or benefit to the energy sector. The total methane emissions from Danish livestock in 1990 were about 160 million kg (from 11.7 million animals, mainly cattle and pigs). This amount is the contribution from enteric fermentation. A similar amount, 125 million kg, is emitted from manure (Fenhann and Kilde, 1994).

parameter	Value	Source	Remarks
Technical data			
Specific investment	2750 US$/kW	Danish Energy Agency, 1995	45.3 million DKr total
O&M	9.6 % p.a.	Danish Energy Agency, 1995	4.6 million DKr annually
Net capacity	2.7 MW	Nielsen & Holm-Nielsen, 1996	10 000 m^3 per day
Annual load	8700 h	Nielsen & Holm-Nielsen, 1996	
Lifetime	20 y	Nielsen & Holm-Nielsen, 1996	
Lifetime generation	469.8 GWh		
Overall net efficiency	around 35%	Danish Energy Agency, 1995	30–60%, dep. on feedstock
Input and composition			
Biomass	410 t d^{-1}	Nielsen & Holm-Nielsen, 1996	60% cow manure, 20% pig slurry, rest is other waste
Biomass transport			
Average total	32 km	Nielsen & Holm-Nielsen, 1996	
Average animal slurry	22 km	Nielsen & Holm-Nielsen, 1996	
Biogas composition			
CH$_4$	64.8%	Nielsen & Holm-Nielsen, 1996	
CO$_2$	35%	Nielsen & Holm-Nielsen, 1996	
Rest (H$_2$, N$_2$, H$_2$S) taken as H$_2$S	0.2%	Fritsche & Rausch, 1993	Generic database
Combustion value (MJ m^{-3})	23.4	Nielsen & Holm-Nielsen, 1996	
Material demand			Energy used for manufacture
Steel	5 t	Fritsche & Rausch, 1993	Generic data, 22.2 GJ/t
Concrete	10 t	Fritsche & Rausch, 1993	Generic data, 4.6 GJ/t
Transport of materials			
Steel by truck	150 km	European Commission, 1995c	For construction
Steel by railway	50 km	European Commission, 1995c	For construction
Concrete by truck	50 km	European Commission, 1995c	For construction
Process demands	per MJ biogas		
Process heat	0.12	Nielsen & Holm-Nielsen, 1996	
Electricity	0.01	Fritsche & Rausch, 1993	Generic data
Emissions	per MJ biogas		
CH$_4$ from storage at plant	0.4 g	Nielsen & Holm-Nielsen, 1996	
CH$_4$ avoided at farm tanks	1.6 g	Nielsen & Holm-Nielsen, 1996	Impacts to be subtracted
Miscellaneous			
Area demand	1 ha		Estimate

Table 7.14. Ribe Biogas plant technical data (Kuemmel *et al.*, 1997).

These emissions have until now been considered in the greenhouse balance of the agricultural sector. They may be divided between the agricultural and the energy sectors, e.g. in proportion to the revenues from the two types of operation. However, they should, of course, not appear as an additional impact. This may serve as a warning against doing LCA only for one sector and neglecting impacts on other sectors. A meaningful LCA must either include all sectors of society affected or must treat what happens outside the system boundaries as indirect impacts for the compartments included in the LCA.

Environmental impacts	Impact type: emissions (g/kWh)	Uncer-tainty*	Monetised value US cents/kWh	Uncertainty & ranges*
From fossil energy currently used in plant construction and operation: CO_2 equiv. (leading to greenhouse effect)				
Plant and truck construction:	23	L	0.7	0.5–1.1
Transportation of feedstock/residues	90	L	2.9	2.0–4.2
Methane leaks (incurred minus avoided)	−285**	M	−9.1	−6 to −13
SO_2 (leading to acid rain and aerosols)	0.25	L	0.07	H,r,n
NO_x (possibly aerosols and health impacts)	0.36	L	0.14	H,r,n
particulates (lung diseases)	0.03	L	0.01	H,r,n
Land use			NQ	
Social impacts				
Occupational health damage	Cases per TWh:			
(manuf. & operation): death	1.7	L	0.55	L,l,n
major injury	2.2	L	0.02	L,l,n
minor injury	0.7	M	0.01	M,l,n
reduced span of life	6.0	M	1.95	M,l,n
Economic impacts				
Direct costs			5–20	
Resource use (energy pay-back time given)	2.1 y	L	NQ	
Labour requirements (manufacture)	17 person y/MW	L	NQ	
Import fraction (for Denmark)	0.1	L	NQ	
Benefits from energy sold			4–11	
Other impacts				
Supply security (variability in wind is high, entry based on plant availability)	High		NQ	
Robustness (up-front investment binds, entry based on technical reliability)	High		NQ	
Global issues (non-exploiting)	Compatible		NQ	
Decentralisation & choice (less with large size)	Good		NQ	
Institution building (collection management)	Modest		NQ	

Table 7.15. Impacts from large biogas plant at Ribe, Denmark (per kWh of biogas) (Kuemmel *et al.*, 1997). See notes below Table 7.6.
** The negative impact is due to a reduction of impacts outside the energy sector that would not otherwise be counted (see text).

Emissions from manure spread on the soil are low, and for the biogas plant residues returned to the fields nearly zero.

Some Danish biogas plants are currently accepting household refuse and waste from food industries. General use of industrial biomass waste is permitted only if there are no significant residues of heavy metals in the slurry.

The biogas chain impacts shown in Table 7.15 are estimated on the basis of emissions and energy production given above. The methane emissions from the cattle metabolism are not included, as they are considered to be the same as without the diversion of manure to the biogas plants. The avoided methane emissions from farm storage of manure are seen to be the dominating contribution to the LCA costs. It varies from installation to installation and has to be assessed for the actual conditions.

An LCA for biofuel chains may be found in Kaltschmitt *et al.* (1996).

LCA of road traffic

The example given in this section deals with passenger transportation by road in Denmark, for which several LCA studies are available. Danish road traffic is characterised by a low rate of deaths by road accidents as compared with other European countries. We are using the same valuation of an accidental death as in the previous examples, and the high number arrived at indicates that road accidents will be the most important externality in many other countries, while in Denmark they turn out not to be dominant.

The transportation sector has, in addition to fuel-related impacts, a large number of other externalities related to the infrastructure needed for road transportation systems. Table 7.16 is based on a recent review of several studies, with some new additions for embracing all identified impacts of owning and using a motor vehicle (Sørensen, 1997a,c). Figure 7.21 summarises these findings. Details of the assumptions are as follows:

The evaluation uses an average-size car, assumed to drive 200 000 km in 10 years with an average efficiency of 13.5 km per litre of gasoline (corresponding to mixed urban and highway driving and equivalent to 7.4 litres per 100 km). The greenhouse warming externality of 9 US c km^{-1} comes from the evaluation made in section 3.1. The health effect caused by air pollution from car exhaust is taken as 4 US c km^{-1}, a number arrived at in several of the Scandinavian studies quoted below Table 7.16. Also, the accident statistics give a firm basis for estimation (although the rate of accidents varies considerably between regions), but the actual value attributed to police and rescue team efforts, hospital treatment, lost workdays and lives all have to be estimated. As mentioned above, the SVL underlying Table 7.16 is 3 million US$, and for time loss a figure of 9.2 US$ h^{-1} is used (based on an interview study on perceived values of waiting time; Danish Technology Council, 1993). This is a "recreational" value in the sense that it corresponds to unemployment compensation rather than to average salary in Denmark. The "stress and inconvenience" entry takes into account the barrier effect of roads with traffic, causing e.g. pedestrians to have to wait (e.g. at red lights) or to walk a larger distance to circumvent the road barrier. This may be valued as time lost.

The noise impact is estimated at 3.1 US c km^{-1} based on hedonic pricing [i.e. the reduction in the value of property exposed to noise (e.g. houses along major highways as compared to those in secluded suburban locations); Danish Transport Council, 1993]. A similar approach is taken to estimate the visual degradation of the environment by roads, signs, filling stations, parking lots and so on. Property values were collected in 1996 (from newspaper advertisements), for detached houses of similar standard located at the same distances from the Copenhagen city centre (but outside the high rise area, at distances of 10–25 km from the centre), but exposed to different levels of visual and noise impact from traffic. The externality is then taken as the number of people exposed times the sum of property losses. The property loss found is 25–45%, and the total damage for 0.5 million people with 0.2 million houses and cars is 20 GUS$ or 52 US c km^{-1}, of which half is assumed to derive from visual impacts. A further reduction by a factor of 2 is introduced by going from a suburban environment near Copenhagen to a country average. The value arrived at is 14 US c km^{-1}.

Environmental impacts	Type of impact: emissions (g per kWh of fuel)	Monetised value (US cents per vehicle-km)	Monetised value (US cents per kWh of fuel)	uncertainty & ranges
Environmental emissions:				
Car manufacture and decommissioning	Average industry	2	3	H
Car maintenance	NQ			
Road construction and maintenance	NQ			
Operation:				
CO_2	277			L
NO_x (may form aerosols)	2.9			M
CO	17			M
HC	3.0			M
particles	0.06			M
Health effects from air pollutants		4	6	H,l,n
Greenhouse warming		9	14	10–21
Noise(variations are large)	Av. increase 1.5 dB(A)	3	5	H,l,n
Environmental & visual degradation				
(from roads, signs, filling stations, etc.)		14	20	H,l,m
Health and injury	*Cases:*			
Occupational (car/road construction				
and maintenance)	NQ			
Traffic accidents (incl. material				
damage, hospital and rescue costs):		10.4	16	M,l,n
Based on deaths (SVL = 3 M US$)	2.4×10^{-8} per kWh-fuel			
heavy injury	24×10^{-8} per kWh-fuel			
light injury (when reported)	16×10^{-8} per kWh-fuel			
Stress and inconvenience (e.g.				
to pedestrian passage)		3.3	5	H,l,n
Economic impacts				
Direct economy (cars, roads, gasoline,				
service and maintenance)	Taxes excluded	27	41	L,l,n
Resource use	Significant		NQ	
Labour requirements and	About 50% of direct			
import fraction (Denmark)	costs are local		NQ	
Benefits (valued at cost of public				
transportation)		35	53	M,l,n
Time use (contingency valuation)		15	23	H,l,n
Other impacts	NA			

Table 7.16. Impact from average Danish 1990 passenger car (Sørensen, 1997c). Data sources: Danish Technology Council (1993), Danish Road Directorate (1981), Danish Statistical Office (1993), Danish Department of Public Works (1987), Christensen and Gudmundsen (1993), and own estimates.

The direct economic impacts include capital expenses and operation for cars and roads, as well as property value of parking space in garages, carports or open parking spaces, but omitting any taxes and duties, and the benefits from driving are taken at the value of public transport (considering that differences in convenience and inconveniences such as not being able to read when driving even out). Time use is, as mentioned above, derived from a contingency valuation (i.e. interviews). The "fairly calculated" cost of driving a passenger car, i.e. all direct costs and indirect LCA impact items except benefits and time use (the most uncertain impacts), is then 0.72 US$ km^{-1}, of which 0.07 US$ km^{-1} are related to owning the car (purchase price without taxes plus environmental impacts of car manufacture) and the remaining 0.65 US$ km^{-1} are related to driving the car. A fair tax level, reflecting external costs, would then be divided into a vehicle tax of around 4000 US$, and a kilometres-driven component that levied onto the fuel would amount to about 5 US$ per litre. Figure 7.21 shows this division between ownership and usage impacts, and compares the "fair" taxation with the current Danish tax levels for car ownership and gasoline purchase.

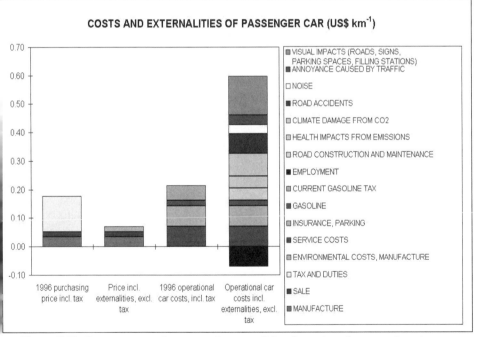

Figure 7.21. Components of costs and externalities for privately owned motor cars. Taxes included in car and gasoline purchasing prices pertain to 1996 levels in Denmark, and the item termed "employment" includes all jobs in the car service sector, but not e.g. in manufacturing. (Kuemmel *et al.*, 1997).

The LCA study includes impacts of both car ownership and driving, with the following types of impacts included in the analysis:

- Health effects from pollutants
- Traffic accidents
- Contributions to enhance the greenhouse effect
- Noise
- Visual aspects
- Barrier effects and inconvenience from road installations
- Road construction
- Car manufacture and decommissioning

The cost of road construction and maintenance is here taken as an externality, again because of the borders selected for the study, which focuses on energy but includes other impacts of car transportation as indirect contributions (rather than attempting a total LCA of society, in which case as mentioned in section 7.4.1 everything could have been calculated separately and all impacts summed at the end). The specific case study includes only the use of roads for passenger automobile traffic, which is, of course, only part of reality. Transportation of goods has a considerable role to play in road planning and contributes significantly to maintenance requirements, accidents, visual impacts, etc. If in the future entirely different modes of transportation were introduced, the infrastructure solutions would also be different, and a new analysis should be made.

7.4.13 LCA of energy systems

In this section, examples of applying the LCA methodology are sketched. Three energy systems are considered, of which one represents the current Danish system (using data from the early part of the 1990s) and the other two are future scenarios for the middle of the 21st century, both with strong penetration of renewable energy sources and the assumption of no come-back for nuclear power. The scenarios and their origins are more fully described in Kuemmel *et al.* (1997).

One future scenario is ecologically driven, assuming a majority of the Danish population prefer a strict environmental policy, where choices involving potentially large but uncertain dangers to the environment are not made even if they appear economically attractive (the basic scenario first appeared in Sørensen *et al.*, 1994). The other scenario is market driven, but assumes externalities to be considered explicitly, either by including their monetised value in costs or by implementing taxes and regulations that will limit the use of the undesirable products or practices. This scenario is the Danish part of the European fair market scenario described in Nielsen and Sørensen (1998). It tries to correct the current omission of environmental and social costs from free market economics by some average or agreed value of the known impacts. This is in opposition to the precautionary principle built into the ecologically driven scenario, which may turn out to overestimate externalities by effectively excluding options that are too uncertain.

From a methodological point of view, the ecological scenario is the simplest, as one just has to calculate the LCA impacts of a prescribed system. For the fair market scenario, there is a coupling between the scenario choices and the LCA calculation, as discussed in section 7.4.1, so the LCA in principle has to be made first, in order to select the precise scenario assumptions regarding the use of different technologies, and then one has to recycle through an iterative optimisation procedure until consistency is achieved.

As discussed in section 7.4.2, there are two ways of treating import and export: either to include impacts from imported goods, e.g. energy, materials or equipment, and neglect impacts associated with goods to be exported ("product view") or vice versa ("economy view"). Let these two views be labelled "P" and "E" type calculations. Of the three scenarios considered, the Danish 1992 system had little import of electricity and no export. The environmentally sustainable scenario does not allow import of energy, whereas the fair market scenario has large amounts of wind-produced electricity and hydrogen exported to Germany. The way in which the two views are explored here is to do the LCA calculations in the following two ways:

- Include impacts along the entire chains of energy conversions, whether they occur in Denmark or abroad, in a full energy chain analysis (cf. Dones *et al.*, 1994). For this the suffix "F" is used.
- Only include impacts occurring as a result of activities within Denmark or from Danish off-shore activities. This will be denoted a "domestic" analysis and labelled by the suffix "D".

The full energy chain analysis is equivalent to the product view, provided that energy and other goods produced for export are omitted from the impact analysis. The domestic analysis is, in principle, identical to the economy view, but not in practice, because it is assumed that all equipment and non-energy materials used in the Danish energy sector are imported (in this way, one avoids having to reduce industrial energy demand by what is used to service the energy industry, for reasons of not double counting, as discussed in section 7.4.1). It should be noted that emissions from domestic activities may impact on foreign countries, e.g. through trans-boundary pollution transport or global dispersion of greenhouse gases.

Available data make it simple to exclude energy produced in Denmark for export from a P-type analysis, but it is not possible to exclude the impacts from other products exported, since their production uses the Danish energy system (for impacts are included) and statistical data does not allow the energy content of exported non-energy products to be singled out in any simple way. This is the reason for not performing a P-type calculation, but replacing it by the F-type full energy chain analysis. The results of the F-type analysis will thus be an upper bound for the correct product view analysis. It makes no difference that it is assumed that equipment and non-energy materials for the energy sector are imported, as their impacts should be, and are, included no matter whether produced in Denmark or abroad.

This is also the reason that one has to make a distinction between the D-type analysis, the one carried out, and the E-type calculation, which should have included the fraction of impacts from equipment and materials produced within Denmark for use in the energy sector. The results of the D-type analysis will thus be a lower bound for the correct economy view analysis. In this way it is possible to state a set of lower and upper bounds for the LCA impacts of the energy system that span and perhaps exaggerate the difference between the two ways of accounting. If the two resulting values are not too far apart, the job of quantifying the life-cycle impacts and exhibiting their uncertainty is finished. It should be noted that with some additional work, the D-type analysis could be transformed into a proper E-type analysis, whereas the upgrading of an F-type calculation to a P-type would require collection of data not presently available.

The remarks made above illustrate the type of practical problems that often arise in implementing the methodology of life-cycle analysis owing to inadequacies of available statistical data.

Sector	Delivered energy	Net energy	End-use energy	Energy quality
Electric heating	9	9	70	Space heating
Fuels for heating	190	153	25	Stationary mechanical energy
Electrical appliances	44	44	10	Electrical appliances
Cooling and refrigeration	15	15	5	Cooling and refrigeration
Electricity for process heat	41	41	18	Process heat under 100°C
Fuels for process heat	111	91	4	Process heat 100–500°C
Electricity for transportation	1	1	3	Process heat over 500°C
Fuels for transportation	173	173	26	Transport work
			20	Energy in food
Total	**584**	**527**	**181**	**Total**

Table 7.17. Danish energy demand 1992 in PJ per year (Sørensen *et al.*, 1994).

The current demand is shown in Table 7.17 and the energy system is shown in Figs. 7.22 and 7.23. These are based on 1992 data, because these were used as the starting point in the ecologically sustainable scenario, including the attempt to distribute the end-uses on categories of energy quality. From 1992 to 1996, the Danish oil and gas production in the North Sea increased, although this is not expected to continue to 2030 (the scenario year for the future scenarios), because of the limited resources expected in the

Danish part of the North Sea. In 1992, coal was still the major source of electric power production, and renewable energy use was increasing although still modest.

The delivered energy in Table 7.17 is the energy delivered to the end-users. The net energy subtracts conversion losses in local boilers and furnaces, but does not correct for conversion losses in the transport sector. The estimated end-use energy is the true net energy needed to deliver the services of the actual system, had all been delivered by the most efficient devices available today (note that the subdivisions are different from those of the left-hand side of the table). The net energy for heating includes hot water used in the domestic and service sectors, whereas this has been listed as low-temperature process heat in the end-use column. The net energy for electrical appliances includes heating of water and low-temperature cooking, which in the end-use column is included under process heat below 100°C, and some stationary mechanical energy used in the service industry. The net process energy includes not only true process heat used in agriculture and industry, but also space heating, hot water, stationary mechanical energy and energy for electric appliances used in these sectors. In the end-use columns, an attempt has been made to categorise these correctly and again to adjust to the efficiency of the best conversion equipment. For the transportation sector, the end-use energy is obtained by applying an average conversion efficiency of current vehicles.

Figures 7.22 and 7.23 give a picture of the 1992 conversions between the primary and end-use energy in the Danish system, starting in Fig. 7.22 with a picture of the agricultural sector in energy units. This illustrates the large magnitude of agricultural energy manipulations and the small fraction of these presently contributing to energy supply, i.e. wood, straw, biogas and residues, not counting the energy provided by food. Each device is described by three numbers: energy input, loss and energy output. In this way, the assumed efficiency of each conversion can be seen.

The energy conversions illustrated in Fig. 7.23 indicate a fairly efficient range of intermediate conversions, followed, however, by a relatively inefficient range of end-use conversions, particularly as regards the combustion engines used in the transportation sector. Efficiencies of electrical appliances have not been estimated here, but may be found together with efficiency analyses of other conversion equipment in Sørensen (1991) and in a condensed form in Sørensen (1992a). Non-energy uses of fuels (e.g. oil as feedstock for the plastics industry and tar for road construction) have been omitted from the overview.

In Table 7.18 the results of LCA calculations for the current Danish energy system are given. The two variations of the calculation are based on the product view (labelled "F" for full chain calculation) and the economy view

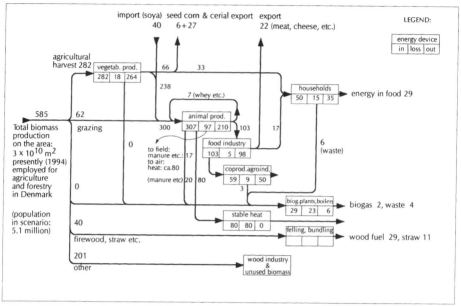

Figure 7.22. Overview of 1992 Danish biomass sector, with energy links but excluding indirect energy inputs for fertilisers, farm machinery, etc. Units are in PJ y^{-1} (Sørensen et al., 1994; Sørensen, 1996c).

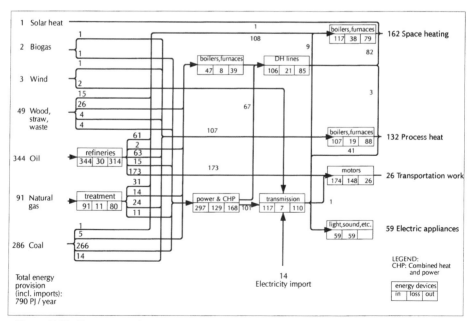

Figure 7.23. Overview of Danish energy system 1992. Units are in PJ y^{-1} (Sørensen et al., 1994a; Sørensen, 1996c).

(labelled "D" for calculation based on impacts arising from domestic activities), as explained previously. As there are impacts not evaluated, this is a partial analysis, and the sum of all calculated impacts cannot be taken as representing the totality of impacts.

In the year 1992 used as basis, Denmark imported some hydro-based electricity from Norway (cf. Fig. 7.23). This gives rise to an externality cost of 13.3 M euro y^{-1}, which has been included in both calculations. The calculation also estimates the greenhouse warming impacts of materials used in the energy sector, assuming that they are all imported and of international transport. These are included in the F-calculations, but not in the D-calculations, where they constitute $10\ 800 \times 10^6$ kg of CO_2-equivalent or slightly more than the difference between the F- and D-calculations, owing to the net energy import in the year considered. For occupational impacts, the accidental deaths are small numbers that exhibit statistical fluctuations from year to year, whereas the larger numbers of injuries are more statistically significant and less fluctuating between years.

Impacts from traffic are taken as entirely domestic impacts except for emission-related impacts. For impacts such as visual, noise, and barrier effects of traffic infrastructure, this is evident. In case of greenhouse warming, the F-type calculation includes international traffic on Danish carriers, whereas the D-type calculation considers only domestic transportation. For air pollution impacts other than greenhouse warming, only impacts from domestic traffic were included in both calculations, because the data on dispersal and dose–response available to us assumed emission over land. For international sea or air transportation, the emissions would reach populated areas only after some atmospheric travel. This could be calculated using existing models, but we did not attempt such calculations.

It is seen that greenhouse warming impacts and a range of other impacts from the transportation sector dominate the picture of impacts included. It is believed that this would also be the case if impacts such as biodiversity and other impacts listed in section 7.4.3 were considered. However, if the valuation of greenhouse warming impacts were altered in the downward direction, e.g. by changing the statistical value of life or making it dependent on economic activity in different regions of the world, then the relative important of other impacts would increase, and it is no longer certain that impacts not included in this evaluation would have a negligible effect.

The evaluation of LCA impacts uses the Table 7.6 example for coal-fired power plants. This means that the atmospheric dispersion calculated for emissions from this particular power plant is considered representative for all Danish power stations. For other fuel combustion, the impacts are scaled up or down from those of the power plant calculations, using data such as those of section 7.4.12 for scaling. For fuel combustion in the transportation sector, the low emission height is considered in estimating the health impacts of land transportation. Also the several non-emission-related impacts identified in the transportation sector are included for this sector only.

Environmental impacts and public health	Emissions: (10^6 kg/y)	Monetised value (Meuro/y)	Range of uncertainty (Meuro/y)
Emission impacts from all sectors except traffic:			
SO$_2$	107 / 90	246 / 206	uncertainty of
NO$_x$ (air pollution impacts in this row)	303 / 285	635 / 577	air pollution
particulates	12.6 / 8.8	48 / 34	impacts almost
HCl	3.8 / 3.6		entirely from
HF	0.2 / 0.2		monetising
CO$_2$	55 095 / 48 695		
CH$_4$	84 / 25		
Other volatile organic compounds	142 / 130		
N$_2$O	2 / 2		
CO	849 / 813		
Above greenhouse gases as CO$_2$ equivalents	73 676 / 65 098	19 156 / 16 925	11 718 to 27 997
Emission impacts from transportation sector:			
Air pollution from car manufacture		875 / 0	H,r,n-m
Air pollution from domestic traffic		1 752 / 1 752	H,l,n-m
Greenhouse gas emissions, CO$_2$ equivalents*	15 485 / 8 105	4 026 / 2 107	1 459 to 5 884
Other impacts from sectors other than traffic:			
Visual impacts		0.1 / 0.1	H,l,n
Noise impacts		0.5 / 0.5	M,l,n
Other impacts from transportation sector:			
Visual impacts		6 497 / 6 497	H,l,n
Noise impacts		1 531 / 1 531	H,l,n
Traffic accidents		5 029 / 5 029	L,l,n
Stress and inconvenience		1 562 / 1 562	H,l,n
Occupational health and injury	Cases:		
All sectors except transportation:	(per year)		
Death	11 / 2	29 / 5	M,l,n
Major injury	298 / 177	24 / 14	Factor of 3 up/down
Minor injury	1 728 / 380	2.1 / 0.5	Factor of 3 up/down
Other impacts			
Road construction and maintenance		1 291 / 1 291	L,r,n
Other infrastructure, labour and import requirements, social benefits, resource depletion		NQ	

Table 7.18. LCA impacts from 1992 Danish energy system. The product and economy views are represented by F- and D-calculations described in the text. The values for the full cycle (F) and the domestic (D) calculations are separated by a "/" for each entry. NA = not analysed, NQ = not quantified. Values are aggregated and rounded (to zero if below 0.1 m euro kWh^{-1}). The greenhouse impact uncertainty estimate is as discussed in section 7.4.12. (L,M,H): low, medium and high uncertainty. (l,r,g): local, regional and global impact. (n,m,d): near, medium and distant time frame (Kuemmel et al., 1997).

* For the transportation sector, the entry before the "/" includes fuels for international transport and stockpiling (in order to fulfil international requirements of minimum energy stores: these "bunkers" appear in the annual energy statistics, when their size changes), the one after the slash includes only domestic. Impacts do not include those from electric trains, which are included (through power plants) in the other sectors.

Environmentally sustainable scenario

The ecologically driven scenario for Denmark (Sørensen *et al.*, 1994) is a normative scenario, assuming broad concern in the Danish population over issues such as greenhouse warming. The study first defines the energy demand at the end-user, by the method described in section 6.2.2: analysing all activities of society, starting from coverage of basic needs for food, shelter and human relations, and continuing into a range of secondary needs that are left open ended in realisation of the futility of trying to guess what possibilities will be explored 35 years from now. However, they are broadly characterised by energy requirements based on the trend of increased introduction of devices using electrical energy, as well as increased efficiency of any energy form used

In order to satisfy the energy need, the energy system layout shown in Fig. 7.25 was constructed, with an agricultural sector providing biomass energy as shown in Fig. 7.24. The primary side of the Danish energy scenario has a large proportion of electric power, primarily due to abundant wind resources, and therefore electricity is supplied to uses of other energy qualities, such as high-temperature process heat as well as low-temperature heat, but in that case using heat pumps to ensure efficient use of the electricity. The cold reservoirs of the heat pumps may under Danish conditions be the ground or aquifers storing absorbed solar heat. The intermittent supply of wind power and solar heat and power is dealt with in the following ways.

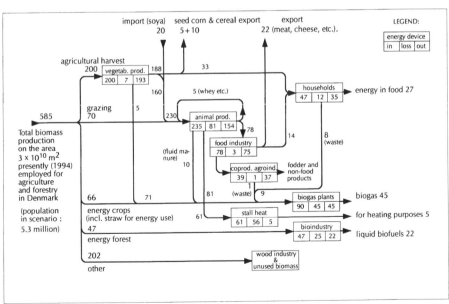

Figure 7.24. Danish 2030 agricultural sector according to the ecologically sustainable scenario. Units: PJ y^{-1} (Sørensen *et al.*, 1994; Sørensen, 1996c).

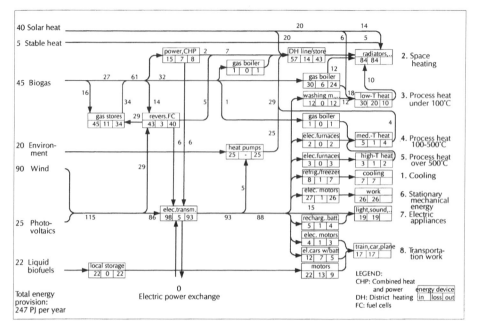

Figure 7.25. Danish 2030 energy system according to the ecologically sustainable scenario. Units: PJ y^{-1} (Sørensen *et al.*, 1994; Sørensen, 1996c).

For electricity, the use of assumed to be developed reversible fuel cells (centrally or dispersed into buildings) allows surplus electricity to be converted into gas, which is stored (Denmark has identified a number of underground caverns and aquifers suitable for this purpose) and later reused for generating electricity and associated heat (distributed through already existing district heating lines). In the case of solar heat, the strong seasonality is dealt with by intelligent use of heat storage and the existing district heating lines: these lines, distributing low-temperature heat, will accept solar heat collected during the sunny months of March to October, but will make use of coproduced heat from fuel cells and combined heat and power plants during winter and in case of insufficient solar heat supply. The source of the alternative heat (except from the surplus power mentioned above) is biogas, which again can be stored. Large biogas plants are already operating in Denmark.

The results of performing a systems LCA calculation for the 2030 ecologically sustainable energy scenario for Denmark are presented in Table 7.19 for both the product and economy views. As for the present system assessment, this is a non-exhaustive list of impacts.

The impacts found are seen to be dominated by those of the transportation sector, while the greenhouse warming impacts of the present system have practically disappeared. In fact, they should go away completely, provided

Environmental impacts and public health	Emissions: (10^6 kg/y)	Monetised value (Meuro/y)	Range of uncertainty (Meuro/y)
Emission impacts from all sectors except traffic:			
SO_2	10.7 / 8.6	25 / 20	Uncertainty of
NO_x (air pollution impacts in this row)	9.8 / 4.7	31 / 15	air pollution
particulates	0.8 / 0.5	3 / 2	impacts almost
HCl	0.004 / 0		entirely from
HF	0.005 / 0		monetising
CO_2	455 / 0		
CH_4**	−52 / −54		
Other volatile organic compounds	14 / 13		
N_2O	2.3 / 1.9		
CO	58 / 48		
Above greenhouse gases as CO_2 equivalents	2472 / 1614	643 / 420	291 to 939
*Emission impacts from transportation sector**			
Air pollution from car manufacture		662 / 0	H,r,n-m
Air pollution from domestic traffic:		192 / 192	H,l,n-m
NO_x from biofuel engines	4.8 / 3.2		
CO from biofuel engines	1.3 / 0.9		
Above two greenhouse emissions in CO_2 eq.	196 / 131	51 / 34	24 to 74
Other impacts from sectors other than traffic:			
Visual impacts		2 / 2	H,l,n
Noise impacts		15 / 15	M,l,n
Other domestic impacts from transportation:			
Visual impacts		4920 / 4920	H,l,n
Noise impacts		1159 / 1159	H,l,n
Traffic accidents		3808 / 3808	L,l,n
Stress and inconvenience		1183 / 1183	H,l,n
Occupational health and injury	Cases: (per year)		
All sectors except transportation:			
Death	4 / 4	10 / 10	M,l,n
Major injury	160 / 156	13 / 12	Factor of 3 up/down
Minor injury	424 / 420	0.5 / 0.5	Factor of 3 up/down
Other impacts			
Road construction and maintenance		978 / 978	L,r,n
Other infrastructure, labour and import requirements, social benefits, resource depletion		NQ	

Table 7.19. LCA impacts from the 2030 Danish energy system according to the ecologically sustainable scenario. The product and economy views are represented by F- and D-calculations described in the text. The values for the full cycle (F) and the domestic (D) calculations are separated by a "/" for each entry (Kuemmel *et al.*, 1997). NA = not analysed, NQ = not quantified. Values are aggregated and rounded (to zero if below 0.1 m euro kWh^{-1}). The greenhouse impact uncertainty estimate is discussed in section 7.4.12. (L,M,H): low, medium and high uncertainty. (l,r,g): local, regional and global impact. (n,m,d): near, medium and distant time frame.

* For the transportation sector, the entry before the "/" includes international transport, the one after includes only domestic and, impacts do not include emission impacts associated with electric trains, which are included (through power plants) in the other sectors. Also the net CO_2 greenhouse warming impacts from biofuels are assumed to be zero.

** The methane emission reduction is relative to the much larger emissions from the agricultural sector.

that the entire world went along with an ecologically sustainable energy supply system, because then the greenhouse gas emissions would be less than the sinks (oceans, plants) could accommodate. However, as we use a fixed greenhouse impact per unit of initiating emission, we do not catch any threshold effect in the calculation. This is not important for the present case, because the greenhouse warming impact is in any case very small.

Occupational impacts are of the same order of magnitude as today, but their origin is shifted to the production, installation and maintenance of renewable energy equipment.

For the transportation sector, the emission-related impacts are practically absent, whereas other impacts are only reduced by the factor of reduced traffic assumed in the ecologically sustainable scenario. We have simply scaled the impacts from Table 7.16 by the difference in kilometres driven, usually denoted the *transport work*. This is adequate for traffic accidents, noise and certain of the visual and barrier effects, while for the impacts associated with road infrastructure the scaling with transport work may for the environmentally sustainable scenario underestimate those impacts associated with the standing traffic structures (roads, filling stations, parking spaces, garages, etc.), because the extent of such structures is not necessarily regulated up or down with traffic intensity.

A comparison of the D- and F-type LCA evaluations will be made below for all the scenarios studied.

Denmark	1990	2000	2010	2030	2050
All sectors	514	522	426	348	252
Households	171	153	127	97	61
electricity	33	19	13	16	18
space heat	113	109	95	69	36
hot water	25	26	19	13	6
fuel for cooking	0	0	0	0	0
Service	40	30	22	24	25
electricity	28	19	12	17	21
fuels	11	11	10	7	4
Industry	116	107	98	76	48
electricity	31	30	28	28	30
fuels	84	77	70	58	48
Transport	188	231	178	150	119
road	134	164	98	72	49
rail	5	5	5	5	4
air	29	40	50	52	43
waterway	20	23	25	21	22

Table 7.20. Danish energy demand development in fair market scenario; units are in PJ y^{-1} (Nielsen and Sørensen, 1996, 1998; Kuemmel *et al.*, 1997).

Fair market scenario

The Danish part of the European fair market scenario (described in Nielsen and Sørensen, 1998; Kuemmel *et al.*, 1997) has a higher demand than the sustainable scenario (shown in Table 7.20) and more international trade of energy, such as Danish export of wind energy to Germany where the demand for renewable energy is higher than the supply. The supply, conversion system and demand are summarised in Fig. 7.26.

The results of performing a systems LCA calculation for the 2050 fair market energy scenario for Denmark are presented in Table 7.21, for both the product and the economy views. As for the corresponding assessments of the sustainable scenario, this is a partial list of impacts owing to areas that we were unable to include in the calculations.

The impacts found are seen to be dominated by those of the transportation sector, just as in the ecologically sustainable scenario, while the greenhouse warming impacts are very small.

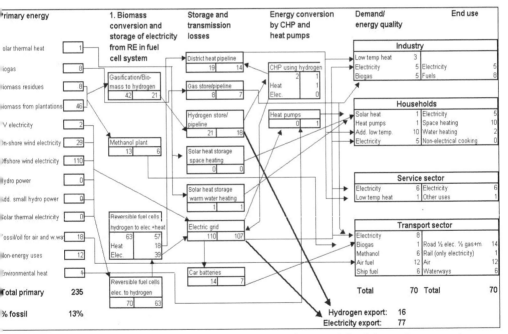

Figure 7.26. Energy conversion system of the Danish part of the fair market scenario for the year 2050. The units are in TWh y^{-1}, and the total primary energy supply of 235 TWh y^{-1} corresponds to 775 PJ y^{-1}. The transport demand listed as "biogas" includes hydrogen (Nielsen and Sørensen, 1996, 1998).

The fair market scenario differs from the ecologically sustainable by having considerably more transport and by using a fair amount of fossil fuels in this sector. However, as the bulk of the oil products are used for international transport (i.e. transport out of Denmark on ships, lorries and airplanes), they show up only in the calculation including such uses. However, for the not emission-related traffic impacts, the higher amount of transport work in the fair market scenario is directly reflected in the externality costs of the transportation sector.

Environmental impacts and public health	Emissions: (10^6 kg/y)	Monetised value (Meuro/y)	Range of uncertainty (Meuro/y)
Emission impacts from all sectors except traffic:			
SO_2	8.1 / 4.6*	19 / 11	Uncertainty of
NO_x	7.3 / 9.2	23 / 29	air pollution
particulates	0.5 / 0.2	2 / 1	impacts almost
HCl	0.002 / 0		entirely from
HF	0 / 0		monetising
CO_2 (from imported equipment/materials only)	893 / 0		
CH_4 HH	−623 / −635		
Other volatile organic compounds	18 / 30		
N_2O	1.5 / 4.7		
CO	11 / 4		
Above greenhouse gases as CO_2 equivalents	1 228 / 1 438	319 / 374	221–546
Emission impacts from transportation sector*:			
Air pollution from car manufacture		1427 / 0	H,r,n-m
Air pollution from domestic traffic		204 / 204	H,l,n-m
CH_4 from biogas	−4.4 / −4.4		
NO_x from methane engines	4.8 / 4.8		
CO from methane engines	1.3 / 1.3		
Oil product combustion CO_2	4 749 / 427		
Total net greenhouse gas emissions, CO_2 eq.	4 854 / 532	1262 / 138	1845 / 96
Other impacts from sectors other than traffic:			
Visual impacts		4 / 14	H,l,n
Noise impacts		27 / 85	M,l,n
Other domestic impacts from transportation:			
Visual impacts		10593 / 10593	H,l,n
Noise impacts		2496 / 2496	H,l,n
Traffic accidents		8200 / 8200	L,l,n
Stress and inconvenience		2547 / 2547	H,l,n
Occupational health and injury	Cases:		
All sectors except transportation:	(per year)		
Death	3 / 9	8 / 23	M,l,n
Major injury	159 / 512	13 / 40	5–35 / 14–113
Minor injury	412 / 1 205	0.5 / 1.4	0.2–1.4 / 0.5–4
Other impacts			
Road construction and maintenance		2 106 / 2 106	L,r,n
Other infrastructure, labour and import requirements, social benefits, resource depletion		NQ	

Table 7.21. LCA impacts from the 2050 Danish energy system according to the fair market scenario. The product and economy views are represented by F- and D-calculations as described in text. The values for the full cycle (F) and the domestic (D) calculations are separated by a "/" for each entry. See notes below Table 7.19 (Kuemmel *et al.*, 1997).

In the several cases where domestic impacts (D-calculation) exceed the full chain impacts (F-calculation), the reason is that in the fair market scenario Denmark exports large quantities of energy, produced within the country at some environmental cost, which is included in the D-calculation but by definition not in the F-calculation.

As in the ecologically sustainable scenario, air pollution and occupational impacts are modest by 2050.

The discussion on scaling of traffic impacts made in the ecologically sustainable scenario discussion is equally valid for the fair market scenario, where the amount of transport work was also used to scale non-emission-type impacts. Efficiency improvements in vehicles are about the same in the two scenarios, and the analysis of passenger transportation by road may only in a rough way be applicable to other types of transport. This adds more uncertainty to the evaluation of impacts in the fair market scenario, because of its higher proportion of international travel and transport, notably by air. Further work on the impacts of airports, airport infrastructure, noise from aircraft and geographical distribution of emissions is clearly needed.

CO_2 from additional fossil fuels required for load matching is not included. It is of the order of 15% of the 1992 fossil externality, according to an estimate made for the entire European scenario.

Comparison of scenario impacts

The LCA calculations described above for the present and two future scenarios for Denmark's energy system are summarised and compared in Figs. 7.27 and 7.28, both for the variant including indirect impacts in imports but neglecting those in exports (F-calculation) and for the variant including impacts from goods and energy produced for export but neglecting indirect impacts from goods imported (D-calculation).

Figure 7.27 shows the results for total monetised LCA impacts for the present system and the ecologically sustainable and fair market scenarios pertaining to a period some 30–60 years into the future. The impacts included are in the future scenarios considerably smaller than those of the present system. The ecological scenario has the smallest impacts, whereas the fair market scenario is somewhere between the present and the ecological scenario. This is primarily due to the higher activity in the transportation sector, which in both future scenarios gives the dominant externality contribution. The greenhouse warming contribution that dominates the present energy system is practically gone and would be completely absent if the rest of the world would cut its greenhouse gas emissions as much as the scenarios assume for Denmark. This is because of the non-linear threshold effect of a certain amount of injected greenhouse gases being absorbed by the oceans and by plants and thus not contributing to increased atmospheric concentrations.

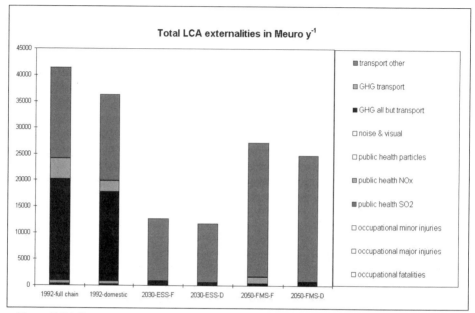

Figure 7.27. Overview of system LCA calculations for present and future Danish energy systems. Total impacts are monetised for full chain (F) and domestic (D) calculations (Kuemmel *et al.*, 1997).

The first observation to make is that there is fairly little difference between the F- and D-calculations. This is true for the totals, but it should be remembered that because of the way the transportation impacts were estimated most of them are the same in the two calculations. Looking instead at the greenhouse warming externalities, it is seen that there is indeed a difference, which for the present energy system is dominated in part by international transportations, and in part by the fossil energy inputs to imported goods used in construction of energy conversion equipment, whereas for the fair market scenario it is dominated alone by the impacts of fossil energy used in international traffic out of Denmark and the contribution from producing excess wind and hydrogen energy for export (almost entirely from equipment) is very small. In both future scenarios, there is a further contribution from additional greenhouse warming impacts associated with imported materials, included only in the F-type calculation.

The message transmitted by this analysis, showing the largest externality coming from greenhouse warming, has a number of qualifications related to the way lives are valuated in different parts of the world. This was discussed in section 7.4.11.

Figure 7.28 summarises the alternative impressions given by presenting the results as totals or relative to the quantities of energy used.

The case studies considered above are intended to illustrate the use of LCA for decision-support in energy planning. They have been based upon scenario work, using time horizons of 30–60 years.

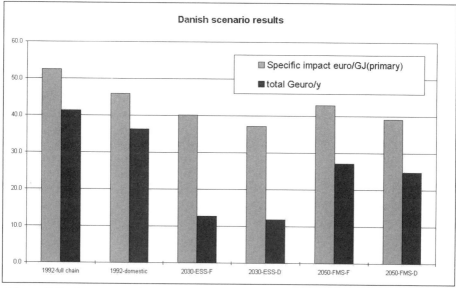

Figure 7.28. Overview of LCA calculations for present and future Danish energy systems. Total impacts and impacts per unit of energy used are given for each scenario. (F): monetised full chain, (D): domestic evaluation (Kuemmel *et al.*, 1997).

The reason for basing the examples of Chapters 6 and 7 on scenario work is explained in the following. For the purpose of assisting decision-makers, a way must be found of describing a given energy system in its social context that is suitable for assessing systems not yet implemented. Simple forecasts of demand and supply based on economic modelling cannot directly achieve this, as economic theory deals only with the past and occasionally the present structure of society. In order to deal with the future, economic modellers may attempt to invoke the established quantitative relations between components and assume that they stay valid in the future. This produces a "business-as-usual" forecast. Because the relations between the ingredients of the economy, e.g. expressed through an input–output matrix, vary with time, one can improve the business-as-usual forecast by taking into account trends already present in the past development. However, even such trend forecasts cannot be expected to retain their validity for very long periods, and it is not just the period of forecasting time that matters, but also changes in the rules governing society. These may change as a result of abrupt changes in technology used (in contrast to the predictable, smooth improvements of technological capability or average rate of occurrence of novel technologies), or they may be changed by deliberate policy choices. It is sometimes argued that econometric methods could include such non-linear behaviour, e.g. by replacing the input–output coefficients by more complex functions. However, prediction of what these should be cannot be based on studies of past or existing societies, because the whole point in human choice is that options are available that are different from past trends. The non-linear, non-predictable relations that may prevail in the future, given certain policy interventions at appropriate times, must therefore be postulated on normative grounds. This is what the scenario method does, and any attempt to mend economic theory also implies invoking a scenario construction and analysis, so in any case this is what has to be done.

It is important to stress that scenarios are not predictions of the future. They should be presented as policy options that may come true only if a pre-scribed number of political actions are indeed carried out. In democratic so-cieties, this can only happen if preceded by corresponding value changes af-fecting a sufficiently large fraction of the society. Generally, the more radi-cally the scenario differs from the present society, the larger must the sup-port of a democratically participating population be. For non-democratic so-cieties, a scenario future may be implemented by decree, with the negative implications of such a procedure.

The actual development may comprise a combination of some reference scenarios selected for analysis, with each reference scenario being a clear and perhaps extreme example of pursuing a concrete line of political preference. It is important that the scenarios selected for political consideration be based on values and preferences that are important in the society in question. The value basis should be made explicit in the scenario construction. Although all analysis of long-term policy alternatives is effectively scenario analysis, particular studies may differ in the comprehensiveness of the treatment of future society. A simple analysis may make normative scenario assumptions only for the sector of society of direct interest for the study (e.g. the energy sector), assuming the rest to be governed by trend rules similar to those of the past. A more comprehensive scenario analysis will make a gross scenario for the development of society as a whole, as a reference framework for a deeper investigation of the sectors of particular interest. One may say that the simple scenario is one that uses trend extrapolation for all sectors of the economy except the one focused upon, whereas the more radical scenario will make normative, non-linear assumptions regarding the development of society as a whole. The full, normative construction of future societies will come into play for scenarios describing an ecologically sustainable global so-ciety, whereas scenarios aiming only at avoiding or coping with one par-ticular problem, such as the climate change induced by greenhouse warm-ing, are often of a simpler kind.

Figure 7.29 illustrates the above discussion by an example (which is in-dicative of the renewable energy systemic transitions), where the total system cost initially goes up, having to cross a barrier before reaching the lower minimum characterising the new system. This would illustrate a very common behaviour, both when only present market costs are considered and also if cost is taken to comprise LCA externalities. The cheaper system will never be selected by a marginal appraisal, but the scenario method correctly identifies it and further enables the transition costs to be calculated using a trajectory method.

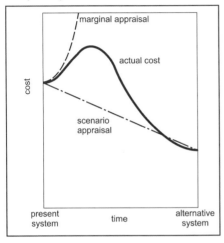

Figure 7.29. Schematic illustration of the difference between marginal appraisal methods and the scenario method, in describing a transition between two local cost minima separated by a cost barrier. The cost may or may not comprise LCA externalities (based upon Kuemmel *et al.*, 1997).

7.5 Examples of break-even price estimations

7.5.1 Wind power system without energy storage

The economic discussion in this section will be based on the wind energy systems considered in section 6.3.2, and first it will be assumed that neither short- nor long-term storage facilities are included in the system.

Consider first a situation where wind energy supplies a marginal contribution to a large, fuel-based electricity generating system. Here the wind energy system acts purely as a fuel-saver, and the assumption that its supplement is marginal implies that the presence of wind generators has no influence on the capacity requirements or on the number of fuel-based units in operation at any time. The fluctuating contribution from wind generators is reflected in small downward adjustments of the generation at the fuel-based units, being completely equivalent to a small drop in load relative to the expected level. The fuel saved at the fuel units, which are regulated, is determined by the marginal cost of generation, i.e. the cost of the last kWh produced. This cost and other parameters used in a reference calculation of the savings introduced by adding a small contribution of wind energy converters to a fossil fuel-based energy system are summarised in Table 7.22.

Parameter	Reference value	Sensitivity
Average wind power production	150 W m^{-2}	1
Operation/maintenance, wind system	$ 0.0033 (kWh)$^{-1}$	-0.29
of which the labour fraction	0.7	-0.061
Depreciation period, wind system	25 y	0.62
Interest rate, corrected for inflation	0.03 y^{-1}	-0.25
Real terms increases of wages	0.03 y^{-1}	-0.060
Real terms increases in raw materials cost	0 y^{-1}	(-0.64)
Conversion efficiency of fuel units	0.316	1.23
Initial fuel price	$0.006 (kWh(t))$^{-1}$	1.23
Real term increases in fuel prices	0 y^{-1}	(14.43)
Fuel-related working costs of fuel units	$0.0007 (kWh(e))$^{-1}$	0.057
of which the labour fraction	0.7	0.012

Table 7.22. Reference parameters assumed for mixed wind/fuel-based electricity system with marginal wind energy contribution. The relative sensitivity of the break-even price to a change of one parameter is given on the right. kWh(t) are thermal energy units, i.e. the fuel energy value, whereas kWh(e) are units of electricity generated from the fuels. The sensitivities are relative ones, except for the two entries in parentheses. They represent the change in break-even price caused by an absolute change of the parameter in question by one unit (cf. text) (based on Sørensen, 1988).

Based on these reference parameters, the break-even price of the wind energy converters can be calculated, i.e. the lowest capital cost for which the accumulated present value of total costs does not exceed the accumulated present value of the fuel expenditures (including cost of fuel handling) saved over the assumed depreciation period, which is taken as the expected lifetime of the converters. For the reference parameters, the derived break-even price is $360 per m^2 swept by the rotor (this being a reasonable "size parameter" for a horizontal axis converter, rather than the fairly arbitrary rating of the electricity generator, cf. Sørensen, 1975c).

In order to determine to which of the parameters the break-even price is more sensitive, the derivatives of the break-even price, $C(b.e.)$, with respect to the ith parameter, π_i, may be used to construct a relative sensitivity, s_i, defined by

$$s_i = \frac{\pi_i}{C(b.e.)} \frac{\partial C(b.e.)}{\partial \pi_i}, \tag{7.10}$$

Close to the reference parameters, this linear expression may be used (evaluated with the reference parameters inserted) to predict the dependence of the break-even price on the different parameters. These sensitivities are given in the right-hand column of Table 7.22, for the example of a marginal wind supply system added to a fuel-based electricity generating system. When the parameter reference value, $\pi_i(ref)$, is zero, the expression (7.10) is not useful, and in these cases $s'_i = C(b.e.)^{-1} \, \partial \, C(b.e.) / \, \partial \, \pi_i$ is given instead.

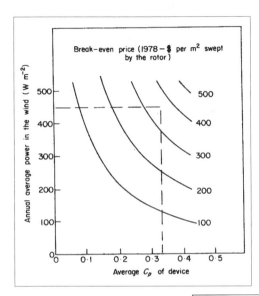

Figure 7.30. Two-parameter study of break-even price of wind energy system supplying a marginal contribution to an electricity system. Other parameters are fixed at the values given in Table 7.22, and the dashed lines correspond to the full reference parameter set (based on Sørensen, 1976d).

Figure 7.31. See legend to Fig. 7.30. An annual inflation rate of 10% has been assumed in deriving the scales corresponding to running prices (based on Sørensen, 1976d).

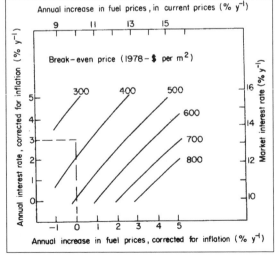

Focusing on the parameters which are found to be the most important in the linear situation close to the reference parameters, a direct parameter study may be made, allowing these parameters to deviate substantially from the reference values. Figures 7.30–7.33 show examples of such studies, in which pairs of parameters are varied while all the remaining parameters are left at the reference values. Figure 7.30 shows the rather trivial but important proportionality between break-even price and average power production, which again is the product of the average power in the wind and the average C_p or average conversion efficiency (including all system losses) of the wind energy converters. The reference parameters roughly correspond to the conditions at a height of 50 m at Danish coastal sites (cf. Figs. 3.34, 3.40, 6.60).

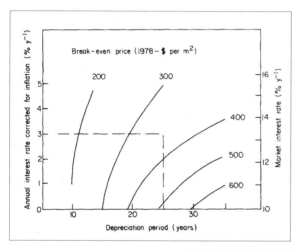

Figure 7.32. See legends to Figs. 7.30 and 7.31 (based on Sørensen, 1975c).

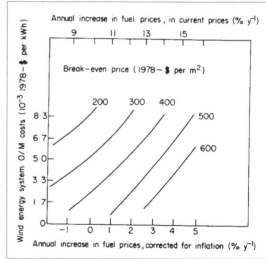

Figure 7.33. See legends to Figs. 7.30 and 7.31 (based on Sørensen, 1975c).

In Fig. 7.31, the parameters focused upon are the fuel price escalation (upper scale in current prices, lower scale corrected for an assumed inflation rate of 0.1 per year) and the interest rate (left-hand scale corrected for 10%

annual inflation, right-hand scale directly giving market interest). Generally speaking, Fig. 7.31 indicates that the break-even price decreases less than linearly with interest rate in fixed prices and increases more than linearly with fuel price escalation in real terms (i.e. corrected for inflation). In Fig. 7.32, the parameters focused upon are the depreciation period (for which an investor may want a value shorter than the physical lifetime of the equipment) and the interest rate. Again, the break-even price is a substantially non-linear function of the parameters. In Fig. 7.33, the parameters chosen are the fuel price escalation and the operation and maintenance cost of the wind energy system, with the latter parameter probably being at present the most uncertain one pertaining to the wind energy system, and also one that is related to the physical lifetime, in that spending more on maintenance may increase the useful life of the system.

It is interesting that the calculation described above was made in 1974-1975 when the first 25-kW Danish wind turbines were just under construction. The notion of megawatt machines operating at heights above 50 m turned out to be quite close to present mature wind turbine technology, and the estimated break-even cost is actually very close to the price today, if the $-values are interpreted as current dollars (European Commission, 2001).

Additional costs of regulation

If the contribution of wind energy is more than marginal, and still no energy storage is possible, then the regulation of the fuel-based back-up system must be considered. This was done in section 6.3.2 in physical terms. The result was the suggestion of a regulation procedure based on extrapolative predictions of wind power production. The penalties of having a wind energy contribution were described in terms of an extra number of start-ups of base and intermediate load facilities and a number of unplanned peak-unit calls for which the energy and maximum power requirements were calculated. Finally, some of the wind energy produced could not be used and became lost. These results were summarised in Figs. 6.64–6.66.

In order to discuss the economy of such a mixed wind/fuel electricity generating system, the cost of start-ups of fuel-based units, as well as the cost of providing peak power through special units or by import, must be known. Independently of this, it can immediately be said that the economy of adding the wind energy converters in this fuel-saving way deteriorates in proportion to the wind energy production that is lost. If the economy is discussed in terms of the break-even price of the wind energy converter, relative to the fuel saved, as done above with the assumption of only a marginal wind energy capacity, I_{wind} [cf. (6.6)], then the effect of not being able to utilise all the energy produced will be to multiply the break-even price by the factor C_{wind}/I_{wind} [where C_{wind} was defined in (6.7)]. Another factor which is evident from Fig. 6.66 is that for different amounts of installed wind capacity, I_{wind}, the mix of fuel-based units being replaced by wind will be different.

In order to illustrate how some of these relations may be incorporated into the direct economic evaluation, the calculation of the wind converter break-even price is repeated as a function of I_{wind} by making the following assumptions. The useful wind power production has an annual average of 150 W m^{-2} × C_{wind}/I_{wind}. The initial fuel price, which in Table 7.22 was $0.006/0.316 = 0.019 per kWh(e), is taken as $0.020/kWh(e) for base load units, $0.024 /kWh(e) for intermediate load units and $0.060/kWh(e) for peak units or equivalent imports. No similar differentiation is made for the fuel-related working costs, but the operation and maintenance cost of the wind energy system is kept at $0.0033 per kWh produced, implying a cost of $0.0033 I_{wind}/C_{wind} per kWh delivered. All other parameters of Table 7.22 are kept at their reference value, but an additional term of the form (7.8) is included. It has the cost of the start-ups and peak-unit calls as p_0, and the present value factor is calculated with a price escalation rate, e, equal to that of the fuel. The start-up costs for base and intermediate load units depend on the size of such units. If the annual average load E_{av} is around 2.3×10^9 W (corresponding to 2×10^{10} kWh per year), then according to the assumptions made in section 6.3.2, the size of each fuel-based unit is 230 MW (rated capacity). The start-up cost comprises fuel to pre-heat the boiler, etc., but also some costs closely related to the organisation and size of the utility system (e.g. depending upon whether the additional start-ups can be accomplished by the personnel already on duty or whether extra labour or overtime expenses have to be added). Cost ranges of $1000–10 000 per start-up, as an average for units with a size average of 230 MW or higher, have been quoted by utility companies (private communication). For the calculations made here, two cases have been considered, viz. a start-up cost of $2000 and $10 000 per start, both for base and intermediate load units. The peak calls may also be characterised by a fixed start-up cost or, if import through grid connections is used, by a fixed charge for an unplanned call (in addition to the higher cost of electricity, which in both cases is reflected in the assumed threefold increase in fuel price for peak load). The actual cost structure for exchange of electricity, through international grid connections, for example, depends on previous agreements specifying the conditions for planned or unplanned exchange.

For simplicity, the present calculation will use a fixed charge per call identical to the charge for start-up of a base or intermediate load unit, i.e. $2000 or $10 000 per call (the number of peak-unit calls being given in Fig. 6.67), which may then also be assumed to include the additional costs of long-distance transmission (grid losses, conversion losses if DC transmission is used for part of the distance, etc.). If the start-up cost per start or per call is

p_u, and the number of starts and peak calls is $(n_u + n_p)$, then the annual average cost of extra fuel-based unit starts or calls is

$$p_0 = \frac{p_u(n_u + n_p)}{E_{av} C_{wind}},$$

per average unit of power delivered by the wind energy system.

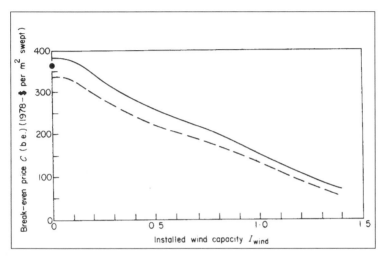

Figure 7.34. Break-even price of wind energy system without storage as a function of installed wind capacity. The contribution is not assumed marginal, and the back-up system regulation procedure based on wind production extrapolations (cf. section 6.3.2) has been used. The solid line corresponds to an assumed $2000 cost per start of back-up or call of peak-power units. The dashed line assumes $10 000 for the corresponding cost. The dot represents the result of the marginal supply calculation with the parameters of Table 7.22.

The results of the break-even cost calculation are presented in Fig. 7.34. For the lower value of start-up costs, these have an extremely small influence on the break-even price, and the drop in C(b.e.) as a function of I_{wind} is primarily due to the loss of wind-produced energy (shaded region in Fig. 6.66), either directly through the above-mentioned factor C_{wind}/I_{wind} or indirectly through the increased operation and maintenance cost per energy unit actually delivered to load areas. The cost influence of the changes in mix of base, intermediate and peak load units operated within the back-up system has a relatively small influence on the break-even cost, with a maximum change in fuel cost per average unit replaced by wind which reaches 14% over the interval of I_{wind} considered. In this connection it may be noted that the break-even cost of Fig. 7.34 does not approach exactly the same value for installed capacity I_{wind} approaching zero as that found for the reference parameters under the assumption of marginal wind energy contribution (Figs. 7.30–7.33, dashed lines). This is partly because of the modifications in fuel prices mentioned, but also due to the start-up rules fixed in section 6.3.2, which imply a modified pattern of operation of the fuel-based system, also in the limit of a very small wind energy contribution.

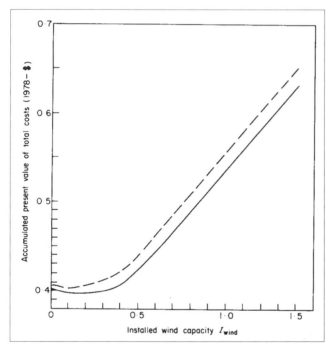

Figure 7.35. Accumulated present value of total costs of supplying 1 kWh of electricity for 25 years, for a mixed wind/fuel system without storage (as in Fig. 7.34), assuming the wind energy converters to cost $300 per m^2 swept by the rotor.

In order to illustrate the optimisation of a fuel/wind energy system mix, from a direct economic point of view, Fig. 7.35 gives the total costs (accumulated present value) as a function of system mix, I_{wind}, for the assumed capital cost of the wind energy converters, $300 per m^2 swept by the rotor (the 1998 wind turbine price is slightly lower for a 1500-kW machine and expected to fall as the market for megawatt machines becomes larger). With the assumptions made so far, the addition of wind energy converters to an installed capacity of 25–40% is seen to be competitive according to the accumulated present value criterion. The lower value corresponds to the assumption of $10 000 start-up costs, the higher value of I_{wind} to the assumption of $2000 start-up costs. On the other hand, the savings relative to zero wind energy contribution are minimal, indicating that wind energy systems would not be introduced unless parameters were changed (i.e. if the fuel price was expected to rise in fixed prices) or indirect economic factors were considered favourable for wind energy.

7.5.2 Wind power systems with storage

In the discussion of a mixed electricity generating system, comprising fuel-based as well as wind energy units (section 7.5.1), no credits were given to the wind energy system for replacing fuel-based installed capacity, i.e. it was assumed that the fuel-based units should be capable of delivering the maximum expected power, irrespective of the installed wind energy capacity. This would seem very reasonable for an isolated utility system, since the peak demand may occur at a time of zero wind power production, and no storage of energy was assumed to take place. However, there are two factors that may modify the situation for the types of utility systems presently found. One is that a surplus of fuel-based generating capacity is usually installed, because of the possibility of failure and repair of individual plants. Thus, the probability of finding a given fuel-based unit available when needed is not unity, although it is likely to be substantially higher than the probability of finding the wind energy converter capable of delivering the desired power at a given time. The other important factor is the presence of grid connections, often linking very large total systems, which geographically may cover areas of different climate as well as different load distributions (particularly if the connections extend in a longitudinal direction, so that the time difference makes the absolute times of peak demand non-overlapping). Generally speaking, the larger the combined grid system is, the greater the probability of finding available capacity somewhere in the system. In many cases, utilities take advantage of strong grid connections to reduce the reserve capacity within their own system to a value lower than that necessary for an isolated system. It also means that units with quite low reliability can be accepted in the system.

Typical *availability factors* (annual fraction of time during which the plant is operating or is ready to go into operation) are above 90% for fossil fuel units if planned shut-downs for maintenance are disregarded and about 80% if they are included. The corresponding figures for nuclear fission plants are a little lower. For an individual wind energy generator placed at a good site, the availability factor may be 80%; for a system of dispersed generators, it may be nearly 100% (cf. Fig. 6.60), but this represents just the availability of a non-zero production. The availability of the annual average power may be in the range 40–60% of time, for different locations. Maintenance shut-downs of wind energy generators are not expected to have a significant influence on the availability factor, since a wind power plant will consist of a large number of individual units that may be serviced independently at different times.

It follows from the above that it may be possible to reduce additional installed capacity to a certain degree when wind energy converters without storage are added to a grid system. By how much depends on the grid connections and reserve capacity. If a time availability of 70% is judged acceptable, it follows from Fig. 6.60 that nearly 20% of the average wind power capacity may be credited with the cost of installing a similar capacity of fuel-based units.

If wind turbines or other intermittently generating renewable energy converters are to provide more firm power, energy storage facilities or special fuel-based back-up units have to be added in order to guarantee firm power availability. If the purpose of introducing the renewable energy systems is to reduce the use of fuels, only storage can provide a long-term solution.

A number of storage types were briefly mentioned in Chapter 5, and the behaviour of wind energy systems including storage facilities of varying size was considered in section 6.3.2.

Two modifications have to be made in the direct economic evaluation in order to take into account the presence of a storage facility. One is to include the direct costs of the storage installation, which may be achieved by adding the cost of the wind energy generator and the cost of the storage accompanying a given installed wind capacity. The other is to add the cost of operating and repairing the storage facility, which may be treated along with the operation and maintenance cost of the converter itself. Furthermore, the inclusion of a storage facility makes it likely that the wind energy system will be able to replace not just fuel but also conventional capacity, as described above. This implies that a term must be included in the break-even price calculation which reflects the capacity credit fraction A, i.e. the fraction of the capital cost of the full back-up system which is rendered superfluous by the wind energy system.

Working entirely in prices referred to a fixed moment in time (i.e. corrected for inflation), the terms successively described above in the calculation of the break-even price may be combined to give the expression

$$\frac{C_{system}(b.e.)}{8.76\,E_{system}^{net}} + \sum_{i} p_{system,i}\,N(n,r,e_i) = \sum_{j} \frac{A_j C_{alt,j}}{8.76\,E_{alt,j}^{net}} + \sum_{j} p_{alt,j}\,N(n,r,e_j). \quad (7.11)$$

The expression is designed to express every term in monetary value for the net production of one kWh of energy every year during the depreciation period of n years. In order to ensure this, the costs C must be in, say, \$ or euro for a given size system (e.g. expressed as m² swept by wind energy generator or as kW rated power), and E must be in watts of net power production for the same size system. The annual payments p must be in units of \$ (or other currency as used in the other terms) per kWh of net energy production. The subscript "*alt*" represents the alternative (e.g. fuel-based) system being compared with, and $p_{system,i}$ for the (renewable energy) system includes terms giving the cost of operation and maintenance of both converter and storage and terms describing eventual penalties in terms of increased costs of regulating the fuel-based back-up system, whereas the $p_{alt,j}$ on the alternative system side includes operation and maintenance related to both equipment and fuels, as well as terms giving the fuel costs themselves, for each of the units forming the alternative system. The capacity credit of the renewable energy system is split on each of the alternative units j, with a fraction A_j specifying the capacity of a particular unit (or unit type, such as base, intermediate or peak unit), which is being replaced. The present value factor $N(n,r,e_i)$ is given as in (7.8), where n is the depreciation time in years, r is the annual interest rate, and e_i is the annual price escalation above inflation, for the ith component of payments (fuels, raw materials, labour, etc.).

In order to illustrate the calculation of the system break-even price for a wind energy system with a certain size of storage, a relation between the storage size and the capacity credits A_j must be established. The storage sizes are described in section 6.3.2 in terms of the number of hours during which the storage is able to absorb the average production of the wind energy converters, E_0. As discussed above, based upon Fig. 6.60 a 20% credit may be given to the current Danish network of wind energy converters in the absence of storage. The single-turbine study of the effect of adding a storage facility (shown in Fig. 6.72 for a loss-free storage) suggests as a first approximation to relate the capacity credit fraction in a linear fashion to the percentage of time during which the average power is available and in such a way that the credit is near zero in the absence of storage. Since, on the other hand, the capacity credit must be 100% if the average power (which is the proper system rating for a wind energy system with storage) is available 100% of time, it is simply assumed that the capacity credit fraction A increases linearly from 0 to 100% as the availability of the average power (eventually multiplied by the load-variation factor $C_1 C_2$ as in the load-following mode also illustrated in Fig. 6.72) increases from 40 to 100%. With this model, A is now known as a function of storage size, and the wind system break-even price can be calculated from (7.11) as a function of storage size, as shown in Fig. 7.36.

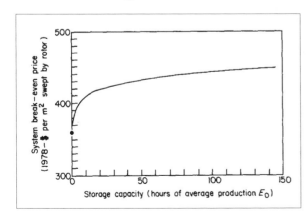

Figure 7.36. Break-even price for wind energy system with energy storage, relative to fuel-based system, as a function of storage capacity (cf. text for assumptions made). The dot on the left-hand axis corresponds to the case of not giving any credits for capacity displaced by the wind energy converters.

Only a single fuel-based unit type has been assumed, as in the reference calculation based on Table 7.22 (the parameters of which have been assumed), implying that there is only a single capital cost term on the right-hand side of (7.11) (and a single $A_j = A$). If the capital cost C_{alt} of the fuel-based system is given in $ per W (rated power), then E_{alt}^{net} is simply the load factor (watts produced divided by watts rated, or annual energy production divided by the energy production corresponding to constantly being at the rated power level). Based on information in the literature (Little, 1975; Blegaa et al., 1976), the capital cost of oil- and coal-based units lies in the range of $300–660 per kW rated power (depending primarily on the level of pollution control), and that of light-water fission units at about $730 per kW or higher (without initial core, but including present value of interest payments over the extended construction period). Load factors are in the range 0.4 to 0.8, depending on the use as intermediate or base load facilities. For the present crude estimate of capacity credits for a wind energy system, the round figures of C_{alt} = $0.5 per W and E_{alt}^{net} = 0.7 have been used.

Figure 7.36 shows that the break-even price of the converter/storage system initially increases steeply with storage capacity, but then evens out. The increase in system break-even price due to the capacity credits amounts to some 15–20%, which is the capital that would be available for the storage facility, provided that the converter itself "uses up" the break-even price found in the absence of capacity credits (the $360 per m^2 swept by the rotor, which was derived for a marginal contribution, using the reference parameters of Table 7.22). This is even assuming that no losses are introduced by the storage system, and since storage facility prices are expected to rise with increasing storage capacity, one expects the existence of the maximum storage size which can be afforded.

Using a 10-h storage as an example, Fig. 7.36 shows an excess of break-even price amounting to $50 (over the $360 for no storage), which would be available for the storage with each square metre of rotor swept converter area. The average power production from 1 m^2 is 150 W, according to Table 7.22, and thus the energy that can be held in a 10-h storage is 1.5 kWh. The capital available to the 10-h storage is thus about $33 per kWh.

Estimates of actual storage costs of different types have indicated probable values (in most cases assuming further industrial development efforts) of about $300 per kW or $30 per kWh for a 10-h storage (cf. Sørensen, 1977, 1978a). The maximum storage within the break-even excess price would thus be about 11 h of average power production. If the cost of the wind energy converter itself is not $360 m^{-2}, but $300 m^{-2} as assumed in Fig. 7.35, then a storage size of about 24 h will still ensure system break-even with the fuel-based alternative considered.

The structure of the system cost C_{system} may be specified as

$$C_{system} = C_{converter} + C_{s,in} + C_s + C_{s,out},$$

where $C_{s,in}$ is the cost of the equipment transferring surplus energy to the storage (e.g. electrolysis unit for hydrogen storage), C_s is the cost of the storage itself (e.g. the hydrogen containers) and $C_{s,out}$ is the cost of retrieving the energy as electricity (e.g. fuel cell unit in the hydrogen example). For some types of storage, the equipment for inserting and withdrawing energy from the storage is the same (the two-way turbines of a pumped hydro installation, the electro-motor/generator of a flywheel storage), and in some cases all three components may be physically combined (some battery types). If the three costs are distinct, only C_s will be expected to increase with the storage capacity. $C_{s,in}$ will be for a device rated at the maximum surplus power expected from the wind energy converter, and $C_{s,out}$ corresponds to a device rated at the maximum power deficit that is expected between current wind power production and load.

If finite losses occur in the storage and retrieval processes, the wind energy system will supply less energy over the year than the production by the converters indicates. The losses associated with a given storage type are usually known (cf. Table 5.5), but the average system loss further depends on the magnitude of the fraction of energy passing by way of the storage. For a load-following system subjected to the conditions of the simulation examples of section 6.3.2, including a 10-h storage with cycle efficiency of 90%, plus a 1250-h storage with cycle efficiency of 50%, the system efficiency is found to be 74% (Sørensen, 1978a), so that E_{system}^{net} in (7.11) should be taken as 0.74 times the average production (of 150 W m^{-2} in the reference case considered above). Since the reduced system efficiency due to storage losses also affects the converter part of the break-even price, the direct economic assessment in terms of accumulated present values leaves little room for introduction of storage types with low cycle efficiency and large capacity. Yet such systems are absolutely necessary in permitting large fractions of total load to be covered by wind or other renewable sources of intermittent nature. This again makes a case for seriously including indirect economic factors or extending the planning perspective to include periods of expected serious difficulties in access to fuel resources.

7.6 Suggested topics for discussion

7.6.1
Compare direct cost evaluations for a definite renewable energy system (such as the one considered in section 7.5.1), in which the annual interest rate has been taken as 15, 10, 5 and 0%.

7.6.2

Discuss differences and similarities between economic models based on dynamic simulation models (e.g. the simplified one presented by Forrester, 1971), models based on quasi-static assumptions (such as the one outlined in section 7.1.2 or the textbook input–output models, see e.g. Cockrane *et al.*, 1974), and finally, models based on scenarios and life-cycle assessment, as described in section 7.4.

7.6.3

Supplement the assumptions made in section 7.5.2 by a relation between the installed wind energy capacity as a fraction I_{wind} of the average load and the size of storage most appropriate (this in particular implies deciding on the capacity credit factor A to aim for, as a function of I_{wind} and coupled with expectations of future escalation of fuel prices). With this at hand, construct accumulated present value curves as functions of installed wind capacity, as done in Fig. 7.35 for the example without energy storage.

7.6.4

Compare centralised and decentralised generation of electricity by means of solar cells. For the centralised system, one may assume support structures made of concrete or of steel frames. The projected cost for support structures and land acquisition is responsible for roughly half the total cost. Consider alternatively decentralised units mounted on individual roofs of existing buildings. The cost of support structures and land would be saved. On the other hand, mounting and controls may be more expensive.

Try to collect costs for the relevant items that may be different for the two systems, and discuss the indirect economic factors relevant to the issue.

Extend the economic discussion to include the requirements for energy storage, based on data for variations in expected solar energy production at a location of your choice.

If wind converter production data for the same region are available, discuss the advantages that may be obtained by having the solar cells and wind energy converters feed into the same energy storage system.

7.6.5

Discuss the economy of central energy conversion followed by transmission, relative to decentralised conversion at the load areas, separately for heat and for electricity.

As an example, construct a small model including possible differences in costs of conversion equipment, depending on unit size, as well as possible differences in conversion efficiency. To this add assumptions regarding the cost of and losses associated with energy transmission of heat or of electricity.

Identify factors of indirect economy that may have particular relevance for the discussion of centralised and decentralised conversion (e.g. supply security in various cases of system failure).

7.6.6

Try to carry through a life-cycle analysis of some energy system of interest to you. First make an inventory of the possible types of social and environmental impacts and the pieces of equipment responsible for these impacts, then try to quantify each in physical terms, and finally, discuss the possible monetarisation of impacts. Did the results become less accurate by monetising?

CHAPTER

8

WINDING UP

The book so far has tried to give the background for renewable energy sources and to describe their occurrence and devices for using them, leading up to a discussion of the role of renewable energy in current and future energy systems, depending on social values and the economic rules used to judge the viability of competing solutions. In this concluding note, I shall make some observations regarding the status of renewable energy development and generally give my personal evaluation of the further requirements, technically as well as institutionally.

Conversion of renewable energy sources

For hydropower, development is expected to be largely restricted to efforts to deal better with environmental and social impacts. This could lead to emphasis on smaller, cascading schemes instead of the very large dam installations seen in the past. Still, the basic technology must be considered as fully developed, and as the cost of environmentally sound hydro schemes is rarely lower than that of fossil plants, the expansion rate of hydropower globally has slowed to nearly zero. There are still possible new sites available, and if social acceptance can be regained after the disasters of the past (flooding the homes of hundreds of thousands people and destroying treasures of cultural heritage – most recently in Turkey), then environmentally integrated hydro could add a further substantial amount to the global renewable energy supply.

In the case of wind turbines, a further development along current trends, particularly in materials, will allow somewhat larger horizontal axis turbines to be built. Although other wind conversion techniques could become viable, the success of the horizontal axis concept is likely to keep it in a leading position. Adjustments in blade profile and regulation features of the blades are likely to follow the widespread interest in off-shore wind. Currently, essentially the same turbines are sold for deployment on land and at sea. However, the wind conditions at sea are so different that the land design, typically chosen to optimise total annual production at good sites on land, far from does so when the same turbine is placed off-shore (cf. the difference between the two patterns in Figs. 4.91 and 4.92 or in Figs. 6.58 and 6.59, as regards the number of hours where the cut-off ceiling in power is reached). But these concerns still involve only small adjustments to a technology already close to maturity: the possibility of working with different blade designs for different locations, and control of pitch, e.g. by being able to turn the entire blade along its axis. The latter has been tried in commercial turbines and gives a few per cent improvement in average efficiency. Thus, the main challenge in wind turbines is improvement of materials, particularly for blades but also for cost reduction of foundations (particularly relevant for off shore turbines, where foundation costs constitute a fairly large fraction of the total). Such advances in blade construction will allow turbines with rated power somewhat larger than the 2–3 MW typical at present. Furthermore, as with any mature technology, continued small improvements in cost per produced kWh with time can be expected.

Entirely new concepts could be imagined for harvesting the still higher wind power levels found at open sea, i.e. much further off-shore than exploited by current machines standing on the sea floor. These could be combined wave- and wind-capturing devices, floating but kept within a restricted area by some advanced navigation system. The wind turbines could be shrouded or ducted constructions, while the wave energy part is likely to use dual air turbines. No developments in this direction have been successful so far. Wave energy devices for near-shore employment have not successfully stricken the balance between the necessary strength and economic power production, and no large mid-ocean wave converters have reached interesting sizes for energy production (only niche devices such as Masuda's air pump, Fig. 4.96, for buoy light generators have showed acceptable performance).

A number of biomass conversion methods are close enough to direct economic viability for the inclusion of indirect economics in cost comparisons to make them acceptable. This is true of biogas plants, particularly on a communal scale and as part of general waste management schemes. Gasification schemes followed by hydrogen or liquid biofuel production are also becoming viable, depending on the valuation used for the externalities of current fuel-based energy systems. Without considerations of indirect economics, such schemes are not currently viable, as demonstrated by the attempts to produce gaseous and liquid fuels from coal and sugar crops. However, in addition to altering the economic valuation, there is certainly room for technology improvements in gasification, cleaning and chemical reaction steps, involving device construction and possibly use of new pathways.

The entire bio-energy field may be in for a radical re-structuring if the current trends of valuing high food quality continues and spreads to regions not currently being concerned. This may alter agricultural practices, both for plant and for animal production, and lead to new concepts of integrated food and energy production based upon ecological principles, as distinct from the concept of dedicated energy crops.

Geothermal resources used in a sustainable way can be a stable source of low-temperature heat, and costs already appear acceptable with some consideration of externalities for the alternatives. New projects of district heating by geothermal energy are coming on-line, e.g. the Amager plant serving the existing Copenhagen heat network. The market for such geothermal solutions depends on the viability of district heating. If the full potential for making existing and new buildings highly energy efficient is realised, there will be very few locations in the world where district heating is economically attractive (probably only city centres with dense, high-rise buildings).

As regards solar energy, the thermal applications are in some, not too high-latitude regions economically viable in an assessment including externalities for conventional solutions. But this also depends on whether the building is energy efficient in the first place. A large heat requirement in winter is little compatible with solar coverage, but if the passive and active efficiency features of the building are optimal, the heat load may be dominated by hot water use, which is more amenable to solar supply. This is particularly so for niche markets, such as vacation areas (e.g. in the Mediterranean region or similar places) where there may be lots of building space being inhabited only during the summer season.

For solar electricity, the situation is different, as the current costs of photovoltaic power systems are still too high even with generous consideration of indirect costs. However, the technical development is rapid in this area, and novel solutions under current development for conventional solar cells may bring about economically attractive solutions within the next one or two decades. In addition, there is a multitude of novel techniques for converting solar energy that could play a decisive role in the future. The use of organic dyes, currently in its infancy, with inherent low efficiency and short lifetimes could gain new momentum by replacing TiO_2 materials and ruthenium-based dyes with concepts based upon polymers and organic dyes, or the emerging three-dimensional photonic crystalline carbon structures, which would potentially allow an engineered establishment of suitable properties (Zakhidov *et al.*, 1998).

Above all, both direct and even more a life-cycle approach to economic valuation will generally make efficient use of energy more attractive in all areas of use, and this will generally improve the conditions for all renewable

energy types by making it possible to supply all required energy from renewables with a minimum strain on resources.

Energy storage and auxiliaries

The intermittency of several renewable energy flows makes it imperative to
include features in systems relying on renewable energy inputs that can as-
sure a match between demand and supply in time and place. For less than
100% renewable energy systems, the non-renewable units can be used for
back-up, but for pure renewable energy systems, storage of energy is necessary. To some extent, the storable biofuels can serve this purpose, but it is
highly likely that additional components will be required.

This brings into focus bulk storage development, as well as means for
transmission to and from such devices, and between load locations and supply locations, also when the closest ones are unavailable. Possible candidates
for a flexible conversion between electric power (i.e. the output energy form
of variable renewable sources from devices such as wind turbines or photovoltaic converters) and storable energy are (preferably reversible) fuel cells
combined with hydrogen storage. In certain areas, pumped hydro (or managed operation of combined hydro and intermittent renewable energy systems) can be used. Other storage forms such as flywheels and batteries are
not evidently suited for bulk or long-term storage. This leaves storage of hydrogen in pressure containers or much more economically in underground
caverns and aquifers as a central candidate for a general solution to the supply–demand mismatch problem. Considerable hopes were placed in utility
uses of advanced batteries, but the last 40 years of development have been
disappointing in this respect. During the last decade, advanced batteries
have finally emerged, albeit on a small scale suited for consumer appliances
(portable computers, cameras, mobile telephones, light sources, etc.).
Whether there is a road from these items of high unit (energy) price to bulk
applications in the energy sector remains to be seen.

A range of hydrogen storage techniques are emerging as suitable for medium-term energy storage, in combination with fuel cell conversion. These
include underground storage of hydrogen (in aquifers or flushed-out salt
domes). Small-scale compressed hydrogen storage in containers for use in
the transportation sector may in the future be replaced by storage in media
such as metal hydrides capable of easily incorporating and releasing hydrogen. Use of hydrogen as an energy carrier is linked to the development of
dependable and affordable fuel cell energy converters.

Fuel cell technology for a while rested with the alkali-based cells originally developed for space applications (until photovoltaic panels took over).
Later, molten carbonate cells were prematurely marketed and then withdrawn. Recently, proton-exchange membrane cells for automotive applications, and high-temperature, solid oxide cells for utility applications have

entered a very active phase of development and planning for market intro-duction, and the expectations of substantial price reductions underlie the considerable investments made (cf. e.g. Daimler-Chrysler-Ballard, 1998). Clearly, there is no guarantee that the substantial price reductions needed will be forthcoming. Fuel cell technology has many similarities to battery technology, so one may fear encountering similar disappointments. The further requirement of substantial infrastructure changes in order to use hy-drogen both in vehicles and for stationary uses is a further impediment for the development of economically viable fuel cell-based energy supply sys-tems (cf. Sørensen *et al.*, 2003; Sørensen, 2004a).

Market penetration of renewable energy

The ongoing slow process of market take-over by renewable energy tech-nologies has been linked to political debates and concrete actions to speed up or slow down the transition process. Political initiatives to facilitate intro-duction have been based on the belief that fossil resources are close to their peak production and will gradually have to be marketed at higher and higher prices, as they physically get more difficult to extract or in some re-gions simply get exhausted. In recent decades, a substantial part of the world has acknowledged the reality of increased greenhouse warming as a result of fossil fuel combustion and used this an the key argument for a rapid intro-duction of renewable energy sources.

One should also take notice of the strong vested interests opposing re-newable energy development: the fossil and nuclear fuel lobbies, large utility preservation organisations and economic advisors opposing any kind of fair market establishment that involves consideration of the different levels of environmental and social indirect costs associated with different energy technologies. It is not enough to have found the overall best solution. There must also be created a political will to implement it despite the possible ef-forts of loud minority opposition groups. This is a basic challenge to demo-cratic decision-making.

Looking at the economically most viable renewable energy technologies, such as wind converters, the market introduction has been characterised by stop-and-go policies in individual countries. Wind energy is an economically viable solution at favoured sites, which can be found in most countries of the world. However, for various reasons the development has been slower than a proper economic evaluation would suggest. Foremost among these reasons is perhaps the liberalisation of the power utility market, which in some countries have been carried through not on the basis of up-to-date, life-cycle assessment type economic thinking, but rather according to the primitive economic theories prevailing in the 18th century in complete disregard of non-direct cost contributions. Private companies are given the licence to harvest the direct economic benefits, while society picks up the bill for externalities.

It should be said at once that life-cycle assessment does not offer four-digit economic comparisons relieving policy-makers from having to reflect upon their actions. Most life-cycle assessments of energy systems are indicative only, but this is also sufficient to state that wind power and certain other renewable energy systems in many places are competitive today, relative to fossil and nuclear alternatives. It is then the job of the responsible government to regulate the marketplace in such a way that individual customers see relative prices the same way as does the life-cycle assessment.

In addition to wind power, a number of biomass-derived energy forms are cost competitive in a life-cycle perspective, including solutions for the otherwise elusive individual transportation sector. Yet, the preference today is to use biomass for simple combustion, an unworthy application with often severe pollution effects.

The political dimension

The conditions for energy transitions away from fossil fuels are different in different parts of the world. In some countries, there is broad political consensus on not redistributing wealth through taxation, This implies a need to keep the price of basic goods, including energy, at such a low level that a large fraction of the population is not pushed below the poverty line. In other parts of the world, there is rather a tradition for significant income redistribution through taxes, and consumers have for many years been accustomed to energy prices far above the direct cost, made possible because the redistribution policy allows such prices to prevail without marginalising large groups of society. In recent years, the general taxation has increasingly been replaced by "fair taxation", meaning that the governments levy taxes on goods such as energy according to estimated *de facto* externality costs of pollution, climate impact, supply insecurity etc. The countries committed to keeping energy prices low are more likely to consider extreme measures, such as going to war, in order to secure access to cheap energy sources from politically unstable parts of the world. Similar remarks can be made in reference to the efforts to encourage more efficient use of energy, although the extra cost of efficiency improvements are generally much lower than the cost of introducing new energy sources.

Seen in this perspective, there is little reason to be optimistic based on the rather spectacular results obtained so far by quite modest investments in renewable energy technology. Technical feasibility, decent price and general public acceptance seem insufficient to persuade policy makers and their industrial basis in many parts of the world. What additional arguments should be brought forward? Air pollution, greenhouse gas emissions, fossil resource exhaustion, nuclear accident and radioactive waste unacceptability have all failed to impact sufficiently on actual political decisions, although few would disagree that all the mentioned ones are problems "in the long run". The wicked circle behind this behaviour is connected with political governance being increasingly short-range motivated, probably at least in part due to the influence of media treatment, by media no longer independent and unbiased in their analysis and criticism, but owned by and serving special interests. So it seems tempting to conclude that while several renewable energy solutions are already ready for the market, the playing rules in the marketplace have to be changed to include long-term human interests if the marketplace is to become a level playing field. This again calls for political action, which will be taken only when and if the rules of the political game have also been modified to include long-term human interests.

REFERENCES

Aarkrog, A. (1971). *Health Physics* **20**, 297–311.

Aastrand, C., Mose, O., Sørensen, B. (1996). Wind power: valuation and finance, in "1996 European Union Wind Energy Conference", pp. 138–139. Stephens and Assoc., Bedford.

Achard, P., Lecomte, D., Mayer, D. (1981). Characterization and modelling of test units using salt hydrates, in "Proc. Int. Conf. on Energy Storage", Vol. 2, pp. 403–410. BHRA Fluid Engineering, Cranfield, UK.

Ackermann, T. (2002). Transmission systems for offshore wind farms. *Renewable Energy World*, **5**, No. 4, 49–61.

Adolfson, W., Mahan, J., Schmid, E., Weinstein, K. (1979). In "Proc. 14th Intersociety Energy Conversion Engineering Conf." pp. 452–454. American Chemical Society, Washington, DC.

Alakangas, E., Hillring, B., Nikolaisen, L. (2002). Trade of solid biofuels and fuel prices in Europe, in "12th European Biomass Conf.", pp. 62–65. ETA Firenze & WIP Munich.

Alcamo, J., Shaw, R., Hordijk, L. (eds.) (1990). "The RAINS Model of Acidification: Science and Strategies in Europe". Kluwer Academic Publ., Dordrecht.

Alexandratos, N. (1995). "World Agriculture: Towards 2010". An FAO study. Wiley, Chichester.

Alfven, H. (1965). *Rev. Mod. Phys.* **37**, 652–665.

Alich, J., Inman, R. (1976). "Energy", Vol. 1, pp. 53–61. Pergamon Press, London.

Allen, C. W. (1973). "Astrophysical Quantities". The Athlone Press, London.

Almquist, E. (1974). *Ambio* **3**, 161–167.

Alsema, E. (1997). Appendix to "Environmental Aspects of PV Power Systems", Utrecht Universiteit VNS Report no. 97072. (irreg. pag.)

Alsema, E., Engelenburg, B. van (1992). Environmental risks of CdTe and CIS solar cell modules. in "Proc. 11th EC Photovoltaics Solar Energy Conference", Montreux, pp. 995–998. Harwood Academic Publ., Newark, NJ.

Alyea, F., Cunnold, D., Prinn, R. (1975). *Science* **188**, 117–121.

Amann, M., Dhoondia, J. (1994). Regional Air Pollution Information and Simulation (RAINS–Asia), User's manual, 34 pp. World Bank and IIASA, Laxenburg.

Andersen, B., Eidorff, S., Lund, H., Pedersen, E., Rosenørn, S., Valbjørn, O. (1974). "Meteorological Data for Design of Building and Installation: A Reference Year", Danish Govt. Build. Res. Inst., Report No. SBI–89.

Andersen, J. (1997). Private communication of data from NORDPOOL ASA.

André, H. (1976). Institute of Electrical and Electronics Engineers, *Transactions* **PAS–95**, 4, 1038–1044.

Andreen, H., Schedin, S. (eds.) (1980). "Den nya energin," Centrum for Tvärvetenskap/Forlaget Tvartryk, Göteborg.

Angrist, S. (1976). "Direct Energy Conversion", 3rd edn. Allyn and Bacon, Boston.

Anonymous (1980). *New Scientist* 11 September, p. 782.

Anonymous (2002). Electricity store. ABB Review No. 4, pp. 62–65.

APACE (1982). Technical Information Bulletin PA/111/1, Apace Research Ltd., Hawkesbury, NSW, Australia.

Appropriate Technology Development Organization (1976). "Gobar Gas", Govt. of Pakistan, Islamabad (undated).

Aqua-Media (1997). Hydropower & Dams, Map prepared for *The Int. J. on Hydropower & Dams.* Aqua-Media Int. Ltd., Sutton UK.

Armstrong, R., Bruce, P. (1996). Synthesis of layered $LiMnO_2$ as electrode for rechargeable lithium batteries. *Nature* **381**, 499–500.

Arnason, G. (1971). In "Man's Impact on the Terrestrial and Oceanic Ecosystems" (W. Matthews, F. Smith, E. Goldberg, eds.). MIT Press, Cambridge, MA.

Asmussen, J. (1975). In "Proc. 2. Workshop on Wind Energy C.onversion Systems" (F. Eldridge, ed.), pp. 112–120, MITRE Corp., USA.

Athey, R. (1976). *Solar Energy* **18**, 143–147.

ATS (2003). Spheral solar technology, website http://www.spheralsolar.com

ATV (1975). *Wind power.* Prepared by C. Estrup, J. Fischer, F. Hvelplund, M. Jensen, M. Johansson, S. Kofoed, K. Lykkegaard, N. Meyer, T. Myhre, B. Pedersen and B. Sørensen for the Danish Academy of Technical Sciences, Lundtofte.

ATV (1976). *Wind power 2.* Prepared by F. Hvelplund, H. Jels, M. Johansson, S. Kofoed, N. Meyer, B. Pedersen, H. Petersen and B. Sørensen for the Danish Academy of Technical Sciences, Lundtofte.

Baader, W., Dohne, Brenndorfer (1978). "Biogas in Theoric und Praxis". Landwirtschaftsverlag, Darmstadt.

Bahadori, M. (1977). "Solar Energy Utilization for Developing Countries", Int. Solar Build. Technology Conf., London, July.

Bahcall, J. (1969). *Sci. Am.* July, 28–37.

Bahcall, J., Davis, R. (1976). *Science* **191**, 264–267.

Baker, A. (1987). Tidal power. *UK IEE Proceedings* **134**, 392–398.

Bakken, K. (1981). System Tepidus, high capacity thermochemical storage/heat pump, in "Proc. Int. Conf. on Energy Storage, Vol. I", pp. 23–28. BHRA Fluid Engineering, Cranfield, UK.

Ballhausen, C., Gray, H. (1965). "Molecular Orbital Theory". Benjamin, New York.

Balling, N. (1976). *J. Geophys.* **42**, 237–256.

Banas, J., Sullivan, W. (1975). Sandia Laboratories Energy Report, SAND75–0530.

Banas, J., Sullivan, W. (eds.) (1976). "Sandia Vertical-Axis Wind Turbine Program", Quarterly Report Oct.–Dec. 1975, SAND76–0036.

Bandel, W. (1981). A review of the possibilities of using alternative fuels in commercial vehicle engines, in "Int. Conf. on Energy Use Management, Berlin, 1981", Session H–3, Daimler-Benz AG, Stuttgart.

Banner, M., Melville, W. (1976). *J. Fluid Mech.* **77**, 825–842.

Barbier, E. (1999). Geothermal energy: a world overview. *Renewable Energy World* **2**, 148–155

Barnes, C. (1971). In "Advances in Nuclear Physics" (M. Baranger and E. Vogt, eds.), Vol. 4, pp. 133–204. Plenum Press, New York.

Barnett, T., Kenyon, K. (1975). *Rep. Prog. Phys.* **38**, 667–729.

Barnett, T., Wilkerson, J. (1967). *J. Mar. Res.* **25**, 292–328.

Barnola, J., Raynauld, D., Korotkevich, Y., Lorius, C. (1987). Vostok ice core provides 160,000-year record of atmospheric CO_2. *Nature* **329**, pp. 408–414.

REFERENCES

Bartels, J. (1957). In "Handbuch der Physik" (Geophysik II) (S. Fliügge, ed.), Vol. 48, pp. 734–774. Springer Verlag, Berlin.

Basore, P. (1991). PC1D v.3 Manual and User Guide. Report 0516/rev. UC–274. Sandia National Laboratory, Albuquerque.

Baumgartner, T. (1993). Product life-cycle analysis. Current practice and methodological implications. In "Life-cycle Analysis of Energy Systems", pp. 172–181. Workshop Proceedings. OECD Publications, Paris.

Bay of Fundy Tidal Power Review Board (1977). Reassessment of Fundy tidal power. New Brunswick.

Beale, W. (1976), as quoted in Hughes (1976).

Bechinger, C., et al. (1996). Nature 383, 608–610

Beck, E. (1975). Science 189, 293–294.

Becke, A. (1993). Density-functional thermochemistry. III: The role of exact exchange. J. Chem. Phys. 98, 5648.

Beckman, W., Schoffer, P., Hartman, W., Lof, G. (1966). Solar Energy 10, 132–136.

Beesch, S. (1952). Ind. Eng. Chem. 44, 1677–1682.

Bell, B. (1998). Looking beyond the internal combustion engine: the promises of methanol fuel cell vehicles. Paper presented at "Fuel Cell Technology Conference, London, September". IQPC Ltd., London.

Benemann, J., Weare, N. (1974). Science 184, 174–175.

Berezin, I., Varfolomeev, S. (1976). Geliotekhnika 12, 60–73.

Berger, W. (1982). Climate steps in ocean history–lessons from the Pleistocene. In "Climate in Earth History" (W. Berger, J. Crowell, eds.), pp. 43–54. National Academy Press, Washington DC.

Berglund, M., Börjesson, P. (2002). Energy efficiency in different types of biogas systems, in "12th European Biomass Conf.", pp. 219–222. ETA Firenze & WIP Munich.

Berkner, L., Marshall, L. (1970). "The Encyclopedia of Geochemistry and Environmental Sciences" (R. Fairbridge, ed.), Vol. 4A, pp. 845–861. Van Nostrand, New York.

Bertness, K., Kurtz, S., Friedman, D., Kibbler, A., Kramer, C., Olsson, J. (1994). 29.5% efficient GaInP/GaAs tandem solar cells. Appl. Phys. Lett. 65, 989–991.

Besant, R., Dumont, R., Schoenau, G. (1979). Solar Age May, 18–24.

Betz, A. (1959). "Stromunglehre", G. Brown Verlag, Karlsruhe, BRD.

Beyer, H., et al. (1994). Journal Wind Engineering Ind. Aerodyn. 51, 111.

Bilton, T., Flowers, E., McCormick, R., Kurfis, K. (1974), pp. 61–67 in Turner (1974).

Biomass Energy Institute (1978). "Biogas Production from Animal Manure", Winnipeg, Manitoba.

Birck, C., Gormsen, C. (1999). Recent developments in offshore foundation design, in "Proc. European Wind Energy Conference, Nice", pp. 365–368 (Petersen, Jensen, Rave, Helm, Ehmann, eds.), James & James, London

Biswas, D. (1977). Solar Energy 19, 99–100.

Bjørnholm, S. (1976). "Energy in Denmark 1990, 2005", Report No. 7, The Intern. Fed. of Institutes for Advanced Study, Copenhagen.

Blackwell, B., Reis, G. (1974). Sandia Laboratories Report SLA–74–0154, Albuquerque, USA.

Blaxter, K. (1974). New Scientist 14 February, 400–403.

REFERENCES

Blegaa, S., Hvelplund, F., Jensen, J., Josephsen, L., Linderoth, H., Meyer, N., Balling, N., Sørensen, B. (1976). "Skitse til Alternativ Energiplan for Danmark", OOA/OVE, Copenhagen (English summary in *Energy Policy*, June 1977, pp. 87–94).

Bligh, T. (1976). *Building Systems Design* October/November, 1–22.

Blomquist, C., Tam, S., Frigo, A. (1979). In "Proc. 14th Intersociety Energy Conversion Engineering Conf.", pp. 405–413. American Chemical Society, Washington, DC.

Bockris, J. (1975). "Energy: The Solar-hydrogen Alternative". Australia and New Zealand Book Co., Brookvale, Australia.

Bockris, J., Reddy, A. (1973). "Modern Electrochemistry". Plenum Press, New York.

Bockris, J., Shrinivasan, S. (1969). "Fuel Cells: Their Electrochemistry". McGraw-Hill, New York.

Boeck, W. (1976). *Science* **193**, 195–198.

Boeniger, M., Briggs, T. (1980). Potential health hazards in the manufacture of photovoltaic solar cells. Ch. 43 in "Health Implications of New Energy Technologies" (Rom, Archer, eds.), Ann Arbor Science Publ., Ann Arbor, MI.

Bohr, A., Mottelson, B. (1969). "Nuclear Structure", Vol. 1. Benjamin, New York.

Boldt, J. (1978). "Solar powered water pump for the rural third world", Danish Technical University, Lab. for Energetics, internal report.

Bolin, B. (1970). *Sci. Am.* **223**, September, 124–135.

Bolin, B. (1974). *Ambio* **3**, 180–188.

Bolle, H.-J. (1972). "Topics in Atmospheric Structure and Composition", Lecture notes, The Int. School of Applied Physics (G. Fiocco, ed.), Frascati, Italy.

Bonefeld, J., Jensen, J., Anom. (2002). Horns Rev 160 MW offshore wind. *Renewable Energy World* **5**, No. 3, 77–87.

Bosanac, M., Sørensen, B., Katic, I., Sørensen, H., Badran, J. (2001). Photovoltaic/thermal solar collectors and their potential in Denmark. Report EFP 1713/00-0014 to the Danish Energy Agency, Copenhagen.

Boussinesq, J. (1877*). Mémoires présentées par div. savants a l'Acad. des Sciences de Paris* **23**, 46 (as quoted by Hinze, 1975).

Brehm, N., Mayinger, F. (1989). A contribution to the phenomenon of the transition from deflagration to detonation. VDI–Forschungsheft No. 653/1989, pp. 1–36. (website: http://www.thermo–a.mw.tu–muenchen.de/lehrstuhl/foschung /eder_gerlach.html).

Broda, E. (1975). *Naturwissenschaftliche Rundschau (Stuttgart)* **28**, 365–372.

Brooks, L., Niiler, P. (1977). *J. Mar. Res.* **35**, 163–175.

Brown, J., Cronin, J. (1974). "Proc. 9th Intersociety Energy Conversion Engineering Conference" (USA), Paper No. 749139.

Brown, C., Warne, D. (1975). Wind Power Report, Ontario Res. Found., Report P.2062/G.

Brown, S. (1998). The automakers' big-time bet on fuel cells. *Fortune Mag.*, 30 March, 12 pages (http://www.pathfinder.com/fortune/1998/980330).

Brüel, P., Schiøler, H., Jensen, J. (1976). "Foreløbig redegørelse for solenergi til boligopvarmning", Project report. Copenhagen.

Bryan, K. (1969). *Monthly Weather Rev.* **97**, 806–827.

Bryan, K., Cox, M. (1967). *Tellus* **19**, 54–80.

Bryson, R. (1971). "Climate Modification by Air Pollution", as quoted by Wilson and

Matthews (1971).

BTM (2001). International Wind Energy Development, March 2001, BTM Consult Aps., http://www.btm.dk

BTM (2002). International Wind Energy Development, March 2002, BTM Consult Aps.

Buch, H. (1954). "Hemispheric Wind Conditions during the Year 1950", Final Report, part 2 of contract AF–19(122)–153, Dept. of Meteorology, Mass. Inst. of Technology, Cambridge, USA.

Buchmann, I. (1998). Understanding your batteries in a portable world. Cadex Inc., Canada, http://www.cadex.com/cfm

Buchner, H. (1980). Thermal energy storage using metal hydrides. In "Energy Storage" (J. Silverman, ed.). Pergamon Press, Oxford.

Budyko, M. (1974). "Climate and Life". Academic Press, New York and London.

Burbridge, E., Burbridge, G., Fowler, W., Hoyle, F. (1957). *Rev. Mod. Phys.* **29**, 547–650.

Butler, D. (1997). Cogema's arrogance adds to la Hague's problems. *Nature* **387**, 839.

Butler, D., Maurice, J., O'Brien, C. (1997). Time to put malaria control on the global agenda. *Nature* **386**, 535–541.

Cacciola, G., Giodano, N., Restuccia, G. (1981). The catalytic reversible (de) hydrogenation of cyclohexane as a means for energy storage and chemical heat pump. In "Proc. Int. Conf. on Energy Storage", pp. 73–89. BHRA Fluid Engineering, Cranfield, UK.

Caldwell, M. (1975). In "Photosynthesis and Productivity in Different Environments" (J. Cooper, ed.), pp. 41–73. Cambridge University Press, Cambridge, UK.

Callen, H. (1960). *Thermodynamics*, Wiley, New York.

Calvin, M. (1974). *Science*, **184**, 375 – 381.

Calvin, M. (1977). Chemistry, population and resources, in "Proc. 26th Meeting of the Int. Union of Pure Applied Chemistry", Tokyo. Also Lawrence Berkeley Laboratory Report LBL–6433.

Canny, M. (1977). *Ann. Rev. Fluid Mech.* **9**, 275–296.

Carslaw. H. and Jaeger, J. (1959). "Conduction of Heat in Solids". Clarendon Press, Oxford.

Casida, M., Casida, K., Jamorski, C., Salahub, D. (1998). *J. Chem. Phys.* **108**, 4439.

Cavanagh, J., Clarke, J. and Price. R. (1993). Ocean energy systems. In "Renewable Energy Sources for Fuels and Electricity" (T. Johansson, H. Kelly , A. Reddy and R. Williams, eds.), pp. 513–547. Island Press, Washington,. DC.

CECSMUD (1982). Sacramento Municipal Utility District 100 MW Photovoltaic Plant. Draft Environmental Impact Report, Californian Energy Commission, State Clearing House # 81111253, Sacramento.

Ceder, G., Chiang, Y-M., Sadoway, D., Aydinol, M., Jang, Y.-I. and Huang, B. (1998). Identification of cathode materials for lithium batteries guided by first-principles calculations. *Nature* **392**, 694–696.

Chapman, P. (1974). *New Scientist* 19 December, 866–869.

Chapman, S. (1943). *Rep. Prog. Phys.* **9**, 92–100.

Chapman, D., Pollack, H. (1975). *Earth and Planetary Sci. Lett.* **28**, 23–32.

Charney, J., Stone, P. (1975). *Science* **187**, 434–435.

Chartier, P., Meriaux, S. (1980). *Recherche* **11**, 766–776.

Chen Li (1951). *Kho Hsueh Thung Paro* ("Science Correspondent") **2**, No. 3, 266.

REFERENCES

Cheremisinoff, N., Cheremisinoff, P., Ellerbusch, F. (1980). "Biomass: Applications, Technology, and Production". Marcel Dekker, New York.

China Reconstructs (1975). December issue, pp. 24– 27 (anonymous).

Chiu. H.-Y. (1964). *Ann. Phys.* **26**, 364–410.

Christensen, P. (1981). "Kemiske Varmelagre", Danish Dept. of Energy, Heat Storage Project Report No. 10, Copenhagen.

Christensen, L., Gudmundsen, H. (1993). "Bilismens fremtid". Danish Technology Council Report 93/3, Copenhagen.

CIESIN (1997). Gridded population of the world. David Simonett Center for Space Studies at University of Santa Barbara, California. NCGIA Technical Report TR-95-6; Consortium for International Earth Science Information Network website http://www.ciesin.org/datasets/gpw/globdem.doc.html

Claesson, S. (1974). Jordbrukstekniska Inst., Sweden, Medd. No. 357.

Clarke, F. (1981). Wave energy technology. In "Long-term Energy Sources" (R. Mayer and J. Olsson, eds.), pp. 1269–1303. Pittman Publ., Boston.

Claude, G. (1930). *Mech. Eng.* **52**, 1039–1044.

Clayton, D. (1968). "Principles of Stellar Evolution and Nucleosynthesis". McGraw-Hill, New York.

Clayton, R. (1965). "Molecular Physics in Photosynthesis". Blaisdell, New York.

Cline, W. (1992). "Global Warming: the Economic Stakes". Inst. Int. Economics. Washington, DC, 399 pp.

Clot, A. (1977). *La Recherche* **8**, March, 213–222.

Cloud, P., Gibor, A. (1970). *Sci. Am.* September, 111–123.

Cockrane, J., Gubins, S., Kiker, B. (1974). "Macro-economics". Scott, Foresman and Co., London.

Cokelet, E. (1977). *Nature* **267**, 769–774.

Collet, P., Frederiksen, E., Hoffmann, T., Madsen, G. (1976). "Boligers Luftskifte", Danish Technological Institute, Copenhagen.

Consoli, F., *et al.* (eds.) (1993). "Guidelines for Life-cycle Assessment: A Code of Practice". Society of Environmental Toxicology and Chemistry (SETAC).

Cooper, P. (1969). *Solar Energy* **12**, 3–8.

Coste, W. (1976). "Evaluation of Wind Power with Energy Storage", Northeastern Utilities (M. Lotker) internal report (Hartford, Connecticut).

Coty, U., Dubey, M. (1976). "The high potential of wind as an energy source", paper presented at the "2nd Annual Energy Symp.", Los Angeles Council of Eng. and Scientists.

Coulson, K. (1975). "Solar and Terrestrial Radiation". Academic Press, New York and London.

Cox, J., Giuli, R. (1968). "Principles of Stellar Structure", Vol. l, Gordon and Breach, New York.

Crafoord, C. (1975). "An estimate of the interaction of a limited array of windmills", Dept. of Meteorology, Univ. of Stockholm, Report DM–16.

Craig, R. (1965). "The Upper Atmosphere". Academic Press, New York and London.

CRES (2002). Wave energy utilization in Europe, Project report for the European Commission, Centre for Renewable Energy Sources, Pikermi.

Cultu, N. (1989a). Energy storage systems in operation. In "Energy Storage Systems" (B. Kilkis, S. Kakac, eds.), pp. 551–574. Kluwer, Dordrecht.

Cultu, N. (1989b). Superconducting magnetic energy storage. In "Energy Storage

Systems" (B. Kilkis, S. Kakac, eds.), pp. 551–574. Kluwer, Dordrecht.

Cutchis, P. (1974). *Science* **184**, 13–19.

Daimler–Chrysler–Ballard (1998). Fuel-cell development programme. Details: http://www.daimler–benz.com/research/specials/necar/necar_e.htm

Damen, K., Faaij, A., Walter, A., Souza, M. (2002). Future prospects for biofuel production in Brazil, in "12th European Biomass Conf.", Vol. 2, pp. 1166–1169, ETA Firenze & WIP Munich.

Danish Association of Boiler Owners ("Kedelforeningen") (1977). "Danmarks Energibalance 1976", Copenhagen.

Danish Department of Energy (1979). "Sæsonlagring af varme i store vandbassiner," Heat Storage Project Report No. 2, Copenhagen.

Danish Department of Environment and Energy (1996). "Danmarks Energifremtider", Copenhagen.

Danish Department of Public Works (1987). "Trafikundersøgelse 1986", Copenhagen.

Danish Electricity Supply Undertakings (1971). "Elværks-statistik 1970/71", Copenhagen.

Danish Energy Agency (1992). "Update on Centralized Biogas Plants". Danish Energy Agency, Copenhagen, 31pp.

Danish Energy Agency (1993). "District Heating in Denmark", Copenhagen, 58 pp.

Danish Energy Agency (1995). "Progress Report on the Economy of Centralized Biogas Plants", Copenhagen, 34 pp.

Danish Energy Agency (1996). "Biomass for energy – Danish solutions", Danish Energy Agency, Copenhagen, 34 pp.

Danish Energy Agency (1999). "Windpower in Denmark", available from website http://www.ens.dk

Danish Energy Agency (2002). Energy Statistics 2001. Danish Energy Agency, Copenhagen, also on http://www.ens.dk

Danish Meteorological Institute (1973). "Oceanographic Observations from Danish Light-vessels and Coastal Stations, 1972", Charlottenlund, Denmark.

Danish Road Directorate (1981). "Trafikulykker 1980", Copenhagen.

Danish Statistical Office (1993). "Miljø", Report 93/1, Copenhagen.

Danish Statistical Survey (1968). "Statistical Inquiries, No. 22" (Agricultural Statistics 1900–1965, Vol. 1), Copenhagen.

Danish Statistical Survey (1972). "Statistical Information No. 9" (Agricultural Statistics 1971), Copenhagen.

Danish Technology Council (1993). "Trafikkens Pris". Copenhagen.

Danish Transport Council (1993). Externaliteter i transportsektoren. Report 93–01, Copenhagen.

Danish Windpower Industry Association (2003). Web: http://www.windpower.dk

Danish wind turbine Price Lists and Overviews (1981–2002). Alternately published by Danish Energy Agency, Renewable Energy Information Secretariat at Danish Institute of Technology, and Energi- og Miljødata Inc., http://www.emd.dk

Dansgaard, W. (1977). *Naturens Verden*, 17–36 (Rhodos Publ. Co., Copenhagen).

Dansgaard, W., Johnsen, S., Reeh, N., Gundestrup, N., Clausen, H., Hammer, C. (1975). *Nature* **255**, 24– 28.

Darmstadter, J., Teitelbaum, P., Polach, J. (1971). "Energy in the World Economy". John Hopkins Press, Baltimore.

REFERENCES

d'Arsonval, J. (1881). *Revue Scientifique* **17**, September, 370–372.

DATV (1995). "Life Cycle Screening of Food Products". Danish Academy of Technical Sciences, Lundtofte.

Davidson, B., *et al.* (1980). *IEE Proc.* **127**, 345–385.

Davis, N. (1974). *New Scientist* 9 May, 319–320.

DEA Wave Program (2002). Final report from Danish Wave Power Committee (in Danish). Rambøll Inc., Virum, website http://www.waveenergy.dk

Deb, S. (1998). Recent developments in high efficiency photovoltaic cells. *Renewable Energy* **15**, 467–472.

Defant, A. (1961). "Physical Oceanography", 2 Vols. Pergamon Press, Oxford.

Delwicke, C. (1970). *Sci. Am.* September, 136–147.

Demler, E., Zhang, S.-C. (1998). Quantitative test of a microscopic mechanism of high-temperature superconductivity. *Nature* **396**, 733–735.

Dennison, B., Mansfield, V. (1976). *Nature* **261**, 32–34.

De Renzo, D. (1978). "European Technology for Obtaining Energy from Solid Waste". Noyes Data Corp., Park Ridge, NJ.

Desrosiers, R. (1981). In "Biomass Gasification" (T. Reed, ed.). Noyes Data Corp., Park Ridge, NJ, pp. 119–153.

Dietz, A. (1954). "Diathermarous Materials and Properties of Surfaces" (as quoted by Duffie and Beckman, 1974).

Digby, S. (1954). In "A History of Technology" (C. Singer, E. Holmyard, A. Hall, eds.), Vol. I, p. 735. Clarendon Press, Oxford.

Dones, R., Frischknecht, R. (1997). Life cycle assessment of photovoltaic systems: Results of Swiss studies on energy chains. In Nieuwlaar and Alsema (1997), 8 pp.

Dones, R., Hirschberg, S., Knoepfel, I. (1994). Greenhouse gas emission inventory based on full energy chain analysis. IAEA Workshop 4–7 October, Beijing.

Douce, R., Joyard, J. (1977). *La Recherche* June, 527–537.

DONG (2003). Gas stores. http://www.dong.dk/dk/publikationer/lagerbrochure/

Douglas, D. (1974*). Trans. IEEE* pp. 37–41.

DR (1997). Testimony of K. Chekerov, Kurchatov Institute, Moscow, in the documentary "Den skjulte faktor", Danish TV1, 18 June.

Draper, L., Squire, E. (1967). *Trans. Inst. Naval Architects* **109**, 85.

Dreicer, M. (1996). "Evaluation of the impacts from the production of electricity: a method for the nuclear fuel cycle". Thesis, l'Ecole des Mines de Paris, December.

Drewes, P. (2003). Spheral solar – a completely different PV technology. In "Proc. PV in Europe Conf. 2002", Roma.

Drift, A. van der (2002). An overview of innovative biomass gasification concepts. in "12th European Biomass Conf.", pp. 381–384. ETA Firenze & WIP Munich.

DTI (2000). 1996 Danish Reference Year, obtained from Danish Technological Institute, Tåstrup.

Duffie, J., Beckman, W. (1974). "Solar Energy Thermal Processes". Wiley, New York.

Duffie, J., Beckman, W. (1991). "Solar Energy Thermal Processes", 2nd edition, Wiley, New York.

DuRy, C. (1969). "Völker des Alten Orient". Holle Verlag, Baden–Baden.

Dutta, J., Wu, Z., Emeraud, T., Turlot, E, Cornil, E., Schmidt, J. and Ricaud, A. (1992). Stability and reliability of amorphous silicon pin/pin encapsulated modules. in "11th EC PV Solar Energy Conference, Montreux" (Guimaraes, L., Palz, W.,

Reyff, C, Kiess, H. and Helm, P., eds.) pp. 545–548. Harwood Academic Publ., Chur.

Dwayne-Miller, R., *et al.* (2001). High density optical data storage and information retrieval using polymer nanostructures, In "Proc. 10 th Int. Conf. On Unconventional Photoactive Systems", Diablerets, p. I-27.

EC (1994). "Biofuels" (M. Ruiz-Altisent, ed.), DG XII Report EUR 15647 EN, European Commission.

EC-ATLAS (2003). European Commission DG Energy: ATLAS programme, http://europa.eu.int/energy_transport/atlas/htmlu/lbpot2.html

Eckart, C. (1958). *Am. J. Sci.* **256**, 225–240.

Eckert, E. (1960). Tidskrift for Landøkonomi (Copenhagen), No. 8.

Ehricke, K. (1971). *Bull. Atom. Sci.* November, 18–26.

Einstein, A. (1905). *Ann. der Physik* **17**, 549–560.

Eisenstadt, M., Cox, K. (1975). *Solar Energy* **17**, 59–65.

Ekman, V. (1902). *Nyt Magasin for Naturvidenskab* **40**, 1 (as discussed by Sverdrup, 1957).

El-Hinnawi, E., El-Gohary, F. (1981). In "Renewable Sources of Energy and the Environment" (E. El-Hinnawi and A. Biswas, eds.), pp. 183–219. Tycooly International Press, Dublin

Ellehauge, K. (1981). Solvarmeanlæg til varmt brugsvand, Danish Department of Energy Solar Heat Program, Report No. 16, Copenhagen.

Elliott, D., Elliott, R. (1976). "The Control of Technology". Wykeham Publ. Co., London.

Ellis, H., Pueschel, R. (1971). *Science* **172**, 845–846.

Eltra/Elkraft (2001). Time series of Danish wind power production 2000, available at websites http://www.eltra.dk and http://www.elkraft-system.dk

Eltra (2003). Søkabel, Danish power utility webpage http://www.eltra.dk

Emiliani, C. (1978). The cause of the ice age. *Earth and Planetary Sci. Lett.* **37**, 349–352.

Engelenburg, van and Nieuwlaar, E. (1993). "Methodology for the Life-cycle Assessment of Energy Technologies." Report 93011. Dept. Science, Technology and Society, Utrecht University.

England, M., Godfrey, J., Hirst, A., Tomczak, M. (1993). The mechanism of Antarctic intermediate water renewal in a world ocean model. *J. Phys. Oceanography* **23**, 1553–1560.

Epstein, E., Norlyn, J. (1977). *Science* **197**, 249–251.

Erdman, N., *et al.* (2002). The structure and chemistry of the TiO_2-rich surface of SrTiO3(001). *Nature* **419**, 55–58.

European Commission (1985). "Test reference years TRY", DGXII: Science, Research & Development, Report EUR 9765, Brussels. 48pp.

European Commission (1995a). "ExternE: Externalities of Energy, Vol. 1: Summary", prepared by ETSU for DGXII: Science, Research & Development, Study EUR 16520 EN, Luxembourg, 179 pp.

European Commission (1995b). "ExternE: Externalities of Energy, Vol. 2: Methodology", prepared by ETSU and Metronomica for DGXII: Science, Research & Development, Study EUR 16521 EN, Luxembourg, 571 pp.

European Commission (1995c). "ExternE: Externalities of Energy, Vol. 3: Coal and lignite", prepared by ETSU and IER for DGXII: Science, Research & Development, Study EUR 16522 EN, Luxembourg, 571 pp.

European Commission (1995d). "ExternE: Externalities of Energy, Vol. 4: Oil and gas", prepared by ETSU and IER for DGXII: Science, Research & Development, Study EUR 16523 EN, Luxembourg, 426 pp.

European Commission (1995e). "ExternE: Externalities of Energy, Vol. 5: Nuclear", prepared by CEPN for DGXII: Science, Research & Development, Study EUR 16524 EN, Luxembourg, 337 pp.

European Commission (1995f). "ExternE: Externalities of Energy, Vol. 6: Wind and hydro", prepared by EEE and ENCO for DGXII: Science, Research & Development, Study EUR 16525 EN, Luxembourg, 249 pp.

European Commission (1997). Energy in Europe, 1997 – Annual Energy Review, DGXII: Science, Research & Development, Special Issue, 179 pp.

European Commission (2001). Concerted Action on Offshore Wind Energy in Europe. Final Report from EC project NNE5-1999-562. Delft University Wind Energy Research Institute, Delft.

Eurostat (1975). "Energy Statistics, 1970–1974", European Community Publ. Office, Luxembourg.

Evans, D. (1976). *J. Fluid. Mech.* **77**, 1–25.

Faaij, A., Hamelinck, C. (2002). Long term perspectives for production of fuels from biomass; integrated assessment and R&D priorities, in "12th European Biomass Conf." vol. 2, pp. 1110—1113. ETA Firenze & WIP Munich.

Fabritz, G. (1954). Wasserkraftmaschinen. In "Hiitte Maschinenbau", Vol. IIA, pp. 865–961. Wilhelm Ernst and Sohn. Berlin.

Fairhall, A. (1973). *Nature* **245**, 20–23.

FAO (2003). UN Food and Agricultural Organisation Statistics, http://apps.fao.org

FAO-Asia (2003). Biomass Energy Technology: Wood Energy Database, Regional Wood Energy Dev. Programme in Asia, http://www.rwedp.org/technobc.html

Federal Energy Administration (1974). "Project Independence, Solar Energy Report", U.S. National Science Found., Washington, DC.

Feld, B. (1976). *Bull. Atom. Sci.* June, 10–13.

Fenhann, J., Kilde, N. (1994). Inventory of emissions to the air from Danish Sources. Systems Analysis Dept., Risø National Laboratory, Roskilde, 111 pp.

Fernandes, R. (1974). "Proc. 9th Intersociety Energy Conversion Engineering Conference", US, Paper No. 749032.

Fittipaldi, F. (1981). Phase change heat storage. In "Energy Storage and Transportation" (G. Beghi, ed.), pp. 169–182. D. Reidel, Dordrecht, Holland.

Flack, H., Morikofer, W. (1964). In "Proc. UN Conf. on New and Renewable Energy Sources", Rome 1961, Paper S/74. UN Printing Office, New York.

Fleagle, R. and Businger, J. (1963). "An Introduction to Atmospheric Physics". Academic Press, New York and London.

Formidlingsrådet (1989). "Forbehandling af biobrændsel. Decentrale kraftvarmeværker". Report prepared by Krüger Engineers Inc., Industry and Trade Board, Danish Department of Commerce, Copenhagen, 93 pp. + maps.

Forrester, J. (1971). "World Dynamics". Wright-Allen Press, Cambridge, MA.

Fowler, W., Hoyle, F. (1964). *Astrophys. J. Suppl.* **9**, No. 91.

Frankhauser, S. (1990). "Global warming damage costs – some monetary estimates". CSERGE GEC Working paper 92/29. Univ. East Anglia, Norwick.

Frankl, P., Masini, A., Gamberale, M., Toccaceli, D. (1997). Simplified LCA of PV systems in buildings, Appendix to "Environmental Aspects of PV Power Systems", Utrecht Universiteit VNS Report no. 97072. (irreg. pag.)

Freeman, M., Haveman, R., Kneese, A. (1973). "The Economics of Environmental Policy". Wiley, New York.

Frenkiel, S. (1964). In "Proc. UN Conf. on New and Renewable Energy Sources", Rome 1961, Paper W/33. UN Printing Office, New York.

Fried, M., Nosten, F., Brockman, A., Brabin, B., Duffy, P. (1998). Maternal antibodies block malaria. *Nature* **395**, 851–852.

Friedli, H., Lotscher, H., Oeschger, H., Siegenthaler, U., Stauffer, B. (1986). Ice core record of the $^{13}C/^{12}C$ ratio of atmospheric carbon dioxide in the past two centuries. *Nature* **324**, 237.

Frisch, M., *et al.* (1998). Gaussian 98 software, Revision A.9. Gaussian, Inc., Pittsburgh PA (1998).

Fritsche, H. (1977). In "Proc. 7th Int. Conf. on Amorphous and Liquid Semiconductors" (W. Spear, ed.), pp. 3–15. University Edinburgh Pres, Edinburgh.

Fritsche, U., Rausch, L. (1993). "TEMIS: Total Emission Model for Integrated Systems", User's guide for version 2.0. Oeko–Institute, Darmstadt.

Frost, W. (1975). In "Initial Wind Energy Data Assessment Study" (M. Changery, ed.), pp. 80–106. Nat. Science Found. (USA), Report No. NSF–RA–N–75–020.

Frost, W., Maus, J., Fichtl, G. (1974). *Boundary Layer Met.* **1**, 123–456.

Fujiwara, L., Nakashima, Y., Goto, T. (1981). *Energy Conversion and Management* **21**, 157–162.

Furbo, S. (1982). Communication No. 116 from Thermal Insulation Laboratory, Technical University, Lyngby, Denmark.

Gale, R. (1987). The environmental movement comes to town: A case study of an urban hazardous waste controversy. pp. 233–250 in Johnson and Covello (1987).

Gates, D. (1966). *Science* **151**, 523–529.

Gates, D. (1968). *Ann. Rev. Plant Physiol.* **19**, 211–238.

Gates, W., Henderson–Sellers, A., Boer, G., Folland, C., Kitoh, A. McAvaney, B., Semazzi, F, Smith, N., Weaver, A., Zeng, Q.-C. (1996). Climate models – evaluation. "Climate Change 1995: The Science of Climate Change", Ch. 5. IPCC working group I report. Cambridge University Press, Cambridge.

Geiger, R. (1961). "Das Klima der bodennahen Luftschicht", Vieweg and Sohn, Braunschweig.

Gerlach, L. (1987). Protest movements and the construction of risk. pp. 103–145 in Johnson and Covello (1987).

Giebel, G. (1999). Effects of distributing wind energy generation over Europe, in "European wind energy conference, Nice 1999", pp. 417-420. James & James, London.

Glaser, P. (1977). *J. Energy* **1**, 75–84.

Glauert, H. (1935). In "Aerodynamic Theory" (W. Durand, ed.), Vol. 4, div. L, pp. 169–360. J. Springer, Berlin.

Goedheer, J., Hammans, J. (1975). *Nature* **256**, 333–334.

Goguel, J. (1976). "Geothermics". McGraw-Hill, New York.

Gøbel, B., Bentzen, J., Hindsgaul, C., Henriksen, U., Ahrenfeldt, J., Houbak, N., Qvale, B. (2002). High performance gasification with the two.stage gasifier, in "12th European Biomass Conf.", pp. 289—395. ETA Firenze & WIP Munich.

Gold, T., Soter, S. (1980). *Sci. Amer.* pp. 130–137

Golibersuch, D., Bundy, F., Kosky, P., Vakil, H. (1976). Thermal energy storage for utility applications. In "Proc. of the Symp. on Energy Storage" (J. Berkowitz, H. Silverman, eds.). The Electrochemical Society, Inc., Princeton, NJ.

Govindjee (ed.) (1975). "Bioenergetics of Photosynthesis". Academic Press, New York and London.

Gramms, L., Polkowski, L., Witzel, S. (1971). *Trans. of the Am. Soc. Agricult. Eng.* **1**, 7.

Granqvist, C., *et al.* (1998). *Solar Energy* **63**, 199-216.

Grasse, W. (1981). *Sunworld* **5**, 68–72.

Gray, T. and Gashus, O. (1972). "Tidal Power". Plenum Press, New York.

Green, M. (1992). Crystalline and polycrystalline silicon solar cells. In "Renewable Energy Sources For Fuels And Electricity" (Johansson *et al.*, eds.), pp. 337–360. Island Press, Washington, DC.

Green, M. (1994). "Silicon Solar Cells: Advanced Principles and Practice". PV Special Research Centre, University of New South Wales, Sydney.

Green, M. (2002). Third generation photovoltaics: recent theoretical progress. In "17th European Photovoltaic Solar Energy Conf., Munich 2001" (McNelis, Palz, Ossenbrink, Helm, eds.), vol. I, p. 14-17.

Green, M., Zhao, J., Wang, A. (1998). 23% module and other silicon solar cell advances. In "Proc. 2nd World Conf. on PV Energy Conversion, Vienna" (J. Schmid *et al.*, eds.), pp. 1187–1192. JRC European Commission EUR 18656 EN, Luxembourg.

Gregg, M. (1973). *Sci. Am.* February, 65–77.

Gregory, D., Pangborn, J. (1976). *Ann. Rev. Energy* **1**, 279–310.

Grimson, J. (1971). "Advanced Fluid Dynamics and Heat Transfer". McGraw-Hill, London.

Gringarten, A., Wintherspoon, P., Ohnishi, Y. (1975). *J. Geophys. Res.* **80**, 1120–1124.

Grove, W. (1839). On voltaic series and the combination of gases by platinum. *Phil. Mag.* **14**, 127–130.

Grubb, M. (1996). Purpose and function of IPCC. *Nature* **379**, 108; response to news items in *Nature* **378**, 322 (1995) and *Nature* **378**, 119 (1995).

Gutenberg, G. (1959). "Physics of the Earth's Interior" (J. Mieghem, ed.). International Geophysics Series, Vol. 1. Academic Press, New York and London.

Hadley, G. (1735). *Phil. Trans.* **29**, 58–62.

Hagen, D., Erdman, A., Frohrib, D. (1979). In "Proc. 14th Intersociety Engineering Conf.", American Chemical Society, Washington, DC, pp. 368–373.

Hagenmuller, P. (1977). *La Recherche* **81**, 756–768.

Hahn, D., Manabe, S. (1975). *J. Atmos. Sci.* **32**, 1515–1541.

Hall, C., Swet, C., Temanson, L. (1979). In "Sun II, Proc. Solar Energy Society Conf., New Delhi, 1978", pp. 356–359. Pergamon Press, London.

Hamakawa, Y. (1998). A technological evolution from bulk crystalline age to multilayers thin film age in solar photovoltaics. *Renewable Energy* **15**, 22–31.

Hamakawa, Y., Tawada, Y., Okamoto, H. (1981). *Int. J. Solar Energy* **1**, 125.

Hambraeus, G. (ed.) (1975). "Energilagring". Swedish Academy of Engineering Sciences, Report No. IVA–72, Stockholm.

Hammond, A. (1976). *Science* **191**, 1159–1160.

Hammond, A., Metz, W., Maugh II, T. (1973). "Energy and the Future". American Assoc. for the Advancement of Science, Washington.

Handley, D., Heggs, P. (1968). *Trans. Econ. & Eng. Rev.* **5**, 7.

Hanneman, R., Vakil, H., Wentorf Jr., R. (1974). Closed loop chemical systems for energy transmission, conversion and storage. In "Proc. 9th Intersociety Energy Conversion Engineering Conf.". American Society of Mechanical Engineers, New York.

Hansen, H. (1985). Poul la Cour (biography). Askov Højskoles Forlag, 516 pp.

Hansen, J. (1975). *J. Atmos. Sci.* **26**, 478–487.

Hara, K., Sayama, K., Ohga, Y., Shinpo, A., Suga, S., Arakawa, H. (2001). A coumarin-derivative dye sensitised nanocrystalline TiO_2 solar cell having a high solar-energy conversion efficiency up to 5.6%. *Chem. Commun.* 569-570

Harth, R., Range, J., Boltendahl, U. (1981). EVA–ADAM System: A method of energy transportation by reversible chemical reactions. In "Energy Storage and Transportation" (G. Beghi, ed.), pp. 358–374. Reidel, Dordrecht, Boston, London.

Hartline, F. (1979). *Science* **206**, 205–206.

Hasselmann, K. (1962). *J. Fluid Mech.* **12**, 481–500.

Hasselmann, K. (1967). *Proc. R. Soc. (Lond.)* **A299**, 77–100.

Hasselmann, K., Barnett, T., Bouws, E., Carlson, H., Cartwrigt, D., Enke, K., Ewing, J., Gienapp, H., Hasselmann, D., Kruseman, P., Meerburg, A., Miiller, P., Olbers, D., Richter, K., Sell, W., Walden, H. (1973). *Deutschen Hydrographischen Zeitschrift*, Erganzungsheft, Reihe A(8°), Nr. 12 (Hamburg).

Hauch, A., *et al.* (2001). New photoelectrochromic device, website describtion: http://www.fmf.uni-freiburg.de/~biomed/FSZ/anneke2.htm

Haug, G., Sigman, D., Tiedemann, R., Pedersen, T., Sarntheim, M. (1999). Onset of permanent stratification in the subarctic Pacific Ocean, *Nature* **401**, 779–782.

Havelock, T. (1919). *Proc. R. Soc. (Lond.)* **A95**, 38–51.

Hay, H., Yellot, J. (1972). *Mech. Eng.* **92**, No. 1, 19–23.

Hayashi, C. (1966). *Ann. Rev. Astronomy and Astrophysics* **4**, 171–192.

Hayes, D. (1977). Worldwatch Paper No. 11. Worldwatch Institute, Washington, DC.

Hays, J., Imbrie, J., Shackleton, N. (1976). *Science* **183**, 1121–1131.

Hein, R. (1974). *Science* **185**, 211–222.

Heppenheimer, T., Hannah, E. (1975). "R & D Requirements for Initial Space Colonization", Contribution No. 2664 from Div. Geol. and Planetary Sci., California Institute of Technology, Pasadena.

Herbig, G. (1967). *Sci. Am.* August, 30–36.

Hermes, J., Lew, V. (1982). Paper CF9/V III/9 in "Proc. UNITAR Conf. on Small Energy Resources, Los Angeles 1981", United Nations Inst. for Training and Research, New York.

Heronemus, W. (1975). Proposal to US Nat. Science Found., Report NSF/ RANN/S E/GI–34979/TR/75/4.

Herrick, C. (1982). *Solar Energy* **28**, 99–104.

Hespanhol, I. (1979). *Energia*, **5**, November–December, quoted from El-Hinnawi and El-Gohary (1981).

Hesstvedt, E. (1973). *Water, Air, and Soil Pollution* **2**, 49–60.

Hewson, E. (1975). *Bull. Am. Met. Soc.* **56**, 660–675.

Hide, R. (1958). *Phil. Trans. R. Soc. (Lond.)* **A250**, 441–478.

Hildebrandt, A., Vant–Hull, L. (1977). *Science* **197**, 1139–1146.

Hind, G., Olson, J. (1968). *Ann. Rev. Plant Physiol.* **19**, 249–282.

Hinze, J. (1975). "Turbulence". McGraw–Hill, New York.

Hippel, F. von, Fels, M., Krugmann, H. (1975). *FAS Professional Bulletin* (Washington

D.C.) **3**, April, 5–6.

Hirsch, R., Gallagher, J., Lessard, R., Wesselhoft, R. (1982). *Science* **215**, 121–127.

Hjalte, K., Lidgren, K., Stahl, I. (1977). "Environmental Policy and Welfare Economics". Cambridge University Press, Cambridge, UK.

Hobbs, P., Harrison, H., Robinson, E. (1974). *Science* **183**, 909–915.

Hoffmann, P. (1998a). ZEVCO unveils fuel cell taxi. *Hydrogen and Fuel Cell Letter*, feature article, August (http://www.mhv.net/~hfcletter/letter).

Hoffmann, P. (1998b). Fuel processors. Record attendance highlight fuel cell seminar in November at Palm Springs. *Hydrogen and Fuel Cell Letter*, feature article, December (http://www.mhv.net/~hfcletter/letter).

Hofmeister, A. (1999). Mantle values of thermal conductivity and the geotherm from phonon lifetimes. *Science* **283**, 1699–1706.

Hohmeyer, O. (1988). "Social Costs of Energy Consumption". Springer-Verlag, Berlin, 126 pp.

Holdren, J. (1974). *Bull. Atom. Sci.* October, 14–23.

Holloway, J., Manabe, S. (1971). *Monthly Weather Rev.* **99**, 335–370.

Holten, T. van (1976). In "Int. Symp. on Wind Energy Systems, Cambridge 1976", Paper E3, British Hydromech. Res. Assoc., Cranfield.

Honsberg, C. (2002). A new generalized detailed balance formulation to calculate solar cell efficiency limits. In "17th European Photovoltaic Solar Energy Conf., Munich 2001" (McNelis, Palz, Ossenbrink, Helm, eds.), vol. I, p. 3-8.

Hooker, C., Hulst, R. van (1980). Institutionalizing a high quality conserver society. *Alternatives* **9**, 25–36.

Hordijk, L. (1991). Use of the RAINS models in acid rain negociations in Europe. *Environmental Science Technology* **25**, 596–603.

Horn, M., *et al.* (1970). *J. Am. Ceram. Soc.* **53**, 124.

Hoven, I. van der (1957). *J. Meteor.* **14**, 160–164.

Hoyle, F. (1975). "Astronomy and Cosmology". W. Freeman and Co., San Francisco.

Hughes, W. (ed.) (1976). "Energy for Rural Development". US Nat. Acad. Sci., Washington, DC.

Hütte (a Berlin-based association, ed.) (1954). "Des Ingenieurs Taschenbuch", Vols. I and II. Wilhelm Ernst, Berlin.

Hütter, U. (1976). In "Proc. Workshop on Advanced Wind Energy Systems", Stockholm, 1974, Swedish Development Board/Vattenfall, Stockholm.

Hütter, U. (1977). *Ann. Rev. Fluid Mech.* **9**, 399–419.

Hvelplund, F. (1975). "Indexing, resource utilization and capital crisis", Internal Report from School of Economics and Business Administration, Aarhus, Denmark.

Hvelplund, F., Illum, K., Jensen, J., Meyer, N., Nørgård, J., Sørensen, B. (1983). "Energy for the future – alternative energy plan" (in Danish). Borgen Publisher, Copenhagen, 96 pp.

Iben, I. (1970). *Sci. Am.* July, 26–39.

Iben, I. (1972). In "Stellar Evolution" (H.-Y. Chiu, A. Muriel, eds.), pp. 1–106. MIT Press, Cambridge, MA.

Ichikawa, Y. (1993). Fabrication technology for large-area a–Si solar cells. In "Technical Digest of 7th Int. PV Science and Engineering Conf. Nagoya", pp. 79–95. Nagoya Institute of Technology.

IEA (1999). Photovoltaic power systems in selected IEA member countries. 3rd survey report of the Power Systems Programme Task 1, International Energy

Agency, Paris.

IEA (2002). Key world energy statistics, International Energy Agency, Paris, Available at http://www.iea.org

IEA-PVPS (2001). Trends in photovoltaic applications, Report T1-10:2001, International Energy Agency Photovoltaic Power Systems Programme.

IEA-PVPS (2002). Trends in photovoltaic applications, Report T1-11:2002, International Energy Agency Photovoltaic Power Systems Programme.

Inaba, A., Shimatani, T., Tabata, S., Kawamura, S., Shibuya, S., Iwase, Y., Kato, K., Kakumoto, T., Kojima, T., Yamada, K., Komiyama, H. (1993). *J. Chem. Eng. Japan* **19**, 809–817.

Ion, D. (1975). "Availability of World Energy Resources". Graham and Trotman, London.

IPCC (1992). "Climate Change 1992: the Supplementary Report to the IPCC Scientific Assessment" (Houghton, J., Callander, B. And Varney, S., eds.). Cambridge University Press, Cambridge, 198 pp.

IPCC (1994). "Climate Scenarios and Socioeconomic Projections for IPCC WGII Assessment" (Greco, S., Moss, R., Viner, D. And Jenne, R., eds.). Consortium for Int. Earth Science Info. Network, Washington, DC (limited distribution working document).

IPCC (1995). "Climate Change 1994. Radiative Forcing of Climate Change and An Evaluation of the IPCC IS92 Emission Scenarios" (Houghton, J. T., Meira Filho, L.G., Bruce, J., Hoesung Lee, Callander, B.A., Haites, E., Harris, N. and K. Maskell, eds.). Cambridge University Press, Cambridge, 339 pp.

IPCC (1996a). "Climate Change 1995: Impacts, Adaptation and Mitigation of Climate Change: Scientific–Technical Analysis". Contribution of Working Group II to the Second Assessment Report of the Intergovernmental Panel on Climate Change (Watson, R.T., Zinyowera, M.C., Moss, R.H., Dokken, D.J., eds.). Cambridge University Press, Cambridge, 878 pp.

IPCC (1996b). "Climate Change 1995: The Science of Climate Change". Contribution of Working Group I to the Second Assessment Report of the Intergovernmental Panel on Climate Change (Houghton, J.T., Meira Filho, L.G., Callander, B.A., Harris, N., Kattenberg, A., and Maskell, K., eds.). Cambridge University Press, Cambridge, 572 pp.

IPCC (1996c). "Climate Change 1995. Economic and Social Dimensions of Climate Change". Contribution of Working Group III to the Second Assessment Report (Bruce, J., Lee, H. and Haites, E.., eds.). Cambridge University Press, Cambridge, 448 pp.

IPCC Data Distribution Centre (1998). Global climate model archive maintained by Deutsches Klimarechenzentrum at http://www.dkrz.de/ipcc/ddc/html/dkrzmain.html

Isaacs, J., Castel, D., Wick, G. (1976). *Ocean Eng.* **3**, 175–187.

ISO (1997). *Life Cycle Assessment- Principles and Framework*, Standard 14040, International Standard Organisation.

ISO (1999). *Life Cycle Assessment- Life Cycle Impact Assessment*, Standard 14040, International Standard Organisation.

Jacobsen, T. (1973), as quoted in "World Almanac 1974", Newspaper Enterprise Ass., New York (p. 427).

Jamshidi, M., Mohseni, M. (1976). In "System Simulation in Water Resources" (G.

VanSteenkiste, ed.), pp. 393–408. North-Holland Publ. Co., Amsterdam.

Jensen, J. (1981). Improving the overall energy efficiency in cities and communities by the introduction of integrated heat, power and transport systems. In "Proc. IEA Inl. New Energy Conservation Technologies Conf." (J. P. Millhom and E. H. Willis, eds.). Springer-Verlag, Berlin, p. 2981.

Jensen, J., Sørensen, B. (1984). "Fundamentals of Energy Storage". Wiley, New York, 345 pp.

Jensen, M. (1962). In "Vindkraftudvalgets Betaænkning", Danish Electricity Supply Undertakings, Copenhagen, pp. 54–66, and personal communication.

Jensen, N. (1994). PARK computer program, described in Beyer *et al.* (1994).

Johansson, T. (2002). Energy for sustainable development – a policy agenda for biomass, in "12th European Biomass Conf.", Vol. 1, pp. 3-6. ETA Firenze, WIP, Munich.

Johnson, B., Covello, V. (eds.) (1987). "The Social and Cultural Construction of Risk". Reidel, Dordrecht.

Johnson, C., Holland, M., Amonier, S. (1994). Assessment of incinerating or recycling newspaper. In "SETAC Workshop on Integrating Impact Assessment into LCA, Brussels".

Johnson, J., Hinman, C. (1980). *Science* **208**, 460–463.

Johnston, H. (1977). *J. Geophysical Research.*

Joukowski, N. (1906). *Bull. de l'Inst. Aeronaut. Koutchino*, Fasc. I, St. Petersburg.

Jørgensen, L., Mikkelsen, S., Kristensen, P. (1980). Solvarmeanlæg i Greve, Danish Department of Energy Solar Heat Program, Report No. 6, Copenhagen (with follow up: Report No. 15, 1981).

Junge, C. (1963). "Atmospheric Chemistry and Radioactivity". Academic Press, New York and London.

Juul, J. (1961). In "Proc. UN Conf. on New and Renewable Energy Sources of Energy", Rome, Vol. 7, paper W/21 (Published 1964 as paper E/Conf. 35 by UN Printing Office, New York).

Kahn, J. (1996). Fuel cell breakthrough doubles performance, reduces cost. Berkeley Lab. Research News, 29. May (http://www.lbl.gov/science–articles/archive/fuel–cells.html).

Kalinin, G., Bykow, V. (1969). *Impact of Science on Society* **19**, No. 2 (as quoted by Lindh, 1972).

Kalkstein, L. (1993). Health and climate change: direct impacts in cities. *The Lancet* **342**, 1397–1399.

Kalkstein, L., Smoyer, K. (1993). The impact of climate change on human health: some international implications. *Experientia* **49**, 969–979.

Kalnay, E., Kanamitsu, M., Kistler, R., Collins, W., Deaven, D., Gandin, L., Iredell, M., Saha, S., White, G., Woollen, J., Zhu, Y., Leetmaa, A., Reynolds, R., Chelliah, M., Ebisuzaki, W., Higgins, W., Janowiak, J., Mo, K., Ropelewski, C., Wang, J., Jenne, R., Joseph, D. (1996). The NCEP/NCAR 40-year reanalysis project. *Bull. Am. Met. Soc.* (March).

Kaltschmitt, M., Reinhardt, G., Stelzer, T. (1996). LCA of biofuels under different environmental aspects, in "Biomass for Energy and the Environment" (Chartier, P., Ferrero, G., Henius, U., Hultberg, S., Sachau, J. and Wiinblad, M., eds.), Vol. 1, pp. 369-386. Pergamon/Elsevier, Oxford.

Kamat, P., Vinodgopal, K. (1998). Environmental photochemistry with semiconductor nanoparticles. Ch. 7 in *Organic and Inorganic Photochemistry* (V. Ramamurthy & K. Schanze, eds.).

Kasahara, A., Washington, W. (1967). *Monthly Weather Rev.* **95**, 389–402.

Kasten, F. (1977). *Solar Energy* **19**, 589–593.

Kato, K., Yamada, K., Inaba, A., Shimatani, T., Tabata, S., Kawamura, S., Shibuya, H., Iwase, Y., Kakumoto, T., Kojima, T. and Komiyama, H. (1993). *J. Chem. Eng. Japan* **20**, 261–267.

Kaye, G., Laby, T. (1959). "Tables of Physical and Chemical Constants", 12th edn. Longmans, London.

Keenan, J. (1977). *Energy Conversion* **16**, 95–103.

Kertz, W. (1957). In "Handbuch der Physik" (S. Fliigge, ed.), Vol. 48, pp. 928–981 (Geophysik II). Springer-Verlag, Berlin.

Keynes, J. (1936). "The General Theory of Employment, Interest and Money". McMillan, New York.

Khaselev, O., Turner, J. (1998). *Science* **280**, 425–427.

Kiilerich, A. (1965). "Oceanografi", Gjellerup, Copenhagen.

King Hubbert, M. (1969). In "Resources and Man" (US Nat. Acad. Sci., and Nat. Sci. Found.), Ch. 8. W. Freeman, San Francisco.

King Hubbert, M. (1971). *Sci. Am.* September, 60–87.

Kinsman, B. (1965). "Wind Waves". Prentice-Hall, Englewood Cliffs, NJ.

Kira, T. (1975). In "Photosynthesis and Productivity in Different Environments" (J. Cooper, ed.), pp. 5–40. Cambridge University Press, Cambridge, UK.

Kitaigorodskii, S. (1970). "The Physics of Air–Sea Interaction" (Israel Program for Scientific Translation, Jerusalem 1973).

Klein, S. (1975). *Solar Energy* **17**, 79–80.

Knöepfel, I. (1993). The Swiss national research project "Total Pollution of Energy Systems". In "Life-cycle Analysis of Energy Systems", pp. 238–247. Workshop Proceedings. OECD Publications, Paris.

Kohn, W., Sham, L. (1965). Self-consistent equations including exchange and correlation effects. *Phys. Rev.* **140** (1965) A1133.

Kok, B., Radmer, R., Fowler, C. (1975). In "Proc. 3rd Int. Congr. on Photosynthesis, Rehovoth" (M. Avron, ed.), Vol. I, pp. 485–496.

Kondratyev, K., Podolskaya, E. (1953). *Izvestiya USSR Acad. Sci. Nauk*, Ser. *Geophys.*, No. 4.

Kondratyev, K., Fedorova, M. (1976). In "Proc. UNESCO/WMO Solar Energy Symp.. Genève 1976". paper ENG.S/Doc. 2.

Koppen, C. van, Fischer, L., Dijkamns, A. (1979). In "Sun II, Proc. Int. Solar Energy Society Conf., New Delhi", pp. 294–299. Pergamon Press, London.

Korn, J. (ed.) (1972). "Hydrostatic Transmission Systems". Int. Text Book Co., London.

Kraemer, F. (1981). "En model for energiproduktion og økomomi for centrale anlæg til produktion af biogas", Report from Physics Laboratory 3, Danish Technical University, Lyngby.

Krasovec, O., *et al.* (2001). Nanocrystalline WO_3 layers for photoelectrochromic and energy storage dye sensitised solar cells. In "Proc. 4 th Int. Symp. New Materials", Montréal, pp. 423-425.

Kreider, J. (1979). "Medium and High Temperature Solar Processes". Academic Press, New York.

REFERENCES

Kuemmel, B. (2000). A global clean fossil scenario. *Int. J. Global Energy Issues*, **13**, 181-195.

Kuemmel, B., Nielsen, S. K., Sørensen, B. (1997). "Life-cycle Analysis of Energy Systems". Roskilde University Press, Copenhagen, 216 pp.

Kussmaul, K., Deimel, P. (1995). Materialverhalten in H_2-Hochdrucksystemen. *VDI Berichte* **1201**, 87-101.

Kutta, W. (1902). "Auftriebkrafte in stromende Flüssigkeiten", *Ill. aeronaut. Mitteilungen*, July.

La Cour, P. (1900). *The experimental mill*, Vol. 1. Nordiske Forlag, Copenhagen; Vols. 2-4 (1901-3).

Ladisch, M., Dyck, K. (1979). *Science* **205**, 898–900.

Ladisch, M., Flickinger, M., Tsao, G. (1979). *Energy* **4**, 263–275.

Landau, L., Lifshitz, F. (1960). "Mechanics". Pergamon Press, Oxford.

Landberg, L., Joensen, A., Giebel, G., Watson, S., Madsen, H., Nielsen, T., Laursen, L., Jørgensen, J., Lalas, D., Tøfting, J., Ravn, H., MacCarthy, E., Davis, E., Chapman, J. (1999). Implementing short-term prediction at utilities. Final report of EC JOULE project JOR3–CT95–0008; a preliminary report dated 1996 was used by Meibom *et al.* (1997; 1999).

Laquer, H. (1976). In "Energy Storage, Compression and Switching" (W. Bostick, V. Nardi and O. Zucker, eds.), pp. 279–305. Plenum Press, New York.

Laskin, A., Gaspar, D., Wang, W., Hunt, S., Cowin, J., Colsom, S., Finlayson-Pitts, B. (2003). Reactions at interfaces as a source of sulphate formation sea-salt particles. *Science*, **301** (2003). 340-344.

Lau, K. (1987). Electricity forecasting in Denmark: Conflict between ministries and utilities. Ch. 8 in "The Politics of Energy Forecasting" (T. Baumgartner and A. Midttun, eds.). Oxford University Press, Oxford.

Leakey, M. (1975), as quoted in *World Almanac* (1977), p. 432.

Lees, L. (1970), as quoted in "Man's Impact on the Global Environment" (C. Wilson and W. Matthews, eds.), p. 63. MIT Press, Cambridge, MA.

Lehmann, J. (1981). Air storage gas turbine power plants: a major distribution for energy storage. In "Proc. Int. Conf. Energy Storage", pp. 327–336. BHRA Fluid Engeneering, Cranfield, UK.

Leichman, J., Scobie, G. (1975). "The Development of Wave Power – a Techno-economic Study", part 2, UK Dept. of Industry, and Nat. Eng. Lab., East Kilbride, Scotland.

Levitus, S., Boyer, T. (1994). "World Ocean Atlas 1994, vol. 4: Temperature", NOAA Atlas NESDIS 4, US Dept. Commerce, Washington, DC.

Lewis, J. (1974). *Sci. Am.*, March. 51–65.

Libowitz, G. (1974). In "Proc. 9th Intersociety Energy Conversion Conference", US, Paper No. 749025.

Linden, S. van der (2003). The commercial world of energy storage: a review of operating facilities. Presentation for "1st Ann. Conf. Energy Storage Council", Houston, Texas.

Lindh, G. (1972). *Ambio* **1**, 185–201.

Lissaman, P. (1976). In "Proc. Int. Symp. on Wind Energy Systems, Cambridge". Paper C2: BHRA Fluid Engineering, Cranfield, UK.

Little, Arthur D. (Inc.) (1975). "Economic Comparison of Base-load Generation Alternatives for New England Electric". S. M. Stoller Corp., March.

Ljungström, O. (1975). "Det svenska vindenergiprogram", Swedish Development Board (STU), Report No. VES–1975: 33.

Locquet, J.–P., Perret, J., Fompeyrine, J., Mächler, E., Seo, J., Tendeloo, G. (1998). Doubling the critical temperature of $La_{1.9}Sr_{0.1}CuO_4$ using epitaxial strain. *Nature* **394**, 453–456.

Loferski, J. (1956). *J. Appl. Phys.* **27**, 777–784.

Longair, M., Sunyaev, R. (1969). *Astrophys. Lett.* **4**, 65–70.

Loomis, R., Gerakis, P. (1975). In "Photosynthesis and Productivity in Different Environments" (J. Cooper, ed.), pp. 145–172. Cambridge University Press, Cambridge, UK.

Lorenz, E. (1967). "The Nature and Theory of the General Circulation of the Atmosphere". World Meteorological Organization, WMO publ. No. 218TP115.

Lorenz, E. (1968). *Meteorol. Monographs* **8**, No. 30, 1–3.

Losciale, M. (2002). Technical experiences and conclusions from introduction of biogas as a vehicle fuel in Sweden, in "12th European Biomass Conf.", Vol. 2, pp. 1124-1127. ETA Firenze & WIP Munich.

Lotker, M. (1974). In "Proc. 9th Intersociety Energy Conversion Engineering Conference", US, Paper No. 749033.

Lovins, A. (1977). '"Soft Energy Paths". Ballinger, New York.

Lund, H., *et al.* (1985). *Test renerence year TRY*. Report EUR 9765 to DG12, European Commission from Cenergia, Ballerup.

Luzzi, A. (1999). IEA Annual Report from Agreement on Production and Utilization of Hydrogen, pp. 35-42.

Lvovitch, M. (1977). *Ambio* **6**, pp. 13–21.

Ma, W., Saida, T., Lim, C., Aoyama, S., Okamoto, H., Hamakawa, Y. (1995). The utilization of microcrystalline Si and SiC for the efficiency improvement in a–Si solar cells. In "1994 First World Conf. on Photovoltaic Energy Conversion, Kona" pp. 417–420. IEEE, Washington, DC.

Maag, W. (1993). *Sonnenenergie und Wärmetechnik*, No. 5/93, 16–19.

MacGregor, I., Basu, A. (1974). *Science* **185**, 1007–1011.

Machta, L. (1971), as quoted in Wilson and Matthews (1971), p. 235.

Machta, L. (1976). In "Proc. UNESCO/WMO Solar Energy Symp., Genève 1976", paper ENGELS./Doc. 1.

Makhijani, A. (1977). *Economic and Political Weekly* (Bombay), Special issue August, pp. 1451–1464.

Manabe, S. (1969). *Monthly Weather Rev.* **97**, 739–774; 775–805.

Manabe, S. (1971). In "Man's Impact on Climate" (W. Matthews, W. Kellogs and G. Robinson, eds.), pp, 249–264. MIT Press, Cambridge, Mass.

Manabe, S., Holloway, J. (1975). *J. Geophys. Res.* **80**, 1617–1649.

Manabe, S., Stouffer, R. (1988). Two stable equilibria of a coupled ocean-atmosphere model. *J. Climate* **16**, 185–192.

Manabe, S., Strickler, R. (1964). *J. Atmos. Sci.* **21**, 361–385.

Manabe, S., Wetherald, R. (1967). *J. Atmos. Sci.* **24**, 241–259.

Manabe, S., Wetherald, R. (1975). *J. Atmos. Sci.* **32**, 3–15.

Manabe, S., Hahn, D., Holloway, J. (1974). *J. Atmos. Sci.* **31**, 43–83.

Manassen, J., Cahen, D., Hodes, G., Sofer, A. (1976). *Nature* **263**, 97–100.

Marchetti, C. (1973). *Chem. Econ. & Eng. Rev.* **5**, 7.

Mardon, C. (1982). High-rate thermophilic digestion of cellulosic wastes. Paper presented at the 5th Australian Biotechnology Conference, Sydney, August 1982.

Margen, P. (1980). *Sunworld* **4**, 128–134.

Marks, S. (1983). *Solar Energy* **30**, 45–49.

Maron, S., Prutton, C. (1959). "Principles of Physical Chemistry". Macmillan, New York.

Marsh, G. (2002). RE storage – the missing link. Elsevier Advanced Technology website: http://www.re-focus.net/mar2002_4.html

Martens, W., Rotmans, J., Niessen, L. (1994). Climate change and malaria risk. GLOBO Report 3–461502003. Global Dynamics & Sustainable Development Programme, RIVM, Bilthoven.

Marx, K. (1859). "A Critique of Political Economy", London.

Masterson, K., Seraphin, B. (1975). "Inter-Laboratory Comparison of the Optical Characteristics of Selective Surfaces for Photo-Thermal Conversion of Solar Energy", Nat. Sci. Found. (USA) Report NSF/RANN–GI–36731X.

Masuda, Y. (1971). Paper presented at "Int. Colloquium on Exploitation of the Oceans", Bordeaux, France.

Mathew, X., *et al.* (2001). In "Proc. 4 th Int. Symp. New Materials", Montréal, 420-421.

Maugh II, T. (1979). *Science* **206**, 436.

McAdams, W. (1954). "Heat Transmission". McGraw-Hill, New York.

McBride, J. (1981). Chemical heat pump cycles for energy storage and conversion. In "Proc. Int. Conf. on Energy Storage", Vol. 2, pp. 29–46. BHRA Fluid Engineering, Cranfield.

McCormick, M. (1973). "Ocean Engineering Wave Mechanics". Wiley, New York.

McCormick, M. (1976). *Ocean Eng.* **3**, 133–144.

McCoy, E. (1967). *Trans. Am. Soc. Agricultural Engineers* No. 6, 784.

McGowan, J. (1976). *Solar Energy* **18**, 81–92.

McGuire, A., Melillo, J., Kicklighter, D., Pan, Y., Xiao, X., Helfrich, J., Moore-III, B., Vorosmarty, C., Schloss, A. (1997). Equilibrium responses of global net primary production and carbon storage to doubled atmospheric carbon dioxide: sensitivity to changes in vegetation nitrogen concentrations. *Global Biogeochem. Cycles* **11**, 173–189.

McKay, R., Sprankle, R. (1974). In "Proc. Conf. on Res. for the Devt. of Geothermal Energy Resources", pp. 301–307. Jet Propulsion Lab., California Inst. of Technology, Pasadena, California.

Meibom, P., Svendsen, T., Sørensen, B. (1997). Import/eksport-politik som redskab til optimeret udnyttelse af el produceret på VE–anlæg. IMFUFA Texts No. 343, 84 pp.

Meibom, P., Svendsen, T., Sørensen, B. (1999). Trading wind in a hydro-dominated power pool system. *Int. J. of Sustainable Development* **2**, 458–483.

Meinel, A., Meinel, M. (1972). *Phys. Today* **25**, No. 2, 44–50.

Meinel, A., Meinel, M. (1976). "Applied Solar Energy". Addison-Wesley, Reading, MA.

Mekel, O., Huppes, G. (1990). "Environmental effects of different packaging systems for fresh milk". Centrum voor Milieukunde, Rijksuniversitet Leiden, CML–meeddelingen # 70.

Melillo, J., Helfrich, J. (1998). NPP database created under NASA and EPRI grants, kindly put at the author's disposal.

Melillo, J., McGuire, A., Kicklighter, D., Moore–III, B., Vorosmarty, C., Schloss, A. (1993). Global climate change and terrestrial net primary production. *Nature* **363**, 234–240.

Merzbacker, E. (1970). "Quantum Mechanics", 2nd edn. Wiley International, New York.

Mesarovic, M., Pestel, E. (1974). "Mankind at the Turning Point". Dutton and Co., New York.

Meyer, H., Morthorst, P., Schleisner, L., Meyer, N., Nielsen, P., Nielsen, V. (1994). Costs of environmental externalities of energy production (in Danish), Report R–770. Risø National Laboratory.

Mie, G. (1908). *Ann. der Phys.* **25**, 377.

Milankovich, M. (1941). *K. Serb. Akad. Beogr.*, Spec. Publ. No. 132 (translation: Israel Program for Scientific Translations, Jerusalem, 1969).

Miles, J. (1957). *J. Fluid Mech.* **3**, 185–204.

Miller, L., Douglas, B. (2004). Mass and volume contributions to twentieth-century global sea level rise. *Nature*, **428**, 406-409.

Millner, A. (1979). *Technology Review* November, 32–40.

Mills, D. (1995). A climate water budget approach to blackfly population dynamics. *Publications in Climatology* **48**, 1–84.

Minami, E., Kawamoto, H., Saka, S. (2002). Reactivity of lignin in supercritical methanol studied with some lignin model compounds, in "12th European Biomass Conf.", pp. 785—788. ETA Firenze & WIP Munich.

Mishan, E. (1969). "Growth: The Price We Pay". Staples Press, London.

Mitchell, J., John, T., Gregory, J., Trett, S. (1995). Climate response to increasing levels of greenhouse gases and sulphate aerosols. *Nature* **376**, 501–504.

Mitchell, J., Johns, T. (1997). On modification of global warming by sulphate aerosols. *J. Climate* **10**, 245–267. Model output (12 months by 240 years) has since 1998 been available at IPCC Data Distribution website http://www.dkrz.de/ipcc/ddc/html/gfdlrun2.html

Mitsui, A., Kumazawa, S. (1977). In "Biological Solar Energy Conversion" (A. Mitsui, S. Miyachi, A. San Pietro and S. Tamura, eds.), pp. 23–51. Academic Press, New York and London.

Mitsuyasu, H. (1968). *Rep. Res. Inst. Appl. Mech.*, Kyushu University, **16**, 459–482.

Mock, J., Tester, J., Wright, P. (1997). Geothermal energy from the earth: its potential impact as an environmentally sustainable ressource. *Ann. Rev. Energy Environ.* **22**, 305–356.

Möller, F. (1957). In "Handbuch der Physik" (S. Flügge, ed.), Vol. 48 (Geophysik III), pp. 155–253. Springer-Verlag, Berlin.

Mollison, D., Buneman, O., Salter, S. (1976). *Nature* **263**, 223–226.

Molly, J. (1977). In "Proc. Int. Symp. on Wind Energy Systems, Cambridge 1976", British Hydromech. Res. Assoc., Cranfield, and *Wind Engineering* **1**, 57–66 (1977).

Monin, A. and Obukhov, A. (1954). *Trudy Geofizicheskova Instituta, Akademiya Nauk, SSSR*, No. 24 (151).

Monin, A. and Yaglom, A. (1965). "Statistical Fluid Mechanics", part I, Izdatel'skvo Nauka, Moscow.

Moore, W. (1972). "Physical Chemistry", 5th edn. Longman Group, London.

Morishima, M. (1976). "The Economic Theory of Modern Society". Cambridge University Press, Cambridge, UK.

Morse, R. (1977). *Ambio* **6**, 209–215.

Moser, J. (1887). Notitz über Verstärkung photoelektrischer Ströme durch optische Sensibilisierung. *Monatshefte für Chemie und verwandte Teile anderer Wissenschaften* **8**, 373.

Moskowitz, L. (1964). *J. Geophys. Res.* **69**, 5161–5179.

Moskowitz, P., Buchanan, W., Shafarman, W. (1994). "Lessons learned from a hydrogen explosion at a photovoltaic research facility". Brookhaven National Laboratory (unpublished report).

Moskowitz, P., Steinberger, H., Thumm, W. (1995). Health and environmental hazards of CdTe photovoltaic module production, use and decommissioning. In "1994 IEEE First World Conf. on Photovoltaic Energy Conversion, Kona", pp. 115–118. IEEE, Washington,DC.

Müller, C., Falcou, A., Reckefuss, N., Rojahn, M., Wiederhirn, V., Rudati, P., Frohne, H., Nuyken, O., Becker, H., Meerholz, K. (2003). Multi-colour organic light-emitting displays by solution processing, *Nature* **421**, 829–833.

Müller, S., Kells, K., Litsios, J., Kumbein, U., Schenk, A., Fichtner, W. (1992). SIMUL v.1 Manual. Integrated Systems Lab., Technical Highschool, Zürich.

Munk, W., MacDonald, G. (1960), as quoted by King Hubbert (1969).

Munk, W. (1980). Affairs at the sea. *Ann. Rev. Earth and Planetary Sci.* **8**, 1–16.

Murphy, D., Christian, P. (1979). *Science* **205**, 651.

Musgrove, P. (1976). *Nature* **262**, 206–207.

Nagaura, T. (1990). Paper for "3rd Int. Battery Seminar, Dearfield Beach, FA".

Nakicenovic, N., Grübler, A., Ishitani, H., Johansson, T., Marland, G., Moreira, J. Rogner, H. (1996). Energy primer. pp. 75–92 in IPCC (1996b).

NASA (1971). Report No. R–351 and SP–8005, May.

NASA (1997). Surface solar energy data set v1.0, NASA Langley Research Center EOSDIS Distributed Active Archive Center, available at website http://eosweb.larc.nasa.gov/

NASA (1998). Ozone data from ADEOS satellite, available at http://jwocky.gsfc.gov

NASA (2004). Sea surface height anomalies from TOPEX/POSEIDON spacecraft measurements, date series at http://sealevel.jpl.nasa.gov/science

Nathan, G., Rajasooria, G., Tan, K., Tan, T. (1976). In "Proc. Int. Symp. on Wind Energy Systems, Cambridge 1976", Paper A5, British Hydromech, Res. Assoc., Cranfield, UK.

Naturlig Energi (1998). Wind power production statistics, pp. 14–17, June issue (in Danish).

Nazeeruddin, M., *et al.* (1993). Conversion of light to electricity by cis-X2Bis(2,2'-bipyridyl-4,4'-dicarboxylate) ruthenium-(II) charge transfer sensitizers (X=Cl-, Br-, I-, CN- and SCN-) on nanocrystalline TiO_2 electrodes, *J. Am. Chem. Soc.* **115**, 6382–6390

Nazeeruddin, M., *et al.* (2001). Engineering of efficient panchromatic sensitizers for nanocrystalline TiO_2-based solar cells. *J. Am. Chem. Soc.* **123**, 1613–1620.

NCAR (1997). The NCAR Community Climate Model CCM3 with NCAR/CSM Sea Ice Model. University Corporation for Atmospheric Research, National Center for Atmospheric Research, and Climate and Global Dynamics Division, http://www.cgd.ucar.edu:80/ccr/bettge/ice

NCEP–NCAR (1998). The NOAA NCEP-NCAR Climate Data Assimilation System I, described in Kalnay *et al.* (1996), data available from Univ. Columbia at http://ingrid.ldgo.columbia.edu

Neftel, A., Moor, E., Oeschger, H., Stauffer, B. (1985). Evidence from polar ice cores for the increase in atmospheric CO_2 in the past two centuries. *Nature* **315**, 45–47.

Neiburger, M., Edinger, J., Bonner, W. (1973). "Understanding our Atmospheric Environment". W. Freeman, San Francisco.

Nelson, J., Haque, A., Klug, D., Durran, J. (2001). Trap-limited recombination in dye-sensitised nanocrystalline metal oxide electrodes. *Phys. Rev.* **B 63**, 205321-9.

Neumann, G. (1949). *Deutscher Hydrogr. Zeitschrift* **2**, 187–199.

Neumann, G. (1956). *Deutscher Hydrogr. Zeitschrift* **9**, 66–78.

Neumann, G., Pierson, W. (1966). "Principles of Physical Oceanography". Prentice-Hall, Englewood Cliffs, NJ.

Newell, R., Vincent, D., Dopplick, T., Ferruzza, D., Kidson, J. (1969). In "The Global Circulation of the Atmosphere" (G. Corby, ed.), pp. 42–90. Royal Met. Soc., London.

Newman, M., Rood, R. (1977). Internal report.

Niel, C. van (1941). *Adv. Enzymol.* **1**, 263.

Nielsen, P.S., Holm-Nielsen, J.B. (1996). CO_2 balance in production of energy based on biogas, poster presented at the European conference: "Environmental Impact of Biomass for Energy", November 1996, Nordwijkerhout.

Nielsen, S., Sørensen, B. (1996). Long-term planning for energy efficiency and renewable energy. Paper presented at "Renewable Energy Conference, Cairo April 1996"; revised as: Interregional power transmission: a component in planning for renewable energy technologies, *Int. J. Global Energy Issues* **13**, No. 1-3 (2000), 170–180.

Nielsen, S., Sørensen, B. (1998). A fair market scenario for the European energy system. In "Long-Term Integration of Renewable Energy Sources into the European Energy System" (LTI–research group, ed.), pp. 127–186. Physica– Verlag, Heidelberg.

Nielsen, S. (1977). Records from Nautical Dept. of the Danish Met. Inst., personal communication.

Nieuwlaar, E., Alsema, E. (1997). Environmental Aspects of PV Power Systems, Utrecht Universiteit VNS Report no. 97072. (irreg. pag.)

Nihoul, P. (1977). *Applied Math. Modelling* **1**, 3–8.

Niiler, P. (1977). *La Recherche* **79**, 517–526.

NOAA (1998). "Global Vegetation Index", weekly, monthly, seasonal and annual data from NOAA-NESDIS-NCDC-SDSD collaboration, available from UNEP (1998).

Nordel (Nordic Electricity Supply Undertakings) (1975). "Drifttekniska specifikationer for varmekraft", Oslo.

Nordhaus, W. (1990). "To Slow or Not to Slow: The Economics of the Greenhouse Effect". Yale University Press, New Haven, CT.

Nordhaus, W. (1994). "Managing The Global Commons: The Economics of Climate Change". MIT Press, Boston, MA, 213 pp.

Norman, R. (1974). *Science* **186**, 350–352.

Norris, R., Röhl, U. (1999). Carbon cycling and chronology of climate warming during the Palaeocene/Eocene transition, *Nature* **401**, 775–778.

NSES (2001). Novator Solar Energy Simulator v1.2, software from Novator Advanced Technology Consulting, Gilleleje.

Obasi, G. (1963). *J. Atmos. Sci.* **20**, 516–528; "Atmospheric momentum and energy calculations for the Southern hemisphere during the IGY", Report No. 6, contract AF19(604)–6108, Dept. of Meteorol., MIT, Cambridge, MA.

Odum, E. (1972). "Ecology". Holt-Reinhardt and Winston, New York.

OECD (2002). OECD Agricultural Outlook 2002–2007. OECD, Paris

OECD/IEA (2002a). Energy Balances of OECD Countries, 1999–2000. International Energy Agency, Paris

OECD/IEA (2002b). Energy Balances of Non-OECD Countries, 1999–2000. International Energy Agency, Paris.

Oetjen, R., Bell, E., Young, J., Eisner, L. (1960). *J. Opt. Soc. Am.* **50**, No. 12, 1308.

Offshore Windenergy Europe (2003). News item dated December 2002, Technical University Delft, posted on the website http://www.offshorewindenergy.org

Ogilvie, T. (1963). *J. Fluid Mech.* **16**, 451–472.

Olsen, H. (1975). Jordbrugsteknisk Institut Medd. No. 24, Royal Veterinary and Agricultural University, Copenhagen.

O'Neill, G. (1975). *Science* **190**, 943–947.

Onsager, L. (1931). *Phys. Rev.* **37**, 405–426.

Oort, A. (1964). *Monthly Weather Rev.* **92**, 483–493.

O'Regan, B., Grätzel, M. (1991). *Nature* **353**, 737.

Osterle, J. (1964). *Appl. Sci. Res.* section A, **12**, 425–434.

Oswald, W. (1973). Progress in Water Technology, *Water Qual. Mgt. Pollut. Contr.* **3**, 153.

Ottinger, R. (ed.) (1991). "Environmental Costs of Electricity". Oceana Publ., New York.

Ovshinsky, S. (1978). *New Scientist* 30 November, 674–677.

Pacific Solar Inc. (1997). Annual Company Review, Sydney, 20pp.

Parker, E. (1964). *Sci. Am.* April, 66–76.

Parry, M., Rosenzweig, C. (1993). Food supply and ths risk of hunger. *The Lancet* **342**, 1345–1347.

Pasachoff, J. (1973). *Sci. Am.* October, 68–79.

Patil, P. (1998). The US DoE fuel cell program. Investing in clean transportation. Paper presented at "Fuel Cell Technology Conference, London, September", IQPC Ltd, London.

Patten, B. (ed.) (1971). "Systems Analysis and Simulation in Ecology", Vol. I. Academic Press, New York and London.

Patten, B. (ed.) (1972). "Systems Analysis and Simulation in Ecology", Vol. II. Academic Press, New York and London.

Patterson, M. (1997). The management of wastes associated with thin film PV manufacturing. Paper presented at "IEA Workshop on Environmental aspects of PV power systems", Utrecht, 25–27 June.

Pattie, R. (1954). *Nature* **174**, 660.

Peebles, P. (1971). "Physical Cosmology". Princeton University Press, Princeton, NJ.

Peixoto, J., Crisi, A. (1965). Dept. of Meteorology, Mass. Inst. of Technology, Sci. Report No. 6, contract AF19(628)–2408, as quoted by Lorenz (1967).

Pelser, J. (1975). In "Proc. 2nd Workshop on Wind Energy Conversion Systems" (F. Eldridge, ed.), pp. 188–195. The Mitre Corp., McLean, Virginia, Report NSF–RA–N–75–050.

Penman, H. (1970). *Sci. Am.* September, 98–109.

Penndorf, R. (1959). Air Force Cambridge Res. Center, AFCRC–TN–59–608, Report No. 1.

Perrin, G. (1981). *Verkehr und Technik*, issue no. 9.

Petersen, E. (1974). Risø Report No. 285, Danish Atomic Energy Commission, Copenhagen; and personal communication.

Petersen, H. (2001). Evaluation of wind turbine performance 2000, HP Consult Report from project ENS-51171/00-0016 of the Danish Energy Agency.

Petersen, H., Lundsager, P. (1998). Comparison of wind turbines based on power curve analysis. Report for the Danish Energy Agency, HP Consult and Darup Associates Ltd.

Petersen, J. (1972). Statens Byggeforskningsinstitut (Danish Building Res. Inst.), Notat No. 20.

Petersen, P. (1991). "Life Cycle Analysis of Local CHP-plants – Energy and Environmental Assessment from Cradle to Grave". Danish Ministry of Environment.

Pfister, G. and Scher, H. (1977). In "Proc. 7th Int. Conf. on Amorphous and Liquid Semiconductors (W. Spear, ed.), pp. 197–208. University of Edinburgh Press, Edinburgh.

Phillips, O. (1957). *J. Fluid Mech.* **2**, 417–445.

Phillips, O. (1966). "The Dynamics of the Upper Ocean". Cambridge University Press, Cambridge, UK.

Pierson, W. (1955). In "Advances in Geophysics" (H. Landsberg, ed.), Vol. 2, pp. 93–178. Academic Press, New York and London.

Pigford, T. (1974). In "Energy, Ecology and the Environment" (R. Wilson and W. Jones, eds.) pp. 343–349. Academic Press, New York.

Pines, D. (1994). Understanding high-temperature superconductivity, a progress report. *Physica B* **199–200**, 300–309.

Pinker, R., Laszlo, I. (1992). Modelling surface solar irradiance for satellite applications on a global scale. *J. Applied Meteorology* **31**, 194–211.

Piper, A. (1975). *Nature* **258**, December 4, news column.

Pippard, J. (1966). "The Elements of Classical Thermodynamics". Cambridge University Press, Cambridge, UK.

Pollack, H., Chapman, D. (1977). *Earth and Planetary Sci. Lett.* **34**, 174–184.

Pommer, K. *et al.* (1991). "Miljømæssig vurdering af mælkeemballage". Miljøprojekt 168. Danish Environmental Agency.

Pond, S. (1971). *Eos (Trans. Am. Geophys. Soc.)* **52**, 389–394.

Popel, F. (1970). *Landtechnische Forschung*, Heft 5, BRD.

Post, R., Post, S. (1973). *Sci. Am.* December, 17–23.

Post, R., Ribe, F. (1974). *Science* **186**, 397–407.

Potter, G., Ellsaesser, H., MacCracken, M., Luther, M. (1975). *Nature* **258**, 697–698.

Pöpperling, R., Schwenk, W., Venkateswarlu, J. (1982). Abschätzung der Korrosionsgefärdung von Behältern und Rohrleitungen aus Stahl für Speicherung von Wasserstoff und wasserstofhältigen Gasen unter hohen Drücken. VDI Zeitschriften Reihe 5, No. 62.

Prandtl, L. (1932). *Beitrage Physik Freien Atmos.* **19**, No. 3 (Bjerknes Festschrift).

Prengle, H., Sun, C.-H. (1976). *Solar Energy* **18**, 561–567.

Priestley, C. (1951). *Australian J. Sci. Res.* **A4**, 315–328.

Prigogine, I. (1968). "Introduction to Thermodynamics of Irreversible Processes".

REFERENCES

Wiley-Interscience, New York.

Pringle, H. (1998). The slow birth of agriculture. *Science* **282**, 1446–1450.

Proton Energy Systems (2003). Unigen. Website http://www.protonenergy.com

Putnam, P. (1953). "Energy in the Future". Van Nostrand, New York.

Rabenhorst, D. (1976). In "Wind Energy Conversion Systems" (J. Savino, ed.). US Nat. Sci. Found., Report NSF/RA/W–73–006.

Rabl, A., Nielsen, C. (1975). *Solar Energy* **17**, 1–12.

Raich, J., Rastetter, E., Melillo, J., Kicklighter, D., Steudler, P., Peterson, B., Grace, A., Moore-III, B. and Vörösmarty, C. (1991). Potential net primary productivity in South America: application of a global model. *Ecological Applic.* **1**, 399–429.

Rajeswaran, G., *et al.* (2000). In *SID 00 Digest*, No. 40 (4 pp)

Ralph, E. (1966). *Solar Energy* **10**, 67–71.

Rambøll Inc. (1999). Wave energy conditions in Danish North Sea, Report (in Danish) from Danish Energy Agency Project 51191/97-0014, Rambøll Inc., Virum.

Raschke, E., Haar, T., Bandeen, W., Pasternak, M. (1973). *J. Atmos. Sci.* **30**, 341–364.

Rasool, S., Schneider, S. (1971). *Science* **173**, 138.

Ravn, H. (2001). Communication of preliminary time series from Danish Utility Public Service Obligation project.

Ratcliffe, J. (ed.) (1960). "Physics of the Upper Atmosphere". Academic Press, New York and London.

Reck *et al.* (1974). as quoted by Manabe and Holloway (1975).

Reed, T. (ed.) (1981). "Biomass Gasification", Noyes Data Corp., Park Ridge, NJ.

Reeves, H. (1965). In "Stellar Structure" (L. Aller and D. McLaughlin, eds.), Vol. 8, pp. 113–193. University of Chicago Press, Chicago.

Reeves, H. (1975). *La Recherche* **60**, 808–817.

Reeves R., Lom, E., Meredith, R. (1982). "Stable Hydrated Ethanol Distillate Blends in Diesels". Apace Res. Ltd., Hawkesbury, NSW, Australia.

Reid, J. (1971). In "Man's Impact on Terrestrial and Oceanic Ecosystems" (W. Matthews, F. Smith and E. Goldberg, eds.). MIT Press, Cambridge, MA.

Reilly, J., Wiswall, R. (1974). *Inorg. Chem.* **13**, 218.

Ren, Y., Zhang, Z., Gao, E., Fang, S., Cai, S. (2001). A dye-sensitized nanoporous TiO_2 photochemical cell with novel gel network polymer electrolyte, *J. Appl. Electrochemistry* **31**, 445-447.

Ricaud, A. (1999). Economic evaluation of hybrid solar systems in private houses and commercial buildings. In "Photovoltaic/Thermal Solar Systems, IEA Solar Heating & Cooling/Photovoltaic Power Systems Programmes, Amersfoort", 13 pp., Ecofys, Utrecht.

Rijnberg, E., Kroon, J., Wienke, J., Hinsch, A., Roosmalen, J., Sinke, W., Scholtens, B., Vries, J., Koster, C., Duchateau, A., Maes, I., Hendrickx, H. (1998). Long-term stability of nanocrystalline dye-sensitized solar cells, in "2nd World Conference on PV Solar Energy Conversion, Vienna" (Schmid, J., Ossenbrink, H., Helm, P., Ehman, H. and Dunlop, E, eds.), pp- 47–52. European Commission, Luxembourg.

Robinson, J. (ed.) (1980). "Fuels from Biomass". Noyes Data Corp., Park Ridge, NJ.

Robinson, N. (ed.) (1966). "Solar Radiation". Elsevier, Amsterdam.

Roll, H. (1957). In "Handbuch der Physik" (S. Flügge, ed.), Vol. 48, pp. 671–733. Springer-Verlag, Berlin.

Roseen, R. (1978). "Central Solar Heat Station in Studsvik", AB Atomenergi Report

ET–78/77, Studsvik.

Rosenzweig, C., Parry, M., Fischer, G., Frohberg, K. (1993). "Climate change and world food supply". Report 3, Environmental Change Unit, Oxford University, Oxford.

Ross, M., Williams, R. (1975). "Assessing the potential for fuel conservation", Inst. Public Policy Alternatives, State University of New York.

Ross, M., Williams, R. (1976). *Bull. Atom. Sci.* November, 30–38.

Rossby, C.-G. (1932). *MIT Meteorological Papers*, Vol. 1, No. 4.

Rowley, J. (1977). *Phys. Today* January, 36–45.

Rubey, W. (1951). *Bull. Geol. Soc. Am.* **62**, 111.

Rubins, E., Bear, F. (1942). *Soil Sci.* 54, 411.

Russell, F., Chew, S. (1981). In "Proc. Int. Conf. on Energy Storage", pp. 373–384. BHRA Fluid Engineering, Cranfield.

Saeki, T. (1975). In "Photosynthesis and Productivity in Different Environments" (J. Cooper, ed.), pp. 297–322. Cambridge University Press, Cambridge, UK.

Sagan, C., Mullen, G. (1972). *Science* **177**, 52–56.

Sah, C., Noyce, R., Shockley, W. (1957). In "Proc. IRE", Vol. 45, September, pp. 1228–1243.

Sakai, H. (1993). Status of amorphous silicon solar cell technology in Japan. In "Technical Digest of 7th Int. PV Science and Engineering Conf. Nagoya", pp. 169–172. Nagoya Institute of Technology.

Salter, S. (1974). *Nature* **249**, 720–724.

Sandwell, D., Smith, W., Smith, S., Small, C. (1998). Measured and estimated seafloor topography. Available together with land topography at the website: http://ingrid.ldgo.columbia.edu/sources/

Sariciftci, N., Smilowitz, L., Heeger, A., Wudl, F. (1992). *Science* **258**, 1474.

Savonius, S. (1931). *Mech. Eng.* **53**, 333–338.

Schlieben, E. (1975). In "Proc. 1975 Flywheel Technology Symp., Berkeley, CA", pp. 40–52. Report ERDA 76.

Schmidt, A., Christiansen, K., Pommer, K. (1994). Livscyklusmodel til vurdering af nye materialer – Metoder, vurdering og fremgangsmåde, dk–TEKNIK Report, Søborg, Denmark, 202 pp.

Schneider, S. and Dennett, R. (1975). *Ambio* **4**, 65–74.

Schramm, D. (1974). *Sci. Am.* January, 69–77.

Schulten, R., Decken, C. van der, Kugeler, K., Barnert, H. (1974). Chemical Latent Heat for Transport of Nuclear Energy over Long Distances. In "Proc. British Nuclear Energy Society Int. Conf., The High Temperature Reactor and Process Applications". BNES, London.

Schumpeter, J. (1961). "Capitalism, Socialism and Democracy". Allen and Unwin, London.

Schwarzschild, M. (1958). "Structure and Evolution of Stars". Princeton University Press, New Jersey.

Scrosati, B. (1995). Challenge of portable power. *Nature* **373**, 557–558.

Seeger, P., Fowler, W., Clayton, D. (1965). *Astrophys. J. Suppl.* **11**, No. 97.

Seitel, S. (1975). *Solar Energy* **17**, 291–295.

Sekera, Z. (1957). In "Handbuch der Physik" (S. Flügge, ed.), Vol. 48 (Geophysik II), pp. 288–328. Springer-Verlag, Berlin.

Seliger, H., McElroy, W. (1965). "Light: Physical and Biological Action". Academic

Press, New York and London.

Sellers, P., Dickinson, R., Randall, D., Betts, A., Hall, F., Berry, J., Collatz, G., Denning, A., Mooney, H., Nobre, C., Sato, N., Field, C., Henderson-Sellers, A. (1997). Modelling the exchanges of energy, water, and carbon between continents and the atmosphere. *Science* **275**, 502–509.

Sellers, W. (1965). "Physical Climatology". University of Chicago Press, Chicago.

Service, R. (1998). Superstrong nanotubes show they are smart, too. *Science* **281**, 940–942.

Sforza, P. (1976). In "Proc. Int. Symp. on Wind Energy Systems, Cambridge, 1976", paper E1. British Hydromech. Res. Assoc., Cranfield.

Shaheen, S., *et al.* (2001). 2.5% efficient organic plastic solar cells. *Appl. Phys. Lett.* **78**, 841–843.

Shakleton, N., Opdyke, N. (1973). *Quaternary Res.* (New York) **3**, 39.

Shelton, J. (1975). *Solar Energy* **17**, 137–143.

Shigley, J. (1972). "Mechanical Engineering Design", 2nd ed., McGraw-Hill, New York.

Shirland, F. (1966). *Advanced Energy Conversion* **6**, 201–221.

Shklover, V., Ovchinnikov, Y., Braginsky, L., Zakeeruddin, S., Grätzel, M. (1998). Structure of organic/inorganic interface in assembled materials comprising molecular components. Crystel structure of the sensitizer bis[4,4'-carboxy-2,2'-bipyridine)(thiocyana-to)]ruthenium(II). *Chem. Mater.* **10**, 2533–2541.

Shockley, W. (1950). "Electrons and Holes in Semiconductors". Van Nostrand, New York.

Shubik, M. (ed.) (1991). "Risk, Organizations and Society". Kluwer Academic Publ., Boston.

Siegel, R., Howell, J. (1972). "Thermal Radiation Heat Transfer". McGraw-Hill, New York.

Siegenthaler, U., Oeschger, H. (1987). Biospheric CO_2 emissions during the past 200 years reconstructed by deconvolution of ice core data. *Tellus* **39B**, 140–154.

Silverstein, S. (1976). *Science* **193**, 229–231.

Sjoblom, C.-A. (1981). Heat storage in phase transitions of solid electrolytes, in "Proc. of the 16th Intersociety Energy Conversion Engineering Conf.", Vol. I, paper 8 19441. The American Society of Mechanical Engineers, New York.

Skyllas-Kazacos, M. (2003). Novel vanadium chloride/polyhalide redox flow battery. *J. Power Sources*, **124**, 299–302.

Slama-Schwok, A., Ottolenghi, M., Avnir, D. (1992). Long-lived photoinduced charge separation in a redox system trapped in a sol–gel glass. *Nature* **355**, 240–242.

Slavîk, B. (1975). In "Photosynthesis and Productivity in Different Environments" (J. Cooper, ed.), pp. 511–536. Cambridge University Press, Cambridge, UK.

Slesser, M. (1975). *Nature* **254**, 170–172.

Slotta, L. (1976). In "Workshop on Wave and Salinity Gradient Energy Conversion" (R. Cohen and M. McCormick, eds.), Paper H, US ERDA Report No. C00–2946–1.

SMAB (1978). "Metanol sam drivmedel," Annual Report, Svensk Metanol-utveckling AB, Stockholm.

Smagorinsky, J., Manabe, S., Holloway, J. (1965). *Monthly Weather Rev.* **93**, 727–768.

Smil, V. (1977). *Bull. Atom. Sci.* February, 25–31.

Smith, D. (1974). *Earth and Planetary Sci. Lett.* **23**, 43–52.

Snyder, R. and Cox, C. (1966). *J. Mar. Res.* **24**, 141–178.

Sørensen, B. (1975a). Computer simulation of ^{131}I transfer from fallout to man. *Water, Air, and Soil Pollution* **4**, 65–87.

Sørensen, B. (1975b). Energy and resources. *Science* **189**, 255–260; and in "Energy: Use, Conservation and Supply" (P. Abelson and A. Hammond, eds.), Vol. II, pp. 23–28. Am. Ass. Advancement of Science, Washington, DC (1978).

Sørensen, B. (1975c). "Vindkraft" (Appendix, pp. 41–67), Danish Academy of Technical Sciences, Lyngby (Summary report *Wind power* available in English).

Sørensen, B. (1976a). A simple model of economic growth or decline under the influence of ressource depletion. *Appl. Math. Modelling* **1**, 24–28.

Sørensen, B. (1976b). Solar heat systems for use at high latitudes. In "Proc. UNESCO-WMO Solar Energy Symp., Geneve, 1976", WMO Paper No. 477, 1977.

Sørensen, B. (1976c). Dependability of wind energy generators with short-term energy storage. *Science* **194**, 935–937.

Sørensen, B. (1976d). Wind energy. *Bull. Atom. Sci.* **32**, September, 38–45; and in "Toward a Solar Civilization" (R. Williams, ed.). MIT Press, Cambridge, MA (1978).

Sørensen, B. (1977). Direct and indirect economics of wind energy systems relative to fuel based systems. In "Proc. Int. Symp. on Wind Energy Systems, Cambridge, 1976", Paper D2, British Hydromech. Res. Assoc., Cranfield; and *Wind Engineering* **1**, 15–22.

Sørensen, B. (1978a). On the fluctuating power generation of large wind energy converters with and without storage facilities. *Solar Energy* **20**, 321–331.

Sørensen, B. (1978b). The regulation of an electricity supply system including wind energy generators. In "Proc. 2nd Int. Symp. on Wind Energy Systems, Amsterdam 1978", Paper G1, British Hydromech. Res. Assoc., Cranfield.

Sørensen, B. (1978c). Wind – large scale utilisation. In "Proc. Conf. on Wind, Wave and Water, Dublin 1978", Industrial Devt. Authority of Ireland, IIRS and Solar Energy Society, Dublin.

Sørensen, B. (1979). *Renewable Energy*, Academic Press, 683 pp., London.

Sørensen, B. (1981a). A combined wind and hydro power system. *Energy Policy* March, 51–55.

Sørensen, B. (1981b). Wind energy. In "Renewable sources of energy and the environment" (E. El-Hinnawi and A. Biswas, eds.), pp. 97–116. Tycooli International Press, Dublin.

Sørensen, B. (1982). Comparative risk assessment of total energy systems. In "Health Impacts of different sources of energy", pp. 455–471. IAEA Publ. SM–254/105, Vienna.

Sørensen, B. (1983). Physics in Society, Lecture notes. IMFUFA Texts 129, Roskilde University, 56 pp. Reprinted as Part I of "Physics revealed", cf. Sørensen (2001a).

Sørensen, B. (1986). A study of wind-diesel/gas combination systems. Energy Authority of New South Wales, EA86/17. Sydney (82 pp).

Sørensen, B. (1987a). Chernobyl accident: assessing the data. *Nuclear Safety* **28**, 443–447.

Sørensen, B. (1987b). "Superstrenge – en teori om alt og intet". Munksgaard Publisher, Copenhagen (184 pp).

Sørensen, B. (1988). Optimisation of wind-diesel systems. In "Proc. Asian and Pacific

REFERENCES

Area Wind Energy Conference, Shanghai", pp. 94–98.

Sørensen, B. (1991). Energy conservation and efficiency measures in other countries. *Greenhouse Studies Series* No. 8, Australian Department of the Arts, Sport, the Environment, Tourism and Territories, Canberra, 118 pp.

Sørensen, B. (1992a). The impact of energy efficiency on renewable energy options. In "Renewable energy technology and the environment" (Sayigh, ed.), pp. 2654–2664. Pergamon Press, Oxford.

Sørensen, B. (1992b). Methods and models for estimating the global circulation of selected emissions from energy conversion. Proc. Technical Committee meeting, Vienna 11–15. May, Int. Atomic Energy Agency, Wien (also "Texts from IM-FUFA", No. 226, Roskilde University, 47 pp.).

Sørensen, B. (1992c). The future of renewable energy. *Ecodesicion* 4, 54-56.

Sørensen, B. (1993a). Technology change: the actor triangle. *Philosophy and Social Action* 19, 7–12.

Sørensen, B. (1993b). What is life-cycle analysis? In "Life-cycle Analysis of Energy Systems", pp. 21–53. Workshop Proceedings, OECD Publications, Paris.

Sørensen, B. (1994a). Strategy for a rich, fulfilling and sustainable society. In "Expanding environmental perspectives" (Lundgren, Nilsson and Schlyter, eds.), pp. 83–101. Lund University Press, Lund.

Sørensen, B. (1994b). Life-cycle analysis of renewable energy systems. *Renewable Energy* 5, part II, 1270–1277.

Sørensen, B. (1994c). Model optimization of photovoltaic cells. *Solar Energy Materials and Solar Cells* 34, 133–140.

Sørensen, B. (1995a). History of, and recent progress in, wind-energy utilization. *Annual Review of Energy & Environment* 20, 387–424.

Sørensen, B. (1995b). Life-cycle analysis. In "Conservation and Environmentalism – An Encyclopedia" (R. Paehlke, ed), pp. 413–415. Garland Publishers, New York.

Sørensen, B. (1995c). Life-cycle approach to assessing environmental and social costs of photovoltaics. *Applied Energy* 52 (suppl.), 357–374.

Sørensen, B. (1996a). Non-linear modelling of integrated energy supply and demand matching systems. Text 327 from IMFUFA (Roskilde University), 12 pp., and *Int. J. Global Energy Issues* 12, 131–137 (1999).

Sørensen, B. (1996b). Life-cycle approach to assessing environmental and social externality costs. In "Comparing energy technologies", Ch. 5, pp. 297–331. International Energy Agency, IEA/OECD, Paris.

Sørensen, B. (1996c). The use of life-cycle analysis to address energy cycle externality problems. In "Comparison of Energy Sources in Terms of Their Full Energy–chain (FENCH) Emission Factors of Greenhouse Gases", pp. 115–131. IAEA Tecdoc–892, Vienna.

Sørensen, B. (1996d). Does wind energy utilization have regional or global climate impacts? In "1996 European Union Wind Energy Conference, Göteborg" (Zervos, A., Ehmann, H. and Helm, P., eds.), pp. 191–194. H. Stephens & Ass., Felmersham.

Sørensen, B. (1996e). The role of life-cycle analysis in risk assessment. *Int. J. Environment and Pollution* 6, 729–746.

Sørensen, B. (1996f). Issues of data collection and use for quantifying the impacts of energy installations and systems. In "Electricity, health and the environment:

Comparative assessment in support of decision making", pp. 123–137. IAEA Report SM–338/21, Vienna.

Sørensen, B. (1997a). Impacts of energy use. In "Human ecology, human economy" (Diesendorf and Hamilton, eds.), pp. 243–266. Allen and Unwin, New South Wales.

Sørensen, B. (1997b). Renewable energy and environmental policy. In "Rural and Renewable Energy: Perspectives from Developing Countries" (Venkata, R., ed.), pp. 49–66. TERI, Tata Energy Research/Rajkamal Press, Delhi.

Sørensen, B. (1997c). Externality estimation of greenhouse warming impacts, *Energy Conversion and Management* **38**, S643–S648.

Sørensen, B. (1999a). Wave energy. In "Encyclopedia of Desalination and Water Resources", UNESCO project. EOLSS Publ., Oxford, by subscription at website http://www.desware.net

Sørensen, B. (1999b), Long-term scenarios for global energy demand and supply: Four global greenhouse mitigation scenarios. Final Report from a project performed for the Danish Energy Agency, IMFUFA Texts 359, Roskilde University, pp. 1–166.

Sørensen, B. (2000a). Role of hydrogen and fuel cells in renewable energy systems. In "Renewable energy: the energy for the 21st century", Proc. World Renewable Energy Conference VI, Reading, Vol. 3, pp. 1469–1474. Pergamon, Amsterdam.

Sørensen, B. (2000b). PV power and heat production: an added value, in "16th European Photovoltaic Solar Energy Conference", vol. 2, pp. 1848-1851. (H. Scheer *et al.*, eds), James & James, London.

Sørensen, B. (2001a). Physics revealed. The methods and subject matter of physics. IMFUFA Texts 129bis, 392 and 394, Roskilde University, 284 pp. Downloadable from http://mmf.ruc.dk/~boson under Fysik E.

Sørensen, B. (2001b). The need for global modelling, Editorial, *Int. J. Global Energy Issues* **13**, 1-3.

Sørensen, B. (2001c). see NSES (2001).

Sørensen, B. (2002a). Handling fluctuating renewable energy production by hydrogen scenarios. In "14th World Hydrogen Energy Conference", Montreal, CD published by CogniScience Publ. for l'Association Canadienne dw l'Hydrogène, Revised CD issued 2003, 9 pp.

Sørensen, B. (2002b). Modelling of hybrid PV-thermal systems, in "Proc. 17th European Photovoltaic Solar Energy Conference", Munich 2001, Vol. 3, pp. 2531-2538. WIP and ETA, Florence.

Sørensen, B. (2002c). Biomass for energy: how much is sustainable?, in "12th European Biomass Conf.", Amsterdam 2002, Vol. 2, pp. 1394-1397. WIP-Munich and ETA, Florence.

Sørensen, B. (2003a). Understanding photoelectrochemical solar cells, in "PV in Europe", Rome 2002, pp. 3-8. WIP-Munich and ETA, Florence.

Sørensen, B. (2003b). Progress in nanostructured photoelectrochemical solar cells. In "Third World Conference on Photovoltaic Energy Conversion, Osaka 2003" (to be published).

Sørensen, B. (2003c). Scenarios for future use of hydrogen and fuel cells. In "Towards a greener world - Hydrogen and Fuel Cells Conference", Vancouver, CD published by Fuel Cells Canada, Vancouver, 12pp.

Sørensen, B. (2003d). Time-simulations of renewable energy plus hydrogen systems.

In "Proc. Hypothesis V Int. Symp.", Porto Conte, Italy, to be published, 8 pp.

Sørensen, B. (2004a). *Hydrogen and Fuel Cells: Emerging Technologies and Applications.* Elsevier/Academic Press Sustainable World Series, Boston (to appear in 2004 or 2005).

Sørensen, B. (2004b). Surface reactions in photoelectrochemical cells (in preparation).

Sørensen, B., Meibom, P. (1998). A global renewable energy scenario. IMFUFA Texts 354, Roskilde University, 112 pp.

Sørensen, B., Meibom, P. (2000). A global renewable energy scenario. *Int. J. Global Energy Issues* **13**, No. 1-3, 196-276, based on Sørensen and Meibom (1998).

Sørensen, B., Petersen, A., Juhl, C., Ravn, H., Søndergren, C., Simonsen, P., Jørgensen, K., Nielsen, L., Larsen, H., Morthorst, P., Schleisner, L., Sørensen, F., Petersen, T. (2001). Project report to Danish Energy Agency (in Danish): Scenarier for samlet udnyttelse af brint som energibærer i Danmarks fremtidige energisystem, *IMFUFA Texts* No. 390, 226 pp., Roskilde University; report download site: http://mmf.ruc.dk/energy

Sørensen, B., Petersen, A., Juhl, C., Ravn, H., Søndergren, C., Simonsen, P., Jørgensen, K., Nielsen, L., Larsen, H., Morthorst, P., Schleisner, L., Sørensen, F., Petersen, T. (2004). Hydrogen as an energy carrier: scenarios for future use of hydrogen in the Danish energy system. *Int. J. Hydrogen Energy* **29**, 23–32; based upon Sørensen *et al.*, (2001).

Sørensen, B., Watt, M. (1993). Life-cycle analysis in the energy field. In Proc. 5th Int. Energy Conf. "Energex '93", Vol. 6, pp. 66–80. Korea Inst. of Energy Research, Seoul.

Sørensen, B., Nielsen, L., Petersen, S., Illum, K., Morthorst, P. (1994). "Future renewable energy system – light green or dark green?" Danish Technology Council, Report 1994/3 (in Danish), available at website http://www.tekno.dk/udgiv/943/943all.htm

Sorensen, K., MacLennan, C. (1974). *IEEE Trans.* 62–68, US Institute of Electrical and Electronics Engineers.

Spear, W., Le Comber, P. (1975). *Solid State Commun.* **17**, 1193–1196.

Spitzer, H. (1954). In "Hütte, Maschinenbau A", pp. 556–566. Wilhelm Ernst and Sohn, Berlin.

Sproul, A., Shi, Z., Zhao, J., Wang, A., Tang, Y., Yun, F., Young, T., Huang, Y., Edmiston, S., Wenham, S., Green, M. (1995). Characterization and analysis of multilayer solar cells. In "1994 First World Conf. on Photovoltaic Energy Conversion, Kona", Vol. 2, pp. 1410–1412. IEEE.

Squires, A. (1974). *Ambio* **3**, 2–14.

Stafford, D., Hawkes, D., Horton, R. (1981). "Methane Production from Waste Organic Matter", CRC Press, Boca Raton, FL.

Stambolis, C. (1976). In "Proc. UNESCO/WMO Solar Energy Symposium, Geneva 1976", Paper ENG.S/Doc. 3, WMO Paper No. 477 (1977).

Stansell, J. (1978). *New Scientist* 16. February, 420–421.

Starr, C. (1969). Social benefit versus technological risk. *Science* **165**, 1232–1238.

Steinberg-Yfrach, G., Rigaud, J.–L., Durantini, E., Moore, A., Gust, D., Moore, T. (1998). Light–driven production of ATP catalysed by F_0F_1-ATP synthase in an artificial photosynthetic membrane. *Nature* **392**, 479–482.

Steinhart, J., Steinhart, C. (1974). *Science* **184**, 306–316.

Stewart, D., McLeod, R. (1980). *New Zealand Journal of Agriculture* Sept., 9–24.

Stewart, G., Hawker, J., Nix, H., Rawlins, W., Williams, L. (1982). "The Potential for Production of Hydrocarbon Fuels from Crops in Australia". Commonwealth Scientific and Industrial Research Organization, Melbourne.

Stewart, R. (1969). *Sci. Am.* September, 76–105.

STI (2002). Sustainable Technology International, Queanbeyan, NSW Australia; website: http://www.sta.com.au

Stock, M., Carr, A., Blakers, A. (1996). Texturing of polycrystalline silicon, *Solar Energy Materials and Solar Cells* **40**, 33.

Stokes, G. (1880). *Math. and Phys. Papers*, Vol. 1, p. 225, as quoted by Wehausen and Laitone (1960).

Stolwijk, J., Canny, P. (1991). Determinants of public participation in management of technological risk. pp. 33–47 in Shubik (1991).

Stommel, H. (1965). "The Gulf Stream: A Physical and Dynamical Description", 2nd edn. University of California Press, Berkeley, California.

Street, R., Mott, N. (1975). *Phys. Rev. Lett.* **39**, 1293–1295.

Strickland, J. (1975). Sandia Laboratories Energy Report SAND75–0431, Albuquerque, New Mexico.

Svendsen, S. (1980). Effektivitetsprøvning af solfangere, Danish Technical University, Laboratory for Heat Insulation, Communication No. 107, Lyngby.

Sverdrup, H. (1957). In "Handbuch der Physik" (S. Flügge, ed.), Vol. 48 (Geophysik II), pp. 608–670. Springer-Verlag, Berlin.

Sverdrup, H., Johnson, M., Fleming, F. (1942). "The Oceans, Their Physics, Chemistry and General Biology". Prentice–Hall, Englewood Cliffs, NJ.

Sverdrup, H., Munk, W. (1947). Wind, Sea and Swell, US Hydrogr. Office Publ. No. 601 (also No. 604, 1951).

Swanson, R., Smith, R., Johnson, C. (1975). In "Proc. 2nd Workshop on Wind Energy Conversion Systems" (F. Eldridge, ed.), pp. 92–97. The Mitre Corp., McLean, Virginia, Report No. NSF–RA–N–75–050.

Swedish Association of Boiler Owners (Ångpanneföreningen) (1973). "Verksamhedsberättelse", Stockholm.

Swedish Ministries of Foreign Affairs and of Agriculture (1971). "Air Pollution across National Boundaries", case study for the UN Conf. on the Human Environment, Stockholm.

Swisher, R. (1995). Wind power a strong contender in US marketplace. In "The World Directory of Renewable Energy Suppliers and Services", pp. 190–196. James & James, London.

Tabor, H. (1967). *Solar Energy* **7**, 189.

Tafdrup, S. (1993). Environmental impact of biogas productiom from Danish centralized plants. Paper presented at "IEA Bioenergy Environmental Impact Seminar, Elsinore, 1993".

Taiganides, E. (1974). *Agricult. Eng.* (*Am. Soc. Agric. Eng.*) **55**, No. 4.

Takahashi, K. (1998). Development of fuel cell electric vehicles. Paper presented at "Fuel cell technology conference, London, September", IQPC Ltd., London.

Takamoto, T., Ikeda, E., Kurita, H., Yamaguchi, M. (1997). Over 30% efficient InGaP/GaAs tandem solar cells with InGaP tunnel junction. In "14th European PV Solar Energy Conf." pp. 970–975. HS Stephens & Assoc., Bedford.

Talbert, S., Eibling, J., Löf, G. (1970). "Manual on Solar Distillation of Saline Water", US Dept. Int., Office of saline water, R&D Report No. 546; cf. also *Solar Energy*

13, 263–276.

Tang, C. (2001). Organic light emitting diodes, In "Proc. 10 th Int. Conf. On Unconventional Photoactive Systems", Diablerets, p. I-23

Tarascon, J., Armand, M. (2001). Issues and challenges facing rechargeable lithium batteries. *Nature*, **414**, 359-367.

Taylor, G., *et al.* (1993). Full scale measurements in wind turbine arrays, In "EC Wind Energy Conference, Travemünde", pp. 755–758. H. Stephens & Ass., Felmersham.

Telegadas, K. (1971). In "Fallout Program Qua. Summary Rep., Mar.–Jun.", pp. I–2 to I–88. Health and Safety Lab. Report HASL–243, Stanford.

Telkes, M. (1952). *Ind. Eng. Chem.* **44**, 1308.

Telkes, M. (1976). In "Critical Materials Problems in Energy Production" (C. Stein, ed.). Academic Press. New York and London.

Templin, R. (1976). In "Proc. Workshop on Advanced Wind Energy Systems, Stockholm 1974" (O. Ljungstrom, ed.). Swedish Development Board/Vattenfall, Stockholm.

Teplyakov, D. and Aparisi, R. (1976). *Geliotekhnika* **12**, No. 3, 35–48.

Thekaekara, M. (1976). *Solar Energy* **18**, 309–325.

Thorpe, T. (2001). Current status and development in wave energy. Proc. "Conf. Marine Renewable Energies", pp. 103-110, Institute of Marine Engineers, UK.

Tideman, J. and Hawker, J. (1981). *Search* **12**, 364–365.

Tol, R. (1995). The damage costs of climate change: towards more comprehensive calculations. *Environmental & Resource Economics* **5**, pp. 353–374.

Toland, R. (1975). In "Proc. 1975 Flywheel Technology Symp.", pp. 243–256. Berkeley, CA, Report ERDA 76–85.

Tomaschek, R. (1957). In "Handbuch der Physik" (S. Flügge, ed.), Vol. 48 (Geophysik II), pp. 775–845. Springer-Verlag. Berlin.

Tonne, F., Normann, W. (1960). *Zeitschrift für Meteorologie* **14**, 166.

Toyota (1996). High-performance hydrogen-absorbing alloy, http:// www.toyota.co.jp/e/november_96/electric_island/press.html

Trebst, A. (1974). *Ann. Rev. Plant Physiol.* **25**, 423–447.

Trenberth, K., Solomon, A. (1994). The global heat balance: heat transfer in the atmosphere and ocean. *Climate Dynamics* **10**, 107–134.

Trenkowitz, G. (1969). "Die Värmepumpe", Verein Deutscher Ingenieurs Berichte Nr. 136.

Trinidade, S. (1980). Energy Crops – the Case of Brazil, in "Int. Conf. on Energy from Biomass", Brighton, UK, 1980, Centro de Tecnologia Promon, Rio de Janeiro

Troen, I., Petersen, E. (1989). "European Wind Atlas". Risø National Laboratory, Roskilde.

Trombe, F. (1973). Centre Nationale de Récherche Scientifique, Report No. B–1–73–100.

Tsang, C., Lippmann, M., Wintherspoon, P. (1979). In "Sun II, Proc. Solar Energy Society Conf., New Delhi 1978", pp. 349–355. Pergamon Press, London.

Tsubomura, H., Matsumura, M., Nomura, Y., Amamiya, T. (1976). Dye sensitised zinc oxide: aqueous electrolyte: platinum photocell. *Nature* **261**, 402–403

Turkenburg, W., *et al.*, (2000). Renewable energy technologies, Ch. 7 in "World Energy Assessment", United Nations Development Programme, New York, pp.

219–272

Turner, C. (ed.) (1974). "Solar Energy Data Workshop", US Nat. Sci. Found. and Nat. Oceanic and Atm. Adm., Silver Spring, Report No. NSF–RA–N–74–062.

Tuttle, J., Ward, J., Duda, A., Berens, A., Contreras, M., Ramanathan, K., Tennant, A., Keane, J., Cole, E., Emery, K., Noufi, R. (1996). The performance of Cu(In,Ga)Se$_2$-based solar cells in conventional and concentrator applications. *Proc. Material Research Society Symposium* **426**, 143.

Twomey, S., Wojciechowski (1969). *J. Atmos. Sci.* **26**, 684–688.

UK Meteorological Office (1997). Unified model (User guide and support documents), London.

Ulrich, R. (1973). In "Explosive Nucleosynthesis" (D. Schramm and W. Arnett, eds.), pp. 139–167. University of Texas Press, Austin.

Ulrich, R. (1975). *Science* **190**, 619–624.

UMIP (1996). "Miljøvurdering af produkter", Project on the development of environmentally sound industrial products (Wenzel *et al.*), Institut for produktudvikling, Report, Danish Technical University.

UN (1981). "World Energy Supplies", United Nations, New York.

UN (1996). "Populations 1996, 2015, 2050". United Nations Population Division and UNDP: available at the website http://www.undp.org/popin/wdtrends/pop/fpop.htm

UN (1997a). "1995 World Energy Supplies", United Nations, New York.

UN (1997b). "UN urban and rural population estimates and projections as revised in 1994". United Nations Population Division and UNDP, Washington. Website: http://www.undp.org/popin/wdtrends/urban.html

UNEP (1998). GRID data website: http://www.unep.ch/noaagnv22.html

United Kingdom Energy Technology Support Unit (1976). "Wave Energy on U.K. Coasts and North Sea". UKAEA Harwell.

United States Bureau of Mines (1975), as quoted by Ross and Williams (1975).

University of Columbia (1998). Selected climate data sets, available at http://ingrid.ldgo.columbia.edu

UNSCEAR (1993). *Sources and Effects of Ionizing Radiation*. United Nations Scientific Committee on the Effects of Atomic Radiation, Report to the General Assembly, with scientific annexes, E.94.IX.2. United Nations, New York.

US DoE (1979). "Peat Prospectus", United States Department of Energy, Washington, DC.

US EPA (1980). US Environmental Protection Agency Report EPA–600/7–80–040 (D. deAngelis *et al.*), Washington, DC; also earlier reports EPA–600/2–76–056 and EPA–600/7–77–091.

US Government (1962). US Standard Atmosphere. Guideline for meteorological observational methodology, Washington DC.

Vermeulen, L., Thompson, M. (1992). Stable photoinduced separation in layered viologen compounds. *Nature* **358**, 656–658.

Vindeløv, S. (1994). Research activities in wave energy, *Sustainable Energy News,* No. 7 (December), 12–13.

Vohra, K. (1982). Rural and urban energy scenario of the developing countries and related health assessment. In "Proc. Int. Symp. on Health Impacts of Different Sources of Energy, Nashville, 1981", pp. 79–96. Int. Atomic Energy Agency, Vienna, Paper No. IAEA–SM–254/102.

REFERENCES

Volfin, P. (1971). *La Récherche* September, 741–755.

Vollmer, R. (1977). *Nature* **270**, 144–147.

Wagner, U., Geiger, B., Schaefer, H. (1998), Energy life cycle analysis of hydrogen systems. *Int. J. Hydrogen Energy* **23**, 1–6.

Walsh, J., Davis, J., Garms, R. (1981). Further studies on the reinvasion of the onchocerciasis control programme by simulus damnosum s.l. *Tropical Medicine & Parasitology* **32**, No. 4, 269–273.

Wan, E., Simmins, J., Nguyen, T. (1981). In "Biomass Gasification" (T. Reed, ed.), pp. 351–385. Noyes Data Corp., Park Ridge, NJ.

Wang, J., Wang, D., Smith, K., Hermes, J. (1982). In "The Future of Small Energy Resources", pp. 465–472. McGraw-Hill, New York.

Ward, R. (1982). *Solar Energy* **29**, 83–86.

Warming, S. (1996). Valuation of environmental impacts from coal–based power production. ELSAM Power Utility, 17 pp. (unpublished note in Danish).

Wassink, E. (1975). In "Photosynthesis and Productivity in Different Environments" (J. Cooper, ed.), pp. 675–687. Cambridge University Press, Cambridge, UK.

Watt, M. (1993). Environmental & Health Considerations in the Production of Cells and Modules. Centre for Photovoltaic Devices & Systems Report # 1993/02, University of New South Wales, Sydney.

WEA (2000). Energy and the challenge of sustainability. *World Energy Assessment*, UNDP, UNDESA, WEC, New York.

Webb, W. (1976). In "Handbuch der Physik" (S. Flügge, ed.), Vol. 49/5 (Geophysik III, K. Rawer, ed.), pp. 117–176. Springer-Verlag, Berlin.

Weber, O. (1975). *Brown Boveri Mitt.* **62**, No. 7/8, 332–337.

Webster, A. (1974). *Sci. Am.* August, 26–33.

WEC (1991). "District heating/combined heat and power". World Energy Council, London.

Wehausen, J., Laitone, E. (1960). In "Handbuch der Physik" (S. Flügge, ed.), Vol. 9 (Stromungsmechanik III), pp. 446–778. Springer-Verlag, Berlin.

Weiner, S. (1977). The Sodium–Sulphur Battery: Problems and Promises. Chapter 12 in "Solid State Chemistry of Energy Conversion and Storage" (J. Goodenough and M. Whittingham, eds.), Advances in Chemistry Series 163, American Chemical Society, Washington, DC.

Weinstein, J., Leitz, F. (1976). *Science* **191**, 557–559.

Weissglas, P. (1977). "Solcellskraftverk i Sverige", Report No. E77–5011, Inst. f. Mikrovågsteknik, Stockholm.

Wenham, S., Green, M., Edminston, S., Campbell, P., Koschier, L., Honsberg, C., Sproul, A., Thorpe, D., Shi, Z., Heiser, G. (1995a). Limits to efficiency if silicon multilayer thin film solar cells. In "1994 First World Conf. on Photovoltaic Energy Conversion, Kona", Vol. 2, pp. 1234–1241. IEEE.

Wenham, S., Robinson, S., Dai, X., Zhao, J., Wang, A., Tang, Y., Ebong, A., Hornsberg, C., Green, M. (1995b). Rear surface effects in high efficiency silicon solar cells. In "1994 First World Conf. on Photovoltaic Energy Conversion, Kona", Vol. 2, pp. 1278–1282. IEEE, Washington, DC.

Wentink, T. (1976). Study of Alaskan Wind Power and its possible Applications. Report No. NSF/RANN/SE/AER–00239/FR–76/1.

Wentorf, R., Hanneman, R. (1974). *Science* **185**, 311–319.

West, C. (1974). "Fluidyne Heat Engine", Harwell Report AERE–R6775, U.K. Atomic Energy Agency.

Wetherald, R., Manabe, S. (1972). *Monthly Weather Rev.* **100**, 42–59.

Wetherald, R., Manabe, S. (1975). *J. Atmos. Sci.* **32**, 2044–2059.

Whittingham, M. (1976). *Science* **192**, 1126–1127.

Wilbur, P., Mancini, T. (1976). *Solar Energy* **18**, 569–576.

Williams, R. (1975). Working paper No. 21, Center for Environmental Studies, Princeton University, NJ.

Wilson, C., Matthews, W. (1970). "Man's Impact on the Global Climate", Report of the Study of Critical Environmental Problems (SCEP). MIT Press, Cambridge, MA.

Wilson, C., Matthews, W. (1971). "Inadvertent Climate Modification", Report of the Study of Man's Impact on Climate (SMIC). MIT Press, Cambridge, MA.

Wilson, R., Lissaman, P. (1974). "Applied Aerodynamics of Wind Power Machines". Oregon State University, Report No. NSF–RA–N–74–113.

Windpower Monthly (2003). "Wind market status 2003". Editorial, March issue, http://www.wpm.co.nz

Winsberg, S. (1981). *Sunworld* **5**, 122–125.

Wise, D. (1981). *Solar Energy* **27**, 159–178.

Wittenberg, L., Harris, M. (1979). In "Proc 14th Intersociety Energy Conversion Engineering Conf." pp. 49–52. American Chemical Society, Washington, DC.

Wizelius, T. (1998). Potential for offshore transmission. *Windpower Monthly* December, p. 25.

Wofsy, S., McElroy, M. and Sze, N. (1975). *Science* **187**, 535–537.

Wolf, M. (1963). *Proc. Inst. Electr. Electron. Eng.* **51**, 674–693.

Wolf, M. (1971). *Energy Conversion* **11**, 63–73.

Wolfbauer, G. (1999). The electrochemistry of dye sensitised solar cells, their sensitisers and their redox shuttles, Ph. D. thesis, Monash University, Victoria, Australia

World Almanac (1977). (G. Delury, ed.), Newspaper Enterprise Ass., New York.

World Energy Conference (1974). "Survey of Energy Resources", privately published in London.

World Energy Conference (1995). *Survey of Energy Resources*, 17th ed., London.

World Meteorological Organization (1964). In "Proc. UN Conf. on Alternative Energy Sources, Rome 1961", Paper W/11. UN Printing Office, New York.

Wrighton, M., Ellis, A., Kaiser, S. (1977). Conversion of visible light to electrical energy: stable cadmium selenide photoelectrodes in aqueous electrolytes. In "Solid State Chemistry of Energy Conversion and Storage" (J. B. Goodenough and M. S. Whittingher, eds.), pp. 71–92. Advances in Chemistry Series 163. American Chemical Society.

Wronski, C., Carlson, D. (1977). In "Proc. 7th Int. Cont; on Amorphous and Liquid Semiconductors" (W. Spear, ed.), pp. 452–456. University of Edinburgh, UK.

Wulff, H. (1966). "The Traditional Crafts of Persia". MIT Press, Cambridge, MA.

Wurster, R. (1997). PEM fuel cells in stationary and mobile applications. Paper for Biel Conference (http://www.hyweb.de/knowledge).

Wynne, B. (1984). The institutional context of science, models, and policy. *Policy Sciences* **17**, 277–320.

Wysocki, J., Rappaport, P. (1960). *J. Appl. Phys.* **31**, 571–578.

Yamada, K., Komiyama, H., Kato, K., Inaba, A. (1995). "Evaluation of photovoltaic energy systems in terms of economic, energy and CO_2 emissions". University of Tokyo, report.

Yamaguchi, M., Horiguchi, M., Nakanori, T., Shinohara, T., Nagayama, K., Yasuda, J. (2000). Development of large-scale water electrolyzer using solid polymer electrolyte in WE-NET, in "Hydrogen energy progress XIII" (Mao and Veziroglu, eds.), Vol. 1, pp- 274-281. IAHE Beijing.

Yamamoto, K., Yoshimi, M., Tawada, Y., Okamoto, Y., Nakajima, A. (1999). Cost effective and high performance thin film Si solar cell towards the 21st century, in "Technical digest of the international PVSEC-11, Sapporo" (Saitoh, T., ed.), pp. 225-228. Tokyo Univ. of Agriculture and Technology, Tokyo.

Yartym, J., *et al.* (2001). In "Proc. 4 th Int. Symp. New Materials", Montréal, pp. 417–419.

Yasukawa, S., Tadokoro, Y., Sato, O., Yamaguchi, M. (1996). Integration of indirect CO_2 emissions from the full energy chain. In "Comparison of Energy Sources in Terms Of Their Full Energy-Chain (FENCH) Emission Factors of Greenhouse Gases", pp. 139–150. IAEA Tecdoc–892, Vienna.

Yazawa, Y., Tamura, K., Watahiki, S., Kitatani, T., Minemura, J., Warabisako, T. (1996). GaInP single-junction and GaInP/GaAs two-junction thin-film solar cell structures by epitaxial lift-off. In "Technical Digest of 9th PV Science & Eng. Conf., Miyazaki", p. 865. Tokyo Inst. of Technology.

Yen, J. (1976). In "Proc. Int. Symp. on Wind Energy Systems, Cambridge 1976", Paper E4. British Hydromech. Res. Ass., Cranfield.

Yoneda, N., Ito, S., Hagiwara, S. (1980). Study of energy storage for long term using chemical reactions, 3rd Int. Solar Forum, Hamburg, Germany, June 24–27.

Yu, G., Gao, J., Hummelen, J., Wudl, F., Heeger, A. (1995). Polymer photovoltaic cells. *Science* **270**, 1789–1791.

Zakeeruddin, S., *et al.* (1997). Molecular engiinering of photosensitizers for nanocrystalline solar cells: synthesis and characterization of Ru dyes based on phosphonated terpyridines. *Inorg. Chem.* **36**, 5937–5944.

Zakhidov, A., Baughman, R., Iqbal, Z., Cui, C., Khayrullin, I., Dantas, S., Marti, J., Ralchanko, V. (1998). Carbon structures with three-dimensional periodicity at optical wavelengths. *Science* **282**, 897–901.

Zener, C. (1973). *Phys. Today* January, 48–53.

Zener, C., Fetkovich, J. (1975). *Science* **189**, 294–295.

Zhang, Y., Suenaga, K., Colliex, C., Iijima, S. (1998). Coaxial nanocables: silicon carbide and silicon oxide sheathed with boron nitride and carbon. *Science* **281**, 973–975.

Zhao, J., Wang, A., Altermatt, P., Wenham, S., Green, M. (1995). 24% efficient solar cells. In "1994 First World Conf. on Photovoltaic Energy Conversion, Kona", Vol. 2, pp. 1477–1480. IEEE, Washington, DC.

Zhao, J., Wang, A., Green, M. (1998). 19.8% efficient multicrystalline silicon solar cells with "honeycomb" textured front surface. In "Proc . 2nd World Conf. on PV Energy Conversion, Vienna" (J. Schmid *et al.*, eds.), pp. 1681–1684. JRC European Commission EUR 18656 EN, Luxembourg.

Zittel, W., Wurster, R. (1996). "Hydrogen in the energy sector". Ludwig-Bölkow-ST Report: http://www.hyweb.de/knowledge/w–i–energiew–eng

REFERENCES

Zlatev, Z., Thomsen, P. (1976). Numerical Inst. at Danish Technical University, Internal Reports No. 76–9/10, Lyngby, Denmark.

Zuidema, G., Born, J. van der, Alcoma, J. Kreileman, G. (1994). Simulating changes in global land cover as affected by economic and climatic factors, *Water, Air and Soil Pollution* **76**, 163–198.

SUBJECT INDEX